GROWTH AND GROWTH HORMONE

GROWTH AND GROWTH HORMONE

PROCEEDINGS OF THE SECOND INTERNATIONAL SYMPOSIUM
ON GROWTH HORMONE
MILAN, MAY 5-7, 1971

Editors
A. PECILE and E. E. MÜLLER
Institute of Pharmacology
University of Milan

1972

EXCERPTA MEDICA, AMSTERDAM

INTERNATIONAL CONGRESS SERIES NO. 244

ISBN 90 219 0154 4

EXCERPTA MEDICA OFFICES

Amsterdam Herengracht 362-364
London Chandos House, 2 Queen Anne Street
Princeton Nassau Building, 228 Alexander Street

Printed in the Netherlands by N.V. Drukkerij Trio, The Hague

SECOND INTERNATIONAL SYMPOSIUM ON GROWTH HORMONE

PRESIDENT

C. H. LI, U.S.A.

LOCAL CHAIRMAN

E. TRABUCCHI, Italy

PROGRAMME COMMITTEE

J. C. BECK, Canada
W. H. DAUGHADAY, U.S.A.
M. M. GRUMBACH, U.S.A.
A. KORNER, U.K.
E. KNOBIL, U.S.A.
C. H. LI, U.S.A.
R. LUFT, Sweden
A. PRADER, Switzerland
A. V. SCHALLY, U.S.A.
E. TRABUCCHI, Italy
A. E. WILHELMI, U.S.A.

EXECUTIVE SECRETARIES

A. PECILE and E. E. MÜLLER, Milan, Italy

ASSISTANT SECRETARY

MARIA LUISA PECILE, Milan, Italy

ACKNOWLEDGEMENT

The President of the Symposium and the Editors gratefully acknowledge the valuable assistance extended by Professor F. J. Ebling and Dr. I. Henderson of Sheffield in the linguistic revision of the papers written by authors whose mother tongue is not English.

FOREWORD

This book contains the complete texts of invited papers presented at the Second International Symposium on Growth Hormone in May of this year. In addition to the discussion on the latest developments on the chemistry and biology of growth hormone, the Symposium included two morning sessions on sulfation factor and human chorionic somatomammotropin (HCS). It has been known for some time that the action of growth hormone on cartilage is partly due to the production and function of sulfation factor. HCS is a single-chain protein consisting of 190 amino acids with a primary structure closely related to human pituitary growth hormone (HGH). Biologically, HCS possesses both growth-promoting and lactogenic activities, as does HGH. Besides 42 invited papers, 122 short communications were presented; abstracts of these communications have already been published elsewhere.

Since the first symposium held four years ago, research activities on growth hormone have been increased enormously. The amino acid sequence of bovine growth hormone is almost completely known. The synthesis of a protein having biological activities of HGH has been achieved. Regulation of growth hormone secretion in both experimental animals and human subjects has largely been clarified. The effectiveness of HGH in treatment of hypopituitary dwarfism is now securely established in clinical medicine.

We wish to thank 325 registrants from 29 countries for their attendance. We are grateful to the members of the programme committee for their enthusiastic advice and to Professors E. Trabucchi, A. Pecile and E. E. Müller for their tremendous task in organizing the Symposium.

Bodega Bay, California
December, 1971

Choh Hao Li

CONTENTS

I. Evolution and Hormones

II. Chemistry, Immunochemistry and Comparative Aspects

III. In Vivo and in Vitro effects

I. Evolution and hormones

EVOLUTION AND HORMONES

E. J. W. BARRINGTON

Department of Zoology, University of Nottingham, Nottingham, United Kingdom

I must try to reduce a potentially vast subject to manageable proportions, and as a contribution to this end I shall refer mainly, although not exclusively, to the vertebrates. This decision is readily justified, for we are much better informed about vertebrate endocrine systems than we are about those of other groups. Moreover, I suspect that whatever principles we can establish for vertebrate hormones will eventually prove to be applicable to those of the invertebrates as well. This follows from current concepts of the essential unity of the organization of living material. It follows, too, from those common principles that have already been demonstrated, one illustration being the widespread occurrence of neurosecretion throughout the animal kingdom (Highnam and Hill, 1969). This, with the associated development of neurohaemal release organs, has led to the quite independent establishment in invertebrates (polychaete worms, for example, and arthropods) of endocrine complexes bearing remarkable resemblances to the pattern of organization of the pituitary gland (see, for example, Golding, 1970).

One principle readily extractable from vertebrate studies is that the molecular structure of some hormones, once they have been incorporated into the physiological organization of a group, can remain entirely untouched by evolutionary change. Thyroxine, triiodothyronine, noradrenaline and adrenaline are examples of hormones that are distributed throughout the vertebrates without any variation. This does not mean, of course, that they are confined to the vertebrates and their immediate ancestors. Triiodothyronine, for example, has been identified in the mucus of a nemertine worm (Major *et al.*, 1969) and trace amounts of this molecule and of thyroxine are found in gorgonid coelenterates (Roche *et al.*, 1964). The physiological significance of this distribution is obscure, for there is certainly no evidence that the molecules have any endocrine function in these particular groups. In gorgonids, as in some other invertebrates, their synthesis is thought to be a by-product of the formation of scleroproteins and of the dependence of this process on the presence of an oxidase system. Whatever the explanation, however, the situation is of great evolutionary interest, for, as has often been pointed out (see, for example, Gorbman, 1955; Major *et al.*, 1969; Barrington, 1972), it carries the suggestion that in this instance, and in others as well, hormones may have originated as metabolic intermediates or by-products. One obvious example of this having occurred is the incorporation of progesterone into the hormonal system of the female mammal.

Many aspects of steroid hormones are, indeed, of evolutionary interest. Those found in vertebrates provide another example of hormones being largely untouched by evolutionary change, for there is a general uniformity of molecular output throughout the group. Certain taxonomic features are detectable (an emphasis upon 1α-hydroxycorticosterone in elasmobranchs, for example), but in general the biosynthetic pathways of these molecules seem to have been established during the earliest stages of vertebrate history, their products thereafter showing very little variation (Sandor, 1969). But here again neither the pathways, nor their molecular products, are confined to the vertebrates. For example, the ovotestis of the slug

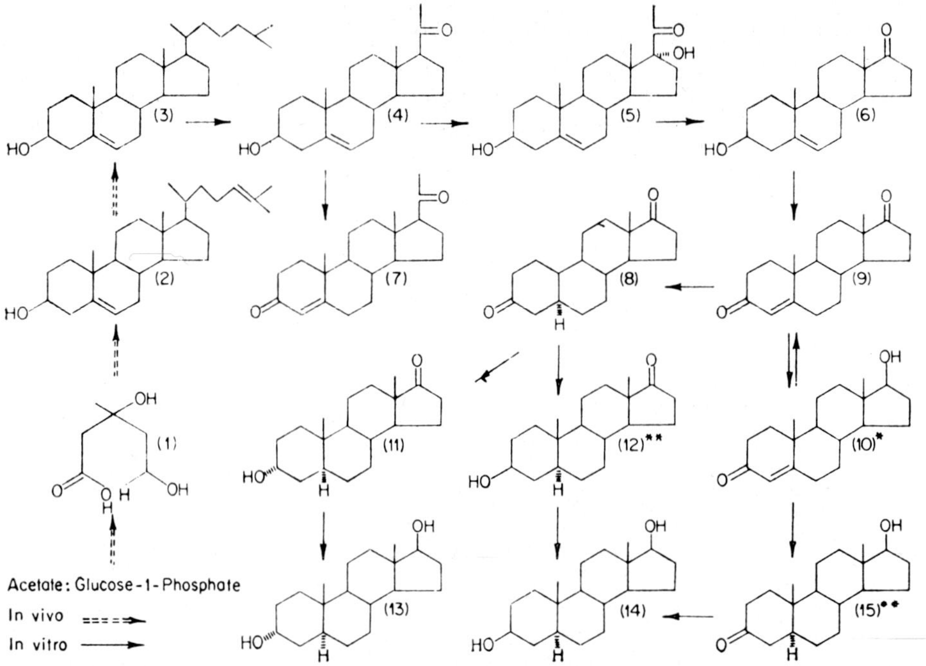

Fig. 1. *In vivo* whole animal cholesterol biosynthesis (left) and a possible interpretation of steroid transformations that occurred during *in vitro* incubation of male-phase ovotestis of *Ariolimax californicus*. (1) mevalonic acid; (2) desmosterol (24-dehydrocholesterol); (3) cholesterol; (4) pregnenolone; (5) 17α-hydroxypregnenolone; (6) dehydroepiandrosterone; (7) progesterone; (8) 5α-androstandione; (9) androstenedione; (10)* testosterone substrate, but not detected; (11) androsterone; (12)** epi-androsterone, not detected, but possible intermediate; (13) 3α-androstandiol; (14) 3β-androstandiol; (15)** 5α-dihydrotestosterone, not detected, but possible intermediate. (From Gottfried and Dorfman, 1970, *Gen. comp. Endocr.*, *15*, 120, by kind permission of the editors).

Ariolimax californicus has a capacity for steroid metabolism very similar to that of the vertebrate gonad (Fig. 1), and it seems likely that steroids participate in some way, still undefined, in the spermatogenesis of this mollusc (Gottfried and Dorfman, 1970). This wide distribution of steroid biosynthetic pathways accounts for the existence in invertebrates of hormonal molecules which we should regard as being primarily characteristic of vertebrates. Examples are oestradiol in the scallop *Pecten henricius* (see Botticelli *et al.*, 1961), 11-deoxycorticosterone in the water-beetle, *Dytiscus marginalis* (Schildknecht *et al.*, 1966) and testosterone in another water-beetle, *Ilybius fenestratus* (Schildknecht *et al.*, 1967). As with thyroxine and triiodothyronine, it is not clear whether the presence of these particular molecules is of physiological significance, although it has been suggested in this instance that the steroids may serve in the two species of beetle as defense substances against vertebrate predators.

The distribution of these biologically active molecules, then, may transcend taxonomic boundaries. Nevertheless, there have been times in evolutionary history when firm decisions were taken; times, that is, when a particular product, or family of products, were selected as part of the chemical characteristics of a major group of animals. The exploitation of thyroxine and triiodothyronine by the vertebrates is one example of this. Another is the regulation of growth and moulting in insects by a group of steroid molecules, the ecdysones (Fig. 2), which are quite distinct from the steroid hormones of vertebrates. In this respect, then, natural selection has established a chemical distinction between the two groups. But ecdysones are not restricted to the insects alone. One of them, ecdysterone (β-ecdysone, or crustecdysone), has been isolated also from crustaceans, and has been shown to promote moulting in these animals, and in arachnids as well. Krishnakumaran and Schneiderman (1969), in discussing this evidence, suggest that the ecdysones evolved in arthropods simultaneously with their chitinous exoskeleton, and must therefore be at least 600 million years old, antedating, perhaps, the steroid hormones of vertebrates, and rivalling them in their stability. It is, however, too early to attach great phylogenetic weight to these findings. In fact, it is still necessary to demonstrate that the ecdysones are indeed the true moulting hormones of arachnids and crustaceans. Moreover, it is to be remembered that the evolution of arthropods must, on any interpretation, have involved much convergence, and that the chitinisation of the exoskeleton may well have developed independently along more than one line (for references, see Barrington, 1967). Conceivably, then, the hormonal use of the ecdysones (supposing that this is their use) may have evolved more than once within the arthropods. If this seems improbable to the invertebrate endocrinologist (and I am not unsympathetic) let it be remembered also that the compound eye has probably evolved more than once, and nothing could seem much more improbable than that!

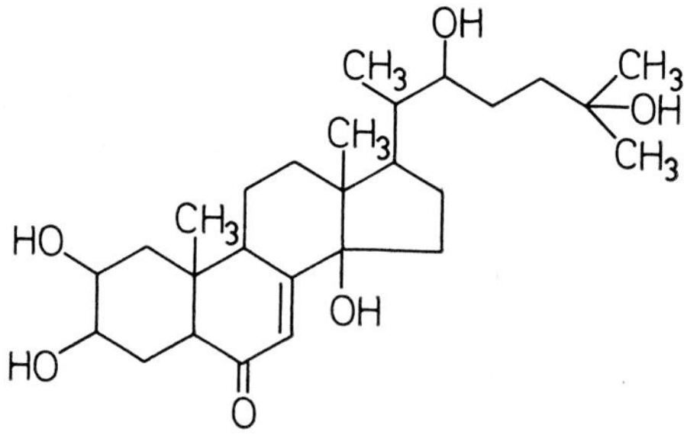

Fig. 2. Structural formula of α-ecdysone.

I have said that these stable hormones, once established, have remained largely untouched by evolution. This stability, however, poses in itself a very important evolutionary problem, for how can such hormones contribute to the continuously changing patterns of adaptive regulation which all groups of animals must achieve if they are to survive during their evolutionary history? The answer to this question is that it is an abstraction to consider hormones as though they can be isolated functionally from their target organs, or from other hormones. The targets and their hormones constitute an integrated complex, and it is evolution of the targets that provides for continuous modulation of the effects of the hormones. In this way, modification of one function can be achieved without disturbing the general relationships of the hormone with other targets or with other hormones.

The operation of this principle is clearly exemplified by the thyroid hormones. These molecules are already present in the protochordates (Salvatore, 1969; Barrington, 1972), which survive today as representatives of an early prevertebrate stage of vertebrate history. Presumably the biosynthesis of biologically active iodothyronines was established in the common ancestors of protochordates and vertebrates, and these substances remained thereafter unaltered in their molecular structure. Yet, despite this stability, they can evoke a wide diversity of physiological responses throughout the vertebrates, the subtlety of their action being well shown (Table I) in the striking range of the transformations initiated and controlled by them during the metamorphosis of the amphibian tadpole (Eaton and Frieden, 1969). There has been more than one view of the way in which these molecules produce their effects (Tata, 1969). It may be that they act upon the mitochondria. Or it may be that they act as gene repressors or derepressors, thereby influencing protein synthesis, and thus the production of

TABLE I

Biochemical systems extensively modified under thyroid control during anuran metamorphosis

Tissue, Organ		Biochemical system	Change	Comments
1.	Whole animal	Respiration	No increase, decrease in certain species	Calorigenic responses still possible
2.	Erythrocytes	Hemoglobin (Hb)	Repression of tadpole Hb synthesis, induction of frog Hb synthesis	Adaptive oxygen binding
3.	Serum proteins	Serum protein biosynthesis (in liver)	Induction of biosynthesis of serum albumin, ceruloplasmin	Probably necessary for homeostasis
4.	Liver	RNA biosynthesis	Increased RNA turnover	Thyroxine-mediated genetic expression via DNA
5.	Liver	Urea production	Induction of urea cycle enzymes	Transition from ammonotelism to ureotelism
6.	Tail	Synthesis of hydrolytic enzymes of lysosomal type	Stimulation of cathepsin, phosphatase, β-glucuronidase synthesis	Leads to tail resorption
7.	Skin	Collagen biosynthesis	Collagenolysis in tail; deposition in back, head	Skin strengthening
8.	Eye	Light-sensitive pigments	Shift to rhodopsin	Repression of porphyropsin (retinene$_2$) synthesis
9.	Intestine	Digestive enzymes	Shift from carbohydrases to proteases	Change from herbivorous to carnivorous
10.	Limb buds	Proteins, nucleic acid	Development and growth of tissues (skin, nerve)	Locomotion on land

(From Eaton and Frieden, 1969, *Gen. comp. Endocr.*, Suppl. 2, 398, by kind permission of the editors).

specific intracellular enzymes. However this may be, studies of amphibian metamorphosis show that the effects of the thyroid hormones on any particular target are ultimately determined by the ways in which the cells are genetically programmed to respond to the hormonal stimulus.

Not all hormones, however, are as stable as those that I have so far mentioned. The polypeptide hormones are very different, for they are known in vertebrates to be much subject to molecular variation by amino acid substitution, and we may expect this to be true of invertebrates as well. This creates a more complex situation in which evolution of the hormones can become linked to a variable extent with evolution of their target organs. I shall take as a model of this situation the results of recent studies of a highly characteristic feature of vertebrates, the hypothalamic polypeptide hormones (for references, see Sawyer, 1966; Heller and Spickett, 1967; Acher, 1969*a*, 1969*b*). Extreme scarcity of information calls for caution in attempting even the simplest generalisation in comparative endocrinology, and the hypothalamic hormones are no exception to this, for so far they have been studied in only some 30 of the 40,000 or more vertebrate species (Acher *et al.*, 1970*a*). Within these limits, however, they can be said to constitute a family of hormones of which seven have so far been characterised (Fig. 3). All show a similar primary molecular structure, but they differ in respect of amino acid substitutions. These can be derived by a theoretical scheme of base changes in the appropriate codons (Geschwind, 1969), provided that we assume in some instances the occurrence of intermediate stages which have either been lost, or at least not yet identified.

Cys–Tyr–Ile–Glu NH_2–Asp NH_2–Cys–Pro–Arg —— Gly NH_2
1 Arginine vasotocin (in all vertebrates, except adult mammals)
Cys–Tyr–Phe–Glu NH_2–Asp NH_2–Cys–Pro–Arg —— Gly NH_2
2 Arginine vasopressin (in most mammals)
Cys–Tyr–Phe–Glu NH_2–Asp NH_2–Cys–Pro–Lys —— Gly NH_2
3 Lysine vasopressin (in a few mammals: Suiformes)
Cys–Tyr–Ile—Ser —— Asp NH_2–Cys–Pro–Glu NH_2–Gly NH_2
4 Glumitocin (in elasmobranchs)
Cys–Tyr–Ile–Glu NH_2—Asp NH_2–Cys–Pro–Ile —— Gly NH_2
5 Mesotocin (in lung fish, amphibians, reptiles, birds)
Cys–Tyr–Ile–Ser —— Asp NH_2–Cys–Pro–Ile —— Gly NH_2
6 Isotocin (in actinopterygians)
Cys–Tyr–Ile–Glu NH_2–Asp NH_2—Cys–Pro–Leu —— Gly NH_2
7 Oxytocin (in mammals)

Fig. 3. The hypothalamic polypeptide hormones of vertebrates. (Data from Acher *et al.*, 1970*a*, *b*).

Study of the taxonomic distribution of these variants (Figs. 3, 4) shows that each major group of vertebrates has one (or, more typically, two) of these molecules, either peculiar to it, or shared with groups that are known to be closely related to it (Acher *et al.*, 1970*a*, *b*). While, therefore, these hormones certainly show some degree of variability, they still show a remarkable stability, as Acher *et al.* emphasise. To give just one example, arginine vasotocin persists throughout almost the entire vertebrate series; only in mammals is it replaced by arginine or lysine vasopressin.

So great is this stability that adaptive specialisation of the targets has been of demonstrable importance. For example, the passage of vertebrates from water to dry land resulted in the establishment in amphibians of the Brunn, or water-balance, response, which is a highly adaptive reaction providing for the conservation and economic utilization of water. This response is hormonally regulated, but no new hormone was evolved for this purpose. Instead, use was made of arginine vasotocin, a hypothalamic hormone that is also present in lampreys and hagfish. These animals are the surviving representatives of the great group of agnathans or jawless vertebrates, which constitute an ancient level of vertebrate organization that preceded the appearance of the more advanced and typical jawed forms. We may conclude that arginine vasotocin had existed for 150,000,000 years or more before it became incorporated in this amphibian adaptation. Its ability to mediate the water-balance response depends upon specialisation of the target organs which respond to it: the skin, the kidney, and the bladder.

Evidence of the importance of target specialisation in establishing the selective advantages of molecular variation emerges also from studies of synthetic analogues of the hypothalamic hormones, of which over 200 have been prepared. Berde and Boissonas (1966) conclude that diversity in the physiological actions of certain of these molecules is due to the individuality

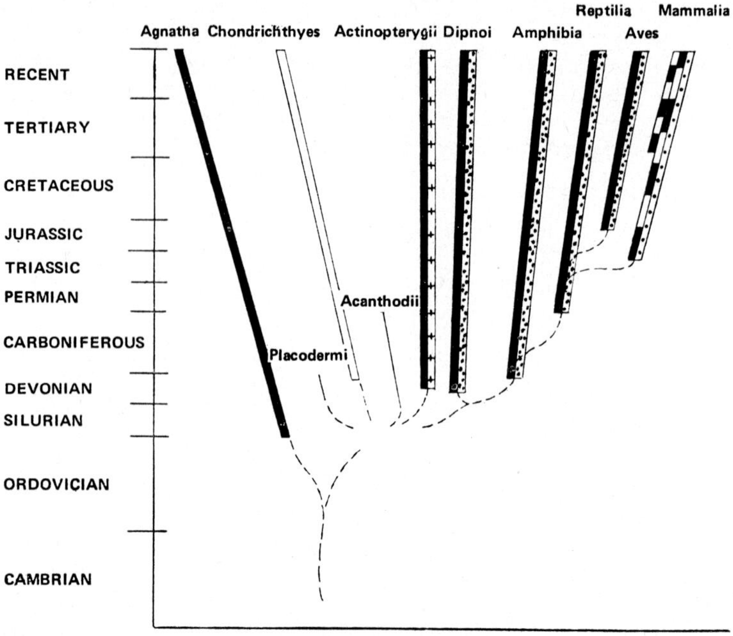

Fig. 4. A phylogenetic tree of the main vertebrate groups, to which has been added the distribution of the hypothalamic polypeptide hormones (data from Acher *et al.*, 1970*a*, *b*). The distribution indicated is an extrapolation from the information available for a small number of living species. Key: arginine vasopressin, long bars; arginine vasotocin, black; glumitocin, white; isotocin, crosses; lysine vasopressin, short bars; mesotocin, close dots; oxytocin, spaced dots.

of the various receptor sites rather than to differences in the primary modes of action of the peptides themselves. But one would expect that the properties of the variant molecules would also be of some importance, and it has indeed been shown that the degree of difference in the physiological actions of these various polypeptides is to some extent correlated with the degree of difference in the physico-chemical properties of the amino acids involved (Sneath, 1966).

All of this presents us with some fascinating evolutionary problems, to which attention has been drawn by Acher *et al.* (1970*a*). They point out that some amino acid substitutions have only small effects upon the biological activity of these molecules, and they suggest that selection pressure would therefore have been unlikely to have operated against such substitutions. Why, then, they ask, should there have been so little molecular variation in this family of hormones? They suggest that either the molecular structure of these hormones may be strictly adapted to functions that are still undiscovered, or else that selection pressure is not the prime cause of their stability.

It cannot, of course, be assumed that small differences in activity would necessarily be immune from selection pressure, but in any case I prefer to argue that hormonal stability can be regarded as a predictable result of the action of natural selection. The validity of this view can, I believe, be illustrated at two levels of analysis, for it can be shown to flow from our knowledge of the mode of action of endocrine systems, as well as from current genetical theory. At the endocrine level, we must take account of hormonal interaction, and of the diverse effects of individual hormones. In such a system there is a strong probability that any mutation will be disadvantageous, for even if it improves one parameter of action, it may disturb others. This is evident from the studies of the structural analogues of the hypothalamic polypeptides to which I have already referred. These have shown that minor modifications of the molecules (for example, the replacement of tyrosine by phenylalanine, which involves the replacement of a phenolic group by a hydrogen atom) may affect many of the biological activities of the molecules, and will then affect them in different ways. If such variants were to arise by mutation, they would have to express themselves in well-adapted phenotypes which must themselves be the result of the long-term action of natural selection. It follows that the complex and disharmonious consequences of such mutations would commonly be disadvantageous and would be rejected by natural selection. This, then, is one factor contributing to stability and conservatism in endocrine systems.

As for the genetical analysis, natural selection acts by bringing genes together into a balanced and efficient gene pool. This is the principle called by Mayr (1963) 'integration' or 'coadaptation' of the genes. The result is an inevitable tendency to conservatism in the gene pool, for the only new mutants that can be incorporated in it will be those that can 'coadapt' harmoniously. Thus we have another factor contributing to the maintenance of stability in endocrine systems.

If, then, we accept that natural selection will favour stability, it follows that departures from it, expressed in the appearance of amino acid substitutions, must themselves have clear selective advantages in order to override this strong tendency to conservatism. In theory, these substitutions might facilitate the action of a hormone in internal environments which must themselves be always evolving. But such minor adjustments, in conjunction with gene duplication, might also lead to molecular diversification and the appearance of new hormones. Admittedly, the application of this view to the interpretation of molecular variation is a matter of current controversy, for it has been argued (Kimura, 1968; Kimura and Ohta, 1971) that the chief cause of molecular evolution is the random fixation of selectively neutral mutations, as a result of genetic drift in small populations. I shall not discuss this controversy here. It must be sufficient to point out that there is still vigorous support for the essentially classical and Darwinian view, that natural selection is of pre-eminent importance in molecular evolution, with genetic drift playing only a minor part (see, for example, Clarke, 1970*a*, *b*).

Future research will have to show how far this selectionist interpretation is adequate to account for the establishment of the active sequences and the amino acid substitutions of growth hormone and the other large polypeptide molecules of endocrine systems. The problem is both biological and chemical, for it will be necessary to determine how far the supposedly 'silent' sequences are truly 'silent' throughout the life history of these molecules, from their synthesis to their final metabolic destruction. It might be thought that it would be easier to evaluate the adaptive significance of the well-defined variations in the small molecules of the hypothalamic hormones. Unfortunately, however, this is not yet so, for we know too little of their functions in the lower vertebrates. There is some evidence of a certain consistency of action throughout the whole vertebrate series, in so far as they can be shown to influence ion transport and the osmotic flow of water, but, as Follett (1970) points out, the actual physiological significance of the hypothalamic polypeptides must remain uncertain until more is known of their levels and half-lives in the blood stream.

A further difficulty is that we cannot be sure of the exact phylogenetic history of these molecules, and authors have understandably differed in their interpretation of this (Vleigenthart and Versteeg, 1967; Peyrot, 1968; Acher *et al.*, 1970*a*). I share the view of Simpson (1964), who contends that if we wish to propose a phylogeny for a particular group of molecular variants, we must be prepared to cross-check our proposition by arranging the molecular data in an entirely independent framework based upon non-molecular evidence. This is just what we cannot do for these molecules, for the precise relationships of the lower vertebrates to each other are very uncertain (Fig. 4). For example, it may be that the three surviving groups of fish (the Chondrichthyes, the Actinopterygii and the Crossopterygii) originated from extinct placoderms or acanthodians (see Barrington, 1968), but we cannot define their ancestry with any precision. We may feel that at least the bony fish (Actinopterygii and Crossopterygii) originated from a common stock. But we have no evidence at all to help us

TABLE II

Partial and complete amino acid sequences of adenohypophysial and synthetic peptides which have melanocyte-stimulating activity

1									H	Ser	Tyr	Ser	Met	Glu	His	Phe	Arg	Trp	Gly	Lys	Pro	Val	Gly	Lys	Lys Arg . . . Phe OH^{39}
2									Acetyl	Ser	Tyr	Ser	Met	Glu	His	Phe	Arg	Trp	Gly	Lys	Pro	Val	NH_2		
3											H	Ser	Met	Glu	His	Phe	Arg	Trp	Gly	Lys	Pro	Val	OH		
4											H	Ser	Met	Glu	His	Phe	Arg	Trp	Gly	Lys	Pro	Met	OH		
5											H	Ser	Met	Glu	His	Phe	Arg	Trp	Gly	Lys	Pro	Met	NH_2		
6												H	Met	Glu	His	Phe	Arg	Trp	Gly	OH					
7														H	His	Phe	Arg	Trp	Gly	OH					
8						H	Asp	Ser	Gly	Pro	Tyr	Lys	Met	Glu	His	Phe	Arg	Trp	Gly	Ser	Pro	Pro	Lys	Asp	OH
9		H	Ala	Glu	Lys	Lys	Asp	Glu	Gly	Pro	Tyr	Arg	Met	Glu	His	Phe	Arg	Trp	Gly	Ser	Pro	Pro	Lys	Asp	OH
10	H Glu . . .	Ala	Ala	Glu	Lys	Lys	Asp	Ser	Gly	Pro	Tyr	Lys	Met	Glu	His	Phe	Arg	Trp	Gly	Ser	Pro	Pro	Lys	Asp	NH_2
11	H Glu . . .	Ala	Ala	Glu	Lys	Lys	Asp	Ser	Gly	Pro	Tyr	Lys	Met	Glu	His	Phe	Arg	Trp	Gly	Ser	Pro	Pro	Lys	Asp^{58}	Lys Arg . . . Glu NH_2^{90}

1, corticotropin; 2, α-MSH; 3, synthetic; 4, dogfish MSH PI; 5, dogfish MSH PII; 6, synthetic; 7, synthetic; 8, ovine β-serine^{2}-MSH; 9, human β-MSH; 10, ovine γ-LPH; 11, ovine β-LPH.
(From Lowry and Chadwick, 1970, *Nature (Lond.)*, *226*, 219, by kind permission of the editors).

decide whether the mesotocin of the crossopterygian lung fish evolved from the isotocin of actinopterygians, or whether the reverse occurred, or whether both evolved separately from an unknown ancestral molecule. Nor can we overlook the possibility that the same molecule might arise independently in different groups by convergent evolution. This has, indeed, been envisaged as one possible explanation for the supposed distribution of oxytocin (Sawyer *et al.*, 1967), and it cannot be lightly dismissed, for structural and functional analysis at the level of whole animal organization is continually producing evidence for the widespread and sometimes quite unpredictable occurrence of convergence.

Obviously, therefore, our analysis of the evolutionary history of the hypothalamic polypeptides is beset with uncertainties, which can only be clarified by much further research, particularly in the field of sequence analysis. Fortunately, however, there are other polypeptide hormones in which the association of molecular diversification with selective advantage is more readily apparent. One example (Li, 1968, 1969*c*) is given by the remarkable structural resemblances between corticotropin and the melanocyte-stimulating hormones (MSH), which share a common heptapeptide sequence (Table II). This resemblance, which is associated with the existence of some melanocyte-stimulating activity in the corticotropin molecule, is strongly suggestive of a common origin of the two molecules. In accord with this is the production of these hormones in two parts of the adenohypophysis which are derived from a common embryonic rudiment, and in cells of the so-called APUD series, which share a number of histochemical and other characters, including mechanisms of synthesis, storage and secretion (Pearse, 1969; Dawson, 1970). The analysis has been complicated by the identification and chemical characterisation of mammalian lipotropin (for references, see Li, 1969*c*). This pituitary secretion (LPH), which is named for its capacity to mobilise lipids, is chemically distinct from corticotropin and all other known adenohypophysial hormones. Yet it contains the heptapeptide sequence that is characteristic of MSH and of corticotropin, and clearly belongs to the same family of molecules.

An important further advance has been the chemical characterisation of the MSH of a dogfish, *Squalus acanthias*. Two active peptides have been identified (Table II), both of them so similar to α-MSH of mammals that there can be little doubt, according to Lowry and Chadwick (1970), of the common origin of these hormones. These authors suggest that this resemblance, taken in conjunction with the high potency of α-MSH and the relative shortness of its molecules, indicates that it is more primitive than β-MSH. The absence of the latter from *Squalus* is in accord with this view, but, as Lowry and Chadwick rightly emphasise, there is need for further elucidation of the MSH sequences of fish before this evidence can be properly evaluated.

It is well known that the pars intermedia of the lamprey secretes a melanocyte-stimulating hormone (Young, 1960). This type of molecule, then, is clearly a very primitive feature of vertebrates, presumably because of the high adaptive value of protective colour change in early agnathans, which are known from their fossil remains to have been sluggishly moving and heavily armoured animals. Whether corticotropin and LPH activities are also present in agnathans is less certain: evidence suggestive of corticotropin secretion in myxinoids has been reported (Fernholm and Olsson, 1969), but so far the only activity that has been convincingly demonstrated in the agnathan pituitary is the secretion of gonadotropin, MSH and arginine vasotocin (Ball and Baker, 1969). It would be premature, therefore, to attempt any further analysis of the molecular evolution which is clearly implicit in this accumulating evidence.

One feature, however, which has received less emphasis than it deserves in this connection is the functional organization of the pituitary (Wingstrand, 1966). In lampreys (Fig. 5A) there is a capillary bed lying between the posterior region of the infundibulum and the pars intermedia. This creates, both in the larva and the adult, a possible path for direct communication between the pars intermedia and the neurosecretory fibres of the hypothalamus, and it is significant that there is a substantial concentration of neurosecretion in this particular region of the infundibulum. There is no obvious sign of such a path between the hypothalamus and the pars distalis, which are separated by relatively avascular connective tissue. Indeed, the secretion of pituitary gonadotropin in male lampreys can continue after the gland has been separated from the hypothalamus and transplanted to the pharyngeal muscle (Larsen, 1969). There is in this respect a marked difference from the teleost *Poecilia*, where the ectopic grafting of the pituitary in the female arrests the maturation of the eggs (Ball *et al.*, 1965). The situation in lampreys thus suggests to me that the association of the adenohypophysis with the floor of the infundibulum, which must have been established very early in vertebrate history, may have been connected with the advantage of securing central regulation of the output of MSH. This suggestion clearly accords well with the conclusion, arising from the quite different evidence of molecular studies, that MSH is a very primitive polypeptide. It must be added, however, that the situation in the myxinoids (the other group of surviving agnathans) is different, for there is no indication in them of a similar functional link between the hypothalamus and either the pars intermedia or the pars distalis (Fig. 5B).

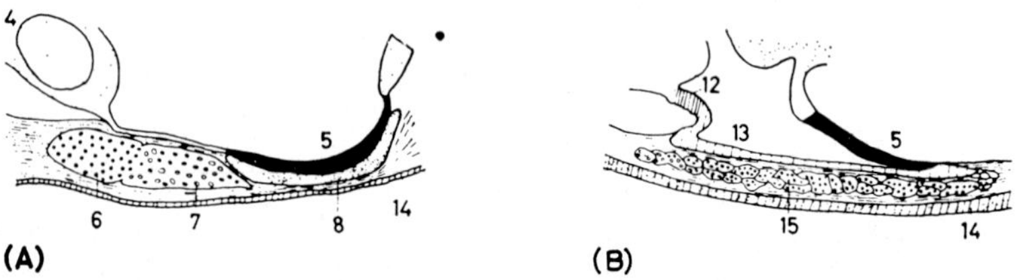

Fig. 5. Median sections of the pituitary gland of adult *Petromyzon* (lamprey) and *Myxine* (hagfish). (A) *Petromyzon.* 4, optic chiasma; 5, neural lobe; 6, 7 and 8, pro-, meso-, and meta-adenohypophysis. (B) *Myxine.* 12, median eminence; 13, infundibulum; 14, epithelium of naso-hypophysial duct; 15, adenohypophysis. (From Wingstrand, 1966, in *The Pituitary Gland, Vol. 1*, pp. 58–126, Butterworths, London; by kind permission of the publishers).

Another example of a suspected evolutionary relationship between polypeptide hormones concerns thyrotropin and the gonadotropins. The evidence, as evaluated by Fontaine (1969*a*, *b*, *c*, *d*,), is too complex to analyse here. It is in large measure biological, and includes the demonstration that mammalian gonadotropins have a thyrotropin-like action in certain lower vertebrates, an effect termed by Fontaine the heterothyrotropic action. His supposition is that the various thyrotropins and gonadotropins evolved from a common ancestral molecule, their divergence being accompanied by a parallel evolution of their target cells (1969*d*). In this case it seems possible that the primary separation may have been particularly associated with (or at least encouraged by) the separation of gnathostomes from agnathans. It is not clearly established that either gonadotropic or thyrotropic activities exist in myxinoids (Fernholm and Olsson, 1969). Gonadotropic activity, however, is present in agnathans (Larsen, 1969), as already mentioned, but it is doubtful whether they secrete thyrotropin. This, at least, is the conclusion drawn in my own laboratory from experimental study of the larva, while studies of the adult by other workers have revealed only traces of thyrotropic activity (Dodd and Dodd, 1969) or else no evidence for its action at all (Larsen and Rosenkilde, 1971). But whatever the precise history of thyrotropin, its appearance (on Fontaine's interpretation) marked a phase of molecular evolution which must have been strongly favoured by natural selection, for it increased the efficiency of action of the thyroid gland without requiring any change in the thyroid hormones themselves.

The biology of growth hormone and prolactin presents problems which also, as I have just suggested, deserve careful attention from this point of view (Li, 1969*a*, *b*; Wilhelmi and Mills, 1969), particularly in view of the remarkable diversity in the actions of prolactin in the main vertebrate groups (Bern and Nicoll, 1969). I should like to conclude, however, by discussing another example, which we have recently been studying in my laboratory. I refer to the hormones of the alimentary tract; a system which has been greatly neglected by comparative endocrinologists. The relevant facts have been discussed in some detail elsewhere (Barrington, 1969, 1971; Barrington and Dockray, 1970), and it will be sufficient now to make two main points. First, the evolution of gnathostomes from the agnathan level of organization involved a considerable increase in the complexity of the alimentary tract (Table III). In agnathans, as exemplified today by lampreys and hagfish, there is an intestine, a liver and a gall-bladder, but there is neither a stomach nor a morphologically identifiable pancreas. Pancreatic zymogen cells are present in the intestinal epithelium, while pancreatic islet tissue is represented by groups of cells in the intestinal wall. These cells, however, consist only of the insulin-secreting B cells. Glucagon-secreting A cells first appear in gnathostomes, as also does the stomach.

TABLE III

Some features of the alimentary tracts of agnathans and gnathostomes

Organ	Functionally related hormone (or hormone-like action) present in alimentary tract	Agnatha	Gnathostomes
Stomach		—	+
	Gastrin		+
Liver		+	+
	Chloretic action		+ (secretin)
Gall bladder		+	+
	Cystokinetic action	+	+ (CCK-PZ)
Pancreas		—	+
Zymogen cells		+	+
	Ecbolic action	+	+ (CCK-PZ)
	Hydrelatic action	+	+ (secretin)
Islet (A) cells		—	+
	Glucagon secretion		+
Islet (B) cells		+	+
	Insulin secretion	+	+

(For explanation, see also text).

The alimentary tract of gnathostomes is, of course, a rich source of polypeptide hormones. These include not only insulin and glucagon, but also secretin, cholecystokinin-pancreozymin (CCK-PZ) and gastrin, which contribute to the regulation of alimentary activity. My second point is that the capacity for secreting biologically active polypeptides is already well-established in agnathans. Insulin is an ancient hormone. It is certainly secreted in agnathans (for references see Barrington, 1968), and there is evidence that it may also be produced in some invertebrates (Davidson *et al.*, 1971). Evidence regarding the other alimentary hormones is less certain, but recent work in my own laboratory and Sweden (Barrington, 1969; Barrington and Dockray, 1970; Nilsson and Fänge, 1970) has shown that the alimentary tract of agnathans contains active materials which can evoke, in the pancreas and gall-bladder of mammals, responses similar to those evoked by secretin (hydrelatic action, or flow of pancreatic fluid) and by CCK-PZ (ecbolic action, or discharge of pancreatic enzymes, and cystokinetic action, or contraction of the gall-bladder). This may seem surprising, considering that agnathans do not possess a fully differentiated pancreas, but, as I have explained, they do have a well developed liver and gall-bladder. It may therefore be that the alimentary hormones were initially evolved to regulate these latter organs rather than the pancreas. This, however, is speculation. Moreover, it must be emphasised that the active materials in the agnathan intestine have not been chemically characterised, nor have their functions been established in the animals from which they have been extracted. All that the present evidence permits one to say is that the ability to secrete biologically active substances, capable of hormone-like activity, seems to be a fundamental and ancient feature of the vertebrate alimentary tract.

The importance of this conclusion in the present context, and in the light of the argument which I have been developing, is that it indicates a situation highly favourable to the diversification of a complex of hormones, evolved through selection pressure to provide for the regulation of the alimentary tract as it advanced in organization. Evidence for such an evolutionary history is, I believe, to be found in structural resemblances between certain of the hormonal molecules. Insulin stands very much on its own, with no structural resemblance to any of the others, an isolation which possibly reflects its supposedly ancient history. Admittedly Schuster (1966) has ingeniously demonstrated the existence of certain matching points on the insulin and glucagon molecules, but it would be straining the evidence to regard this as indicating a common origin of the two, nor does he himself suggest this. As he points out, if there is any significance here, it is likely to be physiological, related perhaps to the existence of common binding sites in the target cells.

Glucagon

His-	Ser-	GLN-	Gly-	Thr-	Phe-	Thr-	Ser-	*Asp*-	TYR-	Ser-	*Lys*-	TYR-	*Leu*-	Asp-	Ser-	ARG-	Arg-	ALA-	Gln-	ASP-	*Phe*-	*Val*-	Gln-	*Trp*-	Leu-	*Met*-	Asn-	Thr
1	2	3	4	5	6	7	8	9	10	11	12	13	14	15	16	17	18	19	20	21	22	23	24	25	26	27	28	29

Secretin

His-	Ser-	ASP-	Gly-	Thr-	Phe-	Thr-	Ser-	*Glu*-	LEU-	Ser-	*Arg*-	LEU-	*Arg*-	Asp-	Ser-	ALA-	Arg-	LEU-	Gln-	ARG-	*Leu*-	*Leu*-	Gln-	*Gly*-	Leu-	*Val*-NH_2
1	2	3	4	5	6	7	8	9	10	11	12	13	14	15	16	17	18	19	20	21	22	23	24	25	26	27

Fig. 6. The primary structure of glucagon (porcine) and secretin (porcine). For those positions in which no substitutions have occurred, ordinary characters are used; where "conservative" substitutions have been observed (*i.e.*, involving residues with similar chemical structure), the residues are italicized; where "radical" substitutions have been found (*i.e.*, involving amino acids with side-chains that are entirely different from a chemical viewpoint), the residues are all in capital letters. (From Weinstein, 1968, *Experientia (Basel)*, *24*, 406, by kind permission of the editors).

There is, however, a striking resemblance between secretin and glucagon (Fig. 6). Secretin has 27 amino acids, and glucagon 29. Of these, the two molecules have no less than 14 in common, including 7 out of 8 in the N-terminal group. This seems more than a coincidence. It indicates rather that glucagon and secretin could well have shared a common origin. Weinstein (1968), who has also discussed this possibility, suggests that glucagon might have appeared in the early Mesozoic. He bases this proposition upon the estimated time for the random occurrence of the necessary mutations, but he does not relate his argument to the morphological evidence to which I have referred. In any case, the early Mesozoic would be much too late, for glucagon is probably present in fish, which means that it must have appeared at least as early as the Devonian. Weinstein, indeed, concedes that the evolution of glucagon from secretin might have occurred in the late Palaeozoic.

Less striking, but still suggestive, are resemblances between gastrin (Fig. 7) and CCK-PZ. The full spectrum of activity of the gastrin molecule requires only the C-terminal tetrapeptide amide (Gregory, 1968*a*, *b*). This group is found also in the cholecystokinin-pancreozymin molecule, a resemblance which is probably associated with the partial immunological identity of this hormone and gastrin I (McGuigan, 1969). Moreover, the cholecystokinetic activity of CCK-PZ depends upon a sulphated tyrosyl residue at the N-terminal end of its C-terminal heptapeptide. This sulphated tyrosyl is also present in one of the two forms of gastrin (gastrin II) of the pig, man, sheep and dog. Here, too, is at least some indication of a possible common ancestry.

I must emphasise that the molecular data are no more than suggestive, and they are certainly not free of complication. One that is quite unexplained is the existence in the skin of certain amphibians of the decapeptide caerulein and phyllocaerulein (Erspamer, 1970), which closely resemble gastrin II and CCK-PZ in their chemical structure and pharmacological activity (Fig. 7). It is impossible to suggest any reason why these particular vertebrates should show this curious feature. It is, however, obvious that if we are to make a satisfactory distinction between chance and design in this complex field of biological organization, we need to know more about the ways in which these various molecules exert their effects on their targets, and, indeed, the structure of the circulating molecules. Is it possible, for example, as Erspamer (1970) suggests, that the molecules extracted from the secretory tissues may be carrier polypeptides, from which smaller polypeptides are split off prior to their release? However, there is already some supplementary information which is helpful to our analysis. Particularly interesting are reports of overlapping in certain of the physiological actions of the hormones of the alimentary tract. For example, gastrin, while primarily a regulator of gastric acid secretion, stimulates also the flow of fluid and enzymes from the pancreas, and of bile from the liver. CCK-PZ, while primarily acting upon the pancreatic enzyme output and gall-bladder contraction, also has some influence upon the output of gastric acid and of pancreatic fluid. Similarly, the structural resemblances between secretin and glucagon are paralleled by some overlap in action.

Such overlapping is to be expected in hormones showing common patterns of molecular structure, just as we find it in corticotropin and MSH. But it must be added that this overlapping of activity in the alimentary hormones is not always associated with common features of structure. Moreover, some of the overlapping responses are very weak, very variable, and, like so much of our endocrinological data, known from only a very few species. Their physiological significance is therefore still obscure, especially as we do not know the normal plasma levels of the circulating hormones. Harper (1967) rightly warns that an additional function should not be attributed to a hormone unless it can be demonstrated within the same dose range as that producing the primary effect of the hormone. But he concedes, following up some of my own thoughts, that the alimentary endocrine system may be at a primitive level of development, and that this may be reflected in an incomplete chemical and functional differentiation. Thus any one of the hormones may show characteristics of other members which, even if only demonstrable at unphysiological dose levels, may yet be of profound evolutionary significance. For myself, I believe that even in this very imperfectly studied field, there is already enough evidence, structural, anatomical, molecular, and functional, to give some weight to the suggestion that the evolution of hormones by molecular diversification must have played a part in the establishment of the endocrine system of the vertebrate alimentary tract.

You will have gathered that my approach to the analysis of endocrine organization is a thoroughgoing Darwinian one, for I am convinced that it is important to view endocrine

(1) Pyr-Gln-Asp-Tyr(SO_3H)-Thr-Gly-Trp-Met-Asp-Phe-NH_2

(2) Pyr-Glu-Tyr(SO_3H)-Thr-Gly-Trp-Met-Asp-Phe-NH_2

(3) -Asp-Tyr(SO_3H)-Met-Gly-Trp-Met-Asp-Phe-NH_2

(4) -Tyr(SO_3H)-Gly-Trp-Met-Asp-Phe-NH_2

Fig. 7. Structural formulae of (1) caerulein (skin of certain anurans), (2) phyllocaerulein (skin of the frog *Phyllomedusa*), (3) C-terminal octapeptide of cholecystokinin-pancreozymin (pig), and (4) C-terminal hexapeptide of gastrin II. (From Erspamer, 1970, *Gut*, *11*, 79, by kind permission of the editors).

systems as complexes of hormones and targets which have been jointly subject to the force of selection pressure. In Darwin's study, which is still preserved in his house as it was during his life, a portrait of Charles Lyell hangs in a prominant position. It is a tribute to the importance which Darwin attached to the proposition, initially formulated by James Hutton, and expounded further by Lyell, that the past and present constitute an interrelated continuum. The present, argued Hutton, is the key to the past, and to this dictum I would add that the past is also the key to the present. It is the peculiar privilege of the biologist to be free to range over vast periods of time in the search for these keys, building up in the process a framework of knowledge in which past and present are inextricably linked. I believe that the exercise of this privilege is essential for securing a full understanding of the endocrine organization of any group of animals, and not least of those most complicated creatures: the mammals, and man.

REFERENCES

ACHER, R. (1969*a*): Neurohypophysin and neurohypophysial hormones. *Proc. roy. Soc. Ser. B, 170*, 7.

ACHER, R. (1969*b*): Evolution des structures des hormones neurohypophysaires. *Coll. int. Cent. nat. Rech. sci., 177*, 3.

ACHER, R., CHAUVET, J. and CHAUVET, M.-T. (1970*a*): Phylogeny of the neurohypophysial hormones. The avian active peptides. *Europ. J. Biochem., 17*, 509.

ACHER, R., CHAUVET, J. and CHAUVET, M.-T. (1970*b*): Molecular evolution of the neurohypophysial hormones: the active peptides of a primitive bony fish *Polypterus bichir. FEBS Letters, 11*, 332.

ASSAN, R., TCHOBROUTSKY, G. and ROSSELIN, G. (1969): Caractérisation radio-immunologique de glucagon dans les tissus digestifs de diverses espèces animales. *Path. Biol., 17*, 747.

BALL, J. N. and BAKER, B. I. (1969): The pituitary gland: anatomy and physiology. In: *Fish Physiology, Vol. 2*, Chapter 1, pp. 1–110. Editors: W. S. Hoar and D. J. Randall. Academic Press, London and New York.

BALL, J. N., OLIVEREAU, M., SLICHER, A. M. and KALLMAN, K. D. (1965): Functional capacity of ectopic pituitary transplants in the teleost *Poecilia formosa* with a comparative discussion on the transplanted pituitary. *Phil. Trans. B, 249*, 69.

BARRINGTON, E. J. W. (1967): *Invertebrate Structure and Function.* Nelson, London.

BARRINGTON, E. J. W. (1968): Phylogenetic perspectives in vertebrate endocrinology. In: *Perspectives in Endocrinology*, Chapter 1, pp. 1–46. Editors: E. J. W. Barrington and C. B. Jørgensen. Academic Press, London and New York.

BARRINGTON, E. J. W. (1969): Unity and diversity in comparative endocrinology. *Gen. comp. Endocr., 13*, 482.

BARRINGTON, E. J. W. (1971): Evolution of hormones. In: *Biochemical Evolution and the Origin of Life*, pp. 174–190. Editor: E. Schoffeniels. North-Holland Publishing Co., Amsterdam.

BARRINGTON, E. J. W. (1972): Biochemistry of primitive deuterostomians. In: *Chemical Zoology.* Editors: M. Florkin and B. Scheer. In press.

BARRINGTON, E. J. W. and DOCKRAY, G. J. (1970): The effect of intestinal extracts of lampreys *(Lampetra fluviatilis* and *Petromyzon marinus)* on pancreatic secretion in the rat. *Gen. comp. Endocr., 14*, 170.

BERDE, B. and BOISSONAS, R. A. (1966): Synthetic analogues and homologues of the protein pituitary hormones. In: *The Pituitary Gland, Vol. 3*, pp. 624–661. Editors: G. W. Harris and B. T. Donovan. Butterworths, London.

BERN, H. A. and NICOLL, C. S. (1969): The taxonomic specificity of prolactins. *Coll. int. Cent. nat. Rech. sci., 177*, 193.

BOTTICELLI, C. R., HISAW, F. L. and WOTIG, H. H. (1961): Estrogens and progesterone in the sea urchin *(Strongylocentrotus franciscanus)* and pecten *(Pecten henricius). Proc. Soc. exp. Biol. (N.Y.), 106*, 887.

CLARKE, B. C. (1970*a*): Darwinian evolution of proteins. *Science, 168*, 1009.

CLARKE, B. C. (1970*b*): Selective constraints on amino-acid substitutions during the evolution of proteins. *Nature (Lond.), 228*, 159.

DAVIDSON, J. K., FALKMER, S., MEHROTA, B. K. and WILSON, S. (1971): Insulin assays and light mi-

croscopical studies of digestive organs in Protostomian and Deuterostomian species and in Coelenterates. *Gen. comp. Endocr.*, *17*, 388.
DAWSON, I. (1970): The endocrine cells of the gastrointestinal tract. *Histochem. J.*, *2*, 527.
DODD, J. M. and DODD, M. H. I. (1969): Phylogenetic specificity of thyroid stimulating hormone with special reference to the Amphibia. *Coll. int. Cent. nat. Rech. sci.*, *177*, 277.
EATON, J. E. and FRIEDEN, E. (1969): Primary mechanisms of thyroid hormone control of amphibian metamorphosis. *Gen. comp. Endocr.*, *Suppl. 2*, 398.
ERSPAMER, V. (1970): Progress report: Caerulein. *Gut*, *11*, 79.
ETKIN, W. (1968): Hormonal control of amphibian metamorphosis. In: *Metamorphosis*. Editors: W. Etkin and L. I. Gilbert. Appleton-Century-Crofts, New York.
FERNHOLM, B. and OLSSON, R. (1969): A cytopharmacological study of the *Myxine* adenohypophysis. *Gen. comp. Endocr.*, *13*, 336.
FOLLETT, B. K. (1970): Effects of neurohypophysial hormones and their synthetic analogues on lower vertebrates. *Int. Enc. Pharmacol. Ther.*, *Section 41*, *Vol. 1*, 321.
FONTAINE, Y. A. (1969*a*): La spécificité zoologique des protéines hypophysaires capables de stimuler la thyroïde. *Acta endocr.*, *Suppl. 130*, 1.
FONTAINE, Y. A. (1969*b*): Studies on the heterothyrotropic activity of preparations of mammalian gonadotropins of teleost fish. *Gen. comp. Endocr.*, *Suppl. 2*, 417.
FONTAINE, Y. A. (1969*c*): La spécificité d'action des hormones thyrétropes. *Coll. int. Cent. nat. Rech. sci.*, *177*, 267.
FONTAINE, Y. A. (1969*d*): Thyrotropins and related glycoproteins. In: *Progress in Endocrinology*, pp. 453–457. Editor: C. Gual. ICS 184, Excerpta Medica, Amsterdam.
GESCHWIND, I. I. (1969): The main lines of evolution of the pituitary hormone. *Coll. int. Cent. nat. Rech. sci.*, *177*, 385.
GOLDING, D. W. (1970): The infracerebral gland in *Nephtys* – a possible neuroendocrine complex. *Gen. comp. Endocr.*, *14*, 114.
GORBMAN, A. (1955): Some aspects of the comparative biochemistry of iodine utilization and the evolution of thyroidal function. *Physiol. Rev.*, *35*, 336.
GOTTFRIED, H. and DORFMAN, R. I. (1970): Steroids of invertebrates. V. *Gen. comp. Endocr.*, *15*, 120.
GREGORY, R. A. (1968*a*): Recent advances in the physiology of gastrin. *Proc. roy. Soc. Ser. B*, *170*, 81.
GREGORY, R. A. (1968*b*): The chemistry of gastrin. In: *The Physiology of Gastric Secretion*, pp. 280–281. Editors: L. S. Semb and J. Myren. Universitets Forlaget, Oslo.
HARPER, A. A. (1967): Hormonal control of pancreatic secretion. In: *Handbook of Physiology*, Section 6. The Alimentary Canal, Vol. II, 969. American Physiological Society, Washington, D.C.
HAYASHIDA, T. and LAGIOS, M. D. (1969): Fish growth hormone: a biological, immunochemical and ultrastructural study of sturgeon and paddlefish pituitaries. *Gen. comp. Endocr.*, *13*, 403.
HELLER, H. and SPICKETT, S. G. (1967): The polymorphism of the neurohypophysial hormones. *Mem. Soc. Endocr.*, *15*, 89.
HIGHNAM, K. L. and HILL, L. (1969): *The Comparative Endocrinology of the Invertebrates*. Arnold, London.
KIMURA, M. (1968): Evolutionary rate at the molecular level. *Nature (Lond.)*, *217*, 624.
KIMURA, M. and OHTA, T. (1971): Protein polymorphism as a phase of molecular evolution. *Nature (Lond.)*, *229*, 467.
KLOSTERMEYER, H. and HUMBEL, R. E. (1966): The chemistry and biochemistry of insulin. *Angew. Chem. int. Ed. Engl.*, *5*, 807.
KRISHNAKUMARAN, A. and SCHNEIDERMAN, H. A. (1969): Induction of molting in Crustacea by an insect molting hormone. *Gen. comp. Endocr.*, *12*, 515.
LARSEN, L. O. (1969): Hypophyseal functions in river lampreys. *Gen. comp. Endocr., Suppl. 2*, 522.
LARSEN, L. O. and ROSENKILDE, P. (1971): Iodine metabolism in normal, hypophysectomized, and thyrotropin-treated river lampreys *(Lampetra fluviatilis)*. *Gen. comp. Endocr.*, *17*, 94.
LI, C. H. (1968): Current concepts on the chemical biology of pituitary hormones. *Perspect. Biol. Med.*, *11*, 498.
LI, C. H. (1969*a*): Comparative chemistry of pituitary lactogenic hormones. *Gen. comp. Endocr.*, *Suppl. 2*, 1.
LI, C. H. (1969*b*): Recent studies on the chemistry of human growth hormone. *Coll. int. Cent. nat. Rech. sci.*, *177*, 175.
LI, C. H. (1969*c*): β-Lipotropin, a new pituitary hormone. *Coll. int. Cent. nat. Rech. sci.*, *177*, 93.

LOWRY, P. J. and CHADWICK, A. (1970): Interrelations of some pituitary hormones. *Nature (Lond.)*, *226*, 219.
MCGUIGAN, J. E. (1969): Studies of the immunochemical I specificity of some antibodies to human gastrin. *Gastroenterology*, *56*, 429.
MAJOR, C. W., HANEGAN, J. L. and ANOLI, L. (1969): Organic binding of iodide in nemertean mucus, *in vivo* and *in vitro*. *Comp. Biochem. Physiol.*, *28*, 1153.
MAYR, E. (1963): *Animal Species and Evolution*, pp. 295–296. Oxford University Press, London.
NILSSON, A. and FÄNGE, R. (1970): Digestive proteases in the cyclostome *Myxine glutinosa* (L.). *Comp. Biochem. Physiol.*, *32*, 237.
PEARSE, A. G. E. (1969): The cytochemistry and ultrastructure of polypeptide hormone-producing cells of the APUD series, and the embryologic, physiologic and pathologic implications of the concept. *J. Histochem. Cytochem.*, *17*, 303.
PEYROT, A. (1968): Alcuni aspetti filogenetici dell'endocrinologia comparata dei vertebrati. *Boll. Zool.*, *35*, 257.
ROCHE, J., RAMETTA, G. and VARRONE, S. (1964): Métabolisme de l'iode et formation d'iodothyronines (T_3 and T_4) au cours de la régeneration de la tunique chez une Ascidie, *Ciona intestinalis* L. *Gen. comp. Endocr.*, *4*, 277.
SALVATORE, G. (1969): Thyroid hormone biosynthesis in Agnatha and Protochordata. *Gen. comp. Endocr.*, *Suppl. 2*, 535.
SANDOR, T. (1969): A comparative survey of steroids and steroidogenic pathways throughout the vertebrates. *Gen. comp. Endocr.*, *Suppl. 2*, 284
SAWYER, W. H. (1966): Neurohypophysial principles of vertebrates. In: *Pituitary Gland, Vol. 3*, pp. 307–329. Editors: G. W. Harris and B. T. Donovan. Butterworths, London.
SAWYER, W. H., FREER, R. J. and TSENG, TSUI-CHIN (1967): Characterization of a principle resembling oxytocin in the pituitary of the holocephalan ratfish *(Hydrolagus colliei)* by partition chromatography on sephadex columns. *Gen. comp. Endocr.*, *9*, 31.
SCHILDKNECHT, H., SIEWARDT, R. and MASCHWITZ, V. (1966): A vertebrate hormone as defense substance of the water-beetle *(Dytiscus marginalis)*. *Angew. Chem. int. Ed. Engl.*, *5*, 421.
SCHILDKNECHT, H., BIRNINGER, H. and MASCHWITZ, V. (1967): Testosterone as a protective agent of the water-beetle *Ilybius*. *Angew. Chem. int. Ed. Engl.*, *6*, 558.
SCHUSTER, T. (1966): Possible structural similarity between insulin and glucagon. *Nature (Lond.)*, 209, 302.
SIMPSON, G. C. (1964): Organisms and molecules in evolution. *Science*, *146*, 1535.
SNEATH, P. H. A. (1966): Relations between chemical structure and biological activity in peptides. *J. theor. Biol.*, *12*, 157.
TATA, J. R. (1969): The action of thyroid hormones. *Gen. comp. Endocr.*, *Suppl. 2*, 385.
VLIEGENTHART, J. F. G. and VERSTEEG, D. H. G. (1967): The evolution of the vertebrate neurohypophysial hormones in relation to the genetic code. *J. Endocr.*, *38*, 3.
WEINSTEIN, B. (1968): On the relationship between glucagon and secretin. *Experientia (Basel)*, *24*, 406.
WILHELMI, A. E. and MILLS, J. B. (1969): The chemistry of the growth hormone of several species. *Coll. int. Cent. nat. Rech. sci.*, *177*, 165.
WINGSTRAND, K. G. (1966): Comparative anatomy and evolution of the hypophysis. In: *The Pituitary Gland, Vol. 1*, pp. 58–126. Editors: G. W. Harris and B. T. Donovan. Butterworths, London.
YOUNG, J. Z. (1960): *The Life of Vertebrates, 2nd ed.* Clarendon Press, Oxford.

II. Chemistry, immunochemistry and comparative aspects

ASPECTS OF THE COMPARATIVE CHEMISTRY OF HUMAN PITUITARY GROWTH HORMONE AND CHORIONIC SOMATOMAMMOTROPIN*

CHOH HAO LI

The Hormone Research Laboratory, University of California,
San Francisco, Calif., U.S.A.

Many contributions for the next three days in this Symposium will center on the chemistry and biology of human pituitary growth hormone (HGH) and human chorionic somatomammotropin (HCS). By way of introducing the scientific programme, I take this opportunity to present some of our recent studies on the chemistry of these two molecules. In addition, ovine prolactin (LTH) is known to be a growth hormone in reptiles and amphibians, and the following discussion will also include this molecule.

Primary structures

The presence of lactogenic activity in the human placenta was first described by Ehrhardt in 1936. Later studies of Josimovich and MacLearen (1962), Kaplan and Grumbach (1964), Friesen (1965) and Florini *et al.* (1966) showed that the human placenta contains a protein hormone possessing biological properties in common with those of HGH. The hormone has since been designated (Li *et al.*, 1968) as human chorionic somatomammotropin (HCS).

HCS is a protein of molecular weight 21,000 with 190 amino acids. The complete amino acid sequence of HCS has only recently been elucidated (Li *et al.*, 1971) as shown in Figure 1. In a direct comparison of the proposed structure of HCS with that of HGH (Li *et al.*, 1969; Li and Dixon, 1971), it is striking that 160 of the 190 amino acid residues occupy identical positions. On the basis of chemical similarity, the 30 remaining residues contain 19 highly acceptable, 4 relatively acceptable and 7 unacceptable replacements. The total homology, including all acceptable replacements, between HCS and HGH molecules is thus 96%.

At the time (December 1970) of the completion of the amino acid sequence of HCS, it was observed that the remarkable similarity between its primary structure and that of HGH could be enhanced by rearrangement of a peptide segment in the HGH structure. This led us to reinvestigate our proposed (Li *et al.*, 1969) amino acid sequence of HGH. Results of the reinvestigation (Li and Dixon, 1971) gave rise to a revised structure shown in Figure 2. It may be noted that a segment (pentadecapeptide) of the sequence containing the single tryptophan residue was misplaced and that two amino acid residues (Leu, Arg) must be added to the original 188 amino acids in the HGH molecule. In addition, the proline (position 130) and glycine (position 132) residues were incorrectly positioned, having been shown in reversed order to that presented in Figure 2. Moreover, Asn is at position 47 instead of Asp, Gln at position 49 instead of Glu, Gln at position 90 instead of Glu and Gln at position 121 instead of Glu.

* These investigations were supported in part by the American Cancer Society, the Allen Foundation and the Geffen Foundation.

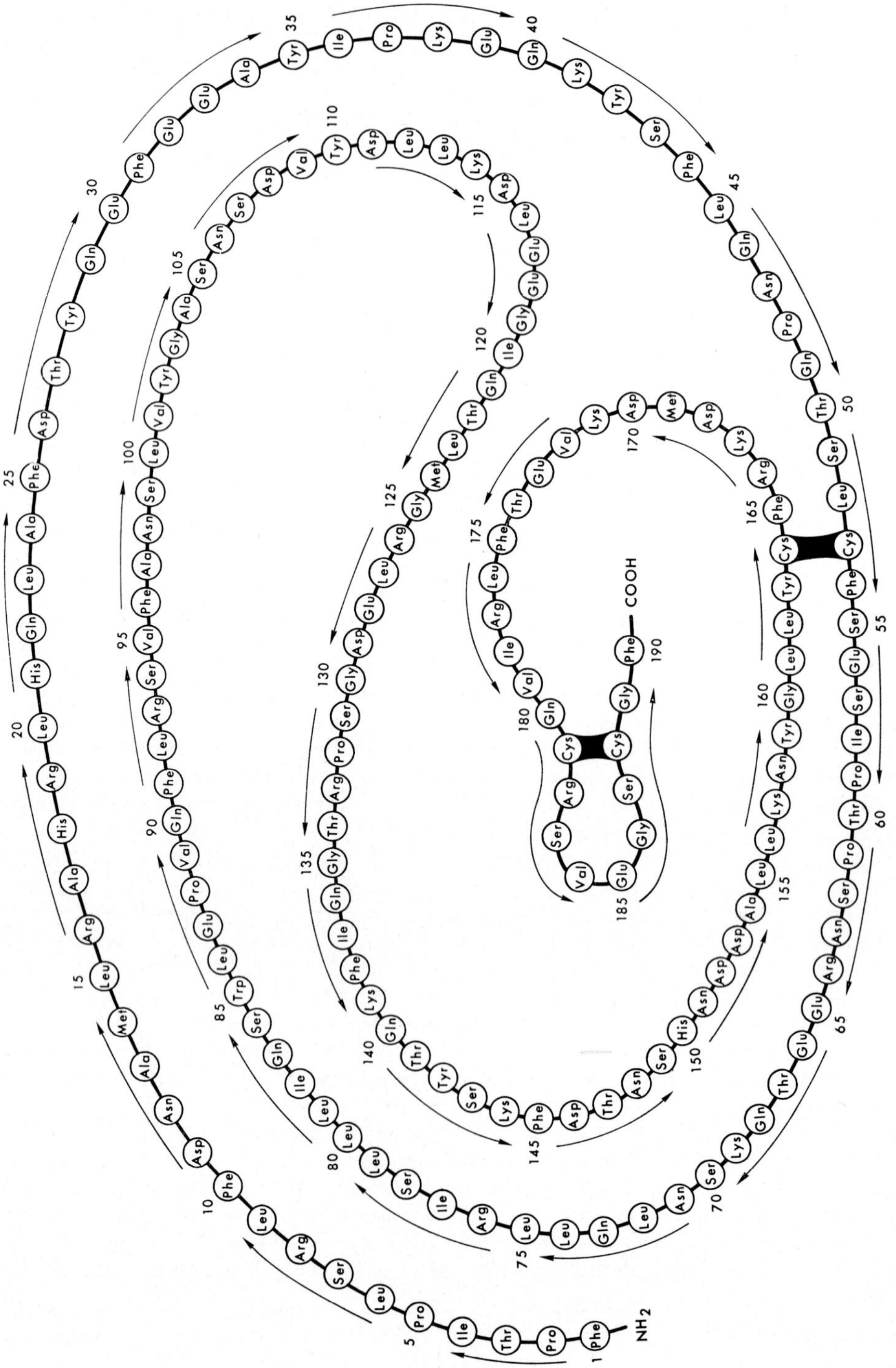

Fig. 1. The amino acid sequence of HCS.

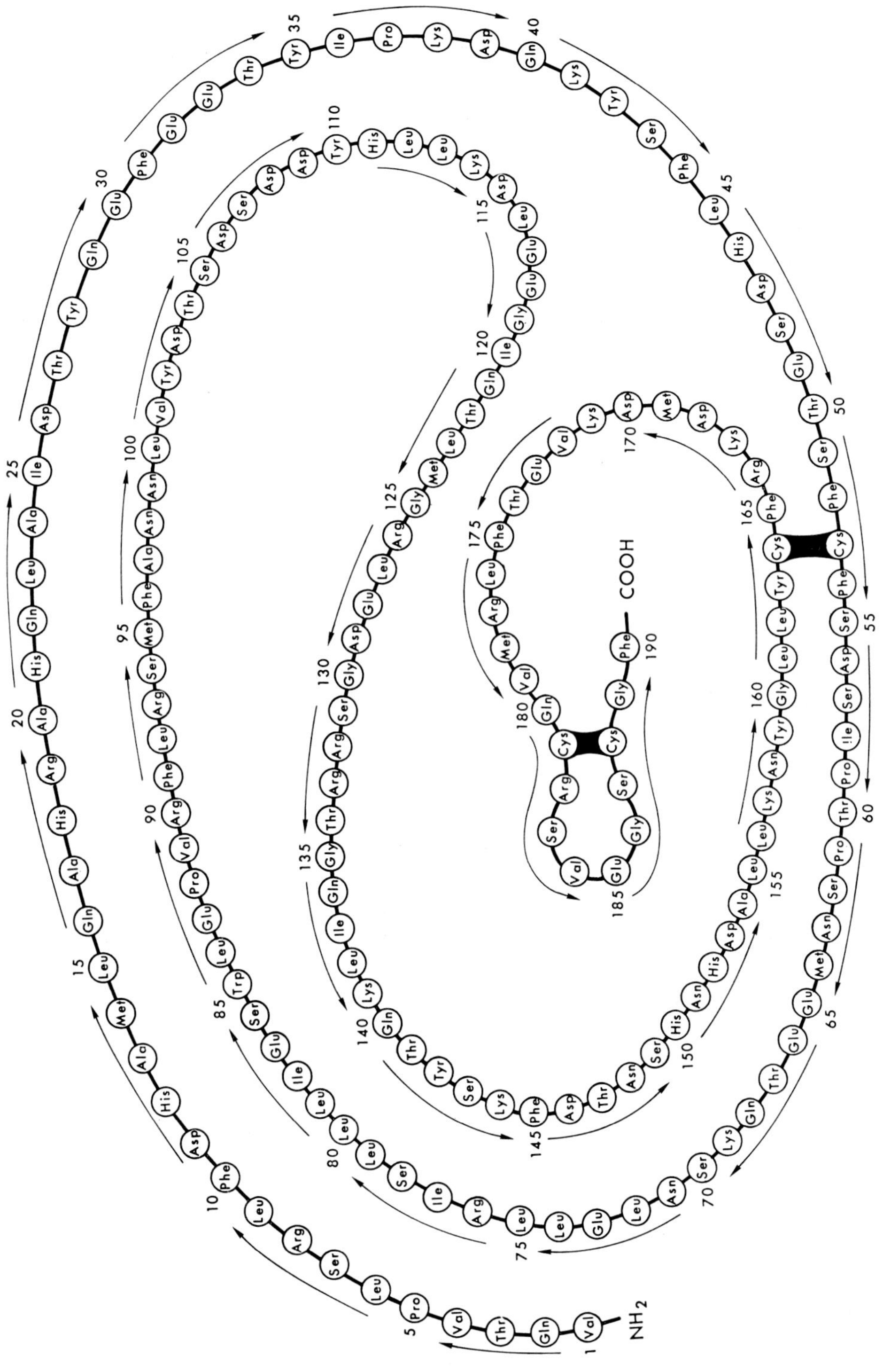

Fig. 2. The amino acid sequence of HGH.

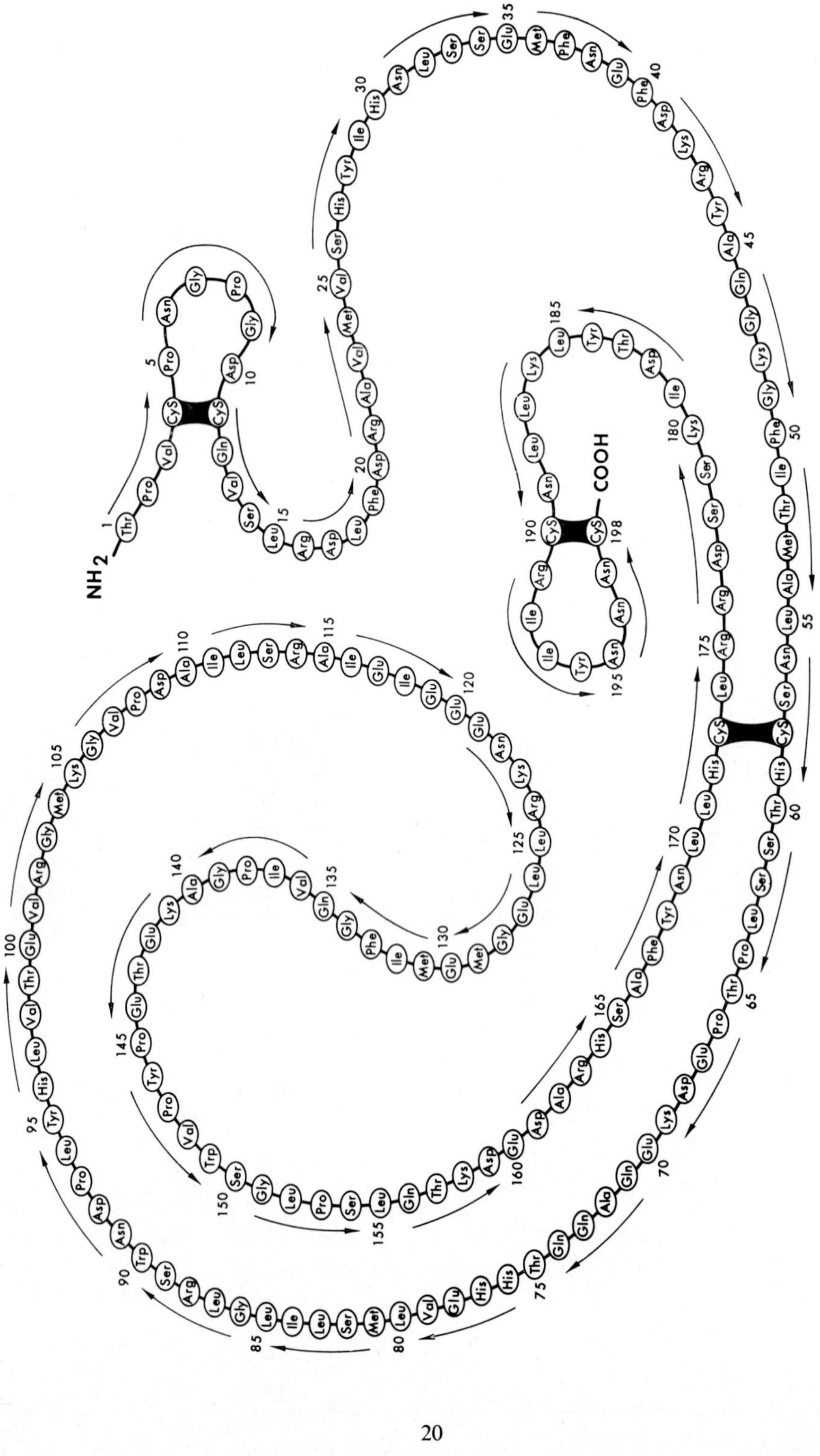

Fig. 3. The amino acid sequence of ovine LTH.

Ovine prolactin is a protein of molecular weight (Li *et al.*, 1957) 23,300 and isoelectric point (Li *et al.*, 1940) at pH 5.7. It consists of a single polypeptide chain (Li, 1957; Li and Cummins, 1958) with two tryptophan residues and three disulfide bridges (Li, 1949). The complete amino acid sequence of LTH has been elucidated (Li *et al.*, 1969, 1970) as shown in Figure 3. It contains 198 amino acids with threonine at the NH_2-terminus and half-cystine the COOH end. The three disulfide bridges are formed between residues 4 and 11, betwe residues 190 and 198 and between 58 and 173. The two tryptophan residues are in posit 90 and 149.

Circular dichroism spectra

The circular dichroism (CD) spectra (Bewley and Li, 1971*a*, *b*) of HGH, HC pH 8.4, in the region of amide bond absorption, give two strong bands with n at 221 and 209 nm. Calculations from these spectra indicate that HGH and α-helix whereas HCS has a lower value (45 %).

Figure 4 gives CD spectra in the region of side chain absorption (Bewley and Li, 1971*a*, *b*). Although the spectra of HGH and LTH shows considerable similarity in the conformations of these two molecules, they are not completely identical. Both proteins show an asymmetric positive peak above 290 nm. This asymmetry is probably due to overlapping with negative bands below 266 nm. The shape of the LTH spectrum suggests that the positive band in this protein overlaps with an additional negative band, also above 290 nm, which is not found in HGH. Simple graphical subtraction of the HGH spectrum from that of LTH between 288 and 310 nm produces a weak negative band with a maximum at 291–292 nm. Since this 'hidden' band in LTH lies above 290 nm, it may tentatively be assigned to tryptophan, along with the positive band. This would be consistent with the fact that LTH contains two tryptophan residues, one showing a positive CD band and the other a negative one, while the single tryptophan in HGH gives rise only to a positive band. The spectrum of HCS in the same region of side chain absorption is quite different from those of HGH and LTH. The two negative peaks

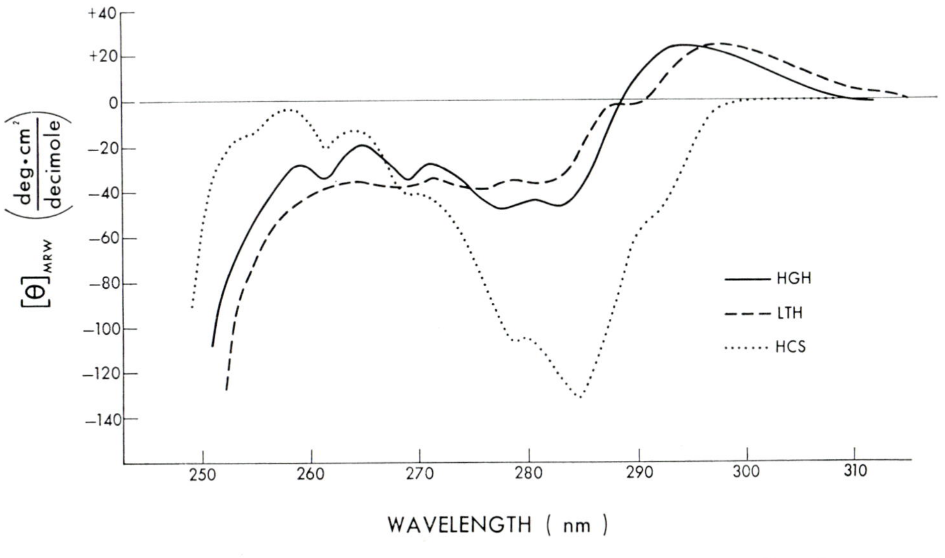

Fig. 4. Side chain circular dichroism spectra of HGH, HCS and LTH in buffers of pH 8.4.

2 5 15 20 29

LTH: -Pro-Val-Cys-Pro Leu-Arg-Asp-Leu-Phe-Asp-Arg-Ala-Val-Met-Val-( )-Ser-His-Tyr-Ile-

HGH: NH_2-Phe-Pro-Thr-Ile-Pro-Leu-Ser-Arg-Leu-Phe-Asp-Asn-Ala-( )-Met-Leu-Arg-Ala-His-Arg-Leu-

HCS: NH_2-Val-Gln-Thr-Val-Pro-Leu-Ser-Arg-Leu-Phe-Asp-His-Ala-( )-Met-Leu-Gln-Ala-His-Arg-Ala-

1 10 20

30 40 41

LTH: -His-Asn-Leu-Ser-Ser-Glu-Met-Phe-Asn-Glu-Phe- -Asp-Lys-

HGH: -His-Gln-Leu-Ala-Phe-Asp-Thr-Tyr-Gln-Glu-Phe-Glu-Glu-Ala-Tyr-Ile-Pro-Lys-Glu-Gln-

HCS: -His-Gln-Leu-Ala-Ile-Asp-Thr-Tyr-Gln-Glu-Phe-Glu-Glu-Thr-Tyr-Ile-Pro-Lys-Asp-Gln-

30 40

45 50 60 65

LTH: -Arg-Tyr-Ala Phe-Ile-Thr-Met-Ala-Leu-Asn-Ser-( )-Cys-His-Thr-Ser-Ser-Leu-Pro-Thr-

HGH: -Lys-Tyr-Ser-Phe-Leu-Gln-Asn-Pro-Gln-Thr-Ser-Leu-Cys-Phe-Ser-Glu-Ser-Ile-Pro-Thr-

HCS: -Lys-Tyr-Ser-Phe-Leu-His-Asp-Ser-Glu-Thr-Ser-Phe-Cys-Phe-Ser-Asp-Ser-Ile-Pro-Thr-

50 60

70 80 84

LTH: -Pro-Glu-Asp-Lys-Glu-Gln-Ala-Gln-Gln-Thr-His-His-Glu-Val-Leu-( )-Met-Ser-Leu-Ile-

HGH: -Pro-Ser-Asn-Arg-Glu-Glu-Thr-Gln-Lys-Ser-Asn-Leu-Gln-Leu-Leu-Arg-Ile-Ser-Leu-Leu-

HCS: -Pro-Ser-Asn-Met-Glu-Glu-Thr-Gln-Lys-Ser-Asn-Leu-Glu-Leu-Leu-Arg-Ile-Ser-Leu-Leu-

70 80

85 90

LTH: -Leu-Gly-Leu-Arg-Ser-Trp-Asn-Asp-Pro-Leu-

HGH: -Leu-( )-Ile-Gln-Ser-Trp-Leu-Glu-Pro-Val-Gln-Phe-Leu-Arg-Ser-Val-Phe-Ala-Asn-Ser-Leu-

HCS: -Leu-( )-Ile-Glu-Ser-Trp-Leu-Glu-Pro-Val-Arg-Phe-Leu-Arg-Ser-Met-Phe-Ala-Asn-Asn-Leu-

90 100

95 100

LTH: -Tyr-His-Leu-Val-Thr-Glu-Val-

HGH: -Val-Tyr-Gly-Ala-Ser-Asn-Ser-Asp-Val-Tyr-Asp-Leu-Leu-Lys-Asp-Leu-Glu-Glu-Gly-Ile-

HCS: -Val-Tyr-Asp-Thr-Ser-Asp-Ser-Asp-Asp-Tyr-His-Leu-Leu-Lys-Asp-Leu-Glu-Glu-Gly-Ile-

110 120

125 135

LTH: -Arg-Leu-Leu-Glu-Gly-Met- -Gly-Gln-Val-Ile-

HGH: -Gln-Thr-Leu-Met-Gly-Arg-Leu-Glu-Asp-Gly-Ser-Pro-Arg-Thr-Gly-Gln-Ile-Phe-Lys-Gln-

HCS: -Gln-Thr-Leu-Met-Gly-Arg-Leu-Glu-Asp-Gly-Ser-Arg-Arg-Thr-Gly-Gln-Ile-Leu-Lys-Gln-

130 140

160 169

LTH: -Gln-Thr-Lys-Asp-Glu-Asp-Ala-Arg-His-Ser-Ala-Phe-Tyr-Asn-

HGH: -Thr-Tyr-Ser-Lys-Phe-Asp-Thr-Asn-Ser-His-Asn-Asp-Asp-Ala-Leu-Leu-Lys-Asn-( )-Tyr-Gly-

HCS: -Thr-Tyr-Ser-Lys-Phe-Asp-Thr-Asn-Ser-His-Asn-His-Asp-Ala-Leu-Leu-Lys-Asn-( )-Tyr-Gly-

150 160

170 180 189

LTH: -Leu-Leu-His-Cys-Leu-Arg-Arg-Asp-Ser-Ser-Lys-Ile-Asp-Thr-Tyr-Leu-Lys-Leu-Leu-Asn-

HGH: -Leu-Leu-Tyr-Cys-Phe-Arg-Lys-Asp-Met-Asp-Lys-Val-Glu-Thr-Phe-Leu-Arg-Ile-Val-Gln-

HCS: -Leu-Leu-Tyr-Cys-Phe-Arg-Lys-Asp-Met-Asp-Lys-Val-Glu-Thr-Phe-Leu-Arg-Met-Val-Gln-

170 180

190 198

LTH: -Cys-Arg-Ile-Ile-Tyr-Asn-Asn-Asn-Cys-COOH

HGH: -Cys-Arg-Ser-Val-( )-Glu-Gly-Ser-Cys-Gly-Phe-COOH

HCS: -Cys-Arg-Ser-Val-( )-Glu-Gly-Ser-Cys-Gly-Phe-COOH

190

shown by HCS at 269 and 261.5 nm are almost identical to corresponding peaks in the HGH spectrum which have been assigned to phenylalanine residues. These peaks are almost entirely absent from the spectrum of LTH. At the present time we cannot make definite chromophore assignments to the negative dichroism in the HCS spectrum between 270 and 290 nm. The two peaks at 279 and 284.5 nm are probably due largely to tyrosine residues although both tryptophan and the disulfide bonds undoubtedly contribute to this region also. It may be pointed out that it is in this region of the spectrum that HCS shows the greatest differences from the other two hormones. The negative shoulder above the 290 nm in the spectrum of HCS is almost certainly due to the protein's single tryptophan residue. It corresponds very closely to the 'hidden' tryptophan band in LTH, while differing from the positive tryptophan in HGH.

Structural comparison of LTH, HCS and HGH

To examine the structures for areas of homology, we have aligned the three sequences according to the best possible fit of certain reference residues (Bewley and Li, 1971*c*). Half-cysteine, tryptophan, tyrosine, histidine and proline were used as references because of their limited content in these proteins and their low relative mutability. Occasionally, the introduction of a gap into one or the other sequence was required to obtain the best alignment.

In Figure 5, an asterisk has been placed above those residue positions which contain the same amino acid in all three structures. There are 50 such positions. These common positions would seem to be more or less randomly distributed, there being no particular area(s) in which they are clearly concentrated. They are also about equally distributed between hydrophilic and hydrophobic residues. However, in terms of the total content in all three proteins, some residue types appear in the common positions much more than others. For example, with the exception of the small disulfide ring near the amino terminal of LTH, each half-cysteine residue is homologous with a corresponding half-cysteine in the other two molecules. Thus, 12 of the 14 cysteine residues, or 86% of the total cysteine content, appear in one or another homologous position in the three hormones. Similarly, 75% of the tryptophan residues (1 in each sequence, or 3 out of a total content of 4) appear in a single homologous position. Other residues whose positions appear to be conserved by identity are: proline, 50%; leucine, 45%; tyrosine, 39%; histidine, 33%; phenylalanine, 30%; alanine and arginine, 27%; serine, aspartic acid and glycine, each 24%. Although glutamic acid, methionine, threonine, glutamine and lysine also occur in common positions in all three sequences, the percentages of their total contents, conserved in this fashion are all well below 25%.

The fact that the amino acid sequences of HGH, HCS and LTH are similar is especially of interest in view of the fact that these three molecules are active as growth-promoting and lactogenic hormones in spite of their differences in origin. It may be that these three hormones are derived through evolution from a common ancestor molecule.

←

Fig. 5. Comparison of the structures of HGH, HCS and LTH. The residue position numbers for HGH and HCS are the same, and appear below the HCS sequence. The residue position numbers for LTH appear above the LTH sequence. Homology is indicated by: identical pairs, vertical bar; highly acceptable replacements, three dots; and acceptable replacements, single dot. Unacceptable replacements are indicated by X. An asterisk has been placed over those residues which are common to all three proteins.

REFERENCES

BEWLEY, T. A. and LI, C. H. (1971*a*): Molecular weight and circular dichroism studies of bovine and ovine pituitary growth hormones. *Biochemistry*, in press.

BEWLEY, T. A. and LI, C. H. (1971*b*): Circular dichroism studies on human chorionic somatomammotropin. *Arch. Biochem.*, *144*, 589.

BEWLEY, T. A. and LI, C. H. (1971*c*): Sequence comparison of human pituitary growth hormone, human chorionic somatomammotropin and ovine pituitary lactogenic hormone. *Experientia (Basel)*, in press.

EHRHARDT, K. (1936): Über das Laktationshormon des Hypophysenvorderlappens. *Münch. med. Wschr.*, *83*, 1163.

FLORINI, J. R., TONELLI, G., BREUER, C. B., COPPOLA, J., RINGLER, I. and BELL, P. H. (1966): Characterization and biological effects of purified placental protein (human). *Endocrinology*, *79*, 692.

FRIESEN, H. (1965): Purification of a placental factor with immunological and chemical similarity to human growth hormone. *Endocrinology*, *76*, 369.

JOSIMOVICH, J. B. and MACLEAREN, J. A. (1962): Presence in the human placenta and term serum of a lactogenic substance immunologically related to pituitary growth hormone. *Endocrinology*, *71*, 209.

KAPLAN, S. L. and GRUMBACH, M. M. (1964): Studies of a human and simian placental hormone with growth hormone-like and prolactin-like activities. *J. clin. Endocr.*, *24*, 80.

LI, C. H. (1949): Studies on pituitary lactogenic hormone XIII. The amino acid composition of the hormone obtained from whole sheep pituitary glands. *J. biol. Chem.*, *178*, 459.

LI, C. H. (1957): Studies on pituitary lactogenic hormone XVII. Oxidation of the ovine hormone with performic acid. *J. biol. Chem.*, *229*, 157.

LI, C. H., COLE, R. D. and COVAL, M. J. (1957): Studies on pituitary lactogenic hormone XVI. Molecular weight of the ovine hormone. *J. biol. Chem.*, *229*, 153.

LI, C. H. and CUMMINS, J. T. (1958): Studies of pituitary lactogenic hormone XVIII. Reduction of disulfide bridges in the ovine hormone and the nature of the C-terminus. *J. biol. Chem.*, *233*, 73.

LI, C. H. and DIXON, J. S. (1971): Human pituitary growth hormone XXXII. The structure of the hormone. Revision. *Arch. Biochem.*, *146*, 233.

LI, C. H., DIXON, J. S. and CHUNG, D. (1971): Primary structure of the human chorionic somatomammotropin (HCS) molecule. *Science*, *173*, 56.

LI, C. H., DIXON, J. S. and LIU, W.-K. (1969): Human pituitary growth hormone XIX. The primary structure of the hormone. *Arch. Biochem.*, *133*, 70.

LI, C. H., DIXON, J. S., LO, T.-B., PANKOV, Y. A. and SCHMIDT, K. D. (1969): Amino acid sequence of ovine lactogenic hormone. *Nature (Lond.)*, *224*, 695.

LI, C. H., DIXON, J. S., LO, T.-B., SCHMIDT, K. D. and PANKOV, Y. A. (1970): Studies on pituitary lactogenic hormone XXX. The primary structure of the sheep hormone. *Arch. Biochem.*, *141*, 705.

LI, C. H., GRUMBACH, M. M., KAPLAN, S. L., JOSIMOVICH, J. B., FRIESEN, H. and CATT, K. J. (1968): Human chorionic somato-mammotropin (HCS), proposed terminology for designation of a placental hormone. *Experientia (Basel)*, *24*, 1288.

LI, C. H., LYONS, W. R. and EVANS, H. M. (1940): Studies on pituitary lactogenic hormone. II. A comparison of the electrophoretic behavior of the lactogenic hormone as prepared from beef and from sheep pituitaries. *J. Amer. chem. Soc.*, *62*, 2925.

COMPARATIVE IMMUNOCHEMICAL STUDIES OF PITUITARY GROWTH HORMONES*

TED HAYASHIDA

Department of Anatomy and the Hormone Research Laboratory, University of California, San Francisco Medical Center, San Francisco, Calif., U.S.A.

I wish to describe to you the comparative immunochemical studies of pituitary growth hormones (GHs) from various vertebrate species which we have been conducting over the past few years, employing monkey antiserum to rat GH (RGH) (Hayashida and Contopoulos, 1967; Hayashida and Lagios, 1969; Hayashida, 1969, 1970, 1971). When RGH first became available to us a few years ago through the efforts of Dr. Stanley Ellis, we were confronted with the problem of selecting an appropriate species of animal for the production of good antisera to this hormone of rat origin. We recalled the results of an earlier study conducted for a different purpose by a group of workers at the Hormone Research Laboratory (Li *et al.*, 1959) which provided us with a clue. In that study relative growth promoting effects of purified GHs from different species including the ox, whale, human, and monkey were being tested in hypophysectomized rats on the basis of changes induced in body weight. Although these hormones were administered daily in very small amounts in saline solution, the rats being injected with either monkey or human GH, became refractory and stopped growing after 10 days of treatment although initially showing a good response, whereas the animals injected with the bovine or whale GH continued to grow at a good rate over a prolonged period of time. It was subsequently demonstrated that the development of refractoriness to primate GH by the rat was in all likelihood due to the facility with which this species formed antibodies against primate GH (Moudgal and Li, 1961). These findings suggested to us that primate GH must be significantly different from RGH in antigenic structure. It appeared therefore, that the primate should be an appropriate species to immunize for the production of a good antiserum to RGH. Fortunately, this turned out to be true. On the basis of precipitin reactions in agar and antiserum features considered to be desirable in radioimmunoassay, antisera produced against RGH in monkeys have proven to be superior to those produced in guinea pigs, which in turn were clearly superior to those produced in rabbits (Hayashida and Contopoulos, 1967; Hayashida, unpublished observations). In fact, we were somewhat surprised to find that none of several rabbits we had immunized with RGH produced any detectable precipitating antibodies against this hormone, and that all of the antisera from these rabbits showed the poorest binding affinity for labelled RGH. On the other hand all of the antisera from the majority of monkeys which had been immunized with the same RGH with dosages equated to those given the rabbits on a body weight basis, showed a moderate to strong precipitin reaction and the highest binding affinity for RGH. I would like now to point out to you

* This investigation was supported by NIH Grants AM-03550 and HD-04063 and also by a grant from the Committee on Research of the University of California School of Medicine. Part of the investigation was performed during the tenure of a USPHS Research Career Development Award (K3-GM-6710).

some of the highlights of the series of studies we have been conducting. The details regarding all of the procedures we have employed have been previously described (Hayashida and Contopoulos, 1967; Hayashida and Lagios, 1969; Hayashida, 1969, 1970).

Preparation of antiserum and pituitary extracts

Antisera were prepared in young adult rhesus monkeys with a total dose of 6.5 mg RGH per animal being administered subcutaneously in Freund's adjuvant in four divided doses given 3 weeks apart, with the last injection being given in saline. The antisera were then absorbed to remove any antibodies to serum proteins and to rat prolactin. For the preparation of pituitary extracts (PEs) a special effort was made to obtain fresh pituitary glands from several species representing different orders within each of the major vertebrate classes. In almost all cases pituitaries were taken from healthy animals which were quickly sacrificed, and the glands were usually frozen immediately after removal. All glands were kept frozen prior to the preparation of extracts, except for those obtained from the lungfish *(Protopterus)* which were acetone-dried specimens. Pituitary extracts were prepared in phosphate-buffered saline at pH 7.2 to which was added 1% by volume of n-butyl alcohol to minimize foaming and to serve as a preservative. Trasylol (FBA Pharmaceuticals, Inc., New York City), a proteinase inhibitor, was added at a concentration of 20 K.I. units per 10 mg pituitary tissue, to minimize possible hormone degradation due to proteolytic enzyme action.

Precipitin reactions in agar gel

I would like first to discuss the results of precipitin reactions in agar employing the double diffusion technique of Ouchterlony (1953). All agar plates used in these studies contained 0.15 ml of the monkey antiserum to RGH in the center well. All solutions placed in the peripheral wells were in a volume of 50 μl each. Figure 1 shows the results seen in an agar plate when we

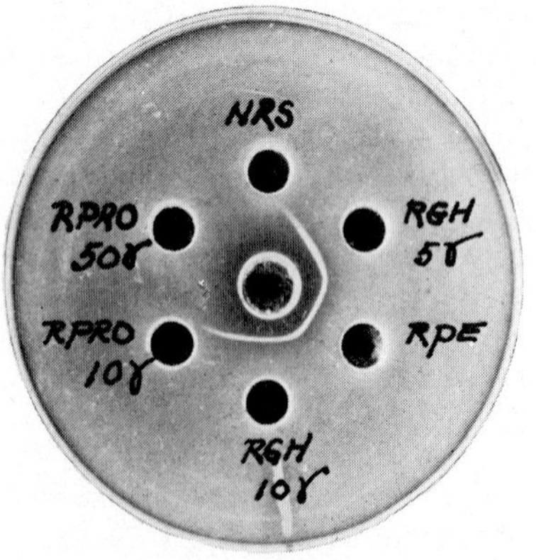

Fig. 1. Ouchterlony plates showing results of precipitin reactions between monkey antiserum to rat GH and pituitary extracts (PEs) from various vertebrate species. Antiserum (0.15 ml) was placed in the center well and 50 μl of the respective PEs were placed in the peripheral wells as indicated. Highly purified rat GH (RGH) of Ellis (2.7 USP units/mg) was used as a reference in all plates, employing 7.5 μg/50 μl, unless otherwise indicated.
RPE denotes an extract representing 1/16 of an adult male rat anterior pituitary. RPRO, rat prolactin (Ellis, 17 IU/mg); NRS, normal rat serum, diluted 1 : 10 in saline. Note the absence of a detectable reaction with RPRO and with rat serum proteins. Photographed at 24 hours. (From Hayashida and Contopoulos, 1967, *Gen. comp. Endocr.*, *9*, 217, by kind permission of the editors).

tested the reaction of the antiserum against the following substances placed in the peripheral wells: highly purified RGH, purified rat prolactin, rat pituitary extract, and normal rat serum. As may be seen in this figure the results of this plate showed a good, single precipitin line with RGH (5 or 10 μg), which in turn showed a line of identity with a substance in the rat PE, presumably due to its RGH content. There was no detectable reaction with rat prolactin (10 or 50 μg) nor with rat serum proteins in the uppermost well. When similar plates were set up with highly purified GHs from several different mammalian species including the ox, sheep, pig, whale, rabbit and rat, it was observed that all of these mammalian GHs gave reactions of qualitative identity with each other and with RGH itself. We were somewhat surprised at the consistency of these results, since species from four different mammalian orders were represented by these GHs. Ellis *et al.*, (1968) have reported essentially similar findings employing immunoelectrophoresis with monkey antiserum to RGH.

We went on then to testing PEs from a number of mammalian species including those representing the same mammalian orders as well as additional ones for which purified GHs were not available. The next two figures (Figs. 2 and 3) are representative of the results ob-

Fig. 2. The HGH well contains 10 μg of a highly purified preparation (Li). Each of the peripheral wells contains 0.25–0.5 mg equivalent of pituitary tissue from various mammalian species. Note reactions of identity shown by all PEs with the highly purified RGH and the absence of a detectable reaction with HGH (or monkey GH). Photographed at 24 hrs. (From Hayashida, 1970, *Gen. comp. Endocr.*, *15*, 432, by kind permission of the editors).

tained with all mammalian PEs. We can see that all PEs showed qualitative reactions of identity with each other and with RGH, the only negative reaction being that of human GH in Figure 2. The various mammals, except for the primates, could not be distinguished from each other by this test. Again, the reactions with the PEs is presumably due to their GH content.

The next plate shows the results we obtained with PEs from another class of vertebrates, the birds. Four different birds from four different orders are represented here, namely, the pigeon, chicken, duck and penguin as indicated (Fig. 4). The PEs of all birds have given a reaction of identity with each other, but with respect to RGH they have given a reaction of partial identity, as indicated by the spur formed by the RGH line which extends beyond the point of juncture. This type of reaction suggests that although certain immunochemical determinants are shared by avian PEs (GHs) in common with RGH, there are other determinants on the RGH molecule which are not shared. The findings indicate that avian PEs (GHs) may therefore be classified immunochemically at a level below that of mammals.

Fig. 3. This plate demonstrates the reactions of identity with RGH shown by PEs (GHs) from additional mammalian species. Each of the extracts represents 0.25–0.5 mg pituitary tissue. Photographed at 24 hrs. (From Hayashida, 1969, *Nature (Lond.)*, *222*, 294, by kind permission of the editors).

Fig. 4. This plate demonstrates the reaction of identity shown by avian PEs (GHs) with each other, but a reaction of partial identity with respect to RGH indicated by the spur formation. All PEs represent 0.5 mg pituitary tissue except for that of the penguin which represents 1.0 mg. Photographed at 42 hrs. (From Hayashida, 1970, *Gen. comp. Endocr.*, *15*, 432, by kind permission of the editors).

The relation of reptilian PE (GH) to avian PE (GH) was examined in the next plate (Fig. 5). The results of this plate indicated that the two reptilian PEs represented by the crocodile and turtle gave a reaction of identity with each other as well as with the avian PE (duck). This reaction of qualitative identity between reptilian and avian PEs was also observed for other birds, indicating the very close immunochemical relationship between these two groups, in agreement with the close phylogenetic relationship believed to exist between them. Thus the reptiles could not be distinguished from birds on this basis and may, therefore, be grouped together into an immunochemical classification that is located at a level below that of the mammalian class. The PE of a urodele amphibian *(Necturus)* was tested in the same plate, and it gave a reaction of partial identity with both the avian (duck) and reptilian (turtle) PEs, placing its immunochemical classification at a level even below that for birds and reptiles.

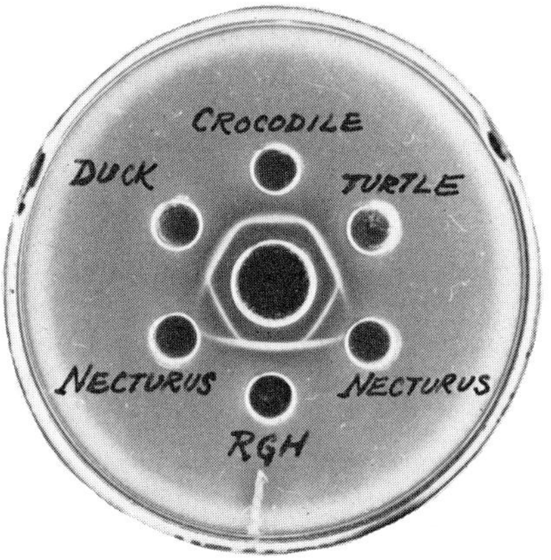

Fig. 5. This plate demonstrates reactions of identity with an avian PE (duck) shown by two reptilian PEs (GHs), the turtle and a crocodilian *(Caiman sclerops)*. Each PE represents 1.0 mg of tissue except duck PE which represents 0.5 mg tissue. Note the reactions of partial identity shown by the amphibian PE *(Necturus)* with respect to the avian and reptilian PEs as well as with RGH. Photographed at 46 hours.

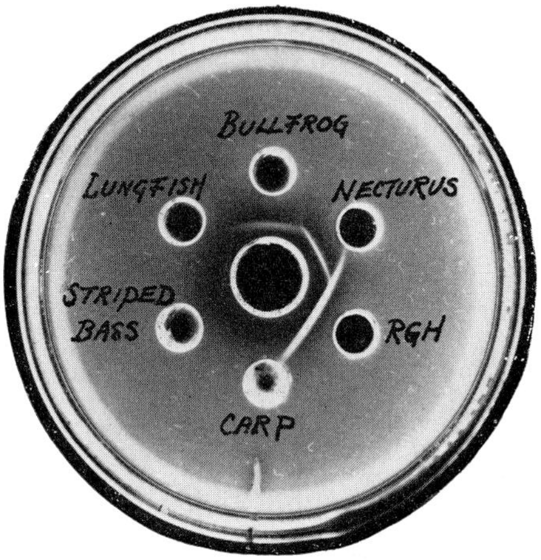

Fig. 6. This plate demonstrates the reaction of partial identity of amphibian PEs *(Necturus* and bullfrog) with respect to RGH. The amphibian PEs contain the equivalent of 1.0 mg tissue per well, while the fish PEs represent 2.0 mg tissue per well. Note that the two amphibian PEs are almost showing a reaction of identity with each other, except for a very slight reaction of partial identity. Photographed at 46 hours.

The amphibian PE also gave a reaction of partial identity with mammalian GH (RGH), as one might have predicted.

In the last plate in this series we see the results of tests with PEs of two amphibians *(Necturus* and bullfrog), the African lungfish *(Protopterus)*, and two modern bony fishes (teleosts), namely the striped bass and carp (Fig. 6). We can readily see that the amphibian PEs almost gave a reaction of identity with each other, with a slight suggestion of a spur at the point of juncture, but both PEs were clearly at a level below that of mammals from the immunochemical standpoint. The PEs of the lungfish and modern bony fishes have been consistently negative by this test as illustrated here.

Radioimmunoassay studies

All of the PEs were also subjected to study by radioimmunoassay, employing the double antibody procedure as described by Schalch and Reichlin (1966) with slight modifications (Hayashida, 1969, 1970). Highly purified RGH (Ellis, 2.7 USP units/mg) was utilized for labelling with ^{131}I and for preparation of standards. All iodinations were performed for us through the courtesy of Drs. Selna Kaplan and Melvin Grumbach. The same monkey antiserum that was used in all agar gel diffusion studies was employed in all radioimmunoassays at a final dilution of 1 : 80,000.

Before commencing the series of studies with the PEs, highly purified GH preparations from several different mammalian species were tested in this system. Figure 7 shows the results of a typical assay. One can see from this figure that the porcine (Wilhelmi), whale (Papkoff) and rabbit (Ellis) growth hormones competed extremely well with the labelled RGH, giving curves with slopes that were very similar to that for RGH itself. However, bovine (Wilhelmi) and ovine (Papkoff) growth hormones gave slopes that were similar to each other, but distinctly different from that of RGH or the other GHs tested. This difference has been noted consistently. When rat prolactin (Ellis, 17 I.U./mg) was tested, the curve remained flat until approximately the 32 mμg level was reached, from which point the prolactin (PRO) preparation began to compete effectively. When a more highly purified rat PRO (Ellis) was subse-

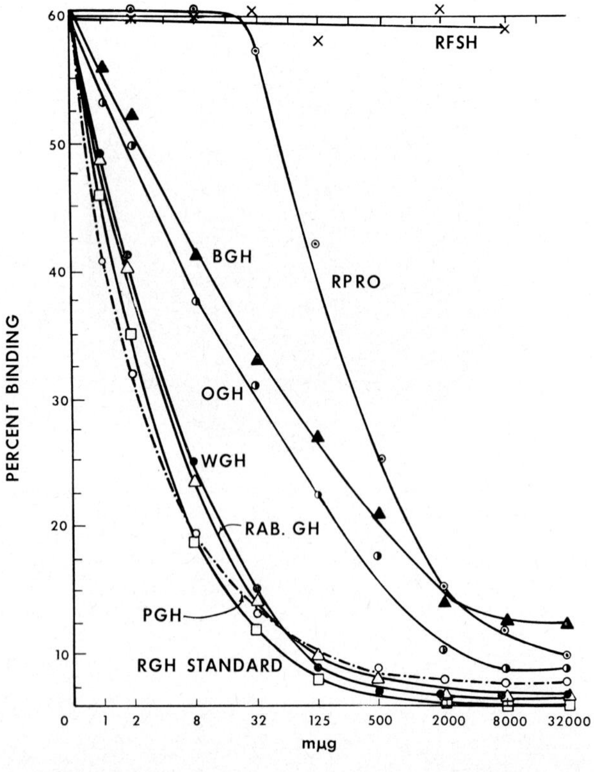

Fig. 7. Cross-reactions of highly purified mammalian GHs with monkey antiserum to RGH by the double antibody radioimmunoassay procedure. Growth hormone from the following species are represented: bovine (BGH), ovine (OGH), whale (WGH), rabbit (RAB. GH) and porcine (PGH). RPRO and RFSH represent purified rat prolactin and follicle-stimulating hormone, respectively. (From Hayashida, 1969, *Nature (Lond.)*, *222*, 294, by kind permission of the editors).

quently tested, the slope remained parallel to the PRO curve shown here, but merely shifted significantly to the right. It was later demonstrated that the curves shown by the PRO preparations were actually due to small amounts of RGH contamination. When the antiserum was tested against a rat FSH preparation (Ellis) there was no detectable reaction.

In testing the reaction of the antiserum with the various PEs, the extracts were diluted in such a way that they paralleled the dilutions of the RGH standards (Fig. 8). The amount of extract representing 0.063 mg of rat pituitary tissue was considered as being equivalent to 1 μg of RGH, on the basis of preliminary immunoassay of the stock rat PE employed as a reference along with the RGH standard. Thus the curves obtained with the various PEs could be directly compared with the RGH standard curve. Pituitary extracts from 14 mammalian species, 4 avian, 3 reptilian, 3 amphibian and 5 piscine species have been tested in this system. Most of the extracts were assayed on more than one occasion and for the majority of species represented, assays were performed on 2 or more extracts prepared from different pools of pituitaries. Figure 8 is a composite drawing of the results obtained in 3 separate radioimmunoassays with PEs from various species representing the major classes of vertebrates. The rat

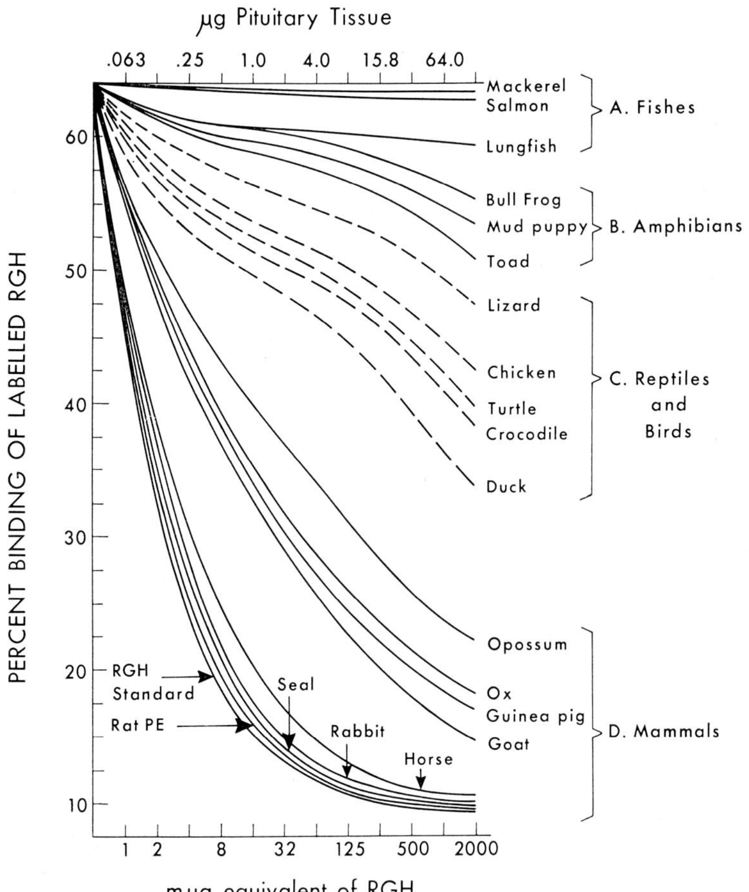

Fig. 8. Composite diagram showing immunochemical relatedness of GH in pituitary extracts from representatives of various vertebrate classes based on radioimmunoassay with antiserum to RGH. (From Hayashida, 1970, *Gen. comp. Endocr.*, *15*, 432, by kind permission of the editors).

PE gave a curve that was essentially identical in its characteristics to that of the standard RGH. It is assumed that the curves for all of the PEs most likely represented the ability of the GH contained in the various extracts, to compete with the labelled RGH for binding with the antiserum to RGH. It may be noted that the mammalian PEs (GHs) competed or cross-reacted more effectively than the PEs of any other vertebrate class. It is of interest to note that the PE of the ox(bovine) as well as that of the sheep (ovine) which is not shown in the drawing, gave slopes that were very similar to those previously observed with highly purified bovine and ovine GHs in Figure 7. The PE of the opossum which is considered to be one of the most primitive forms of living mammals, has always given the lowest degree of cross-reaction among all mammalian PEs tested. The curves for all reptilian and avian PEs fell into an entirely separate zone from that of the mammals, showing a lesser degree of cross-reaction than did the members of the mammalian group in agreement with the results obtained in the agar gel diffusion studies. Amphibian PEs showed an even lesser degree of cross-reaction, constituting yet another group again in accordance with the results of agar diffusion studies. The PEs of all modern bony fishes tested, exemplified by the mackerel and salmon shown in this figure, gave negligible cross-reactions. The PE of the lungfish which is considered to be a form of fish with features similar to those which gave rise to the land vertebrates, has usually given a low but probably significant degree of cross-reaction in this system. In other words, its PE(GH) did show a small degree of immunochemical relatedness to mammalian GHs, or more specifically, to RGH.

It should be added that as far as mammalian species are concerned, Garcia and Geschwind (1968) have also reported evidence for the relatively close immunochemical relationships for GH among several mammalian species, employing radioimmunoassay with monkey antiserum to RGH. They also obtained essentially similar results with the use of monkey antiserum to rabbit GH. The studies reported earlier by Tashjian *et al.* (1965) and by Trenkle and Li (1964), had indicated the immunochemical relatedness of GH among several mammalian species, by micro-complement-fixation (Wasserman and Levine, 1961) in the former investigation, employing rabbit antiserum to HGH, and in the latter study through use of precipitin reactions with guinea pig antiserum to porcine GH.

Bioassays of pituitary extracts

Most of the PEs which were subjected to immunochemical studies were also tested for their ability to stimulate growth in the mammal by use of the standard tibia assay (Greenspan *et al.*, 1949) in immature hypophysectomized rats which were injected daily for 4 days beginning 2 weeks postoperatively. The animals were sacrificed on the fifth day at which time the tibias were taken for processing and subsequent examination. The PEs for each species studied were in most instances injected at two or three different dose levels, usually three; and in the majority of cases a repeat assay was performed utilizing an extract prepared from a different pool of pituitaries. A bovine GH preparation (Li, 1.2 USP unit/mg) was employed as a reference standard. The results of the bioassays indicated that all mammalian PEs were considerably more potent than PEs of species representing other vertebrate classes on a basis of wet tissue weight, with the exception of those of certain amphibians such as the bullfrog and the toad which showed relative potencies that were equivalent to those of several mammals. The mean potency for either the avian or reptilian PEs amounted to approximately one-tenth the mean potency for all mammalian PEs. The tailed amphibian *(Necturus)* PE, in contrast to the results obtained with the bullfrog and toad PEs, showed a potency similar in magnitude to those at the lower range of values for birds and reptiles. The lungfish PE which was prepared from acetone-dried glands, when equated on a wet tissue basis, showed a potency which was again similar to those obtained with PEs of birds and reptiles. The modern bony fishes (teleosts) such as the striped bass, salmon and mackerel, represented the only group whose PEs showed no significant stimulation of the rat tibia, which is in general accord with the observations of earlier investigators (see reviews by Knobil and Hotchkiss, 1964; and Geschwind, 1967).

A rather surprising observation made during the study of various fish PEs was the finding that the PEs of relatively primitive bony fishes (chondrosteans) such as the sturgeon *(Acipenser)* and the paddlefish *(Polyodon)* (Hayashida and Lagios, 1969) as well as that of the bowfin *(Amia)*, a holostean fish (Hayashida, 1971), are capable of inducing significant stimulation of the rat tibia. In the latter study it was also noted that PEs of the gar *(Lepisosteus)* which is another holostean, and the *Polypterus* (chondrostean), were also capable of showing some stimulation of the rat tibia, although because of the lower degree of stimulation the results must still be considered equivocal. The findings of precipitin reactions in agar employing the above primitive fishes showed a general correlation between the relative degree of immunochemical relatedness of a particular PE (GH) to RGH and the ability of that PE to stimulate growth in the mammal based on the rat tibia assay. For example, as may be seen in the agar plate illustrated by Figure 9, both the paddlefish and sturgeon PEs (GHs) which show positive

Fig. 9. Ouchterlony plate showing immunochemical relatedness of GH in PEs of primitive bony fishes to RGH, and the lack of detectable relatedness of GH in PEs of modern bony fishes to RGH. RPE indicates 0.3 mg equivalent of an adult male rat pituitary. The sturgeon and paddlefish PEs represent 1.0 mg and 1.5 mg of tissue, respectively, while the salmon and mackerel PEs represent 2.0 mg of tissue each. Photographed at 24 hours. (From Hayashida, 1970, *Gen. comp. Endocr.*, *15*, 432, by kind permission of the editors).

rat tibial stimulation, gave precipitin reactions with the antiserum to RGH, showing a reaction of identity with each other, and a reaction of partial identity with RGH (Hayashida, 1970). The PEs of the modern bony fishes (salmon and mackerel) have always given a negative precipitin test as shown in this figure, correlating with the inability of these PEs to stimulate the rat tibia. Radioimmunoassay with the PEs from the sturgeon and paddlefish and the two modern bony fishes, yielded results that paralleled those of the agar diffusion studies (Fig. 10). In the case of the holostean fishes, the bowfin and the gar, the PEs (GHs) again showed clear precipitin lines, giving reactions of partial identity with RGH. The results obtained by radioimmunoassay with the latter PEs have again been entirely in accord with those of the agar diffusion studies and preliminary findings suggest that these results show a positive correlation with those of tibia assays (Hayashida, 1971).

In addition to the correlations observed between (1) the ability of a PE of GH from a particular species of fish to promote growth in the mammal (rat tibia) and (2) the immunochemical relatedness to RGH, there appears to be yet a third positive correlation, namely, the extent of development of the median eminence of the brain which is concerned with the trans-

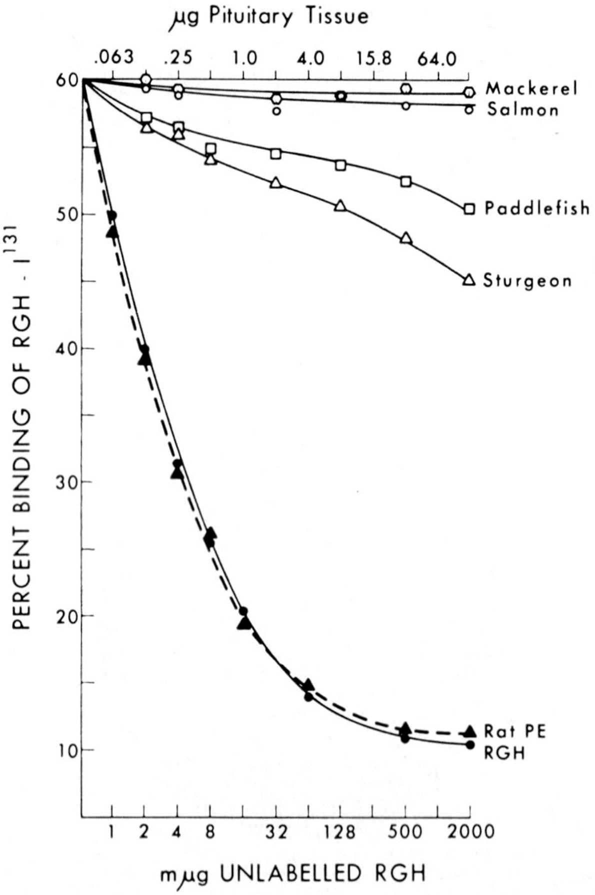

Fig. 10. Cross-reactions of GH in PEs of primitive and modern bony fishes based on radioimmunoassay with monkey antiserum to RGH. The relative ability of GH in the various fish PEs to competitively inhibit the binding of labelled RGH is represented. (From Hayashida and Lagios, *Gen. comp. Endocr., 13*, 403, by kind permission of the editors).

mission of releasing factors from the hypothalamus to the anterior pituitary (Hayashida and Lagios, 1969; Lagios, 1970; Hayashida, 1970, 1971). In other words, it appears that the modern bony fishes do not show a positive correlation with regard to any of these three parameters when compared to the same parameters for mammals; on the other hand, the findings of our studies have indicated that such a positive correlation appears to exist in the case of all of the more primitive fishes we have examined. At least from the standpoint of the above cited references, the results thus far suggest that the primitive fishes we have studied (chondrosteans and holosteans) are less divergent from the main line of vertebrate evolution leading to the land vertebrates, than are the modern bony fishes.

Specificity of the antiserum for growth hormone

The specificity of the reaction of the antiserum for GH in rat PE has been demonstrated in several ways, such as by the reaction of identity demonstrated between highly purified RGH

TABLE I

Neutralization of the tibia-stimulating activity of purified RGH or a saline extract of rat pituitary with monkey antiserum to RGH

Group	No. rats	Treatment[1]	Tibial plate width ($\mu \pm$ S.E.)	Significance[3]
1.	6	Hypophysectomized controls	158 ± 6.9	
2.	3	RGH 15 μg/day	226 ± 7.7	1 vs. 2 $p < .001$
3.	6	RGH 15 μg/day + NMS (0.1 ml)	226 ± 11.6	1 vs. 3 $p < .001$
4.	8	RGH 15 μg/day + MA/S (0.1 ml)	162 ± 6.3	3 vs. 4 $p < .001$
5.	4	RGH 15 μg/day + MA/S (.05 ml)	206 ± 8.4	3 vs. 5 $p < .01$
6.	3	RPE (1/16 pit.)[2] + NMS (0.1 ml)	249 ± 13.3	1 vs. 6 $p < .001$
7.	4	RPE (1/16 pit.) + MA/S (0.1 ml)	175 ± 14.1	6 vs. 7 $p < .001$

[1] RGH, rat growth hormone, Ellis, II-35-C4; NMS, normal monkey serum; MA/S, monkey antiserum to RGH.
[2] RPE, a saline extract of adult male rat pituitary, equivalent to 1/16 of anterior lobe/day.
[3] Significance in terms of *p* values of Fisher.

(From Hayashida and Contopoulos, 1967, *Gen. comp. Endocr.*, *9*, 217, by kind permission of the editors).

TABLE II

Cross-neutralization of GH in duck and sturgeon pituitary extracts with monkey antiserum (MAS) to RGH in the tibia assay[1]

Group	No. rats	Serum (ml)	Tibial plate width (mi ± S.E.)	Significance
1. Hypophysectomized controls	8	—	154.4 ± 6.1	
2. Normal monkey serum (NMS)	3	0.3	160.1 ± 6.4	
3. Duck pit. extract + NMS	4	0.1	203.2 ± 9.4	1 vs 3 $p < .001$
4. Duck pit. extract + NMS	4	0.2	209.1 ± 13.9	
5. Duck pit. extract + NMS	5	0.3	201.8 ± 5.5	
6. Duck pit. extract + MAS (RGH)	5	0.1	184.4 ± 15.5	3 vs 6 n.s.
7. Duck pit. extract + MAS (RGH)	4	0.2	167.7 ± 11.0	3 vs 7 $p < .05$
8. Duck pit. extract + MAS (RGH)	5	0.3	157.6 ± 9.0	3 vs 8 $p < .01$
9. Sturgeon pit. extract + NMS	5	0.2	201.5 ± 4.2	1 vs 9 $p < .001$
10. Sturgeon pit. extract + NMS	5	0.3	206.2 ± 10.2	
11. Sturgeon pit. extract + MAS (RGH)	3	0.1	204.2 ± 6.7	
12. Sturgeon pit. extract + MAS (RGH)	5	0.2	184.8 ± 7.8	9 vs 12 n.s.
13. Sturgeon pit. extract + MAS (RGH)	4	0.3	160.5 ± 10.8	9 vs 13 $p < .01$
14. Sturgeon pit. extract + MAS (RPRO)	4	0.3	202.5 ± 5.7	

[1] Pituitary extracts and serum were injected subcutaneously at different sites, daily for 4 days. The daily dosage of pituitary extract was equivalent to 1.8 mg wet weight of pituitary per rat. The tibial response to this dosage, as indicated by tibial plate widths of Groups 3 and 9, is approximately equivalent to that obtained with 5 μg per day of purified BGH of 1 USP unit/mg potency. RPRO, monkey antiserum to rat prolactin; n.s., not significant. Significance of difference between mean values, in terms of *p* values of Fisher.

(From Hayashida, 1970, *Gen. comp. Endocr.*, *15*, 432, by kind permission of the editors).

and a substance in rat PE in agar diffusion studies, by the demonstration of a single component in the PE which shows the same mobility as purified RGH after immunoelectrophoresis; and by radioimmunoassay, the demonstration of parallelism between RPE and purified RGH standard, as well as the lack of reaction with rat prolactin and FSH. Also, in comparing the radioimmunoassay slopes of bovine and ovine PEs with the respective purified GHs, it has been consistently observed that the slopes for these purified hormones which are so distinctly different from that of the RGH standard, are also essentially identical to the slopes observed with the PEs. Additional evidence for the specificity of the antiserum for GH was demonstrated with PEs of submammalian species in a radioimmunoassay study. It was found that the preincubation for 48 hours of the amount of antiserum (0.1 ml of 1 : 8000) ordinarily used per tube in the assay system, with a precise amount (125 mμg) of highly purified RGH (Ellis, 3 USP units/mg) (see Fig. 8), which should be just sufficient to block all combining sites on the anti-RGH antibodies, resulted not only in an almost complete inhibition of the development of a RGH standard curve, but also in blocking any detectable cross-reaction with PEs of the duck. Likewise, studies involving the neutralization of the biological activity of GH with antiserum have also added supporting evidence for the reaction of the antiserum with the GH in PEs of submammalian species. For example, it was previously shown that the rat tibial plate stimulating activity of either purified RGH or of rat PE could be completely blocked with simultaneous injections of monkey antiserum to RGH (Table I). It was subsequently demonstrated that the same antiserum was also capable of completely blocking the rat tibial plate stimulating activity of both duck and sturgeon PEs, as may be observed from the data shown in Table II. The data in the last line of the table suggests that this blocking or neutralization is not due to a non-specific or toxic effect of the antiserum, since equivalent amounts of monkey antiserum to rat prolactin showed no neutralization of the tibial stimulation induced by sturgeon PE. All of the findings presented above provide supporting evidence that the observed reactions between the antiserum and the various PEs, are due to reactions with the GH contained in the extracts.

CONCLUDING REMARKS

The concept of the possible existence of a common core in the GH molecule from various vertebrate species has been entertained for a number of years (Wilhelmi, 1955; Li, 1957), since GH from all species tested up until that time, with the exception of fish GH, had been shown to be active in promoting growth based on the tibia assay in the hypophysectomized rat. On the basis of the evidence obtained in the current series of investigations, this concept has been considerably strengthened with the use of immunochemical as well as biological procedures. The applicability of this concept has now been extended to include several primitive fishes and a number of additional vertebrate species representing each of the major vertebrate classes.

In general summary it may be said that with the use of monkey antiserum to RGH employing precipitin reactions in agar and radioimmunoassay, it has been demonstrated that all mammalian GHs must share a number of antigenic determinants with each other and that with increased phylogenetic distance from the mammalian class, the fewer such determinants appear to be shared with mammalian GH by the GH from any submammalian species. It appears that GH from the various vertebrate species are separable into general immunochemical categories which closely correspond to their respective phylogenetic classes from which the species originate. The immunochemical approach has therefore shown a general correlation with evolutionary development and was more reliable in this regard than the biological approach under the conditions of our investigations.

An interesting area of research should be the extraction and purification of GH from the pituitaries of appropriate submammalian species and subjecting such products to the same kind of immunochemical and biological studies as have been reported here, in an attempt to test the validity of the generalizations made. This should eventually provide us with insight into the molecular determinants of immunochemical and biological activities, the relation of such determinants to each other, and how they have become modified during the process of evolution.

ACKNOWLEDGEMENTS

The author is particularly grateful for the skilled technical assistance provided by Miss Eleanor Lasky during the entire period of this investigation and wishes to thank Miss Irma Chang for her assistance during part of the period. He is indebted to Dr. Alfred E. Wilhelmi for making available to us a number of highly purified growth hormone preparations, and is also indebted to Drs. Stanley Ellis, Choh Hao Li and Harold Papkoff for supplying us with various other purified growth hormone preparations. The author wishes to express his appreciation to the various individuals who have facilitated the procurement of the large numbers of pituitary glands employed in this investigation.

REFERENCES

Ellis, S., Grindeland, R. E., Nuenke, J. M. and Callahan, P. X. (1968): Isolation and properties of rat and rabbit growth hormones. *Ann. N. Y. Acad. Sci., 148*, 328.

Garcia, J. F. and Geschwind, I. I. (1968): Investigation of growth hormone secretion in selected mammalian species. In: *Growth Hormone*, pp. 267–291. Editors: A. Pecile and E. E. Müller. ICS 158, Excerpta Medica, Amsterdam.

Geschwind, I. I. (1967): Molecular variation and possible lines of vertebrate evolution of peptide and protein hormones. *Amer. Zool., 7*, 89.

Greenspan, F. S., Li, C. H., Simpson, M. E. and Evans, H. M. (1949): Bioassay of hypophyseal growth hormone: the tibia test. *Endocrinology, 45*, 455.

Hayashida, T. (1969): Relatedness of pituitary growth hormone from various vertebrate classes. *Nature (Lond.), 222*, 294.

Hayashida, T. (1970): Immunological studies with rat pituitary growth hormone (RGH). II. Comparative immunochemical investigation of GH from representatives of various vertebrate classes with monkey antiserum to RGH. *Gen. comp. Endocr., 15*, 432.

Hayashida, T. (1971): Biological and immunochemical studies with growth hormone in pituitary extracts of holostean and chondrostean fishes. *Gen. comp. Endocr., 17*, 275.

Hayashida, T. and Contopoulos, A. N. (1967): Immunological studies with rat pituitary growth hormone. I. Basic studies with immunodiffusion and antihormone tests. *Gen. comp. Endocr., 9*, 217.

Hayashida, T. and Lagios, M. D. (1969): Fish growth hormone: A biological, immunochemical, and ultrastructural study of sturgeon and paddlefish pituitaries. *Gen. comp. Endocr., 13*, 403.

Knobil, E. and Hotchkiss, J. (1964): Growth hormone. *Ann. Rev. Physiol., 26*, 47.

Lagios, M. D. (1970): The median eminence of the bowfin, *Amia calva* L. *Gen. comp. Endocr., 15*, 453.

Li, C. H. (1957): Properties of and structural investigations on growth hormones isolated from bovine, monkey and human pituitary glands. *Fed. Proc., 16*, 775.

Li, C. H., Papkoff, H. and Jordan, C. W. (1959): Difference in biological behavior between primate and beef or whale pituitary growth hormones. *Proc. Soc. exp. Biol. (N.Y.), 100*, 44.

Moudgal, N. R. and Li, C. H. (1961): Production of antibodies to human pituitary growth hormone in the rat. *Endocrinology, 68*, 704.

Ouchterlony, O. (1953): Antigen-antibody reactions in gels; types of reactions in coordinated systems of diffusion. *Acta path. microbiol. scand., 32*, 231.

Schalch, D. S. and Reichlin, S. (1966): Plasma growth hormone concentration in the rat determined by radioimmunoassay; influence of sex, pregnancy, lactation, anesthesia, hypophysectomy and extrasellar pituitary transplants. *Endocrinology, 79*, 275.

Tashjian Jr, A. H., Levine, L. and Wilhelmi, A. E. (1965): Immunochemical relatedness of porcine, bovine, ovine and primate pituitary growth hormones. *Endocrinology, 77*, 563.

Trenkle, A. and Li, C. H. (1964): An immunological investigation of porcine pituitary growth hormone. *Gen. comp. Endocr., 4*, 113.

Wasserman, E. and Levine, L. (1961): Quantitative micro-complement fixation and its use in the study of antigenic structure by specific antigen-antibody inhibition. *J. Immunol., 87*, 290.

Wilhelmi, A. E. (1955): In: *Hypophyseal Growth Hormone, Nature and Actions*, p. 69. Editors: R. W. Smith Jr, O. H. Gaebler and C. N. H. Long. McGraw-Hill (Blakiston), New York.

STUDIES ON THE PRIMARY STRUCTURE OF PORCINE GROWTH HORMONE* **

J. B. MILLS † and A. E. WILHELMI

Department of Biochemistry, Emory University, Atlanta, Georgia, U.S.A.

We have undertaken the study of the primary structure of porcine growth hormone for a number of reasons. First, the hormone is interesting in itself. Second, certain of its properties suggested that it might be more closely related to human growth hormone than bovine growth hormone is. Third, the work is intended to be complementary to work already accomplished, with human growth hormone, as described in the first of these Symposia by C. H. Li (1968), and work in progress on human growth hormone and human placental lactogen (Niall, 1971; Niall *et al.*, 1971), on bovine growth hormone (Fellows and Rogol, 1969; Dellacha *et al.*, 1968) and on ovine growth hormone (Peña *et al.*, 1970). Finally, the work already cited, and our own work, have indicated that the mammalian growth hormones are a much more uniform family of molecules, especially in regard to the size of their minimum covalent units, than could have been supposed a decade ago. The basis for their differences in physical and biological behavior is therefore more subtle than it at first appeared to be, and it is more useful, therefore, to study as many different kinds of these molecules as possible.

The porcine growth hormone used in these studies was prepared by the method of Chen *et al.* (1970), using as starting material the filtrate left after the adsorption of ACTH onto oxidized cellulose. A partially purified fraction is derived in good yield by a series of simple adjustments of pH and fractionations with ammonium sulfate. This fraction is readily brought to a high state of purity and homogeneity by ion exchange chromatography on DEAE-cellulose equilibrated with 0.01 M Tris-formate, pH 8.0. The active principle is eluted by stepwise increase of the buffer concentration to 0.0375 M. This is the latest and most refined of several systems that have been applied to the purification of the hormone. The product, subjected to gel filtration on Sephadex G-100 equilibrated in 0.04 M ammonium formate, pH 8.0, emerges as a single symmetrical peak. By disc gel electrophoresis (Ornstein, 1964) it migrates as a single band. The carboxyl-terminal amino acid is phenylalanine, and the amino-terminal residue is mainly phenylalanine, but occasional preparations also contain a significant amount of amino-terminal alanine. The amino acid composition of porcine GH is unremarkable, and it is similar to that of a number of other mammalian growth hormones as we have determined them in our laboratory.

Porcine growth hormone contains three molecules of methionine per mole, and it seemed

---

* Publication No. 1029 of the Division of Basic Health Sciences, Emory University.

** Abbreviations used in the text are: GH, growth hormone; PGH, BGH, OGH, HGH, porcine, bovine, ovine and human growth hormone, respectively; ACTH, adrenocorticotrophic hormone; HPL, human placental lactogen.

† Dr. Mills is supported by a Research Career Development Award from the National Institute of Child Health and Human Development (HD29346) and the work itself has been generously supported by grants HD 01231 and AM 03598 of the National Institutes of Health.

probable, therefore, that an attack on the molecule with cyanogen bromide might make a profitable first approach to the determination of the amino acid sequence. This proved to be so, and a report of the first stage of this work, establishing the sequence of the 67 residues at the carboxyl-terminal end of the molecule, as well as of the amino-terminal tetrapeptide, has already appeared (Mills *et al.*, 1970). The fragments derived by the reaction of PGH with cyanogen bromide are readily separated by gel filtration on Sephadex G-75 in 20% formic acid. If the fully reduced and amino-ethylated molecule is used, four peaks are obtained. The least retarded peaks, one and two, have the same amino acid composition and are thought to be the same peptide in different states of aggregation. They are combined and designated Fragment A. Peak three, which emerges next, appears to be a single large peptide (Fragment B). If the hormone is treated with CNBr prior to reduction, peaks 1, 2, and 3 emerge as a single peak. Fragment B is generated by subsequent reduction of this product. The most retarded peak four contains three peptides which can be separated by paper electrophoresis. One of these (Fragment C) is the carboxyl-terminal dodecapeptide. The other two (D and E) are clearly related to the amino-terminal end of the molecule. The tetrapeptide, with the sequence Phe-Pro-Ala-HSer-, confirms the amino-terminal sequence reported by Papkoff *et al.* (1962) for porcine GH. But we find in addition a small amount of the dipeptide, Ala-HSer-, which may be derived by partial degradation of the native molecule. We have not observed the pentapeptide found at the amino-terminal of bovine GH by Wallis (1969) and by Fellows and Rogol (1969). There is a marked resemblance of the terminal dodecapeptides of several of the growth hormones. Those of the pig and horse, and probably of the dog, are identical (Mills *et al.*, 1970; Oliver and Hartree, 1968); those of the ox and sheep are reported to be identical (Dellacha *et al.*, 1968; Peña *et al.*, 1970). All of them closely resemble the sequence of HGH in this region.

The sequence of Fragments B and C, as they have been determined by conventional procedures, is illustrated in Figure 1, in comparison with HGH and BGH as reported by Li (1968) and by Fellows and Rogol (1969). A very high degree of homology is seen between all three hormones. The porcine hormone, as we suspected originally, resembles HGH somewhat more closely than the bovine hormone does.

We are presently at work on the sequencing of the large remaining Fragment A. A number of the constituent tryptic peptides have been separated, purified, and sequenced, but the work has not gone far enough to put them in order. We must therefore be content, at this stage, to present our information merely on a 'matching' basis with the sequence of HGH as recently

```
HUMAN           X              Pro     Gly     Thr             Phe
PORCINE    Arg-Glu-Leu-Glu-Asp-Gly-Ser-Pro-Arg-Ala-Gly-Gln-Ile-Leu-Lys-Gln-Thr-
BOVINE                             Thr

HUMAN          Ser                         Ser-His-Asn
PORCINE    Tyr-Asp-Lys-Phe-Asp-Thr-Asn-Leu-Arg-Ser-Asp-Asp-Ala-Leu-Leu-Lys-Asn-
BOVINE                                 Met

HUMAN                      Tyr         Arg         Met-Asp     Val         Phe
PORCINE    Tyr-Gly-Leu-Leu-Ser-Cys-Phe-Lys-Lys-Asp-Leu-His-Lys-Ala-Glu-Thr-Tyr-
BOVINE                                 Arg                     Thr

HUMAN              Ile-Val-Gln     X       Ser         Gly         Gly
PORCINE    Leu-Arg-Val-Met-Lys-Cys-Arg-Arg-Phe-Val-Glu-Ser-Ser-Cys-Ala-PheCOOH.
BOVINE                                         Gly     Ala
```

Fig. 1. Sequence of 67 residues at the C-terminal end of porcine growth hormone as compared to that for HGH (Li, 1968) and BGH (Fellows and Rogol, 1969). The close similarity of the three sequences is readily apparent.

```
PGH              Ala-Met                                              /  Val      /
HGH  Phe-Pro-Thr-Ile-Pro-Leu-Ser-Arg-Leu-Phe-Asp-Asn-Ala-Met-Leu-Arg-Ala-

PGH  Gln-His                          Ala                Lys/             Arg/
HGH  His-Arg-Leu-His-Gln-Leu-Ala-Phe-Asp-Thr-Tyr-Glu-Glu-Phe-Glu-Glu-Ala-

PGH                                        Gln-Ala
HGH  Tyr . . . Asp-Leu-Glu-Glu-Gly-Ile-Glu-Thr-Leu-Met-
               113                                122
```

Fig. 2. Tentative sequence of a portion of porcine Fragment A. The sequence of the corresponding portion of HGH as revised by Niall (1971) is presented from the amino-terminal end; the numbered portion is taken from the sequence as presented by Li (1968). The dots indicate the section of Fragment A for which the peptides have not been sequenced. Strokes indicate the separate tryptic peptides of porcine Fragment A, and the underlining indicates the peptides of the porcine hormone for which the sequence has been established. The ordering of these peptides is assumed from the "match" of the sequence of HGH.

revised by Niall (1971) in Figure 2. In this Figure, the sequence of HGH is continuous, and the points at which PGH differs are placed separately. The underlining denotes regions in which the sequences of the porcine fragment have been accounted for. The strokes mark off separate peptides of the porcine fragment. The numbers at the carboxyl-terminal portion are those of the sequence as reported by Li (1968). The three dots indicate a region about which we as yet know nothing. Unfortunately it includes the very interesting region in which the tryptophan is thought to be, as well as the last half-cystine residue, but we cannot add anything to that story. Although we may be too bold in our alignment of peptides in Figure 2, it is both agreeable and a cause for hope and confidence to see the very great extent to which the sequences appear to agree. Dr. Fellows will doubtless add more firm information of this kind in his report, which follows this one.

As our view of the primary structure of HGH, BGH, OGH, PGH and HPL rapidly broadens in detail, the similarities in sequence seem greatly to outweigh the differences. It raises the question, whether the center of activity is to be found in some relatively simple sequence, or shall we finally learn that the biological differences between these molecules are dependent upon their tertiary structure? This question has not really been seriously considered, but I am glad to know that there are now at least three groups of X-ray crystallographers who have begun to take an interest in crystallizing growth hormones and in taking intimate pictures of their crystals. I hope that their work may be a subject for the Third International Symposium on Growth Hormone.

ACKNOWLEDGEMENTS

We express our continuing gratitude to our former colleagues, Dr. H. C. Chen and R. Ashworth and our present ones, Miss S. Scapa and Mr. S. Howard. A great burden of the work was carried by a group of loyal, willing and skilled technical assistants: Misses M. L. Willcox, M. Yager and M. Pappy, Mrs. L. Chang and Mrs. S. Wagner.

REFERENCES

CHEN, H. C., WILHELMI, A. E. and HOWARD, S. C. (1970): Purification and characteristics of porcine growth hormone. *J. biol. Chem.*, *245*, 3402.

DELLACHA, J. M., SANTOMÉ, J. A. and PALADINI, A. C. (1968): Physicochemical and structural studies on bovine growth hormone. *Ann. N. Y. Acad. Sci.*, *148*, 313.

FELLOWS Jr, R. E. and ROGOL, A. D. (1969): Structural studies on bovine growth hormone. I. Isolation and characterization of cyanogen bromide fragments. *J. biol. Chem.*, *244*, 1567.

LI, C. H. (1968): The chemistry of human pituitary growth hormone: 1956–1966. In: *Growth Hormone*, pp. 3–28. Editors: A. Pecile and E. E. Müller. ICS 158, Excerpta Medica, Amsterdam.

LI, C. H., EVANS, H. M. and SIMPSON, M. E. (1945): Isolation and properties of the anterior hypophyseal growth hormone. *J. biol. Chem.*, *159*, 353.

MILLS, J. B., HOWARD, S. C., SCAPA, S. and WILHELMI, A. E. (1970): Cyanogen bromide cleavage and partial amino acid sequence of porcine growth hormone. *J. biol. Chem.*, *245*, 3407.

NIALL, H. D. (1971): A revised primary structure for human growth hormone. *Nature (Lond.)*, *230*, 90.

NIALL, H. D., HOGAN, M. L., SAUER, R., ROSENBLUM, I. Y. and GREENWOOD, F. L. (1971): Sequences of pituitary and placental lactogenic and growth hormones; evolution from a primordial peptide by gene reduplication. *Proc. nat. Acad. Sci. (Wash.)*, *68*, 866.

OLIVER, L. and HARTREE, A. S. (1968): Amino acid sequences around the cystine residues in horse growth hormone. *Biochem. J.*, *109*, 19.

ORNSTEIN, L. (1964): Disc electrophoresis-I Background and theory. *Ann. N. Y. Acad. Sci.*, *121*, 321.

PAPKOFF, H., LI, C. H. and LIU, W. R. (1962): The isolation and characterization of growth hormone from porcine pituitaries. *Arch. Biochem.*, *96*, 216.

PEÑA, C., PALADINI, A. C., DELLACHA, J. M. and SANTOMÉ, J. A. (1970): Structural studies on ovine growth hormone. Cyanogen bromide fragments: N- and C-terminal sequences. *Europ. J. Biochem.*, *17*, 27.

WALLIS, M. (1966): A C-terminal sequence from ox growth hormone. *Biochim. biophys. Acta (Amst.)*, *115*, 423.

WALLIS, M. (1969): The N-terminus of ox growth hormone. *FEBS Letters*, *3*, 118.

STRUCTURAL STUDIES ON BOVINE GROWTH HORMONE*

ROBERT E. FELLOWS JR, ALAN D. ROGOL and ANNE MUDGE

Department of Physiology and Pharmacology,
Duke University Medical Center, Durham, N. C., U.S.A.

In the 25 years or more that relatively pure preparations of bovine growth hormone (BGH) have been available, a large body of information has been obtained concerning its physiological activity. However, the physical and chemical properties of the bovine hormone present many problems which have delayed progress in understanding its molecular structure. Among these are the prominent amino terminal heterogeneity (Ellis, 1961), which originally led to the suggestion that BGH might be a branched chain structure (Li, 1956), the tendency to aggregate in solutions of high ionic strength (Dellacha *et al.*, 1968) resulting in initial estimates of molecular weight as high as 46,000 and the demonstration of species specificity, suggesting

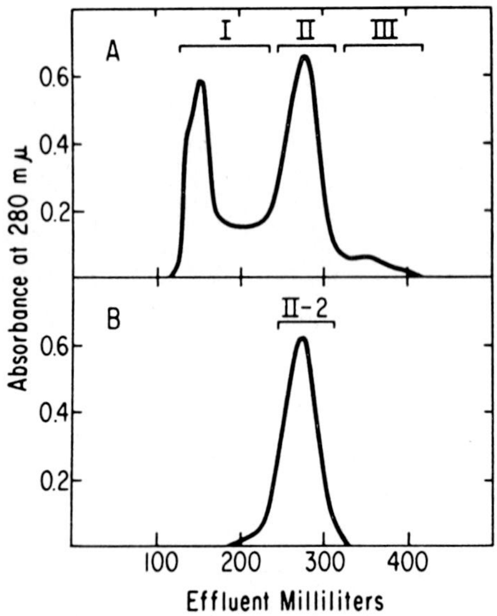

Fig. 1. Chromatographic purification of BGH.
A. Chromatography of NIH-GH-B11 on Sephadex G-100 in 1% formic acid.
B. Rechromatography of fraction II from A. Fraction II-2 was pooled as indicated. (From Fellows and Rogol, 1969. Courtesy of the Editors of *The Journal of Biological Chemistry*.)

* Supported in part by Grant No. AM12861 from the National Institute of Arthritis and Metabolic Diseases.

TABLE I

Amino terminal analysis of NIH-GH-B12 fraction II-2

Amino acid	Mole ratio[1] Observed	Blank	Net
Alanine	0.46	0.09	0.37
Phenylalanine	0.32	0.00	0.32
Methionine	0.26	0.00	0.26
Glutamic acid	0.27	0.24	0.03
Valine	0.03	0.02	0.01
Total			0.99

Values were obtained by the hydantoin method of Stark and Smyth (1963) and are based on a molecular weight of 20,800. (From Fellows and Rogol, 1969. Courtesy of the Editors of *The Journal of Biological Chemistry.)*

[1] Corrected to 100% recovery.

TABLE II

Amino acid composition of bovine growth hormone

Amino acid	NIH-GH-B12 fraction II-2	Free and Sonenberg (1966) Component I	Mills and Wilhelmi (1965)
Lysine	11.10 (11)	10.50	10.70
Histidine	2.55 (3)	2.96	2.81
Arginine	11.90 (12)	11.90	11.90
Aspartic acid	16.20 (16)	15.20	15.00
Threonine	12.30[1] (12)	11.40	11.20
Serine	11.60[1] (12)	11.70	11.50
Glutamic acid	23.90 (24)	21.70	23.40
Proline	6.10 (6)	6.10	6.88
Glycine	9.52 (10)	9.56	9.84
Alanine	12.80 (13)	13.50	13.30
Half-cystine	3.99[2] (4)	4.60	3.23
Valine	5.87 (6)	6.37	5.13
Methionine	3.98[3] (4)	3.87	3.70
Isoleucine	6.24 (6)	6.51	5.50
Leucine	24.40 (24)	24.30	23.70
Tyrosine	5.93 (6)	6.01	5.59
Phenylalanine	11.70 (12)	11.60	11.60
Tryptophan	1.03[4] (1)		
No. of residues	182		

Values for amino acids are expressed as residues per 20,800 g. Nearest integral residues of fraction II-2 are given in parenthesis. Component I of Free and Sonenberg (1966) was recalculated from the original data based on a molecular weight of 45,700. (From Fellows and Rogol, 1969. Courtesy of the Editors of *The Journal of Biological Chemistry.)*

[1] Extrapolated to zero time of hydrolysis.

[2] Determined as cysteic acid.

[3] Determined as methionine sulfone.

[4] Determined by the method of Edelhoch (1967).

that the primary structure of the bovine hormone might vary extensively from that of human growth hormone (HGH). While the amino acid sequence of BGH is still not completely established, recent progress in several laboratories has shed much light on its primary structure and its structural relationship to HGH and growth hormone from other species (Fellows and Rogol, 1969; Santome *et al.*, 1966; Wallis, 1969; Seavey *et al.*, 1971).

The sequence analysis of BGH has been pursued in our laboratory with material obtained from the Endocrinology Study Section of the National Institutes of Health. Preparations B11, B12, B14, and B16, having mean relative potencies of .83 to 1.04 USP units per mg, were initially chromatographed on columns of Sephadex G-100 in 1% formic acid or .05 M Tris HCL, pH 9.0 to obtain the monomeric fraction II (Fig. 1). This was rechromatographed on the same column to obtain fraction II-2 used for subsequent chemical studies. Material from fraction II-2 shows a single major and two minor bands by disc gel electrophoresis at pH 9.0 and a single band by vertical starch gel electrophoresis at pH 8.9. End group analysis of fraction II-2 (Table I) by the method of Stark and Smyth (1963) indicated that this fraction had, in addition to the usual amino terminal alanine and phenylalanine, a significant amount of methionine, as originally seen by Ellis (1961). The amino acid composition of fraction II-2 (Table II) was found to be very similar to that reported for highly purified preparations of BGH prepared in other laboratories. From 24, 48 and 72 hour hydrolysates it was calculated that there were 11 lysines and 12 arginines so that a maximum of 24 peptides should be produced by cleavage with trypsin. Among other residues which are important as structural markers are 3 histidines, 4 methionines, 4 cysteines and 1 single tryptophan.

Our initial strategy for the sequence analysis of BGH has involved isolation and characterization of five fragments produced by cleavage of the native hormone at methionine residues with cyanogen bromide according to the method of Steers *et al.* (1965). When the cyanogen

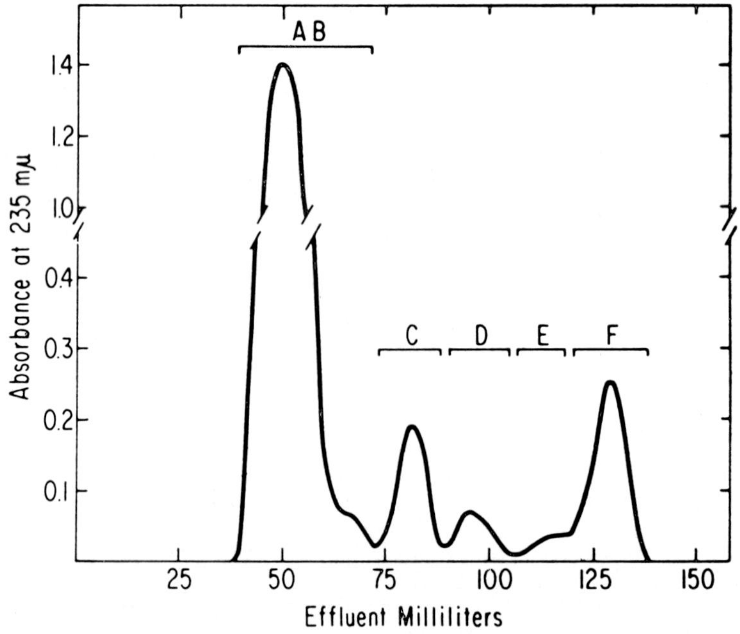

Fig. 2. Chromatographic separation of cyanogen bromide fragments of fraction II-2. Fragments were separated on Sephadex G-50 in 10% acetic acid and pooled as indicated. Pool F is a salt peak. (From Fellows and Rogol, 1969. Courtesy of the Editors of *The Journal of Biological Chemistry.*)

bromide fragments were initially separated by chromatography on Sephadex G-50 in 10% acetic acid (Fig. 2), four peaks were observed. The initial fraction, AB, contains two fragments connected by an intact disulfide bond. These were subsequently separated into fragment A and fragment B by chromatography on Sephadex G-75 in 50% acetic acid after reduction and alkylation of the cystine residue with either iodoacetic acid, iodoacetamide or ethyleneimine. Fragments C and D were purified by rechromatography on the same column and the mixture of small peptides found in fragment E were isolated by chromatography on Dowex 50 × 8 resin. Amino acid analysis of the individual fragments (Table III) showed that fragment A contains approximately 109 amino acids. Fragments B, C, and D have 30, 35 and 12 residues respectively and the major constituent of pool E is a pentapeptide. The sum of amino acids in the five fragments closely approximates the composition of the starting material. The order in which these fragments occur in the native molecule (Fig. 3) was deduced from data on the

TABLE III

Amino acid composition of cyanogen bromide fragments of bovine growth hormone

Amino acid	Fragment						NIH-GH-B12 fraction II-2
	A	B	C	D	E	Total	
Lysine	5.02 (5)	3.01 (3)	2.07 (2)	0.95 (1)		11	11
Histidine	1.88 (2)	0.99 (1)				3	3
Arginine	6.07 (6)	2.99 (3)	2.05 (2)	2.05 (2)		13	12
Aspartic acid	7.87 (8)	4.04 (4)	4.07 (4)			16	16
Threonine	7.12[1] (7)	2.04[1](2)	2.56 (3)			12	12
Serine	8.12[1] (8)	2.03[1](2)		0.96 (1)		11	12
Glutamic acid	16.70 (17)	1.44 (1)	4.23 (4)	1.20 (1)		24	24
Proline	4.01 (4)		1.18 (1)		0.97 (1)	6	6
Glycine	5.77 (6)	1.25 (1)	1.91 (2)	1.06 (1)		10	10
Alanine	8.00 (8)	1.26 (1)	1.14 (1)	2.03 (2)	2.00 (2)	14	13
Cystine	0.89[2] (1)	1.00[2](1)		1.79[2](2)		4	4
Valine	4.32 (4)	1.10 (1)	0.40			5	6
Methionine	0.92[3] (1)	0.80[3](1)	0.74[3](1)		0.83[3](1)	4	4
Isoleucine	4.93[4] (5)		0.96 (1)			6	6
Leucine	15.60[4](16)	5.97[4](6)	1.94 (2)			24	24
Tyrosine	3.18 (3)	2.05 (2)	0.84 (1)			6	6
Phenylalanine	6.98 (7)	1.26 (1)	1.04 (1)	1.89 (2)	1.00 (1)	12	12
Tryptophan	1.04[5] (1)					1	1
Number of residues	109	30	25	12	5	182	182
Amino terminal residue (% yield)	none detected	Arg (99)	Arg (67)	Lys (98)	Ala (95)		Ala (37) Phe (32) Met (26)

Values are expressed as residues per mole. Sums of the assumed number of residues, in parentheses, are compared with integral values for the whole molecule. Amino terminal residues of all fragments were determined by the subtractive Edman procedure and for fragment A and fraction II-2 by the method of Stark and Smyth (1963). (From Fellows and Rogol, 1969. Courtesy of the Editors of *The Journal of Biological Chemistry.)*

[1] Extrapolated to zero time of hydrolysis.
[2] Determined as carboxymethylcysteine.
[3] Estimated as the sum of homoserine and homoserine lactone.
[4] Taken from 72 hour hydrolysate.
[5] Determined by the method of Barman and Koshland (1967).

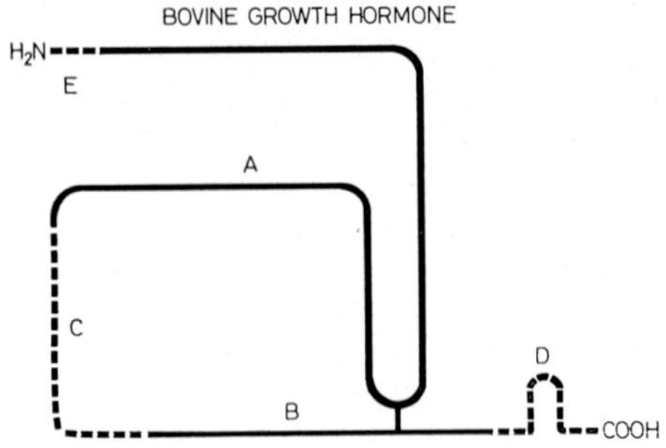

Fig. 3. The location of cyanogen bromide fragments in BGH. Fragments have been placed in the order E-A-C-B-D starting from the amino terminus.

structure of methionine containing tryptic peptides of BGH, the composition of fragments produced by cleavage at the single tryptophan residue and by direct sequence analysis of the native molecule. The pentapeptide fragment E is first, followed by the large fragment A. Fragments C, B, and D, in that order, make up the rest of the molecule.

Our investigation of primary structure has been directed initially to fragments C, B, and D, then to fragment E, and finally to fragment A. The sequence of fragments C, B, and D has been determined by classical methods of protein chemistry involving the isolation and subtractive Edman degradation (Konigsberg and Hill, 1962) of sets of small peptides. The sequence of fragment C (Table IV) was deduced by sequential degradation of peptides obtained after cleavage with trypsin, thermolysin, chymotrypsin and dilute hydrochloric acid. The order of the three tryptic peptides was apparent from amino and carboxyl terminal analysis of each peptide and of the whole fragment. Two peptides were obtained which contain the

TABLE IV

Summary of proof of sequence of cyanogen bromide fragment C of bovine growth hormone

Peptide	Sequence of peptides
Tp-1	Arg
Tp-2	(Arg,Glu,$^{\text{Leu}}_{\text{Val}}$,Glu,Asp,Gly,Thr,Pro)Arg
Tp-2 Th-1	Arg-Glu
Tp-2 Th-2	$^{\text{Leu}}_{\text{Val}}$-Glx-Asx-Gly-Thr-Pro-Arg
Tp-2 A-3	Gly-Thr-Pro-Arg
Tp-3	Ala-Gly-Gln-Ile-Leu-Lys
Sequence	Arg-Glu-$^{\text{Leu}}_{\text{Val}}$-Glu-Asp-Gly-Thr-Pro-Arg-Ala-Gly-Gln-Ile-Leu-Lys
Tp-4	(Glx,Thr,Tyr,Asx,Lys, Phe,Asx,Thr,Asx,Met)
Tp-4 C-1	Gln-Thr-Tyr
Tp-4 C-2	Asp-Lys-Phe
Tp-4 C-3	Asp-Thr-Asn-Met
Sequence	Gln-Thr-Tyr-Asp-Lys-Phe-Asp-Thr-Asn-Met

leucine-valine microheterogeneity originally found in this fragment and since confirmed by Seavey *et al.* (1971). The sequence of fragment B was also deduced by isolation and sequential degradation of tryptic and thermolytic peptides (Table V). The cysteine residue which forms the disulfide bond with the cysteine in fragment A is in position 15 of this fragment, which also contains one of the three histidine residues of BGH.

The sequence of fragment D, the carboxyl terminal fragment, was determined by amino and carboxyl terminal degradation of the carbamidomethylated fragment, and of a tryptic peptide and three thermolytic peptides (Table VI). Our findings are similar to those of Santome *et al.* (1966) and confirm the presence of the Arg-Arg sequence which is not present in human growth hormone (HGH). This fragment also contains the second and smaller disulfide loop of BGH entirely within it. The sequence of the carboxyl terminal 68 residues of BGH established by studies on fragments C, B, and D, are shown in Table VII. The points of cya-

TABLE V

Summary of proof of sequence of cyanogen bromide fragment B of bovine growth hormone

Peptide	Sequence of peptides
Tp-1	Arg-Ser- Asp-Asp-Ala-Leu-Leu-Lys
Tp-2	Asx-Tyr(Gly, Leu, Leu, Ser, Cys)
Th-3	Leu-Lys-Asn
Th-4	Leu-Lys-Asx-Tyr-Gly
Th-6	Leu-Leu-Ser-Cys
Sequence	Arg-Ser -Asp-Asp-Ala-Leu-Leu-Lys-Asn-Tyr-Gly-Leu-Leu-Ser-Cys
Tp-2	Phe-Arg
Tp-4	Asp-Leu-His -Lys
Tp-5	Thr-Glx-Thr-Tyr-Leu-Arg
Tp-6	Val-Met
Th-7	Phe-Arg-Lys-Asp-
Th-8	Leu-His -Lys(Thr, Glu, Thr, Tyr)
Th-9	(Leu, Arg, Val)Met
Th-10	(Leu, Arg)
Th-11	Val-Met
Sequence	Phe-Arg-Lys -Asp-Leu-His -Lys-Thr-Glu-Thr-Tyr-Leu-Arg-Val-Met

TABLE VI

Summary of proof of sequence of cyanogen bromide fragment D of bovine growth hormone

Peptide	Sequence of peptides
Fragment S-CAM D	Lys-Cys-Arg-Arg-Phe(Gly, Glx, Ala, Ser)Cys-Ala-Phe
Th-2	Phe-Gly(Glx, Ala, Ser-Cys, Ala, Phe)
Th-2B	Phe-Gly-Glu
Th-2C	Ala-Ser-Cys
Th-2D	Ala-Phe
Sequence	Lys-Cys-Arg-Arg-Phe-Gly-Glu-Ala-Ser-Cys-Ala-Phe

TABLE VII

The carboxyl terminal sequence of bovine growth hormone

CnBr
↓
----Met-Arg-Glu-(Leu/Val)-Glu-Asp-Gly-Thr-Pro-Arg-Ala-Gly-Gln-

CnBr
↓
Ile-Leu-Lys-Gln-Thr-Tyr-Asp-Lys-Phe-Asp-Thr-Asn-Met-Arg-

|
Ser-Asp-Asp-Ala-Leu-Leu-Lys-Asn-Tyr-Gly-Leu-Leu-Ser-Cys-

Phe-Arg-Lys-Asp-Leu-His-Lys-Thr-Glu-Thr-Tyr-Leu-Arg-Val-

CnBr
↓
Met-Lys-Cys-Arg-Arg-Phe-Gly-Glu-Ala-Ser-Cys-Ala-Phe-COOH

TABLE VIII

Peptides isolated from crude fragment E by chromatography on Dowex 50 × 8

Amino acid	Chromatographic fraction			
	IV	VI	VII	X
Proline	1.00	1.00		
Alanine	2.14	1.03		1.00
Methionine	0.85	0.93	1.00	0.95
Phenylalanine	1.12	1.11		
No. of residues	5	4	1	2

Values are expressed as mole ratios. Methionine was determined as homoserine and homoserine lactone.

TABLE IX

The amino terminus of growth hormones

Bovine:	Ala-Phe-Pro-Ala-Met-
	Phe-Pro-Ala-Met-
	Ala-Met-
	Met-
Porcine:	Phe-Pro-Ala-Met-
	Ala-Met-
Human:	Phe-Pro-Thr-Ile-

nogen bromide cleavage, the location of the valine-leucine heterogeneity and the location of cysteine residues participating in the major disulfide loop and forming the carboxyl terminal loop are evident.

After completion of the sequence analysis of this part of the molecule, our attention was directed to the amino terminus and the problem of amino terminal heterogeneity. From pool E (Fig. 2), a series of small fragments containing 5, 4, 2, and 1 amino acids and having the compositions shown in Table VIII were isolated by ion exchange chromatography. By subtractive Edman degradation, those proved to be members of an homologous series related to the parent sequence Ala-Phe-Pro-Ala-Met. Although we had expected to find the pentapeptide and tetrapeptide, the sequences of which have been independently reported by Wallis (1969), and also free methionine in the form of homoserine, the dipeptide Ala-Met was obtained as well. These findings, summarized in Table IX, indicate that the amino terminal structure of BGH is somewhat more complex than is seen in the obviously homologous regions of porcine growth hormone (Mills *et al.*, 1970) and HGH (Li *et al.*, 1969; Niall, 1971).

Our studies on cyanogen bromide fragment, A, which follows fragment E in BGH, have involved a different strategy based on use of the automatic Sequencer and direct Edman degradation of larger peptides and fragments of BGH. Table X shows the amino terminal

TABLE X

NH_2-terminal sequence of cyanogen bromide fragment A of bovine growth hormone

```
                    10
Ser -Leu-Ser -Gly-Leu-Phe-Ala-Asn-Ala-Val-Leu-Arg-Ala-
     20                                   30
Gln-His -Leu-His-Gln-Leu-Ala-Ala-Asp-Thr-Phe-Lys-Glu-
                                         40
Phe-Glu-Arg-Thr-Tyr-Ile -Pro-Glu-Gly-Gln-Arg-Tyr-Ser-
```

sequence of fragment A which was obtained with the automatic Sequencer, initially at the Massachusetts General Hospital with the help of Dr. Hugh Niall, and subsequently repeated in part at Duke University. The finding of amino terminal serine, also indicated by the data of Wallis (1969), explains our original inability to identify the end group of this fragment by the method of Stark and Smyth in which there is severe destruction of serine. The remaining two histidine residues are found in positions 20 and 22 of BGH, and again there is evidence suggesting subspecies microheterogeneity at position 24 involving a Leu-Gln substitution. The latter, however, remains to be proved by isolation and characterization of small peptides corresponding to each variant. The sequence obtained by automated methodology stops shortly before the cysteine in fragment A which, with the cysteine in fragment B, forms the major disulfide loop of BGH. In order to obtain additional sequence data on fragment A, large peptides have been produced by cleavage of native and succinylated BGH at the single tryptophan residue by a modification of the method of Atassi (1967) and by tryptic digestion of fragment A after blocking lysine residues with maleic anhydride (Butler *et al.*, 1969). Two tryptophan cleavage fragments were separated by gel filtration and ion exchange chromatography. From their composition it has been determined that the tryptophan occurs approximately at position 86 in BGH, a location which is similar to that described for human placental lactogen and the revised structure of HGH (Niall *et al.*, 1971). The sequence following the tryptophan in BGH, Leu-Glu-Pro-Leu, is the same as the sequences of the analogous regions of HPL and HGH. From data on the lysine blocked arginine peptides of fragment A we have deduced a preliminary structure for the remainder of fragment A. Although this must be confirmed, it appears that all of fragment A will prove to be homologous to the revised sequence of HGH.

While the sequence analysis of fragment A was in progress, we focused our attention on several features of BGH which could be explained on the basis of the primary sequence data available.

In order to study possible methods by which the amino terminal heterogeneity of BGH might be generated, we have synthesized by solid phase methodology the tetrapeptide, Ala-Phe-Pro-Ala, corresponding to the first four residues of BGH. When this peptide was incubated at 40° C with a high speed supernatant of a bovine pituitary homogenate (Table XI),

TABLE XI

Degradation of synthetic Ala-Phe-Pro-Ala

Enzyme	pH	Alanine released[1] time (hours)			
		1	2	3	4
Bovine pituitary homogenate	7.3	0.04	0.09	0.14	0.18
Plasmin	7.3	0.00	0.00	0.00	0.00

[1] Values for free alanine are μmoles per 1.0 μmole of substrate.

the time dependent release of free alanine (but not phenylalanine or proline) was demonstrated by thin layer chromatography and by direct amino acid analysis. The generation of amino terminal phenylalanine in amounts equal to free alanine released was also shown by direct Edman degradation. Because it has been suggested that the enzyme that degrades BGH is plasmin (Ellis *et al.*, 1968), this experiment was repeated using highly purified human plasminogen which was activated for 30 minutes with urokinase. Since free alanine was not released we conclude that the amino terminal heterogeneity in BGH is caused by a pituitary enzyme, but it is not plasmin. It seems quite reasonable to expect, since 18% of amino terminal alanine was cleaved from the synthetic tetrapeptide in 4 hours, that generation of the Phe end group in BGH occurs by a similar mechanism before or during its extraction from the pituitary, as we and others have suggested (Fellows and Rogol, 1969; Wallis, 1969; Ellis, 1961). Our findings are also compatible with data obtained from single pituitaries indicating that the amino terminal heterogeneity is non-allelic in origin (Peña *et al.*, 1969).

A second area of particular interest concerns the biochemical evolution of growth hormone and related proteins. Using a computer program which identifies regions of statistically significant homology based on minimum nucleotide base changes per codon, we have observed apparent internal homology in BGH (Table XII). In this and the following tables of sequence comparisons, a solid line indicates identical residues and a broken line signifies one step mutations of the nucleotide code. In this comparison, 27% of positions have identical residues and 46% have single step mutations. In addition to the three regions shown (one near the amino terminus at residues 16–33 and two in the carboxyl terminal region which are homologous with the positions of HGH indicated in parentheses) we have also identified a fourth homologous region following residue 90 in BGH. This pattern suggests that the BGH molecule evolved from a small ancestral peptide which underwent two successive reduplications to produce the present molecule, a mechanism similar to that which has been proposed for the immunoglobulins (Hill *et al.*, 1966). A similar pattern of internal homology has been

TABLE XII

Internal homology in bovine growth hormone

		20 30
BGH (16–33):	Leu-Arg-	Ala-Gln-His-Leu-His-Gln-Leu-Ala-Ala-Asp-Thr-Phe-Lys-Glu-Phe-Glu-Arg-
	(130)	(140)
BGH (C1):	Leu-Glu-Asp-Gly-Thr-Pro-Arg-Ala	-Gly-Gln-Ile-Leu-Lys-Gln-Thr-Tyr-Lys-Lys-Phe-Asp-Thr-
		(170) (180)
BGH (C2):	Leu-Ser-	Cys-Phe-Arg-Lys-Asp-Leu-His-Lys-Thr-Glu-Thr-Tyr-Leu-Arg-Val-Met-Lys-

Identical residues are indicated by a solid line and single step mutations by a broken line. Numbers in brackets indicate homologous positions in human growth hormone.

TABLE XIII

Amino terminal homology of growth hormones and prolactin

	10 20
BGH:	Ala-Phe-Pro-Ala-Met-Ser-Leu-Ser-Gly-Leu-Phe-Ala-Asn-Ala-Val-Leu-Arg-Ala-Gln-His-Leu-His-
	10 20
OGH:	Ala-Phe-Pro-Ala-Met-Ser-Leu-Ser-Gly-Leu-Phe-Ala-Asn-Ala-Val-Leu-Arg-Ala-Gln-His-Leu-His-
	10 20
HGH:	Phe-Pro-Thr-Ile-Pro-Leu-Ser-Arg-Leu-Phe-Asp-Asn-Ala-Met-Leu-Arg-Ala-His-Arg-Leu-His-
	20 30
OP:	Thr-Pro-Val/Val-Ser-Leu-Arg-Asp-Leu-Phe-Asp-Arg-Ala-Val-Met-Val-Ser-His-Tyr-Ile-His-
	30 40
BGH:	Gln-Leu-Ala-Ala-Asp-Thr-Phe-Lys-Glu-Phe-Glu-Arg-Thr-Tyr-Ile-Pro-Glu-Gly-Gln-Arg-Tyr-
	30 40
OGH:	Gln-Leu-Ala-Ala-Asp-Thr-Phe-Lys-Glu-Phe-Glu-Arg-Thr-Tyr-Ile-Pro-Glu-Gly-Gln-Arg-
	30 40
HGH:	Gln-Leu-Ala-Phe-Asp-Thr-Tyr-Gln-Glu-Phe-Glu-Glu-Ala-Tyr-Ile-Pro-Lys-Glu-Gln-Lys-Tyr-
	40 50
OP:	Asn-Leu-Ser-Ser-Glu-Met-Phe-Asn-Glu-Phe-Asp-Lys-Arg-Tyr- Ala-Gln-Gly-Lys-Gly-Phe-

Identical residues are indicated by a solid line and single step mutations by a broken line.

found in HGH and HPL (Niall *et al.*, 1971), indicating that the gene reduplications occurred before the subsequent evolution of growth hormones to produce contemporary species differences.

By contrast to the moderate degree of homology observed when BGH is compared with itself, very strong homology is seen when the amino terminal segments of growth hormones and prolactin are compared (Table XIII). Here, the amino terminal 43 residues of BGH are compared with similar regions of ovine growth hormone (OGH), HGH and ovine prolactin (OP) (Li *et al.*, 1970). The preliminary animo terminal sequence shown for OGH was derived from data obtained in our laboratory on native OGH and cyanogen bromide fragment A of OGH by manual and automatic Edman degradation and from peptide data of Dr. Terry Bellair (personal communication). The HGH sequence is the revised sequence reported by

TABLE XIV

Carboxyl terminal homology of growth hormones and prolactin

	Leu
BGH:	Glu-Val -Glu-Asp-Gly -Thr -Pro -Arg-Ala-Gly -Gln-Ile -Leu-Lys-Gln-Thr-Tyr-Asp-Lys -Phe -Asp-Thr-
	130 140
HGH:	Arg-Leu -Glu-Asp-Gly -Ser -Pro -Arg-Thr-Gly -Gln-Ile -Phe-Lys-Gln-Thr-Tyr-Ser -Lys -Phe -Asp-Thr-
	140 150
OP:	Val -Ile -Pro -Gly -Ala -Lys -Glu-Thr -Glu-Pro -Tyr -Pro-Val-Trp-Ser -Gly-Leu-Pro -Ser -Leu -Gln-Thr-
BGH:	Asn-Met-Arg-Ser -Asp-Asp-Ala -Leu-Leu-Lys -Asn-Tyr-Gly-Leu-Leu-Ser -Cys -Phe-Arg -Lys -Asp-Leu-
	150 160
HGH:	Asn-Ser -His -Asn-Asp-Asp-Ala -Leu-Leu-Lys -Asn-Tyr-Gly-Leu-Leu-Tyr-Cys -Phe-Arg -Lys -Asp-Met-
	160 170
OP:	Lys - Asp-Glu-Asp-Ala -Arg-His -Ser -Ala -Phe-Tyr-Asn-Leu-Leu-His-Cys -Leu-Arg -Arg-Asp-Ser-
BGH:	His -Lys -Thr-Glu-Thr -Tyr -Leu-Arg-Val -Met-Lys -Cys-Arg-Arg-Phe-Gly-Glu-Ala -Ser -Cys-Ala -Phe
	170 180
HGH:	Asp-Lys -Val -Glu-Thr -Phe -Leu-Arg-Ile -Val -Gln-Cys-Arg- Ser -Val-Glu-Gly -Ser -Cys-Gly -Phe
	180 190
OP:	Ser -Lys -Ile -Asp-Thr -Tyr -Leu-Lys-Leu-Leu -Asn-Cys-Arg-Ile -Ile -Tyr-Asn-Asn-Asn-Cys

Identical residues are indicated by a solid line and single step mutations by a broken line.

Niall (1971). The amino terminal disulfide loop (residue 4–12) has been deleted from prolactin to facilitate the comparison. Here 71 % of positions compared are identical and 26 % involve a single base change. In contrast to what had been suggested from comparisons of partial sequences of the amino terminal region of BGH reported by others with the amino terminal sequence of HGH, it is clear that the homology among all four proteins, as well as with the corresponding sequence of HPL (Niall *et al.*, 1971) which is not included in this comparison, is very strong. OGH is clearly most closely and prolactin least closely related to BGH in an evolutionary sense. A comparison of the carboxyl terminal structure of BGH, HGH and OP demonstrates the same strong homology in this part of the molecule (Table XIV). Among the positions compared 49 % are identical and 29 % are homologous. BGH is clearly more closely related to HGH than to prolactin.

Although the sequence of the middle portion of BGH is provisional and it is therefore premature to attempt a detailed comparison with HGH, our data derived from tryptophan fragments and arginine peptides of BGH fragment A suggest that the homology with HGH is consistent through this region as well. It now seems most unlikely that a nonhomologous segment exists in BGH which might be responsible for the species specificity of the hormone; rather, it appears that the molecular determinants of species specificity must lie in the relatively few nonhomologous residues occurring in otherwise homologous regions of primary structure or in conformational changes produced by amino acid substitutions occurring throughout the molecule.

With the common evolutionary origin of growth hormones and prolactin clearly established by comparisons such as these, it is then of interest to look for evidence of homology between growth hormone and other protein hormones which are not closely related in a functional sense. When the partial sequence of BGH and the revised sequence of HGH were

TABLE XV

Homology between growth hormone and alpha subunit of LH and TSH

	40
BGH:	Glu-Gly-Gln-Arg-Tyr-Ser -Ile -Gln-Asp-
	40 50
HGH:	Lys-Glu-Gln-Lys-Tyr-Ser-Phe-Leu-Gln-Asp-Pro-Glu-Thr-Ser-Leu-Cys-
LHα :	20 30
TSHα:	Lys-Glu-Asn-Lys-Tyr-Phe-Ser-Lys-Pro-Asp-Ala-Pro-Ile-Tyr-Gln-Cys-
	60
HGH:	Leu-Cys-Phe-Ser-Glu-Ser-Ile-Pro-Thr-Pro-Ser-Asn-Arg-Glu-Glu-Thr
LHα :	40 50
TSHα:	Cys-Cys-Phe-Ser-Arg-Ala-Tyr-Pro-Thr-Pro-Ala-Arg-Ser-Lys-Lys-Thr

Identical residues are indicated by a solid line and single step mutations by a broken line.

compared with the primary structure of the subunits of luteinizing hormone (LH) (Liu *et al.*, 1971) and thyroid stimulating hormone (TSH) (Liao and Pierce, 1971) homology was observed with a limited area of the alpha subunits of the two glycoprotein hormones (Table XV). In the comparison between HGH and the alpha subunit of LH or TSH, 41 % of positions have identical residues and 41 % involve one step mutations. Since LH and TSH appear also to share a common alpha subunit with human chorionic gonadotrophin (Pierce *et al.*, 1971) it is likely that the homology extends to this hormone as well. This suggests that both the glycoprotein hormones and the simple protein hormones, growth hormone, prolactin and placental lactogen, have evolved from some even more primitive common precursor. Clearly a more extensive search for homology among these hormones will be necessary in order to determine whether this conclusion is indeed valid.

ACKNOWLEDGEMENTS

We are indebted to Dr. Hugh Niall, Dr. Thomas Vanaman, Robert Vanzant and Delores Johnson for assistance in various phases of this work. Purified human plasminogen was a gift of Mr. Sal Pizzo and bovine growth hormone was generously supplied by the Endocrinology Study Section, NIH.

REFERENCES

Atassi, M. Z. (1967): Specific cleavage of tryptophyl peptide bonds with periodate in sperm whale myoglobin. *Arch. Biochem.*, *120*, 56.

Barman, T. E. and Koshland Jr, D. E. (1967): A colorimetric procedure for the quantitative determination of tryptophan residues in proteins. *J. biol. Chem.*, *242*, 5771.

Butler, P. J. G., Harris, J. I., Hartley, B. S. and Leberman, R. (1969): The use of maleic anhydride for the reversible blocking of amino groups in polypeptide chains. *Biochem. J.*, *112*, 679.

Dellacha, J. M., Enero, M. A. and Paladini, A. C. (1968): Physiochemical behaviour and biological activity of bovine growth hormone in acidic solution. *Biochim. biophys. Acta (Amst.)*, *168*, 95.

Edelhoch, H. (1967): Spectroscopic determination of tryptophan and tyrosine in proteins. *Biochemistry*, *6*, 1948.

Ellis, S. (1961): Studies on the serial extraction of pituitary proteins. *Endocrinology*, *69*, 554.

Ellis, S., Nuenke, J. M. and Grindeland, R. E. (1968): Identity between the growth hormone degrading activity of the pituitary gland and plasmin. *Endocrinology*, *83*, 1029.

FELLOWS, R. E. and ROGOL, A. D. (1969): Structural studies on bovine growth hormone. *J. biol. Chem.*, *244*, 1567.

FREE, C. A. and SONENBERG, M. (1966): Separation and properties of multiple components of bovine growth hormone. *J. biol. Chem.*, *241*, 5076.

HILL, R. L., DELANEY, R., FELLOWS, R. E. and LEBOVITZ, H. E. (1966): The evolutionary origins of the immunoglobins. *Proc. nat. Acad. Sci. (Wash.)*, *56*, 1762.

KONIGSBERG, W. and HILL, R. J. (1962): The structure of human hemoglobin. *J. biol. Chem.*, *237*, 2547.

LI, C. H. (1956): Hormones of the anterior pituitary gland. Part I. Growth and adrenocorticotropic hormones. *Advanc. Protein Chem.*, *11*, 101.

LI, C. H., DIXON, J. S. and LIU, W. K. (1969): Human pituitary growth hormone. XIX. The primary structure of the hormone. *Arch. Biochem.*, *133*, 70.

LI, C. H., DIXON, J. S., LO, T. B., SCHMIDT, K. D. and PANKOV, Y. A. (1970): Studies on pituitary lactogenic hormone. XXX. The primary structure of the sheep hormone. *Arch. Biochem.*, *141*, 705.

LIAO, T. H. and PIERCE, J. G. (1971): The primary structure of bovine thyrotropin. *J. biol. Chem.*, *246*, 850.

LIU, W. K., NAHM, H. S., SWEENEY, C. M., BAKER, H. N., LAMKIN, W. M. and WARD, D. N. (1971): The amino acid sequence of the S-aminoethylated ovine luteinizing hormone S-subunit (LH-a). *Res. Commun. chem. Path. Pharm.*, *2*, 168.

MILLS, J. B., HOWARD, S. C., SCAPA, S. and WILHELMI, A. E. (1970): Cyanogen bromide cleavage and partial amino acid sequence of porcine growth hormone. *J. biol. Chem.*, *245*, 3407.

MILLS, J. B. and WILHELMI, A. E. (1965): Sulfitolysis of bovine growth hormone. *Endocrinology*, *76*, 522.

NIALL, H. D. (1971): Revised primary structure for human growth hormone. *Nature new Biol.*, *230*, 65.

NIALL, H. D., HOGAN, M. L., SAUER, R., ROSENBLUM, I. Y. and GREENWOOD, F. C. (1971): Sequences of pituitary and placental lactogenic and growth hormones: Evolution from a primordial peptide by gene reduplication. *Proc. nat. Acad. Sci. (Wash.)*, *68*, 866.

PEÑA, C., PALADINI, A. C., DELLACHA, J. M. and SANTOME, J. A. (1969): Evidence for nonallelic origin of the two chains in ox growth hormone. *Biochim. biophys. Acta (Amst.)*, *194*, 320.

PIERCE, J. G., BAHL, O. P., CORNELL, J. S. and SWAMINATHAN, N. (1971): Biologically active hormones prepared by recombination of the a chain of human chorionic gonadotropin and the hormone-specific chain of bovine thyrotropin or of bovine luteinizing hormone. *J. biol. Chem.*, *246*, 2321.

SANTOME, J. A., WOLFENSTEIN, C. E. M., BISCOGLIO, M. and PALADINI, A. C. (1966): Sequence of thirteen amino acids in the C-terminal end of bovine growth hormone and localization of a disulfide bridge. *Arch. Biochem.*, *116*, 19.

SEAVEY, B. K., SINGH, R. N. P. and LEWIS, U. J. (1971): Bovine growth hormone: evidence for two allelic forms. *Biochem. biophys. Res. Commun.*, *43*, 189.

STARK, G. L. and SMYTH, D. G. (1963): The use of cyanate for the determination of NH_2-terminal residues in proteins. *J. biol. Chem.*, *238*, 214.

STEERS, E., CRAVEN, G. R., ANFINSEN, C. B. and BETHUNE, J. J. (1965): Evidence for nonidentical chains in the B-galactosidase of Escherichia coli K12. *J. biol. Chem.*, *240*, 2478.

WALLIS, M. (1969): The N-terminus of ox growth hormone. *FEBS Letters*, *3*, 118.

SEPARATION OF THE PHENYLALANYL AND ALANYL CHAINS OF BOVINE AND OVINE GROWTH HORMONES BY ELECTROFOCUSING

S. ELLIS, M. LORENSON, R. E. GRINDELAND and P. X. CALLAHAN

Environmental Biology Division, Ames Research Center, National Aeronautics and Space Administration, Moffett Field, Calif., U.S.A.

The growth hormones which have been isolated from human, monkey, pig (Li and Liu, 1964) or rat pituitary glands (Ellis *et al.*, 1968) contain only phenylalanine at the NH_2-terminus. On the other hand, bovine and ovine growth hormones contain approximately equimolar quantities of NH_2-terminal phenylalanine and alanine. Contrary to previous suppositions (Li *et al.*, 1955) recent studies have shown that the minimum molecular weight of the bovine and ovine hormones is about 21,000 rather than 45,000 (Andrews, 1966; Dellacha *et al.*, 1966) and that there are two COOH-termini rather than one (Wallis, 1966). Presumably, the discrepant molecular weights which have been reported arise from the strong tendency of these hormones to dimerize under appropriate conditions of pH, ionic strength and protein concentration which as yet have not been fully defined.

The foregoing findings indicate that the bovine and ovine hormones are comprised of two subunits which differ at their NH_2-termini and perhaps in other portions of their peptide chains as well. With a view to learning whether the subunits differ significantly in biological properties and composition, separation of the subunits was undertaken in the hope that the high resolution afforded by the technique of isoelectric focusing (Vesterberg and Svensson, 1966) would accomplish a useful resolution.

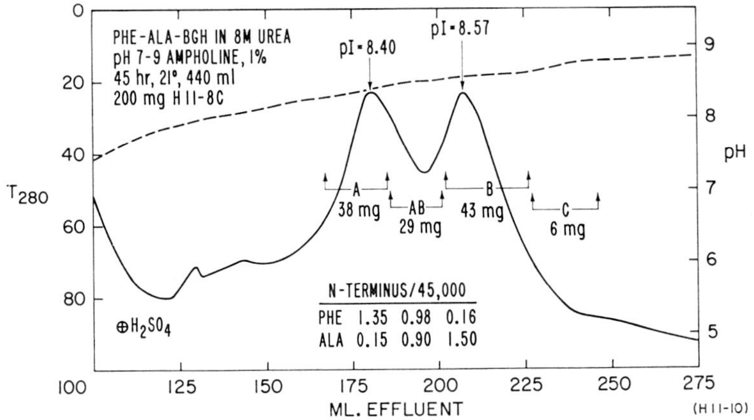

Fig. 1. Isoelectric focusing of (Ala,Phe)-bovine growth hormone in the presence of 8 M urea. A gradient of 0-36% sucrose, 8 M urea, 1% Ampholine (pH 7-9) was prepared with 200 mg of hormone solubilized in fraction No. 18 (14% sucrose). Conditions: 440 ml column, 45 hours of focusing at 21°; total power constant at 3 watts (final voltage = 695); pH measured on 5 ml fractions of effluent. A turbid fraction was noted near the anode chamber within 5 hours after focusing was begun.

Our first experiments (Ellis *et al.*, 1970) employed pH 7–9 Ampholine containing 8 M urea throughout the column for the purpose of solubilizing the hormone which is poorly soluble at the isoelectric region in solvents of low ionic strength. The hormone was dissolved in a single 9 ml volume of pre-mixed Ampholine-sucrose-8 M urea and applied to a 440 ml LKB column (Haglund, 1967) at about the midpoint of the gradient. The profile of the protein distribution which was obtained after 45 hours of electrofocusing at 21° is shown in Figure 1. Two major peaks were discernible, one of which showed an apparent pI of 8.57 and the other, a pI of 8.40. After recovering the pooled effluent by dialysis and lyophilization, NH_2-terminal analysis was performed by the dinitrofluorobenzene method. The component having a pI of 8.57 showed predominantly alanine at the NH_2-terminus while the component with a pI of 8.40 yielded mainly phenylalanine. The fraction intermediate between the two peaks had the same NH_2-terminal composition as the starting hormone, namely, 0.98 moles of phenylalanine and 0.90 moles of alanine as calculated on a molecular weight of 45,000. The growth potency of both peaks was estimated to be 1.4 to 1.5 USP units per mg by the tibial assay as compared to 2.6 for the original hormone. These results indicate that the phenylalanyl and alanyl peptide chains of bovine growth hormone are separable by electrofocusing in 8 M urea and that both chains are equipotent in their biological activity as measured by tibial assay.

The reproducibility of the protein profiles was examined by resubmitting the isolated peaks to isoelectric focusing under the original conditions. In Figure 2 are shown the absorbancy profiles of an initial separation of the two components as well as that of the derived components. The refocused components yielded the expected pI values and absorbancy profiles. The growth hormone used in these experiments was isolated from near neutral pituitary extracts (Ellis, 1961) where partial proteolytic cleavage of an NH_2-terminal peptide

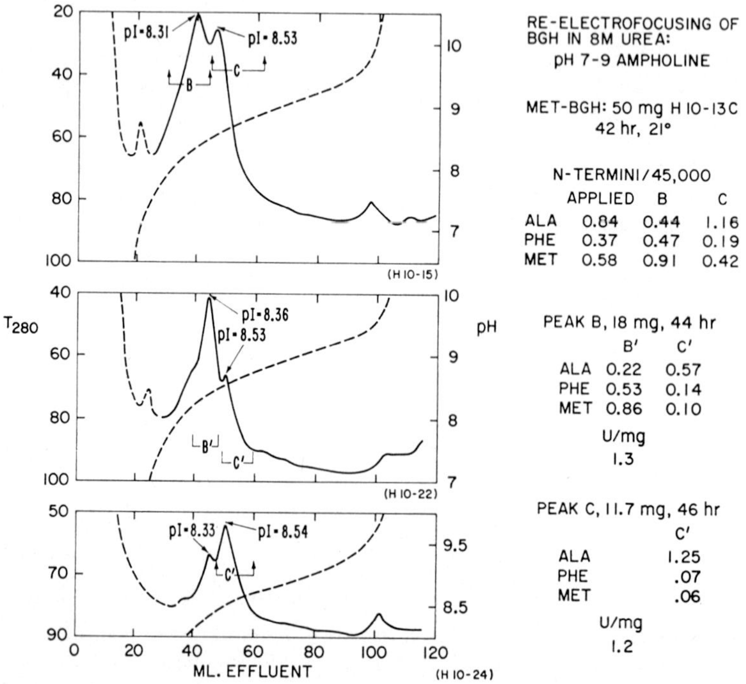

Fig. 2. Effect of re-electrofocusing. Conditions are as those given in Figure 1 except that a 110 ml column and the (Ala,Met)-hormone were used.

is rapidly incurred in the case of bovine extracts. Figure 2 therefore shows the distribution of three NH_2-terminal residues which were present in the applied hormone in the amounts of 0.84 moles of alanine, 0.37 moles of phenylalanine and 0.58 moles of methionine per 45,000 mol.wt. The association of NH_2-terminal phenylalanine and methionine in the refocused B peak and their virtual absence from the C peak shows that the phenylalanyl and methionyl chains have the same pI and that the methionyl chain is probably derived only from the phenylalanyl chain by proteolytic cleavage.

Since the dimer of bovine growth hormone appears to undergo a concentration-dependent dissociation in neutral solutions (Andrews, 1966), it was of interest to determine if a separation of the peptide chains of bovine and ovine growth hormones could be achieved in the absence of urea. Fortunately, the solubilizing influence of the sucrose density gradient used in the electrofocusing column prevents precipitation of the isoelectric zones. For application to the column the hormone was dissolved in the 40% sucrose-Ampholine solution and distributed throughout the electrofocusing column in a density gradient. The absorbancy profile of bovine hormone electrofocused in pH 3–10 Ampholine for 69 hours at 10° is shown in Figure 3. Three major peaks were obtained with pI at 8.06, 7.71 and 6.85. As will be shown

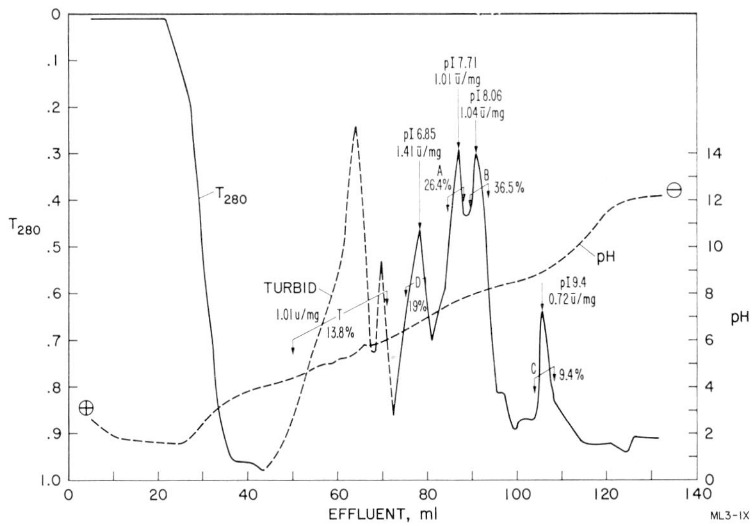

Fig. 3. Isoelectric focusing of bovine growth hormone (Ala, Met) in a pH 3–10 Ampholine gradient. A gradient of 0–45% sucrose: 1% Ampholine (pH 3–10) was prepared with 30 mg of hormone (1.5 USP U/mg) solubilized in the concentrated sucrose-Ampholine solution. Conditions of electrofocusing: 110 ml column with the anode at the bottom; 69 hours at 10° with the power maintained at 1 watt (final voltage = 800); pH measured on 2.5 ml fractions. A region of precipitation appeared approximately 5 cm above the anode electrode block within 3 hours after focusing was begun.

below, peak B (pI 8.06) consisted of the alanyl chain while peak A (pI 7.71) contained the methionyl chain. Peak D (pI 6.85) as well as the turbid region T (pI 6 to 6.5) appeared to be deamidated forms of the hormone as judged by their increased mobilities in gel electrophoresis (Fig. 7). In spite of wide differences in pI, each component displayed a biological activity of approximately 1 USP unit per mg as compared to the applied hormone which had 1.5 units per mg. These results show that the two chains of bovine growth hormone are separable by isoelectric focusing even in the absence of urea. In passing, it should be noted

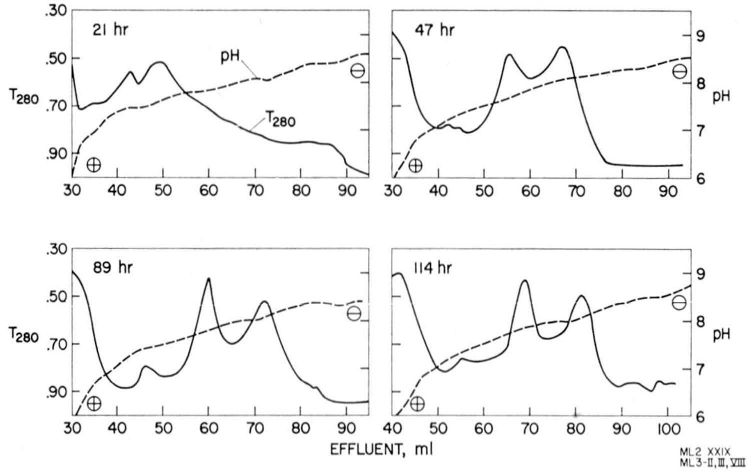

Fig. 4. The effect of time of isoelectric focusing on the separation of (Ala, Met)-bovine growth hormone. Electrofocusing columns were prepared as in Figure 3 except that pH 7–9 Ampholine was used, and the time of electrofocusing was 21, 47, 89, or 114 hours.

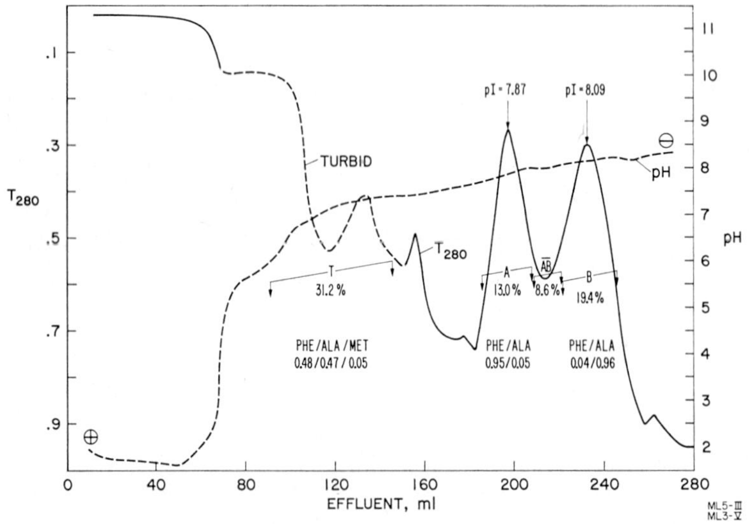

Fig. 5. Large-scale electrofocusing of (Ala, Phe)-bovine growth hormone. A gradient of 0–45% sucrose: 1% Ampholine (pH 7–9) was prepared using a gradient mixer with the hormone (167 mg) solubilized in the concentrated sucrose-Ampholine solution. After 3 hours of electrofocusing, a band of precipitation appeared 15 cm above the anode chamber. Conditions of electrofocusing: 440 ml gradient with 18 ml of dense Ampholine without hormone above anode-electrode block; total time 92 hours at 10°, maintaining the power at 3 watts (final voltage = 800); pH measured on 3 ml fractions of effluent. The protein peaks were dialyzed, lyophilized, and purified on a G-25 Sephadex column (2.5 × 40 cm) equilibrated in 0.1 M Tris-1 M NH_4HCO_3-3% sucrose (pH 9.1). The NH_2 termini were determined by Edman degradation.

that peak C is an artifact formed by the adhesion to the vessel walls of a small amount of the turbid fraction which was eluted by the alkaline pH when the column was drained.

Since improved resolution can be obtained in a narrower pH gradient, pH 7–9 Ampholine without urea was used in subsequent studies. Figure 4 shows that maximum resolution is achieved at about 90 hours of electrofocusing. Longer periods of equilibrations had no effect on pI, yield of protein or biological activity.

A large scale separation of the two chains of the bovine hormone in pH 7–9 Ampholine without urea is shown in Figure 5. Peak B having a pI of 8.09 consisted of the alanyl chain contaminated with about 5 % of the phenylalanyl chain while the reverse proportions prevailed in peak A. It is noteworthy that the acidic region of turbid protein contained equimolar amounts of NH_2-terminal phenylalanine and alanine. Preliminary studies of this fraction have revealed the following changes in properties as compared to the native hormone: (1) a loss of 3 moles of amide nitrogen, (2) a markedly reduced solubility below pH 7, (3) an increased electrophoretic mobility at pH 9, and (4) no change in amino acid composition from that of the native hormone. These properties are suggestive of a rather extensive denaturation which, for reasons unknown as yet, appears to occur during the initial hours of electrofocusing. In Table I are shown the tibial activities of the fractions derived from a typical separation of

TABLE I

Isoelectric focusing of (Ala,Phe)-bovine growth hormone

Fraction	pI	Protein recovery (%)	NH_2-termini Phe/Ala/Met (moles/22,000)	Biological activity (U/mg) (95 % C.L.)	[No. of rats]
T	6–6.5	25.6	.48/.47/.05	1. 1.01 (0.88–1.17)	[12]
D	6.8–7.2	9.8	.47/.47/.06	— —	—
A	7.87	13.0	.96/.04	1. 1.58 (1.30–1.93)	[11]
				2. 1.95 (1.68–2.22)	[11]
B	8.09	19.4	.05/.95	1. 1.74 (1.48–2.30)	[12]
				2. 1.66 (1.43–1.94)	[12]
Native	—	—	.45/.50/.05	1. 1.85 (1.58–2.16)	[12]
				2. 1.80 (1.55–2.09)	[12]

200 mg of growth hormone (Li, 1954), pH 7–9 Ampholine (1 %), 88 hours, 10°, 440 ml column.

bovine growth hormone prepared by the method of Li (1954). Whereas the tibial potencies of the phenylalanyl (A) and alanyl (B) chains were equal to that of the native hormone, the denatured acidic protein (T) showed a 30 % reduction in potency.

When ovine growth hormone (Ellis, 1961) was submitted to electrofocusing in a urea-free gradient of pH 7–9 Ampholine as shown in Figure 6, resolution into four peaks and an acidic turbid fraction was obtained. As in the case of the bovine hormone, the component having the most alkaline pI contained mainly NH_2-terminal alanine while the adjacent component of pI 7.95 contained NH_2-terminal phenylalanine. The peaks with more acidic pI values including the turbid fraction (pI 6.6 to 7.0) were mixtures of both chains. Some properties of these fractions are summarized in Table II. In contrast to the results with the bovine hormone, all the fractions derived from the ovine hormone, including the denatured fraction I from the turbid acidic region of the gradient, exhibited potencies which were not different from the original hormone. Moreover, the fractions did not show differences in amide N content or in minimum molecular weight and, as in the case of the bovine hormone, the fractions with the two most acidic pI had the same NH_2-terminal composition as the original hormone.

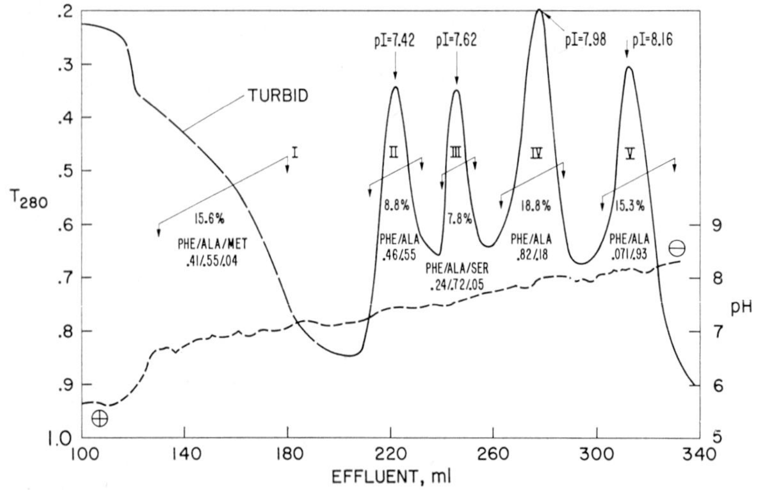

Fig. 6. Isoelectric focusing of ovine growth hormone in a pH 7–9 Ampholine gradient. The conditions were as described in Figure 5 except that ovine growth hormone (160 mg) was electrofocused, the total time was 88 hours, and the isolated fractions were not chromatographed on G–25 Sephadex. The NH_2-terminal residues of native ovine growth hormone were .43Phe, .50 Ala, and .06 Met per 22,000.

TABLE II

Isoelectric focusing of ovine growth hormone

Fraction	pI	Protein recovery (%)	NH_2-termini Phe/Ala/ Met/Ser (moles /22,000)	Minimum molecular weight[a]	Tibial assay (USP U/mg) (95% C.L.)	[No. of rats]
I	6.6–7.0	15.6	.41/.55/.04/0	22,500	1.42 (1.27–1.60)	[19]
II	7.42	8.7	.46/.55/0/0	21,000; 16,700 minor	1.34 (1.16–1.55)	[21]
III	7.62	7.5	.24/.72/0/.05	21,000; 14,700 minor	1.34 (1.19–1.40)	[21]
IV	7.98	18.8	.82/.18/0/0	21,400	1.65 (1.49–1.79)	[21]
V	8.16	15.3	.07/.93/0/0	21,900	1.33 (1.24–1.43)	[32]
Native			.43/.50/.06/0	22,300; 18,300 minor	1.71 (1.55–1.75)	[24]

[a] Determined by SDS-acrylamide gel electrophoresis. According to Weber and Osborn (1969).

On acrylamide gel electrophoresis at pH 9 (Fig. 7), the phenylalanyl and alanyl chains from the bovine (A and B) and the ovine (IV and V) hormones had identical mobilities. When the mobilities of these components were measured as a function of gel concentration (5.5 to 13%), the molecular weights as calculated by the method of Hedrick and Smith (1968) ranged from 36,000 to 47,000. It would appear, therefore, that the isolated chains dimerize as readily as the native hormone. Moreover, if the association constants are approximately equal for both chains, it is likely that the purified native hormone consists of a randomly dimerized population of alanyl and phenylalanyl chains.

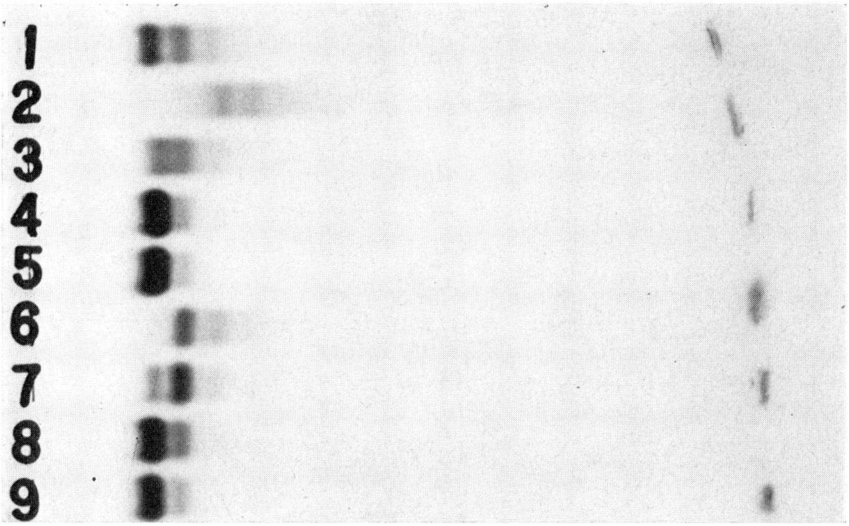

Fig. 7. Electrophoresis of electrofocused bovine and ovine growth hormones in acrylamide gel (7%) buffered with 83 mM Tris-2.7 mM Na_2EDTA-6.7 mM H_3BO_3 at pH 9.1; 2 hrs at 300 v; stained with Amido Black. The numbered channels contained 40 μg of the following growth hormones (1, starting BGH; 2-5, electrofocused bovine fractions from Fig. 3; 6-9, electrofocused ovine fractions from Fig. 6):

2. T, pI 4.6	6. II, pI 7.42
3. D, pI 7.3-7.7	7. III, pI 7.62
4. A (Phe), pI 7.87	8. IV (Phe), pI 7.98
5. B (Ala), pI 8.09	9. V (Ala), pI 8.16

Micro-complement fixation (Wasserman and Levine, 1960) was employed to assess the degree of immunological similarity between the alanyl and phenylalanyl chains of the bovine

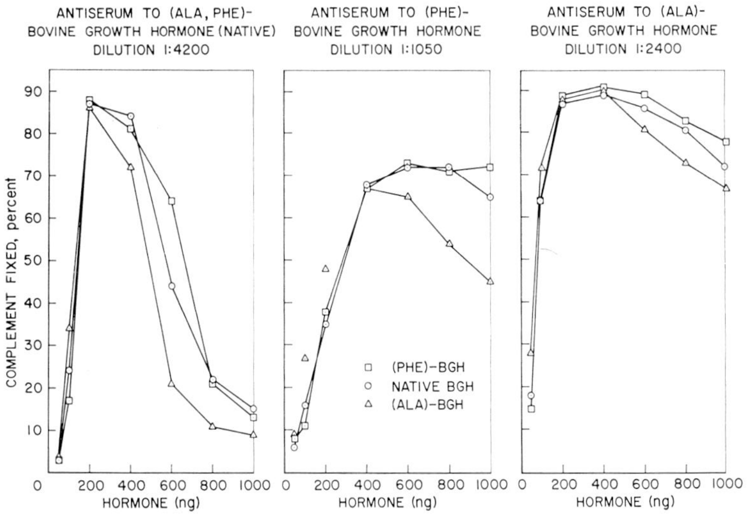

Fig. 8. Complement fixation by monkey antisera to native and electrofocused bovine growth hormones.

hormone (Fig. 8). Antisera against the native hormones (Phe, Ala-BGH) as well as the individual chains were obtained by immunization of Rhesus monkeys with the appropriate protein in Freund's complete adjuvant. When each of the three types of growth hormone was tested against each of the three different antisera, there was no significant difference in the equivalence points between any of the fixation curves. The absence of significant immunological differences was also evident from radioimmunoassay in which the phenylalanyl and alanyl chains were found to be identical with the native hormone in the displacement of labeled native hormone from native hormone antisera. These results indicate that the immunogenic and antigenic sites of the two peptide chains of bovine growth hormone are indistinguishable.

Isoelectric focusing of the bovine and sheep hormones in acrylamide gel is shown in Figure 9. Two intense bands were evident in the gel segment having a pH of 7–8. The acidic minor bands would appear to correspond to the components with acidic pI which were observed in isoelectric focusing with sucrose density gradients. Human, porcine and rat growth

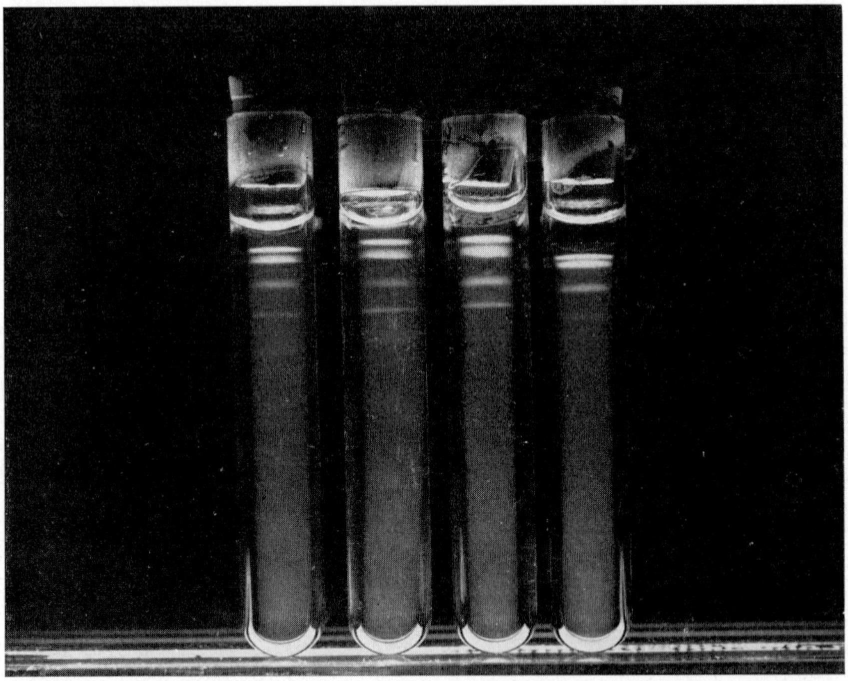

Fig. 9. Isoelectric focusing of ovine growth hormone (left pair) and bovine growth hormone (right pair) in 1 M urea -1% pH 3–10 Ampholine at 5% for 4 hours; 100 μg of protein in each tube. Precipitin bands formed with 20% sulfosalicylic acid.

hormones yielded only a single intense band as well as several minor contaminants also having acidic pI. Thus, the two chains of bovine and ovine growth hormone are demonstrable by isoelectric focusing in acrylamide gel as well as in free solution.

When rat GH was submitted to electrofocusing in a pH 5–7 gradient of Ampholine containing 8 M urea, only one major peak having a pI of 7.05 was obtained (Fig. 10). The small amount of protein at pH 6.69 (B) and the turbid material appearing at A consisted of denatured hormone similar to that observed with the bovine and ovine hormones. The electrofocused product as well as the applied hormone, which had been further purified by

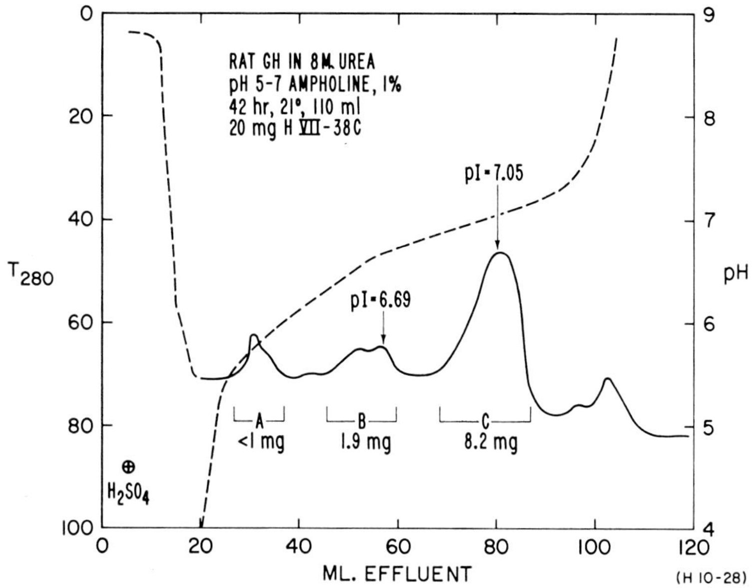

Fig. 10. Isoelectric focusing of rat growth hormone in 8 M urea, pH 5–7 Ampholine (1 %) for 42 hours, at 21°; 20 mg of purified rat GH were applied to a 110 ml column.

TABLE III

Amino acid compositions of the Ala- and Phe-chains of bovine growth hormone

Residue	Ala-BGH (pI 8.09)	Phe-BGH (pI 7.87)	Native BGH 1	Native BGH 2
Lys	10.8	10.4	11.1	10.5
His	3.4	3.1	2.6	3.0
Arg	12.6	12.2	11.9	12.2
Asp	15.2	15.0	16.2	15.1
Thr	11.1	11.3	12.3	11.3
Ser	12.4	12.5	11.6	11.7
Glu	23.1	23.2	23.9	21.6
Pro	5.6	6.0	6.1	6.1
Gly	9.6	9.8	9.5	9.8
Ala	13.2	12.9	12.8	13.5
½ Cys	4.8	4.9	4.0	4.6
Val	6.0	6.1	5.9	6.4
Met	2.5	3.3	4.0	4.4
Ileu	6.3	6.5	6.2	6.5
Leu	26.1	26.1	24.4	24.3
Tyr	6.1	5.8	5.9	6.0
Phe	12.4	12.1	11.7	11.7
Trp	—	—	1.0	—
Residues	181		182	

1 Fellows and Rogol (1969)
2 Yamasaki, Kikutani and Sonenberg (1970)

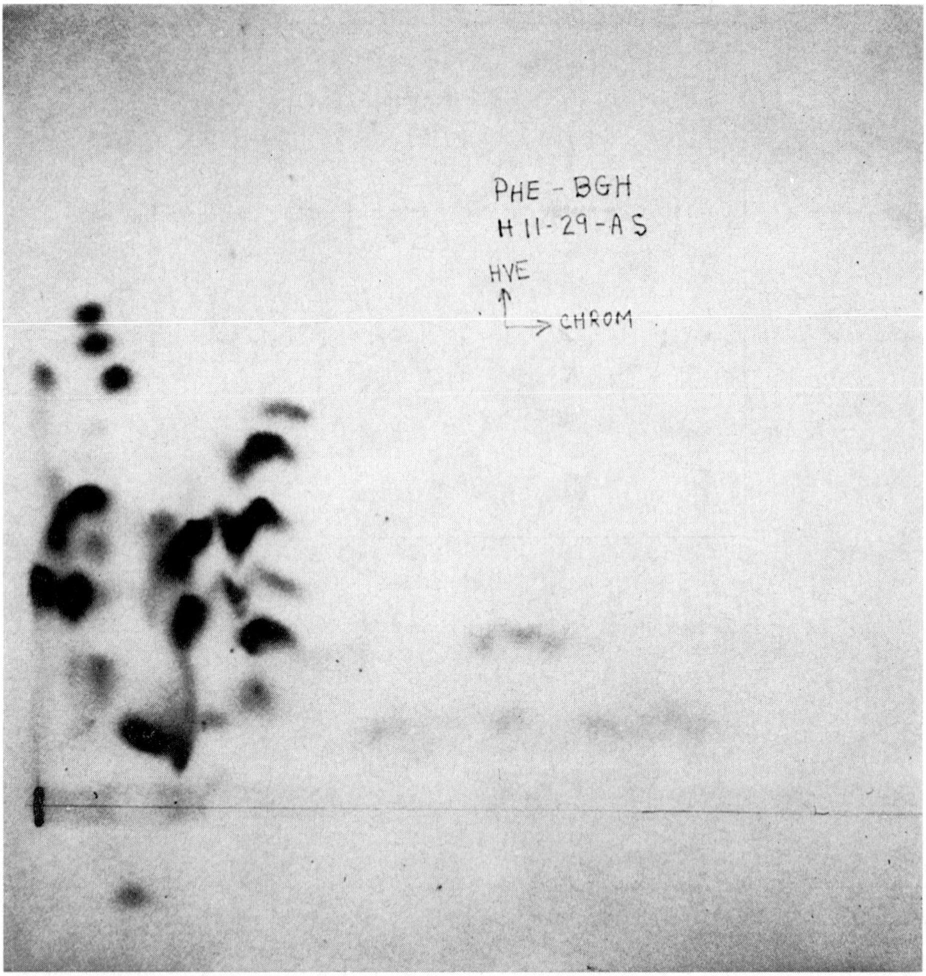

Fig. 11a. Peptide map of trypsin-digested Phe-chain of bovine growth hormone. Conditions: 0.75 mg of digested protein applied to Whatman No. 1 paper; first dimension, descending chromatography in n-butanol:acetic acid:water (450:50:125) for 20 hours; second dimension, pyridine:acetic acid:water (100:10:1000), pH 3.6.

DEAE-cellulose chromatography, contained 1 mole NH_2-terminal phenylalanine per mole of protein (mol.wt. 21,000) rather than 0.6 moles as previously reported (Ellis *et al.*, 1968). Hydrolysis with DFP-treated carboxypeptidase yielded 0.9 moles of COOH-terminal phenylalanine. Thus, in contrast to the growth hormones of bovine and ovine origin, only a single NH_2-terminal phenylalanyl chain is demonstrable in rat growth hormone.

The resolution of the bovine and sheep hormones into two peptide chains which differ in pI raises a question regarding the basis for this difference. While amide analyses have shown that the denatured bovine components with a pI of 5–6 are reduced in amide content by 3 moles as compared to that found in peptides of pI 7.87 and 8.09, no significant difference in amide content could be detected between the phenylalanyl and alanyl chains themselves. It therefore appears unlikely that the difference in pI can be attributed to a difference in the degree of amidation of the acidic residues of the two chains.

Fig. 11b. Peptide map of trypsin-digested Ala-chain of bovine growth hormone. Conditions as given in Fig. 11a.

Amino acid analyses of the bovine chains are given in Table III from which it is apparent that the composition of both chains is essentially identical. However, certain borderline differences were consistently evident in the analyses of several different preparations of the respective peptide chains. Thus the more basic alanyl chain (pI 8.09) consistently showed from 0.3 to 0.5 additional residues of arginine, lysine, alanine and tyrosine, whereas the remaining residues varied within $\pm$ 0.1 from one preparation to another. The inability to achieve integral differences in residues may be ascribed in part to the presence of trace peptide impurities. This is supported by the observations that dansylation and chromatography of the unhydrolyzed peptide revealed variable amounts of many small peptides or amino acids which were present in the order of 0.02 nanomoles each per nanomole of peptide. These impurities, which persist even after the hormone has been purified by gel filtration, DEAE-chromatography and isoelectric focusing, may obscure integral differences in residue composition between the two chains. If substantiated, the apparently higher content of basic residues in the alanyl chain would be consistent with its more alkaline pI value.

Further evidence for differences in the composition of the two bovine chains is suggested from a comparison of the tryptic peptide maps of the non-reduced chains. The results shown in Figure 11 suggest that there are 7 to 8 differences in peptide composition between the two chains. This degree of difference has also been observed by column chromatography on sulfonated cation-exchange resins eluted with pyridine buffer gradients (Benson *et al.*, 1966). On the other hand, when tryptic digests were prepared from the dithiothreitol reduced, carboxymethylated phenylalanyl and alanyl chains (Fig. 12), the absorbancy profiles of the two chains were essentially identical, except for two minor shoulders. Why the tryptic digests of the reduced hormones show little difference while the non-reduced digests display marked differences is not clear and requires further study.

That compositional differences in the peptide chains do in fact exist has been reported

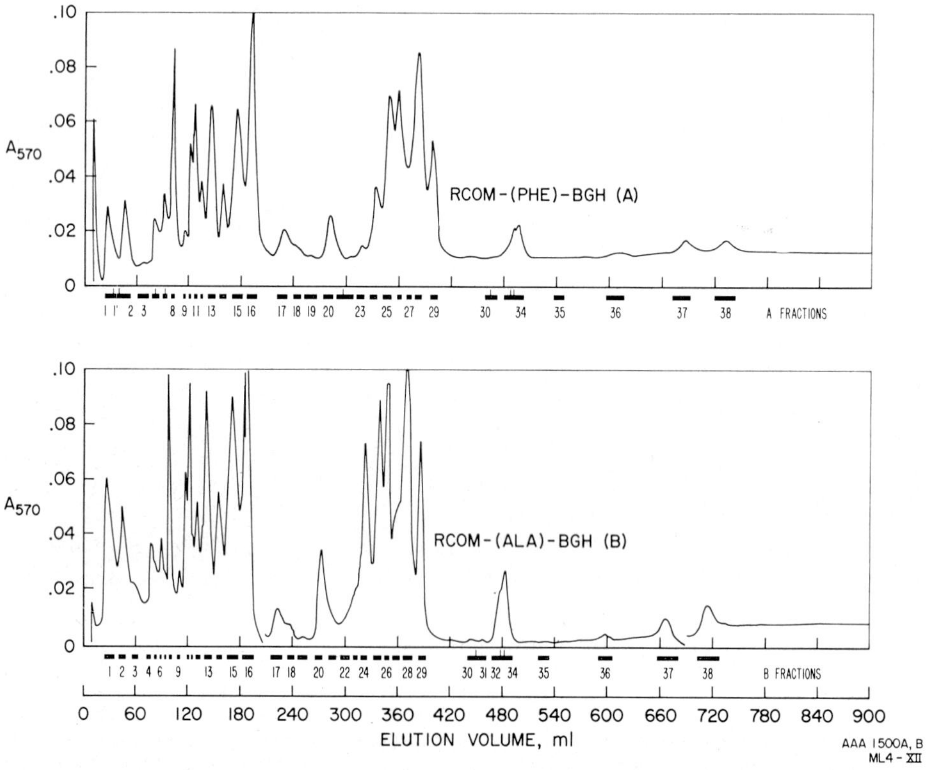

Fig. 12. Column chromatography of trypsin-digested, reduced and carboxymethylated (Ala)- and (Phe)-growth hormones from Figure 5. The method of Bewley *et al.* (1968) was used for the preparation of the reduced tetra-S-carboxymethylated (RCOM) hormones. 15 mg of each fraction was reduced with dithiothreitol (DDT : BGH = 25 : 1) and carboxymethylated with recrystallized iodoacetate (IOAc : DTT = 10 : 1). The reaction was stopped with excess DDT and dialyzed. Tryptic digestion was carried out at pH 9 and 25° for 6 hours at a hormone: enzyme ratio of 50 : 1. The tryptic peptides were separated on a (0.9 × 60 cm) column of Beckman-Spinco PA–35 (Spherical) resin with a linear gradient of pyridine acetate (600 ml: 0.2 M (pH 3.1) to 2 M (pH 4.98)). This was followed by 250 ml of 2 M (pH 4.98) buffer. A splitstream technique was used whereby 80% of the eluate was collected into fractions and 20% was used for color development in a Beckman-Spinco Model 120B amino acid analyzer.

recently by Seavey *et al.* (1971), who have observed an allelic interchange of leucine and valine. However, the distribution of the allelic forms was found to be independent of the NH_2-terminal residues which are non-allelic according to Peña *et al.* (1969). In addition, Fellows (1971) has observed a partial interchange between leucine and glutamine in the native hormone. It therefore appears possible that further sequence studies may reveal additional amino acid substitutions which may explain the isoelectric difference between two chains of bovine and ovine growth hormones.

REFERENCES

ANDREWS, P. (1966): Molecular weights of prolactins and pituitary growth hormones. *Nature (Lond.)*, *209*, 155.

BENSON, J. V., JONES, R. T., CORMICK, J. and PETERSON, J. A. (1966): Accelerated automatic chromatographic analyses of peptides on a spherical resin. *Analyt. Biochem.*, *16*, 91.

BEWLEY, T. A., DIXON, J. S. and LI, C. H. (1968): Human pituitary growth hormone. XVI. Reduction with dithiothreitol in the absence of urea. *Biochim. Biophys. Acta (Amst.)*, *154*, 420.

DELLACHA, J. M., ENERO, M. A. and FAIFERMAN, I. (1966): Molecular weight of bovine growth hormone. *Experientia (Basel)*, *15*, 16.

ELLIS, S. (1961): Studies on the serial extraction of pituitary proteins. *Endocrinology*, *69*, 554.

ELLIS, S., GRINDELAND, R. E. and CALLAHAN, P. X. (1970): Subunits cf bovine and ovine growth hormones: separation by electrofocusing. In: *Abstracts, 52nd Endocrine Society Meeting, St. Louis, Mo.*, p. 113.

ELLIS, S., GRINDELAND, R. E., NUENKE, J. M. and CALLAHAN, P. X. (1968): Isolation and properties of rat and rabbit growth hormones. *Ann. N. Y. Acad. Sci.*, *148*, 328.

FELLOWS Jr, R. E. (1971): Presented at National Institutes of Health Workshop on Prolactin, Bethesda, Md., 1971 (unpublished).

FELLOWS Jr, R. E. and ROGOL, A. D. (1969): Structural studies on bovine growth hormone. I. Isolation and characterization of cyanogen bromide fragments. *J. biol. Chem.*, *244*, 1567.

HAGLUND, H. (1967): Isoelectric focusing in natural pH gradients – a technique of growing importance for fractionation and characterization of proteins. *Sci. Tools*, *14*, 17.

HEDRICK, J. L. and SMITH, A. J. (1968): Size and charge isomer separation and estimation of molecular weights of proteins by disc gel electrophoresis. *Arch. Biochem.*, *126*, 155.

LI, C. H. (1954): A simplified procedure for the isolation of hypophyseal growth hormone. *J. biol Chem.*, *211*, 555.

LI, C. H., CLAUSER, H., FONSS-BECH, P., LEVY, A. L., CONDLIFFE, P. G. and PAPKOFF, H. (1955): Hypophyseal growth hormone as a protein. In: *The Hypophyseal Growth Hormone, Nature and Actions*, Chapter 1, pp. 70–98. Editors: R. W. Smith Jr, O. H. Gaebler and C. N. H. Long. McGraw-Hill, New York.

LI, C. H. and LIU, W. H. (1964): Human pituitary growth hormone. *Experientia (Basel)*, *20*, 169.

PEÑA, C., PALADINI, A. C., DELLACHA, J. M. and SANTOME, J. A. (1969): Evidence for nonallelic origin of the two chains in ox growth hormone. *Biochim. Biophys. Acta (Amst.)*, *194*, 320.

SEAVEY, B. K., SINGH, R. N. P., LEWIS, U. J. and GESCHWIND, I. I. (1971): Bovine growth hormone: evidence for two allelic forms. *Biochem. Biophys. Res. Commun.*, *43*, 189.

VESTERBERG, O. and SVENSSON, H. (1966): Isoelectric fractionation, analysis, and characterization of ampholytes in natural pH gradients. *Acta. chem. scand.*, *20*, 820.

WALLIS, M. (1966): The C-terminal sequence from ox growth hormone. *Biochim. biophys. Acta (Amst.)*, *115*, 423.

WALLIS, M. (1969): The N-terminus of ox growth hormone. *FEBS Letters*, *3*, 118.

WASSERMAN, E. and LEVINE, L. (1960): Quantitative micro-complement fixation and its use in the study of antigenic structure by specific antigen-antibody inhibition. *J. Immunol.*, *87*, 290.

WEBFR, K. and OSBORN, M. (1969): The reliability of molecular weight determinations by dodecyl sulfate-polyacrylamide gel electrophoresis. *J. biol. Chem.*, *244*, 4406.

YAMASAKI, N., KIKUTANI, M. and SONENBERG, M. (1970): Peptides of a biologically active tryptic digest of bovine growth hormone. *Biochemistry*, *9*, 1107.

RELATION OF THE STRUCTURE OF HUMAN GROWTH HORMONE TO THE CONTROL OF CARBOHYDRATE AND FAT METABOLISM

J. BORNSTEIN

Department of Biochemistry, Monash University, Victoria, Australia

Previously reported work from this laboratory has described techniques of hydrolysis of human and other growth hormones yielding two polypeptides with direct action on metabolism of glucose and fat in various systems (Bornstein *et al.*, 1968*a*, *b*; 1969*a*, *b*).

These peptides originally code named Ac-G (suggested new name 'cataglykin')* and In-G ('somantin')** are obtained by enzymic hydrolysis by pituitary endopeptidases which can be found contaminating most growth hormone preparations.

By applying the actions of these peptides as seen in Table I, to the metabolic pathways, it is possible to account for the known metabolic actions of growth hormone as inhibition of the

TABLE I

Comparison of biological activity of natural and synthetic somantins

Test	Natural	Synthetic S-S	Synthetic SH
Glucose uptake by muscle	+	+	0
Fat synthesis by liver	+	+	0
Fat mobilisation from fat pad	+	+	0
Inhibition of			
Glyceraldehyde 3-P dehydrogenase	+	+	0
Glycerol 1-P dehydrogenase	+	+	0
Acetyl CoA carboxylase	+	+	0
Hexokinase	0	0	0
Glucose 6-P dehydrogenase	0	0	0
Fructose 6-P kinase	0	0	0
Lactic dehydrogenase	0	0	0
Malic dehydrogenase	0	0	0
Transaminases	0	0	0
Pyruvic carboxylase	0	0	0
Fatty acid synthetase	0	0	0

In the case of G3PD, Gl1PD and Acetyl CoA carboxylase, both synthetic and natural cataglykin reversed the inhibition of these enzymes by the above somantins.

* The name Cataglykin was chosen for this polypeptide as it lowers blood glucose in man and increases the utilisation of glucose.

** The name Somantin is a contraction of *Som*atotrophic *Ant*i *In*sulin polypeptide.

triose phosphate dehydrogenases results in an accumulation of glycolytic intermediates from glyceraldehyde-3-phosphate up to glucose-6-phosphate and thus inhibition of phosphorylation of glucose by product inhibition of hexokinase (Crane and Sols, 1953) and lipolysis as a result of increased fructose-1, 6-phosphate concentration (Chlouverakis, 1968). Inhibition of hexokinase results in β-oxidation, thus lowering the respiratory quotient. It has also been found that low concentrations of somantin accelerate protein synthesis by muscle, skin and liver tissues. High concentrations inhibit protein synthesis and no explanation is at present available for this finding.

The pituitary endopeptidases are at least three in number and some of their properties are summarised in Table II.

It is seen that the pepsin-like enzyme is non-specific, yielding both cataglykin and somantin as well as other peptides, but that the other enzymes, cataglykin releasing enzyme (CRE) and somantin releasing enzyme (SRE), are more specific as only one or other active sequence is released by purified preparations. These active sequences have been obtained from human growth hormone (HGH), ovine growth hormone (OGH) and human placental lactogen (HPL), each of which has yielded both cataglykin and somantin.

We would point out that one of the major problems in the preparation of these polypeptides is the destruction of the released polypeptide either by further action of the enzyme or enzymes used or by other contaminating polypeptidases.

Following the hydrolysis of two samples of HGH with CRE and SRE, the cataglykin and

TABLE II

Properties of pituitary endopeptidases releasing cataglykin- and somantin-like peptides

Enzyme	pH optimum	Peptide released		Amino acid sequence attacked
		Cataglykin	Somantin	
Pepsin-like	2–3	Yes	Yes	Unknown
CRE[1]	5.4	Yes	No	X-Ileu ↓
SRE[2]	7.8	No	Yes	Cys-Phe-Arg-Lys ↓ Cys-Ala-Lys-Lys

Hydrolysis of substrates by CRE and SRE

Enzyme	Substrate hydrolysed	Peptide released	
		Cataglykin	Somatin
CRE	Human GH	Yes	No
	Ovine GH	Yes	No
	Human PL	Yes	No
	Lysozyme	No	No
SRE	Human GH	No	Yes
	Ovine GH	No	Yes
	Human PL	No	Yes

[1] Cataglykin Releasing Enzyme.
[2] Somantin Releasing Enzyme.

TABLE III

Properties of cataglykin and somantin obtained by CRE and SRE hydrolysis

	Cataglykin		Somantin	
	Max.	Min.	Max.	Min.
Spectrum Å	2580–2620	2420	2580–2620	2460
N-terminal sequence	H_2N Phe-Pro-Thr		H_2N Arg-Lys-Asp-Met	
C-terminal sequences	-Ser-Leu-Leu-Leu COOH		-Gly-Phe COOH	

No tyrosine or tryptophan found in either on amino acid analysis, thus confirming spectral characteristics.

TABLE IV

Active polypeptides prepared by solid-state synthesis

Cataglykin

H_2N Phe-Pro-Thr-Ileu-Pro-Leu-Ser-Arg-Leu-Phe-Asp-Asn-Ala-Met-Leu-Arg-Ileu-Ser-Leu-
1 2 3 4 5 6 7 8 9 10 11 12 13 14 15 16 17 18 19
Leu-Leu COOH
20 21

Somantin

H_2N Arg-Lys-Asp-Met-Asp-Lys-Val-Glu-Thr-Phe-Leu-Arg-Ileu-Val-Glu-
1 2 3 4 5 6 7 8 9 10 11 12 13 14 15
16 Cys = Cys – Gly – Phe COOH
| | 23 24 25
17 Arg Ser 22
| |
18 Ser Gly 21
| |
19 Val – Glu 20

somantin peptides were isolated as previously described and analysed with a view to determination of sequence and synthesis (Table III).

When these part sequences were applied to the proposed structure of Li *et al.* (1969), it appeared that the part sequences were unique to amino acids 1–21 and 164–188 of that structure (Table IV). These sequences were then synthetised by the solid state technique and were found to be active (Bornstein *et al.*, 1971).

However, the proposed sequence for cataglykin is not possible in the revised structure proposed by Niall (1971). As this revision appears to be correct, the activity of cataglykin must lie within the structure between amino acids 1 and 16, and the C-terminal sequence reported by us be due to an impurity the N-terminal of which was not detected. This problem is now being re-examined*, and preliminary work on a urinary polypeptide with cataglykin-like activity (Table V) suggests that the natural sequence is shorter.

* In order to test this hypothesis, the synthetic molecule was digested with Carboxypeptidase A and the sequence Ileu-Ser-Leu-Leu-Leu removed, leaving the sequence from 1–16. This material was then tested in the Glyceraldehyde-3-phosphate dehydrogenase assay and found to be active in reversing ovine somantin.

TABLE V

Structural resemblances between N- and C-terminal ends of human growth hormone and human placental lactogen

	N-term. HGH	N-term. HPL	C-term. HGH and HPL	
			Phe	190
1	Phe	Val	Gly	189
2	Pro	Glu	Glu-Gly-Ser-Cys	188
3	Thr	Thr	Val-Ser-Arg-Cys	181
4	Ileu	Val	Gln	180
5	Pro	Pro	Met or Val	179
6	Leu	Leu	Ileu	178
7	Ser	Ser	Arg	177
8	Arg	Arg	Leu	176
9	Leu	Leu	Phe	175
10	Phe	Phe	Thr	174
11	Asp	Asp	Glu	173
12	Asn	His	Val	172
13	Ala	Ala	Lys	171
14	Met	Met	Asp	170

When the relevant part sequences of the C-terminal and N-terminal areas of HGH and HPL are compared, it is seen that close similarity exists between hexapeptide sequences at C- and N-terminal ends of both molecules as HGH and HPL share a common sequence from 6–11, and a similar sequence occurs between 173 and 178 of the C-terminal end of HGH and HPL, the substitutions being isoleucine for leucine, threonine for serine and glutamic for aspartic.

When models of the sequence 1–14 and the sequence 170–190 of HGH are built, it is seen that the two hexapeptides are able to form a markedly similar tertiary structure (Fig. 1), and we would suggest that the activity of cataglykin in reversing the action of somantin is due to competition by these part sequences for appropriate enzyme binding sites, the actual inhibitory activity of somantin being probably associated with the ring structure, as the S-S bond must be present for activity (Bornstein *et al.*, 1971).

PRESENCE OF SOMANTIN AND CATAGLYKIN IN BIOLOGICAL FLUIDS

In order to determine whether these peptides were hydrolysed from growth hormone *in vivo*, blood and urine were processed under conditions identical with those used in isolation of

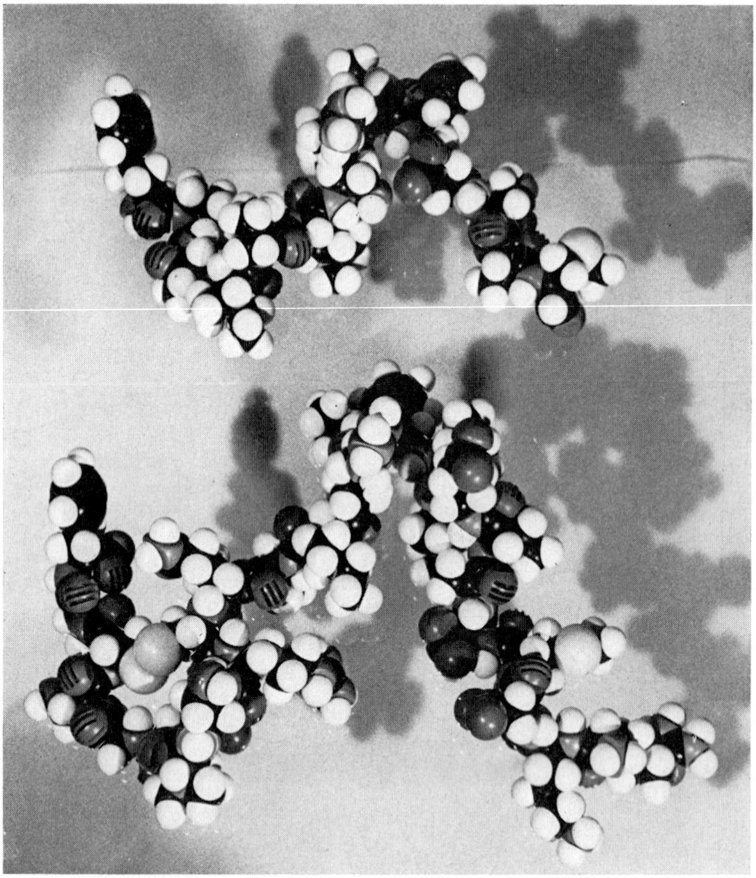

Fig. 1. Cataglykin (upper) and somantin (lower).

the two compounds from a growth hormone hydrolysate. The results are shown in Table VI.

It is seen that the somantin-like fragment from blood and urine has the same spectrum, chromatographs identically, has the same enzyme specificity and N-terminus as that obtained from HGH. It is reversed by ovine cataglykin, urinary cataglykin and synthetic human cataglykin.

The urinary cataglykin, however, has a major difference from the original structure proposed, *i.e.*, the C-terminal amino acids being -ala-met, suggesting strongly that the natural active peptide is 1–14 and thus accounts for the activity of the synthetic material despite the problems arising as to the correct sequence between 17 and 21.

Although absolute proof of the identical nature of the blood and urinary polypeptides with those derived from human growth hormone must await complete chemical analysis, the close biological and physical resemblance between the peptides from biological fluids and those hydrolysed from growth hormone suggests a close relationship and led to an examination of the possible mechanisms of release of the active peptides from growth hormone.

RELEASE OF CATAGLYKIN AND SOMANTIN BY INTACT MUSCLE

Two real possibilities exist for the release of the peptides from growth hormone. The first,

TABLE VI

Properties of cataglykin- and somantin-like fractions isolated from blood and urine

	Cataglykin not isolated	Somantin isolated
Blood		
Spectrum Å	—	Max. 2580–2620 Min. 2460
Effect on isolated tissues	—	as for HGH and OGH prep.
Enzymic specificity	—	as for HGH and OGH prep.
Chromatography	—	as for HGH and OGH prep.
Urine		
Spectrum Å	Max. 2580–2620 Min. 2410	Max. 2580–2620 Min. 2460
Enzyme specificity	as for HGH and OGH preps.	as for HGH and OGH preps.
Chromatography	as for HGH and OGH preps.	as for HGH and OGH preps.
	OGH	
Sources of somantins reversed (G3PDH assay)	Human urine Human blood 'Human' synthetic	
Structural characteristics		
N-terminus	Phe	Arg
C-terminus	Ala-Met	

supported by the presence of apparently specific polypeptidases in the pituitary, is that hydrolysis occurs in the pituitary, the metabolically appropriate peptide being released. However, a number of workers have observed growth hormone effects on tissues *in vitro*, and growth hormone is metabolically active in hypophysectomised animals. Further, the fragments only react with antibody very poorly and could not account for immune reactive HGH in plasma. Accordingly, it was decided to test the hypothesis that the active peptides are released in the target tissue, release of one or other being actuated by the metabolic state of the cell.

TABLE VII

Hydrolysis of HGH and OGH by intact rat diaphragm

	Tissue	Pituitary hormone[1]	Insulin	Peptide detected
1.	0	HGH	0	0
2.	0	OGH	0	0
3.	Muscle homogenate	OGH	0	0
4.	Liver homogenate	OGH	0	0
5.	Intact fed	HGH	0	Cataglykin, somantin
6.	Intact fed	OGH	0	Cataglykin, somantin
7.	Intact fasted	OGH	0	Somantin
8.	Intact fasted	OGH	0.01 μ/ml	Cataglykin, trace somantin
9.	Intact fed	OFSH	0	0
10.	Ovary	HGH	0	0

[1] Pituitary hormones 0.1 mg/ml, total 20 mg. Treated with DFP to inactivate residual endopeptidases. Cataglykin and somantin detected by glyceraldehyde-3-P dehydrogenase assay. Expts. 7 and 8 paired diaphragms.

The experimental procedure was to incubate the test substance with intact muscle, and muscle or liver homogenates in Krebs-phosphate glucose buffer for 4 hours and to examine the buffer ultrafiltrate for the presence of cataglykin and somantin activity after removal of salts by ultrafiltration on a UM2 filter. The results are seen in Table VII.

Consideration of these data leads to the preliminary conclusion that these peptides are released at the target organ and, as intact cells are required for the release, that the hydrolytic enzymes occur in the cell membrane, thus supporting the findings of Rillema and Kostyo (1971), who found that most of the effects of growth hormone could be demonstrated in intact tissue on prolonged incubation but were abolished if the tissue were treated with collagenase or trypsin.

The finding that insulin inhibits the release of somantin supports the findings of Zimmet *et al.* (1971), who found that there was a fall in plasma somantin during the raised insulin part of glucose tolerance curves in man, the level returning to normal as blood glucose reaches the fasting level.

In view of these findings we would suggest that these peptides could play an important role in the control of glucose and fat metabolism by competitively modulating the rate of glucose utilisation and fat metabolism, that this competition is based on the structural similarity of parts of their molecules, that the release at target organ level is modulated by insulin concentration, and that at least one factor in the genesis of diabetes mellitus may be the failure of insulin to inhibit somantin release from growth hormone, but would emphasise that we have no evidence as to the growth promoting activity of either somantin or cataglykin.

ACKNOWLEDGEMENTS

Some of the work presented in this paper has been carried out by Mr. C. J. I. Driver, as part of studies for the degree of Doctor of Philosophy.

We would thank the National Health and Medical Research Council of Australia for their generous support.

REFERENCES

BORNSTEIN, J., KRAHL, M. E., MARSHALL, L. B., GOULD, M. K. and ARMSTRONG, J. McD. (1968*a*): Pituitary peptides with direct action on the metabolism of carbohydrates and fatty acids. *Biochim. biophys. Acta (Amst.)*, *156*, 31.

BORNSTEIN, J., ARMSTRONG, J. McD. and JONES, M. D. (1968*b*): Effect of a growth hormone fraction on the activity of glyceraldehyde-3-phosphate dehydrogenase. *Biochim. biophys. Acta (Amst.)*, *156*, 38.

BORNSTEIN, J., ARMSTRONG, J. McD., GOULD, M. K., HARCOURT, J. A. and JONES, M. D. (1969*a*): Mechanism of the diabetogenic action of growth hormone. I. Effect of polypeptides derived from growth hormone on glycolysis in muscle. *Biochim. biophys. Acta (Amst.)*, *192*, 265.

BORNSTEIN, J., TAYLOR, W. M., MARSHALL, L. B., ARMSTRONG, J. McD. and GOULD, M. K. (1969*b*): Mechanism of the diabetogenic action of growth hormone. II. Effect of polypeptides derived from growth hormone on fat metabolism. *Biochim. biophys. Acta (Amst.)*, *192*, 271.

BORNSTEIN, J., ARMSTRONG, J. McD., NG, F. M., PADDLE, B. M. and MISCONI, L. (1971): Structure and synthesis of biologically active peptides derived from growth hormone. *Biochem. biophys. Res. Commun.*, *42/2*, 252.

CHLOUVERAKIS, C. (1968): The lipolytic action of fructose-1-6-diphosphate. *Metabolism*, *17*, 708.

CRANE, R. K. and SOLS, A. (1953): Association of hexokinase with particulate fractions of brain and other tissue homogenates. *J. biol. Chem.*, *203*, 273.

LI, C. H., DIXON, J. S. and LIU, W. (1969): Human pituitary growth hormone. XIX. The primary structure of the hormone. *Arch. Biochem.*, *133*, 70.

NIALL, H. D. (1971): Revised structure for human growth hormone. *Nature New Biol.*, *230*, 90.

RILLEMA, J. A. and KOSTYO, J. L. (1971): Studies on the delayed action of growth hormone on the metabolism of the diaphragm. *Endocrinology*, *88*, 240.

ZIMMET, P., NG, F. M., BORNSTEIN, J., ARMSTRONG, J. McD. and TAFT, H. P. (1971): Insulin antagonist of pituitary origin in plasma of normal and diabetic subjects. *Brit. med. J.*, *1*, 203.

STUDIES ON ACTIVE FRAGMENTS OF BOVINE GROWTH HORMONE*

M. SONENBERG[1], N. YAMASAKI[2], M. KIKUTANI[3], N. I. SWISLOCKI[1], L. LEVINE[4] and M. NEW[4]

[1] Division of Endocrinology, Sloan-Kettering Institute for Cancer Research, New York, N. Y., U.S.A.; [2] Laboratory of Biochemistry, Faculty of Agriculture, Ehime University, Matsuyama City, Japan; [3]Faculty of Pharmaceutical Sciences, Nagasaki University, Nagasaki, Japan; [4] Department of Pediatrics, Cornell University Medical College, New York, N.Y., U.S.A.

The consensus is that pituitary growth hormone from species other than man or monkey is without hormonal activity in humans. Some understanding of this species specificity for growth hormone is provided by the chemical differences between primate and non-primate growth hormones (Li and Liu, 1964). Unfortunately, inadequate supplies of the native hormone or synthetic human growth hormone (HGH) (Li and Yamashiro, 1970), have hampered investigations which would lead to an understanding of the role of HGH in health and disease.

The similarities of metabolic effects elicited by various pituitary growth hormones in man and lower species suggest that there may be a common active portion in all pituitary growth hormones. From the primary structures of HGH (Li *et al.*, 1969) and bovine growth hormone (BGH) (Santomé *et al.*, 1971) there do indeed appear to be significant areas of homology.

Earlier studies have suggested that it is possible to degrade proteolytically BGH to varying extents with retention of significant amounts of biological activity, *e.g.* chymotrypsin (Li *et al.*, 1956, 1959; Kolli *et al.*, 1966), trypsin (Li, 1956), carboxypeptidase (Harris *et al.*, 1954), pepsin (Li *et al.*, 1955; Laron *et al.*, 1964; Kolli *et al.*, 1966) and streptomycete extracts (Reusser, 1965). A preparation obtained by chymotryptic digestion of BGH (Forsham *et al.*, 1958; Li *et al.*, 1959) was reported to have growth hormone activity in humans, although this was not confirmed (Bergenstal and Lipsett, 1960). Recently (Elsair *et al.*, 1964; Sonenberg *et al.*, 1965*a*, 1969) the chemical procedure has been modified and preparations of chymotryptic digests have been obtained with anabolic activity in pituitary dwarfs. Similarly, papain digests (Sonenberg *et al.*, 1965*a*, 1967; Levie and Sonenberg, 1970) of BGH have been prepared and found to have anabolic activity in the human.

In addition, it has been possible to prepare cyanogen bromide fragments of BGH (Fellows and Rogol, 1969) and porcine growth hormone (Nutting *et al.*, 1970*a*). Although the cyanogen bromide fragments of BGH were without growth promoting activity, they did possess metabolic activities in adipose tissue *in vitro* (Nutting *et al.*, 1970*b*). The porcine growth hormone fragments were effective in stimulating a small but significant body weight gain in rats as well as ^{14}C-leucine incorporation into diaphragm protein (Nutting *et al.*, 1970*a*).

Tryptic digests of bovine growth hormone

With unfractionated enzymatic digests of BGH it is difficult to be sure that the biological response is not due to some undigested component, although such components from non-primate pituitary growth hormone are without activity in humans. These difficulties have been

* Supported in part by Research Grants CA-08748 and FR-47 of the National Institute of Health, GB-19797 of the National Science Foundation and P-437 of the American Cancer Society.

Fig. 1. Reading from left disc electrophoretic patterns of components A-I and A-II. For preparative and analytical techniques in this and subsequent figures see Yamasaki *et al.* (1970).

answered in part in the case of tryptic digests of BGH that we have prepared (Sonenberg *et al.*, 1965*b*; Nadler *et al.*, 1967; Sonenberg *et al.*, 1968). These digests were found to have most of the metabolic activity of HGH. From such digests with an average of 2 bonds/molecule hydrolysed we have isolated a component (TBGH-d) which was homogeneous by gel filtration, sedimentation equilibrium, and disc electrophoresis. It had the same molecular weight and amino acid composition as the undigested BGH from which it was derived. In addition to its increased electrophoretic mobility, this component had two new amino terminal and two new carboxyl terminal amino acids. Valine and serine as amino terminal acids were found in addition to alanine and phenylalanine of undigested BGH. Similarly, two arginine carboxyl terminal residues were found in addition to the phenylalanine of undigested BGH. This component had growth promoting activity similar to undigested BGH. When administered to humans it produced most of the metabolic effects associated with HGH.

When this homogeneous component (TBGH-d) from a tryptic digest of BGH was subjected to gel filtration in 50% acetic acid, two homogeneous (Fig. 1) components were isolated (Yamasaki *et al.*, 1970), whereas gel filtration of BGH under the same conditions gave only one fraction (Fig. 2). The larger peptide (A-I) of 16,000 molecular weight (Fig. 3) and the smaller peptide (A-II) of 5,000 molecular weight (Fig. 4) accounted for the total weight

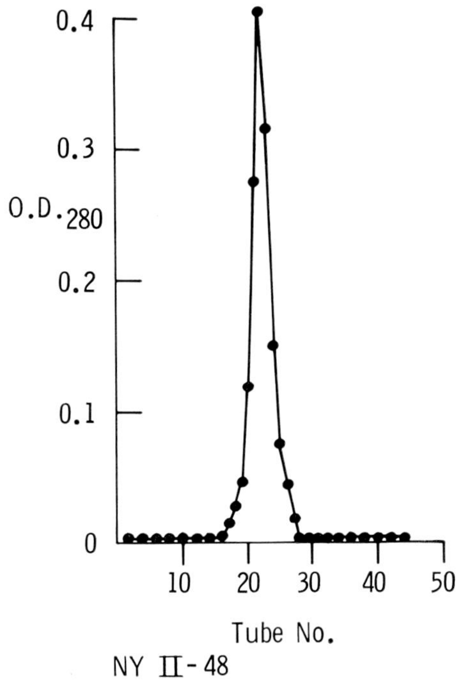

Fig. 2. Gel filtration of 15 mg of BGH in 50% acetic acid on Sephadex G-75. Column size 2.5 × 40 cm.

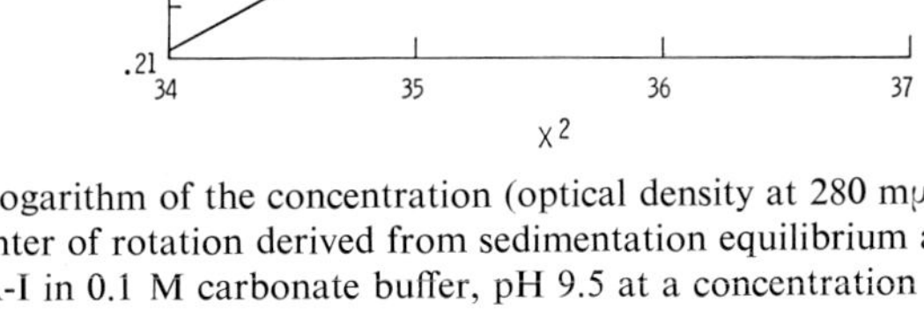

Fig. 3. Plot of the logarithm of the concentration (optical density at 280 mμ) against square of the distance from the center of rotation derived from sedimentation equilibrium at 8,000 r.p.m. of component A-I in 0.1 M carbonate buffer, pH 9.5 at a concentration of 1 mg/ml.

(21,000) and amino acid composition of the component from which they were derived. Reduction and carboxymethylation of TBGH-d and subsequent gel filtration (Fig. 5) yielded two homogeneous fractions, RCM-1 and RCM-2 of 11,000 and 5,000 molecular weight (Figs. 6 and 7). Similarly the 16,000 molecular weight peptide in turn could be fractionated by reduction and carboxymethylation and gel filtration into two peptides, RCM-A-I_1 and RCM-A-I_2 which were homogeneous by disc electrophoresis and sedimentation equilibrium (Figs. 8 and 9). The molecular weights, 11,000 (RCM-A-I_1) and 5,000 (RCM-A-I_2) and amino

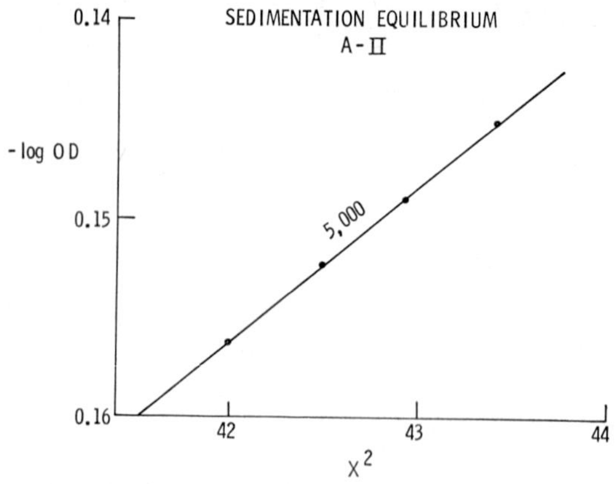

Fig. 4. Plot of logarithm of the concentration (optical density at 280 mμ) against square of the distance from center of rotation derived from sedimentation equilibrium at 8,000 r.p.m. of component A-II in 0.1 M carbonate buffer, pH 9.5 at a concentration of 0.7 mg/ml.

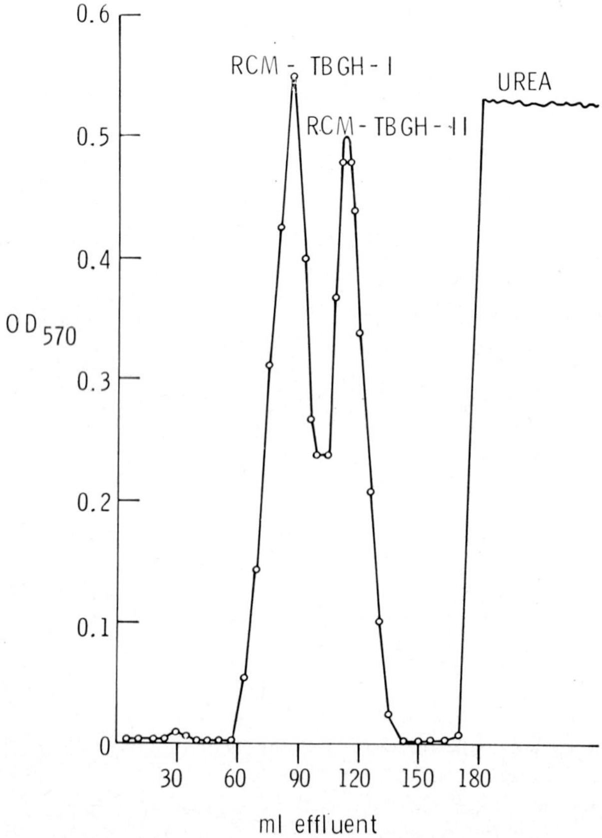

Fig. 5. Gel filtration on Sephadex G-75 in 50% acetic acid of TBGH-d after reduction and carboxymethylation in urea. Column size 2.5 × 40 cm. Color developed with ninhydrin.

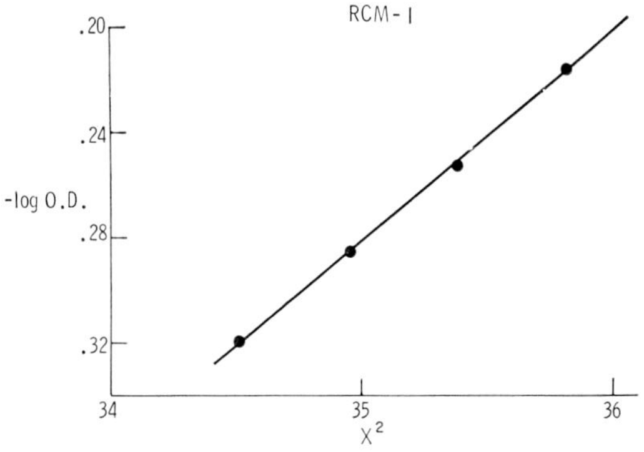

Fig. 6. Plot of the logarithm of the concentration (optical density at 280 mμ) against square of the distance from center of rotation derived from sedimentation equilibrium at 8,000 r.p.m. of first major peak (RCM-TBGH-I) of reduced and carboxymethylated TBGH-d. The component was dissolved in 8 M urea and 0.1 M carbonate buffer, pH 9.5 at a concentration of 1 mg/ml. See Fig. 5.

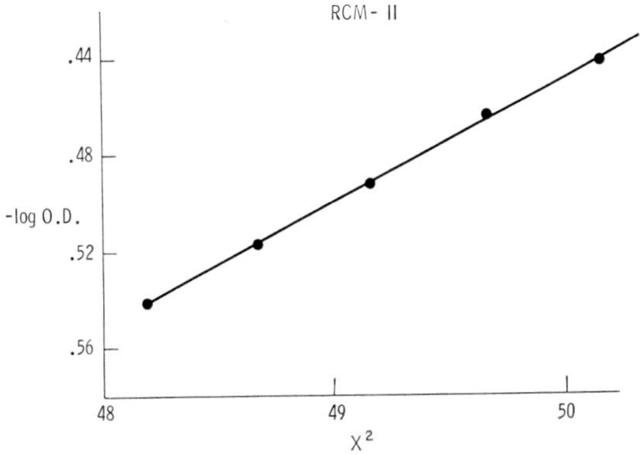

Fig. 7. Plot of the logarithm of the concentration (optical density at 280 mμ) against square of the distance from center of rotation derived from sedimentation equilibrium at 20,000 r.p.m. of second major peak (RCM-TBGH-II) of reduced and carboxymethylated TBGH-d. The fraction was dissolved in 0.1 M carbonate buffer, pH 9.5 at a concentration of 0.5 mg/ml. See Fig. 5.

acid compositions accounted for the entire 16,000 molecular weight peptide (Yamasaki *et al.*, 1970). Thus, the 5,000 molecular weight fraction, RCM-2, seemed to contain two 5,000 molecular weight peptides A-II and RCM-A-I_2.

Chemical characterization of small peptide

As noted above, gel filtration in 50% acetic acid of the homogeneous component with 2 bonds hydrolysed yielded 2 peptides, which on a calibrated Sephadex G-75 column were eluted at volumes corresponding to molecular weights of 16,000 and 5,000. This was verified by sedimentation equilibrium which also indicated monodispersity (Figs. 3 and 4). The more sensitive technique of polyacrylamide disc electrophoresis also indicated homogeneity of these 2 peptides (Fig. 1).

The amino acid composition (Table I) of A-II revealed no disulfide or histidine residues and only single residues of proline, alanine, methionine, isoleucine and tyrosine. Of additional interest for subsequent sequence analysis was the presence of 5 basic residues, 2 lysine and 3 arginine, and 3 aromatic residues, 1 tyrosine and 2 phenylalanine.

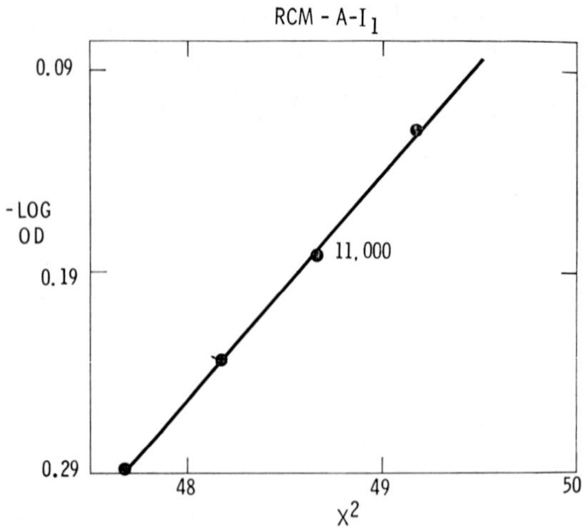

Fig. 8. Plot of the logarithm of the concentration (optical density at 280 mμ) against square of the distance from center of rotation derived from sedimentation equilibrium at 20,000 r.p.m. of component RCM-A-I_1. Sample dissolved in 0.1 M carbonate buffer, pH 9.5 and 8 M urea at a concentration of 0.7 mg/ml.

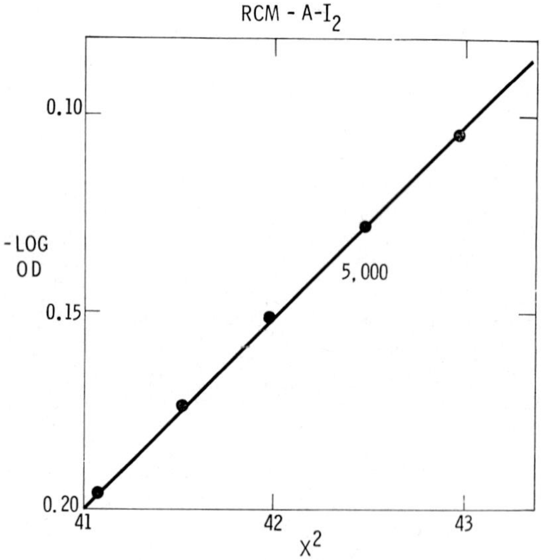

Fig. 9. Plot of the logarithm of the concentration (optical density at 280 mμ) against square of the distance from center of rotation derived from sedimentation equilibrium at 20,000 r.p.m. of component RCM-A-I_2. Sample dissolved in 0.1 M carbonate buffer, pH 9.5 and 8 M urea at concentration of 1.0 mg/ml.

Far ultraviolet circular dichroism of BGH and the homogeneous component from a tryptic digest of BGH revealed spectra suggesting a helical content of approximately 55% (Sonenberg and Beychok, 1971). The spectrum of A-II suggested elements of helix and random coil with the latter predominating (Sonenberg and Beychok, 1971). Near ultraviolet circular dichroic spectra of BGH contained negative bands at 285 nm and 290 nm and positive bands at 275 nm, 265 nm and 260 nm with a shoulder at 270 nm. The component from the tryptic digest of BGH demonstrated fewer circular dichroic bands in the near ultraviolet region. There was no circular dichroism above 253 nm in the smaller peptide, A-II.

TABLE I

Amino acid composition of A-II and its cyanogen bromide fragments. Number of residues

	A-II	AMO CB-1	AMO CB-2
Aspartic acid	4	3	1
Threonine	3	2	1
Serine	2	2	–
Glutamic acid	5	3	2
Proline	1	–	1
Glycine	3	2	1
Alanine	1	1	–
Half cystine	–	–	–
Valine	3	3	–
Methionine	1	–	–
Isoleucine	1	1	–
Leucine	5	4	1
Tyrosine	1	1	–
Phenylalanine	2	2	–
Histidine	–	–	–
Lysine	2	2	–
Arginine	3	1	2

Biological characterization of small peptide

Although BGH and the homogeneous component of a tryptic digest of BGH produced approximately equivalent increases in weight and tibial width of hypophysectomized rats, there was a decreased response to the small and large peptide. In addition, the slope of the response curve of these peptides was not parallel to that of undigested BGH (Fig. 10). Both peptide fragments obtained from the tryptic digest of BGH were effective *in vitro* as well, albeit with considerably less potency (Swislocki *et al.*, 1970). The *in vitro* metabolic activities of BGH noted with the fragments were the promotion of ^{14}C-glucose uptake, its oxidation to CO_2 and incorporation into glyceride glycerol, glycerol release and ^{14}C-histidine incorporation into adipose tissue protein. Thus, the 2 fragments of the tryptic digest of BGH possess BGH activity *in vivo* and *in vitro*. Moreover, the various metabolic effects of BGH were not dissociated by limited tryptic fragmentation of the BGH molecule since each of the 2 fragments of different primary structure exerted similar metabolic effects. It may be that in BGH there is more than one active site which elicits the multiple metabolic effects.

HGH, but not BGH, is able to produce a conformational change in the proteins of the human erythrocyte membrane *in situ* as demonstrated by circular dichroism (Sonenberg, 1969) and intrinsic fluorescence (Sonenberg, 1971). Peptide A-II, like HGH, is also able to produce a similar conformational change in human erythrocyte membrane proteins (Sonenberg, 1971).

The 5,000 molecular weight peptide had metabolic effects in humans similar to HGH. When administered to one growth hormone deficient and one low birth weight dwarf in doses (50 mg/day for 6 days) about ten times that employed with human growth hormone, the anti-insulin effect was of equivalent magnitude (Figs. 11 and 12). There was also a decrease in blood urea nitrogen, urinary nitrogen and urinary creatine and an increase in urinary calcium (Figs. 13 and 14). Smaller or larger doses were not tested.

The marked decrease in biological activity of A-II compared to untreated BGH is of some

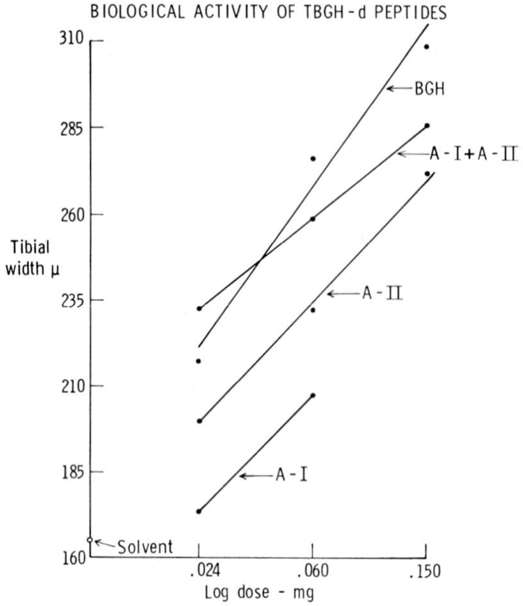

Fig. 10. Tibial width response of growth hormone derivatives. A-I + A-II were present in a ratio of 3 : 1 (w : w).

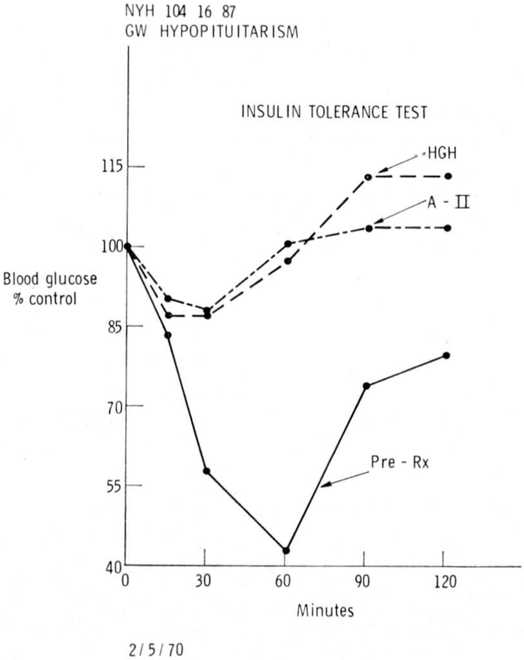

Fig. 11. Effects in hypopituitary dwarf, GW, of 4 days of intramuscular injections of peptide A-II (40 mg daily) on insulin (0.03 units/kg) tolerance test.

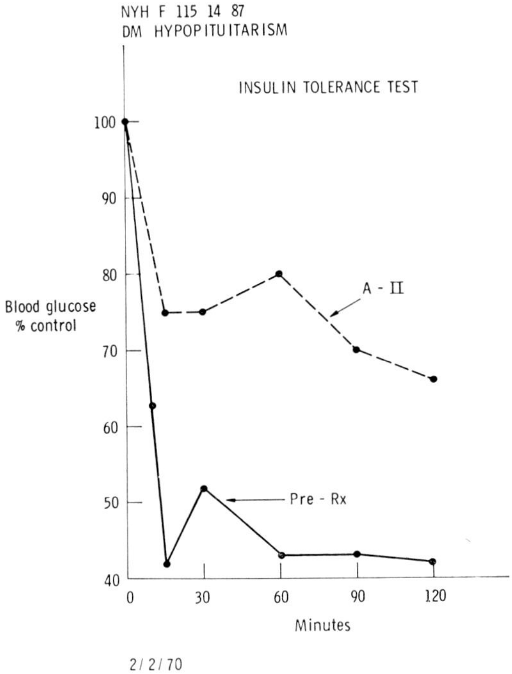

Fig. 12. Effects in low birth weight dwarf, DM, of 2 days of intramuscular injections of peptide A-II (40 mg daily) on insulin (0.03 units/kg) tolerance test.

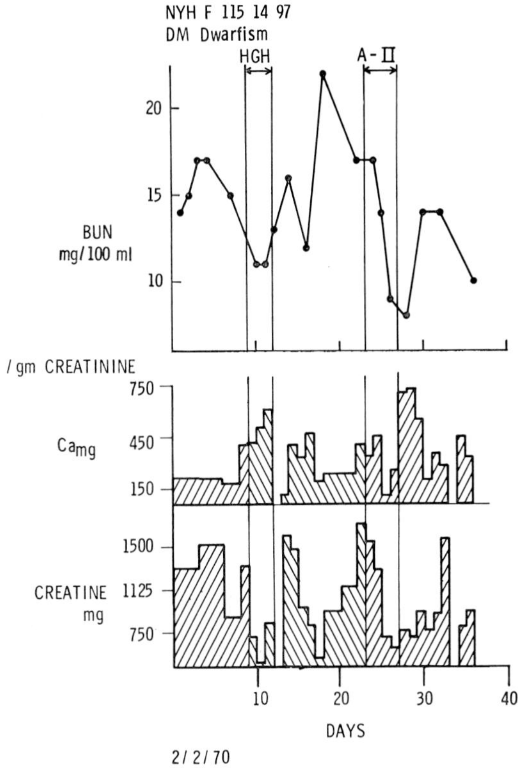

Fig. 13. Effects in low birth weight dwarf, DM, of growth hormone preparations on urinary calcium, urinary creatine, and blood urea nitrogen. Human growth hormone (HGH) was injected in a dose 5 mg daily for 3 days and peptide A-II injected in a dose of 20 mg daily for 4 days. Both were administered intramuscularly.

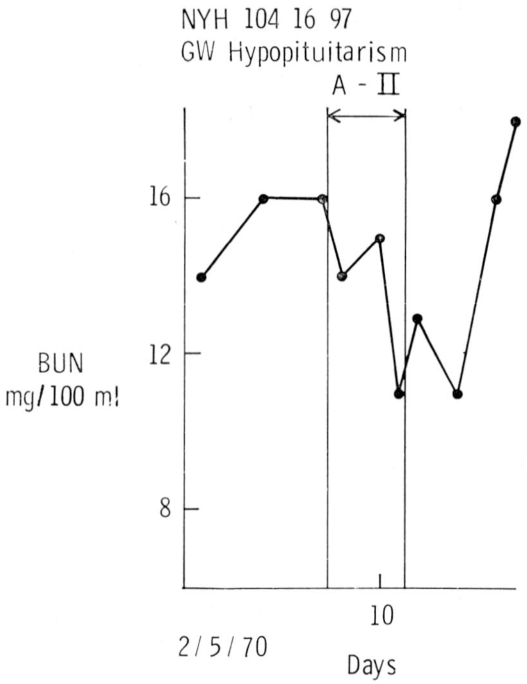

Fig. 14. Effect on blood urea nitrogen (BUN) in hypopituitary dwarf, GW, of 4 days of intramuscular injections of peptide A-II (40 mg daily).

interest (Yamasaki *et al.*, 1970). When it was noted (Table II) that the biological activity of undigested BGH decreased from 1.0 to 0.37 $\pm$ 0.06 I.U./mg when dissolved in 35% acetic acid for 18 hours at 4° C, we considered the possibility of some chemical reaction with acetic acid. Similarly, the biological activity of TBGH-d, the parent component from which peptides A-I and A-II were derived, decreased from approximately 1.5 to 0.13 $\pm$ 0.03 I.U./mg, when dissolved in 35% acetic acid. This is under study.

In view of the lack of parallelism of the growth response of undigested BGH and A-II, it is possible that the decreased growth response to A-II is related to its smaller size and more rapid catabolism. With this in mind, we have prepared a series of polyalanylated growth hormone derivatives (Table III) as was done with HGH (Canfield, 1968). If the size of these growth hormone active peptides could be enlarged, their catabolism might be decreased significantly with a consequent increased biological response. From these experiments (Tables III and IV), it can be seen that despite the fact that BGH has a smaller number, 139, of additional alanine groups per mole of BGH than HGH, polyalanylated BGH was completely without growth promoting activity. HGH, of the same molecular weight as BGH, had introduced 265 additional alanine residues with retention of approximately 50% of the growth promoting activity. This may be related to the smaller helical content of HGH as opposed to BGH (Sonenberg and Beychok, 1971). Similarly, TBGH-d, also of the same molecular weight, polyalanylated with 200 additional alanine groups, retained about 30% of its growth promoting activity. On the other hand, when the larger peptide and the smaller peptides were polyalanylated with 193 and 51 additional alanine groups respectively there was complete loss of growth promoting activity. Thus increasing the size of either the A-I or A-II peptide with this chemical modification did not increase growth hormone activity.

TABLE II

Effect of acetic acid on growth promoting activity

Preparation	Lot	No. of rats	Total dose (μg)	Tibia width (μ)	Potency (I.U./mg ± SE)	p
Solvent		5		160		
BGH	I.S.	5	24	211 ± 13.6	1.0	<.001
		5	60	229 ± 5.0		<.001
		5	150	279 ± 8.5		<.001
BGH (35% HAc)	ANY 6	5	24	193 ± 2.4	0.37 ±	<.001
		5	60	207 ± 6.5	0.06	<.001
		5	150	234 ± 8.1		<.001
Solvent		5		161		
BGH	I.S.	5	24	212 ± 12.9	1.0	<.01
		5	60	255 ± 8.2		<.001
		5	150	308 ± 5.1		<.001
TBGH-d	ANK-B-2	5	60	197 ± 8.3	0.13 ±	<.01
(35% HAc)		5	150	213 ± 7.3	0.03	<.001
		5	375	237 ± 13.8		<.001
Preparation	**Lot**	**No. of rats**	**Daily dose (μg)**	**Weight gain (g)**	**Potency (I.U./mg ± SE)**	**p**
Solvent		8		3.8		
BGH	I.S.	8	25	10.9 ± 3.1	1.0	<.01
		8	100	19.4 ± 2.3		<.001
TBGH-d	AMH	8	17	10.8 ± 1.2	1.48 ±	<.001
		8	67	19.4 ± 3.0	0.28	<.001

I.S. = International standard.

TABLE III

Polyalanylation of growth hormone preparations

Preparation	Polyalanylated	Mol. wt. of parent	Lys + NH_2-terminals	μM Ala / 10 mg	mole Ala / mole parent
BGH	—	21,000	12	6	13
	+			72	152
TBGH-d	—	21,000	12	6	13
	+			101	213
A-I	—	16,000	10	8	12
	+			128	205
A-II	—	5,000	3	2	1
	+			109	52
HGH	—	21,000	10	3	7
	+			130	272

For native preparations of BGH and derivatives of BGH see analyses in Yamasaki *et al.* (1970). For native HGH see analysis in Li *et al.* (1969).

TABLE IV

Effect of polyalanylation on growth promoting activity

Preparation	Lot	No. of rats	Total dose (μg)	Tibia width (μ ± SE)	Biol. act. (I.U./mg ± S.E.)	p
Solvent		5		160 ± 3.6		
BGH	Intern.	5	24	211 ± 13.6	1.0	<.001
	Standard	5	60	229 ± 5.0		<.001
		5	150	279 ± 8.5		<.001
BGH	ANY-1	5	24	161 ± 5.7	<0.16	N.S.
		5	60	186 ± 3.6		<.001
		5	150	197 ± 9.8		<.01
TBGH-d	ANY-2	5	24	182 ± 7.9	<0.32 ± .08	<.05
		5	60	206 ± 7.9		<.001
		5	150	226 ± 13.1		<.01
A-I	ANY-3	5	500	184 ± 5.0	<0.008	<.01
		5	3000	193 ± 8.4		<.01
A-II	ANY-5	5	60	170 ± 4.3	<0.16	N.S.
		5	150	154 ± 4.9		N.S.
		5	375	155 ± 6.5		N.S.
HGH	ANY-4	5	24	195 ± 10.1	0.53 ± 0.9	<.02
		5	60	210 ± 6.7		<.001
		5	150	254 ± 12.3		<.001

TABLE V

Recombination of A-I and A-II

Preparation	Lot	No. of rats	Total dose (μg)	Tibia width (μ)	p
Solvent		6		165	
BGH	Intern.	6	24	218 ± 8.9	<.001
	Standard	6	60	277 ± 19.5	<.001
		6	150	309 ± 12.7	<.001
A-I	AMF-2	6	480	174 ± 5.4	N.S.
		6	1200	208 ± 7.8	<.001
A-II	AMF-5A	6	120	233 ± 15.5	<.01
		6	300	259 ± 8.6	<.001
		6	750	286 ± 9.6	<.001
A-I + A-II	3 AMF-2 / 1 AMF-5A	6	40	200 ± 12.4	<.05
		6	100	232 ± 9.0	<.001
		6	250	272 ± 17.5	<.001

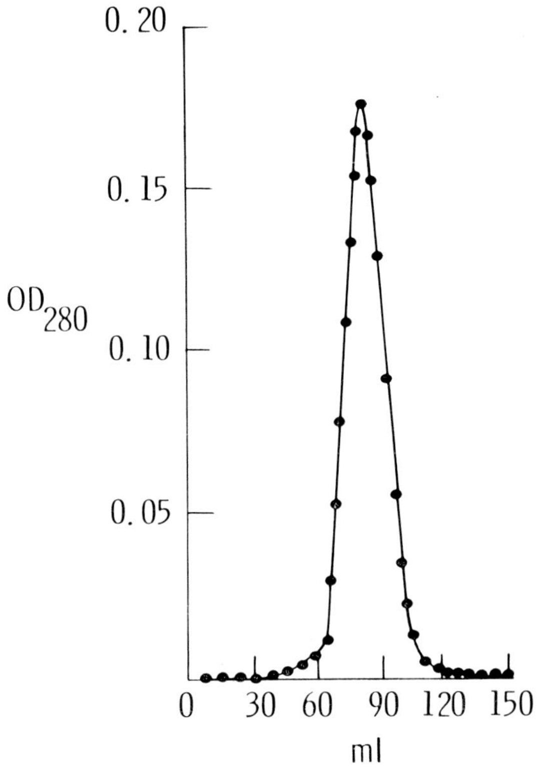

Fig. 15. Gel filtration of 8 mg of component A-I and 2.5 mg of component A-II in N acetic acid on Sephadex G-75. Column size 2.5 × 40 cm.

Recombination of large and small peptide of BGH

The possibility remained that the lower growth promoting potency of the smaller peptide was the consequence of the loss of some contribution of the larger peptide to the conformation of the smaller peptide and possibly the reverse. Evidence for the partial recombination of the large and small peptide was obtained from gel filtration in 1 N acetic acid where a single elution peak with a molecular weight of 21,000 was noted (Fig. 15). This fraction on a calibrated column under the same conditions was eluted at the same volume as BGH and TBGH-d (Figs. 16 and 17). In contradistinction, gel filtration of TBGH-d in 35% (6.15 N) acetic acid yielded two fractions (Yamasaki *et al.*, 1970). In addition, circular dichroism at pH 9.5 of equimolar amounts of small and large peptides placed in separate compartments of a tandem optical cell yielded a spectrum which was unlike the spectrum noted with the peptides in the same compartment of an optical cell, suggesting greater helicity on interaction (Sonenberg and Beychok, 1971). The helical content (23%) of the small and large peptide together was still significantly less than that (50–58%) of the component of a tryptic digest of BGH from which they were derived (Sonenberg and Beychok, 1971). In addition, the growth promoting activity (Table V) of the recombined equimolar amounts of large and small peptide was greater than the sum of the growth promoting activities of the individual peptides.

Degradation of small peptide

The 37 amino acid peptide has been cleaved with cyanogen bromide to yield 2 fragments

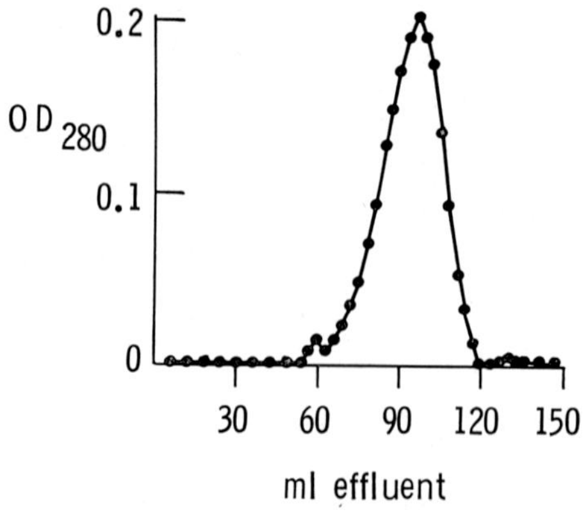

Fig. 16. Gel filtration of 10 mg of BGH in N acetic acid on Sephadex G-75. Column size 2.5 × 40 cm.

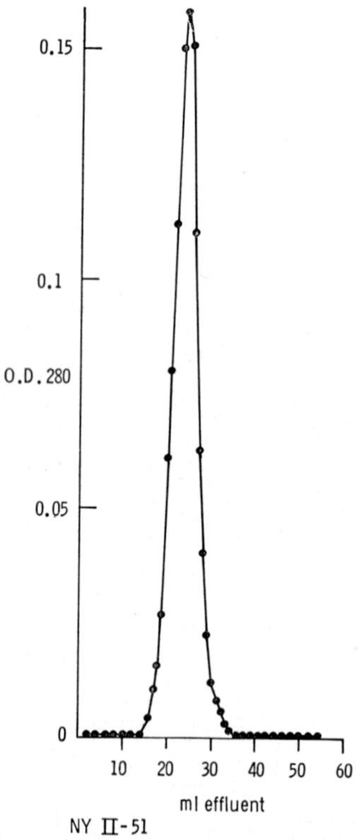

Fig. 17. Gel filtration of 13 mg of TBGH-d in N acetic acid on Sephadex G-75. Column size 2.5 × 40 cm.

(Fig. 18) consistent with the presence of a single methionine residue (Yamasaki *et al.*, 1970). Amino acid analysis of these two cyanogen bromide fragments revealed peptides with 9 and 27 amino acids with the absence of methionine (Table I). The nonapeptide had arginine as

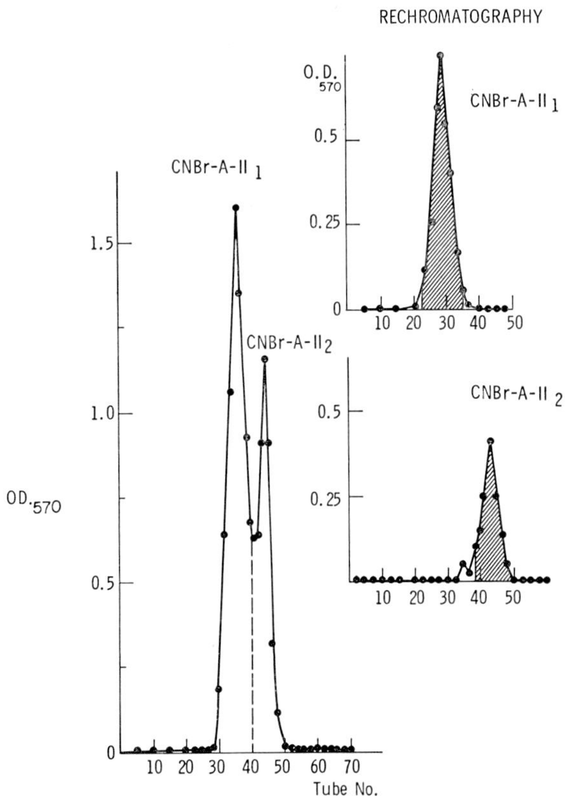

Fig. 18. Gel filtration of cyanogen bromide cleaved A-II peptide in 0.2 N acetic acid on Sephadex G-25. Column size 2.5 × 95 cm. Color developed with ninhydrin.

both amino and carboxyl terminal amino acids. The amino acid composition of the nonapeptide of A-II and the amino terminal nonapeptide of fragment C of Fellows and Rogol (1969) are identical.

REFERENCES

Bergenstal, D. and Lipsett, M. (1960): Metabolic effects of human growth hormone and growth hormone of other species in man. *J. clin. Endocr.*, *20*, 1247.

Canfield, R. E. (1968): Concepts relating to the structure of protein hormones. *Ann. N. Y. Acad. Sci.*, *148*, 289.

Elsair, J., Variel, E., Gerbeaux, S., Dartois, A. M. and Royer, P. (1964): Effects de l'hormone de croissance dans le nanisme hypothalamo-hypophysaire. II. Étude des actions metaboliques de nouvelles preparations d'hormone de croissance bovine partiellement hydrolysée. *Rev. franç. Etud. clin. biol.*, *9*, 614.

Fellows Jr., R. E. and Rogol, A. D. (1969): Structural studies on bovine growth hormone. I. Isolation and characterization of cyanogen bromide fragments. *J. biol. Chem.*, *244*, 1567.

Forsham, P. H., Li, C. H., Di Raimondo, V. C., Kolb, F. O., Mitchell, D. and Newman, S. (1958): Nitrogen retention in man produced by chymotrypsin digests of bovine somatotropin. *Metabolism*, *7*, 762.

Harris, J. I., Li, C. H., Condliffe, P. G. and Pon, N. G. (1954): Action of carboxypeptidase on hypophyseal growth hormone. *J. biol. Chem.*, *209*, 133.

KOLLI, E. A., SINITSYNA, A. L., KEDA, Y. M. and BOGACHEVA, I. D. (1966): Enzyme hydrolysates of growth hormone and the effect of different degrees of hydrolysis on the growth promoting activity and physico-chemical properties of the preparations. *Probl. Endokr. Gormonoter.*, *12*, 89.

LARON, Z., YED-LEKACH, A., ASSA, S. and KOWADLO-SILBERGELD, A. (1964): Immunochemical properties of bovine and human pituitary growth hormone after pepsin digestion. *Endocrinology*, *74*, 532.

LEVIE, H. L. and SONENBERG, M. (1970): Studies on papain digests of bovine growth hormone. *Chem. Engin. News*, August, 83.

LI, C. H. (1956): Hormones of the anterior pituitary gland. Part I. Growth and adrenocorticotropic hormones. *Advanc. Prot. Chem.*, *11*, 101.

LI, C. H., CLAUSER, H., FØNSS-BECH, P., LEVY, A. L., CONDLIFFE, P. G. and PAPKOFF, H. (1955): Hypophyseal growth hormone as a protein. In: *Hypophyseal Growth Hormone, Nature and Action*, p. 70. Editors: R. Smith, O. H. Gaebler and C. N. H. Long. McGraw-Hill, New York.

LI, C. H., DIXON, J. S. and LIU, W. K. (1969): Human pituitary growth hormone. XIX. The primary structure of the hormone. *Arch. Biochem.*, *133*, 70.

LI, C. H. and LIU, W. K. (1964): Human pituitary growth hormone. *Experientia (Basel)*, *20*, 169.

LI, C. H., PAPKOFF, H., FØNSS-BECH, P. and CONDLIFFE, P. G. (1956): Action of chymotrypsin on hypophyseal growth hormone. *J. biol. Chem.*, *218*, 41.

LI, C. H., PAPKOFF, H. and HAYASHIDA, T. (1959): Preparation and properties of beef α-core from chymotryptic digestion of bovine growth hormone. *Arch. Biochem.*, *85*, 97.

LI, C. H. and YAMASHIRO, D. (1970): The synthesis of a protein possessing growth-promoting and lactogenic activities. *J. Amer. chem. Soc.*, *92*, 7608.

NADLER, A., SONENBERG, M., FREE, C. A. and NEW, M. I. (1967): Growth hormone activity in man with components of tryptic digests of bovine growth hormone. *Metabolism*, *16*, 830.

NUTTING, D. F., KOSTYO, J. L., MILLS, J. B. and WILHELMI, A. E. (1970*a*): A cyanogen bromide fragment of reduced and S-aminoethylated porcine growth hormone with anabolic activity. *Biochim. biophys. Acta (Amst.)*, *200*, 601.

NUTTING, D. F., KOSTYO, J. L., GOODMAN, H. M. and FELLOWS, R. E., Jr. (1970*b*): Biologically active cyanogen bromide fragments of bovine growth hormone. *Endocrinology*, *86*, 416.

REUSSER, F. (1965): Enzymatic modification of bovine growth hormone by proteolytic streptomycete extracts. *Acta Endocr. (Kbh.)*, *49*, 578.

SANTOMÉ, J. A., DELLACHA, J. M., PALADINI, A. C., WOLFENSTEIN, C. E. M., PEÑA, C., POSKUS, E., DAURAT, S. T., SESÉ, Z. M., SANGUESA, V. F., BISCOGLIO, M. J. and FERNANDEZ, H. N. (1971): *Atlas of Protein Sequence and Structure*. National Biomedical Research Foundation, Silver Spring, Md.

SONENBERG, M. (1969): Interaction of human growth hormone and human erythrocyte membranes as demonstrated by circular dichroism. *Biochem. biophys. Res. Commun.*, *36*, 450.

SONENBERG, M. (1971): Interaction of human growth hormone and human erythrocyte membranes studied by intrinsic fluorescence. *Proc. nat. Acad. Sci. (Wash.)*, *68*, 1051.

SONENBERG, M. and BEYCHOK, S. (1971): Circular dichroism studies of biologically active growth hormone preparations. *Biochim. biophys. Acta (Amst.)*, *229*, 88.

SONENBERG, M. and DELLACHA, J. M. (1967): Anabolic effects in man of papain digests of bovine growth hormone. *J. clin. Endocr.*, *27*, 1035.

SONENBERG, M., DELLACHA, J. M., FREE, C. A. and NADLER, A. C. (1969): Growth hormone activity in man of chymotryptic digests of bovine growth hormone. *J. Endocr.*, *44*, 265.

SONENBERG, M., FREE, C. A., DELLACHA, J. M. and NADLER, A. C. (1965*a*): In: *Abstracts, VI Pan American Congress of Endocrinology*, Abstr. 200. ICS 99, Excerpta Medica, Amsterdam.

SONENBERG, M., FREE, C. A. DELLACHA, J. M., BONADONNA, G., HAYMOWITZ, A. and NADLER, A. C. (1965*b*): The metabolic effects in man of bovine growth hormone digested with trypsin. *Metabolism*, *14*, 1189.

SONENBERG, M., KIKUTANI, M., FREE, C. A., NADLER, A. C. and DELLACHA, J. M. (1968): Chemical and biological characterization of clinically active tryptic digests of bovine growth hormone. *Ann. N. Y. Acad. Sci.*, *148*, 532.

SWISLOCKI, N. I., SONENBERG, M. and YAMASAKI, N. (1970): In vitro metabolic effects of bovine growth hormone fragments in adipose tissue. *Endocrinology*, *87*, 900.

YAMASAKI, N., KIKUTANI, M. and SONENBERG, M. (1970): Peptides of a biologically active tryptic digest of bovine growth hormone. *Biochemistry*, *9*, 1107.

EVIDENCE FOR THE SEPARATE EXISTENCE OF A HUMAN PITUITARY PROLACTIN – A REVIEW AND RESULTS*

FREDERICK C. GREENWOOD

Department of Biochemistry and Biophysics, University of Hawaii, Honolulu, Hawaii

The problem of human prolactin was posed in 1961. In that year Lyons *et al.* (1961) demonstrated unequivocably that purified human growth hormone possessed intrinsic prolactin bioactivity. This was confirmed and extended by Chadwick *et al.* (1961), Rivera *et al.* (1967) and Hartree *et al.* (1965). An experimental observation is seldom doubted; the interpretation is more frequently questioned. In this instance the results were interpreted to suggest that 'in the human the hormone control of lactation and growth seems to be effected through a single pituitary protein namely human growth hormone' (Li and Bewley, 1970). Further experimental support had been obtained by Lyons *et al.* (1968) from the increase in weight of infants of mothers injected with growth hormone, taking the weight increase of the infants as a measure of lactation performance. The homologies in the amino acid sequences of human growth hormone and ovine prolactin (Li and Bewley, 1970) likewise gave a rational basis for the growth hormone – like activities noted for ovine prolactin on injection into man (Bergenstal and Lipsett, 1958; McCalister and Welbourn, 1962; Beck *et al.*, 1965). The existence of a single growth hormone/prolactin in the human appeared also to apply to the monkey since monkey growth hormone was shown to possess prolactin bioactivities (Peckham *et al.*, 1968).

The opposing view expressed a belief that a human prolactin could exist apart from a human growth hormone (Wilhelmi, 1961), that the prolactin bioactivities of human growth hormone were real but only expressed in species other than man since plasma immunoreactive growth hormone changes were not consistent with a role in lactation (Greenwood, 1967). A partial separation of the two biological activities from human pituitaries was achieved by Chen and Wilhelmi (1964), and by Apostolackis (1965). However, immunological studies by Tashijian *et al.* (1965) and a search in lactation plasma using a radioimmunoassay for these prolactin enriched fractions (Stephenson and Greenwood, 1967; Greenwood, 1967) suggested that they were modified growth hormone preparations. Nevertheless the 'separatist' view persisted as plasma growth hormone levels did not reflect a lactogenic role (Greenwood, 1967; Glick, 1969; Spellacy *et al.*, 1970). Likewise lactation was present in women with an isolated growth hormone deficiency (Rimoin *et al.*, 1968). It may be noted that the separate identity of animal prolactin and growth hormone has never been in dispute. Separate radioimmunoassays and distinctive stimuli for animal prolactins have been uncovered (*cf* Bryant and Greenwood, 1972).

Biological assays per se for prolactin in human plasma showing its presence (Canfield and Bates, 1965) initially could be criticized as measuring prolactin/growth hormone. However, this work was substantiated when prolactin bioactivity of lactation plasma and plasma from

* This work was supported by contract (NIH 69–2190) and research grant NIH AM 13217 and a grant from G. D. Searle and Co.

galactorrhea patients was detected where immunoreactive growth hormone was too low to account for the biological activity (Roth *et al.*, 1968; Forsyth, 1970; Benjamin *et al.*, 1969). Using a sensitive *in vitro* biological assay for prolactin and a conventional radioimmunoassay for growth hormone Frantz and Kleinberg (1970) and Kleinberg and Frantz (1970) have measured high levels of prolactin in lactating women, in subjects on phenothiazines and in patients with galactorrhea. This activity was not neutralized by adding anti-human growth hormone serum and the plasma levels of growth hormone were shown to be low. In plasmas with a high plasma growth hormone (>15 ng/ml), the prolactin bioactivity detected was abolished by antiserum to growth hormone. A significant contribution using biological assays in conjunction with histological, immunological and electron microscopic studies has been presented by Peake *et al.* (1969). Plasma obtained from patients before and after removal of pituitary tumors causing galactorrhea and the tumors themselves were intensively studied. The conclusion was inescapable that a human prolactin exists in its own right. Hence from this series of, what have been termed for convenience, bioassay papers it may be concluded that human prolactin exists as a separate entity from growth hormone. That this is also applicable to the monkey is shown by the ready detection of prolactin bioactivity in monkey plasma, no immunoreactive growth hormone being detected (Dr. C. S. Nicoll, NIH Workshop on Prolactin).

Heterologous radioimmunoassays have been devised to measure a peptide hormone, not available as purified material, using a cross-reacting system. A typical example is the measurement of caprine prolactin in plasma against an ovine prolactin standard in a labelled ovine prolactin/anti-ovine prolactin system (Bryant and Greenwood, 1968). The available radioimmunoassays for HCS and ovine prolactin are logical contenders for heterologous systems for possibly detecting human prolactin. Lactation plasma taken after suckling in women appeared to contain material cross-reacting with HCS which was not HCS, by definition a pre-partum hormone and not growth hormone, by direct measurement (Stephenson and Greenwood, 1969). Subsequent unpublished work using a different anti-HCS serum in a more sensitive assay convinced us that the inhibitions were artefacts generated by plasma damage. Further attempts using an antiserum to chemically modified HCS have been reported (Dr. J. Josimovich, NIH Workshop on Prolactin), showing a lactation antigen cross-reacting in a labelled HCS/anti-modified HCS system. Dr. A. R. Midgley Jr. reported at the same meeting a lactation antigen inhibiting in an anti-ovine prolactin system. The latter has been used by Herbert and Hayashida (1970) in the fluorescent antibody technique to distinguish prolactin from growth hormone cells in the human pituitary. The cross-reactions between HCS, ovine prolactin and a postulated human prolactin in lactation plasma appear surprising until the amino acid sequences of HGH, ovine prolactin and HCS reported by Professor C. H. Li and his colleagues (Li *et al.*, 1966, 1969) are studied. The homologies noted between ovine prolactin and HGH (Li and Bewley, 1970) and between HCS and HGH (Li, 1971) have been extended to the three peptides and internal homologies noted (Niall *et al.*, 1971) and it is postulated that these arose by gene reduplication of a primordial peptide. The remarkable homologies shown by Niall *et al.* (1971) make it surprising that antisera generated to one of these peptides can ever be wholly specific for just that one peptide! By extrapolation human prolactin would have similar four areas of internal homology, be about 200 amino acids in length, have an off-center tryptophanyl residue, a large disulphide loop and a C terminal bridge and by analogy with ovine prolactin an N terminal bridge. Any antiserum generated to HCS, HGH or ovine prolactin with antibodies directed to one of the homologous areas would be expected to react with human prolactin. The test of such heterologous radioimmunoassays for human plasma prolactin would seem to be their physiological validity. This is now possible since suckling and phenothiazine stimuli have been shown to be powerful releasers of prolactin in many animals (*cf* Bryant *et al.*, 1968, 1970). This has already been carried out for a labelled porcine prolactin/anti-ovine prolactin system used to measure human prolactin (Dr. W. H. Daughaday, discussion at this Symposium).

Biosynthesis of human or monkey prolactin in vitro offered a possible solution to the view that a separate prolactin exists in the anterior pituitary but is present in small quantities relative to growth hormone. The studies of Pasteels and his colleagues (Pasteels *et al.*,1963;Pasteels, 1968) demonstrated that anterior pituitary tissue from the human fetus when cultured initially released both prolactin bioactivity and growth hormone activity but in later cultures the prolactin bioactivity only was released. The fall in growth hormone was confirmed by immunoassay and the prolactin bioactivity extracted and used to generate an antiserum. The latter neutralized the prolactin bioactivity of human lactation plasma. Nicoll *et al.* (1970) in culturing the adenohypophyses from the rhesus monkey likewise demonstrated the independence of growth hormone and prolactin release. Using polyacrylamide gel, bioassay and radioimmunoassays a sample of monkey prolactin assaying 30 I.U./mg has been obtained (Dr. C. S. Nicoll, presented at the NIH Prolactin Workshop meeting, January 1971). Friesen *et al.* (1970) have used short-term incubations of small fragments of human anterior pituitary glands with tritiated leucine to study the biosynthesis of pituitary proteins. Using acrylamide gel electrophoresis and antiserum to HGH they obtained a principal radioactive protein, not HGH, and suggested it was human prolactin.

Undoubtedly biosynthetic studies using human and monkey pituitaries have produced evidence difficult to reconcile with a one molecule (growth hormone/ prolactin) view.

Radioimmunoassay for human prolactin. 'Biosynthetic studies have been used to produce sufficient quantities of human and monkey prolactin to produce antisera, to show their distinction by biological assay and by polyacrylamide gel electrophoresis from GH, and to develop radioimmunoassays for prolactin in plasma for the first demonstrations of the stimuli which cause a release of human prolactin.' This is a convenient summary of what we know to be true at this time. In addition it is apparent from the NIH Prolactin meeting and this Symposium that valid heterologous assays for human prolactin are soon to be published. However, a scholarly and fully referenced dissection of the separate parts of the summary is not easy in an area which is of such active interest. It could be predicted from the work of Pasteels' group (*cf* Pasteels, 1968), Nicoll *et al.* (1970) and Friesen *et al.* (1970) that homologous radioimmunoassays would logically follow the isolation of sufficient prolactin. Preliminary reports of a radioimmunoassay based on Pasteel's human prolactin were presented in September 1970 (Bryant *et al.*, 1971*a*) and as an abstract in February 1971 (Bryant *et al.*, 1971*b*) and a detailed paper (Bryant *et al.*, 1971*c*). However it was apparent from unpublished presentations in January 1971 by Drs. Friesen and Guyda (NIH Prolactin Workshop) that radioimmunoassays for both Pasteels' human prolactin and Friesen's monkey prolactin had been developed and applied to human plasma. Guyda *et al.* (1971) have presented evidence for monkey and human prolactin. They used an anti-ovine prolactin serum to precipitate their presumptive monkey prolactin. This was then purified by removal of monkey growth hormone. The final fraction was used to produce an antiserum which cross reacted with ovine prolactin. This allowed a heterologous radioimmunoassay, labelled ovine prolactin/anti monkey prolactin and the measurement of human prolactin in pituitary extracts only at that state. HGH constituted 7.8% and human prolactin 0.9% of a pituitary powder. The purification of adequate amounts of human prolactin for essentially a homologous assay seems probable *viz.* labelled human prolactin and an anti-human prolactin antiserum. We concur with Nicoll *et al.* (1970) and Friesen *et al.* (1970) that ultimate proof of the discrete identity of prolactin will require its isolation and chemical and biological characterization. The independent and essentially simultaneous development of radioimmunoassays for human prolactin are a significant step forwards since they allow routine measurements in plasma and will doubtless be fully exploited in studies of the normal physiology and pathophysiology of human prolactin.

Radioimmunoassay for human prolactin (based on Pasteels' human prolactin) and its application to plasma. A study of Pasteels' human prolactin and an antiserum to it was carried out by determining the binding of the labelled antigen to the homologous and heterologous antisera (anti-HP, anti-HCS and anti-HGH). Conversely the unlabelled antigen was used to inhibit in the labelled HGH/anti-HGH system. From the results it was apparent that the human prolactin contained some 1% of immunoreactive HGH and that the anti-human prolactin serum contained antibodies with specificities directed to HGH and HCS. However antibodies were present directed to the human prolactin material only and a radioimmunoassay was shown to be possible showing little interference by HGH in plasma. Thence a small number of plasma samples were assayed for HGH and for HP, using Pasteels' HP as standard. In situations shown previously to cause a marked release of sheep or goat prolactin – suckling (Bryant *et al.*, 1970) or certain phenothiazines (Bryant *et al.*, 1968) human plasma prolactin levels were shown to increase. Further evidence for the physiological validity of the plasma assays has been produced (Bryant and Greenwood, 1971). In this study the stimulation of prolactin secretion in the human by suckling or by phenothiazine treatment was confirmed and a stress induced release of prolactin obtained by the injection of insulin, oxytocin and hypertonic saline. We have some reservations whether the latter is a specific or non-specific stimulus. Of seven subjects studied after oral glucose one showed a marked increase readily ascribed to stress in this individual akin to the anomalous rises in plasma HGH and cortisol seen after oral glucose tests in some patients with breast cancer (Greenwood *et al.*, 1968). Our first impression of studies on few cases of galactorrhea are that prolactin levels taken at random may be high or low and it is evident that dynamic tests, without a stress element, are required for a definitive study. The imprecision of the present radioimmunoassay, due to the poor antiserum and radioiodination damage makes it comparable, although easier to carry out, to the biological assay of Frantz and Kleinberg (1970). Ten plasmas previously assayed by these authors showed at best 8/10 agreements on ranking in order of relative prolactin levels. Taken with the physiological measurements the comparison adds its increment to the evidence that the radioimmunoassay is a valid estimate of human prolactin in plasma.

The requirement for a small but continuing supply of purified human prolactin for labelling and for standards and the requirement for better antisera led us to repeat Pasteels' tissue culture technique for human fetal pituitaries. In addition this technique offered an opportunity to isolate adequate amounts of human prolactin for characterization and sequence.

Results obtained from some thirty pituitaries in culture since June 1970 presented by Siler *et al.* (1971) confirm the results summarized by Pasteels (1968) that human prolactin release continues and increases with culture time whereas HGH declines but is usually present. Microscale purification and radioiodination of fetal and pituitary tumor prolactin, the latter obtained from pituitary tumors kindly supplied by Dr. B. R. Webster of the Toronto General Hospital, from individual culture dishes have been accomplished and gave similar results when used with anti-Pasteels' human prolactin serum in radioimmunoassay of plasmas previously assayed. In addition the application of radioimmunoassays for LH, FSH, ACTH/MSH to the media of pituitary tissues have demonstrated the sporadic release of these hormones, a sex difference in release of FSH and LH and the presence of these hormones and of HGH and HP in the pituitary tissue from fetuses as early as 5 weeks of gestation.

CONCLUSION

A review of the ten-year-old problem of the separate identity of human and monkey prolactins is presented citing evidence from clinical, biological and biosynthetic studies and from heterologous and homologous radioimmunoassays of human prolactin in plasma. It is concluded that human and monkey prolactins exist as separate peptides in their respective anterior pituitary glands and that they are released by stimuli different from those causing growth hormone release although both are released in stress.

The observations that human growth hormone shows prolactin bioactivities in animals and that ovine prolactin shows growth hormone-like activities in the human and their cross-reactivities can now be more readily visualized from the known sequences and their homologies. Specificity of biological activity and immunospecificity seem to be relative rather than absolute properties of peptide hormones. The role of target tissue receptors in perhaps imposing a higher degree of specificity is not known. However target tissue receptors of difficult species do not seem to distinguish between similar peptides with different major bioactivities as readily as does the immune defense mechanism.

ACKNOWLEDGEMENTS

We acknowledge with pleasure gifts of human prolactin, anti-human prolactin from Professor J. L. Pasteels and of FSH, LH and their antisera from the Hormone Distribution Officer, Dr. R. W. Bates, National Institutes of Health, an anti-ACTH serum from Dr. Lesley Rees and an anti-LH serum from Dr. Richard Donald.

ADDENDUM

The following is a personal and journalistic summary of recent developments in the area covered by the review. Since it is based on notes and discussions at this Symposium and at the subsequent Ciba Foundation Symposium on Lactogenic Hormones the findings will not be attributed. Readers interested in this area are referred to papers in this volume, the volume to be published by the Ciba Foundation, that of the Cornell Symposium on Gonadotropins and to Recent Progress in Hormone Research in 1972. Research groups known to be actively working and publishing in the area of plasma bio- and radioimmunoassays of human and monkey prolactin, their isolation and sequence, include the laboratories of, in alphabetical order, Daughaday, Forsyth, Frantz, Friesen, Greenwood, Josimovich, Kwa, Midgley, Niall, Nicoll, Pasteels, Peckham and Turkington and their colleagues.

Measurements in plasma show that human prolactin is secreted in normal children, adult males and females, and that a pattern in the menstrual cycle has not yet been discerned. Increases in concentration have been shown after stress, immediately and for several days after phenothiazines, during human but not monkey pregnancy, after suckling and in some cases after insulin. Release is inhibited by ergocornine, and no rises after arginine have been observed. Stalk section causes an elevation in plasma levels in the majority of patients whereas hypophysectomy causes the expected fall with a half-life of the rapid phase of about 15 minutes. Plasma levels are high in renal disease and LDOPA decreases normal levels of prolactin. Some but not all patients with galactorrhea have elevated levels; few acromegalics have elevated levels. Likewise some patients with idiopathic hypopituitarism have elevated levels. Patients with myxedema have higher than normal levels of plasma prolactin.

Heterologous radioimmunoassays using porcine or ovine prolactin and antisera to ovine prolactin and HPL may well become the method of choice now that homologous radioimmunoassays and bioassays have shown that a human and monkey prolactin exists as a separate entity from growth hormones. Supplies of human and monkey prolactin are inadequate for general distribution at this time but are sufficient for preliminary amino acid sequence. The synthesis of the first 30 or so amino acids is a reasonable expectation and the sequence of the whole molecules should be completed by late 1972, using prolactin from either tissue culture medium from fetal or tumor pituitaries or from pooled adult pituitaries using current homologous and heterologous radioimmunoassays.

The human prolactin problem has been in the literature for a decade but a method for its solution existed certainly in 1963 from the work of Dr. J. L. Pasteels and his colleagues. Its actual and full solution spans no more than 2 years work. Perhaps there is a lesson to be learned here for other outstanding problems in endocrinology.

REFERENCES

Apostolackis, M. (1965): The extraction of prolactin from human pituitary glands. *Acta endocr.*, *49*, 1.

Beck, J. C., McGarry, E. E., Dawson, K. G., Gonda, A., Hamid, M. A. and Rubinstein, D. (1965): Comparative metabolic actions of primate growth hormone in man. In: *Proceedings of the 2nd International Congress of Endocrinology*, p. 1242. Editor: S. Taylor. ICS 83, Excerpta Medica, Amsterdam.

BENJAMIN, F., CASPER, D. J. and KOLODNY, H. (1969): Immunoreactive human growth hormone in conditions associated with galactorrhea. *Obstet. Gynec., 34*, 34.
BERGENSTAL, D. M. and LIPSETT, M. B. (1958): The anabolic effect of sheep prolactin in man. *J. clin. Invest., 37*, 877.
BRYANT, G. D. and GREENWOOD, F. C. (1968): Radioimmunoassay for ovine, caprine and bovine prolactin in plasma and tissue extracts. *Biochem. J., 109*, 831.
BRYANT, G. D. and GREENWOOD, F. C. (1971): The concentrations of human prolactin in plasma by radioimmunoassay: experimental and physiological modifications. In: *Ciba Foundation Colloquia on Endocrinology.* Editors: G. E. W. Wolstenholme and J. Knight.
BRYANT, G. D. and GREENWOOD, F. C. (1972): Prolactin immunoassay. In: *Methods in Investigative and Diagnostic Endocrinology – Peptide Hormones.* Editors: R. S. Yalow and S. A. Berson. North-Holland Publishing Co., Amsterdam.
BRYANT, G. D., CONNAN, R. M. and GREENWOOD, F. C. (1968): Changes in plasma prolactin induced by acepromazine in sheep. *J. Endocr., 41*, 613.
BRYANT, G. D., LINZELL, J. L. and GREENWOOD, F. C. (1970): Plasma prolactin in goats measured by radioimmunoassay: The effects of teat stimulation, mating behavior, stress, fasting and of oxytocin insulin and glucose injection. *Hormones, 1*, 26.
BRYANT, G. D., SILER, T. M., GREENWOOD, F. C., PASTEELS, J. L., ROBYN, C. and HUBINONT, P. O. (1971*a*): Human prolactin radioimmunoassay. In: *Radioimmunoassay Methods*, p. 218. European Workshop, September, 1970. Editors: K. E. Kirkham and W. M. Hunter. Churchill Livingstone, London.
BRYANT, G. D., SILER, T. M., MORGENSTERN, L. L. and GREENWOOD, F. C. (1971*b*): Radioimmunological studies on a human pituitary prolactin from tissue culture and from human lactation plasma. *Abstracts, Fourth Asia and Oceania Congress of Endocrinology*, Abstract 12.
BRYANT, G. D., SILER, T. M., GREENWOOD, F. C., PASTEELS, J. L., ROBYN, C. and HUBINONT, P. O. (1971*c*): Radioimmunoassay of a human pituitary prolactin in plasma. *Hormones, 2*, 139.
CANFIELD, C. J. and BATES, R. W. (1965): Nonpuerperal galactorrhea. *New. Engl. J. Med., 273*, 897.
CHADWICK, A., FOLLEY, S. J. and GEMZELL, C. A. (1961): Lactogenic activity of human pituitary growth hormone. *Lancet, 2*, 241.
CHEN, H. C. and WILHELMI, A. E. (1964): *Program, 46th meeting of the Endocrine Society, 1964*, Abstr. p. 80. Thomas, Springfield, Ill.
FORSYTH, I. A. (1970): The detection of lactogenic activity in human blood by bioassay. *J. Endocr., 46*, iv.
FRANTZ, A. G. and KLEINBERG, D. L. (1970): Prolactin: evidence that it is separate from growth hormone in human blood. *Science, 170*, 745.
FRIESEN, H., GUYDA, H. and HARDY, I. (1970): Biosynthesis of human growth hormone and prolactin. *J. clin. Endocr., 31*, 611.
GLICK, S. M. (1969): The regulation of growth hormone secretion. In: *Frontiers in Neuroendocrinology*, p. 141. Editors: W. F. Ganong and L. Martini. University Press, Oxford.
GREENWOOD, F. C. (1967): Growth hormone. In: *Hormones in Blood, Vol. 1*, p. 229. Editors: C. H. Gray and A. L. Bacharach. Academic Press, London.
GREENWOOD, F. C., JAMES, V. H. T., MEGGITT, B. F., MILLER, J. D. and TAYLOR, P. H. (1968): Pituitary function in breast cancer. In: *Prognostic Factors in Breast Cancer*, p. 409. Editors: A. D. M. Forrest and P. B. Kunkler. Livingstone, London.
GUYDA, H., HWANG, P. and FRIESEN, H. (1971): Immunologic evidence for monkey and human prolactin. *J. clin. Endocr., 32*, 120.
HARTREE STOCKELL, A., KOVACIC, N. and THOMAS, M. (1965): Growth promoting and luteotrophic activities of human growth hormone. *J. Endocr., 33*, 249.
HERBERT, C. D. and HAYASHIDA, T. (1970): Prolactin localization in the primate pituitary by immunofluorescence. *Science, 169*, 378.
KLEINBERG, D. L. and FRANTZ, A. C. (1970): Measurement of prolactin in human plasma by in vitro bioassay. *Clin. Res., 28*, 363.
LI, C. H. (1971): The amino acid sequence of HCS. Presented at the National Institutes of Health Workshop on Prolactin, January 1971.
LI, C. H. and BEWLEY, T. A. (1970): Primary structure of human pituitary growth hormone and sheep pituitary lactogenic hormone compared. *Science, 168*, 1361.
LI, C. H., DIXON, J. S., LO, T-B., PANKOV, Y. A. and SCHMIDT, K. D. (1969): Amino acid sequence of ovine lactogenic hormone. *Nature (Lond.), 224*, 695.

LI, C. H., LIU, W. K. and DIXON, J. S. (1966): Human pituitary growth hormone. XII. The amino acid sequence of the hormone. *J. Amer. chem. Soc.*, *88*, 2050.

LYONS, W. R., LI, C. H., AHMAD, N. and RICE-WRAY, E. (1968): Mammotrophic effects of human hypophysial growth hormone preparations in animals and man. In: *Growth Hormone*, p. 349. Editors: A. Pecile and E. E. Müller. ICS 158, Excerpta Medica, Amsterdam.

LYONS, W. R., LI, C. H. and JOHNSON, R. E. (1961): Biologic activities of human hypophysial mammotrophin. *Program, 43rd meeting of the Endocrine Society*, p. 4. Thomas, Springfield, Ill.

MCCALISTER, A. and WELBOURN, R. B. (1962): Stimulation of mammary cancer by prolactin and the clinical response to hypophysectomy. *Brit. med. J.*, *1*, 1669.

NIALL, H. D., HOGAN, M. L., SAUER, R., ROSENBLUM, I. Y. and GREENWOOD, F. C. (1971): Pituitary and placental lactogenic and growth hormones: evolution from a primordial peptide by gene duplication. *Proc. nat. Acad. Sci.*, *68*, 866.

NICOLL, C. S., PARSONS, J. A. and FIORINDO, R. P. (1970): Evidence of independent secretion of prolactin and growth hormone *in vitro* by adenohypophyses of rhesus monkeys. *J. clin. Endocr.*, *30*, 512.

PASTEELS, J. L. (1968): Nouvelles recherches sur la structure et le comportement des cellules hypophysaires en culture. *Mém. Acad. roy. Méd. Belg. II*, *7*, 1.

PASTEELS, J. L., BRAUMANN, H. and BRAUMANN, J. (1963): Etude comparée de la secrétion d'hormone somatotrope par l'hypophyse humaine *in vivo* et de son activité lactogénique. *C. R. Acad. Sci.*, *256*, 2031.

PEAKE, G. T., MCKEEL, D. W., JARETT, L. and DAUGHADAY, W. H. (1969): Ultrastructural histologic and hormonal characterization of a prolactin-rich human pituitary tumor. *J. clin. Endocr.*, *29*, 1383.

PECKHAM, W. D., HOTCHKISS, J., KNOBIL, E. and NICOLL, C. S. (1968): Prolactin activity of homogeneous primate growth hormone preparations. *Endocrinology*, *82*, 1247.

RIMOIN, D. L., HOLZMANN, G. B., MERIMEE, T. J., RABINOWITZ, D., BARNES, A. C., TYSON, J. E. H. and MCKUSICK, V. A. M. (1968): Lactation in the absence of human growth hormone. *J. clin. Endocr.*, *28*, 1183.

RIVERA, E. M., FORSYTH, I. A. and FOLLEY, S. J. (1967): Lactogenic activity of mammalian growth hormones. *Proc. Soc. exp. Biol.* (N.Y.) *124*, 859.

ROTH, J., GORDEN, P. and BATES, R. W. (1968): Studies of growth hormone and prolactin in acromegaly. In: *Growth Hormone*, p. 124. Editors: A. Pecile and E. E. Müller. ICS 158, Excerpta Medica, Amsterdam.

SILER, T. M., MORGENSTERN, L. L. and GREENWOOD, F. C. (1971): The release of human prolactin and other peptide hormones from human anterior pituitary tissue culture. In: *Ciba Foundation Colloquia on Endocrinology*. Editors: G. E. W. Wolstenholme and I. Knight. To be published.

SPELLACY, W. N., BUHI, W. C. and BIRK, S. A. (1970): Normal lactation and blood growth hormone studies. *Amer. J. Obstet. Gynec.*, *107*, 244.

STEPHENSON, F. A. and GREENWOOD, F. C. (1967): Radioimmunological studies on a prolactin-rich growth hormone fraction from human pituitaries. In: *First International Symposium on Growth Hormone*, p. 29. Editors: A. Pecile and E. E. Müller. ICS 142, Excerpta Medica, Amsterdam.

STEPHENSON, F. A. and GREENWOOD, F. C. (1969): A new hormone of lactation or an artefact of radioimmunoassay. In: *Protein and Polypeptide Hormones*, p. 28. Editor: M. Margoulies. ICS 161, Excerpta Medica, Amsterdam.

TASHJIAN Jr., A. H., LEVINE, L. and WILHELMI, A. E. (1965): Immunochemical studies with antisera to fractions of human growth hormone which are high or low in pigeon crop gland stimulating activity. *Endocrinology*, *77*, 1023.

WILHELMI, A. E. (1961): Fractionation of human pituitary glands. *Canad. J. Biochem. Physiol.*, *39*, 1659.

III. In vivo and in vitro effects

THE EFFECT OF GROWTH HORMONE ON INDUCIBLE LIVER ENZYMES

ASHER KORNER* and BRIGID L. M. HOGAN

School of Biological Sciences, University of Sussex, Falmer, United Kingdom

Growth hormone administration to hypophysectomized rats increases protein and RNA synthesis in the liver (for review, see Korner, 1970) and changes the activity of certain liver enzymes, notably tyrosine aminotransferase (TAT) and tryptophan oxygenase (TO) (Kenney, 1967; Labrie and Korner, 1968*a*), and ornithine decarboxylase (OD) (Jänne and Raina, 1969; Russell *et al.*, 1970). Growth hormone also modifies the induction of the first two enzymes by cortisol (Harding and Rosen, 1963; Holten and Kenney, 1967; Csányi and Greengard, 1968; Labrie and Korner, 1968*a*; Liberti *et al.*, 1970). This paper describes experiments designed to elucidate further the mechanism(s) through which growth hormone modifies the activity and inducibility of these enzymes and discusses possible interpretations of the results.

EXPERIMENTAL PROCEDURE

Experiments in which TAT and TO activity were measured were performed as described by Labrie and Korner (1968*a*, *b*). Young albino rats (100 g weight) were found to give the best response of ornithine decarboxylase activity to growth hormone administration. Livers were rapidly excised and homogenised at 4°C in 0.25 M sucrose, 50 mM Tris HCl pH 7.1, 2.5 mM β-mercaptoethanol. After centrifugation at 15,000 *g* for 10 min the supernatant was made 0.05 mM with pyridoxal phosphate and assayed immediately, essentially as described by Pegg and Williams-Ashman (1968). The final concentration of ornithine in the incubation medium was 0.25 mM. L-1-C^{14}-ornithine was purchased from the Radiochemical Centre, Amersham, U.K. and diluted to a final specific activity of 2 mc/mmole by the addition of non-radioactive L-ornithine.

RESULTS

Figure 1 confirms the observations of many previous workers that hydrocortisone administration to hypophysectomised as compared to adrenalectomised rats results in a greater and more prolonged increase in tryptophan oxygenase activity. The plateau of enzyme activity occurs later (between 6 and 12 hours depending on the experiment) and the fall in activity occurs more slowly.

This response of hypophysectomised rats to cortisol is shown again in Figure 2, together with the effect of injecting growth hormone, either alone or in combination with hydrocortisone. Combined hormone treatment results in an enhancement of TO activity at 3 hours, but this is followed by a sharp and more rapid fall in activity, compared to animals treated with hydrocortisone alone. Growth hormone itself causes a smaller but significant increase in tryptophan oxygenase at 3 hours but this activity falls back to basal level very quickly. In some experiments the enzyme falls to below basal for some hours after growth hormone

* Prof. Asher Korner deceased in September 1971.

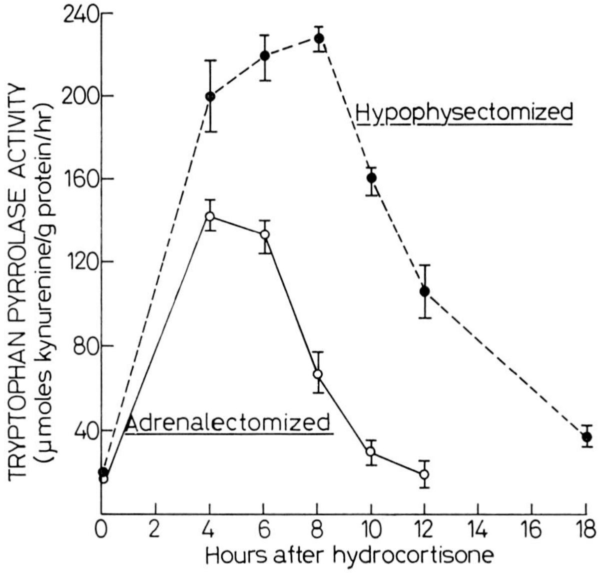

Fig. 1. Time course of induction of tryptophan oxygenase after hydrocortisone injection in adrenalectomised and hypophysectomised rats. 30 mg/kg of sodium succinate hydrocortisone was injected intraperitoneally at zero time.

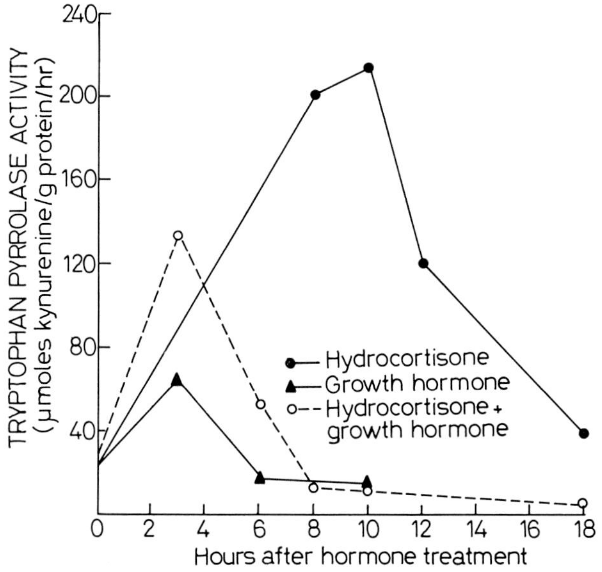

Fig. 2. The effect of intraperitoneal injection of 0.3 mg/kg bovine growth hormone alone, 30 mg/kg hydrocortisone alone, or 30 mg/kg hydrocortisone plus 0.3 mg growth hormone, on tryptophan pyrrolase activity of liver of hypophysectomised rats.

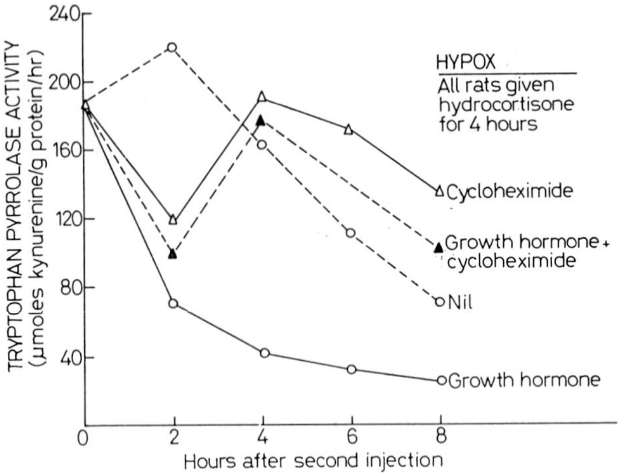

Fig. 3. Effect of 0.3 mg/kg growth hormone, 1.0 mg/kg cycloheximide, growth hormone and cycloheximide given together, or saline given at 4 hours on the tryptophan oxygenase activity of liver of hypophysectomised rats all given 30 mg/kg hydrocortisone at zero time.

treatment. These results suggest that growth hormone has two opposing effects; it enhances the cortisol induction at 3 hours to some extent but inhibits it greatly at times subsequent to about 4 hours, apparently by increasing the rate of fall in enzyme activity. This second effect was investigated further in the experiments described in Figure 3.

In this experiment a group of hypophysectomised rats were injected with cortisol and 4

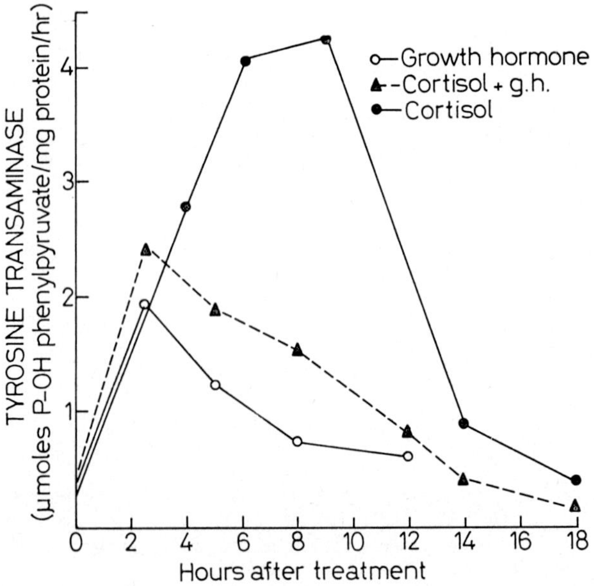

Fig. 4. Effect of growth hormone alone, hydrocortisone alone, and hydrocortisone plus growth hormone, on the tyrosine aminotransferase activity of hypophysectomised rats. Details as for Fig. 2.

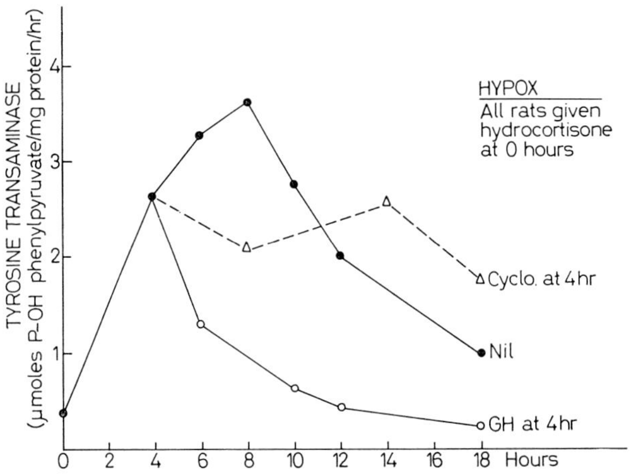

Fig. 5. Effect of growth hormone, cycloheximide, or saline given at 4 hours on the tyrosine aminotransferase activity of liver of hypophysectomised rats all given hydrocortisone at zero time. Details as for Fig. 3.

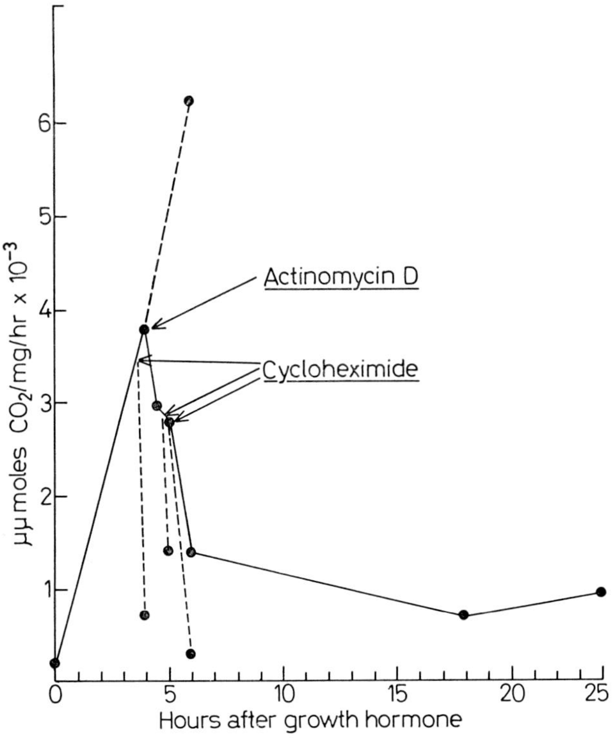

Fig. 6. Induction of ornithine decarboxylase in rats given growth hormone. Young (100 g) rats were injected intraperitoneally with 0.4 mg bovine growth hormone. At various times groups of five rats were killed and the livers removed and assayed for ornithine decarboxylase activity. Cycloheximide (5 mg) was given 15 min or 1 hour, and actinomycin D (1 mg) 2 hours before death at the times indicated.

hours later were given a second injection of either saline, growth hormone, cycloheximide, or cycloheximide and growth hormone together. It can be seen that growth hormone caused a marked and rapid inhibition of enzyme activity and that this effect was itself inhibited by the simultaneous administration of cycloheximide, as was the fall in enzyme activity in animals given hydrocortisone alone.

Figures 4 and 5 show that essentially similar responses of hypophysectomised rats to growth hormone, cortisol and cycloheximide administration were found for TAT. None of the effects of growth hormone administration on TAT or TO activity or inducibility observed by us conflict with previously published experiments, but confirm and extend them (Kenney, 1967; Csányi and Greengard, 1968; Labrie and Korner, 1968*a*; Liberti *et al.*, 1970).

Figure 6 shows the effect of growth hormone administration to young rats on the activity of liver ornithine decarboxylase. In agreement with observations of other workers (Jänne and Raina, 1969; Russell *et al.*, 1970) there is an approximately 20-fold increase during the first 4 hours, followed by a rapid and then a slower decline in activity. However, in contrast to the findings of others, the enzyme level remains somewhat elevated above the basal value even 25 hours after hormone treatment. Figure 6 also shows that injection of cycloheximide into rats at various times after growth hormone treatment resulted in a very rapid fall in enzyme activity. It was found that injection of actinomycin D simultaneously or 2 hours after growth hormone resulted in an inhibition of the growth hormone response as measured at 4 hours. In contrast, as shown in Figure 6, injection of 1 mg actinomycin D 4 hours after growth hormone resulted in a further large increase in enzyme activity.

DISCUSSION

Interpretation of our *in vivo* effects of growth hormone on the activity and inducibility of TAT and TO rests heavily on the hypothesis developed by Tomkins and his associates (Tomkins *et al.*, 1969; Thompson *et al.*, 1970) to account for induction of TAT in rat hepatoma (HTC) cells in tissue culture. According to this theory (outlined in Fig. 7) messenger RNA for TAT is synthesised continuously, but is reversibly complexed with a repressor molecule (R) and made unavailable for protein synthesis and is degraded. Cortisol inactivates the

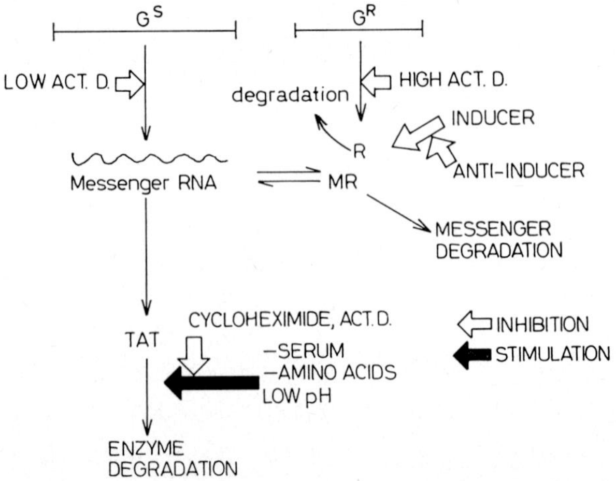

Fig. 7. Theory of tyrosine aminotransferase induction by corticosteroids in hepatoma cells (Tomkins *et al.*, 1969, and see text). The G^S refers to the structural gene for the enzyme, while G^R refers to the regulatory gene. R is the hypothetical repressor.

repressor and allows messenger RNA to accumulate and be translated, thus producing a rise in enzyme activity that plateaus when a steady state is reached between enzyme synthesis and degradation. The repressor is seen to be labile and its synthesis inhibited by high concentrations of actinomycin D, resulting in the breakdown of any inactive messenger-repressor complex and a further increase in messenger RNA available for translation, thus producing a 'superinduction' of TAT activity. Tomkins *et al.* (1969) find that in most of their experiments (see below) TAT degradation is not inhibited by actinomycin D, thus eliminating an alternative explanation of 'superinduction'.

Objections to this theory and the explanation for the 'superinduction' phenomenon upon which it rests have come from Kenney and his associates (Reel and Kenney, 1968; Lee *et al.*, 1970) working with both HTC and H35 hepatoma cells in culture. They find that, under their conditions of culture (see below), degradation of TAT *is* inhibited by actinomycin D.

Work from Tomkins' laboratory (Auricchio *et al.*, 1969; Hershko and Tomkins, 1971) has begun to reconcile these, and other, observations. They find that degradation of TAT is more rapid when cells are transferred to suboptimal conditions (*e.g.* into medium lacking serum (see Reel and Kenney, 1968) or amino acids). This 'enhanced' rate of degradation is reduced almost immediately to the control rate by cycloheximide or, less effectively, by actinomycin D. They argue that these results need not necessarily imply that a protein and its messenger RNA, both with a rapid turnover, are required for enzyme degradation. Instead, by analogy with observations made on the control of protein degradation in bacteria (Goldberg, 1971), they suggest that the level of some metabolite in the cell, *e.g.* the level of charged amino-acyl tRNA, may control the rate of TAT degradation. Inhibition of protein synthesis (either directly, with cycloheximide, or possibly indirectly by actinomycin D) would increase the level of charged tRNA and thus inhibit 'enhanced' degradation. It is not known what role, if any, the phenomenon of 'enhanced' degradation and its control has in the induction of TAT in intact rat liver. However, several workers, including ourselves (Figs. 3 and 5) have observed inhibition of TAT degradation by cycloheximide and actinomycin D *in vivo*, prompting the speculation that it may be a 'normal' component of the metabolism of rat liver. Levitan and Webb (1970) have measured immunologically the rates of synthesis and degradation of TAT following hydrocortisone administration *in vivo* and find during the first 4 hours that there is both an increase in enzyme synthesis and virtual inhibition of degradation. At later times synthesis declines to the basal rate, while degradation is resumed (and can be inhibited by cycloheximide).

Given this background of complexity, not to mention controversy, in the control of TAT (and, by analogy, TO) activity by steroid hormones, it is difficult to interpret the effects of growth hormone on these enzymes in intact animals. The early, transient rise in TAT and TO after growth hormone administration is reminiscent of the early increase in TAT activity obtained in intact (Holten and Kenny, 1967) and perfused (Hagen and Kenny, 1968) rat liver with insulin. Growth hormone failed to stimulate enzyme activity in perfused livers under the conditions used by these workers, leading them to suggest that early growth hormone stimulation *in vivo* was a secondary consequence of insulin secretion (Kenney, 1967). However, Ottolenghi and Cavagna (1968) have reported an increase in TAT in livers perfused with growth hormone, and, unlike growth hormone, insulin does not produce an early increase in TO activity in whole animals (Labrie and Korner, 1969). Experiments in our laboratory using HTC cells have failed to show any effect of bovine growth hormone (50 μg/ml) on basal or dexamethasone induced TAT levels, under conditions (essentially those described by Gelehrter and Tomkins, 1970) in which 5 μg/ml of bovine insulin gave a 2-fold increase in enzyme activity. Insulin stimulation of TAT synthesis is additive to steroid induction and has been attributed to increased utilization of pre-existing messenger RNA (Gelehrter and Tomkins, 1970; Lee *et al.*, 1970) but of course it is not known if such a mechanism applies to the stimulatory action of growth hormone.

Several explanations are possible for the later growth hormone inhibition of cortisol induction of TAT and TO. Among these are an increase in the metabolism of hydrocortisone by the liver, or an increase in the synthesis (or decrease in degradation) of the hypothetical repressor (R in Fig. 7), both resulting (according to the hypothesis outlined in Fig. 7) in an earlier shut-off of enzyme synthesis for a given dose of steroid. Alternatively, growth hormone may in some way promote 'enhanced' degradation of TAT. For example, hydrocortisone administration to hypophysectomised (or even normal) rats may result in a flooding of the liver with amino acids from extrahepatic tissues, where protein degradation is stimulated by cortisol. Growth hormone could presumably prevent or reduce this effect by increasing extrahepatic utilisation of amino acids. In support of this argument it has been observed (Labrie and Korner, 1968*a*) that in force-feeding rats amino acids inhibit the growth hormone repression of cortisol induction of TAT and TO. However, it is obvious that much more work needs to be done before a full explanation for growth hormone control of TAT and TO levels is reached.

The response of ornithine decarboxylase activity to growth hormone administration that we have observed confirms the reports of others (Jänne and Raina, 1969; Russell *et al.*, 1970). The rapid fall in activity following cycloheximide treatment suggests (*a*) that the rise in ornithine decarboxylase is a result of increased net synthesis of the enzyme and not activation of a precursor and (*b*) that the enzyme has a very rapid turnover (Russell *et al.*, 1970). A similar rapid fall ($t_{1/2}$ = approximately 10 min) in OD activity after cycloheximide addition is seen in HTC cells in tissue culture (Hogan, unpublished observations). The apparent 'superinduction' of OD by a high dose of actinomycin D given at 4 hours has not been reported previously.

Jänne *et al.* (Jänne *et al.*, 1968; Jänne and Raina, 1969) have shown an apparent correlation between the increase in OD activity, RNA polymerase activity in isolated nuclei (measured at low ionic strength when most activity is from ribosomal RNA polymerase I), putrescine levels, and spermidine synthesis in liver, following growth hormone administration to young rats.

We have found that the rate of increase in both ribosomal RNA synthesis *in vivo* and RNA polymerase activity measured at low ionic strength in isolated nuclei is greatest during the first 4 to 6 hours after growth hormone administration to hypophysectomised rats (Oravec and Korner, 1971; Pegg and Korner, unpublished observations). However, between 18 and 36 hours after growth hormone there appears to be a preferential increase in the synthesis of DNA-like RNA which does not correlate with an increase in ornithine decarboxylase activity (Oravec and Korner, 1971).

ACKNOWLEDGEMENTS

We are grateful to the Medical Research Council for support in the form of a group, and to Mrs Lauraine Cornwall and Mrs. Susan Murden for excellent technical assistance.

REFERENCES

AURICCHIO, F., MARTIN, D. and TOMKINS, G. (1969): Control of degradation and synthesis of induced tyrosine aminotransferase studied in hepatoma cells in culture. *Nature (Lond.)*, *224*, 806.

CSÁNYI, V. and GREENGARD, O. (1968): Effect of hypophysectomy and growth hormone on the inductions of rat tyrosine aminotransferase and tryptophan oxygenase by hydrocortisone. *Arch. Biochem.*, *125*, 824.

GELEHRTER, T. B. and TOMKINS, G. M. (1970): Post-transcriptional control of tyrosine aminotransferase by insulin. *Proc. nat. Acad. Sci. (Wash.)*, *66*, 390.

GOLDBERG, A. L. (1971): A role of aminoacyl-tRNA in the regulation of protein breakdown in *E. coli*. *Proc. Nat. Acad. Sci. (Wash.)*, *68*, 362.

HAGEN, C. B. and KENNEY, F. T. (1968): Regulation of tyrosine-α-ketoglutarate transaminase in rat liver. VII. Hormonal effects on synthesis in the isolated, perfused liver. *J. biol. Chem.*, *243*, 3296.

HARDING, H. R. and ROSEN, F. (1963): Effects of hypophysectomy and growth hormone on the endogenous levels and induction of two adaptive enzymes. *Fed. Proc.*, *22*, 409.

HERSKO, A. and TOMKINS, G. M. (1971): Studies on the degradation of tyrosine aminotransferase in hepatoma cells in culture. *J. biol. Chem.*, *246*, 710.

HOLTEN, D. and KENNEY, F. T. (1967): Regulation of tyrosine-α-ketoglutarate transaminase in rat liver. VI. Induction by pancreatic hormones. *J. biol. Chem.*, *242*, 4372.

JÄNNE, J. and RAINA, A. (1969): On the stimulation of ornithine decarboxylase and RNA polymerase activity in rat liver after treatment with growth hormone. *Biochim. biophys. Acta (Amst.)*, *174*, 769.

JÄNNE, J., RAINA, A. and SIIMES, M. (1968): Mechanism of stimulation of polyamine synthesis by growth hormene in rat liver. *Biochim. biophys. Acta (Amst.)*, *166*, 419.

JEFFERSON, L. S. and KORNER, A. (1967): A direct effect of growth hormone on the incorporation of precursors into protein and nucleic acids of perfused rat liver. *Biochem. J.*, *104*, 826.

KENNEY, F. T. (1967): Regulation of tyrosine-α-ketoglutarate transaminase in rat liver. V. Repression in growth hormone-treated rats. *J. biol. Chem.*, *242*, 4367.

KORNER, A. (1970): Insulin and growth hormone control of protein biosynthesis. In: *Ciba Foundation Symposium, Control Processes in Multicellular Organisms*, p. 86. Editors: G. E. W. Wolstenholme and J. Knight. J. and A. Churchill, London.

LABRIE, F. and KORNER, A. (1968*a*): Growth hormone inhibition of hydrocortisone induction of tyrosine transaminase and tryptophan pyrrolase and its reversal by amino acids. *J. biol. Chem.*, *243*, 1120.

LABRIE, F. and KORNER, A. (1968*b*): Actinomycin-sensitive induction of tyrosine aminotransferase and tryptophan pyrrolase by amino acids and tryptophan. *J. biol. Chem.*, *243*, 1116.

LABRIE, F. and KORNER, A. (1969): Effect of glucagon, insulin and thyroxine on tyrosine transaminase and tryptophan pyrrolase of rat liver. *Arch. Biochem.*, *129*, 75.

LIBERTI, J. P., LONGMAN, E. S. and NAVON, R. S. (1970): Effects of hydrocortisone and growth hormone on tyrosine aminotransferase and tryptophan oxygenase in hypophysectomised rats. *Endocrinology*, *86*, 1448.

LEE, K.-L., REEL, J. R. and KENNEY, F. T. (1970): Regulation of tyrosine α-ketoglutarate transaminase in rat liver. IX. Studies of the mechanisms of hormonal inductions in cultured hepatoma cells. *J. biol. Chem.*, *245*, 5806.

LEVITAN, I. B. and WEBB, T. E. (1970): Hydrocortisone-mediated changes in the concentration of tyrosine aminotransferase in rat liver: an immunochemical study. *J. molec. Biol.*, *48*, 339.

ORAVEC, M. and KORNER, A.: *J. molec. Biol.*, *58*, 489.

OTTOLENGHI, C. and CAVAGNA, R. (1968): Effect of hypophyseal growth hormone on the activity of tyrosine amino transferase in the isolated perfused rat liver. *Endocrinology*, *83*, 924.

PEGG, A. E. and WILLIAMS-ASHMAN, H. G. (1968): Biosynthesis of putrescine in the prostate gland of the rat. *Biochem. J.*, *108*, 533.

REEL, J. R. and KENNEY, F. T. (1968): "Superinduction" of tyrosine transaminase in hepatoma cell cultures: differential inhibition of synthesis and turnover by actinomycin D. *Proc. nat. Acad. Sci. (Wash.)*, *61*, 200.

RUSSELL, D. H., SNYDER, S. H. and MEDINA, V. J. (1970): Growth hormone induction of ornithine decarboxylase in rat liver. *Endocrinology*, *86*, 1414.

THOMPSON, E. B., GRANNER, D. K. and TOMKINS, G. M. (1970): Superinduction of tyrosine aminotransferase by actinomycin D in rat hepatoma (HTC) cells. *J. molec. Biol.*, *54*, 159.

TOMKINS, G. M., GELEHRTER, T. D., GRANNER, D. K., MARTIN, D., SAMUELS, H. H. and THOMPSON, E.B. (1969): Control of specific gene expression in higher organisms. *Science*, *166*, 1474.

EFFECTS OF HYPOPHYSECTOMY ON PROTEIN AND CARBOHYDRATE METABOLISM IN THE PERFUSED RAT LIVER*

L. S. JEFFERSON, J. W. ROBERTSON and E. L. TOLMAN**

Department of Physiology, College of Medicine, The Milton S. Hershey Medical Center, The Pennsylvania State University, Hershey, Pa., U.S.A.

Hypophysectomy of the rat is known to cause marked alterations in the delicate endocrine balance of this animal that result in many primary and secondary hormonal effects on protein and carbohydrate metabolism in the liver. Attempts to understand the biochemical alterations in the liver cell following hypophysectomy mainly have involved studies on the whole animal or on disrupted and fractionated liver tissue removed from animals following various hormonal treatments. On the basis of these studies, it is well established that hypophysectomy reduces liver growth and the synthesis of hepatic protein and RNA and that these defects are corrected by treatment of animals with growth hormone (Simpson *et al.*, 1949; Talwar *et al.*, 1962; Korner, 1960). These effects are associated with corresponding changes in the activities of liver ribosomes and nuclear RNA-polymerase when these are assayed in cell-free systems (Korner, 1961; Pegg and Korner, 1965; Widnell and Tata, 1966). The protein anabolic effect of growth hormone may be related to the stimulated transport of amino acids into liver cells (Noall *et al.*, 1957), although transport is also increased following hypophysectomy (Riggs and Walker, 1960). Since growth hormone recently has been demonstrated to stimulate directly protein and RNA synthesis in the perfused liver (Jefferson and Korner, 1967) and in liver slices (Clemens and Korner, 1970), it now is possible to investigate *in vitro* the effects of hypophysectomy and growth hormone on hepatic protein metabolism.

Following hypophysectomy, profound changes in carbohydrate metabolism occur, particularly during fasting. Since the liver plays an essential role in the maintenance of blood sugar concentrations due to its ability to produce glucose from glycogen and its capacity to carry out gluconeogenesis, alterations in hepatic carbohydrate metabolism would be expected. It is well known that fasted hypophysectomized (hypox) animals have low levels of blood glucose and liver glycogen, and evidence suggesting that these changes are related to a diminished hepatic gluconeogenesis has been presented (Soskin *et al.*, 1935; DeBodo and Sinkoff, 1953; Weber and Cantero, 1959; Freedland *et al.*, 1968). On the other hand, fed hypox animals maintain normal levels of blood glucose in the face of above normal rates of carbohydrate utilization (Russell and Bennett, 1937) and preliminary studies have indicated that hypophysectomy causes an increased synthesis of glucose from amino acids and lactate in the perfused rat liver (Tolman *et al.*, 1970).

In the studies presented here, the isolated perfused rat liver preparation was used to investigate the effects of hypophysectomy on hepatic protein and carbohydrate metabolism.

---

* This work was supported by Grant No. AM 13499 from the National Institutes of Health, U.S. Public Health Service.

** Postdoctoral fellow of the National Institutes of Health, U.S. Public Health Service (Grant No. HE 43762).

EXPERIMENTAL PROCEDURES

Normal and hypox male Wistar rats maintained on regular laboratory chow and water up to time of sacrifice were used in these studies. Hypox rats were 14–16 days postoperative and all rats weighed 95–115 g at time of use.

Livers were perfused *in situ* by the technique of Mortimore (1961) as described by Jefferson *et al.* (1972). The basic perfusion medium consisted of Krebs-Henseleit bicarbonate buffer containing 3% bovine serum albumin, 8 mM glucose, sufficient washed ovine or bovine erythrocytes to give a hematocrit of 20–22%, and other substrates at varying concentrations. The perfusate was equilibrated with humidified O_2-CO_2 (95% : 5%) at 37° C in rotating reservoirs and pumped through the liver at a constant rate of 7.0 ml/min. Depending on the experiment, the perfusate was either recirculated or passed through the liver a single time and the effluent collected as serial samples. Other details of the experiments as well as methods for analysis of the perfusate and liver samples are given in the legends to the Tables and Figures.

RESULTS AND DISCUSSION

In vivo observations

Some of the effects of hypophysectomy related to protein and carbohydrate metabolism of the liver that were observed *in vivo* are shown in Table I. It can be seen that hypophysectomy impaired liver growth relative to that of the whole body, as evidenced by the diminished liver to body weight ratio. Liver glycogen and RNA contents were reduced by 49% and 22%, respectively, but liver protein concentration, when expressed on the basis of liver weight, appeared to be increased slightly. Plasma urea levels were increased by 100% in the hypox

TABLE I

Effects of hypophysectomy observed in vivo

	Condition of rats	
	Normal	Hypox
Liver weight (g · 100 g body wt^{-1})	5.17 ± .12	4.34 ± .18[1]
Liver glycogen content (μmoles glucose · g^{-1})	312 ± 13	158 ± 30[2]
Liver RNA content (mg · g^{-1})	15.12 ± .47	11.76 ± .44[2]
Liver protein content (mg · g^{-1})	178 ± 4	197 ± 3[1]
Plasma glucose concentration (mM)	8.56 ± .28	8.48 ± .42
Plasma urea concentration (mM)	5.42 ± .33	11.02 ± .65[2]

Rats were anesthetized with an intraperitoneal injection of sodium pentobarbital, their abdominal cavities exposed and aortic blood drawn into chilled heparinized tubes. Immediately following collection of the blood sample, livers were quickly excised and frozen in Wollenberger clamps at the temperature of liquid nitrogen. Blood plasma samples were analyzed for glucose by the method of Huggett and Nixon (1957) and for urea by the method of Searcy *et al.* (1961). Frozen liver samples were analyzed for glycogen by the method of Walaas and Walaas (1950) and for protein by the method of Lowry *et al.* (1951). Glycogen levels are reported as glucose equivalents per g wet liver weight. Samples of liver homogenates that were prepared, as described in the legend to Figure 3, were analyzed for RNA by the alkaline hydrolysis method of Fleck and Munro (1962). Mean values are presented followed by the standard error of the mean.

[1] differs from normal by $P < .005$

[2] differs from normal by $P < .001$

rats, while the plasma levels of glucose were normal. These changes are similar to those reported by other workers (Phillips and Robb, 1939) and suggest the presence of a number of metabolic alterations in the livers of hypox rats.

Protein synthesis in the perfused liver – the ribosome cycle

Before detailing the effects of hypophysectomy and growth hormone, the optimal conditions for studying protein synthesis in the perfused liver will be described. Earlier studies had demonstrated that the incorporation of amino acids into liver protein, the aggregation of ribosomes and the activity of isolated ribosomes in a cell-free system all depended on the total amount and the composition of the amino acid supply (Jefferson and Korner, 1969). In these

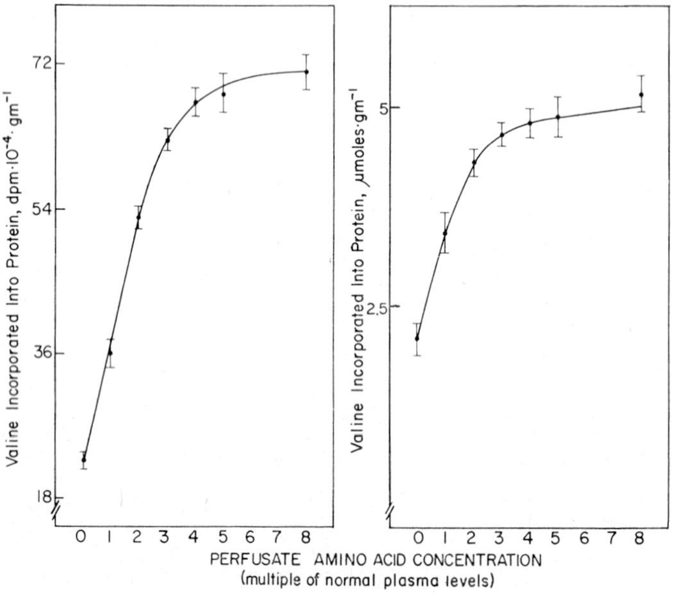

Fig. 1. Effect of perfusate amino acid concentration on incorporation of valine into liver protein. Livers were perfused for 30 min with non-recirculating medium containing mixtures of amino acids in multiples of their concentrations in normal rat plasma. In the experiments denoted zero perfusate amino acid concentration, the medium contained only valine at normal plasma levels. In each case ^{14}C-valine was present at a constant specific activity. At the end of perfusion, livers were quickly removed and frozen in a Wollenberger clamp at the temperature of liquid nitrogen. The frozen liver samples were powdered in a percussion mortar and a portion of this powder used to quantitate the incorporation of radioactive valine into protein after the sample was freed of amino acids, RNA and lipid (Munro *et al.*, 1964). Powdered tissue samples were prepared for amino acid analysis by treatment with 1 % picric acid to extract the free amino acids as described by Tallan *et al.* (1954). Amino acids were quantitated using a Beckman amino acid analyzer. A portion of the effluent from the column of the analyzer was collected and used for the determination of the specific activity of the radioactive valine. The number of μmoles of valine incorporated into liver protein was calculated by the following formula:

$$\text{Incorporation, } \mu\text{moles/g protein} = \frac{\text{Incorporation, d.p.m./g}}{\text{Average specific activity of liver intracellular valine, d.p.m./}\mu\text{mole}}$$

Mean values ± the standard error of the mean are presented. At least three livers were perfused under each condition.

studies in which the perfusate was recirculated, addition of a mixture of all amino acids at initial concentrations ten times their normal plasma levels seemed optimal during a one hour period. However, it is quite difficult to define accurately amino acid requirements under conditions where the perfusate is recirculated since its composition is constantly changing. In order to maintain perfusate amino acid concentrations at constant levels, livers were perfused for 30 min with a non-recirculating medium containing mixtures of amino acids in multiples of their concentrations in normal rat plasma. The results of these experiments are depicted in Figure 1. When expressed as a function of the perfusate amino acid concentration, incorporation of ^{14}C-valine into liver protein increased linearly reaching a maximal rate at 3 to 5 times the normal plasma amino acid levels (left panel, Figure 1). In other experiments, it had been determined that ^{14}C-valine is a suitable amino acid for a study of the rates of protein synthesis in liver since it (1) is not oxidized to an appreciable extent, (2) is not converted to other amino acids, and (3) has a measurable rate of incorporation into liver protein. For more reliable estimates of the rates of protein synthesis it is also necessary to determine the specific activity of the labeled precursor, since degradation of liver proteins occurs during perfusion with consequent dilution of the specific activity of the radioactive amino acid. As shown in a subsequent section, addition of amino acids to the perfusate diminished the extent of protein

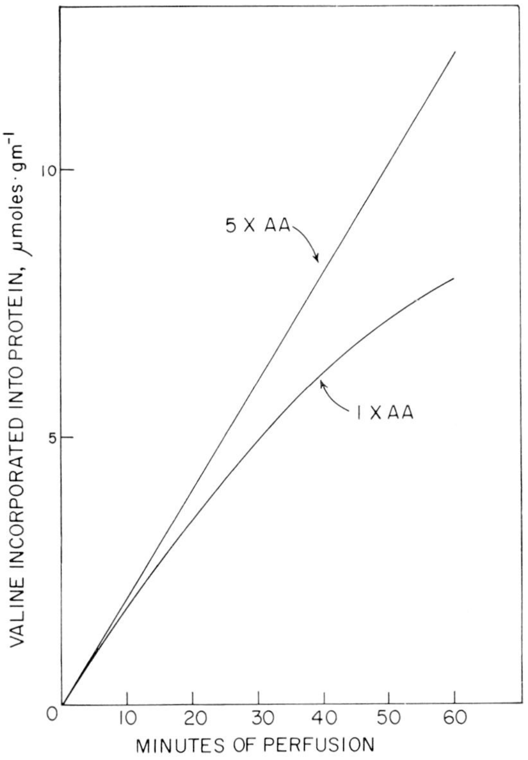

Fig. 2. Time course of protein synthesis in the perfused liver. Livers were perfused for 5, 15, 30 and 60 min with non-recirculating medium containing either all amino acids at concentrations near those found in normal rat plasma (labeled 1 × AA) or all amino acids at concentrations 5 times those found in plasma (labeled 5 × AA). In each case ^{14}C-valine was present at a constant specific activity. The rate of valine incorporation into protein was based on the average specific activity of liver intracellular valine for each time point. At least three livers were perfused under each condition.

degradation which would result in less dilution of the labeled amino acid and lead to the incorporation of more radioactive valine but not necessarily more total valine into liver protein. In the right panel of Figure 1, the rates of protein synthesis are shown based on the intracellular specific activity of valine. The results are similar to those shown in the left panel of the figure, but dilution of the radioactive valine influenced quantitatively the rate of protein synthesis at the lower amino acid levels.

The time course of protein synthesis in the perfused system was determined in livers that were perfused for 5, 15, 30 and 60 min with non-recirculating medium containing either all amino acids at normal plasma concentrations or all amino acids at concentrations 5 times

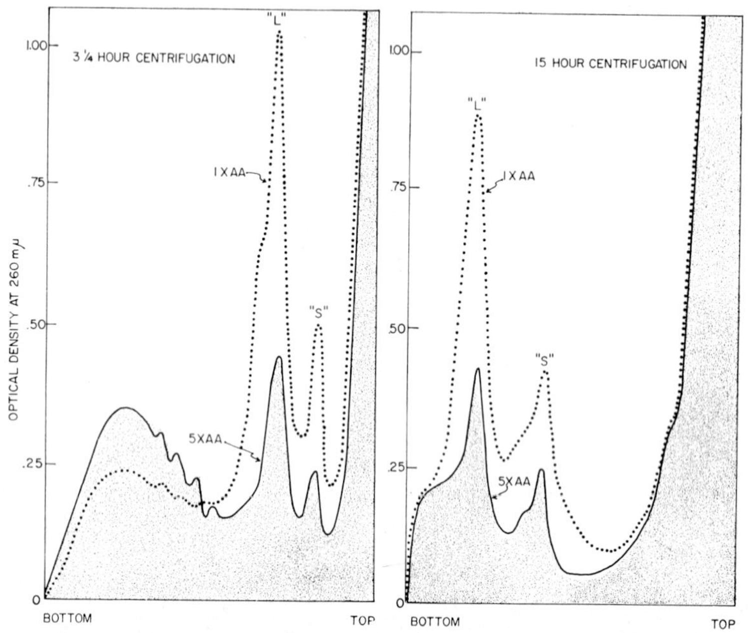

Fig. 3. Effects of perfusion and amino acid concentrations on ribosomal aggregation. Livers that were either unperfused or perfused for 30 minutes with non-recirculating medium containing the amino acid concentrations indicated were rapidly dropped into ice-cold 0.25 M sucrose medium containing 100 mM KCl, 40 mM NaCl, 1 mM $MgCl_2$ and 20 mM tris buffer, pH 7.6. The livers were homogenized in 8 volumes of the same medium and the homogenate centrifuged at 10,000 $\times$ *g* for 10 minutes to remove mitochondria, nuclei and debris. The post-mitochondrial supernatant was aspirated and sufficient 10% Triton X-100 was added to give a final concentration of 1%. Supernatants (0.75 ml) were layered onto 12.5 ml exponential sucrose gradients (15% to 2 M) which were formed in the apparatus of Noll (1967) as described by Morgan *et al.* (1971). The gradients were centrifuged for either 3¼ or 15 hours at 40,000 r.p.m. in an SW-40 rotor of a Spinco ultracentrifuge. Following centrifugation, the gradients were analyzed by pumping the contents of the tube through a flow cell of a Gilford recording spectrophotometer. Optical density at 260 mμ was recorded and in some experiments effluent from the flow cell was collected for RNA analysis. In the perfusion studies, the medium contained either all amino acids at concentrations near those found in normal rat plasma (dotted line, labeled 1 $\times$ AA) or all amino acids at concentrations 5 times those found in plasma (solid line, labeled 5 $\times$ AA). Large (60S) and small (40S) subunit peaks are labeled as 'L' and 'S', respectively. The sedimentation patterns of polysomes and subunits from livers perfused with buffer containing 5 $\times$ AA were identical to those of unperfused livers represented by the shaded area. At least five experiments were carried out for each condition and the distribution of ribosomes and subunits on the gradients in all experiments was similar to that shown above.

those found in plasma (Figure 2). The specific activity of intracellular ^{14}C-valine was determined for each time period and rates of protein synthesis calculated as described in the legend of Figure 1. Protein synthesis was maintained at a linear rate for the entire perfusion period when the medium contained amino acid concentrations at 5 times their normal plasma levels, but the rate of synthesis declined when the medium contained normal plasma levels of amino acids.

Earlier studies had indicated that these effects of amino acids on protein synthesis in the liver were related to changes in the aggregation and activity of ribosomes. Figure 3 illustrates the effects of perfusion and amino acid concentrations on ribosomal aggregation in the presence of 1 mM Mg^{++}. This was found to be the optimal Mg^{++} concentration for studying ribosomal aggregation in liver under these conditions (unpublished observations). Ribosomal subunits could be separated more effectively from heavier aggregates and from each other by centrifugation for 15 hours (Fig. 3, right panel). It can be seen that during perfusion with medium containing amino acids at their normal plasma concentrations, levels of polysomes decreased and ribosomal subunits increased compared to those of unperfused livers which are indicated by the shaded area. Addition of amino acids at concentrations 5 times those found in plasma increased polysomes and decreased subunits to levels identical with those of unperfused livers.

These effects of amino acids on protein synthesis and levels of polysomes and subunits in perfused liver can be interpreted on the basis of a ribosome cycle as outlined in Figure 4.

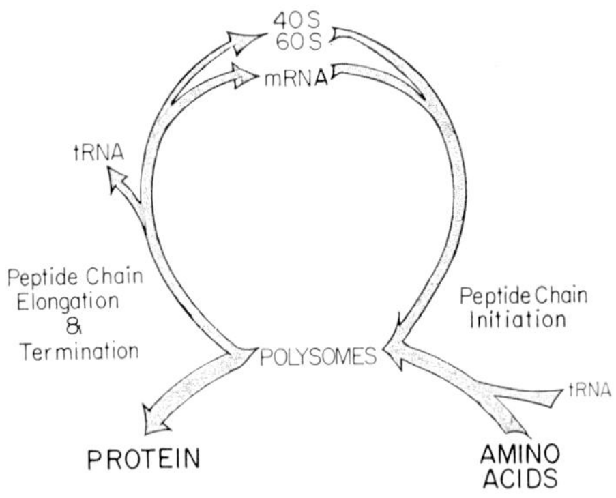

Fig. 4. Model of the ribosome cycle in protein synthesis. Details are discussed in the text.

Recent evidence (Hogan and Korner, 1968; Kabat and Rich, 1969; Hoerz and McCarty, 1969) suggests that protein synthesis in mammalian cells involves a ribosome cycle similar to that described in bacteria (Algranati *et al.*, 1969; Schlessinger, 1969). According to this proposal, ribosomal subunits associate from a free subunit pool to initiate peptide synthesis on messenger RNA. The ribosome units move down the message until, upon peptide chain termination, the subunits dissociate and rejoin the subunit pool. Reactions within the cycle can be divided into two groups, those involved in (1) initiation and (2) elongation of peptide chains. If the rates of protein synthesis are known, levels of subunits and polysomes can be interpreted in terms of the relative rates of these groups of reactions. In the steady state, cycle activity must equal the overall rate of protein synthesis. If the rate of protein synthesis falls in association with an increase in subunits, a restraint on initiation of peptide chains greater than the restraint on elongation is indicated. On the other hand, a fall in the rate of synthesis in conjunction with low levels of subunits indicates a greater inhibition of chain elongation than initiation. Since polysomes decreased and ribosomal subunits increased in association with a decline in the rate of protein synthesis, a block in the initiation of peptide chains appeared to develop during perfusion of livers with buffer containing normal plasma levels of amino acids. Addition of amino acids at concentrations 5 times those found in plasma produced a maximal rate of protein synthesis and restored polysomes and subunits to *in vivo* levels, indicating that peptide initiation had been facilitated. Similar effects of perfusion and amino acid levels on peptide chain initiation were observed in heart (Morgan *et al.*, 1971).

Effects of hypophysectomy and growth hormone on protein synthesis

In parallel studies to those discussed above, effects of hypophysectomy and growth hormone on protein synthesis in the perfused liver were evaluated by investigating effects on ribosome cycle activity. As can be seen in Table II, the rate of protein synthesis in livers of hypox

TABLE II

Effects of hypophysectomy and growth hormone on the incorporation of ^{14}C-valine into liver protein

Perfusate amino acid concentration	Valine incorporated into protein, dpm · mg^{-1}		
	Normal	Hypox	Hypox + GH
0 × AA	226.4 ± 11.1	145.0 ± 10.4	243.6 ± 7.9
1 × AA	358.6 ± 18.2	293.6 ± 13.6	433.9 ± 13.6
3 × AA	581.1 ± 22.1	410.0 ± 20.7	597.1 ± 10.0
5 × AA	679.3 ± 26.8	392.1 ± 31.4	674.3 ± 29.3

Livers from either normal, hypox or growth hormone-treated hypox rats were perfused for 30 min with non-recirculating medium containing either valine only at normal plasma levels (0 × AA), all amino acids at concentrations near those found in normal rat plasma (1 × AA), or all amino acids at concentrations 3 and 5 times those found in plasma (3 × AA and 5 × AA, respectively). At the end of perfusion, the livers were frozen and valine incorporation into liver protein determined as described in the legend to Figure 1. Treated animals were injected intraperitoneally with 100 μg of bovine growth hormone (NIH-GH-B14) 48 and 24 hours prior to the start of perfusion. Mean values are presented followed by the standard error of the mean. From 3 to 6 livers were perfused under each condition.

rats was decreased at every perfusate amino acid concentration studied and the maximal rate was about 50% of that seen in control livers. Treatment of hypox rats with growth hormone (100 μg/100 g) 48 and 24 hours prior to perfusion restored the rate of protein synthesis to normal (Table II). In these studies, the effects of hypophysectomy and growth hormone on ribosomal aggregation were evaluated by collecting the fractions containing the large and small subunit peaks from 15 hour gradients and the levels of subunits quantitated by RNA analysis (Table III). First it should be noted that the levels of ribosomal subunits in unperfused livers of hypox rats and hypox rats treated with growth hormone were unchanged from the levels found in unperfused normal livers (bottom line, Table III). During perfusion with medium containing the lower concentrations of amino acids (0 and 1 × AA), levels of ribosomal subunits increased substantially in the livers of both hypox and hormone treated rats. In accord with the data presented in Figure 3, levels of ribosomal subunits in livers of normal rats showed a substantial increase during perfusion with the lower amino acid concentrations. In all three groups, addition of amino acids at concentrations 5 times those found in plasma restored subunits to the levels found in unperfused livers. Taken together, these data suggest that the initiation process was unimpaired in livers of hypox rats, but that a defect existed in

TABLE III

Effects of hypophysectomy and growth hormone on levels of ribosomal subunits in livers perfused with varying amino acid concentrations

Perfusate amino acid concentration	Subunit RNA, mg RNA · mg homogenate RNA^{-1}					
	Normal		Hypox		Hypox + GH	
	Large subunit	Small subunit	Large subunit	Small subunit	Large subunit	Small subunit
0 × AA	.140 ± .003	.070 ± .003	.103 ± .005	.058 ± .008	.107 ± .009	.056 ± .005
1 × AA	.099 ± .004	.063 ± .003	.059 ± .001	.030 ± .002	.069 ± .001	.037 ± .001
2 × AA	.081 ± .001	.040 ± .002	—	—	—	—
3 × AA	.064 ± .004	.034 ± .004	.044 ± .001	.023 ± .001	.040 ± .004	.023 ± .002
5 × AA	.048 ± .003	.020 ± .004	.045 ± .005	.024 ± .002	.042 ± .004	.021 ± .002
Unperfused	.046 ± .003	.024 ± .002	.048 ± .005	.019 ± .001	.038 ± .004	.022 ± .001

Livers that were either unperfused or perfused for 30 minutes with non-recirculating medium containing the amino acid concentrations indicated were homogenized and samples of the detergent treated post-mitochondrial supernatants layered onto gradients as described in the legend to Figure 3. The gradients were centrifuged for 15 hours and fractions containing the large and small subunit peaks were collected and analyzed for RNA by the alkaline hydrolysis method of Fleck and Munro (1962). Results are expressed per mg of RNA in the whole homogenate. In the perfusion studies, the medium contained either valine only at normal plasma levels (0 × AA), all amino acids at concentrations near those found in normal rat plasma (1 × AA), or all amino acids at concentrations 2, 3 and 5 times those found in plasma (2 × AA, 3 × AA, and 5 × AA, respectively). Treated animals were injected intraperitoneally with 100 μg of bovine growth hormone (NIH-GH-B14) 48 and 24 hours prior to the start of perfusion. Mean values are presented followed by the standard error of the mean.

elongation of peptide chains. It appeared that this alteration was the result of a single hormone deficiency, since growth hormone treatment was effective in correcting the defect. These findings are in agreement with the studies of Korner (1961) which showed that hypophysectomy of the rat decreased, and treatment with growth hormone increased, the ability of liver ribosomes to synthesize protein in a cell-free system.

Effects of hypophysectomy and growth hormone on protein degradation

The decrease in liver growth in hypox rats (Table I) could result from the defect in protein synthesis discussed above, but might also involve increased proteolysis. To test this possibility, the liver protein of normal and hypox rats was prelabeled with ^{14}C-valine and the rate of proteolysis estimated during a subsequent perfusion. As can be seen in the left panel of Figure 5, protein degradation occurred at the same rate in livers of both normal and hypox rats and could be inhibited by 50% in both tissues by addition of 10 times the normal plasma levels of amino acids to the perfusate. The addition of growth hormone (10 μg/ml) to the perfusion medium had no effect on the rate of protein degradation in livers of normal rats (right panel, Fig. 5). The observation that amino acids can prevent degradation of the protein of perfused rat liver is in agreement with the findings of Woodside and Mortimer (1970).

Effects of hypophysectomy on carbohydrate metabolism

As noted earlier, hypox rats maintain a normal level of blood glucose (Table I) while carbohydrate utilization relative to other nutrients is increased (Russell and Bennett, 1937).

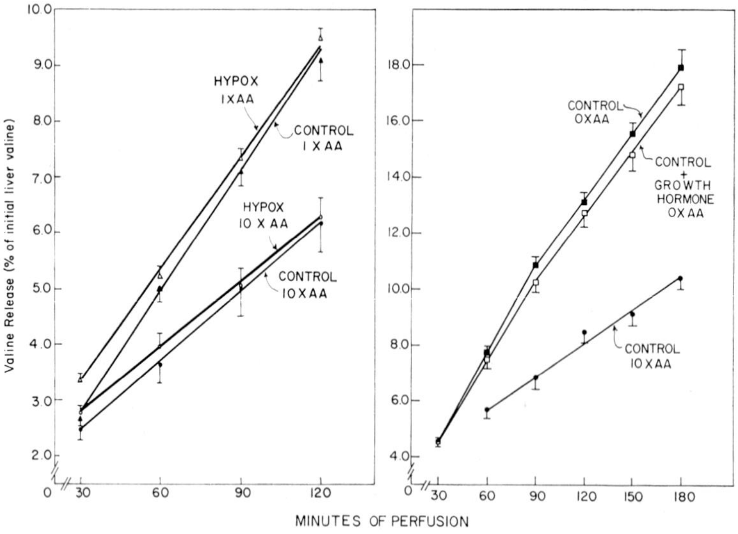

Fig. 5. Effects of hypophysectomy and growth hormone on proteolysis in livers perfused with varying amino acid concentrations. The liver protein of normal and hypox rats was prelabeled by administering intraperitoneal injections of valine-1-^{14}C (10 μCi per dose) at 15 and 3 hours prior to the start of perfusion. Livers were perfused for 2 or 3 hours with recirculating medium containing 15 mM valine to minimize the reincorporation of ^{14}C-valine into protein. The release of ^{14}C-valine was determined on samples of perfusate withdrawn at the indicated times. Results are expressed as the percentage of the initial ^{14}C-valine content of the liver. As indicated, the medium also contained initially either no amino acids except valine (0 × AA), all other amino acids at concentrations near those found in normal rat plasma (1 × AA), or all other amino acids at concentrations 10 times those found in plasma (10 × AA). Where indicated, bovine growth hormone (NIH-GH-B14) was added to the perfusate at an initial concentration of 10 μg per ml. Mean values are presented and the vertical bars represent the standard errors of the mean. At least 6 livers were perfused under each condition.

In addition, the reduction in liver glycogen content by one-half (Table I) and the impaired ability to mobilize free fatty acids (Goodman, 1970) emphasizes the need of the hypox rat for additional glucose to maintain energy metabolism. As also noted in Table I, plasma urea levels were more than doubled following hypophysectomy, suggesting that amino acid catabolism was increased in these animals. Since gluconeogenesis is an important catabolic pathway for many amino acids, it is possible that increased conversion of amino acids to glucose enables the hypox animal to maintain a normal blood sugar level. The data presented in Table IV show the effect of hypophysectomy on plasma amino acid concentrations and support the suggestion that amino acids are an important source for increased hepatic glucose synthesis after removal of the pituitary. It can be seen that the plasma concentrations of eight amino acids, including aspartate, threonine, serine, asparagine, glutamine, glutamate, alanine and lysine, were significantly reduced in the hypox rat. Perfusion studies in other laboratories (Exton and Park, 1967; Ross *et al.*, 1967) have shown that most of these amino acids, and particularly alanine, are readily converted to glucose in the liver. Note that the plasma level of alanine was reduced by 44% by hypophysectomy. The plasma levels of the other amino acids were not changed significantly, but the total plasma amino acid concentration was decreased significantly from 3.6 to 2.7 mM following hypophysectomy.

The reduction in plasma levels of amino acids following hypophysectomy could have resulted from an increase in their transport into liver cells or from an increase in their hepatic

TABLE IV

Effect of hypophysectomy on plasma amino acid concentrations

Amino acid	Condition of rat	
	Normal	Hypox
	plasma concentration, 10^2 mM	
Aspartic acid	1.84	0.51[3]
Threonine	26.97	18.79[2]
Serine	24.34	18.65[1]
Asparagine	6.26	4.48[1]
Glutamine	66.66	44.29[2]
Glutamic acid	7.57	3.87[2]
Alanine	47.05	26.32[2]
Lysine	41.81	27.33[2]
Proline	18.58	15.44
Glycine	40.80	36.73
Valine	17.32	15.30
Cysteine	2.34	2.78
Methionine	4.65	4.64
Isoleucine	9.05	7.62
Leucine	16.07	13.11
Tyrosine	8.33	9.31
Phenylalanine	5.38	5.64
Histidine	6.33	5.01
Arginine	13.22	11.70
Tryptophan	—	—
Total	364.57	271.52[2]

Samples of aortic blood were drawn into heparinized syringes from normal and hypox rats anesthetized by intraperitoneal injection of sodium pentobarbital and the plasma stored at –70° C for subsequent analysis. Plasma amino acids were extracted with 1% picric acid as described by Tallan *et al.* (1954) and quantitated using a Beckman amino acid analyzer. Each value represents the mean of 3 determinations. For clarity of presentation, the standard errors of the means are not included, but differences between the values for the hypox and normal animals were compared by the Student's *t* test.

[1] differs from normal by $P < .05$

[2] differs from normal by $P < .01$

[3] differs from normal by $P < .005$

utilization. As seen in Table V, hypophysectomy significantly decreased intracellular concentrations of most of these amino acids, including aspartate, asparagine, glutamine, alanine, lysine, and histidine. Alanine levels showed the greatest reduction, being only 45% of normal in the livers of hypox rats. The total intracellular amino acid concentration was decreased significantly from 33.7 to 27.5 mM following hypophysectomy. These data suggested that the hepatic utilization of amino acids was increased by hypophysectomy and that this contributed to the observed decrease in plasma amino acid levels.

This possibility was tested by observing the effect of hypophysectomy on amino acid uptake by the isolated perfused rat liver. In these studies livers of both normal and hypox rats were perfused with a recirculating medium that contained initially all amino acids at 10 times the concentrations found in the plasma of normal rats. Amino acid analyses were carried out on samples of perfusate plasma prior to and after 60 min of perfusion. The differences between the initial and final concentrations represented amino acid uptake by the livers. As shown in

TABLE V

Effect of hypophysectomy on liver amino acid concentrations

Amino acid	Condition of rat	
	Normal	Hypox
	liver concentration, mM	
Aspartic acid	2.21	1.67[1]
Asparagine	.44	.27[2]
Glutamine	12.22	8.27[1]
Alanine	2.55	1.15[3]
Lysine	.82	.68[1]
Histidine	1.21	.90[3]
Threonine	.93	.85
Serine	1.77	1.59
Proline	.25	.41
Glutamic acid	6.75	6.45
Glycine	3.29	3.92
Valine	.34	.35
Methionine	.14	.13
Isoleucine	.22	.22
Leucine	.37	.36
Tyrosine	.14	.16
Phenylalanine	.11	.10
Cysteine	—	—
Arginine	—	—
Tryptophan	—	—
Total	33.73	27.48[2]

Normal and hypox rats were anesthetized with an intraperitoneal injection of sodium pentobarbital, their abdominal cavities exposed by a midline incision and the liver excised and rapidly frozen in Wollenberger clamps at the temperature of liquid nitrogen. The frozen livers were pulverized and weighed samples of the powder were homogenized in cold 1% picric acid (Tallan *et al.*, 1954). The extracted amino acids were quantitated on a Beckman amino acid analyzer. Each value represents the mean of 3 determinations and the standard errors of the mean are not shown for the sake of clarity. Significance of differences between the normal and hypox values were tested by the Student's *t* test.

[1] differs from normal by $P < .05$

[2] differs from normal by $P < .01$

[3] differs from normal by $P < .001$

Table VI, the total amino acid uptake by perfused livers of hypox rats was 50% greater than that by control livers. Of all the amino acids, alanine was taken up to the greatest extent in control livers. Uptakes of alanine, glutamine, glycine and lysine were all above 10 μmoles/g/hr. In livers of hypox rats, the uptake of these four amino acids was approximately doubled and marked increases in the uptake of other amino acids, including aspartate, threonine, serine, asparagine, glutamate and arginine, also were observed. At the end of perfusion, intracellular amino acid concentrations in the livers of hypox rats were lower than those in control livers (data not shown), suggesting that the increased uptake resulted, at least in part, from an enhanced amino acid utilization. In these experiments, the production of glucose and urea by the livers of hypox rats was increased above that in normal livers (from 13.76 to 35.11 μmoles/g/hr and from 35.82 to 69.61 μmoles/g/hr, respectively), indicating that gluconeogenesis and ureogenesis were two pathways of amino acid utilization which were stimulated.

TABLE VI

Effect of hypophysectomy on amino acid uptake by the isolated perfused rat liver

Amino acid	Liver donor	
	Normal	Hypox
	amino acid uptake, μmoles $\cdot$ g^{-1} $\cdot$ hr^{-1}	
Aspartic acid	1.69	2.48[1]
Threonine	4.12	7.01[2]
Serine	6.00	11.29[1]
Asparagine	2.80	4.03[1]
Glutamine	12.54	22.57[2]
Proline	6.99	9.11
Glutamic acid	0	3.31[1]
Glycine	14.60	22.55[1]
Alanine	16.59	20.79[1]
Valine	1.67	1.84
Methionine	2.89	3.49
Isoleucine	1.87	1.93
Leucine	3.09	2.72
Tyrosine	4.61	4.62
Phenylalanine	3.38	4.43
Lysine	10.69	18.85[2]
Histidine	3.16	4.32
Arginine	3.73	5.99[2]
Total	100.42	151.33[1]

Livers from normal and hypox rats were perfused for 60 minutes with recirculating medium that initially contained all amino acids at 10 times their normal plasma concentrations. Amino acid concentrations in samples of perfusate plasma were determined prior to and after perfusion by extraction with 1% picric acid (Tallan *et al.*, 1954) followed by quantitation on a Beckman amino acid analyzer. The differences between the initial and final concentrations represented amino acid uptake by the perfused liver. Perfusate samples from 3 different experiments were pooled for analysis and the values represent the means of 3 analyses.

[1] differs from normal by $P < .05$

[2] differs from normal by $P < .01$

To confirm that hypophysectomy increased the conversion of amino acids to glucose, livers were perfused for 30 min with non-recirculating medium containing normal plasma concentrations of amino acids and ^{14}C-alanine. Substrates were maintained at constant levels by this technique and conversion of alanine to glucose could be determined by measuring the rate of ^{14}C-glucose synthesis. As can be seen in Figure 6, the rate of ^{14}C-glucose synthesis by livers of hypox rats was 4-fold greater than that by normal livers. As can be seen in the right panel of this figure, a concomitant 40% decrease in the oxidation of alanine was observed in the hormone-deficient livers. This difference of 20 nmoles/g/min in the rate of alanine oxidation would be sufficient to account for the increase in the rate of conversion of alanine to glucose.

Since carbon flow through the gluconeogenic pathway was increased, the following studies were undertaken to investigate the regulation of gluconeogenesis in livers of hypox rats. In these studies, livers were perfused for 30 minutes with non-recirculating medium containing ^{14}C-lactate as a substrate for gluconeogenesis. The rate of ^{14}C-glucose synthesis from ^{14}C-lactate increased as a function of the perfusate lactate concentration (Fig. 7). Glucose synthesis

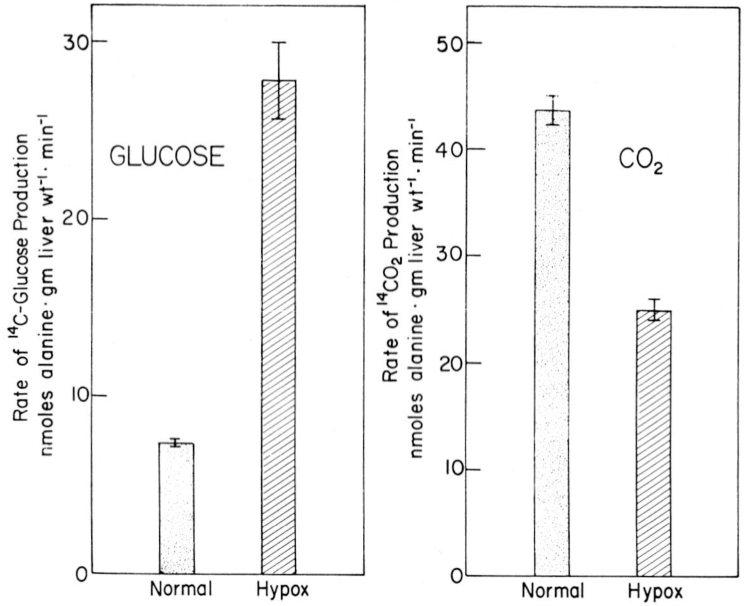

Fig. 6. Conversion of alanine to glucose and CO_2 by perfused livers of normal and hypophysectomized rats. Livers from normal and hypox rats were perfused for 30 min with non-recirculating medium containing all amino acids at their normal plasma concentrations and L-alanine-U-C^{14}. Samples of the perfusate were collected prior to perfusion and at 28 to 30 min into the perfusion. ^{14}C-glucose was determined by the method described by Exton and Park (1967). The amount of ^{14}C-glucose in the samples was taken to represent a measure of the rate of conversion of ^{14}C-alanine to ^{14}C-glucose. To determine the oxidation of ^{14}C-alanine to $^{14}CO_2$, the perfusate was collected under chilled heptane to minimize exposure to ambient air. Aliquots of the whole perfusate were added to rubber-stoppered Kontes flasks fitted with plastic center wells containing hyamine hydroxide 10-X. The medium was acidified by the addition of 0.5 ml 5 N H_2SO_4. After 30 min the center wells were dropped into liquid scintillation counting solution and the radioactivity determined. In all experiments, radioactivity was corrected for quenching and the efficiency of counting was determined. Radioactivity in experimental solutions was converted to molar equivalents by reference to the specific activity of the initial perfusion medium. The values presented represent the mean of 7 determinations for the normal animals and 4 for the hypox rats. The vertical bars represent the standard errors of the mean. The production of ^{14}C-glucose and $^{14}CO_2$ by the livers of the hypox rats both differed significantly from their respective normal values ($P < .001$).

by livers of hypox rats was 3- to 4-fold greater than that by control livers at perfusate lactate concentrations up to 8 mM. Lactate levels were saturating for gluconeogenesis at 6-8 mM in livers of hypox rats and 14 mM in control livers. It should be noted that the level of lactate required to saturate gluconeogenesis in the livers of fed normal rats is much higher than the level (4 mM) which saturated the process in livers of fasted rats (Exton and Park, 1967). At saturating lactate concentrations, similar maximum gluconeogenic rates of 1.05 and 0.90 μmoles/g/min were observed in livers of normal and hypox rats, respectively. These data indicated that within the physiological range of blood lactate levels, around 1 to 4 mM, the gluconeogenic activity of livers from hypox animals was greater than that of control livers, but that the total gluconeogenic capacities of the two tissues were similar. In other experiments it was found that fasting increased the gluconeogenic capacity of livers from normal but not hypox rats and that hypophysectomy prevented the stimulatory effect of glucagon on gluconeogenesis (unpublished observations).

Similar to the data obtained with alanine, the rate of lactate oxidation by livers of hypox

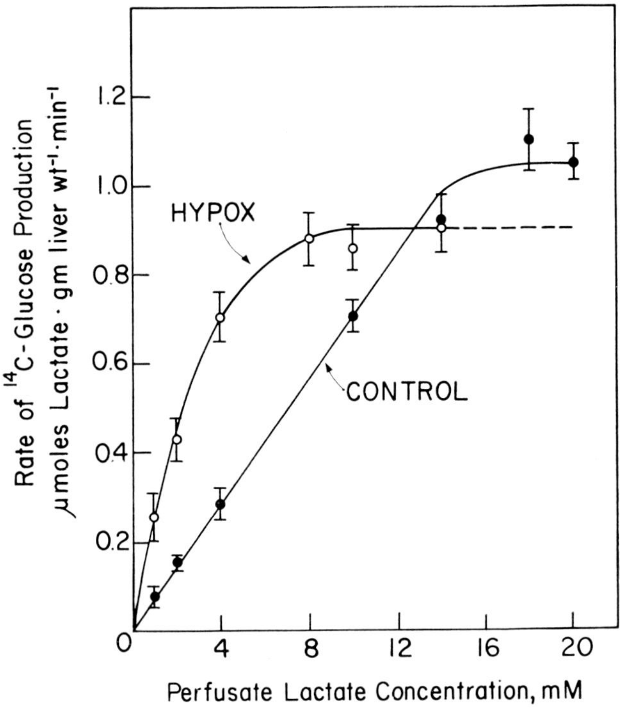

Fig. 7. *^{14}C-glucose synthesis from ^{14}C-lactate as a function of lactate concentrations.* Livers from normal and hypox rats were perfused for 30 min with non-recirculating medium containing the indicated levels of L (+) lactate-^{12}C and sodium DL-lactate-2-^{14}C. Conversion of ^{14}C-lactate to ^{14}C-glucose was measured as described in the legend to Figure 6. Each point represents the mean of 3 to 8 observations. The vertical bars represent the standard errors of the mean.

TABLE VII

Oxidation of ^{14}C-lactate by perfused livers of normal and hypox rats

Perfusate lactate concentration mM		Conversion of ^{14}C-lactate to $^{14}CO_2$ µmoles · g liver wt^{-1} · min^{-1}	
		Normal	Hypox
	DL-Lactate-2-^{14}C		
1		.148 ± .016	.066 ± .004[1]
2		.206 ± .018	.154 ± .016[1]
4		.372 ± .006	.206 ± .018[1]
8		.441 ± .046	.300 ± .010[2]
10		.524 ± .012	.374 ± .028[2]
14		.558 ± .038	.450 ± .022[2]
	L-Lactate-1-^{14}C		
1		.165 ± .009	.129 ± .013[2]
10		.214 ± .018	.100 ± .010[1]

Experiments were performed as described in the legend to Figure 7. In some experiments, sodium-L-lactate-1-^{14}C was used. Values shown represent the means of 3 to 8 determinations ± the standard errors of the mean.

[1] differs from normal by $P < .05$

[2] differs from normal by $P < .001$

rats was significantly less than that by control livers at all perfusate concentrations of lactate that were studied (Table VII). The conversion of lactate-2-^{14}C to $^{14}CO_2$ by the experimental livers was only 44 to 81 % of the control rate, though the difference diminished with increasing perfusate lactate levels. Similar results were obtained if L-lactate-1-^{14}C was used as the radioactive tracer, except that $^{14}CO_2$ formation by the livers of hypox animals did not increase as lactate concentration was raised. The primary site for the oxidation of lactate-2-^{14}C is assumed to be the citric acid cycle, while the production of radioactive CO_2 from lactate-1-^{14}C occurs both in the citric acid cycle and at the pyruvate dehydrogenase step.

A reduction in the oxidation of pyruvate to acetyl CoA might lead to an increase in the availability of substrate for gluconeogenesis by diminishing the flow through alternate metabolic pathways. The data shown in Table VIII support this suggestion. These results demonstrate that fatty acid synthesis from lactate was inhibited by 80–90 % in perfused livers of hypox rats. It is also noted that the incorporation of label from ^{14}C-lactate into protein was diminished by 20–50 % in the livers of the experimental animals. A depressed rate of glycogen synthesis in the livers of the hypox animals was observed as expected from the adrenocortical atrophy known to occur subsequent to hypophysectomy.

The preceding data on the metabolic fate of lactate in livers from normal and hypox rats

TABLE VIII

Conversion of ^{14}C-lactate to fatty acids, protein adn glycogen by perfused livers of normal and hypox rats

Perfusate lactate concentration mM		Conversion of ^{14}C-lactate to fatty acids, protein and glycogen nmoles · g liver wt^{-1} · min^{-1} Normal	Hypox
	Fatty acids		
1		12.53 ± 1.47	1.40 ± .33[1]
4		20.33 ± 1.70	3.27 ± .73[3]
10		73.73 ± 6.47	10.33 ± 1.67[3]
	Protein		
1		26.00 ± .67	14.27 ± .53[3]
10		86.93 ± 4.40	55.60 ± 4.67[3]
14		94.53 ± 4.13	77.07 ± 6.93[1]
	Glycogen		
1		14.20 ± 1.03	10.12 ± .96[1]
10		58.20 ± 3.90	29.34 ± 2.07[1]
14		61.54 ± 5.74	36.34 ± 2.67[2]

Perfusion conditions were the same as described in the legend to Figure 7. At the end of perfusion the livers were quickly frozen in Wollenberger clamps at the temperature of liquid nitrogen and subsequently pulverized. The conversion of ^{14}C-lactate to ^{14}C-fatty acids was determined by the method of Exton (1964). Incorporation of radioactivity into protein was determined as described in Figure 1. The conversion of ^{14}C-lactate to ^{14}C-glycogen was estimated by digesting samples of powdered liver in 30 % KOH for 60 min and precipitating the glycogen with ethanol (final concentration of 65 %). The pellet was dissolved in 1 N H_2SO_4 and the solution hydrolyzed for 3 hr at 100°C. After neutralization with NaOH, aliquots of the solution were dissolved in an aqueous counting solution. Appropriate corrections were made for counting efficiency and quenching. Radioactivity in the test substances was converted to molar equivalents using the specific activity of lactate in the initial perfusate. Each value represents the mean of 3 to 8 observations ± the standard error.

[1] differs from normal by $P < .05$

[2] differs from normal by $P < .01$

[3] differs from normal by $P < .005$

TABLE IX

Metabolic fate of ^{14}C-lactate in perfused livers of normal and hypox rats

Perfusate lactate concentration	Liver donor	Total lactate utilized μmoles · g liver wt^{-1} · 30 min^{-1}	Metabolic fate of ^{14}C-lactate (per cent of total lactate utilized)				
			Glucose	CO_2	Protein	Fatty acids	Glycogen
1 mM	Normal	8.19 ± 0.93	26.4	54.2	9.5	4.8	5.1
	Hypox	10.29 ± 1.12	73.5	19.2	4.1	0.3	2.9
14 mM	Normal	51.27 ± 2.67	53.8	32.7	5.6	4.3	3.6
	Hypox	45.15 ± 3.40	61.9	29.9	5.1	0.7	2.4

Perfusion conditions were the same as described in the legend to Figure 7. Data are presented as a per cent of the total lactate utilized and were calculated by the following formula:

$$\frac{^{14}\text{C-lactate incorporated into specific metabolite per g per min}}{\text{Total } ^{14}\text{C-lactate utilized per g per min}} \times 100$$

are summarized in Table IX. Experiments with near physiologic and saturating levels of lactate are shown and the data are presented as a per cent of the total lactate utilized at each lactate concentration. With 1 mM lactate in the perfusing media, the control livers oxidized over 50% of the substrate while converting only 26% to glucose. Livers of hypox rats, on the other hand, oxidized only 19% of the substrate that was utilized while converting more than 70% to glucose. The per cent of lactate converted to protein, fatty acid and glycogen by the control livers exceeded that by livers of hypox rats. It can be seen that the total amounts of lactate utilized at both low and high lactate levels were similar for livers of both groups.

SUMMARY

During perfusion of livers from either normal or hypox rats, the initiation of peptide chains became rate-limiting for protein synthesis. Addition of amino acids to the perfusion medium facilitated the initiation process equally in both tissues, and revealed a defect in the elongation of peptide chains in the livers of hypox rats. The defect was apparently due to the deficiency of growth hormone, since hormone treatment for 48 hours restored protein synthesis to normal. Since 24 to 48 hours of growth hormone treatment was required to correct the elongation defect, this effect of the hormone may not be related to its direct effect on protein synthesis *in vitro* which has been observed within 30 minutes and which appeared to involve a stimulatory effect on amino acid transport (Jefferson and Korner, 1967, and unpublished observations).

Hypophysectomy produced marked changes in amino acid metabolism which were contributed to in part by increased conversion of amino acids to glucose. Increased ability of the hypox animal to convert amino acids and other precursors to glucose has physiological importance in that it enables the animal to maintain a normal level of blood glucose at a time when the rate of glucose utilization is relatively increased and the availability of free fatty acids as substrates for energy metabolism is decreased. Since amino acid utilization was increased while lactate utilization was normal, it appears that hypophysectomy enhanced the amount of substrate available for gluconeogenesis, possibly because of increased amino acid transport and/or transamination. Increased AIB transport and increased activities of glutamic-pyruvic transaminase, histidase and serine dehydratase have been observed in livers of hypox animals (Riggs and Walker, 1960; Sakuma *et al.*, 1968; Freedland *et al.*, 1968; Feigelson,

1971). The lactate studies demonstrate that hypophysectomy also increased glucose synthesis as a result of alterations in substrate carbon flow. Thus, the amount of substrate available for gluconeogenesis was enhanced in livers of hypox animals as a result of diminished activities of alternative metabolic pathways, including synthesis of fatty acid, protein and glycogen and oxidation to CO_2. It would appear that hypophysectomy led to a restraint of the conversion of pyruvate to acetyl CoA. In support of this suggestion, we observed that acetyl CoA and citrate levels were only 50% of normal in livers of hypox rats perfused with lactate as substrate. Pyruvate dehydrogenase exists in two forms, an active dephosphorylated form and an inactive phosphorylated form (Linn *et al.*, 1969). It has been suggested that interconversion of this enzyme is an important regulatory mechanism. A marked fall in the active form without a change in total activity of this enzyme was observed in heart muscle and kidney following starvation and diabetes (Wieland *et al.*, 1970). Insulin has also been implicated in the regulation of this enzyme in adipose tissue (Jungas, 1970). It is possible that the low plasma levels of insulin that are associated with hypophysectomy led to conversion of pyruvate dehydrogenase to the inactive form.

REFERENCES

ALGRANATI, I. D., GONZALEZ, N. S. and BADE, E. G. (1969): Physiological role of 70S ribosomes in bacteria. *Proc. nat. Acad. Sci. (Wash.)*, *62*, 574.

CLEMENS, M. J. and KORNER, A. (1970): Amino acid requirement for the growth-hormone stimulation of incorporation of precursors into protein and nucleic acids of liver slices. *Biochem. J.*, *119*, 629.

DEBODO, R. C. and SINKOFF, M. W. (1953): Anterior pituitary and adrenal hormones in the regulation of carbohydrate metabolism. In: *Recent Progress on Hormone Research*, Vol. VIII, pp. 511–570. Editor: G. Pincus. Academic Press, New York.

EXTON, J. H. (1964): Metabolism of rat-liver cell suspensions. I. General properties of isolated cells and occurrence of the citric acid cycle. *Biochem. J.*, *92*, 457.

EXTON, J. H. and PARK, C. R. (1967): Control of gluconeogenesis in liver. I. General features of gluconeogenesis in the perfused rat liver. *J. biol. Chem.*, *242*, 2622.

FEIGELSON, M. (1971): Hypophyseal regulation of hepatic histidase during postnatal development and adulthood. I. Pituitary suppression of histidase activity. *Biochim. biophys. Acta (Amst.)*, *230*, 296.

FLECK, A. and MUNRO, H. N. (1962): The precision of ultraviolet absorption measurements in the Schmidt-Thannhausen procedure for nucleic acid estimation. *Biochim. biophys. Acta (Amst.)*, *55*, 571.

FREEDLAND, R. A., AVERY, E. G. and TAYLOR, A. R. (1968): Effect of thyroid hormones on metabolism. II. The effect of adrenalectomy or hypophysectomy on responses of rat liver enzyme activity to L-thyroxine injection. *Canad. J. Biochem.*, *46*, 141.

GOODMAN, H. M. (1970): Permissive effects of hormones on lipolysis. *Endocrinology*, *86*, 1064.

HOERZ, W. and MCCARTY, K. S. (1969): Evidence for a proposed initiation complex for protein synthesis in reticulocyte polyribosome profiles. *Proc. nat. Acad. Sci. (Wash.)*, *63*, 1206.

HOGAN, B. L. M. and KORNER, A. (1968): The role of ribosomal subunits and 80-S monomers in polysome formation in an ascites tumour cell. *Biochim. biophys. Acta (Amst.)*, *169*, 139.

HUGGETT, A. ST G. and NIXON, D. A. (1957): Enzymic determination of blood glucose. *Biochem. J.*, *66*, 12P.

JEFFERSON, L. S. and KORNER, A. (1967): A direct effect of growth hormone on the incorporation of precursors into proteins and nucleic acids of perfused rat liver. *Biochem. J.*, *104*, 826.

JEFFERSON, L. S. and KORNER, A. (1969): Influence of amino acid supply on ribosomes and protein synthesis of perfused rat liver. *Biochem. J.*, *111*, 703.

JEFFERSON, L. S., ROBERTSON, J. W. and CAHN, J. (1972): Regulation of hepatic protein synthesis. I. Effect of amino acid levels on protein synthesis and on ribosomal aggregation. Submitted for publication.

JUNGAS, R. L. (1970): Effect of insulin on fatty acid synthesis from pyruvate, lactate or endogenous sources in adipose tissue: Evidence for the hormonal regulation of pyruvate dehydrogenase. *Endocrinology*, *86*, 1368.

KABAT, D. and RICH, A. (1969): The ribosomal subunit-polyribosome cycle in protein synthesis of embryonic skeletal muscle. *Biochemistry*, *8*, 3742.

KORNER, A. (1960): The effect of hypophysectomy of the rat and of treatment with growth hormone on the incorporation *in vivo* of radioactive amino acids into the proteins of subcellular fractions of rat liver. *Biochem. J.*, *74*, 462.

KORNER, A. (1961): The effect of hypophysectomy and growth hormone treatment of the rat on the incorporation of amino acids into isolated liver ribosomes. *Biochem. J.*, *81*, 292.

LINN, T. C., PETIT, F. H., HUCHO, F. and REED, L. J. (1969): α-Keto acid dehydrogenase complexes. XI. Comparative studies of regulatory properties of the pyruvate dehydrogenase complexes from kidney, heart and liver mitochondria. *Proc. nat. Acad. Sci. (Wash.)*, *62*, 227.

LOWRY, O. H., ROSEBROUGH, N. J., FARR, A. L. and RANDELL, R. J. (1951): Protein measurement with the Folin phenol reagent. *J. biol. Chem.*, *193*, 265.

MORGAN, H. E., JEFFERSON, L. S., WOLPERT, E. B. and RANNELS, D. E. (1971): Regulation of protein synthesis in heart muscle. II. Effect of amino acid levels and insulin on ribosomal aggregation. *J. biol. Chem.*, *246*, 2163.

MORTIMORE, G. E. (1961): Effect of insulin on potassium transfer in isolated rat liver. *Amer. J. Physiol.*, *200*, 1315.

MUNRO, H. N., JACKSON, R. J. and KORNER, A. (1964): Studies on the nature of polyribosomes. *Biochem. J.*, *92*, 289.

NOALL, M. W., RIGGS, T. R., WALKER, L. M. and CHRISTENSEN, H. N. (1957): Endocrine control of amino acid transfer. *Science*, *126*, 1002.

NOLL, H. (1967): Characterization of macromolecules by constant velocity sedimentation. *Nature (Lond.)*, *215*, 360.

PEGG, A. E. and KORNER, A. (1965): Growth hormone action on rat liver RNA polymerase. *Nature (Lond.)*, *205*, 904.

PHILLIPS, R. A. and ROBB, P. D. (1939): Metabolism studies in the albino rat. *Endocrinology*, *25*,187.

RIGGS, T. R. and WALKER, L. M. (1960): Growth hormone stimulation of amino acid transport into rat tissues *in vivo*. *J. biol. Chem.*, *235*, 3603.

ROSS, B. D., HEMS, R. and KREBS, H. A. (1967): The rate of gluconeogenesis from various precursors in the perfused rat liver. *Biochem. J.*, *102*, 942.

RUSSELL, J. A. and BENNETT, L. L. (1937): Carbohydrate storage and maintenance in the hypophysectomized rat. *Amer. J. Physiol.*, *118*, 196.

SAKUMA, M., NAKAGAWA, H., SUDA, M. and NAKOO, K. (1968): Effect of bovine growth hormone on serine dehydratase in rat liver. *Endocrinology*, *83*, 381.

SCHLESSINGER, D. (1969): Ribosomes: Development of some current ideas. *Bact. Rev.*, *33*, 445.

SEARCY, R. L., GOUGH, G. S., KOROTZER, J. L. and BERQUIST, L. M. (1961): Evaluation of a new technique for the estimation of urea nitrogen in serum. *Amer. J. med. Technol.*, *27*, 255.

SIMPSON, M. E., EVANS, H. M. and LI, C. H. (1949): The growth of hypophysectomized female rats following chronic treatment with pure pituitary growth hormone. I. General growth and organ changes. *Growth*, *13*, 151.

SOSKIN, S., MIRSKY, I. A., ZIMMERMAN, L. M. and CROHN, N. (1935): Influence of hypophysectomy on gluconeogenesis in the normal and depancreatectomized dog. *Amer. J. Physiol.*, *114*, 110.

TALLAN, H. H., MOORE, S. and STEIN, W. H. (1954): Studies on the free amino acids and related compounds in the tissues of the cat. *J. biol. Chem.*, *211*, 927.

TALWAR, G. P., PANDA, N. C., SARIN, G. S. and TOLANI, A. J. (1962): Effect of growth hormone on ribonucleic acid metabolism. *Biochem. J.*, *82*, 173.

TOLMAN, E. L., METZGER, L. E. and JEFFERSON, L. S. (1970): A stimulated rate of gluconeogenesis in perfused livers of hypophysectomized rats. *Biochem. biophys. Res. Commun.*, *41*, 294.

WALAAS, O. and WALAAS, E. J. (1950): Effect of epinephrine on rat diaphragm. *J. biol. Chem.*, *187*, 769.

WEBER, G. and CANTERO, A. (1959): Effect of hypophysectomy on liver enzymes involved in glycogenolysis and in glucogenesis. *Amer. J. Physiol.*, *197*, 699.

WIDNELL, C. C. and TATA, J. R. (1966): Additive effects of thyroid hormone, growth hormone and testosterone on deoxyribonucleic acid-dependent ribonucleic acid polymerase in rat liver nuclei. *Biochem. J.*, *98*, 621.

WIELAND, O., SPESS, E. and SCHULZE-WETHMER, F. H. (1970): Der Einfluss von Hunger und Diabetes auf die Interconvertierung von aktiver und inaktiver Pyruvat-Dehydrogenase verschiedener Rattenorgane *in vivo*. *Hoppe-Seyler's Z. physiol. Chem.*, *351*, 326.

WOODSIDE, K. H. and MORTIMORE, G. E. (1970): Control of proteolysis in the perfused rat liver: Influence of amino acids, insulin and glucagon. *Fed. Proc.*, *29*, 379.

SOME OBSERVATIONS CONCERNING SERUM 'THYMIDINE FACTOR'*

M. S. RABEN**, S. MURAKAWA and M. MATUTE***

New England Medical Center Hospitals and Department of Medicine, Tufts University School of Medicine, Boston, Mass., U.S.A.

Stimulation by serum of thymidine incorporation into DNA in hypophysectomized rat costal cartilage *in vitro* was attributed by Daughaday and Reeder (1966) to a factor comparable or identical to 'sulfation factor', termed 'thymidine factor' by Van Wyk *et al.* (1969). Interest in this subject has been stimulated by the possibility that this factor or factors may mediate the growth-promoting action of growth hormone.

'Thymidine factor' activity of hypophysectomized rat serum

Daughaday and Reeder (1966) found that whereas normal rat serum stimulated thymidine incorporation, hypophysectomized rat serum was inhibitory. Using more complex medium (enriched Medium 199) (Esanu *et al.*, 1969), but probably chiefly by virtue of changing the medium after the 18–20 hour preincubation period, we found hypophysectomized serum markedly stimulatory although less potent than normal rat serum.

In the modified procedure, the medium was removed after the preincubation period, replaced with a fresh sample of the same medium, whether control or test medium, ^{3}H-thymi-

TABLE I

Influence of change of medium after the preincubation period on the stimulatory effect of normal rat serum (NRS) on thymidine incorporation

Preinbucation (1st 24 hours)	Incubation (2nd 24 hours)	Change of medium after 24 hours	^{3}H-thymidine uptake % of control
Medium	Medium	yes	100%
4% NRS	Medium	yes	196%
Medium	4% NRS	yes	293%
4% NRS	4% NRS	yes	530%
4% NRS	4% NRS	no	111%

The medium was enriched 199. ^{3}H-thymidine was added at the end of the preincubation period.

* This investigation was supported by Grant AM 01567 from the National Institute of Arthritis and Metabolic Diseases, USPHS.
** Recipient of Research Career Award, National Institutes of Health.
*** Trainee, Graduate Research Training Program, Grant HE 05391 from the National Heart and Lung Institute, USPHS.

dine was added, and incubation continued for 24 hours. Samples were run in quadruplicate, and the cartilage was processed as previously described (Murakawa and Raben, 1968). Assays were run in groups of 7 samples, each in quadruplicate, so that there were 28 vials each containing 6 pieces of costal cartilage, one from each of 6 hypophysectomized rats. The wet weight of cartilage in each vial was approximately 12 mg. The change of medium after the preincubation period caused a marked enhancement of sensitivity of the assay even with normal rat serum (Table I). The dose-response with the modified procedure is shown in Table II with normal human serum. There was an increasing response from 145% of the control value with 0.1% serum in the medium to 6301% of control with 10% serum. The dose-response in this example was better than in the average assay. The sensitivity varied from assay to assay but the dose-response with normal serum was usually satisfactory above 0.5–1%.

The activity of hypophysectomized rat serum was increased by the presence of a small amount of dexamethasone in the medium (Table III). Dexamethasone (0.02 μg/ml) is now usually included in the medium in tests of hypophysectomized serum and of isolated substances.

TABLE II

Effect of normal human serum at various concentrations on the thymidine incorporation in DNA in costal cartilage from hypophysectomized rats

Samples	% Normal human serum	^{3}H-thymidine incorporation % of control
Control	0	100
Normal human serum	0.1	145
Normal human serum	0.5	355
Normal human serum	1.0	828
Normal human serum	2.0	1913
Normal human serum	5.0	3655
Normal human serum	10.0	6301

The tissue was preincubated for 20 hours and then incubated for 24 hours with ^{3}H-thymidine and fresh medium. Each value is the average of 4 samples. All samples contained pieces of cartilage from the same rats. The medium was enriched 199, supplemented as indicated with serum.

TABLE III

Influence of dexamethasone on thymidine factor activity of serum from hypophysectomized rats

Dexamethasone:	0	0.02 μg/ml	0.1 μg/ml
Expt. 1			
4% Hypox serum (6)	184%	399%	410%
Expt. 2			
4% Hypox serum (5)	120%	376%	—

Values are the average of the indicated number of individual serums tested and are expressed as per-cent of control uptake. Control and experimental samples contained the indicated amount of dexamethasone. The medium was enriched 199.

TABLE IV

Comparison of normal and hypophysectomized rat serum

	% Serum in medium		Hypox serum	Normal rat serum
Expt. 1	4%		282%	556%
			330	
			472	
			443	
			354	
		Avr.	376%	556%
Expt. 2	4%		575%	555%
			618	1718
			664	
			646	
		Avr.	626%	1137%
Expt. 3	2%		150%	620%
	10%		1137%	1972%

Thymidine incorporation is expressed as percent of incorporation with basal medium without serum. Basal medium was enriched 199 with dexamethasone 0.02 μg/ml. In Expts. 1 and 2, each value represents an individual rat serum. In Expt. 3, 1 hypox and 1 normal serum were tested at 2 doses.

The data in Table IV indicate that normal rat serum tends to be more active than hypophysectomized serum, perhaps about twice as active, but hypophysectomized serum tested in the range of 2–10% is more impressive for its potency than for its weakness relative to normal serum. There is a good dose response with hypophysectomized serum seen in experiment 3, Table IV.

Effect of growth hormone on thymidine factor activity of serum

Since no data have yet been published to support the hypothesis that the potency of serum as thymidine factor is growth-hormone dependent, this point was studied fairly extensively. In 8 experiments done over several years involving 59 control and 57 growth hormone-treated hypophysectomized rats, the serum was tested after 48 hours of growth hormone treatment (Table V). Thymidine incorporation was greater with treated serum in all experiments, averaging twice the control uptake, but the increase was quite variable from experiment to experiment. A single injection of growth hormone appeared to increase the activity of serum after 17 hours but not after 2 or 7 hours (Table VI). Despite this evidence of growth hormone stimulation, it was found that the potency of serums from 6 patients with hereditary isolated growth hormone deficiency (provided by Dr. T. Merimee) was similar to that of pooled normal human serum (Table VII). It was, however, observed that several serum samples from an acromegalic patient were more potent than the pooled serum.

Active substances in human serum

Attempts to identify the thymidine factor indicated that at least 3 substances in human serum possessed activity: a substance, probably peptide, of small molecular weight, roughly estimated 1,000–1,500 by gel filtration, NSILA-S (the type of non-suppressible insulin-like activity with a molecular weight of 6,000–10,000) and insulin.

TABLE V

Effect of growth hormone treatment on thymidine factor activity of hypophysectomized rat serum

Expt.	Source of serum	No. of rats	Thymidine uptake % of basal medium ± SE	BGH-treated / untreated
1[1]	Untreated 5%	9	648 ± 138	2.7
	BGH-treated 5%	9	1762 ± 378	
2	Untreated 2%	5	738 ± 78	4.1
	BGH-treated 2%	5	2996 ± 578	
3	Untreated 4%	9	289 ± 69	1.53
	BGH-treated 4%	8	443 ± 74	
4	Untreated 4%	6	233 ± 20	1.39
	BGH-treated 4%	6	323 ± 63	
5	Untreated 4%	9	255	1.25
	BGH-treated 4%	7	318	
6	Untreated 4%	8	258 ± 30	2.08
	BGH-treated 4%	8	538 ± 94	
7	Untreated 2%	8	348 ± 87	1.28
	BGH-treated 2%	8	445 ± 60	
8[2]	Untreated 2%	5	177	1.59
	BGH-treated 2%	6	282	
				Avr. 1.99

BGH was administered for 2 days in a dose of 0.25 mg SC twice a day; medium (enriched 199) was changed after preincubation period. Female rats operated at 21 days were used 10 days later.

[1] Medium was Eagle's supplemented with glutamine and 9 amino acids.

[2] BGH 2 mg daily intraperitoneally for 2 days. Dexamethasone 0.02 μg/ml in medium.

TABLE VI

Effect of a single injection of BGH (2 mg intraperitoneally) on the potency of serum in stimulating thymidine uptake

	No. of rats	Percent serum in medium	Thymidine uptake % of control[1]
Expt. 1			
Untreated hypox rat serum	3	2%	144
17 hour BGH-treated	3	2%	228
Expt. 2			
2 hour BGH-treated	2	2%	161
7 hour BGH-treated	2	2%	179
17 hour BGH-treated	2	2%	477

Female rats hypophysectomized 1–2 months earlier at 21 days of age were used. Medium was enriched 199 with dexamethasone 0.02 μg/ml.

[1] Control value is uptake with basal medium without serum.

TABLE VII

Serum from 6 hereditary isolated growth hormone-deficient subjects compared with 2 pools of normal human serum for thymidine factor activity

Source of serum	Test No.: % serum:	(1) 2%	(2) 2%	(3) 2%	(4) 4%	(5) 4%
RS		246	233	340	1066	427
MA		201	247	179	698	401
ES		421	—	—	1333	518
M		213	227	250	841	408
DT		255	146	197	—	275
B		242	200	296	—	481
	Avr.	263	211	252	985	418
Pooled normal (1)		—	292	275	653	496
Pooled normal (2)		—	175	—	776	827
	Avr.		234	275	715	662
Ratio GH deficient / normal		—	0.90	0.92	1.38	0.63 Avr. 0.96

Enriched medium 199 without dexamethasone was used with change after preincubation period.

The small molecular weight substance was found in the supernatant after the addition of 3 volumes of cold ethanol to serum at pH 7.1. The supernatant was evaporated, dissolved in H_2O in a volume 1/10 the original serum, washed 3 times with 3 volumes of ethyl ether, and run through a column of Sephadex G50 in H_2O (Fig. 1). The active fraction, which was markedly delayed, was concentrated and run through a Sephadex G15 column. The active frac-

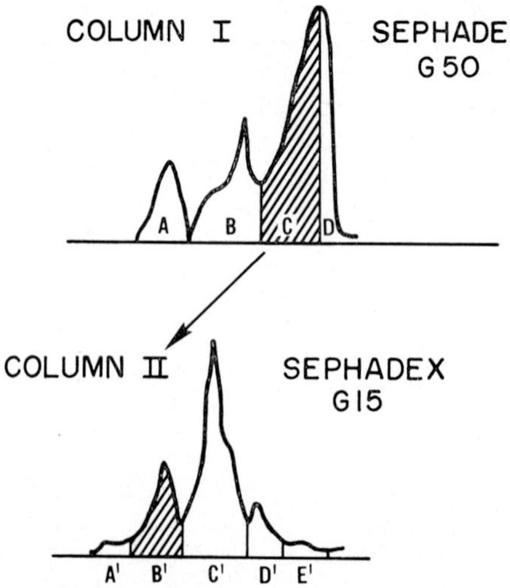

Fig. 1. Fractionation of small molecular weight substance with thymidine factor activity. Alcoholic extract of human serum was applied first to Sephadex G50 column. The active fraction, indicated by cross-hatching, was then applied to Sephadex G15 column. The assays on cross-hatched fraction B′ are presented in Table VIII.

TABLE VIII

Stimulation of thymidine and sulfate incorporation by the active fraction from Sephadex G15

G15 Fraction B′	Sulfate incorporation % of control	Thymidine incorporation % of control
0 (control)	100	100
5%	104	215
10%	124	159
20%	136	212
40%	173	328
80%	181	181
160%	167	—

The basal medium contained Eagle's amino acids supplemented with glutamine. For thymidine uptake, medium was supplemented with 5% normal human serum. Incubation period was 24 hours, preceded for thymidine uptake only by a preincubation period of 18–20 hours, without change of medium. The amount of Fraction B′ is expressed as the equivalent amount of serum from which it was prepared.

tion was slightly delayed on this column. It stimulated both sulfate and thymidine incorporation in cartilage (Table VIII) but appeared to account for only a small portion of the total activity of serum. This substance may be related to that isolated by Liberti (1970) by ultrafiltration of bovine plasma. It will be noted in Table VIII that at the time these tests were done, supplemented Eagle's medium was used, and the medium was not changed after the preincubation in the thymidine factor assay.

NSILA-S, extracted from human serum by the method of Burgi *et al.* (1966), was provided by Dr. E. R. Froesch. The batch tested in the spring of 1968 was contaminated with insulin in about equal insulin-like unitage but the preparation was found to be more potent in thymidine factor activity than insulin (Table IX). More recently a batch of insulin-free NSILA-S was tested and found to be active in the dose range of 1.2 to 12 μU/ml (1 to 10 μg/ml) (Table X). The material isolated by Van Wyk *et al.* (1971) may be related to NSILA-S since the extraction procedure was derived from the procedure used to prepare NSILA-S.

Although insulin added to the medium in relatively large amounts was active as observed by Salmon *et al.* (1968), the activity of serum appeared not to be proportional to its insulin content. There was no increase in activity during glucose tolerance tests in 12 subjects.

TABLE IX

Stimulation of thymidine incorporation in cartilage by insulin and by insulin-contaminated NSILA-S

		% of Control
Insulin	1 mU/ml	169%
NSILA-S	1 mU/ml	307%

Medium was enriched 199 containing dexamethasone (0.02 μg/ml); the medium was changed after preincubation period.

TABLE X

Thymidine factor activity of insulin-free NSILA-S

Expt.	Serum in medium	NSILA-S μU/ml	Thymidine incorporation % of control
1	[1]NHS 0.5%	2.4	127
		12.0	196
2	NHS 0.5%	12.0	219
3		1.2	169
	NHS 0.5%	3.6	173
		12.0	272
4	[2]HRS 2.0%	1.2	167
		12.0	277

The insulin-like activity of the preparation was 1.2 mμ/mg. The medium was enriched 199 containing dexamethasone 0.02 μg/ml and serum as indicated.
[1] Normal human serum (NHS).
[2] Hypophysectomized rat serum (HRS).

In unsuccessful attempts to locate the origin of the serum activity, it was noted that rat serum obtained 18 hours after partial hepatectomy (67%) and 24 hours after bilateral nephrectomy had normal activity.

COMMENTS AND SPECULATION

The effect of serum on thymidine incorporation in cartilage appears to be greater than the sum of the activities of the substances found to be stimulatory. There is much yet to be learned about the nature, source and function of this activity.

The role of thymidine factor as a mediator of the action of growth hormone remains speculative. In view of the considerable activity observed in hypophysectomized rat serum, it would seem that if the factor stimulates growth, it is only when the amount is above a rather high threshold level. Perhaps the amount present in hypophysectomized animals represents the amount of anabolic stimulus that causes slow growth in young animals and is needed to maintain tissue mass constant in rats of about 100 g size. The normal thymidine factor activity of the serum in hereditary isolated growth hormone deficiency is a disconcerting finding in regard to these speculations.

It is also of speculative interest to consider the relations of sulfation and thymidine factor to the serum factors required for cell culture, to such factors as the nerve and the epithelial growth factors, both isolated from mouse submaxillary gland (Cohen, 1962), and to postulated humoral substances involved in tissue regeneration. Hypophysectomized rat serum has been stated to be half as potent as normal rat serum in maintaining the growth of 3T3 cells, a line of mouse fibroblasts (Holley and Kiernan, 1968). This is similar to the relative potency we observed in the thymidine factor test in cartilage. The epithelial growth factor appears not to be as active as thymidine factor. We found material provided by Dr. S. Cohen inactive at a dose of 2μg/ml, an amount far greater than the minimal dose in systems in which it is active.

The rapid regeneration of the liver following partial hepatectomy in hypophysectomized rats is always a good reminder of how much of the regulation of tissue growth is beyond the control of growth hormone. The absence of growth hormone delays the early biochemical events following partial hepatectomy (Russell and Snyder, 1969) but the overall effect on regeneration is slight. There is evidence that this regenerative process may have its own humoral agent (Moolten and Bucher, 1967). In view of these many factors possibly related to growth, perhaps the role of growth hormone will prove to be that of a modulator of a still poorly defined humoral system that controls organ and tissue mass.

ACKNOWLEDGEMENTS

We gratefully acknowledge the expert assistance of Mrs. Vera Grinbergs in these studies. We wish to thank Dr. Rajendra K. Sogani for performing the hepatectomies, Dr. E. R. Froesch for providing the NSILA-S, Dr. S. Cohen for the 'epithelial growth factor' and Dr. T. J. Merimee for the serums of patients with isolated growth hormone deficiency.

REFERENCES

Burgi, H., Muller, W. A., Humbel, R. E., Labhart, A. and Froesch, E. R. (1966): Non-suppressible insulin-like activity of human serum. I. Physicochemical properties, extraction and partial purification. *Biochim. biophys. Acta (Amst.)*, *121*, 349.

Cohen, S. (1962): Isolation of a mouse submaxillary gland protein accelerating incisor eruption and eyelid opening in the new-born animal. *J. biol. Chem.*, *237*, 1555.

Daughaday, W. H. and Reeder, C. (1966): Synchronous activation of DNA synthesis in hypophysectomized rat cartilage by growth hormone. *J. Lab. clin. Med.*, *68*, 375.

Esanu, C., Murakawa, S., Bray, G. A. and Raben, M. S. (1969): DNA synthesis in human adipose tissue *in vitro*. I. Effect of serum and hormones. *J. clin. Endocr.*, *29*, 1969.

Holley, R. W. and Kiernan, J. A. (1968): 'Contact inhibition' of cell division in 3T3 cells. *Proc. nat. Acad. Sci. (Wash.)*, *60*, 300.

Liberti, J. P. (1970): Partial purification of bovine sulfation factor. *Biochem. biophys. Res. Commun.*, *39*, 356.

Moolten, F. L. and Bucher, N. L. (1967): Regeneration of rat liver: Transfer of humoral agent by cross-circulation. *Science*, *158*, 272.

Murakawa, S. and Raben, M. S. (1968): Effect of growth hormone and placental lactogen on DNA synthesis in rat costal cartilage and adipose tissue. *Endocrinology*, *83*, 645.

Russell, D. H. and Snyder, S. H. (1969): Amine synthesis in regenerating rat liver: Effect of hypophysectomy and growth hormone on ornithine decarboxylase. *Endocrinology*, *84*, 223.

Salmon, W. D. Jr., DuVall, M. R. and Thompson, E. Y. (1968): Stimulation by insulin *in vitro* of incorporation of (^{35}S) sulfate and (^{14}C) leucine into protein-polysaccharide complexes, (^{3}H) uridine into RNA and (^{3}H) thymidine into DNA of costal cartilage from hypophysectomized rats. *Endocrinology*, *82*, 493.

Van Wyk, J. J., Hall, K., Van den Brande, J. L. and Weaver, R. P. (1971): Further purification and characterization of sulfation factor and thymidine factor from acromegalic plasma. *J. clin. Endocr.* *32*, 389.

Van Wyk, J. J., Hall, K. and Weaver, W. P. (1969): Partial purification of sulphation factor and thymidine factor from plasma. *Biochim. biophys. Acta (Amst.)*, *192*, 560.

RELATION OF GROWTH HORMONE TO THYMUS AND THE IMMUNE RESPONSE

E. SORKIN, W. PIERPAOLI, N. FABRIS and ELENA BIANCHI

Schweizerisches Forschungsinstitut, Medizinische Abteilung, Davos, Switzerland, and Institute of Normal Anatomy, University of Pavia, Pavia, Italy

This paper is concerned with the interaction between growth hormone (somatotrophic hormone = STH) and the thymus, a primary lymphoid organ which plays a decisive part in the formation of lymphocytes and thereby in the immune response. The number of lymphocytes in an adult human being is about 10^{12}. Each lymphocyte probably makes one single kind of immunoglobulin which it carries at its surface but there is a wide range of different immunoglobulin molecules which are capable of combining with some part of presumably any foreign material (antigens) which penetrates a vertebrate. This fact is of obvious biological value since the immune response is thus a versatile defence mechanism for eliminating microbes, other parasites and for suppressing the host's own antigenically abnormal cells.

The thymus in vertebrates and the bursa of Fabricius in birds are the primary lymphoid organs into which stem cells migrate and where they differentiate presumably under the influence of maturational hormones. Thymus-influenced lymphoprogenitive cells (T-lymphocytes) mediate cell-mediated immune response (*e.g.* transplantation immunity, contact sensitivity, bacterial allergy) by being stimulated to release pharmacologically active agents capable of activating, damaging or killing other cells in their environment (Burnet, 1969).

It is the aim of this paper to provide evidence for our present and previous proposition

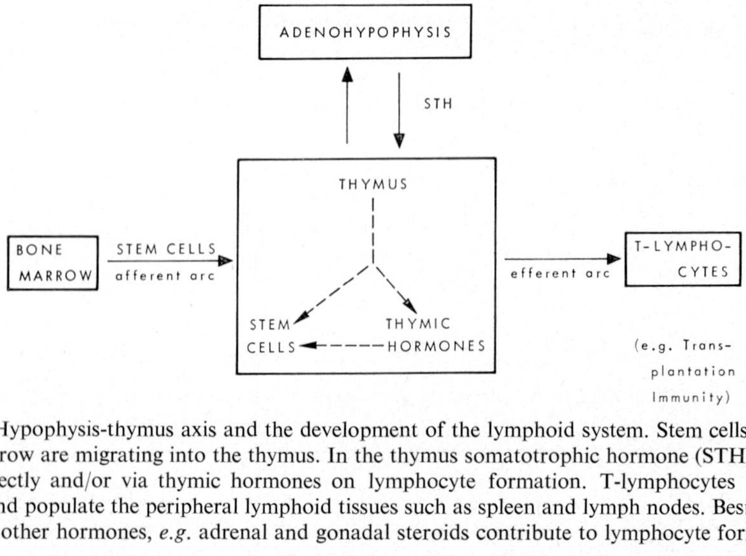

Fig. 1. Hypophysis-thymus axis and the development of the lymphoid system. Stem cells from the bone marrow are migrating into the thymus. In the thymus somatotrophic hormone (STH) is acting either directly and/or via thymic hormones on lymphocyte formation. T-lymphocytes leave the thymus and populate the peripheral lymphoid tissues such as spleen and lymph nodes. Besides STH, several other hormones, *e.g.* adrenal and gonadal steroids contribute to lymphocyte formation.

(Pierpaoli and Sorkin, 1969*b*; Pierpaoli *et al.*, 1970), that growth hormone is involved in the development of the thymus and the peripheral lymphoid tissues. Figure 1 summarizes in an abbreviated form our contention. Whether STH acts directly on the stem cells in the thymus or through some thymic hormones must be left an open question.

We have used several experimental approaches to prove our proposition: (1) neonatal thymectomy in mice and its effect on the adenohypophysis; (2) effect of anti-hypophysis serum and anti-STH serum on the thymus and peripheral lymphoid tissue in mice; (3) study of the thymus and of immunological parameters in STH-deficient Snell-Bagg dwarf mice before and after reconstitution with STH.

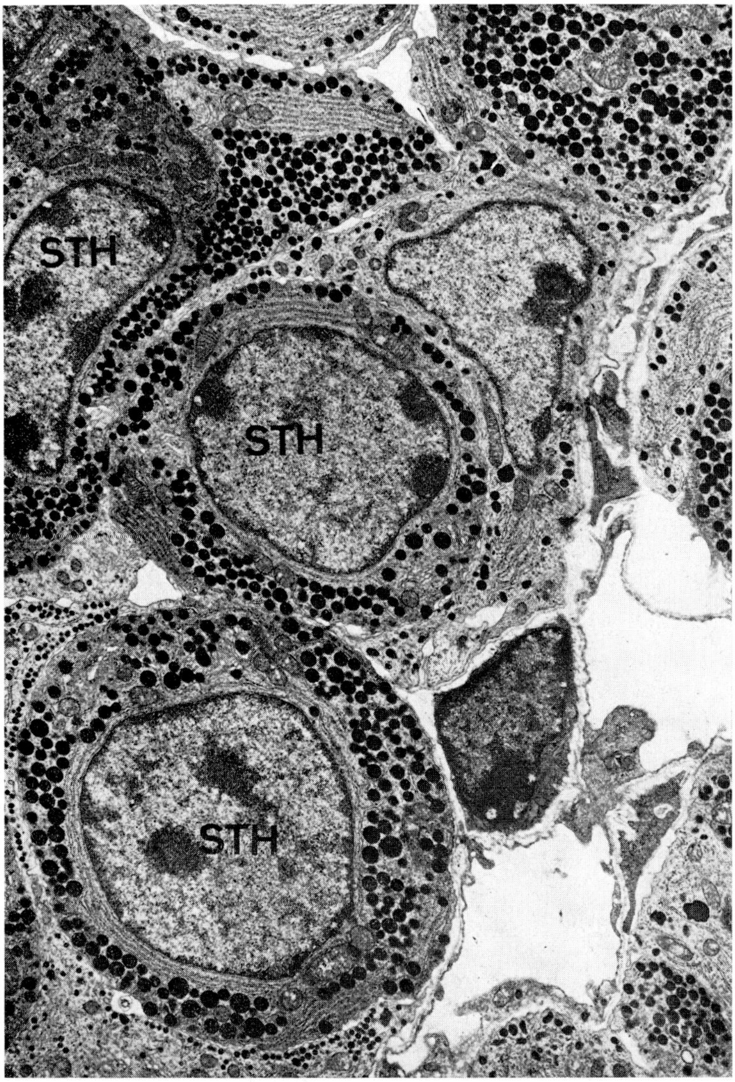

Fig. 2a. Electron micrograph of the anterior pituitary gland of a normal NMRI mouse at 47 days of age. The field is completely occupied by heavily granulated STH cells. Glutaraldehyde-osmium, × 5600.

The results reported previously (Pierpaoli *et al.*, 1970) and the present ones are definite evidence for the view that there is a relation between endocrinological function, especially of growth hormone, and immunological maturation.

1. Neonatal thymectomy in mice and its effect on the adenohypophysis

The first approach to test the possibility that thymus and hypophysis are reciprocally interrelated was a study of the effect of neonatal thymectomy on the adenohypophysis. This investigation was suggested by the evidence that neonatal thymectomy in rodents produces a

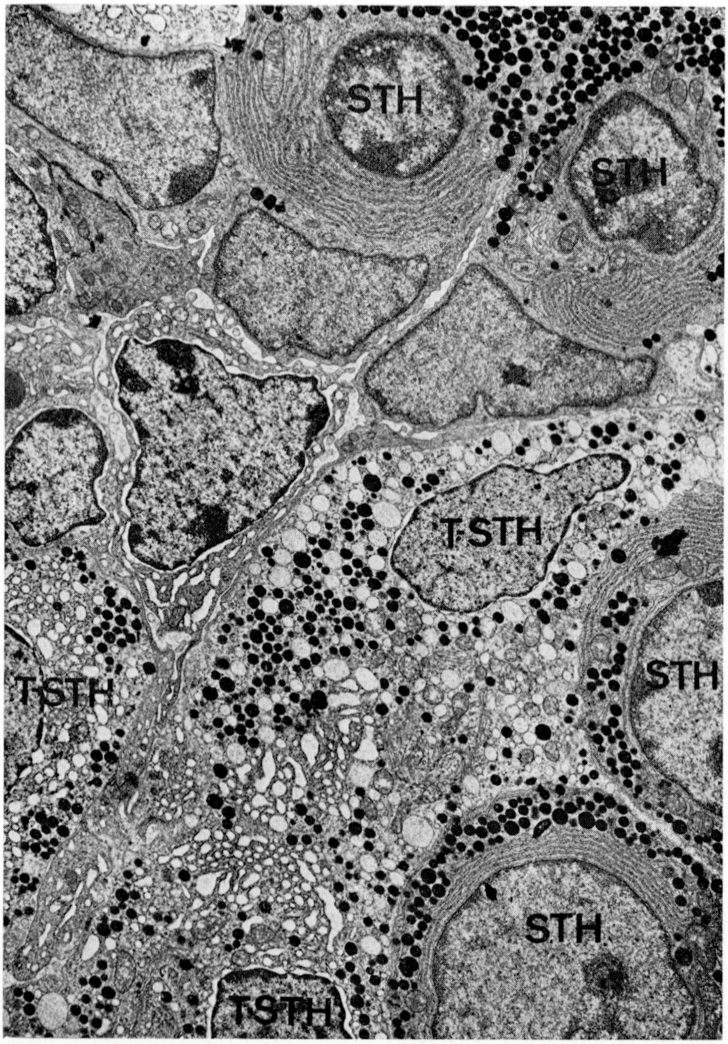

Fig. 2b. Electron micrograph of the anterior pituitary gland of a neonatally thymectomized litter mate of mouse providing section for Fig 2a. The field is occupied by several STH cells; most of them show typical degranulation (thymotropic cells, T-STH). Glutaraldehyde-osmium, × 5600.

severe impairment of their immunological capacity and induces in certain strains of mice the onset of a fatal wasting syndrome with characteristics similar to certain endocrinopathies. Light microscopy studies (Pierpaoli and Sorkin, 1967*a*, *b*) and recent electron microscopy studies (Bianchi *et al.*, 1971) on the adenohypophysis of neonatally thymectomized mice proved that removal of the thymus immediately after birth produces alterations in the acidophilic, growth hormone-producing cells (STH cells) such as degranulation and striking enlargement of the endoplasmic reticulum with formation of cisternae (Figs. 2a and 2b). The number of these changed, degranulated STH cells increased with age after neonatal thymectomy but was not higher when the neonatally thymectomized mouse showed the symptoms of the wasting syndrome. This indicated that this syndrome is not directly determined by the altera-

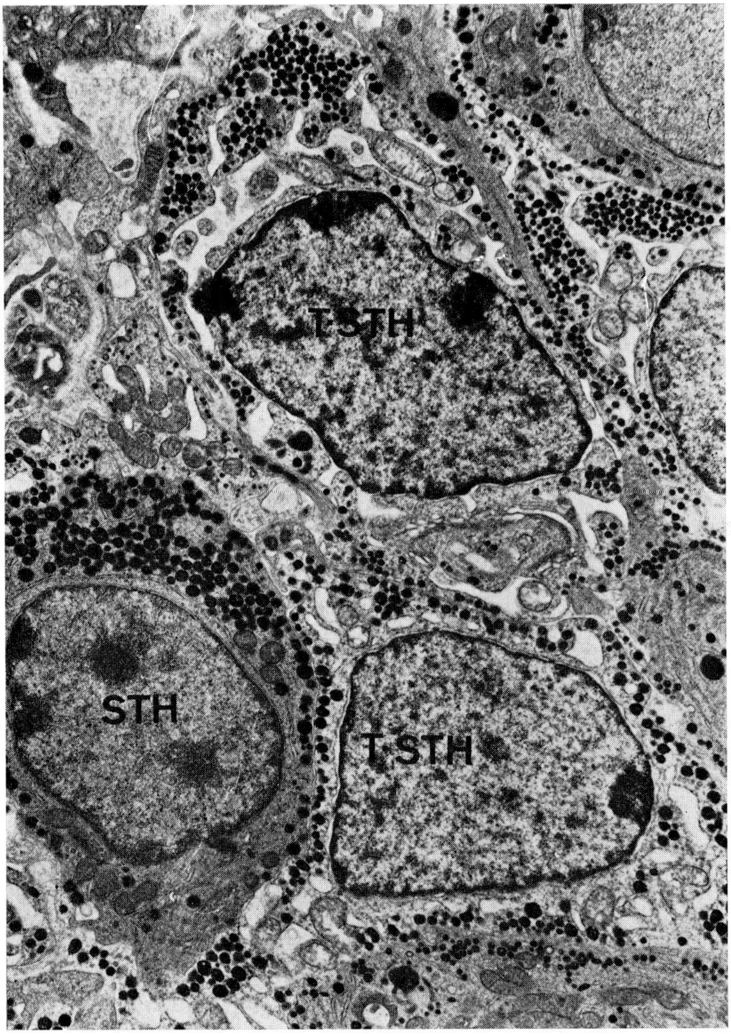

Fig. 3. Electron micrograph of the anterior pituitary gland of an 11-day-old neonatally thymecto mized germ-free NMRI mouse. Precocious degranulation of some of the few differentiated STH cells. Glutaraldehyde-osmium, × 6720.

tion of the hypophysis but rather by lack of STH-dependent, thymus-derived cells (Pierpaoli and Sorkin, 1971). Other recent and yet unpublished results indicate that if mice are neonatally operated in germ-free conditions, where possible effects of infections are prevented, the same type of degranulation of STH cells appears even more precociously than in mice bred under conventional conditions (Fig. 3). Therefore the degranulation of STH cells after neonatal thymectomy cannot be attributed to a secondary effect of bacterial infections in immunologically deficient mice. Neonatal removal of a peripheral lymphatic organ, the spleen, did not result in any adenohypophysial alterations.

These findings suggested that the neonatal thymus behaves as a target gland for the adenohypophysis and that the product of STH cells regulates its activity and possibly its function by acting directly on some cells in the thymus or on thymus factors (Pierpaoli *et al.*, 1970).

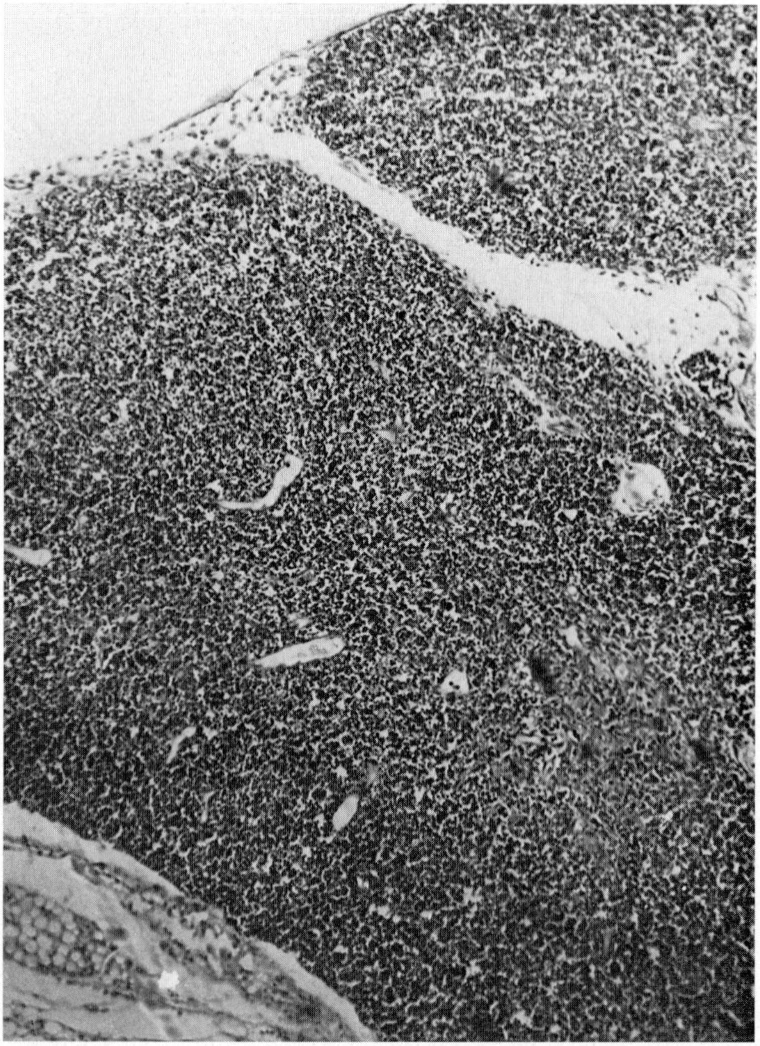

Fig. 4a. Thymus of a 2-week-old Charles River mouse, wasting after 7 days treatment with anti-STH globulins. Note involution of the organ with loss of thymocytes in the cortex.

The degranulation of STH cells in the adenohypophysis after neonatal thymectomy seems to be a proof of this relationship. The immunological and nonimmunological significance of this thymus-hypophysis relationship has been recently analysed (Pierpaoli and Sorkin, 1971).

2. Effect of anti-hypophysis serum and anti-STH serum on the thymus and peripheral lymphoid tissue in mice

A second system was used to evaluate the influence of the hypophysis on the development of the thymus. A rabbit anti-mouse hypophysis serum was produced (Pierpaoli and Sorkin, 1969*a*). Groups of mice were injected once intraperitoneally at different ages, the controls received normal rabbit serum or anti-thymus serum prepared by the same method. One single

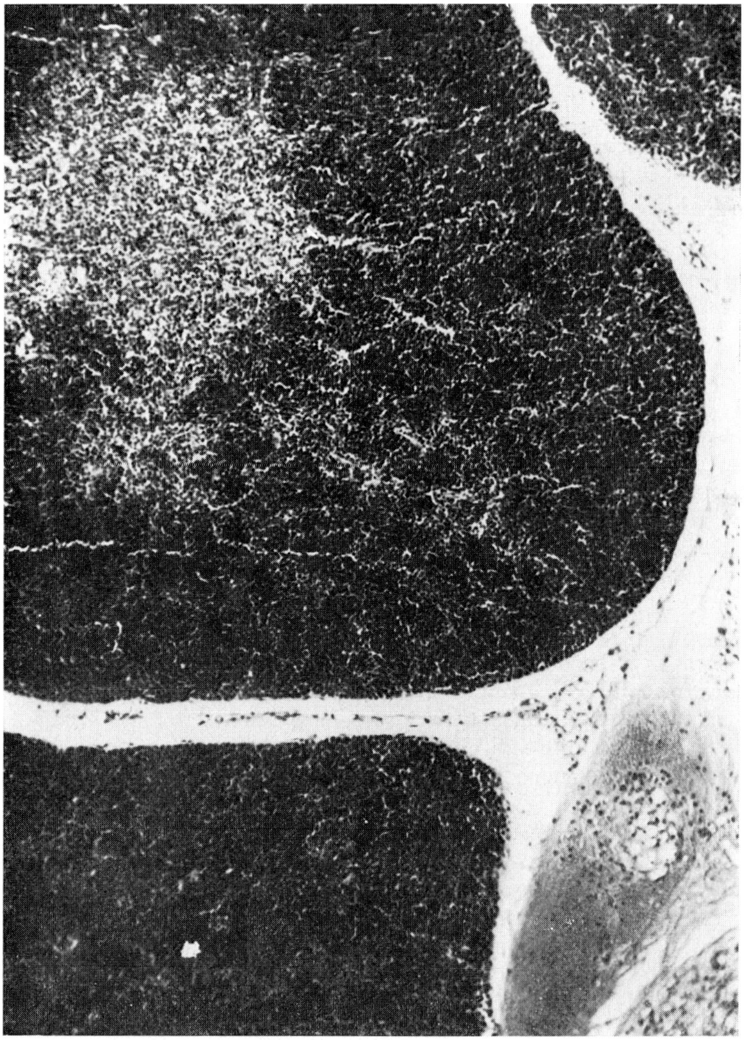

Fig. 4b. Thymus of a litter mate control, injected with the same amount of normal rabbit serum globulins, showing normal structural appearance. Haematoxylin-eosin, × 42.

inoculation of the serum produced inhibition of growth, involution of the thymus and wasting syndrome similar to that observed after neonatal thymectomy. The effect of the anti-hypophysis serum was age-dependent. Its activity in inducing wasting syndrome was maximal at the age when the same strain of mice, if neonatally thymectomized, would develop the wasting disease (Pierpaoli and Sorkin, 1967*a*). These results and the previous finding that degranulation of STH cells follows neonatal thymectomy indicated the possibility that STH might play a major role in thymus development.

Therefore an antiserum against STH was prepared in rabbits by using either the commercially available bovine Raben-type growth hormone or bovine STH supplied by Prof. A. E. Wilhelmi through the courtesy of the Endocrinological Study Section of the National Institutes of Health, Bethesda, Maryland, U.S.A. These antisera or concentrated globulin preparations containing high titers of anti-STH antibody and having no direct toxic effect on thymocytes were injected for several days into young adult mice. Similar results to those obtained by a single injection of anti-hypophysis serum were observed. Body growth was inhibited and wasting syndrome was observed (Pierpaoli and Sorkin, 1968, 1969*b*). The thymus was reduced in size due to depletion of thymocytes in the cortex (Figs. 4a and 4b) and the thymus-dependent areas in the spleen were reduced or absent. A corresponding reduction of antibody synthesis was also detected in the immunized animals during treatment with anti-STH antibody (Pierpaoli and Sorkin, 1969*b*). These data correspond to those obtained in genetically hypopituitary, STH-deficient dwarf mice whose thymus is involuted and whose immune capacity is impaired (see the following section).

3. The thymus and immunological functions in STH-deficient Snell-Bagg dwarf mice before and after reconstitution with STH.

At the same time as the immunological observations carried out on animals whose pituitary function was depressed by immunological blockage (see previous section), experiments have been performed on dwarf mice with naturally occurring hypopituitarism. Their endocrine deficiency is congenital and due to a recessive character, phenotypically present only in homozygous animals.

Endocrinological studies carried out on these dwarf mice (Snell-Bagg strain – genetic symbol dw) have shown that the adenohyphophysis is primarily affected; growth hormone-producing cells are scanty, if not absent (Bartke, 1964). The total content of growth hormone in pituitary glands of dwarf mice as measured by radioimmunoassay is 1 : 1000 of that of normal animals and plasma STH is undetectable (Garcia and Geschwind, 1968). Besides the growth hormone deficiency, dwarf mice show also a reduced synthesis of thyrotropic hormone with consequent hypofunction of the thyroid (Wegelius, 1959).

Tests performed on these animals (Pierpaoli *et al.*, 1969; Fabris *et al.*, 1971*a*) have shown that the thymus of dwarf mice is smaller than would be expected from the impaired growth of the animals. Histologically the thymus cortex, where the bone-marrow derived stem cells undergo maturation to lymphocytes, shows in dwarf mice a poor cellularity (Fig. 5a) whereas in normal littermates the cortex is richly populated (Fig. 5b). As a consequence of this thymus deficiency, dwarf mice have fewer than normal lymphocytes in the peripheral blood and a reduced cellularity in those areas of the peripheral lymphoid tissue, where T-lymphocytes are normally migrating from the thymus. The immunological functions dependent on these thymus-derived lymphocytes are depressed in dwarf mice in comparison with normal littermates. Thus the capacity to reject allogeneic skin-grafts is strongly reduced. The humoral antibody response is slightly depressed, and immunoglobulin levels and types are normal (Wilkinson *et al.*, 1970). These data support the view that dwarf mice are affected by a distinct thymus deficiency syndrome, which is due to their pituitary deficiency.

The proof that STH is the main hormone involved in this immuno-endocrinological syndrome derives from experiments in which 250 γ bovine STH (NIH-GH) was injected daily

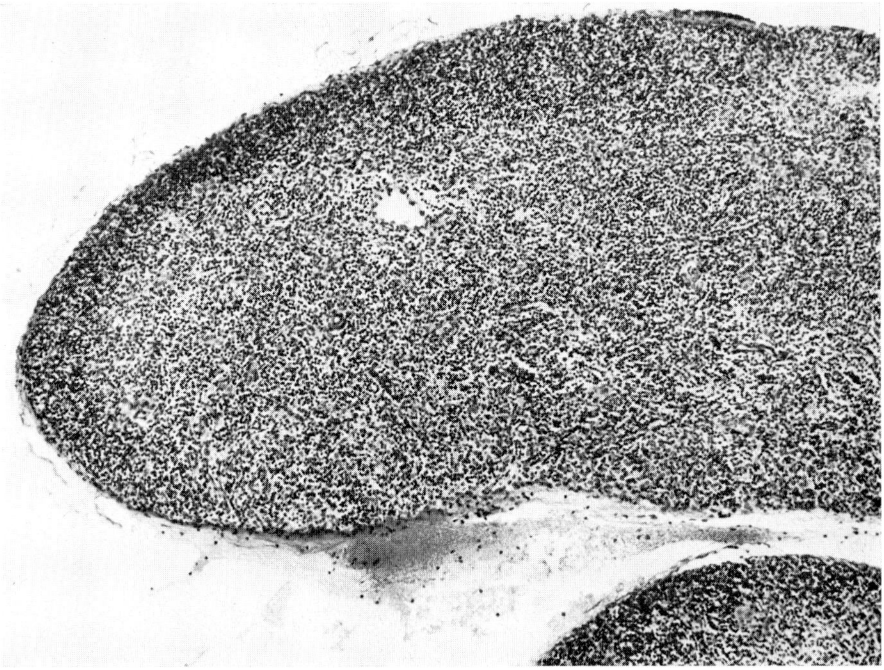

Fig. 5a. Histological section of thymus from a 60-day-old Snell-Bagg mouse. Untreated dwarf mouse (note the reduction in thickness and in cellularity of the cortex). Haematoxylin-eosin, × 42.

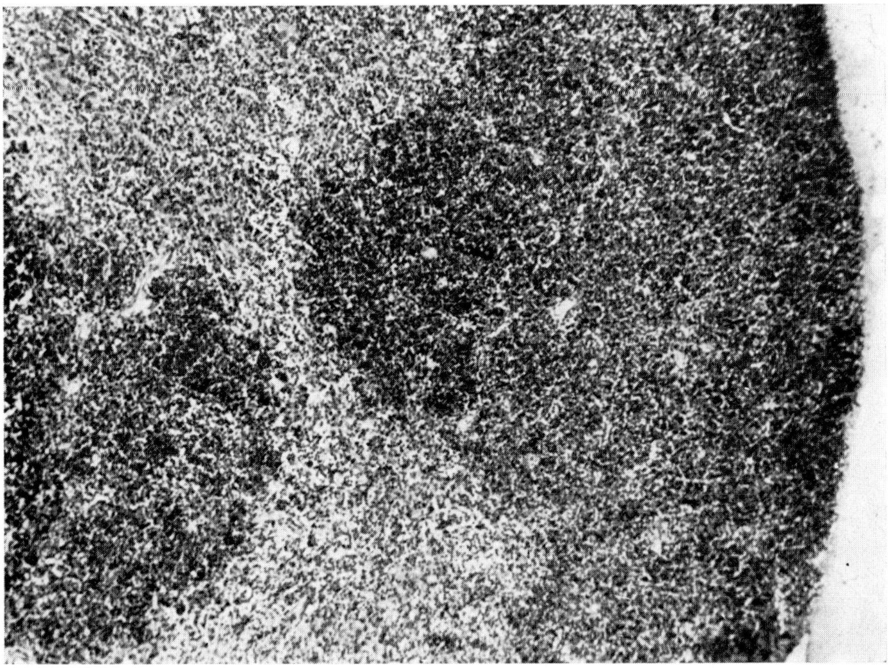

Fig. 5b. Histological section of thymus from a 60-day-old Snell-Bagg mouse. Normal animal (note the thickness of the cortex, populated by lymphoid cells). Haematoxylin-eosin, × 42.

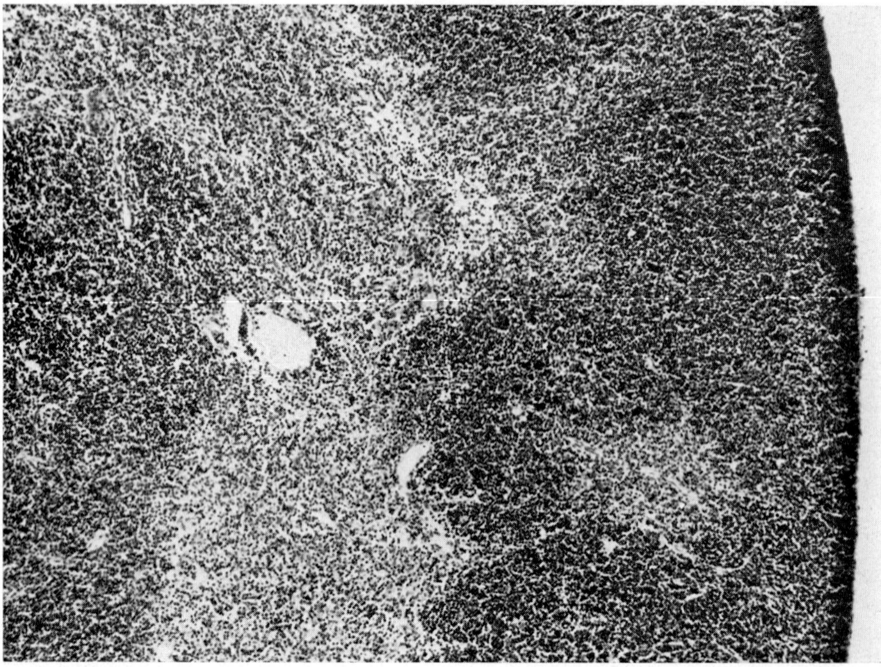

Fig. 5c. Histological section of thymus from a 60-day-old dwarf mouse treated with 250 μg bovine STH for 20 days (note the complete restoration of thymus structure). Haematoxylin eosin, × 42.

into dwarf mice. An increase in body weight together with a complete restitution of their immunological deficiencies was obtained. In particular, the thymus was restored to its normal size and the cortex of the thymus increased in thickness showing lymphoid cells as in normal mice (Fig. 5c). At the same time the thymus-dependent peripheral lymphoid tissues were repopulated by lymphocytes and the immune capacity was completely recovered (Fabris *et al.*, 1971*b*). The specific sensitivity of the thymus to STH has been clearly proven by the observation that complete reconstitution of the peripheral lymphoid tissue and its function in dwarf mice is not achieved by growth hormone therapy if the thymus has been surgically removed before the treatment starts. On the other hand the effect of growth hormone therapy on the immune system of dwarf mice can be fully substituted by the injection of peripheral lymphocytes from mature normal donors (Fabris *et al.*, 1971*b*).

From these data it can be deduced that the main action of STH in these experiments is to promote formation of thymic lymphocytes, which thereafter migrate from the thymus to populate the peripheral lymphoid tissues.

COMMENTS AND CONCLUSIONS

It has been proposed here that growth hormone has important functions for the development of the immune system. Stem cells, presumably of bone marrow origin, can enter the thymus, where they proliferate. The present work has provided evidence that this proliferation is stimulated by growth hormone. The exact mode of action of STH on the thymic cells is not known. It could act directly on the stem cells or through thymic factors (hormones ?), but this needs clarification (Fig. 1).

Three types of experiments have been used to establish our proposition.

1. A detailed light and electron microscopical examination of the anterior pituitary gland of conventional or germ-free mice at different times after neonatal thymectomy revealed a progressive increase in number of degranulated growth hormone-producing cells in the adenohypophysis. Also characteristic changes, namely, extremely distended cisternae of the endoplasmic reticulum and a reduction in number of hormone granules were noted. These modifications were not observed in sham-operated controls or in neonatally splenectomized mice. The observed cytological changes are similar to those which have been described in basophil cells of the anterior pituitary gland after removal of the thyroid or after castration (Farquhar and Rinehart, 1954*a*, *b*). One can assume therefore that some kind of feed-back system has been upset by the neonatal removal of the thymus, thereby having an effect, probably direct, on the STH-producing cells. However, a more direct experimental proof of this hypothesis, *e.g.* by reversing the hypophysial changes by thymic hormones, will be needed.

2. Injection of heterologous antihypophysis serum into young mice resulted in atrophy of thymus and of peripheral lymphoid tissues and in development of a wasting syndrome similar to that observed after neonatal thymectomy in some strains of mice. Anti-growth hormone serum produced similar effects in the cortex of the thymus and the thymus-dependent (paracortical) areas of the spleen. These results demonstrate the interdependence between the perinatal thymus and the STH-producing cells in the hypophysis.

3. In our third experimental situation, Snell-Bagg hypopituitary dwarf mice, which produce only about a thousandth of the normal amount of growth hormone were used. Hypoplasia of the thymus and peripheral lymphoid tissues, reduction of peripheral blood lymphocytes, and impairment of transplantation immunity, are typical of these STH-deficient mice and these data suggest that they show a thymus-deficiency syndrome. Treatment of these dwarf mice with bovine growth hormone completely restores the impaired immune capacity. However, this action was not exerted when the dwarf mice were thymectomized in the adult age (Fabris *et al.*, 1971*b*). It is of interest that lymphocytes, which presumably result from the action of growth hormone on the thymus, also can restore the deficiencies of the dwarf mice.

We believe that taken together these experiments provide good evidence that STH is a thymotropic hormone and that the most critical effect of its action on thymus growth and function is at the time of ontogeny of the immune system in mammals. In view of our data it is proposed that the thymus should be considered as an endocrine gland, albeit one with unusual features.

ACKNOWLEDGEMENT

This work was supported by the Schweizerischer Nationalfonds zur Förderung der wissenschaftlichen Forschung, Grant No. 3.246.69 SR and the Emil Barell-Stiftung, Basel.

REFERENCES

Bartke, A. (1964): Histology of the anterior hypophysis, thyroid and gonads of two types of dwarf mice. *Anat. Rec.*, *149*, 225.

Bianchi, E., Pierpaoli, W. and Sorkin, E. (1971): Cytological changes in the mouse anterior pituitary after neonatal thymectomy: a light and electron microscopical study. *J. Endocr.*, *51*, 1.

Burnet, M. F. (1969): *Cellular Immunology*. Melbourne and Cambridge University Press, Cambridge.

Fabris, N., Pierpaoli, W. and Sorkin, E. (1971*a*): Hormones and the immunological capacity. III. The immunodeficiency disease of the hypopituitary Snell-Bagg dwarf mouse. *Clin. exp. Immunol.*, *9*, 209.

Fabris, N., Pierpaoli, W. and Sorkin, E. (1971*b*): Hormones and the immunological capacity. IV. Restorative effects of developmental hormones or of lymphocytes on the immunodeficiency syndrome of the dwarf mouse. *Clin. exp. Immunol.*, *9*, 227.

FARQUHAR, M. G. and RINEHART, J. F. (1954*a*): Electron microscopic studies on the anterior pituitary gland of castrate rats. *Endocrinology, 54*, 516.

FARQUHAR, M. G. and RINEHART, J. F. (1954*b*): Cytologic alterations in the anterior pituitary gland following thyroidectomy: an electron microscope study. *Endocrinology, 55*, 857.

GARCIA, J. F. and GESCHWIND, I. I. (1968): Investigation of growth hormone secretion in selected mammalian species. In: *Growth Hormone*, pp. 267–291, Editors: A. Pecile and E. E. Müller. ICS 158, Excerpta Medica, Amsterdam.

PIERPAOLI, W., BARONI, C., FABRIS, N. and SORKIN, E. (1969): Hormones and immunological capacity. II. Reconstitution of antibody production in hormonally deficient mice by somatotropic hormone, thyrotropic hormone and thyroxin, *Immunology, 16*, 217.

PIERPAOLI, W., FABRIS, N. and SORKIN, E. (1970): Developmental hormones and immunological maturation. In: *Hormones and the Immune Response.* Ciba Foundation Study Group No. 36, pp. 126–143. Editors: G. E. M. Wolstenholme and Julie Knight. Churchill, London.

PIERPAOLI, W. and SORKIN, E. (1967*a*): Relationship between thymus and hypophsyis. *Nature (Lond.) 215*, 834.

PIERPAOLI, W. and SORKIN, E. (1967*b*): Cellular modification in the hypophysis of neonatally thymectomized mice. *Brit. J. exp. Path., 48*, 627.

PIERPAOLI, W. and SORKIN, E. (1968): Hormones and immunologic capacity. I. Effect of heterologous anti-growth hormone (ASTH) antiserum on thymus and peripheral lymphatic tissue in mice. Induction of a wasting syndrome. *Immunol., 101*, 1036.

PIERPAOLI, W. and SORKIN, E. (1969*a*): A study on anti-pituitary serum. *Immunology, 16*, 311.

PIERPAOLI, W. and SORKIN, E. (1969*b*): Effect of growth hormone serum on the lymphatic tissue and the immune response. *Antibiot. et Chemother. (Basel), 15*, 122.

PIERPAOLI, W. and SORKIN, E. (1972): Physiological significance of the thymus-hypophysis axis for immunological and other lymphocyte function. Submitted for publication.

WEGELIUS, O. (1959): The dwarf mouse, an animal with secondary myxedema. *Proc. Soc. exp. Biol. (N.Y.), 101*, 225.

WILKINSON, P. C., SINGH, H. and SORKIN, E. (1970): Serum immunoglobulin levels in thymus deficient pituitary dwarf mice. *Immunology, 18*, 437.

THE EFFECT OF GROWTH HORMONE ON THE SYNTHESIS AND ACCUMULATION OF POLYAMINES IN MAMMALIAN TISSUES*

A. RAINA and E. HÖLTTÄ

Department of Medical Chemistry, University of Helsinki, Helsinki, Finland

Analyses of a number of systems have demonstrated that rapidly growing tissues, and more specifically, tissues engaged in active RNA and protein synthesis contain high concentrations of polyamines (see Raina and Jänne, 1970). Our interest in the possible significance of polyamines in growth processes was aroused by observations of the high polyamine content in the developing chick embryo (Raina, 1963). Furthermore, analysis of rat tissues at different ages revealed that the concentration of polyamines, expecially that of spermidine, was high in the newborn rats and decreased with age (Jänne *et al.*, 1964). A rapid accumulation of spermidine was also observed in the regenerating rat liver soon after partial hepatectomy (Raina *et al.*, 1966). As a further example, some of the effects of growth hormone on the synthesis and accumulation of polyamines in rat liver will be described here.

It is appropriate to begin by briefly summarizing current knowledge of polyamine biosynthesis in animal tissues. This is schematically presented in Fig. 1. Putrescine synthesis is catalysed by ornithine decarboxylase, a pyridoxal phosphate-requiring enzyme, found in the

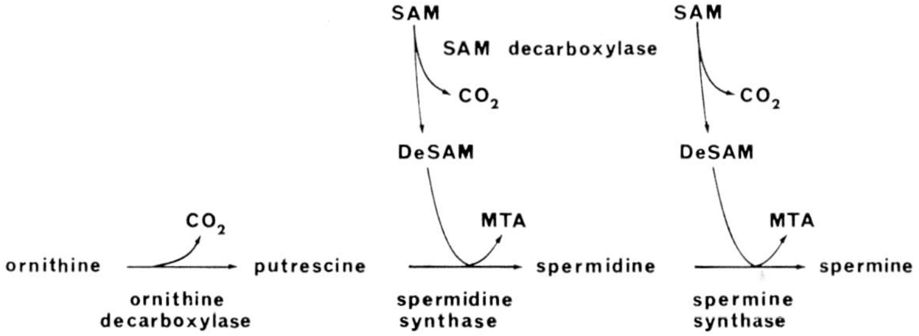

Fig. 1. The biosynthetic pathway of polyamines in animal tissues. SAM: S-adenosyl-L-methionine; DeSAM: decarboxylated S-adenosyl-L-methionine; MTA: methylthioadenosine.

soluble fraction of liver homogenate (Jänne *et al.*, 1968) and having an extremely rapid turnover rate (Jänne and Raina, 1969; Russell and Snyder, 1969). Putrescine then serves as a precursor of spermidine. The propylamine moieties of spermidine and spermine are derived from S-adenosylmethionine. Contrary to earlier results (Pegg and Williams-Ashman, 1970) obtained with partially purified enzyme preparations from rat prostate, it now appears that

* This study was supported by grants from the Sigrid Jusélius Foundation and the National Research Council for Medical Sciences, Finland.

separate enzymes catalyse the synthesis of spermidine and spermine. We recently succeeded in separating three enzyme activities from rat brain (Raina and Hannonen, 1971): one catalysing the decarboxylation of S-adenosylmethionine, and two propylamine transferases, one specific for putrescine (spermidine synthase) and the other specific for spermidine (spermine synthase). A complete separation of S-adenosylmethionine decarboxylase and spermidine synthase was also reported recently by Jänne *et al.* (1971).

The first observation indicating that growth hormone has an effect on the accumulation of polyamines was reported by Kostyo in 1966. The low level of spermidine found in the livers of hypophysectomized rats rapidly returned to normal following growth hormone treatment. Shortly thereafter it was demonstrated by Jänne (1967) in our laboratory that the administration of growth hormone to normal rats stimulated *in vivo* biosynthesis of putrescine and spermidine by the liver. The stimulation of putrescine synthesis appeared to be due to a marked increase in hepatic ornithine decarboxylase activity, which leads to a rapid accumulation of putrescine (Jänne *et al.*, 1968). Our results further suggested that the increase in spermidine synthesis observed after growth hormone treatment resulted from a primary increase in liver putrescine. The following results concern the effect of growth hormone on polyamine synthesis, and on ornithine decarboxylase in particular. The rapid changes in ornithine decarboxylase activity produced by growth hormone treatment may provide a useful tool for the study of mechanisms regulating polyamine synthesis on the one hand, and the mode of growth hormone action on the other.

EXPERIMENTAL

Female rats of the Wistar strain, weighing 85–110 g, were used. Unless otherwise indicated they were fasted for 12–14 hours before sacrifice. A commercial preparation of porcine growth hormone (Somacton®, Ferring, Malmö, Sweden) was used throughout the study. This preparation had a relatively high degree of purity, containing approximately 1 I.U. per 0.5 mg of dry substance (personal communication by Ferring). Growth hormone and α-amanitin were administrated intraperitoneally in 0.5 ml of 0.9 % sodium chloride. Ornithine decarboxylase activity was assayed in the high speed supernatant fraction of liver homogenate as described earlier (Jänne *et al.*, 1968) except that mercaptoethanol in the assay mixture was replaced by 5 mM dithiotreitol (Jänne and Williams-Ashman, 1970). The *in vitro* synthesis of spermidine and spermine was assayed using S-adenosyl-L-methionine as the labelled precursor and the dialysed supernatant fraction as the source of the enzyme (Raina *et al.*, 1970). RNA polymerase activity of isolated nuclei was assayed in a medium of low ionic strength in the presence of Mg^{2+} (Jänne and Raina, 1969).

RESULTS

The basal level of liver ornithine decarboxylase activity of fasted animals is quite low as is shown in Table I. Although administration of 0.1–0.5 I.U. of growth hormone per 100 g body weight significantly increased this enzyme activity, a dose of 2 I.U. or even higher was used in most experiments to detect changes in the activity of other enzymes less sensitive to growth hormone, *i.e.* tyrosine aminotransferase, tryptophan pyrrolase and nuclear RNA polymerase. As demonstrated earlier (Jänne and Raina, 1969; Russell *et al.*, 1970), the rise in liver ornithine decarboxylase activity after a single dose of growth hormone was relatively rapid. It was markedly elevated only 2 hours after the treatment, reached its maximum at 4 hours and then rapidly declined. The sharp decline suggests a rapid turnover rate of liver ornithine decarboxylase, as demonstrated in other systems (Russell and Snyder, 1969). We also noticed that the stimulation of ornithine decarboxylase was accompanied by a parallel increase in nuclear RNA polymerase activity (Jänne and Raina, 1969). In contrast to the marked changes seen in ornithine decarboxylase activity, growth hormone treatment only slightly increased spermi-

TABLE I

Response of liver ornithine decarboxylase activity to different doses of growth hormone

Dose (I.U.)	Ornithine decarboxylase (pmoles $^{14}CO_2$/mg protein/30 min)
Saline	10 ± 5
0.1	81 ± 50
0.5	402 ± 117
2.0	1089 ± 257

Female Wistar rats (95–107 g) were treated intraperitoneally with growth hormone in 0.5 ml of 0.9% NaCl 4 hours before sacrifice. The rats were fasted for 14 hours. The values are means ± standard deviation of 5 animals.

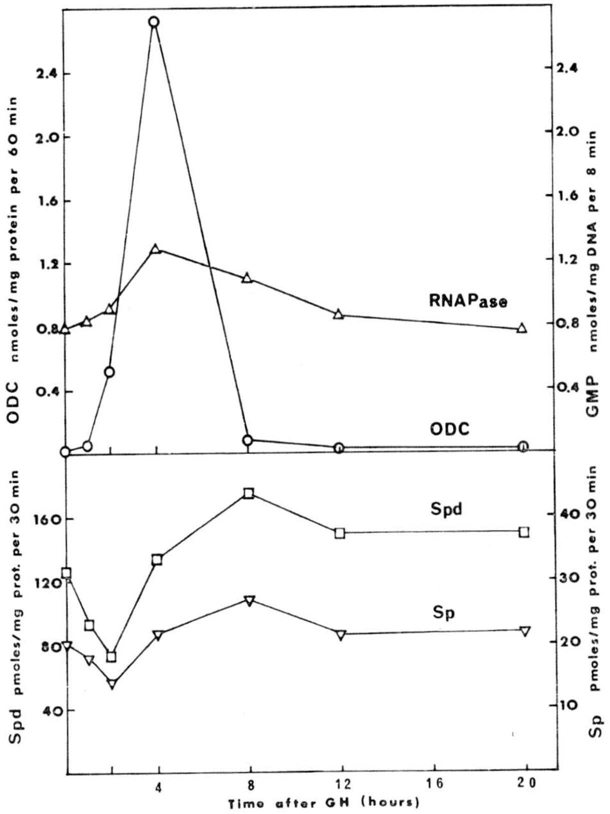

Fig. 2. Effect of a single dose of growth hormone on polyamine synthesizing enzyme activities and nuclear RNA polymerase in rat liver. Analyses were performed at the times indicated after a single dose of growth hormone (8 I.U./ 100 g body weight). Values are means of 3 animals in each group. ODC: ornithine decarboxylase; RNAPase: nuclear RNA polymerase; Spd: spermidine; Sp: spermine.

dine and spermine synthesizing enzyme activities (Fig. 2). Because the assay system used reflects mainly the activity of S-adenosyl-L-methionine decarboxylase, the possibility of more marked changes in the activities of propylamine transferases was not excluded. It would be of interest to determine whether the enzyme activities responsible for the synthesis of spermidine and spermine are co-ordinately regulated. The relatively small changes in the spermidine and spermine synthesizing enzyme activities observed after administering growth hormone or after partial hepatectomy (Raina *et al.*, 1970) seem to indicate that ornithine decarboxylase plays a crucial role in the regulation of the rate of polyamine synthesis.

The following experiments may shed some light on the mode of action of growth hormone in this system. It has been reported that growth hormone increases the level of liver tyrosine aminotransferase and tryptophan pyrrolase in intact but not in adrenalectomized rats (Kenney *et al.*, 1965). Our results (Jänne and Raina, 1969) showed that treatment with growth hormone stimulated ornithine decarboxylase and nuclear RNA polymerase in the adrenalectomized animals also, although less than in the intact animals. In agreement with the data of Kenney *et al.*, the activities of tyrosine aminotransferase and tryptophan pyrrolase were decreased in adrenalectomized rats after growth hormone treatment. It appears therefore that the stimulation of ornithine decarboxylase is independent of the presence of adrenals.

To exclude the possibility that the growth hormone effect might be mediated through the action of some other hormone(s), we studied the effect of growth hormone in isolated perfused liver (Jänne *et al.*, 1969). When growth hormone was added to the perfusion fluid, the ornithine decarboxylase activity was twice the control level after two hours perfusion. This effect was sensitive to cycloheximide and independent of the addition of amino acids to the perfusion medium. Because of the very high turnover rate of the enzyme these data must be interpreted with caution.

One can now ask whether the increase in ornithine decarboxylase activity induced by growth hormone is due to *de novo* synthesis of the enzyme protein or merely due to stabilization or activation of pre-existing enzyme molecules. A further question is whether hormone acts by activating the translation of pre-existing RNA templates or by stimulating transcription. At present, none of these possibilities can definitely be excluded, although our results favour *de novo* synthesis of the enzyme. As shown in Table II, the administration of 2 I.U. of growth hormone at zero and 2 hours prevents the decrease of ornithine decarboxylase activity, normally seen at 6 hours after a single dose of the hormone. Table II further shows that ornithine decarboxylase can be reinduced by a second injection of growth hormone given at 6 hours, although the response was smaller than that observed after the first injection.

TABLE II

Effect of multiple treatment with growth hormone on liver ornithine decarboxylase activity

Time of treatment (hr)	Time of analysis (hr)	Ornithine decarboxylase (pmoles $^{14}CO_2$/mg protein/30 min)
0 (saline)	4	10 ± 5
0	4	1089 ± 257
0	6	38 ± 19
0 and 2	6	1437 ± 538
0 and 6	10	618 ± 128

Female rats received 2 I.U. of growth hormone per 100 g body weight at the times indicated. 5 to 7 rats in each group. Other conditions as in Table I.

In this connection the recent report of Panko and Kenney (1971) deserves some consideration. They reported that in contrast to the results obtained with tyrosine aminotransferase, repeated injections of either hydrocortisone or growth hormone were unable to maintain the peak level of liver ornithine decarboxylase in adrenalectomized animals. These authors therefore concluded that the increase in ornithine decarboxylase activity after treatment with hydrocortisone or growth hormone is not due to a direct action of these hormones, as seen with tyrosine aminotransferase. However, when comparing the overall changes in these two enzyme activities, the differences in the turnover rates of the enzymes must be taken into account. From the data of Panko and Kenney it appears likely that during continuous treatment with hydrocortisone, the rate of synthesis of both enzymes remains elevated, even if not maximal. In regard to the growth hormone effect, it is difficult to compare the results obtained in the present study directly with those reported by Panko and Kenney, because of differences in the age and hormonal status of the animals as well as in the dose of growth hormone used.

TABLE III

Effect of growth hormone on liver ornithine decarboxylase and RNA polymerase activity

Group	Ornithine decarboxylase (pmoles $^{14}CO_2$/mg protein/60 min)	RNA polymerase (pmoles GMP/mg DNA/8 min)
Fasted controls	99	250
Fasted plus GH	2500	727
Fed controls	72	390
Fed plus GH	3010	736
Fed plus cycloheximide	56	100
Fed plus cycloheximide plus GH	72	88

Cycloheximide (5 mg/100 g body weight) was given intraperitoneally 4.5 hours and growth hormone (8 I.U./100 g) 4 hours before sacrifice. The fasted groups had been kept without food for 20 hours. (Data from J. Jänne and A. Raina, 1969, *Biochim. biophys. Acta*, *174*, 769).

Table III shows that growth hormone increases ornithine decarboxylase and nuclear RNA polymerase both in fed and fasted rats. This effect was completely prevented by cycloheximide. Actinomycin D only partially abolished the growth hormone effect when given simultaneously with the hormone (Jänne *et al.*, 1968), but totally blocked the increase if given 30 min before the administration of the hormone. α-Amanitin is probably a more specific inhibitor of nucleoplasmic RNA polymerase (see Fiume and Wieland, 1970). In a dose of 100 μg/100 g body

TABLE IV

Inhibition of growth hormone-induced increase in liver ornithine decarboxylase activity by α-amanitin

Treatment	Ornithine decarboxylase (pmoles $^{14}CO_2$/mg protein/30 min)
GH	1089 ± 257
GH plus α-amanitin	48 ± 46
α-amanitin 60 min before GH	60 ± 43

Female rats received 2 I.U. of growth hormone (GH) per 100 g body weight 4 hours and α-amanitin (100 μg/100 g) 4 or 5 hours before sacrifice. 5 rats in each group. Other conditions as in Table I.

weight this drug completely inhibited the growth hormone-induced increase in ornithine decarboxylase activity (Table IV) whether given simultaneously with the hormone or 60 min earlier. These results seem to suggest that the enhanced enzyme activity may be due to increased synthesis of the enzyme protein, and may depend on RNA synthesis. However, additional mechanisms on other levels of control are by no means excluded. Development of a radioimmunological assay method for ornithine decarboxylase would be helpful in resolving part of the problem.

In conclusion, growth hormone treatment markedly stimulates liver ornithine decarboxylase activity. Furthermore, the increase in ornithine decarboxylase activity seems to be the main factor in the stimulation of spermidine synthesis. Although it appears that high ornithine decarboxylase activity is a characteristic of a rapidly growing tissue, it is by no means specific for growing tissue. A number of stimuli which are known to increase the rate of protein and RNA synthesis, also increase the level of ornithine decarboxylase (see Raina and Jänne, 1970). These include several other hormones, such as thyroid hormone, insulin, glucagon, androgens and estrogens. Compounds such as thioacetamide and folic acid are also known to modify the level of this enzyme activity (Raina and Jänne, 1970). It is tempting to speculate that all these agents might have a common mediator, *e.g.* cyclic AMP. In fact, our recent results show that treatment of the rat with dibutyryl cyclic AMP can markedly increase liver ornithine decarboxylase activity, an effect sensitive to α-amanitin (Hölttä and Raina, 1971). Future work will show, we hope, whether these effects have any physiological significance. As discussed in more detail elsewhere (Raina and Jänne, 1970), polyamines themselves may be important in the regulation of cell metabolism, especially in the synthesis and function of ribonucleic acids.

SUMMARY

Administration of a single dose of growth hormone to intact rats results in a rapid and transient increase in liver ornithine decarboxylase activity. This effect was also seen in the livers of adrenalectomized rats and in isolated perfused livers. After repeated treatments of intact rats with growth hormone the elevated level of the enzyme persisted for at least 6 hours. The growth hormone-induced increase in ornithine decarboxylase activity could be abolished by inhibitors of RNA and protein synthesis. In several systems it was also noticed that the activities of liver ornithine decarboxylase and nuclear RNA polymerase show parallel changes. Some aspects of the regulation of polyamine synthesis are discussed.

ACKNOWLEDGEMENT

We wish to thank Professor Th. Wieland for the generous gift of α-amanitin.

REFERENCES

Fiume, L. and Wieland, Th. (1970): Amanitins. Chemistry and action. *FEBS Letters*, *8*, 1.

Hölttä, E. and Raina, A. (1971): The effect of feeding, glucagon and cyclic AMP on liver ornithine decarboxylase activity. *Scand. J. clin. Lab. Invest.*, *27*, *Suppl. 116*, 54.

Jänne, J. (1967); Studies on the biosynthetic pathway of polyamines in rat liver. *Acta physiol. scand.*, *Suppl. 300*, 1.

Jänne, J., Kekomäki, M., Raina, A. and Hölttä, E. (1969): *Abstracts, 6th FEBS Meeting, Madrid*, p. 195.

Jänne, J. and Raina, A. (1969): On the stimulation of ornithine decarboxylase and RNA polymerase activity in rat liver after treatment with growth hormone. *Biochim. biophys. Acta (Amst.)*, *174*, 769.

Jänne, J., Raina, A. and Siimes, M. (1964): Spermidine and spermine in rat tissues at different ages. *Acta physiol. scand.*, *62*, 352.

JÄNNE, J., RAINA, A. and SIIMES, M. (1968): Mechanism of stimulation of polyamine synthesis by growth hormone in rat liver. *Biochim. biophys. Acta (Amst.)*, *166*, 419.

JÄNNE, J., SCHENONE, A. and WILLIAMS-ASHMAN, H. G. (1971): Separation of two proteins required for synthesis of spermidine from S-adenosyl-L-methionine and putrescine in rat prostate. *Biochem. biophys. Res. Commun.*, *42*, 758.

JÄNNE, J. and WILLIAMS-ASHMAN, H. G. (1970): Mammalian ornithine decarboxylase: Activation and alteration of physical behaviour by thiol compounds. *Biochem. J.*, *119*, 595.

KENNEY, F. T., WICKS, W. D. and GREENMAN, D. L. (1965): Hydrocortisone stimulation of RNA synthesis in induction of hepatic enzymes. *J. cell. comp. Physiol.*, *66*, 125.

KOSTYO, J. L. (1966); Changes in polyamine content of rat liver following hypophysectomy and treatment with growth hormone. *Biochem. biophys. Res. Commun.*, *23*, 150.

PANKO, W. B. and KENNEY, F. T. (1971): Hormonal stimulation of hepatic ornithine decarboxylase. *Biochem. biophys. Res. Commun.*, *43*, 346.

PEGG, A. E. and WILLIAMS-ASHMAN, H. G. (1970): Enzymic synthesis of spermine in rat prostate. *Arch. Biochem. Biophys.*, *137*, 156.

RAINA, A. (1963): Studies on the determination of spermidine and spermine and their metabolism in the developing chick embryo. *Acta physiol. scand.*, *60*, *Suppl. 218*, 1.

RAINA, A. and HANNONEN, P. (1971): Separation of enzyme activities catalysing spermidine and spermine synthesis in rat brain. *FEBS Letters*, *16*, 1.

RAINA, A. and JÄNNE, J. (1970): Polyamines and the accumulation of RNA in mammalian systems. *Fed. Proc.*, *29/4*, 1568.

RAINA, A., JÄNNE, J., HANNONEN, P. and HÖLTTÄ, E. (1970): Synthesis and accumulation of polyamines in regenerating rat liver. *Ann. N. Y. Acad. Sci.*, *171*, 697.

RAINA, A., JÄNNE, J. and SIIMES, M. (1966): Stimulation of polyamine synthesis in relation to nucleic acids in regenerating rat liver. *Biochim. biophys. Acta (Amst.)*, *123*, 197.

RUSSELL, D. H. and SNYDER, S. H. (1969): Amine synthesis in regenerating rat liver: Extremely rapid turnover of ornithine decarboxylase. *Mol. Pharmacol. 5*, 253.

RUSSELL, D. H., SNYDER, S. H. and MEDINA, V. J. (1970): Growth hormone induction of ornithine decarboxylase in rat liver. *Endocrinology*, *86*, 1414.

THE INTERRELATIONSHIP OF THE IN VITRO ACTIONS OF GROWTH HORMONE TO THOSE IN VIVO AND TO EFFECTS OF INSULIN

K. L. MANCHESTER

Departments of Biochemistry, University College London, London, England, and University of the West Indies, Kingston, Jamaica*

The interrelationship of the *in vivo* and *in vitro* actions of growth hormone still constitutes a problem – a problem because the *in vitro* actions of the hormone are not necessarily those to be expected from its apparent *in vivo* functions. Several forms of explanation of this discrepancy have been variously put forward – the most obvious is that growth hormone as normally available is not pure and that some of the *in vitro* effects result from the presence of contaminants, alternatively that the insulin-like effects of growth hormone *in vitro* on muscle from hypophysectomised animals arise because of an ability of the hormone to displace insulin bound to the tissue in an inert form, and thirdly that the different actions of the hormone, *e.g.* the ability to promote growth and to mobilise fat, are indeed all properties of the same molecule but that different parts of the comparatively large structure are responsible for the different effects. The following results bear on these points.

Possible impurities in growth hormone preparations and the association of growth and fat mobilising activities

Growth hormone mobilises fat both *in vivo* and *in vitro*. With adipose tissue *in vitro* however fat mobilisation is produced by a large number of peptide and other materials which of course do not promote growth.

By chromatography on DEAE-cellulose in borate buffer at pH 8.7 it is possible to fractionate Wilhelmi preparations of growth hormone into at least three fractions and a gift of these fractions from Dr Wallis to Dr Salaman and myself enabled us to make a comparison of both the *in vivo* and *in vitro* fat mobilising activity of the fractions which can be compared with their measured growth promoting capacities. We found that significant rise in plasma free fatty acids of fasting rats was observable 6 hours after injection of as little as 45–75 μg of the fractions A and B, which possess growth promoting activities equivalent to the starting material, whereas fraction C (which constituted 15–30% of the recovered material and had little growth promoting activity) only produced significant fat mobilisation *in vivo* in amounts greater than 150 μg (Table I). Similar results were found in separate experiments with a further preparation of growth hormone made by direct chromatography of an aqueous extract of anterior pituitary lobes on DEAE-cellulose and Sephadex (Wallis and Dixon, 1966). This material, which showed high growth promoting potency, also possessed fat mobilising activity when administered in quantities as low as 45 μg per animal. Thus fat mobilising activity *in vivo* correlates with growth promoting capacity (and in parentheses the quantities of growth hormone required here to show significant fat mobilisation are much lower than those frequently quoted).

* Present address.

TABLE I

Fat mobilising activity of fractions of bovine growth hormone in vivo

Assay No.	Dose (μg)	Mean plasma FFA (μmoles/ml)	P value	
			test vs control	test vs test
1	B: 250	0.93 ± 0.05 (6)	$P < 0.001$	$P > 0.05$
	C: 250	0.77 ± 0.07 (6)	$P < 0.05$	
	Control	0.57 ± 0.04 (6)		
2	B: 150	0.94 ± 0.06 (6)	$P < 0.01$	$P < 0.01$
	C: 150	0.61 ± 0.03 (6)	$P > 0.05$	
	Control	0.61 ± 0.07 (6)		
3	A: 75	0.76 ± 0.02 (12)	$P < 0.001$	$P > 0.05$
	B: 75	0.76 ± 0.05 (12)	$P < 0.01$	
	Control	0.57 ± 0.02 (12)		
	A: 45	0.54 ± 0.04 (12)	$P < 0.05$	$P > 0.05$
	B: 45	0.56 ± 0.03 (12)	$P > 0.05$	
	Control	0.50 ± 0.04 (12)		

Female rats ca. 160–185 g were starved overnight and placed in three groups of six. Two groups were injected with the stated quantities of the different fractions dissolved in saline and the control group with saline alone. After 6 hr the rats were bled and samples of plasma taken for determination of FFA (Salaman and Robinson, 1961). Each figure is the mean ± S.E. of the mean of the number of observations in parentheses. The values quoted in assays 3 and 4, where there was a group size of 12 rats, were obtained by combining the data from two replicate assays.

The method of preparation of the three fractions involved application of a sample of Wilhelmi material to a DEAE-cellulose column in borate buffer at pH 8.7. Fraction A is eluted from the column as a retarded peak in the starting buffer, fraction B on application of a NaCl gradient to 0.1 M, fraction C on raising the molarity of the NaCl gradient to 0.5 (Manchester and Wallis, 1963). Any prolactin activity is associated with fraction C (Wallis and Kovacic, 1965).

TABLE II

Fat mobilising activity of fractions of bovine growth hormone in vitro

	Mobilising activity of fractions (μmoles of FFA/fat body/hr) at concentrations in the incubation medium of		
	5.0 μg/ml	1.0 μg/ml	0.5 μg/ml
Fraction A	0.49 ± 0.05 (10)	0.15 ± 0.05 (10)	not tested
Fraction B	1.02 ± 0.08 (8)	1.15 ± 0.10 (8)	0.60 ± 0.10 (6)
Fraction C	1.04 ± 0.10 (8)	0.42 ± 0.08 (8)	0.26 ± 0.06 (6)

Each figure is the mean ± S.E. of the mean and derives from a series of replicate tests with the paired fat bodies of rats starved overnight. In each test one fat body was incubated at 37° C for 3.5 hr in 2.5 ml of Krebs-Ringer bicarbonate buffer containing 5% (w/v) bovine serum albumin, and the paired fat body was incubated under the same conditions in the medium containing a given concentration of fraction. After incubation, samples of the media were taken for determination of FFA. Activity is expressed as the mean of the differences between paired fat bodies in the rate of release of FFA into the incubation medium. The mean rate of FFA release in control systems was approximately 1.1 μmoles/fat body/hr. This figure is corrected for FFA added to the medium in the albumin component, an amount equivalent to a release of 0.28 μmoles/fat body/hr.

When tested *in vitro* a different pattern emerged. Whereas fraction B still retained potent activity, that of C was at least as great as A and significant increase in release of FFA was achieved with as little as 0.5 μg/ml in the absence of added glucocorticoid (Table II). Since puromycin can block mobilisation of FFA *in vitro* inducible by growth hormone (Fain *et al.*, 1965), it was of interest to find that, although the drug blocked the fat mobilising action of fraction A, it did not block that of B – a somewhat unexpected result.

Thus ability to mobilise fatty acid *in vitro* does not necessarily correlate with mobilising capacity *in vivo*.

The nature of the insulin-like actions of growth hormone

The ability of growth hormone to enhance lipolysis by fat tissue is in obvious contrast to the capacity of insulin to suppress this process. With muscle from hypophysectomised animals growth hormone *in vitro* mimics a number of actions of insulin such as in promoting uptake of glucose and of certain amino acids and in stimulating incorporation of amino acids into protein. Ottaway (1953) suggested that growth hormone might be displacing inertly bound insulin already present in the tissue which became active when released.

Many of the effects of insulin have been characterised with rat diaphragm muscle. The ability of this tissue to respond to insulin decreases following denervation (Buse and Buse, 1959, 1961). In the first week following nerve section the tissue undergoes a transient hypertrophy during which it becomes unresponsive to insulin (though adrenaline still produces glycogenolysis). A possible way in which an insulin-like effect of growth hormone would manifest itself as not being due to release of bound insulin would be if growth hormone were still effective with the hypertrophied denervated muscle. The data in Table III however show that sensitivity of the muscle to growth hormone, like that to insulin, is lost during the hypertrophy following nerve section. It could be argued that, since during hypertrophy the basal rate of uptake of aminoisobutyrate and incorporation of amino acids into protein is enhanced

TABLE III

Effect of denervation and addition of growth hormone on the uptake of glucose, accumulation of aminoisobutyrate and incorporation of glycine into protein by isolated rat diaphragm

	Uptake of glucose (mg/g wet wt)	Accumulation of aminoisobutyrate (Ratio $\frac{^{14}C/g\ tissue}{^{14}C/ml\ medium}$)	Incorporation of glycine into protein (c.p.m./mg protein)
Innervated tissue	2.32 ± 0.06 (8)	0.97 ± 0.01 (8)	230 (4)
+ growth hormone	3.41 ± 0.07 (8)	1.05 ± 0.03 (8)	328 (4)
Denervated tissue	3.24 ± 0.10 (8)	1.90 ± 0.09 (8)	361 (4)
+ growth hormone	3.29 ± 0.08 (8)	1.92 ± 0.09 (8)	383 (4)

Each figure is the mean ± S.E. of the mean of the number of observations shown in parentheses. Specific-pathogen-free rats of the Sprague-Dawley strain, weight ca. 160 g were hypophysectomised and after not less than 14 days were subject to left unilateral phrenectomy. Three or four days later the diaphragm muscles, both denervated and innervated portions, were removed and incubated with shaking for 1.5 hr in Krebs-Ringer bicarbonate containing glucose (1 mg/ml) and [^{14}C]aminoisobutyrate or [^{14}C]glycine. At the end of the period the glucose consumption of the tissue, the accumulation of aminoisobutyrate and the incorporation of glycine into protein were measured as previously described (Manchester, 1966). Growth hormone (NIH-GH-B12) when present had a concentration of 100 μg/ml.

(Buse *et al.*, 1965; Harris and Manchester, 1966), a 'ceiling' has been reached which allows no further room for increase, but this is demonstrably not so in the case of glucose uptake and it must be concluded that hormonal sensitivity under these conditions is in some way lost. The literature in fact contains several other examples of this sort of change. Thus heart muscle in failure ceases to respond to the inotropic effect of glucagon (Gold *et al.*, 1970) because of an inability of the hormone to activitate adenyl cyclase – yet response to adrenaline persists. The regenerating liver loses sensitivity to the protein catabolic influence of glucagon (Miller *et al.*, 1970), though in this case the glycogenolytic effects of the hormone persist. There is thus seemingly a change in hormone sensitivity during the 'stress' of induced growth. Mammary gland tissue also shows varying sensitivity towards insulin (Friedberg *et al.*, 1970), but in this case it is highest during the hyperplasia of pregnancy.

The loss of sensitivity of the hypertrophying diaphragm to growth hormone, as to insulin, does not therefore help to differentiate between the mechanism of action of the two hormones under these conditions. Rather it serves to illustrate how the effects of either hormone on the various aspects of metabolism are likely to arise from a common point of action. The loss of sensitivity to growth hormone is in a sense the more striking in view of the apparent diversity of growth hormone fragments which retain the essential activities of the hormone (Nutting *et al.*, 1970; Swislocki *et al.*, 1970). The apparent lack of extreme specificity would indeed be consistent with a rather unspecific ability to dislodge other proteins (including insulin) bound to the tissue. It is however possible to distinguish between insulin-like effects of growth hormone as being due to growth hormone *per se* as opposed to insulin in at least two ways. First, the *in vitro* actions are not neutralised by antiserum to insulin (Manchester and Young, 1959), despite the ability of such antiserum to neutralise the effects of insulin when added to muscle prior to the antiserum (Pastan *et al.*, 1966). Secondly, growth hormone added to normal muscle depresses incorporation of labelled acetate into fat (Manchester, 1963), which is quite the reverse of the influence of insulin.

It would be particularly interesting to look at the influence of some of the fragments of growth hormone which retain growth promoting activity on this last parameter, since the effect is unusual in being visible with muscle from normal animals. The influence of sulphation factor which, though dependent on growth hormone in origin, seems to exemplify characteristics similar in many ways to those of insulin (Salmon and DuVall, 1970), would also bear investigation in this respect.

CONCLUDING COMMENTS

Recent investigation of the structure of insulin (Blundell *et al.*, 1971) suggests that the three-dimensional conformation of the molecule, as much as the amino acids available on its surface, is the most important determinant for activity. Likewise some conservation of structure of glucagon is necessary for function (Rodbell *et al.*, 1971) and variation in amino acid sequence in different species is not seen (Bromer *et al.*, 1971). The close sequence homologies between growth hormone, placental lactogen and prolactin, particularly for the hydrophobic amino acids (Niall *et al.*, 1971), despite different biological activities, suggest by contrast that some of the pituitary hormones may have similar conformations and that surface groups will therefore be of critical importance. This possibility is made the more likely with the finding that growth promoting activity is retained by a variety of fragments and derivatives of the molecule in a manner quite different from what appears to be the case for insulin (Li and Yamashiro, 1970; Nutting *et al.*, 1970). It is not surprising, therefore, that the *in vitro* effects of the pituitary hormones should prove so complex.

REFERENCES

BLUNDELL, T. L., DODSON, G. G., DODSON, E., HODGKIN, D. C. and VIJAYAN, M. (1971): X-ray analysis and the structure of insulin. In: *Recent Progress in Hormone Research.* Editor: E. B. Astwood, *27*, 1.

BROMER, W. W., BOUCHER, M. E. and KOFFENBURGER, J. E. (1971): Amino acid sequence of bovine glucagon. *J. biol. Chem.*, *246*, 2822.

BUSE, M. G. and BUSE, J. (1959): Glucose uptake and response to insulin of the isolated rat diaphragm. *Diabetes*, *8*, 218.

BUSE, M. G. and BUSE, J. (1961): The effect of denervation and insulin on the penetration of D-xylose into rat hemidiaphragms. *Diabetes*, *10*, 134.

BUSE, M. G., MCMASTER, J. and BUSE, J. (1965): The effect of denervation and insulin on protein synthesis in the isolated rat diaphragm. *Metabolism*, *14*, 1220.

FAIN, J. N., KOVACEV, V. P. and SCOW, R. O. (1965): Effect of growth hormone and dexamethasone on lipolysis and metabolism in isolated fat cells of the rat. *J. biol. Chem.*, *240*, 3522.

FRIEDBERG, S. H., OKA, T. and TOPPER, Y. J. (1970): Development of insulin sensitivity by mouse mammary gland *in vitro*. *Proc. nat. Acad. Sci. (Wash.)*, *67*, 1493.

GOLD, H. K., PRINDLE, K. H., LEVEY, G. S. and EPSTEIN, S. E. (1970): Effects of experimental heart failure on the capacity of glucagon to augment myocardial contractility and activate adenyl cyclase. *J. clin. Invest.*, *49*, 999.

HARRIS, E. J. and MANCHESTER, K. L. (1966): The effects of potassium ions and denervation on protein synthesis and the transport of amino acids in muscle. *Biochem. J.*, *101*, 135.

LI, C. H. and YAMASHIRO, D. (1970): The synthesis of a protein possessing growth-promoting and lactogenic activities. *J. Amer. chem. Soc.*, *92*, 7608.

MANCHESTER, K. L. (1963): Effects of pituitary growth hormone and corticotropin on fat metabolism in isolated rat diaphragm. *Biochim. biophys. Acta (Amst.)*, *70*, 531.

MANCHESTER, K. L. (1966): Some factors affecting the response of muscle to insulin. *Biochem. J.*, *98*, 711.

MANCHESTER, K. L. and WALLIS, M. (1963): Comparison of *in vivo* and *in vitro* actions of various ox pituitary growth hormone fractions. *Nature (Lond.)*, *200*, 888.

MANCHESTER, K. L. and YOUNG, F. G. (1959): Hormones and protein biosynthesis *J. Endocr.*, *18*, 381.

MILLER, L. L., MUTSCHLER NAISMITH, L. and CLOUTIER, P. F. (1970): Protein catabolism in the isolated perfused regenerating rat liver. Loss of protein catabolic response to glucagon. In: *Homologies in Enzymes and Metabolic Pathways*, pp. 516–529. Editors: W. J. Whelan and J. Schultz. North Holland Publ. Co., Amsterdam.

NIALL, H. D., HOGAN, M. L., SAUER, R., ROSENBLUM, I. Y. and GREENWOOD, F. C. (1971). Sequences of pituitary and placental lactogenic and growth hormones: Evolution from a primordial peptide by gene reduplication. *Proc. nat. Acad. Sci. (Wash.)*, *68*, 866.

NUTTING, D. F., KOSTYO, J. L., MILLS, J. B. and WILHELMI, A. E. (1970): A cyanogen bromide fragment of reduced and S-aminoethylated porcine growth hormone with anabolic activity. *Biochim. biophys. Acta (Amst.)*, *200*, 601.

OTTAWAY, J. H. (1953): The insulin-like effect of growth hormone. *Biochim. biophys. Acta (Amst.)*, *11*, 443.

PASTAN, I., ROTH, J. and MACCHIA, V. (1966): Binding of hormone to tissue: the first step in polypeptide hormone action. *Proc. nat. Acad. Sci. (Wash.)*, *56*, 1802.

RODBELL, M., BIRNBAUMER, L., POHL, S. L. and SUNDBY, F. (1971): The reaction of glucagon with its receptor: Evidence for discrete regions of activity and binding in the glucagon molecule. *Proc. nat. Acad. Sci. (Wash.)*, *68*, 909.

SALAMAN, M. R. and ROBINSON, D. S. (1961): The effect of fasting on the clearing factor lipase activity of rat adipose tissue and plasma. In: *Enzymes of Lipid Metabolism*, p. 218. Editor: P. Desnuelle. Pergamon Press, Oxford.

SALMON, W. D. and DUVALL, M. R. (1970): *In vitro* stimulation of leucine incorporation into muscle and cartilage protein by a serum fraction with sulfation factor activity: Differentiation of effects from those of growth hormone and insulin. *Endocrinology*, *87*, 1168.

SWISLOCKI, N. I., SONENBERG, M., and YAMASAKI, N. (1970): *In vitro* metabolic effects of bovine growth hormone fragments in adipose tissue. *Endocrinology*, *87*, 900.

WALLIS, M. and DIXON, H. B. F. (1966): A chromatographic preparation of ox growth hormone. *Biochem. J.*, *100*, 593.

WALLIS, M. and KOVACIC, N. (1965): Prolactin activity of ox growth hormone. *J. Endocr.*, *33*, 443.

IV. Sulfation factor

PARTIAL PURIFICATION FROM HUMAN PLASMA OF A SMALL PEPTIDE WITH SULFATION FACTOR AND THYMIDINE FACTOR ACTIVITIES*

JUDSON J. VAN WYK, KERSTIN HALL, J. LEO VAN DEN BRANDE, ROBERT P. WEAVER, KNUT UTHNE, RAYMOND L. HINTZ, JOHN H. HARRISON and PAUL MATHEWSON

Departments of Pediatrics and Chemistry, University of North Carolina, Chapel Hill, N.C., U.S.A.; Department of Endocrinology and Metabolism, Karolinska sjukhuset, Stockholm, Sweden; and Department of Pediatrics, University of Rotterdam School of Medicine, Rotterdam, The Netherlands

Almost a decade and a half has elapsed since Daughaday and Salmon proposed that the action of growth hormone (GH) on skeletal tissue is mediated through a secondary substance, detectable in plasma, which they termed 'sulfation factor' (Salmon and Daughaday, 1957). Since then their observations have been amply confirmed. It has been shown that a GH dependant plasma factor stimulates not only the incorporation of sulfate into chondroitin sulfate, but also the incorporation of thymidine into DNA (Daughaday and Reeder, 1967), proline into collagen (Daughaday and Mariz, 1962), and uridine into RNA (Salmon and DuVall, 1970). GH itself has substantially no *in vitro* effect on cartilage metabolism (Salmon and Daughaday, 1957).

Prior to 1962, many laboratories seized upon the sulfation factor assay as an indirect means of assessing pituitary status. It was found that sulfation factor activity is high in serum from acromegalic patients, whereas hypopituitary patients have subnormal levels which are restored by the administration of human growth hormone (Daughaday *et al.*, 1958; Almqvist *et al.*, 1961). With the advent of the radioimmunoassay for human growth hormone, interest in sulfation factor assays waned, although the phenomenon itself remained unexplained.

The present studies** were undertaken to determine the nature of the factor or factors in serum which are responsible for stimulating sulfate uptake by cartilage. At the outset of these studies it was not clear whether the sulfation promoting activity of plasma could be attributed to an unique molecular species or whether it represented the net effect of multiple chemical and hormonal alterations resulting from GH action in non-skeletal tissue***. It remained to be

* This research was supported by AB Kabi, Stockholm, Sweden, and by the following grants to the University of North Carolina from the U.S. Public Health Service: Research Grant AM01022, Training Grant AM05330, and Career Research Award 5 K06 AM14115 (JJVW). Research by K.H. was additionally supported by grants from the Karolinska Institute and Nordisk Insulin Foundation.

** Studies on sulfation factor were initiated independently by Dr. Kerstin Hall in Stockholm and by Drs. J. L. Van den Brande and J. J. Van Wyk in Chapel Hill, North Carolina. The collaborative effort began during the 1968–1969 academic year while Dr. Van Wyk was a guest in the laboratory of Professor Rolf Luft, Department of Endocrinology and Metabolism at the Karolinska sjukhuset. Dr. Van den Brande is now a member of the University of Rotterdam Medical Faculty, Rotterdam, Holland.

*** Koumans and Daughaday had shown earlier that the sulfation promoting activity of normal serum in their *in vitro* system was accounted for by both dialyzable and non-dialyzable components. The dialyzable component was accounted for, at least in large part, by certain non-essential amino acids, particularly serine, which were lacking from their original incubation medium. The non-dialyzable component was found to be heat stable and excluded by Sephadex G-25 (Koumans and Daughaday, 1963).

shown whether the sulfation factor represented a degraded portion of the growth hormone molecule or whether it was an entirely new substance induced in non-skeletal tissue by GH action. Lastly, it seemed of particular importance to determine whether the same plasma factor (or factors) stimulate the synthesis of both chondroitin sulfate and DNA. This question seemed of particular interest since growth hormone stimulates both the specialized biosynthetic processes which are characteristic of differentiated cartilage cells and cellular replication, actions which are not synchronous within a given cell cycle.

BIOASSAY PROCEDURES

The Daughaday and Reeder assay for measuring thymidine uptake in costal cartilage segments from hypophysectomized rats (Daughaday and Reeder, 1967) was modified by the incorporation of a dual isotope labeling procedure which permitted the simultaneous measurement of sulfation factor activity (PSF) and thymidine factor activity (PTF) (Van den Brande *et al.*, 1971).

Classical bioassay statistics are based on a symmetric 4 or 6 point assay design. All computations are made with an IBM 1130 computer by a program which tests for linearity, parallelity, preparation difference, opposite curvature, and lambda values. Quantitative differences between preparations are expressed as potency ratios with 95% confidence limits. Dose-response curves for PTF are steeper than for PSF, but also have higher variances (Fig. 1).

This cumbersome bioassay system continues to be the rate limiting factor in following purification. Using a 6-point assay design, a maximum of 9 unknown preparations can be tested against a standard preparation in one assay. The results frequently fail to meet the statistical criteria of a valid bioassay because of wide fiducial limits and unexplained deviations from linearity and parallelity. Many of these problems are explained by inhibitors which are normally present in blood or which are accumulated during chemical fractionation.

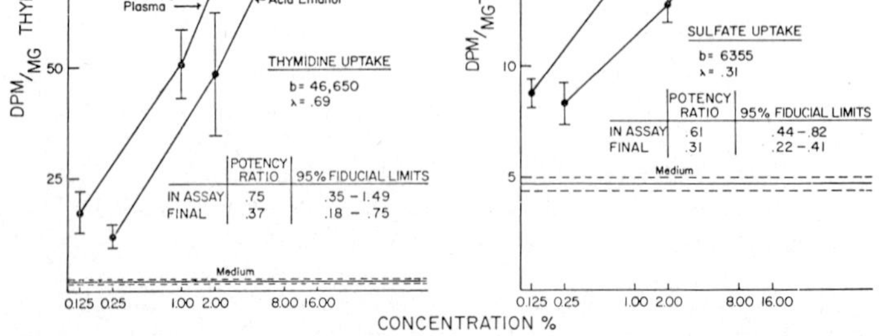

Fig. 1. Dose-response curves obtained with an acid ethanol extract compared with the acromegalic plasma from which it was derived. Uptake of $^{35}SO_4{}^{2-}$ and ^{3}H-methyl thymidine are expressed in DPM per mg dry cartilage. Each point represents the mean of 6 cartilages (1 from each of 6 rats) and the brackets indicate ± S.E.M. Potency ratios 'In Assay' refer to relative responses in the assay irrespective of dose. 'Final' potency ratios reflect the relative activities of the two preparations adjusted for the dosages used in the assay. (From J. J. Van Wyk *et al.*, 1971, *J. clin. Endocr.*, *32*, 389, by kind permission of the editors).

An alternative assay using the pelvic rudiments of 11-day chick embryos has been described by Hall (1970). This method is easier to perform and also has lower variances. However, it cannot measure thymidine incorporation and has only one fourth the sensitivity of the rat assay.

Both laboratories independently confirmed that human growth hormone, like bovine growth hormone, stimulated PSF and PTF activity in the plasma of hypophysectomized rats whereas HGH is itself inactive when added directly *in vitro*. Rat cartilage was found to be much more sensitive to insulin than was embryonic chick cartilage.

STUDIES ON WHOLE ACROMEGALIC PLASMA

A pool of plasma from several untreated acromegalic patients was heated in a Dubnoff shaker at 56° C for 30 min, stored for 15 hours at 4° C and then centrifuged at 2500 r.p.m. This procedure did not alter the biologic activity and preserved the linearity of responses at higher dosage levels. The supernatant was lyophilized and stored at -20° C. The term 'plasma equivalents' (pl. eq.) is used to indicate the original volume of plasma in ml from which each preparation was derived.

CHROMATOGRAPHY ON DEAE CELLULOSE

Whole plasma was adjusted to pH 8.5 and applied to DEAE cellulose at low ionic strength. Elution was performed in a decreasing pH and increasing ionic strength gradient. Most of the sulfation and thymidine factor activity appeared in the 'fall through' volume and there was

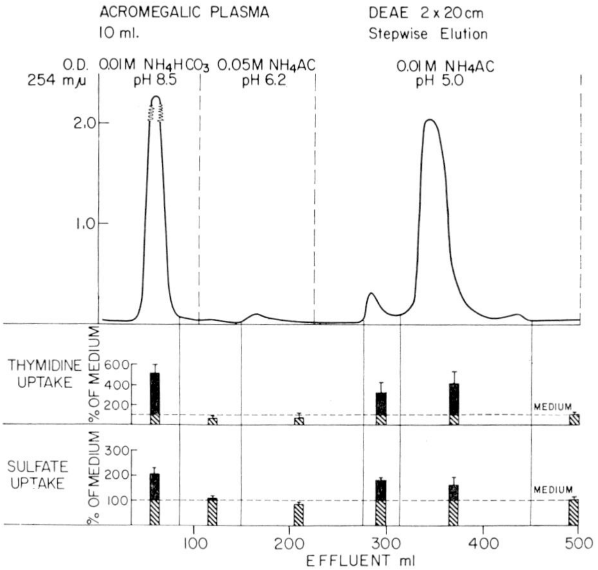

Fig. 2. Chromatography on DEAE-cellulose. 10 ml acromegalic plasma were applied to column in 0.01 M NH_4HCO_3 pH 8.5 and thereafter eluted stepwise as shown with 0.01 M NH_4HCO_3 pH 8.5, 0.05 M NH_4 acetate pH 6.2 and 0.1 M NH_4 acetate pH 5.0. After measuring the absorbance at 254 mμ the fractions were combined into pools as indicated by vertical lines. Bars represent mean uptake of $^{35}SO_4^{2-}$ and ^{3}H-methyl thymidine expressed as percentages of the mean uptake of cartilages incubated in basal medium. The uptakes shown for the first peak were obtained at a 2% concentration (pl. eq. in medium), whereas those in the second peak were obtained at a 16% concentration. (From J. J. Van Wyk *et al.*, 1971, *J. clin. Endocr.*, *32*, 389, by kind permission of the editors).

only a hint of activity associated with subsequent protein peaks. The non-adsorbed protein (termed 'DE Material') was used as starting material in a number of subsequent studies (Van den Brande *et al.*, 1971).

In another experiment, whole acromegalic plasma was applied to DEAE at pH 8.5 and stepwise elution was carried out at decreasing pH and increasing ionic strength increments. Although most of the activity again appeared in the fall-through volume, between 5 and 20% of the original biologic activity was adsorbed at pH 8.5 and eluted at pH 5 with 0.1 M ammonium acetate (Fig. 2). Although not positively identified, it is possible that this minor component may be attributed to the high insulin content of this plasma since insulin is active in the rat assay and insulin behaves similarly on DEAE.

STABILITY STUDIES

The biologic activity of 'DE Material' was stable for 6 hours at 4° C over a pH range between pH 2 and pH 10. All biologic activity was destroyed by incubation with pronase. Active preparations of 'DE Material' were heated for 30 min in a constant temperature water bath at temperatures ranging between 40 and 80° C. They were then cooled, the heat coagulated protein removed by centrifugation, and the supernatant fractions assayed. Heating at 80° C totally removed activity from the residual supernatant and there were partial losses at 70° C. These studies did not distinguish whether the active material was itself inactivated by heat or whether it was coprecipitated with larger proteins.

FRACTIONATION BY COLD ETHANOL

In an attempt to obtain a more concentrated starting preparation than whole plasma, ethanol was added in the cold to whole plasma and the concentration stepped up progressively to 20, 40, and 80%. The precipitate was removed at each step and aliquots of the supernate removed for biologic assay. There was moderate loss of activity at 20% concentration, a greater loss at 40%, and at 80% there was no residual biologic activity (Van den Brande *et al.*, 1971). A number of Cohn fractions were also examined*. None of these preparations offered much advantage since there was no substantial enhancement of biologic activity per mg protein.

STARCH GEL ELECTROPHORESIS OF 'DE MATERIAL'

Starch gel electrophoresis of 'DE Material' at pH 8.5 revealed that both thymidine and sulfate activities moved only slightly anodally from the origin. There was in addition, however, some front-running sulfation factor activity.

MOLECULAR SIEVING

When DE material was applied to Sephadex G-100 or whole plasma to Sephadex G-25 most of the activity was recovered with the major protein peaks. Similar results have been reported by Bala *et al.* (1970). Results obtained when acromegalic plasma was passed serially through graded ultrafiltration membranes in an Amicon cell are shown in Table I. All activity was retained by the UM-05 and XM-50 filters and most of the activity was retained on the XM-100 filter. Recoveries of activity from initial starting plasma were nearly quantitative.

---

* We are indebted to Dr. Björn Holmström of AB Kabi, Stockholm, Sweden, for furnishing these preparations.

TABLE I

Ultrafiltration of acromegalic plasma

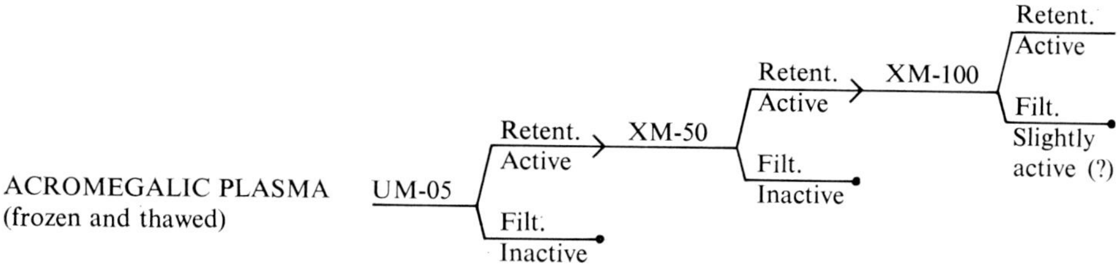

Preparation	Concentration % (pl.eq.) in medium	DPM/mg % of uptake in medium (± S.E.M.) Thymidine	Sulfate
Medium	—	100 ± 13	100 ± 7
Plasma (K.W.)	0.75	464 ± 53	223 ± 11
	1.50	508 ± 56	277 ± 18
	3.00	574 ± 43	264 ± 10
XM-100	1.5	495 ± 65	232 ± 13
Retentate	3.0	561 ± 103	251 ± 10
XM-100	3.0	115 ± 34	123 ± 9
Filtrate	12.0	111 ± 21	136 ± 7

(From J. J. Van Wyk *et al.*, 1971, *J. clin. Endocr.*, *32*, 389, by kind permission of the editors).

STUDIES ON ACID ETHANOL EXTRACTS OF ACROMEGALIC PLASMA

The preceding studies on whole acromegalic plasma, or on that portion of plasma not adsorbed to an anionic exchange resin, suggested that most of the biologic activity in acromegalic plasma is neutral or basic in character and is associated with protein fractions of substantial size. Nevertheless, it seemed quite possible that the active material might be considerably smaller than this and either aggregated or adsorbed to a larger carrier protein. Attempts were therefore made to split non-covalent bonds and to recover activity in a smaller peptide fraction. The technique which finally proved successful was that of acid ethanol extraction (Van Wyk *et al.*, 1969, 1971). The procedure was similar to that described by Jacob *et al.* (1968) in extracting non-suppressible insulin-like activity from plasma. Since non-suppressible ILA is also elevated in acromegaly and low in hypopituitarism, it seemed possible that the active substance(s) might be similar if not identical. The acid ethanol extraction procedure has now been carried out many times on plasma from both normal and acromegalic subjects with recoveries of 30–40% of the biological activity in native plasma. Over 99% of the proteins in native plasma are removed by this procedure. The dose-response curves of the acid ethanol extracts parallel those of native plasma both for thymidine activity and sulfate activity, suggesting that the substances extracted are similar to those in native plasma (Fig. 1).

CHROMATOGRAPHY ON SEPHADEX

When an acid ethanol (AE) extract was chromatographed on Sephadex G-100, most of the biologic activity was recovered in a zone corresponding to a molecular weight somewhat smaller than the cytochrome c marker of 12,400; however, small amounts of activity were also

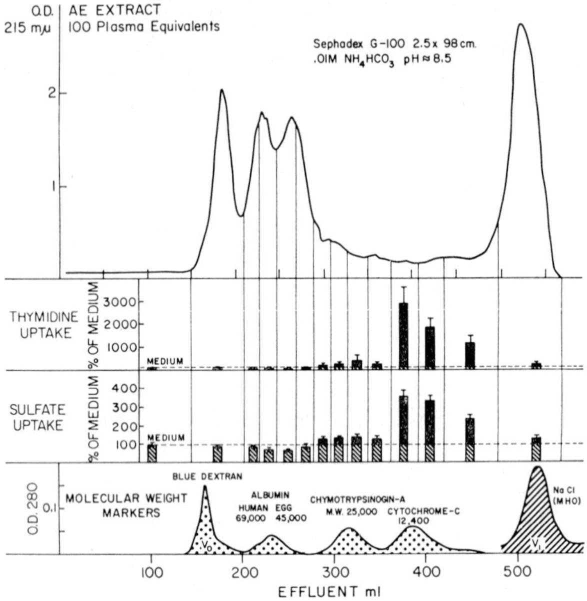

Fig. 3. 100 pl. eq. of an acid ethanol extract of plasma from an acromegalic patient was chromatographed on Sephadex G-100 in 0.01 M NH_4HCO_3. (From J. J. Van Wyk *et al.*, 1971, *J. clin. Endocr.*, *32*, 389, by kind permission of the editors).

recovered in the succeeding smaller molecular weight fractions (Fig. 3). Chromatography on Sephadex G-75 and Sephadex G-50 led to refinement of the estimated molecular weight with an average value of about 8,000.

CHROMATOGRAPHY ON DOWEX 50-H$^+$

Dowex 50 WX 2 (200-400 mesh) H$^+$ has been used in a variety of ways to provide an initial purification step for AE extracts. All of the biologic activity is retained when applied between pH 3.0 and 6.0 at low ionic strength. Elution with either diethyl amine or 0.2 N NH_4OH leads to nearly quantitative recovery with an approximately 10-fold purification (biologic activity/mg protein).

CHROMATOGRAPHY ON CMC CELLULOSE

An AE extract of acromegalic plasma was dissolved in ammonium acetate pH 5.6 and further diluted with water until the conductivity matched that of 0.01 M NH_4 acetate (0.6 MHO). The entire volume was pumped onto a carboxymethyl cellulose column equilibrated with the same buffer and the column thoroughly washed with buffer. Elution was carried out in a linear ionic gradient between 0.01 M and 0.4 molar. The bulk of biologically inactive proteins in the AE extract failed to be adsorbed to the resin under these conditions, whereas the biologic activity was recovered in a small protein peak which was eluted between 2 and 5.6 MHO (Fig. 4). The net recovery of biologic material from acromegalic plasma after chromatography on CMC and Sephadex ranged between 5 and 10% with net purification of 6,200 X for PSF and 15,000 X for PTF. The apparently greater purification of PTF was probably due to loss of inhibitors which affect the PTF assay to a greater extent.

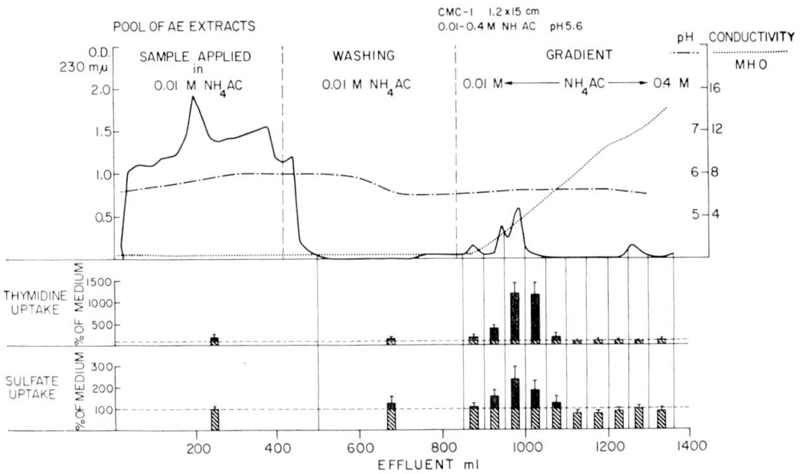

Fig. 4. Chromatography on CMC-cellulose. 500 pl. eq. of acid ethanol extract were applied to the column (1.2 × 15 cm) in 0.01 M NH_4 acetate pH 5.6 and washed extensively with the same buffer. Thereafter, the column was eluted in a linear ionic gradient at pH 5.6 Fractions of 10 ml were collected at a flow rate of 40 ml per hour and absorbancies were measured at 230 mμ. The fractions were pooled as indicated by the vertical lines and assayed for biologic activity. (From J. J. Van Wyk *et al.*, 1971, *J. clin. Endocr.*, *32*, 389, by kind permission of the editors).

HIGH VOLTAGE ELECTROPHORESIS

The diethyl amine eluate from the Dowex 50 column was lyophilized, passed over Sephadex G-25, and thereafter applied to paper in a small volume of buffer. High voltage electrophoresis in pyridine acetate buffer pH 6.5 was carried out at 60 volts/cm for 50 min. Ninhydrin staining revealed 5 protein bands with biologic activity present only in the band migrating with the mobility of glycine. This material was eluted and applied to another paper and electrophoresed at pH 2. Again there was separation into 4 ninhydrin positive bands but again the activity migrated with the velocity of glycine. These results suggested that the biologic activity recovered has a neutral charge.

ISOELECTRIC FOCUSING

Several crude acid ethanol extracts were electrofocused with an ampholyte buffer mixture in the range between pH 5 and pH 8. A small sharp band of activity was recovered with a pI of 5.2 and a larger band with an isoelectric point of 6.5–6.7 (Fig. 5). The isoelectric point of the biologically active material recovered in the more acidic band corresponds closely with that of insulin, and this material may be identical with the minor component recovered from DEAE. This possibility has not been pursued further.

STUDIES ON POOLED NORMAL PLASMA

It was apparent that to achieve greater purification of the biologically active substance (or substances) and to obtain sufficient quantities for meaningful determinations of structure, considerably larger quantities of starting material would be required than could be acquired from acromegalic blood donors. We therefore turned our attention to AE extracts of pooled

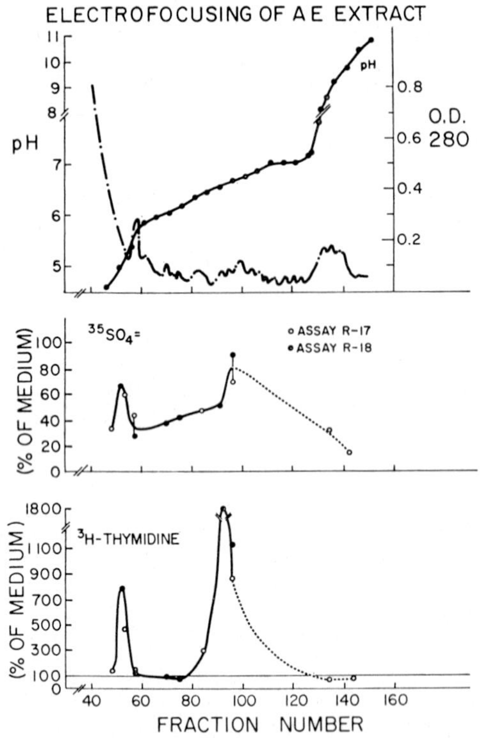

Fig. 5. Electrofocusing on an LKB-8102 Ampholine column (440 ml) at pH 5-8, equilibrated for 93 hours. 8 plasma equivalents of acid ethanol extract were applied to the column. Fractions of 2.1 ml were collected, and absorbance measured at 280 mμ. The eluates were diluted in an equal volume of medium for assay. Each specimen was assayed in triplicate in an assay method utilizing cartilage segments from a single immature male rhesus monkey. Contamination of the fractions with SO_4^{2-} from the electrode lock artifactually lowered the apparent uptake of $^{35}SO_4^{2-}$. (From J. J. Van Wyk *et al.*, 1971, *J. clin. Endocr.*, *32*, 389, by kind permission of the editors).

plasma from normal blood donors*. Results with these extracts have been generally similar to the results obtained with plasma from single acromegalic subjects, although the data obtained to date provide greater suggestion that the active material may be heterogeneous.

CHROMATOGRAPHY ON CMC

The behavior of AE extracts from pooled normal plasma on CMC resembles that of acromegalic plasma, except that the peak of biologic activity is eluted at a slightly higher ionic strength and activity is recovered over a wider range (Fig. 6).

CHROMATOGRAPHY ON SEPHADEX

Chromatography on Sephadex G-75 in ammonium bicarbonate buffer again suggests that

* We are indebted to Professor Bertil Åberg, Director of Research and Development of AB Kabi Pharmaceutical Company for making available for these studies acid ethanol extracts prepared from large quantities of outdated pooled human plasma.

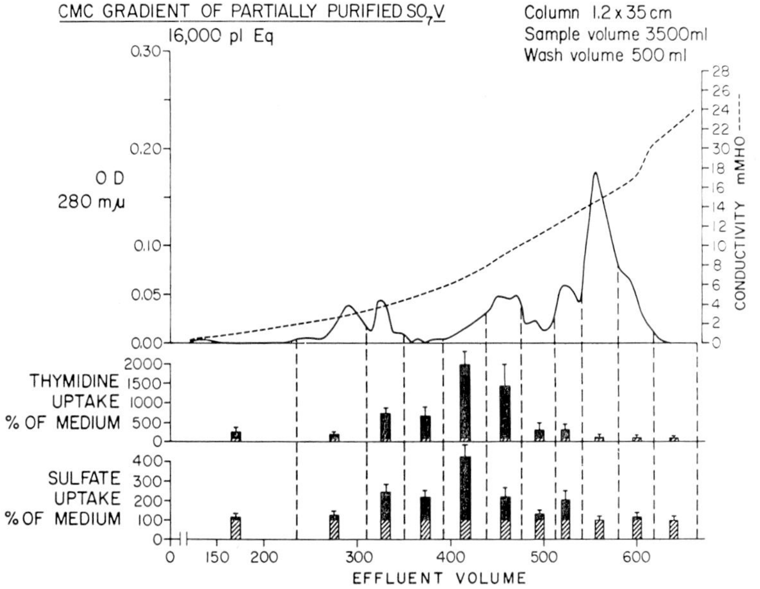

Fig. 6. The acid ethanol extract derived from 16,000 ml of pooled normal plasma was dialyzed against NH_4 acetate 0.01 M pH 5.6 and pumped on a column of CMC equilibrated with the same buffer. Elution was carried out in an exponential ionic gradient as shown. The eluates from the 'pump on' and 'wash' volumes (not shown) contained some biologic activity, probably due to overloading of the column.

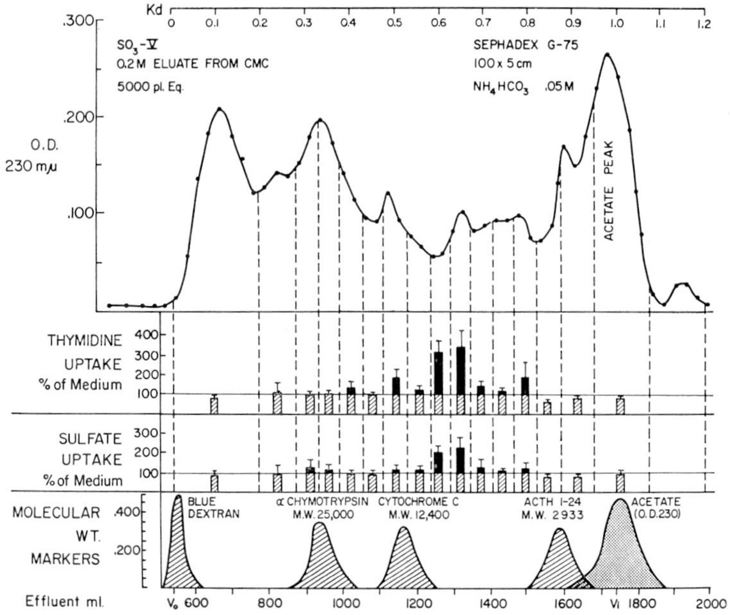

Fig. 7. The acid ethanol extract derived from 5000 ml of pooled normal plasma was first chromatographed on CMC. The 0.2 M NH_4 acetate eluate was lyophilized, applied to Sephadex G-75, and eluted with NH_4HCO_3, 0.05 M. The molecular weight markers were applied in the same volume as the sample.

most of the activity has a molecular weight of approximately 8,000, but again there is a suggestion of activity in smaller molecular weight fractions (Fig. 7).

HIGH VOLTAGE ELECTROPHORESIS

AE extracts from pooled normal serum were further purified on Dowex 50 and Sephadex G-25 and then subjected to HV electrophoresis at pH 6.5. As with acromegalic plasma all of the recoverable activity migrated with the velocity of neutral amino acids. This band was eluted and again subjected to H.V.E. at pH 2.0. Seven ninhydrin positive bands were visible in addition to the albumin marker (Fig. 8). Biologic activity was recovered over 4 adjacent ninhydrin positive bands, with mobilities ranging between the amino acid markers glycine and leucine.

ELECTROFOCUSING

Another extract from the same plasma pool was subjected to preliminary purification on CMC and Sephadex G-75 and then applied to a 110 ml LKB electrofocusing column in a sucrose gradient between pH 3.0 and pH 10.0. Most of the activity was recovered between pH 7.5 and 10 (Fig. 9) with only a small amount around pH 6.5 (where the major component in acromegalic plasma focused). The absence of any activity at pH 5.2, as previously found with acid ethanol extracts of acromegalic plasma, is not surprising since it is likely that any material with such a charge would have been removed by ion exchange chromatography.

Since the ampholyte buffer range was different in this study and there were other important experimental differences (smaller column and a larger quantity of protein), it is not yet

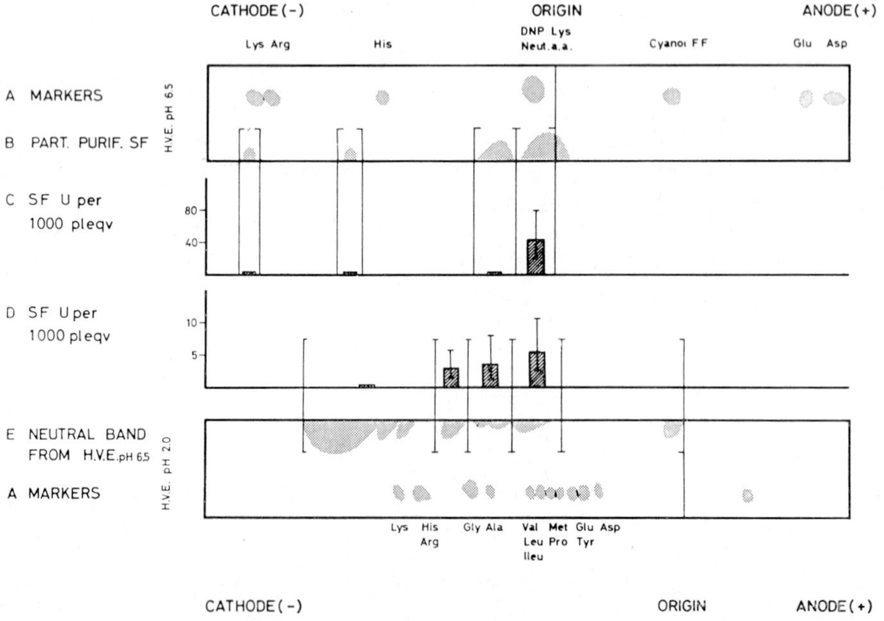

Fig. 8. A partially purified acid ethanol extract from pooled normal plasma was applied to paper and subjected to H.V. electrophoresis at pH 6.5. Ninhydrin positive bands (B) were identified on a small peripheral strip. The biologic activity eluted from areas of paper corresponding to these bands are shown in C. An aliquot of the neutral zone was applied to another paper and subjected to HVE at pH 2. The ninhydrin positive bands are shown in E and the corresponding assay results in D. Results are expressed in sulfation factor units with 95% fiducial limits.

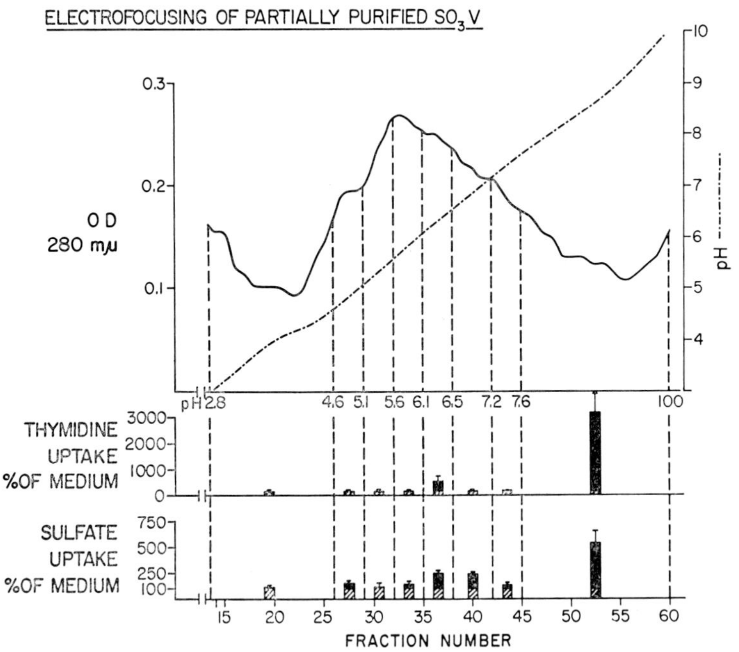

Fig. 9. 4000 pl.eq of a partially purified preparation of pooled normal plasma (acid ethanol, CMC, Sephadex G-75) was added in a sucrose gradient to a 110 ml LKB Ampholine column. Electrofocusing was carried out for 72 hours in ampholyte buffer in a range between pH 3 and pH 10.

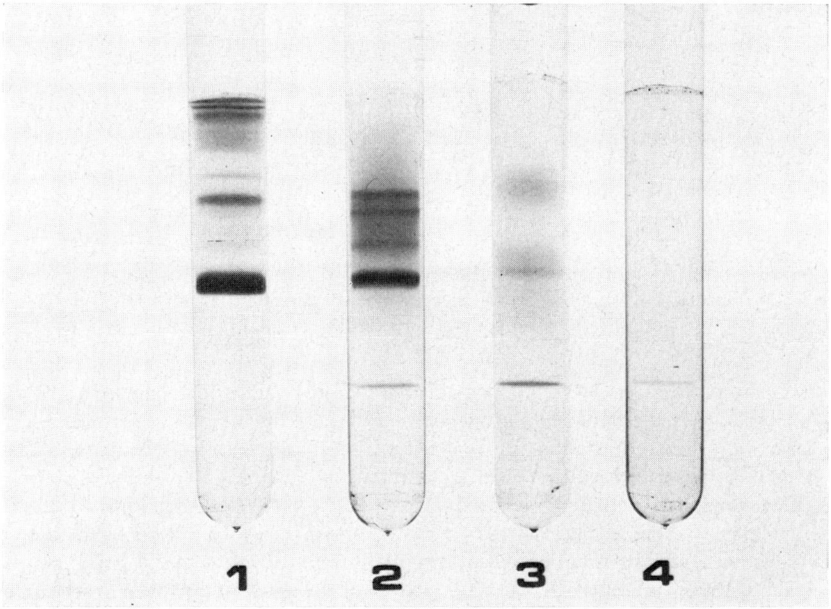

Fig. 10. Polyacrylamide gel patterns of (1) pooled normal plasma, (2) acid ethanol extract, (3) active fraction after chromatography on CMC and (4) active fraction after CMC eluate was chromatographed on Sephadex G-75 in 0.05 N NH_4HCO_3. All were run at pH 8.9 in a gel concentration of 7.5%. Several additional bands in the upper portion of tube 4 were barely visible in the freshly stained gel.

clear whether there is indeed a basic peptide with sulfation factor activity present in pooled normal plasma or whether the apparent high pI represents an artifact. A precedent for peptides yielding discrepant isoelectric points in differing ampholyte mixtures has been provided by Reichert in the case of LH (Reichert, 1971). Presumably this is due to binding of the peptide to components of the ampholyte mixture.

ANALYTICAL ACRYLAMIDE GEL ELECTROPHORESIS

In Figure 10 are the protein patterns observed at different stages of purification after disc electrophoresis on 7.5% acrylamide gel at pH 8.9. In the more highly purified preparations the strongest band stainable with Coomasie Blue runs at the front with multiple barely visible retarded bands (not visible on the photograph). On preparative acrylamide gel electrophoresis the front running band was inactive. Biologic activity was recovered only in fractions with Rf values less than 0.55.

SUMMARY

In none of these purification procedures has it been possible to dissociate thymidine incorporating activity from sulfate incorporating activity and present data suggest that these activities are resident in the same molecule or molecules. Dr. Hall has obtained evidence that non-suppressible insulin-like activity also parallels sulfation factor activity (Hall and Uthne, 1971).

TABLE II

Summary of major purification steps

Material	Recovery of starting activity (%)	Recovery of protein (μg/pl.eq.)	Purification achieved
Native plasma	100	75,000	0
Acid ethanol	35	200	135 X
CMC	20	4	704 X
Sephadex G-75	5	0.3	12,500 X
Isofocus >pH 7.5	3	0.1	22,500 X

Approximate recoveries of biologic activity and protein content per original ml of starting material at successive steps of purification. The starting material was pooled normal plasma.

Table II shows the recovery values, protein content and purification achieved at the various steps starting from pooled normal plasma. Although these figures may reflect quite large errors in assay, it is apparent that an approximately 22,500 fold degree of purification has been achieved with an overall recovery of 3%. Assuming a molecular weight of 8,000, a high degree of *in vitro* potency is demonstrated at 3×10^{-9} molar concentration.

The evidence presented suggests that a large portion of the sulfation factor activity in plasma is accounted for by small peptides. A small amount of the activity, possibly attributable to insulin, is resident in an acidic fraction. Most of the activity, however, is accounted for by an 8,000 MW neutral or slightly basic peptide. The data presented suggest that final purification of the major peptide can be achieved from normal human plasma. Very large quantities of plasma will be required, however, to purify sufficient quantities for complete structural analysis.

REFERENCES

ALMQVIST, S., IKKOS, D. and LUFT, R. (1961): Studies on sulfation factor (SF) activity of human serum. *Acta endocr. (Kbh.)*, *36*, 577.

BALA, R. M., FERGUSON, K. A. and BECK, J. C. (1970): Plasma biological and immunoreactive growth hormone-like activity. *Endocrinology*, *87*, 506.

DAUGHADAY, W. H. and MARIZ, I. K. (1962): Conversion of proline-U-C^{14} to labeled hydroxyproline by rat cartilage in ribs: Effects of hypophysectomy, growth hormone and cortisol. *J. Lab. clin. Med.*, *59*, 741.

DAUGHADAY, W. H. and REEDER, C. (1967): Synchronous activation of DNA synthesis in hypophysectomized rat cartilage by growth hormone. *J. Lab. clin. Med.*, *68*, 357.

DAUGHADAY, W. H., SALMON, W. D. and ALEXANDER, F. (1958): Sulfation factor activity of sera from patients with pituitary disorders. *J. clin. Endocr.*, *19*, 743.

HALL, K. (1970): Quantitative determination of the sulphation factor activity in human serum. *Acta endocr. (Kbh.)*, *63*, 338.

HALL, K. and UTHNE, K. (1971): Human growth hormone and sulfation factor. *This Volume*, pp. 192–198.

JACOB, A., HAURI, Ch. and FROESCH, E. R. (1968): Non-suppressible insulin-like activity in serum. III. Differentiation of two distinct molecules with non-suppressible ILA. *J. clin. Invest.*, *47*, 2678.

KOUMANS, J. and DAUGHADAY, W. (1963): Amino acid requirements for activity of partially purified sulfation factor. *Trans. Ass. Amer. Phycns*, *76*, 152.

REICHERT, L. E., JR. (1971): Electrophoretic properties of pituitary gonadotropins as studied by electrofocusing. *Endocrinology*, *88*, 1029.

SALMON JR., W. D. and DAUGHADAY, W. H. (1957): A hormonally controlled serum factor which stimulates sulfate incorporation by cartilage in ribs. *J. Lab. clin. Med.*, *49*, 825.

SALMON JR., W. D. and DUVALL, M. R. (1970): A serum fraction with 'sulfation factor activity' which stimulates *in vitro* incorporation of leucine and sulfate into protein-polysaccharide complexes, uridine into RNA, and thymidine into DNA of costal cartilage from hypophysectomized rats. *Endocrinology*, *86*, 721.

VAN DEN BRANDE, J. L., VAN WYK, J. J., WEAVER, R. P. and MAYBERRY, H. E. (1971): Partial characterization of sulphation and thymidine factors in acromegalic plasma. *Acta Endocr. (Kbh.)*, *66*, 65.

VAN WYK, J. J., HALL, K. and WEAVER, R. P. (1969): Partial purification of sulphation and thymidine factor from plasma. *Biochim. biophys. Acta (Amst.)*, *192*, 560.

VAN WYK, J. J., HALL, K., VAN DEN BRANDE, J. L. and WEAVER, R. P. (1971): Further purification and characterization of sulfation factor and thymidine factor from acromegalic plasma. *J. clin. Endocr.*, *32*, 389.

THE SULFATION FACTOR HYPOTHESIS: RECENT OBSERVATIONS*

WILLIAM H. DAUGHADAY and JOHN T. GARLAND**

Metabolism Division, Department of Internal Medicine, Washington University School of Medicine, St. Louis, Mo., U.S.A.

We shall try to summarize in this paper some of the basic facts that support the sulfation factor hypothesis. We became aware of the possibility of indirect control of cartilage metabolism by GH after a series of disasterously negative attempts to stimulate sulfate uptake directly with GH in cartilage segments from hypox rats. In 1956 Salmon made the critical observation in my laboratory that normal serum, but not hypox rat serum, could stimulate cartilage directly (Salmon and Daughaday, 1957). Sulfation factor (SF) was defined at the time as shown in Table I. I am happy to state that these basic observations have been extensively confirmed.

TABLE I

Basic definition of sulfation factor

1. Normal rat serum contains a nondialyzable component, *sulfation factor*, absent in the serum of hypophysectomized rats, which greatly stimulates chondroitin sulfate synthesis by cartilage from hypophysectomized rats in *vitro*.
2. Somatotropin treatment of hypophysectomized rats restores *sulfation factor* activity to serum.
3. *Sulfation factor* activity cannot be duplicated by somatotropin, insulin or thyroxine addition *in vitro*.

(Salmon and Daughaday, 1957)

Subsequent work in my laboratory and that of Salmon in Nashville, Tennessee, has clearly established that the effects of SF on cartilage are not limited to promoting the sulfation of chondroitin but include stimulation of many aspects of cell anabolism and cell replication (Table II). Dr. Salmon will have much to say about the basic effects of SF on the chondrocyte and, perhaps, other target cells.

Isolation of SF from plasma has been a difficult task because of the tedious *in vitro* bioassay which is sensitive to alterations in other plasma components. It also is likely that there

* Supported by Research Grants No. AM-01526 and No. AM-05105 from the National Institute of Arthritis and Metabolic Diseases. Clinical studies were conducted in the Clinical Research Center, supported by Grant FR-0036 of the Division of General Medical Sciences, National Institutes of Health.

** Trainee in Metabolism. Supported by Training Grant No. AM-05027 of the National Institutes of Health.

TABLE II

Metabolic processes of cartilage stimulated by sulfation factor

1. Chondroitin sulfate synthesis – ^{35}S-sulfate uptake.
   (Salmon and Daughaday, 1957)
2. Chondromucoprotein synthesis – ^{14}C-leucine incorporation.
   (Salmon *et al.*, 1968)
3. Collagen synthesis – conversion of U-^{14}C-proline to collagen-^{14}C-hydroxyproline.
   (Daughaday and Mariz, 1962)
4. RNA synthesis – ^{3}H-uridine incorporation into RNA.
   (Salmon *et al.*, 1968)
5. DNA synthesis – ^{3}H-thymidine into DNA.
   (Daughaday and Reeder, 1966)

TABLE III

Procedures used in purifications of SF

1. *Koumans and Daughaday, 1963*: pH 5.5, boiled extract, Sephadex G-25.
2. *Salmon and DuVall, 1970:* pH 5.5, boiled extract. Absorption and elution from carboxymethyl cellulose.
3. *Liberti, 1970:* Sequential filtration through Diaflo membrane.
4. *VanWyk et al., 1971:* Acid ethanol extraction, CMC – chromatography, Sephadex G-50, Dowex 50 chromatography.

TABLE IV

Properties of SF

1. *Circulating SF has a large molecular weight:*
   Nondialyzable
   Sedimented in ultracentrifuge
   Excluded by Sephadex G-25
   Retained by Diaflo membranes up to 50,000 MW cutoff
2. *Physical stability in plasma:*
   Thermostable – activity has been recovered after boiling at pH 5.5
   Tolerates pH extremes
   Destroyed by Trypsin
3. *No immunologic relationship to GH:*
   Does not react in radioimmunoassay
   Is not neutralized by anti-GH serum

is only a small amount present. Some of the approaches to isolation are outlined in Table III. The most promising initial extraction procedures are the preparation of active boiled extracts by Salmon and DuVall (1970) and the acid ethanol extraction procedure of VanWyk and his associates (1971). Further purification has been accomplished on carboxymethyl cellulose and on Sephadex gels.

Only general statements can be made about the chemical and physical properties of SF (Table IV). It is clear that it circulates in the plasma as a large molecular weight compound or complex. It should be emphasized that the partially purified product of Liberti (1970) and of VanWyk *et al.* (1971) is of considerably smaller molecular size. SF is thermostable. It retains full activity in plasma at 37° C for 24 hours and may be stored virtually indefinitely in the frozen or lyophilized form. Active extracts have been obtained after boiling at pH 5.5. SF activity is also tolerant of pH extremes. Activity is greatly decreased by trypsin digestion. SF does not share important immunologic determinants with GH because it cannot be neutralized by exposure to anti-GH gamma globulin (Daughaday and Kipnis, 1966).

I would like now to turn to a number of observations which we have recently made on SF.

Action of SF on isolated chondrocytes

Adamson and Anast (1966) and later Hall (1970) have shown that SF stimulates sulfate uptake in chick embryo cartilage. Garland, in our laboratory, has taken this a step further by demonstrating SF-like effects of serum on isolated chick chondrocytes prepared from chick

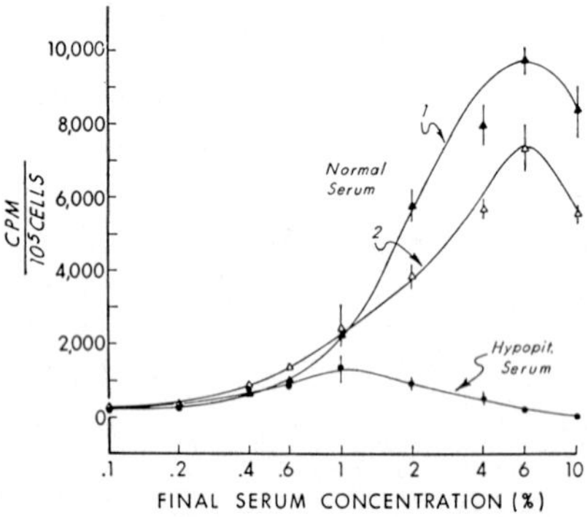

Fig. 1. Stimulation of ³H-thymidine uptake into isolated chick chondrocytes in the presence of varying amounts of normal human serum and hypox serum (Garland *et al.*, 1970).

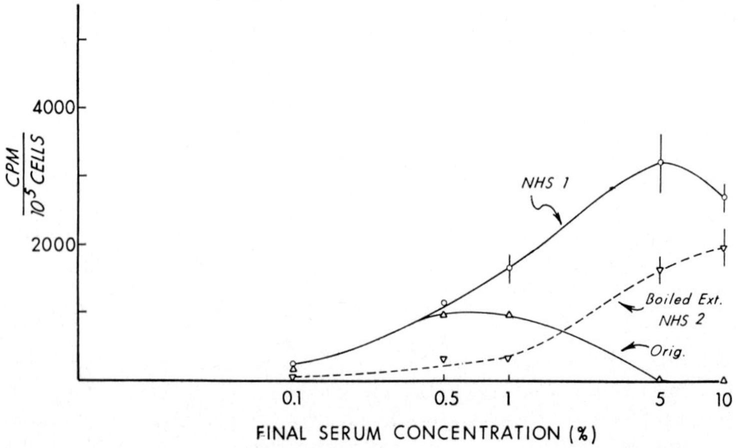

Fig. 2. ³H-thymidine uptake by isolated chondrocytes in the presence of varying amounts of an inhibitory serum and an active boiled extract prepared from this serum (Garland *et al.*, 1970).

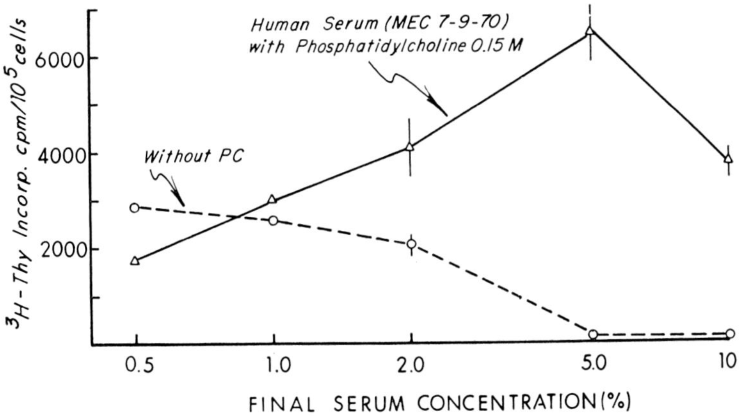

Fig. 3. ^{3}H-thymidine uptake by isolated chondrocytes in the presence of varying concentrations of an inhibitory serum with and without phosphatidyl choline.

pelvic rudiments by trypsin-collagenase treatment (Garland *et al.*, 1970). The isolated chondrocytes have been incubated with ^{3}H-thymidine in dilute serum for 20 hours and thymidine uptake into DNA determined. Most normal human sera cause an impressive stimulation of thymidine uptake (Fig. 1). Unfortunately, many normal human sera and all rat sera are inhibitory at higher concentrations (Fig. 2). Often this inhibition is not observed in boiled extracts of the inhibitory sera. Also, Garland has observed that the inhibitory effects of some sera can be neutralized by phosphatidyl choline (Fig. 3). The isolated chondrocyte preparation is a promising test object for study of SF action. Unfortunately, the frequent occurrence of non-hormonal inhibition renders this an unsuitable test object for routine assays of serum, but it may be of use with purified fractions.

TABLE V

Utilization of rat serum sulfation factor by preincubation with hypox cartilage

Medium	Preincubation	Incubation SO_4 uptake (μM/100 mg wet cartilage/24 hours)
Buffer	None	39.7 ± 5.0[1] (3)[2]
Buffer	Without cartilage	40.3 ± 2.5 (3)
Buffer	With cartilage	48.8 ± 4.1 (3)
Buffer	With boiled cartilage	42.5 ± 2.1 (3)
Serum	None	106.5 ± 8.8 (5)
Serum	Without cartilage	96.1 ± 2.2 (5)
Serum	With cartilage	48.6 ± 1.3 (5)
Serum	With boiled cartilage	100.3 ± 3.2 (5)

All preincubations were 37° C for 24 hours. Wet hypox rat cartilage (65 to 89 mg), either living or killed by boiling, was added for preincubation. The second incubations were carried out in the usual manner, and the incubation medium contained 10% normal rat serum.

[1] The results are expressed as the mean ± standard error.

[2] The figures in parentheses are the number of incubation flasks per group.

(Daughaday *et al.* 1968)

Cartilage binding of SF

We have carried out a limited number of experiments which suggest binding or utilization of SF by cartilage (Daughaday *et al.*, 1968). A double incubation method was used for these studies. In the first incubation 60 to 89 mg of cartilage was added to absorb the SF from a serum containing incubation medium. After 24 hours the first lot of costal cartilages was removed and fresh hypox rat cartilage was added and sulfate uptake determined in the usual manner. The results of such an experiment, along with the appropriate controls, are summarized in Table V. Incubation without cartilage did not affect SF activity and the addition of cartilage to a serum free medium did not inhibit subsequent sulfate uptake. Living cartilage, but not boiled cartilage, removed essentially all sulfation factor activity from the medium. Under comparable conditions of incubation, cartilage did not remove any growth hormone from a medium containing 10% acromegalic serum. The results of these experiments provide evidence of selective uptake of removal of SF, but not GH, by cartilage. Further demonstration of specific SF receptors is being attempted in my laboratory.

Effect of age on SF responsiveness

Intrauterine growth is largely independent of pituitary control. In the rat, growth hormone does not appear in the pituitary until the last few days of gestation and, therefore, could not be expected to exert much influence on fetal growth. Heins *et al.* (1970) wished to determine the age at which cartilage acquires its responsiveness to SF. We were unable to detect any stimulation of the rapid rate of sulfate uptake by fetal cartilage when SF was present in the incubation medium (Fig. 4). SF responsiveness was noted in cartilage from rats soon after

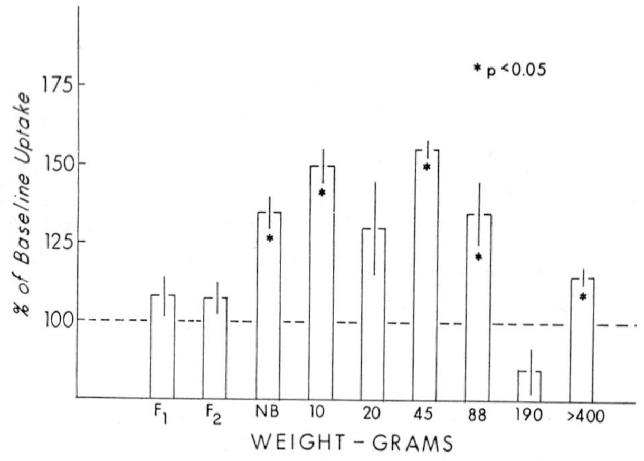

Fig. 4. Comparison of stimulation of sulfate uptake by SF by cartilage of normal rats of different ages (Heins *et al.*, 1970).

birth and reached a peak at about the time of weaning. Sulfate uptake by cartilage from fully mature rats was relatively low and not greatly stimulated *in vitro* by the addition of SF. These observations correlate well with the embryological evidence and suggest that rat fetal chondrocyte cultures probably would not be a satisfactory test object while chondrocytes from the chick appear to be responsive. It is of interest that embryonic chick growth has been shown to be partially dependent on the pituitary (Fugo, 1940).

Generation of SF without GH

The exact relationship between GH and sulfation factor remains unclear. Certainly SF does not share common immunologic determinants of GH. If SF represents an activated GH core, it would be unlikely that substances other than GH would give rise to SF generation. For this reason we have been interested in the mechanism of the remarkable stimulation of growth of hypox rats infected with spargana of *Spirometra mansonoides*, a cat tape worm. The metabolic effects of this worm were first observed in the laboratory of Mueller and Reed (1968) and have subsequently been extensively studied by Steelman *et al.* (1970). Infected hypox rats continue to grow rapidly for a period of about 4 to 6 weeks and then a plateau occurs as immunologic resistance develops. During the period of active growth we have not detected circulating rat growth hormone by radioimmunoassay. The serum from actively growing infected rats can stimulate growth of non-infected hypox rats (Steelman *et al.*, 1970). Garland *et al.* (1971) have demonstrated that serum from such infected rats can stimulate thymidine uptake in cartilage when injected into non-infected hypox rats. This effect is comparable to that induced by GH and cannot be attributed to transfer of SF. Significantly, when serum from actively growing infected rats is mixed with serum from rats who have developed resistance to the worm factor, the ability to stimulate thymidine uptake is neutralized by the inactive serum (Fig. 5).

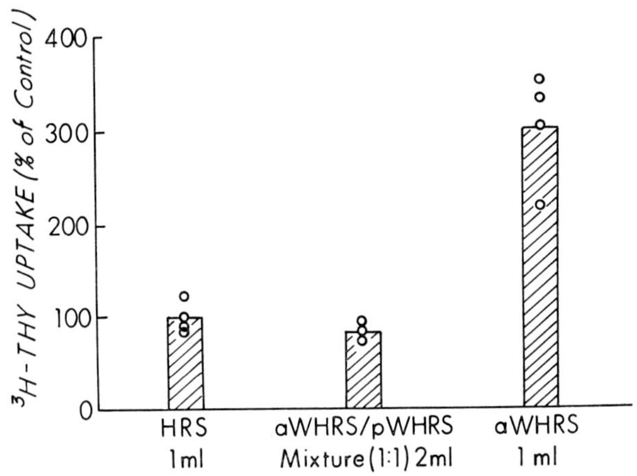

Fig. 5. In this experiment hypox rats were injected with the sera from actively growing hypox rats infected with *Spirometra mansonoides* (aWHRS) by serum from similar rats after growth had plateaued (pWHRS) and mixtures of the two. Forty-eight hours later cartilage was removed and incubated with ^{3}H-thymidine (Garland *et al.*, 1971).

When serum from actively growing infected rats was tested in our standard sulfation factor assay, it was found to contain about six-tenths of the SF activity of normal rat serum (Fig. 6). This moderate SF activity is to be contrasted to the extremely potent worm factor demonstrated by *in vivo* experiments. When we repeated the mixing experiments with serum from actively growing and from plateaued rats using the sulfation factor activity as an end-point, we were unable to detect any neutralization of SF.

Garland *et al.* (unpublished data, 1971) have recently completed experiments with infected normal male rats to determine whether the worm growth factor activates short loop inhibition of growth hormone storage and secretion. During the three weeks that the rats were observed, there was no stimulation of growth over that observed in non-infected control male rats.

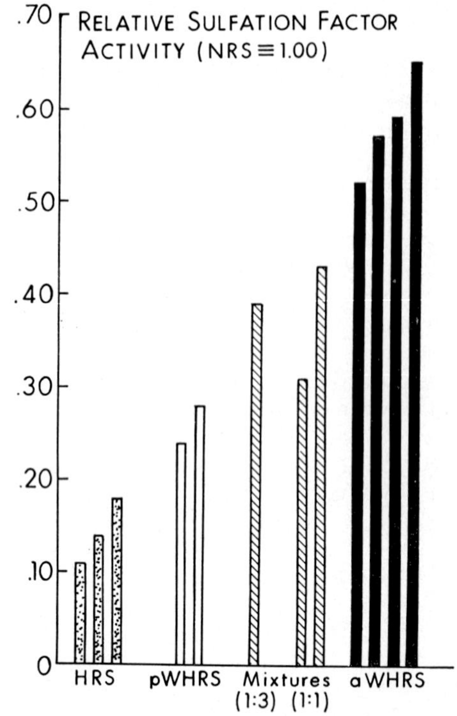

Fig. 6. Addition of aWHRS, pWHRS and mixture of the two directly to hypox rat cartilage *in vitro*. ^{35}S-sulfate uptake compared to normal rat serum stimulation (Garland *et al.*, 1971).

TABLE VI

Effects of Spirometra mansonoides infection on pituitary GH

	Infected	Control	%Δ
Number of rats	7	7	
Pituitary weight, mg	6.34 ± .49	9.61 ± .33	–34%
Pituitary GH, μg	319 ± 34	1001 ± 96	–68%

This is not surprising because in infected hypox rats SF concentration does not exceed normal. We did observe a 34% decrease in pituitary weight (Table VI) and a 68% decrease in pituitary GH concentration. Serum GH content measured after decapitation under pentobarbital anesthesia was somewhat lower in the infected animals, but because of the large variance we could not establish the significance of the difference.

Our concept of growth stimulated by *Spirometra mansonoides* can be summarized as follows: Spargana of this parasite release a potent worm growth factor which is not immunologically related to rat GH. In addition to stimulating growth *in vivo*, this worm factor inhibits GH storage and release, either directly, or through SF. Worm factor does not act on cartilage directly, but does stimulate SF generation. Growth continues until immunologic resistance to the worm factor develops. This resistance is directed against the worm factor and not against the endogenously generated SF.

TABLE VII

Growth without GH

A 9-year-old girl had an operation for craniopharyngioma. Three months post-operatively she was evaluated by Dr. V. Weldon.

Ht 121 cm; *HA* 7 yr; *BA* 8½ yr
PBI 5.2 μg%, *RAI* uptake 2.1% at 24 hours
17 OH 0.2 and 0.4 mg/day
17 KS 0.4 and 0.4 mg/day. No increase with metyrapone.
GH – No values higher than 2 ng/ml, even after arginine and insulin.
SF – 0.5, 0.67, 0.54, 1.0

Height after 21 months was 135 cm. Growth velocity 9.4 cm/yr.

Prolactin concentration: 260 μl equivalent/ml (normal < 5.0 μl equivalent/ml). Roughly 18 mU/ml.

We have been intrigued by the possibility that SF can be induced in human beings in the absence of GH. This thought has been suggested by the finding that normal skeletal growth has been reported in the absence of detectable GH (Holmes *et al.*, 1968). Cases with measurable GH without rises in response to provocative stimuli cannot be considered particularly remarkable. Review of the published reports has convinced us that these patients are usually found to have chromophobe adenomas or craniopharyngiomas. Because these pituitary diseases are often associated with increased levels of prolactin, we suspect that when the secretory rate of prolactin is sufficiently large, normal growth can be sustained by prolactin. Such a case, studied with Dr. Virginia Weldon, is summarized in Table VII. In contrast to GH, prolactin was easily measured by radioimmunoassay and sulfation factor was present in the serum.

It is well known that prolactins and growth hormones of non-primates have amino acid homologies and overlapping biologic properties. Beck *et al.* (1964) have stimulated growth in human pituitary dwarfs with ovine prolactin so that it is not unreasonable to postulate that human prolactin, when present in large amounts, can serve as a growth hormone.

Defective SF generation

Familial dwarfism with elevated serum GH concentrations was first described by Laron *et al.* (1966). My laboratory has been involved in collaborative studies of SF generation in this condition (Daughaday *et al.*, 1969). Laron dwarfs are phenotypically similar to hyposomatotropic dwarfs. Like hyposomatotropic dwarfs, there is a consistent absence or low level of plasma SF in the untreated state. When GH is administered there is no significant rise in SF activity in contrast to the brisk rise of SF in hyposomatotropic dwarfs (Fig. 7).

Recently, through the cooperation of Dr. Elders of Little Rock, we have studied an Arabian boy with this syndrome. No rise in serum SF followed treatment with large doses of GH. It is also noteworthy that 15 mg daily of GH did not induce the anorexia and malaise which would be expected with this dose in normal children or patients with hyposomatotropism.

GH administration will greatly decrease the GH response to arginine in normal individuals (Abrams *et al.*, 1971) and patients with Turner's syndrome (Peake *et al.*, 1971). Arginine infusions were given to our patient before and after GH treatment in a dose of 7.5 mg twice a day (Fig. 8). Despite the loss of some early specimens immediately after arginine infusion, it is evident that plasma GH was elevated compared to the baseline value 60 min and 75 min after arginine.

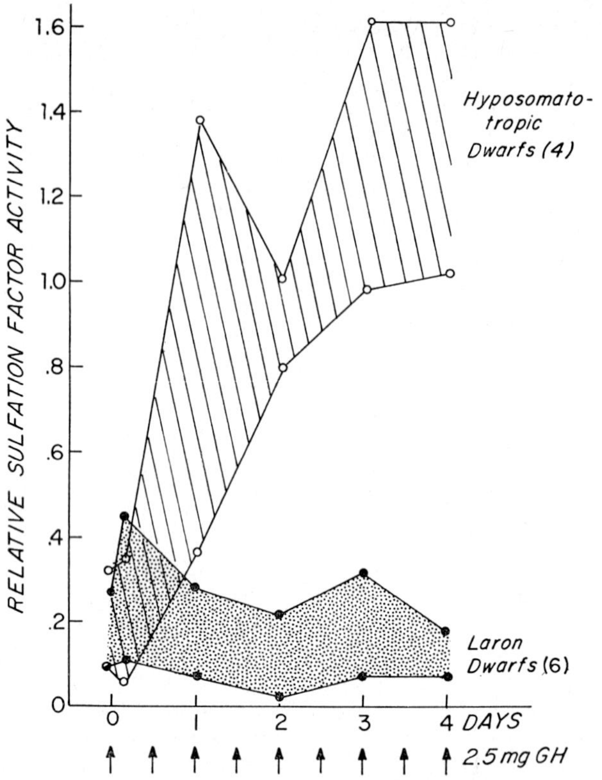

Fig. 7. Relative SF activity of hyposomatotropic and Laron dwarfs before and after GH treatment (Daughaday *et al.*, 1969).

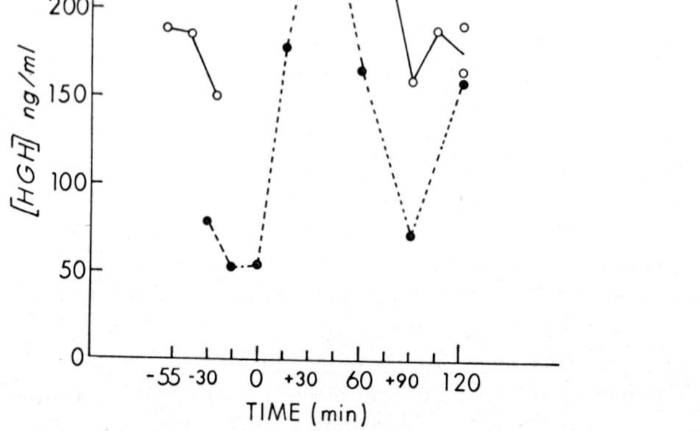

Fig. 8. GH response to arginine before (dotted line) and after (solid line) 15 mg/day of GH for 4 days (Elders *et al.*, 1971).

TABLE VIII

Prolactin activity of circulating GH in Laron dwarfism

	Apparent prolactin activity (mU/ml)
SH's serum	1.84
SH's serum + anti HGH	0.2
HGH standard, equivalent to that present in SH's serum	1.7

Assayed by the *in vitro* mouse mammary gland method of Loewenstein *et al.* (1970). Potency expression only relative as assay curves of HGH and ovine prolactin standard are not parallel.

Laron *et al.* (1971) and more recently, Najjar *et al.* (1971) have found that biochemical indices of response to potent GH in Laron dwarfs are impaired and no sustained stimulation of skeletal growth followed GH treatment.

Initially, Laron *et al.* (1966) suggested that the GH in this syndrome was biologically inactive but immunologically reactive. While we have not been able to establish that the circulating GH has growth promoting activity, we have confirmed its prolactin activity. We assayed prolactin by a modification of the *in vitro* pregnant mouse mammary gland procedure in which the response is a rise in the N-acetyl lactosamine synthetase activity (Loewenstein *et al.*, 1970). When we added serum from our patient to this assay system easily detected prolactin-like activity was found (Table VIII). The prolactin activity per nanogram of GH was closely similar to our reference GH. The prolactin activity in our patients' serum was entirely neutralized by anti-GH gamma globulin.

The syndrome of familial dwarfism with high plasma GH can all be explained by a universal defect in GH receptors, including those involved in SF generation, short loop feedback inhibition of GH secretion, lipolysis, renal handling of phosphate and calcium, insulin sensitivity and β-cytotropic effects.

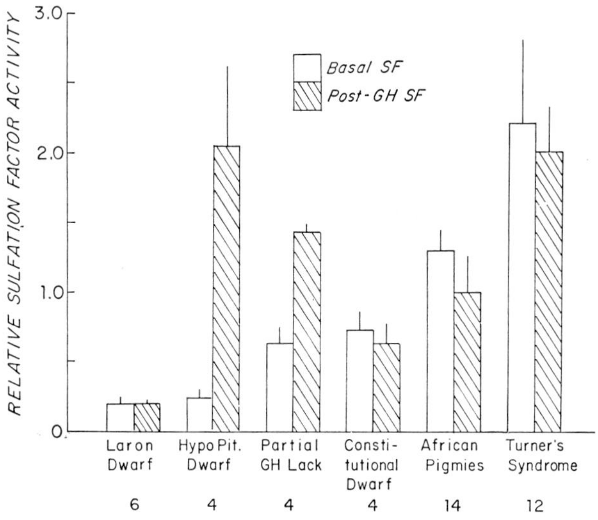

Fig. 9. Comparison of SF levels before and after GH in different types of growth impairment (Daughaday, 1969).

The metabolic defect in Laron dwarfism differs from those recognized in African pigmies by Merimee *et al.* (1968). These individuals have normal GH secretion and sulfation factor and there is no increase in SF with moderate doses of GH, an entirely normal response (Fig. 9) (Daughaday *et al.*, 1969). Pigmies have increased insulin sensitivity and decreased insulin secretion in response to glucose and arginine. Nitrogen retention and lipolysis following GH administration are subnormal. Consideration of these observations leads to the view that the defect in African pigmies could not involve a general defect in GH receptors because short loop feedback is retained and the receptors responsible for sulfation factor generation are also retained. Even a selective defect in some GH receptors is not a sufficient explanation of the growth failure because in addition there is subnormal response to sulfation factor. We suggest that the pigmies have several adaptive changes which lead to short stature.

SUMMARY

Evidence continues to support the hypothesis that growth hormone acts on cartilage through the mediation of a hormonal substance which we have called sulfation factor. ^{3}H-thymidine incorporation into isolated chondrocytes from chick embryos is stimulated by a serum component presumed to be sulfation factor. Embryonic rat cartilage and cartilage from aged rats is unresponsive to sulfation factor. *Spirometra mansonoides* spargana can stimulate sulfation factor generation in the absence of growth hormone. Familial dwarfism with high plasma GH (Laron dwarfism) is characterized by low levels of sulfation factor unresponsive to GH treatment. The GH in this condition has normal lactogenic activity. Exogenous GH does not inhibit endogenous GH secretion in keeping with the hypothesis that this disease is the result of a general disorder of GH receptors. Growth may occur in human beings with markedly impaired GH secretion if prolactin secretion is sufficiently elevated.

REFERENCES

ABRAMS, R. L., GRUMBACH, M. M. and KAPLAN, S. L. (1971): The effect of administration of human growth hormone on the plasma grcwth hormone, cortisol, glucose and free fatty acid response to insulin: Evidence for growth hormone autoregulation in man. *J. clin. Invest.*, *50/4*, 940.

ADAMSON, L. F. and ANAST, C. S. (1966): Amino acid, potassium and sulfate transport and incorporation by embryonic chick cartilage: The mechanism of the stimulatory effects of serum. *Biochim. biophys. Acta (Amst.)*, *121/1*, 10.

DAUGHADAY, W. H., HEINS, J. N., SRIVASTAVA, L. and HAMMER, C. (1968): Sulfation factor: Studies of this removal from plasma and metabolic fate in cartilage. *J. Lab. clin. Med.*, *72/11*, 803.

DAUGHADAY, W. H. and KIPNIS, D. M. (1966): The growth promoting and anti-insulin actions of growth hormone. *Recent Progr. Hormone Res.*, *22*, 49.

DAUGHADAY, W. H., LARON, Z., PERTZELAN, A. and HEINS, J. N. (1969): Defective sulfation factor generation: A possible etiologic link in dwarfism. *Trans. Ass. Amer. Phycns*, *82*, 129.

DAUGHADAY, W. H. and MARIZ, I. K. (1962): Conversion of proline -4-C^{14} to labeled hydroxyproline by rat cartilage *in vitro:* effects of hypophysectomy, growth hormone and cortisol. *J. Lab. clin. Med.*, *59/5*, 741.

DAUGHADAY, W. H. and REEDER, C. (1966): Synchronous activation of DNA synthesis in hypophysectomized rat cartilage by growth hormone. *J. Lab. clin. Med.*, *68/3*, 357.

FUGO, N. W. (1940): Effects of hypophysectomy in the chick embryo. *J. exp. Zool.*, *85/2*, 271.

GARLAND, J. T., LOTTES, M. E., KOZAK, S. and DAUGHADAY, W. H. (1970): Stimulation of DNA synthesis in isolated chondrocytes by sulfation factor. *J. Lab. clin. Med.*, *76/11*, 862.

GARLAND, J. T., RUEGAMER, W. R. and DAUGHADAY, W. H. (1971): Induction of sulfation factor activity by infection of hypophysectomized rats with *Spirometra mansonoides*. *Endocrinology*, *88/4*, 924.

HALL, K. (1970): Quantitative determination of the sulfation factor activity in human serum. *Acta endocr. (Kbh.)*, *63/2*, 338.

HEINS, J. N., GARLAND, J. T. and DAUGHADAY, W. H. (1970): Incorporation of ^{35}S-sulfate into rat cartilage explants *in vitro:* Effects of aging on responsiveness to stimulation by sulfation factor. *Endocrinology, 87/10,* 688.

HOLMES, L. B., FRANTZ, A. G., RABKIN, M. T., SOELDNER, J. S. and CRAWFORD, J. D. (1968): Normal growth with subnormal growth hormone levels. *New Engl. J. Med., 279/11,* 559.

KOUMANS, J. and DAUGHADAY, W. H. (1963): Amino acid requirement for activity of partially purified sulfation factor. *Trans. Ass. Amer. Phycns, 76,* 152.

LARON, Z., PERTZELAN, A., KARP, M., KOWADLO-SILBERGELD, A. and DAUGHADAY, W. H. (1971): Administration of growth hormone to patients with familial dwarfism with high plasma immunoreactive growth hormone: Measurement of sulfation factor, metabolic and linear growth responses. *J. clin. Endocr., 33/332,* 197.

LARON, Z., PERTZELAN, A. and MANNHEIMER, S. (1966): Genetic pituitary dwarfism with high serum concentrations of human growth hormone: A new inborn error in metabolism. *Israel J. med. Sci., 2/1,* 152.

LIBERTI, J. P. (1970): Partial purification of bovine sulfation factor. *Biochem. biophys. Res. Commun., 39/3,* 356.

LOEWENSTEIN, J. E., MARIZ, I. K., PEAKE, G. T. and DAUGHADAY, W. H. (1970): Prolactin bioassay by measurement of N-acetyllactosamine (NAL) synthetase in mouse mammary gland explants. In: *Program of the 52nd Meeting of the Endocrine Society,* p. 127.

LOEWENSTEIN, J. E., MARIZ, I. K., PEAKE, G. T. and DAUGHADAY, W. H. (1971): Prolactin bioassay by induction of N-acetyllactosamine synthetase in mouse mammary gland explants. *J. clin. Endocr., 33/2,* 217.

MERIMEE, T. J., RIMOIN, D. L., CAVALLI-SFORZA, L. C., RABINOWITZ, D., McKUSICK, V. A. (1968): Metabolic effects of human growth hormone in the African pygmy. *Lancet, 2/7501,* 194.

MUELLER, J. F. and REED, P. (1968): Growth stimulation induces infection with *Spirometra mansonoides* spargana in propylthiouracil treated rats. *J. Parasit., 54/1,* 51.

NAJJAR, S. S., KHACHADURIAN, A. K., ILBAWI, M. N. and BLIZZARD, R. M. (1971): Dwarfism with elevated levels of plasma growth hormone. *New Engl. J. Med., 284/15,* 809.

PEAKE, G. T., RIMOIN, D. L., PACKMAN, S. and DAUGHADAY, W. H. (1971): Feedback inhibition of growth hormone (GH) secretion in man. *Clin. Res., 29/1,* 129.

SALMON, W. D., JR. and DAUGHADAY, W. H. (1957): A hormonally controlled serum factor which stimulates sulfate incorporation by cartilage *in vitro. J. Lab. clin. Med., 49/6,* 825.

SALMON, W. D., JR. and DUVALL, M. R. (1970): A serum fraction with 'sulfation factor activity' which stimulates *in vitro* incorporations of leucine and sulfate into protein-polysaccharide complexes, uridine into RNA and thymidine into DNA of costal cartilage from hypophysectomized rats. *Endocrinology, 86/12,* 721.

SALMON, W. D., JR., DUVALL, M. R. and THOMPSON, E. Y. (1968): Stimulation by insulin *in vitro* of incorporation of (^{35}S) sulfate and (^{14}C) leucine into protein-polysaccharide complexes, (^{3}H) uridine into RNA and (^{3}H) thymidine into DNA of costal cartilage from hypophysectomized rats. *Endocrinology, 82/3,* 493.

STEELMAN, S. L., MORGAN, E. R., CUCCARO, A. J. and GLITZER, M. S. (1970): Growth hormone-like activity in hypophysectomized rats implanted with *Spirometra mansonoides* spargana (34453). *Proc. Soc. exp. Biol. (N.Y.), 133/2,* 269.

VANWYK, J. J., HALL, K., VANDENBRANDE, J. L. and WEAVER, R. P. (1971): Further purification and characterization of sulfation factor and thymidine factor from acromegalic plasma. *J. clin. Endocr., 32/3,* 389.

INVESTIGATION WITH A PARTIALLY PURIFIED PREPARATION OF SERUM SULFATION FACTOR: LACK OF SPECIFICITY FOR CARTILAGE SULFATION

WILLIAM D. SALMON JR

Veterans Administration Hospital and Department of Medicine, Vanderbilt University School of Medicine, Nashville, Tennessee, U.S.A.

Sulfate incorporation into the mucopolysaccharide of cartilage in rats is decreased after hypophysectomy and increased in hypophysectomized rats by treatment with growth hormone (Ellis *et al.*, 1953; Denko and Bergenstal, 1955; Murphy *et al.*, 1956). Studies in Dr. Daughaday's laboratory established that the action of growth hormone on cartilage sulfation *in vivo* is mediated by a factor distinct from growth hormone itself (Salmon and Daughaday, 1957). The term 'sulfation factor' has been applied to this non-dialyzable component of serum, which has been found in man (Daughaday *et al.*, 1959) as well as the rat. Subsequently, it has been observed that other components of serum may be inhibitory in the *in vitro* assay for sulfation factor and that a valid estimation of the concentration of the latter in many sera requires at least a partial purification. As an outgrowth of those observations, a simple procedure for the extraction of sulfation factor from serum has been devised. Studies with a serum fraction prepared by this procedure indicate that it has multiple actions on cartilage and also may influence other tissues.

TISSUE INCUBATION METHOD

The experiments to be described involved *in vitro* tests of rat serum, or fractions thereof, in a phosphate-buffered basal incubation medium which contained 14 amino acids, glucose, penicillin and streptomycin. The concentration of serum per ml of incubation medium has been expressed in milliliters. For serum fractions the concentration per ml of incubation medium has been expressed in milliliters of whole serum from which the fraction was prepared. Costal cartilage and diaphragm were obtained from hypophysectomized rats.

NON-DIALYZABLE, HEAT-LABILE INHIBITOR OF SULFATION

Assays for sulfation factor activity suggested that certain pools of serum from normal rats contained one or more substances which inhibited sulfation. Such sera were recognized from the characteristic that increasing the concentration failed to produce the anticipated increase in sulfation factor activity. Serum from starved rats, which was a potent source of inhibitory material, was utilized for studies designed to eliminate inhibitory effects. When tested at low concentrations, serum from normal rats which had been starved for 4 days had no detectable sulfation factor activity (Fig. 1). At higher concentrations such serum inhibited basal incorporation of sulfate, which suggested that the inhibitor therein was not specific for sulfation factor itself. However, sulfation factor activity in serum from fed rats was inhibited by the addition of serum from starved rats. Increasing the concentration of the combined sera resulted in decreasing sulfation factor activity.

Several procedures were successful in enhancing sulfation factor activity of serum from

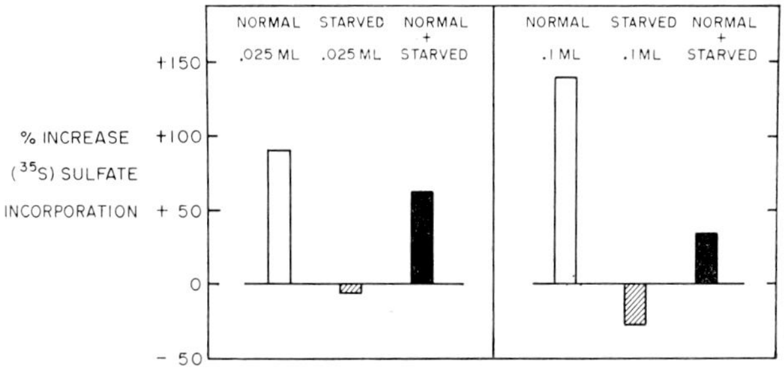

Fig. 1. Inhibition of sulfation factor activity in serum from normal rats by addition of serum from rats which had been starved for 4 days. Rats weighed approximately 90 g at the beginning of starvation. Sulfate incorporation was corrected for differences in sulfate pool of incubation media which resulted from additions of serum. Number of hypophysectomized rats used for assay = 8.

normal rats and diminishing the inhibitory activity of serum from starved rats. A combination of some of these, which included heating at 60° C for 15 to 30 min, passage through a column of an anion-exchange resin (Dowex-1 in bicarbonate form) and dilution, proved useful in additional efforts to fractionate serum. Nevertheless, strongly inhibitory sera were not freed of inhibitory material by these manipulations.

The inhibitor in serum from starved rats was found to be non-dialyzable (Fig. 2). It was

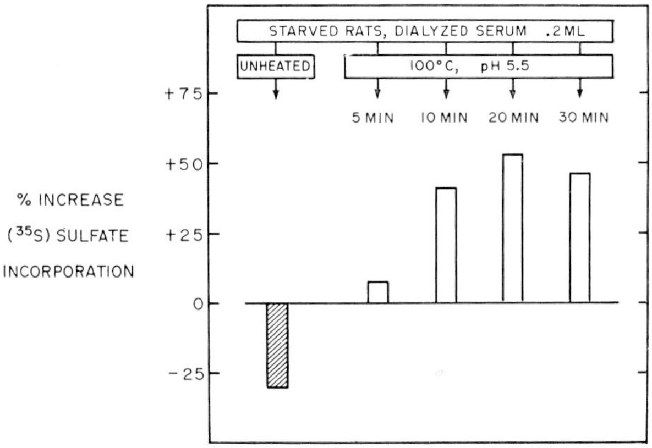

Fig. 2. Persistence after dialysis of inhibitory activity in serum from rats which had been starved for 4 days: effects of acidification and heating at 100° C. Rats weighed approximately 90 g at the beginning of starvation. Serum was dialyzed for 18 hours at 4° C against phosphate-buffered saline, pH 7.4. Dialyzed serum 40% in 0.15 M NaCl was acidified to pH 5.5 with 0.3 N HCl. Portions were heated with mixing in a sealed tube immersed in boiling water for the indicated times and cooled. Unheated, acidified serum and the supernatant solutions from the heated samples after centrifugation were neutralized with 0.3 N NaOH. Number of hypophysectomized rats used for assay = 7.

eliminated in serum acidified to pH 5.5 and heated at 100° C. The residual solution after removal of coagulated protein exhibited sulfation factor activity. This activity, like that in heated, acidified serum from fed rats, was retained after dialysis in boiled cellophane tubing (Fig. 3). The combined effects of individual extracts prepared from serum of starved rats and serum of fed rats were greater than either alone. Extension of these observations led to a simple method for partial purification of sulfation factor.

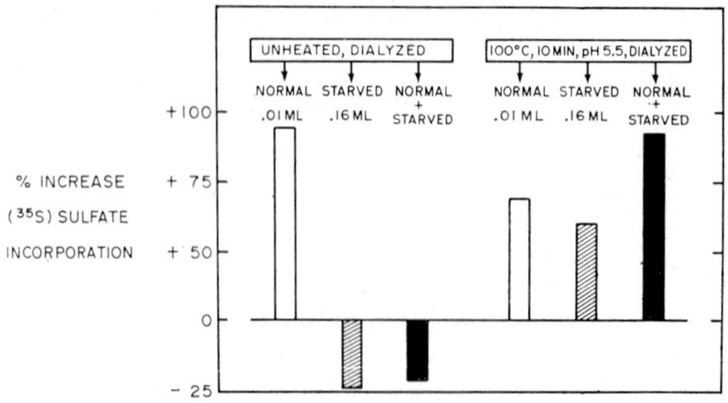

Fig. 3. Elimination by acidification and heating of inhibitory activity in serum from rats which had been starved for 4 days: persistence of sulfation factor activity after dialysis. Rats weighed approximately 90 g at the beginning of starvation. Serum from either normal or starved rats was diluted with an equal volume of 0.15 M NaCl and acidified to pH 5.5 with 0.3 N HCl. A portion was heated with mixing in a sealed tube immersed in boiling water for 10 min and cooled. Unheated, acidified serum and the supernatant solutions from the heated samples after centrifugation were neutralized with 0.3 N NaOH. Each was dialyzed for 18 hours at 4° C against phosphate-buffered saline, pH 7.4. Number of hypophysectomized rats used for assay = 8.

FRACTION WITH SULFATION FACTOR ACTIVITY

Preparation from normal rat serum and characteristics

Serum was separated from pooled blood collected from normal rats after decapitation. Diluted, acidified (pH 5.5) serum was heated at 100° C for 15 min and coagulated protein was discarded. The residual solution (pH 5.5, 0.05 M) was passed through an anion-exchanger, DEAE-Sephadex, and then through a cation-exchanger, CM-Sephadex, which absorbed the active fraction. The eluate (pH 7.5, 0.2 M) was lyophilized, redissolved, placed in cellophane tubing which had been treated with boiling 0.15 M $NaHCO_3$ solution and dialyzed at 4° C against phosphate-buffered saline.

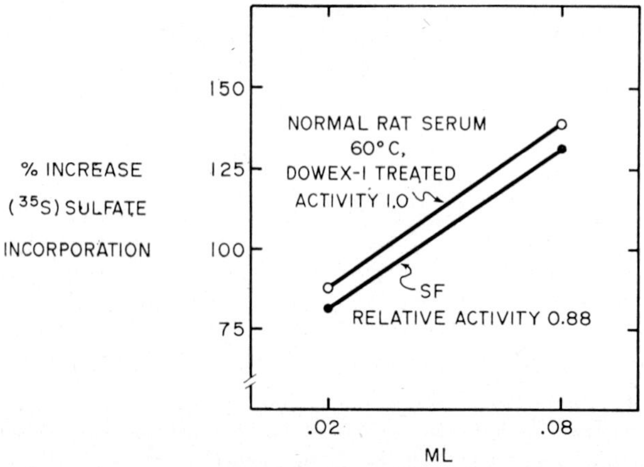

Fig. 4. Assay of serum fraction with sulfation factor activity (SF). The reference serum was the same lot of normal rat serum, which had been heated at 60° C, passed through a column of Dowex-1 resin equilibrated with 0.15 M $NaHCO_3$ and dialyzed for two hours at 4° C against phosphate-buffered saline, pH 7.4. Number of hypophysectomized rats used for assay = 8.

Such a fraction, which was prepared from volumes of serum as small as 3 to 6 ml, exhibited as much as 90% of the apparent sulfation factor activity of serum which had been heated at 60° C and passed through a column of Dowex-1 (Fig. 4). The protein content, as determined by the method of Lowry *et al.* (1951), using bovine serum albumin as standard, was 0.1 to 0.2 mg per ml of the whole serum. When this fraction was subjected to either discontinuous electrophoresis in polyacrylamide gel or column chromatography on Sephadex G-50, most of

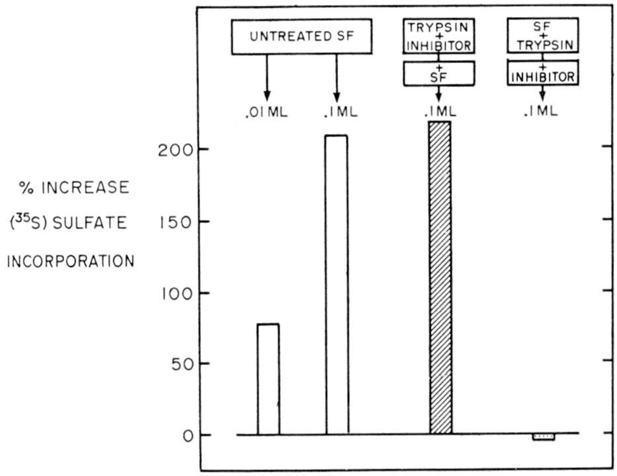

Fig. 5. Activity of SF after treatment with trypsin. Solutions of enzyme and inhibitor were dialyzed against phosphate-buffered saline, pH 7.4, prior to use. A concentrated solution of SF in phosphate-buffered saline, pH 7.4, was treated with 0.005% trypsin for 3 hours at 37° C. Soybean trypsin inhibitor was added to a concentration of 0.05% and the solution was diluted for addition to the incubation medium. Two controls with untreated SF were provided, one in which SF was incubated without additions and another in which the enzyme and inhibitor were mixed prior to addition of SF. Cartilage was incubated for 24 hours in test media supplemented with amino acids, glucose, antibiotics, and 0.1% ovalbumin. Sulfate incorporation was then determined after transfer of tissue samples to the basal medium with [^{35}S] sulfate and further incubation for 2 hours. Number of hypophysectomized rats used for assay = 8.

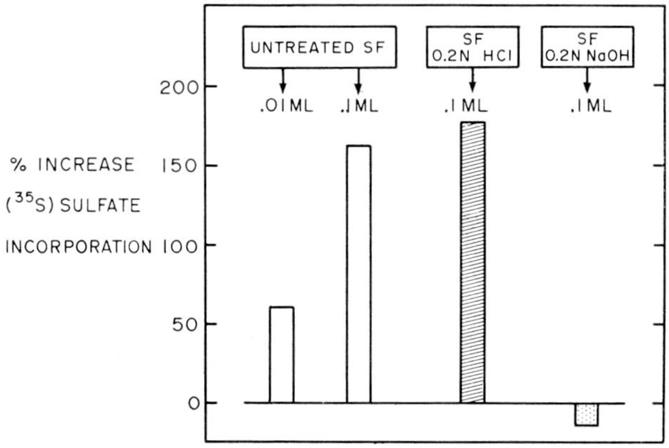

Fig. 6. Activity of SF after treatment with acid or alkali. A concentrated solution of SF was made 0.2 N with either HCl or NaOH, incubated for 3 hours at 37° C, neutralized and diluted for addition to the incubation media. Assay conditions similar to the experiment shown in Figure 5.

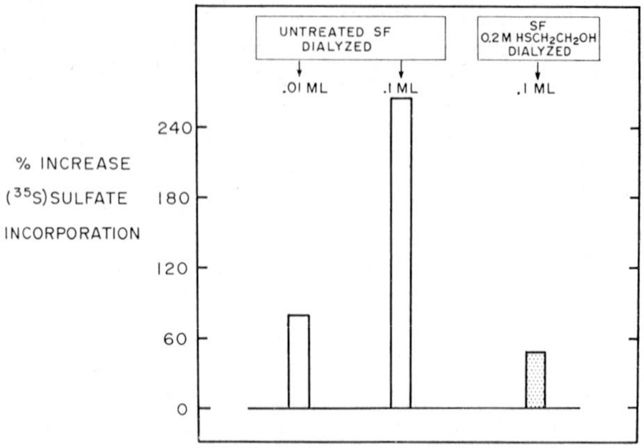

Fig. 7. Activity of SF after treatment with 0.2 M 2-mercaptoethanol. A concentrated solution of SF in phosphate-buffered saline, pH 7.4, was incubated in a nitrogen atmosphere for 4 hours at 37° C with or without mercaptoethanol, diluted and dialyzed for 20 hours at 4° C against repeated changes of phosphate-buffered saline, pH 7.4. Assay conditions similar to the experiment shown in Figure 5.

the protein was separated from the active material. The latter was eluted with 1 M acetic acid from Sephadex in a volume suggesting a molecular weight in the range 7,000 to 10,000.

The protein nature of sulfation factor was supported by loss of activity following treatment with trypsin 0.005% (Fig. 5). Incubation with 0.2 N HCl did not affect activity, while activity was lost on incubation with 0.2 N NaOH (Fig. 6). Because of qualitative similarities in actions of sulfation factor and insulin, the effect of a thiol compound was tested. Reagents of this class have been used in the past to reduce and inactivate both insulin (Lens and Neutelings, 1950; Fraenkel-Conrat and Fraenkel-Conrat, 1950) and serum insulin-like activity (Bürgi *et al.*, 1966; Jakob *et al.*, 1968). Less than 10% of sulfation factor activity was retained after treatment with 2-mercaptoethanol (Fig. 7). Such treatment produced similar destruction of the sulfation-promoting activity of insulin.

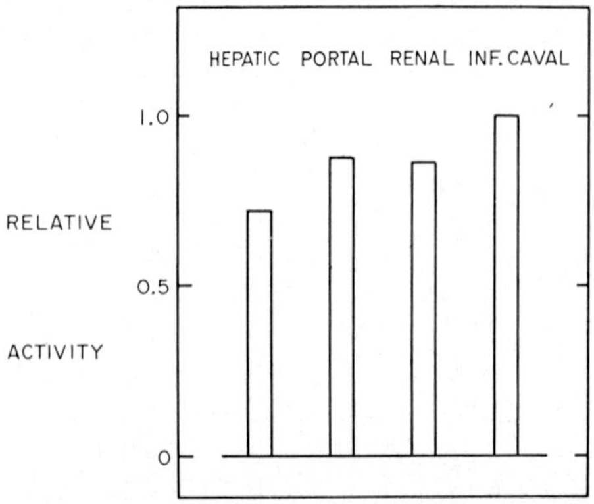

Fig. 8. Assay of fractions prepared from venous plasma from various sites. Blood was collected in heparinized syringes from normal rats anesthetized with pentobarbital. Aliquots of plasma from an individual sampling site of each rat were pooled. The assay of SF was performed at two dilutions. The inferior vena caval fraction was selected as the reference and assigned an activity of 1.0. Number of hypophysectomized rats used for assay = 10.

Other serum and tissue preparations

Fractions which were prepared by the preceding method from plasma from hepatic, portal, renal and inferior caval veins of normal rats had comparable activities (Fig. 8). This was of interest because of the report by McConaghey and Sledge (1970) that sulfation factor was produced by perfused normal rat liver in the presence of bovine growth hormone. While one might have anticipated from their report that a higher concentration would be found in hepatic vein plasma than elsewhere, the relatively long half-life of sulfation factor (3 to 4 hours) found by Daughaday *et al.* (1968) suggests that a small gradient at the source may be adequate to sustain a given level in the general circulation. The detection of such small differences with the present relatively insensitive assay may be impossible.

Fractions were also prepared from a number of tissues (Fig. 9). While those from the kidney, pancreas, heart and pituitary had significant activity, none approached that of serum.

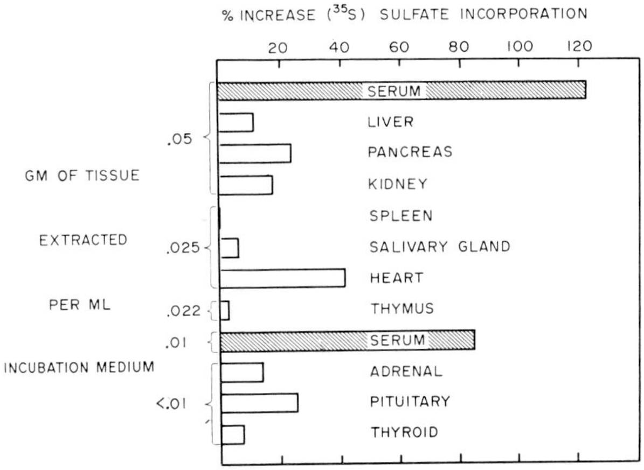

Fig. 9. Sulfation-promoting activity of fractions which were prepared from various tissues of a normal rat by the method used for SF. The concentration of the fraction per ml of incubation medium represents 5% of either the wet weight of tissue sample prior to homogenization or whole adrenal, pituitary and thyroid glands. Number of hypophysectomized rats used for assay = 8. Levels of significance determined from analysis of variance and F test: serum, pancreas and heart $p < .001$; kidney $p < .005$; pituitary $p < .025$; all others $p > 0.1$.

The relationship of such tissue activity to that of serum was not explored further. This study was done prior to the report of Hall *et al.* (1970) that skeletal muscle extracts contain sulfation-promoting activity, and we have not tested a skeletal muscle fraction.

Fractions prepared from pooled serum of hypophysectomized rats had low activity, which appeared to be due to the same substance found in serum from normal rats (Fig. 10). Such fractions contained amounts of protein similar to those prepared from serum of normal rats. This reflects the crude nature of the preparations, which has been indicated above. Other experiments suggested that the low level of sulfation factor was largely inhibited in whole serum. Since the presence of pituitary remnants in the rats providing serum for these studies was not rigidly excluded by microscopic examinations of the sella turcica, the presence of sulfation factor in the pooled serum could not be assumed to have occurred in the complete absence of growth hormone. However, Daughaday (1971) has reported that sulfation factor activity can be induced in serum of hypophysectomized rats by infestation of those animals with *Spirometra mansonoides*, suggesting that the factor does not have an absolute requirement for growth hormone and, therefore, is not an active fragment of the latter.

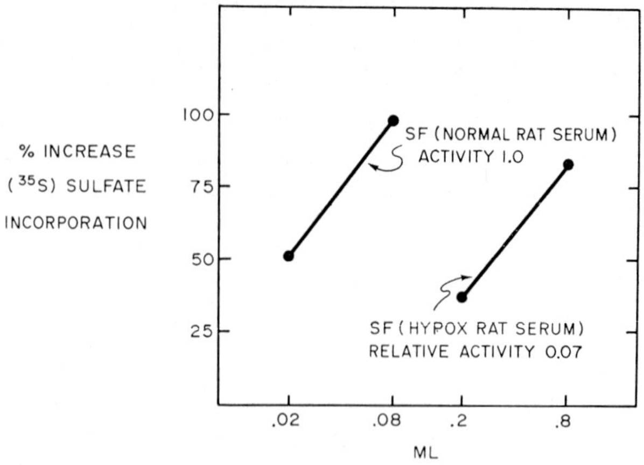

Fig. 10. Assay of SF prepared from pooled serum of 80 hypophysectomized (hypox) rats. Number of hypophysectomized rats used for assay = 8.

Actions on synthesis of protein-polysaccharide complexes, RNA and DNA in cartilage

Early studies with dialyzed serum from normal rats suggested that the action of sulfation factor to increase sulfation in cartilage from hypophysectomized rats involved protein synthesis. Amino acids augmented the effect of the serum factor (Salmon and Daughaday, 1958) and either amino acid imbalance (Salmon, 1960) or puromycin (Salmon *et al.*, 1967) inhibited that effect. Since it had been shown by several groups of investigators that the sulfated mucopolysaccharide of cartilage existed in a complex form with noncollagenous protein (Shatton and Schubert, 1954; Mathews and Lozaityte, 1958; Muir, 1958; Partridge and Davis, 1958)

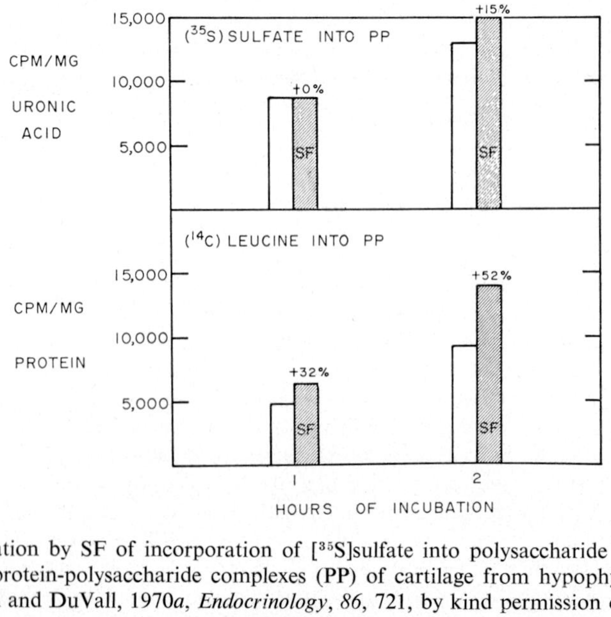

Fig. 11. Stimulation by SF of incorporation of [^{35}S]sulfate into polysaccharide and [^{14}C]leucine into protein of protein-polysaccharide complexes (PP) of cartilage from hypophysectomized rats. (From Salmon and DuVall, 1970*a*, *Endocrinology*, *86*, 721, by kind permission of the editors).

and since Gross *et al.* (1960) found that this complex was metabolized as a unit, it was postulated that the action of sulfation factor on sulfation might be directed primarily at synthesis of protein in protein-polysaccharide complexes (Salmon, 1960). A general action on cartilage protein synthesis seemed likely after the report by Daughaday and Mariz (1962) that serum from normal rats increased incorporation of proline into hydroxyproline of collagen, the other major component of cartilage intercellular substance. Subsequent experiments utilizing a serum fraction with sulfation factor activity have supported such hypotheses.

This fraction increased incorporation of sulfate into polysaccharide and leucine into pro-

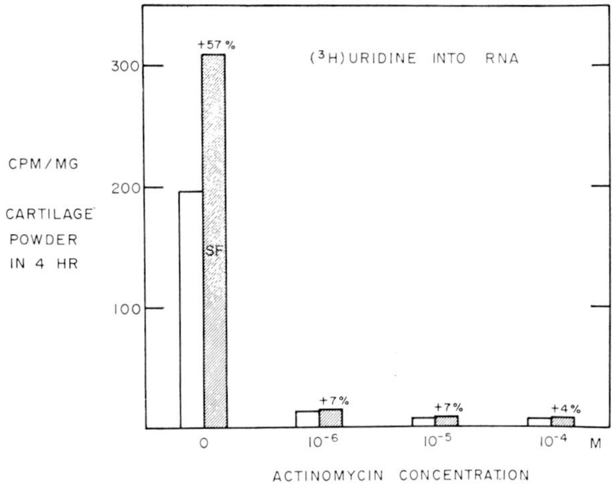

Fig. 12. Stimulation by SF of incorporation of [^{3}H]uridine into RNA of cartilage from hypophysectomized rats: inhibition by actinomycin D. Period of incubation 4 hours. (From Salmon and DuVall, 1970*a*, *Endocrinology*, *86*, 721, by kind permission of the editors).

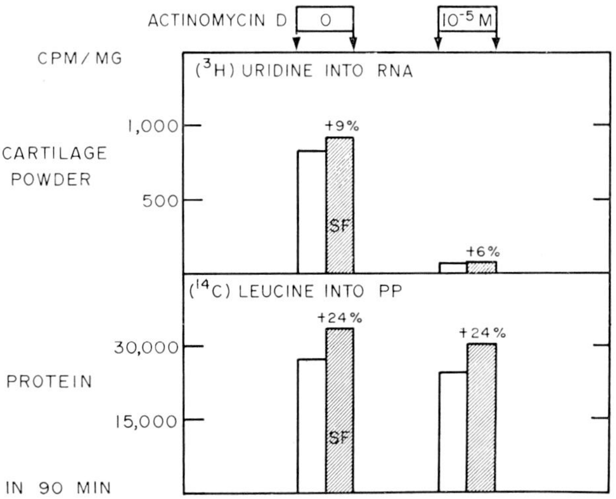

Fig. 13. Stimulation by SF of incorporation of [^{3}H]uridine into RNA and [^{14}C]leucine into protein of PP of cartilage from hypophysectomized rats: dissociation of effects in the presence of actinomycin D. Period of incubation 90 min. (From Salmon and DuVall, 1970*a*, *Endocrinology*, *86*, 721, by kind permission of the editors).

tein of protein-polysaccharide complexes. Incorporation of the labeled compounds was stimulated in roughly parallel manner after several hours, but the first effect was increased incorporation of leucine into protein (Fig. 11). This, together with earlier findings, led to conclusions that the action of sulfation factor on sulfation does reflect synthesis of protein-polysaccharide complexes, that limited sulfation in cartilage of rats following hypophysectomy is a consequence of limited ability to synthesize protein and that the latter is due to deficiencies of growth hormone and the dependent sulfation factor (Salmon and DuVall, 1970*a*).

The serum fraction also stimulated incorporation of uridine into cartilage RNA. This effect was inhibited more than 90% by actinomycin D in concentrations from 10^{-6} to 10^{-4} M (Fig. 12). However, an increase in uridine incorporation into RNA was not detected in any experiment prior to enhancement of leucine incorporation into protein of protein-polysaccharide complexes. Furthermore, the former could be inhibited by actinomycin D without a parallel inhibition of the latter (Fig. 13). Therefore, the action of sulfation factor to initiate increased protein-polysaccharide synthesis did not appear to be dependent upon new RNA synthesis.

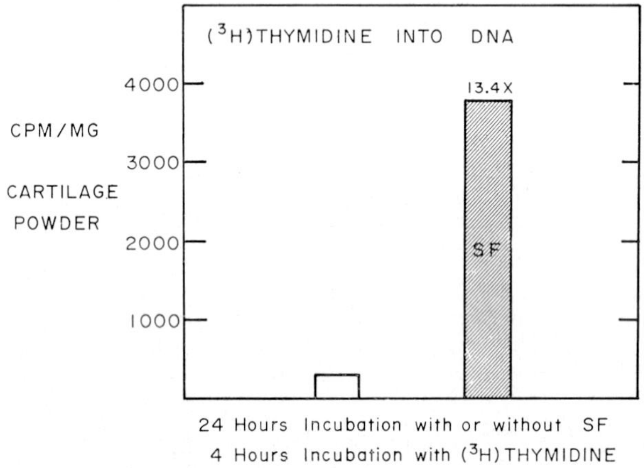

Fig. 14. Stimulation by SF of incorporation of [^{3}H]thymidine into DNA of cartilage from hypophysectomized rats. Cartilage was incubated in the basal medium with or without SF for 24 hours and thymidine incorporation was then determined after transfer of tissue to the basal medium with [^{3}H]thymidine and further incubation for 4 hours. (From Salmon and DuVall, 1970*a*, *Endocrinology*, *86*, 721, by kind permission of the editors).

Daughaday and Reeder (1966) were the first to report that serum from normal rats enhanced *in vitro* incorporation of thymidine into cartilage DNA. The serum fraction with sulfation factor activity had a similar action. An effect was detectable 6 to 10 hours after beginning incubation of cartilage with the fraction and after 24 hours of incubation thymidine incorporation into DNA was greatly increased (Fig. 14).

General actions on protein synthesis in cartilage and muscle

Because it seemed likely that sulfation factor had a general action on protein synthesis in cartilage, studies have been extended to include other tissues and cells. A comparison of the effects of a serum fraction with sulfation factor activity, growth hormone and insulin on leucine incorporation into trichloroacetic acid-insoluble protein of cartilage and muscle from hypophysectomized rats has been made (Salmon and DuVall, 1970*b*). Both tissues were stimu-

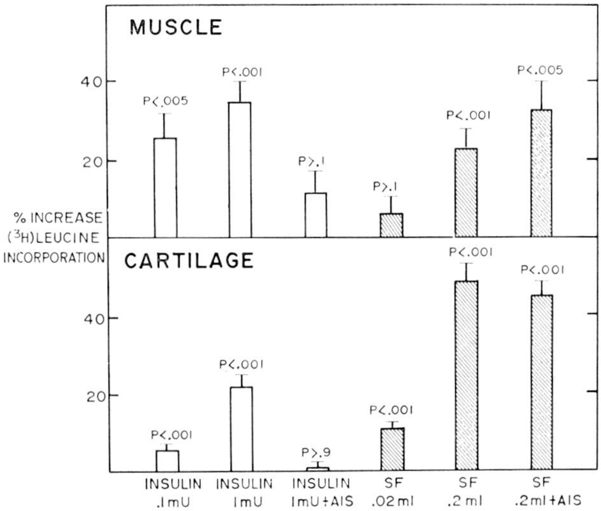

Fig. 15. Stimulation by SF of incorporation of (^{3}H)leucine into trichloroacetic acid-insoluble protein of muscle and cartilage from hypophysectomized rats: comparison with insulin and effect of anti-insulin serum (AIS). Period of incubation 2 hours. (From Salmon and DuVall, 1970*b*, *Endocrinology*, *87*, 1168, by kind permission of the editors).

lated by the serum fraction and those effects were not inhibited by anti-insulin serum (Fig.15). The minimum effective concentration for cartilage was less than for muscle. By contrast, cartilage was less sensitive than muscle to either growth hormone or insulin and often unaffected by even very high concentrations of the former. Nevertheless, cartilage was more sensitive than muscle to growth hormone *in vivo*. Serum from hypophysectomized rats treated with growth hormone in doses which were effective on muscle and cartilage *in vivo* contained sulfation factor sufficient to stimulate both tissues from hypophysectomized rats *in vitro*. These findings suggested that sulfation factor is involved in the *in vivo* action of growth hormone to stimulate protein synthesis in muscle as well as cartilage.

Other actions

Glucose uptake by muscle was also enhanced by the serum fraction, although glucose was not required in the medium for the effect on leucine incorporation into protein by either muscle or cartilage.

A beef serum fraction with sulfation factor activity has also been tested in cell culture. When added to a medium which contained bovine serum albumin and Eagle's minimum essential medium supplemented with other amino acids, that fraction increased both number and total protein of HeLa cells (Salmon and Hosse, 1971).

DISCUSSION

These results provide evidence that actions of the so-called 'sulfation factor' are broad and not limited to the promotion of sulfate incorporation by cartilage. Indeed, the effect on sulfation appears to be the result primarily of stimulation of protein synthesis in protein-polysaccharide complexes.

All of the reported actions of sulfation factor have been produced by insulin in sufficient concentrations. Because of the striking parallel between the qualitative effects of sulfation factor and insulin, the suggestion has been made that the former is represented in the total serum insulin-like activity (Salmon and DuVall, 1970*b*). In this regard the heat stability of sulfation factor is noteworthy. Jakob *et al.* (1968) have reported that insulin-like activity in human serum consists of at least two components, one of which is destroyed by heat at 80° C. They found that only about 5% of the insulin-like activity was heat stable, which would suggest that sulfation factor could be only a small part of the total activity. However, the critical dependence of the heat stability of sulfation factor upon pH suggests caution in any conclusion at this time about the contribution of sulfation factor to serum insulin-like activity.

Finally, it seems possible also, in view of the qualitative similarities in actions of sulfation factor and insulin, that the non-dialyzable, heat-labile inhibitor of cartilage sulfation is the same as or similar to an inhibitor of the action of insulin on glucose uptake by muscle which has been detected by other investigators in serum from either diabetic rats (Bornstein and Park, 1953; Bornstein, 1953; Whitney and Young, 1957) or normal rats fasted for 24 hours (Krahl *et al.*, 1959). While the inhibitory action does not appear to be specific for sulfation factor itself, the inhibitor may profoundly impair the tissue response to sulfation factor. Consequently, sulfation factor activity in whole serum is related to concentrations of both sulfation factor and the inhibitor. The striking increase of this inhibitor in serum of rats during starvation suggests a physiological regulatory role.

REFERENCES

Bornstein, J. (1953): Insulin-reversible inhibition of glucose ultilization by serum lipoprotein fractions. *J. biol. Chem.*, *205*, 513.

Bornstein, J. and Park, C. R. (1953): Inhibition of glucose uptake by the serum of diabetic rats. *J. biol. Chem.*, *205*, 503.

Bürgi, H., Müller, W. A., Humbel, R. E., Labhart, A. and Froesch, E. R. (1966): Non-suppressible insulin-like activity of human serum. I. Physicochemical properties, extraction and partial purification. *Biochim. biophys. Acta (Amst.)*, *121*, 349.

Daughaday, W. H. (1971): Sulfation factor regulation of skeletal growth. A stable mechanism dependent on intermittent growth hormone secretion. *Amer. J. Med.*, *50*, 277.

Daughaday, W. H., Heins, J. N., Srivastava, L. and Hammer, C. (1968): Sulfation factor: Studies of its removal from plasma and metabolic fate in cartilage. *J. Lab. clin. Med.*, *72*, 803.

Daughaday, W. H. and Mariz, I. K. (1962): Conversion of proline-U-C^{14} to labeled hydroxyproline by rat cartilage *in vitro*: Effects of hypophysectomy, growth hormone, and cortisol. *J. Lab. clin. Med.*, *59*, 741.

Daughaday, W. H. and Reeder, C. (1966): Synchronous activation of DNA synthesis in hypophysectomized rat cartilage by growth hormone. *J. Lab. clin. Med.*, *68*, 357.

Daughaday, W. H., Salmon Jr, W. D. and Alexander, F. (1959): Sulfation factor activity of sera from patients with pituitary disorders. *J. clin. Endocr.*, *19*, 743.

Denko, C. W. and Bergenstal, D. M. (1955): The effect of hypophysectomy and growth hormone on S^{35} fixation in cartilage. *Endocrinology*, *57*, 76.

Ellis, S., Hublé, J. and Simpson, M. E. (1953): Influence of hypophysectomy and growth hormone on cartilage sulfate metabolism. *Proc. Soc. exp. Biol. (N.Y.)*, *84*, 603.

Fraenkel-Conrat, J. and Fraenkel-Conrat, H. (1950): The essential groups of insulin. *Biochim. biophys. Acta (Amst.)*, *5*, 89.

Gross, J. I., Mathews, M. B. and Dorfman, A. (1960): Sodium chondroitin sulfate-protein complexes of cartilage. II. Metabolism. *J. biol. Chem.*, *235*, 2889.

Hall, K., Holmgren, A. and Lindahl, U. (1970): Purification of a sulphation factor from skeletal muscle of rat. *Biochim. biophys. Acta (Amst.)*, *201*, 398.

Jakob, A., Hauri, Ch. and Froesch, E. R. (1968): Nonsuppressible insulin-like activity in human serum. III. Differentiation of two distinct molecules with nonsuppressible ILA. *J. clin. Invest.*, *47*, 2678.

KRAHL, M. E., TIDBALL, M. E. and BREGMAN, E. (1959): Preparation and anti-insulin activity of lipoprotein fractions from rat serum. *Proc. Soc. exp. Biol. (N.Y.)*, *101*, 1.

LENS, J. and NEUTELINGS, J. (1950): The reduction of insulin. *Biochim. biophys. Acta (Amst.)*, *4*, 501.

LOWRY, O. H., ROSEBROUGH, N. J., FARR, A. L. and RANDALL, R. J. (1951): Protein measurement with the folin phenol reagent. *J. biol. Chem.*, *193*, 265.

MATHEWS, M. B. and LOZAITYTE, I. (1958): Sodium chondroitin sulfate-protein complexes of cartilage. I. Molecular weight and shape. *Arch. Biochem.* *74*, 158.

McCONAGHEY, P. and SLEDGE, C. B. (1970): Production of 'sulphation factor' by the perfused liver. *Nature (Lond.)*, *225*, 1249.

MUIR, H. (1958): The nature of the link between protein and carbohydrate of a chondroitin sulphate complex from hyaline cartilage. *Biochem. J.*, *69*, 195.

MURPHY, W. R., DAUGHADAY, W. H. and HARTNETT, C. (1956): The effect of hypophysectomy and growth hormone on the incorporation of labeled sulfate into tibial epiphyseal and nasal cartilage of the rat. *J. Lab. clin. Med.*, *47*, 715.

PARTRIDGE, S. M. and DAVIS, H. F. (1958): The chemistry of connective tissues. IV. The presence of a non-collagenous protein in cartilage. *Biochem. J.* *68*, 298.

SALMON JR, W. D. (1960): Importance of amino acids in the actions of insulin and serum sulfation factor to stimulate sulfate uptake by cartilage from hypophysectomized rats. *J. Lab. clin. Med.* *56*, 673.

SALMON JR, W. D. and DAUGHADAY, W. H. (1957): A hormonally controlled serum factor which stimulates sulfate incorporation by cartilage *in vitro*. *J. Lab. clin. Med.*, *49*, 825.

SALMON JR, W. D. and DAUGHADAY, W. H. (1958): The importance of amino acids as dialyzable components of rat serum which promote sulfate uptake by cartilage from hypophysectomized rats *in vitro*. *J. Lab. clin. Med.*, *51*, 167.

SALMON JR, W. D. and DUVALL, M. R. (1970*a*): A serum fraction with 'sulfation factor activity' stimulates *in vitro* incorporation of leucine and sulfate into protein-polysaccharide complexes, uridine into RNA, and thymidine into DNA of costal cartilage from hypophysectomized rats. *Endocrinology*, *86*, 721.

SALMON JR, W. D. and DUVALL, M. R. (1970*b*): *In vitro* stimulation of leucine incorporation into muscle and cartilage protein by a serum fraction with sulfation factor activity: differentiation of effects from those of growth hormone and insulin. *Endocrinology*, *87*, 1168.

SALMON JR, W. D. and HOSSE, B. R. (1971): Stimulation of HeLa cell growth by a serum fraction with sulfation factor activity. *Proc. Soc. exp. Biol. (N.Y.)*, *136*, 805.

SALMON JR, W. D., VON HAGEN, M. J. and THOMPSON, E. Y. (1967): Effects of puromycin and actinomycin *in vitro* on sulfate incorporation by cartilage of the rat and its stimulation by serum sulfation factor and insulin. *Endocrinology*, *80*, 999.

SHATTON, J. and SCHUBERT, M. (1954): Isolation of a mucoprotein from cartilage. *J. biol. Chem.*, *211*, 565.

WHITNEY, J. E. and YOUNG, F. G. (1957): Some hormonal influences on the glucose uptake of normal rat diaphragm *in vitro*. *Biochem. J.*, *66*, 648.

HUMAN GROWTH HORMONE AND SULFATION FACTOR

KERSTIN HALL and KNUT UTHNE

Department of Endocrinology and Metabolism, Karolinska sjukhuset, and Department of Biochemistry, AB KABI, Stockholm, Sweden

It is well established that human serum contains some factor(s) which stimulate the incorporation *in vitro* of labelled sulfate into cartilage (Salmon and Daughaday, 1957; Daughaday *et al.*, 1958; Almqvist 1960, Hall, 1970). The appearance of this so-called sulfation factor (SF) in serum seems to be dependent upon human growth hormone (HGH). Growth hormone itself has no discernible effect on cartilage.

In the present investigation SF activity was determined by measuring the incorporation of labelled sulfate into cartilaginous pelvis from 12-day-old embryonic chicks (Hall, 1970). The serum samples or serum fractions were incubated 4 or 6 hours in a medium containing an optimal concentration of amino acids. Normal human serum added to a final concentration of 2.5–60% gave a dose-response curve with linearity between the logarithm of the dose and the sulfate uptake. Porcine insulin in concentrations of 250–4000 μU/ml caused a slight, but not significant, stimulation of sulfate uptake. HGH in concentrations of 10–1000 ng/ml had no effect on sulfate uptake. The determination of SF activity has usually been made as a 4- or 6-point-assay with 6 cartilaginous leaflets at each dose level. The lambda values of the method are approximately 0.20. One unit of SF activity is defined as the activity of one ml of an arbitrary reference serum.

We have studied the induction of SF activity both *in vivo* and *in vitro*. In eight patients with pituitary dwarfism, SF activity was determined before and after i.v. injection of 2 mg HGH. The low basal value of SF was not changed during the first hour. Three hours after the injection, when the HGH values approached normal levels, there was a significant rise in SF activity. The mean difference in SF activity between 1 and 3 hours after HGH injection was

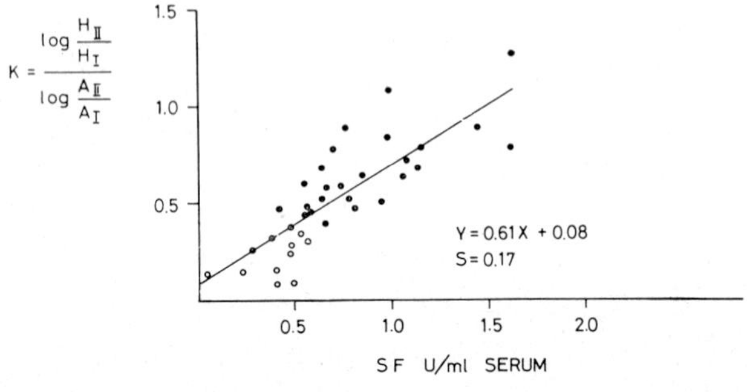

Fig. 1. Sulfation factor activity in serum and corresponding growth rate, expressed as K-value, in patients with pituitary dwarfism before (○) and during (●) treatment with HGH. The line represents the regression equation. (From Hall and Olin, 1971).

0.52 U of SF (Hall, 1971). There was a slight decrease in glucose 30 min after administration, but no significant change in the insulin determined radioimmunologically. The human growth hormone, used in this study and during long-term treatment of patients with pituitary dwarfism, was prepared by AB Kabi-Recip according to the method of Roos *et al.* (1963). One mg of this preparation corresponds to 2 IU of the first international reference standard of HGH for immunoassay.

In collaboration with P. Olin we have followed the SF activity and the growth rate in 20 patients with pituitary dwarfism during treatment with HGH, for 6 months to 2½ years (Hall and Olin, 1971). All these patients had a growth pattern consistent with the diagnosis of pituitary dwarfism and they showed no rise in immunoreactive HGH in serum after insulin-induced hypoglycemia. The age range of the patients at the start of the treatment was 6–25 years. Eleven of the patients had previously been treated with various other HGH preparations for 1 to 5 years. The body height curves of the untreated patients were linear when the logarithms of both body height and age were used. The slopes of the height curves increased during treatment with HGH. The slopes, calculated as a K-value,

$$K = \frac{\log H_{II} - \log H_{I}}{\log A_{II} - \log A_{I}}$$

(where H_I and H_{II} are heights in cm at ages A_I and A_{II}) were used as a measure of growth rate. When the SF activities in serum, before and during HGH treatment, were correlated with the corresponding K-values of growth rate, the following regression equation was obtained (Fig. 1):

$$K = 0.61 \times \text{SF unit} + 0.08$$

The standard deviation around the computed line was 0.17. When the SF activities in serum

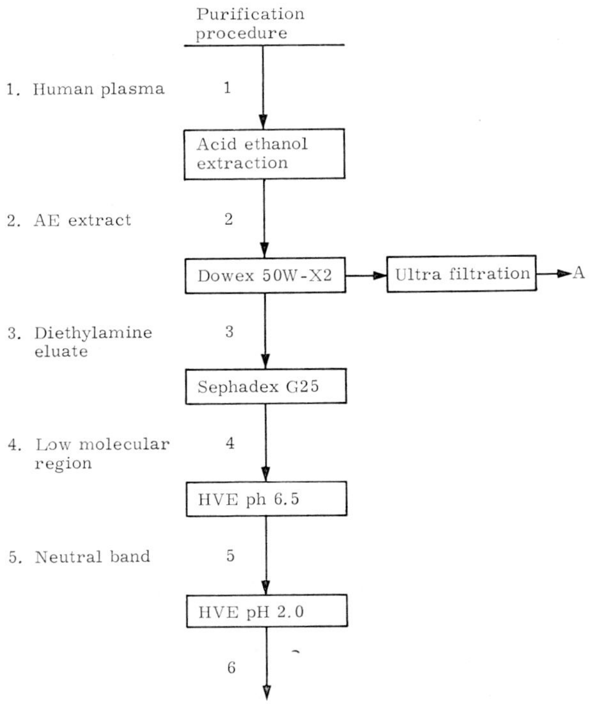

Fig. 2. Purification scheme for sulfation factor from human plasma.

and the K-values of the growth rate were correlated to the dosage of HGH, both decreased with the duration of HGH treatment. We have taken these results in patients with pituitary dwarfism as indirect evidence that HGH induces the SF factor in serum.

We have further studied the effect of HGH on the induction of SF activity *in vitro*. A crude microsomal fraction prepared from rat liver was incubated in the presence of a pharmacological dose of HGH, NaDH, glucose-6-phosphate, nicotinamide and $MgSO_4$ for 6 hours. The lyophilized supernatant after ethanol precipitation contained SF activity. During gel chromatography on Sephadex G25 the biological activity was recovered in the same molecular range as partial purified SF activity from human plasma (Hall and Uthne, 1971).

During purification of the sulfation factor from normal human plasma, we have followed various biological effects on normal and hypophysectomized rats. The acid ethanol extraction procedure was used as an initial step in purification (Van Wyk *et al.*, 1969, 1971*a*). Further purification of SF from human plasma was done in collaboration with J. J. Van Wyk (Van Wyk 1971*b*). The purification procedure in the present study is shown in Figure 2. Since we suspected polymerization of our active component, we treated our lyophilized eluate from Dowex 50 W-X2, with 2-mercaptoethanol, before gel chromatography, using Sephadex G25.

A partially purified SF fraction (A), see Figure 2, containing 2.1 U of SF activity and 46 μU insulin per mg protein, was administered to hypophysectomized rats to study its effect on the width of their epiphyseal lines (Table I). The animals, five in each group, were given

TABLE I

Tibia-assay

Test substance	Daily dose		Epiphyseal width (mean ± S.E.)
	μg protein	Units	
Control	—	—	13.45 ± 0.89
S.F.	47	0.10	16.91 ± 1.60[1]
S.F.	2400	50	17.40 ± 1.75[1]
HGH	40	0.08	21.06 ± 0.43[2]
HGH	80	0.16	20.63 ± 2.82[1]

[1] $p < 0.01$
[2] $p < 0.001$

a daily dose of either 47 or 2400 μg of partially purified SF for one week. In the tibia-assay (Evans *et al.*, 1943) SF caused a significant increase in the width of the epiphyseal lines at a daily dosage of 47 μg (0.1 SF units). We also observed that SF caused a thickening of the skin. Although the dose of SF was increased, we were unable to obtain the maximal increase which was achieved with 40 or 80 μg HGH. This failure may reflect inhibitory material in this SF preparation.

The effect of normal serum on sulfation is the result of a primary action on DNA (Van Wyk *et al.*, 1969), RNA, and on protein synthesis in cartilage (Salmon *et al.*, 1968; Salmon and DuVall, 1970*a*, *b*). As shown by Salmon and DuVall (1970*a*, *b*), a fraction from rat serum, containing SF activity, also stimulates the incorporation of amino acids into the diaphragm of hypophysectomized rats. In experiments on α-aminoisobutyric acid (^{14}C-AIB) transport in diaphragms from hypophysectomized rats, we used a SF fraction, containing 10.5 SF units per mg protein and no detectable radioimmunological insulin. The incubation medium contained amino acids (Salmon and DuVall, 1970*a*, *b*). As shown in Figure 3, a slight increase

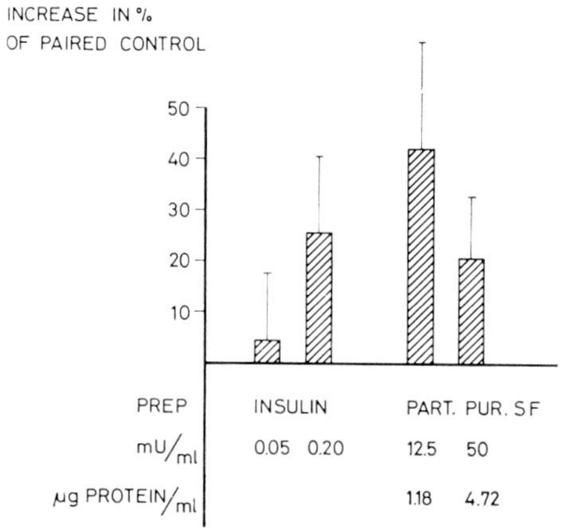

Fig. 3. Transport of ^{14}C-α-aminoisobutyric acid (AIB) into diaphragms from hypophysectomized rats.

in ^{14}C-AIB transport was obtained with 0.2 mU insulin per ml incubation medium. The lowest dose of SF used (1.18 μg protein per ml incubation medium containing 12.5 mU SF activity) caused a 40% increase in ^{14}C-AIB transport compared with the paired hemidiaphragms. By administering a larger dose of SF (50 mU/ml) the increase above that of the control was only 20%.

There are similarities in the *in vitro* effects of insulin and serum fractions containing SF activity. Both stimulate the transport and incorporation of amino acids into proteins of the cartilages and diaphragms of hypophysectomized rats. It has been suspected that SF has a similar stimulating effect as insulin on glucose uptake in hemidiaphragm and epididymal fat tissue from rats but no evidence has so far been presented. Human serum contains insulin-

TABLE II

SF activity and nonsuppressible ILA during purification of 10,000 ml human plasma

Preparation	SF activity, U and and fiducial limits (p = 0.95)	Nonsuppressible ILA, mU and fiducial limits (p = 0.95)
1. Plasma	5650 (3000–9110)	1270 (710–2780)
2. AE extract	1040 (680–1490)	265 (33–609)
3. Diethylamine eluate from Dowex-50W-X2	860 (410–1400)	98 (7–240)
4. Low molecular region from Sephadex G25	840 (480–1280)	239 (96–428)
5. Neutral band from HVE pH 6.5	430 (160–1270)	60 (36–98)
6. HVE pH 2.0	104 (39–179)	33 (1–113)

like activity (ILA), and only 10–20% of this activity is insulin-dependent. ILA in serum increases four hours after i.v. injection of HGH (Zahnd *et al.*, 1960), thus showing similarities with the occurrence of SF in serum.

During purification of SF, insulin, ILA, and nonsuppressible insulin-like activity have been determined at each step. Epididymal fat from normal rats incubated with ^{14}C-glucose, and the conversion of ^{14}C-glucose to $^{14}CO_2$ were used in the determination of ILA. Nonsuppressible ILA was determined after addition of insulin antibodies. The determinations were made as a 4- or 6-point-assay with 6 rats at each dose level, and porcine insulin as reference. The acid ethanol extract contained both nonsuppressible ILA and insulin. This procedure has been used for the purification of nonsuppressible ILA (Bürgi *et al.*, 1966; Jacob *et al.*, 1968; Poffenbarger *et al.*, 1968). Our recovery of nonsuppressible ILA in the acid ethanol extract was 20% and somewhat less in the subsequent steps. During gel chromatography with Sephadex G25, equilibrated with 1-M HAc containing 0.01-M 2-mercaptoethanol, we observed SF and nonsuppressible ILA concentrated in the same low molecular region (Fig. 4). The pooled fractions with SF activity contained 70 U of SF activity and 20 mU of non-

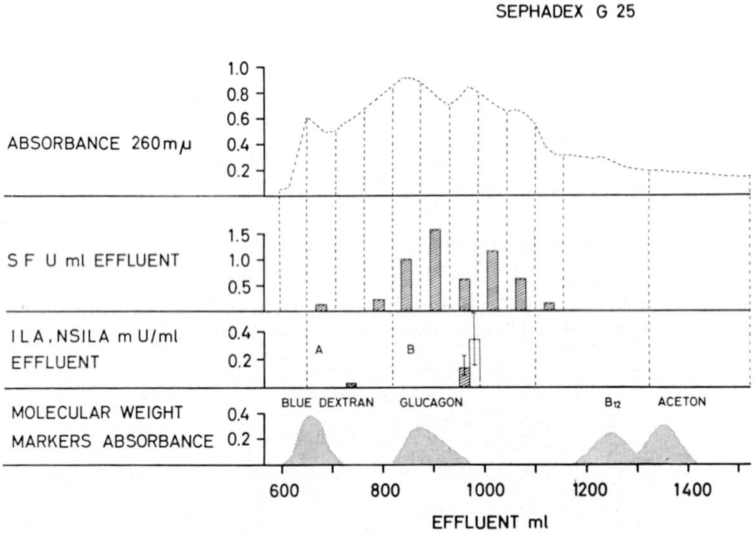

Fig. 4. Gel chromatography on Sephadex G25 (5 × 89 cm) equilibrated with 1-M HAc containing 0.01-M 2-mercaptoethanol. Diethylamine eluate containing 0.37 g protein in a volume of 10 ml was applied. The fractions were pooled as indicated by the dotted vertical lines and assayed for biological activities. SF activity and ILA are shown in the columns shaded-in patch. The open column represents the nonsuppressible ILA. (From Hall and Uthne, 1971).

suppressible ILA per mg protein. This means a purification of about 10,000 times compared with the activity in the original plasma. Treatment with 2-mercaptoethanol abolished the radioimmunological property of insulin. Also during further purification of SF by high voltage electrophoresis on paper at pH 6.4 and 2.0 the nonsuppressible ILA was recovered in the same band as SF.

During the whole purification procedure of SF, the nonsuppressible ILA was present in the same fractions as the SF. The recovery of SF, nonsuppressible ILA, and insulin is shown in Table II. At each purification step there is about 100–300 μU nonsuppressible ILA per SF unit (Table III).

TABLE III

Recovery of SF activity, nonsuppressible ILA, and insulin in per cent of original activity during purification from human plasma

Preparation	SF activity	Non-suppressible ILA	Insulin	Non-suppressible ILA, μU per U of SF activity
1. Plasma	100	100	100	225
2. AE extract	18	21	35	255
3. Diethylamine eluate from Dowex 50W-X2				
4. Low molecular region from Sephadex G25	15	19	0	285
5. Neutral band from HVE pH 6.5	8	5	0	140
6. HVE pH 2.0	2	3	0	310

At present, we do not know if our purified SF is homogeneous. The possibility remains open that SF and nonsuppressible ILA are two polypeptides with similar physical and chemical properties. A tentative hypothesis is that the sulfation factor represents the nonsuppressible insulin-like activity in human serum.

REFERENCES

ALMQVIST, S. (1960): Studies on sulfation factor (SF) activity of human serum. *Acta endocr., (Kbh.), 35/3*, 381.

BÜRGI, H., MÜLLER, W. A., HUMBEL, R. E., LABHART, A. and FROESCH, E. R. (1966): Non-suppressible insulin-like activity of human serum. I. Physico-chemical properties, extraction and partial purification. *Biochim. biophys. Acta (Amst.), 121/2*, 349.

DAUGHADAY, W. H., SALMON, W. D. and ALEXANDER, F. (1958): Sulfation factor activity of sera from patients with pituitary disorders. *J. clin. Endocr., 19/7*, 743.

EVANS, H. M., SIMPSON, M. E., MARX, W. and KIBRICK, E. (1943): Bioassay of the pituitary growth hormone. Width of the proximal epiphyseal cartilage of the tibia in hypophysectomized rats. *Endocrinology, 32/1*, 13.

HALL, K. (1970): Quantitative determination of the sulphation factor activity in human serum. *Acta endocr. (Kbh.), 63/2*, 338.

HALL, K. (1971): Effect of intravenous administration of human growth hormone on sulphation factor activity in serum of hypopituitary subjects. *Acta endocr. (Kbh.), 66/3*, 491.

HALL, K. and OLIN, P. (1972): Sulphation factor activity and growth rate of patients with pituitary dwarfism during long term treatment with human growth hormone. *Acta endocr. (Kbh.), 69/3*, 417.

HALL, K. and UTHNE, K. (1971): Some biological properties of purified sulphation factor from human plasma. *Acta med. scand., 190/1–2*, 137.

JACOB, A., HAURI, C. H. and FROESCH, E. R. (1968): Non-suppressible insulin-like activity in human serum. III. Differentiation of two distinct molecules with non-suppressible ILA. *J. clin. Invest., 47/12*, 2678.

POFFENBARGER, P. L., ENSINCK, J. W., DIETER, K. H. and WILLIAMS, R. H. (1968): The nature of human serum insulin-like activity (ILA): Characterization of ILA in serum and serum fractions obtained by acid-ethanol extraction and adsorption chromatography. *J. clin. Invest., 47/2*, 301.

Roos, P., Fevold, H. R. and Gemzell, C. A. (1963): Preparation of human growth hormone by gel filtration. *Biochim. biophys. Acta (Amst.)*, *74/3*, 525.

Salmon Jr, W. D. and Daughaday, W. H. (1957): A hormonally controlled serum factor which stimulates sulfate incorporation by cartilage *in vitro*. *J. Lab. clin. Med.*, *49/6*, 825.

Salmon Jr, W. D. and DuVall, M. R. (1970*a*): A serum fraction with 'sulfation factor activity' stimulates *in vitro* incorporation of leucine and sulfate into protein polysaccharide complexes, uridine into RNA and thymidine into DNA of costal cartilage from hypophysectomized rats. *Endocrinology*, *86/4*, 721.

Salmon Jr, W. D. and DuVall, M. R. (1970*b*): *In vitro* stimulation of leucine. Incorporation into muscle and cartilage protein by a serum fraction with sulfation factor activity: Differentiation of effects from those of growth hormone and insulin. *Endocrinology*, *87/6*, 1168.

Salmon Jr, W. D., DuVall, M. R. and Thompson, E. Y. (1968): Stimulation by insulin *in vitro* of incorporation of (^{35}S) sulfate and (^{14}C) leucine into protein polysaccharide complexes, (^{3}H) uridine into RNA and (^{3}H) thymidine into DNA of costal cartilage from hypophysectomized rats. *Endocrinology*, *82/2*, 493.

Van Wyk, J. J., Hall, K. and Weaver, R. P. (1969): Partial purification of sulphation factor and thymidine factor from plasma. *Biochim. biophys. Acta (Amst.)*, *192/3*, 560.

Van Wyk, J. J., Hall, K., Van den Brande, J. L. and Weaver, R. P. (1971*a*): Further purificatiou and characterization of sulphation factor and thymidine factor from acromegalic plasma. *J. clin. Endocr.*, *32/2*, 389.

Van Wyk, J. J., Hall, K., Van den Brande, J. L., Weaver, R. P., Uthne, K., Hintz, R. L., Harrison, J. H. and Mathewson, P. (1971*b*): Partial purification from human plasma of a small peptide fraction with sulfation factor and thymidine factor activities. *This Volume*, pp. 155–167.

Zahnd, G., Steinke, J. and Renold, A. E. (1960): Early metabolic effects of human growth hormone. *Proc. Soc. exp. Biol. (N.Y.)*, *105*, 455.

V. Human chorionic somatomammotropin

EFFECT OF CHEMICAL MODIFICATIONS AND TRYPTIC DIGESTION ON BIOLOGICAL AND IMMUNOLOGICAL ACTIVITIES OF HUMAN CHORIONIC SOMATOMAMMOTROPIN (HCS)

P. NERI, C. AREZZINI, G. CANALI, F. COCOLA and P. TARLI

Research Centre, I.S.V.T. Sclavo, Siena, Italy

Among the many problems faced during studies on human chorionic somatomammotropin (HCS), those concerned with the structure-function relationship are particularly interesting, because of their practical and theoretical implications. It is well known that HCS shows a lower growth promoting activity in comparison with HGH (Friesen, 1965; Florini *et al.*, 1966; Li, 1970) despite the fact that both hormones share a number of structural similarities. On the other hand, HCS and HGH have a comparable lactogenic potency (Josimovich and MacLaren, 1962).

The existence in the HGH molecule of separate cores for the lactogenic and somatotropic activities have been postulated on the basis of indirect evidence (Li, 1968). The recent findings of Yamasaki *et al.* (1970) provide direct experimental support for this view, at least in the case of BGH. Thus it could be reasonably assumed that in the molecules of HGH and HCS the active centres for the lactogenic potency are equivalent but not those for the growth promoting activity.

Furthermore, the question may be raised whether the cores of growth promoting activity are different, or identical but not equally accessible to the target cell. If the first alternative is true, a comparison between the amino acid sequences would shed light on the importance of the different amino acid residues; otherwise, one might hope to enhance the activity of HCS by modifying the molecule, *e.g.* by limited proteolysis, in such a way as to render the active core more accessible. A suitable experimental approach to these problems is to study the effects of a number of chemical or enzymatic modifications of the HCS molecule on the immunological and biological activities of HCS and, whenever possible, to compare these effects with those obtained from similar experiments carried out on HGH.

We shall report in this paper some preliminary results of a study to determine which groups are not essential for the HCS activities.

MATERIALS

Hormones and immunochemicals

Human chorionic somatomammotropin (HCS) was isolated from frozen human placenta by the procedure previously described (Neri *et al.*, 1970*a*). Standard sheep prolactin (lot NIH-P-S-8) was generously supplied by the Endocrinology Study Section of the National Institutes of Health. HGH for bioassay was prepared from fresh human pituitaries, according to Raben (1957). HCS labelled by S.O.R.I.N. (Saluggia, Italy) with ^{125}I and anti-HCS serum (SCLAVO, Siena, Italy) were used.

Chemicals

The following reagents were used for the modification of HCS: *o*-methylisourea (NBCo., Cleveland, Ohio, U.S.A.); 1,2-cyclohexanedione, 1-cyclohexyl-3-(2-morpholinyl-(4)-ethyl)-carbodiimide metho-*p*-toluene sulphonate, 2-nitrophenylsulphenyl chloride, iodoacetamide and tetranitromethane (Fluka, Buchs, Switzerland); 2-hydroxy-5-nitrobenzyl bromide (BDH, Poole, England); dithiothreitol (Calbiochem, Los Angeles, California, U.S.A.); maleic anhydride, 2-mercaptoethanol and iodoacetic acid (Merck, Darmstadt, Germany). Trypsin (analytical grade, Boehringer, Mannheim, Germany) and soybean inhibitor (Mann, New York, N.Y., U.S.A.) were used in experiments involving limited proteolysis of HCS.

The following standards were used for amino acid analysis: standard amino acid mixture type 1 (Beckman, Palo Alto, California, U.S.A.), cyclohexanedione-arginine and S-carboxymethylcysteine prepared in our laboratory according to Toi *et al.* (1967) and Armstrong and Lewis (1951) respectively. All other chemicals were reagent grade and used without further purification, with the exception of urea, which was de-ionized on a mixed bed ion exchange resin (Amberlite MB-1; Rohm and Haas Co., Philadelphia, U.S.A.).

Animals

Female rats of the Sprague Dawley strain coming from our colony, were hypophysectomized at 26-28 days of age, by the transauricular method of Falconi and Rossi (1964). The animals were used for the tibia test after a postoperative period of 12-14 days.

Pigeons 4-5 weeks old were obtained from a local source.

METHODS

1. *Gel filtration and electrophoresis*

Gel filtrations were carried out on Sephadex G-100 or G-25 (Pharmacia, Uppsala, Sweden) prepared according to the manufacturer's directions. In general, columns of 2.5 $\times$ 100 cm and 3.5 $\times$ 45 cm were used for G-100 and G-25 gels, respectively. Acrylamide gel electrophoresis was performed using an apparatus as described by Davis (1964), but built in our laboratory with minor modifications.

2. *Amino acid analysis*

Acid hydrolysis of HCS derivatives was performed for 24 hours under vacuum at 110°C in constant-boiling HCl. The hydrolysate was subjected to amino acid analysis according to Moore and Stein (1963) in an amino acid analyser (Beckman, Model Unichrom).

3. *Chemical modifications of HCS*

(a) Reaction with maleic anhydride. HCS was subjected to the reaction of Sia and Horecker (1968) and the derivative characterized as previously described (Neri *et al.*, 1970*b*).

(b) Reaction with o-methylisourea. The reaction was carried out at 4-6°C for 5 days using a protein concentration of 2.5% and 0.5 M *o*-methylisourea. The pH was adjusted daily to 10.5. The reaction product was isolated by dialysis and lyophilization. A further purification was achieved by gel filtration on Sephadex G-100 in 0.1 M $NaHCO_3$-Na_2CO_3 buffer, pH 10.4. The extent of the modification was evaluated by determining both the homoarginine in an amino acid analyser after acid hydrolysis of the protein (Kimmel, 1967), and the residual free amino groups (Mokrasch, 1967).

(c) Modification of arginyl residues. The guanidine group of arginyl residues was condensed with 1,2-cyclohexanedione by the method of Toi *et al.* (1967). The modified HCS was isolated by dialysis and lyophilization. Further purification was achieved by gel filtration on Sephadex G-100 in 0.01 M NH_4HCO_3, pH 8.4. The disappearance of free arginine and the resulting cyclohexanedione-arginine were measured in the acid hydrolysate in order to estimate the number of modified groups.

(d) Amidation of carboxyl groups. The HCS carboxyl groups were made to react with ammonium chloride by a soluble carbodiimide, adapting the conditions described by Armstrong and McKenzie (1967) for β-lactoglobulin A. The amide nitrogen increase estimated in the product of hydrolysis using a micro method (Botti *et al.*, unpublished data; Leach and Parkhill, 1955) was taken as a measure of the extent of the reaction.

(e) Modification of tryptophan residue. Tryptophan was arylated by two different reagents, *e.g.* 2-hydroxy-5-nitrobenzyl bromide (HNB-Br) and 2-nitrophenylsulphenyl chloride (NPS-Cl) using the same procedures as for ribonuclease T1 (Terao and Ukita, 1969) and for HGH (Brovetto-Cruz and Li, 1969), respectively. The completion of the tryptophan modification was checked by two independent methods, *i.e.* with *p*-dimethylaminobenzaldehyde (Spies and Chambers, 1949) and spectrophotometrically at 412 nm for HNB-HCS (Koshland *et al.*, 1964), and 365 nm for NPS-HCS (Scoffone *et al.*, 1968).

(f) Reaction with tetranitromethane. Tyrosil residues were nitrated by tetranitromethane according to Sokolovsky *et al.* (1966). The amount of modified tyrosines was determined, both by the absorption at 428 nm due to the nitrotyrosines formed (Sokolovsky *et al.*, 1966) and by measuring the number of residual tyrosines, after acid hydrolysis, with an amino acid analyser.

(g) Modification of disulphide bonds. HCS disulphide bonds were reduced by dithiothreitol in the presence of iodoacetamide, using the same conditions for the preparation of the corresponding reduced S-carbamidomethylated HGH (Bewley *et al.*, 1968). Reduced and carboxymethylated HCS was prepared by 2-mercaptoethanol and iodoacetic acid in the presence of 8 M urea (Hirs, 1967). Both derivatives were freed of excess reagent by gel filtration on Sephadex G-25 and further purified on Sephadex G-100 in 0.01 M NH_4HCO_2, pH 8.4. The number of reduced and alkylated bridges was estimated by the analysis of residual half-cystine and newly formed S-carboxymethyl-cysteine in the acid hydrolysate.

4. *Preparation of tryptic digests of HCS*

Digestions of HCS with trypsin were carried out at 25°C in water-jacketed vessels under nitrogen, with the pH held constant at 9.5 by adding 0.02 N KOH. HCS (5 mg/ml) in 0.1 M KCl was incubated with crystalline trypsin (0.5 mg/ml) in a ratio of 300 : 1 w/w. Soybean inhibitor (5 mg/ml) was added to stop the digestion after the uptake of 1, 2, 3, 4 alkali equivalents per mole of HCS, in separate experiments. The trypsin and the inhibitor solution were prepared immediately before use, all in 0.1 M KCl. The product of the proteolysis was recovered by dialysis against distilled water at 4°C, and lyophilization. Acrylamide electrophoretic patterns of trypsin-digested HCS show an increasing number of fast moving bands as the proteolysis progresses.

5. *Radioimmunoassays*

Chemical derivatives and tryptic digestion were assayed in a system for HCS (Aubert *et al.*, 1970) using the double antibody technique. For the HCS assay, antigens were incubated at 37°C for 1 hour with anti-HCS rabbit serum at a 1 : 900 dilution. After 1 hour, 10^3 c.p.m. of ^{125}I-HCS were added, followed by a 2-hour incubation period at 37°C. The second antibody (goat) was then added and the tubes kept at 4°C overnight. The labelled complex was membrane-filtered and counted in a well-automatic γ-counter (Autogamma, Packard, Illinois, U.S.A.). The effect of modifications on the immunoreactivity was estimated from the altered ability to displace the label, when an equivalent amount of HCS derivative was substituted for that of native HCS in the HCS system. The percentage of the initial immunoreactivity was calculated by plotting the HCS weight which displaced the same amount of label as did the derivative, against the corresponding weight of the derivative itself. The slope of the curve in the linear range multiplied by 100 gave the percentage of initial immunoreactivity.

6. *Assay of growth promoting activity*

The tibia test of Greenspan *et al.* (1949) was used. Native or modified HCS was dissolved in saline (1 mg/ml) with the addition of a few microlitres of NaOH, diluted to the desired concentration with saline, and injected into groups of 5-6 hypophysectomized rats. The final pH was about 7. The control group was given the solvent mixture used for preparation of the solution, diluted in the same manner. In each experiment, care was taken to use HCS of the batch from which the modified hormones were derived. Thus each rat was injected daily intraperitoneally with 0.5 ml of a solution containing 100 μg/ml of protein, over a period of 4 days. Tibiae taken from animals killed by decapitation were dissected and stained. Growth promoting potency was estimated from the cartilage growth, measured by a micrometer eyepiece (Baush and Lomb, Rochester, N.Y.) mounted on a microscope.

7. *Assay of prolactin-like potency*

The pigeon crop-sac assay according to Nicoll (1967) was used. Control and test groups, each of 5-6 animals, were treated with solvent or hormone solution prepared as described for the bioassay of growth promoting activity. Each pigeon was injected intradermally twice daily with 0.1 ml of a solution containing 200 μg/ml of hormone, over a 2-day period. The increase in the average mucosal dry weight of the treated animals over that of the controls was taken as a measure of the prolactin-like potency.

8. *Calculations*

Statistical analyses and other numerical calculations were carried out with an Olivetti Programma 101 desk top computer.

RESULTS

1. *Characterization of HCS derivatives*

In Table I we have summarized the chemical modifications of HCS considered in this work, together with the changes in the absorption spectrum and the percentage of modified groups. This was established by determining independently the residual and modified groups, both in native and modified HCS.

2. *Effects of chemical and tryptic modifications on immunological activity (Table II)*

(a) Chemical modifications. The chemical modifications carried out on HCS impair the immunological activity. The modification of tyrosil and argynil groups, the maleylation of

TABLE I

Characteristics of modified HCS

Derivative	Modified group	No.of groups in HCS*	% of modified groups	ε (278 nm)	(λ max)**
HCS		—	—	17430	—
Mal – HCS	Free amino	11 (b)	96.9 (a)	22400	—
Guan – HCS	ε-free amino	10 (b)	79.6 (b)	17700	—
CHD – HCS	Guanidyl	10 (b)	91.6 (b)	20300	—
CMC – HCS	Carboxyl	22 (c)	65.4 (c)	17450	—
HNB – HCS	Tryptophanyl	1 (a)	112.0 (a)	18680	18470 (412)
NPS – HCS	Tryptophanyl	1 (a)	142.0 (c)	33700	7300 (365)
TNM – HCS	Tyrosyl	8 (b)	84.2 (a)	49800	27700 (428)
RCAM – HCS	Disulphide bridge	4 (b)	45.5 (b)	17600	—
RCOM – HCS	Disulphide bridge	4 (b)	110.5 (b)	17700	—

Abbreviations: Mal: maleylated; Guan: guanidinated; CHD: cyclohexanedione-treated; CMC: amidated; HNB: 2-hydroxy-5-nitrobenzyl bromide-treated; NPS: 2-nitrophenylsulphenyl chloride-treated; TNM: nitrated; RCAM: reduced and S-carbamido-methylated; RCOM: reduced and S-carboxymethylated.

* Calculated on the basis of M.W. 21000.

** Molar extinction coefficient, at the wavelength (in brackets) of the characteristic maximum in the spectrum of HCS derivative.

(a) Determined spectrophotometrically; (b) determined by amino acid analysis in the hydrolysate of the native or modified protein; (c) determined by specific chemical methods.

All the spectra were carried out in 0.9% NaCl solution, with the exception of that of HNB-HCS and NPS-HCS, which were carried out in 0.1 M KOH and 80% acetic acid, respectively.

TABLE II

Effects of chemical (left) and tryptic (right) modifications on immunological activity of HCS

Derivative	% of initial reactivity in HCS assay*	Base uptake (eq./21000 g)	% of initial reactivity in HCS assay*
HCS	100	0	100
Mal – HCS	3	1	100
Guan – HCS	53	2	68
CHD – HCS	0	3	61
CMC – HCS	35	4	44
HNB – HCS	25		
NPS – HCS	7		
TNM – HCS	4		
RCAM – HCS	14		
RCOM – HCS	27		

See Table I for abbreviations of chemically modified HCS.

* Percentage of initial reactivity = weight of HCS/weight of HCS derivative displacing the same amount of label.

free NH_2 and the arylation of tryptophan by NPS appear to exert the greatest influence on the ability of HCS to bind the specific antibodies. On the other hand, the conversion of lysil into homo-argynil residues and the amidation of carboxyl groups have less effect. Although only one tryptophan is present in the HCS molecule, it appears to play an important role, because its modification drastically reduces the immunological activity. However, the method and/or the type of reaction used in the preparation of HCS derivatives is also of importance. In the case of tryptophan, if the reaction was carried out with HNB, the immunological activity was influenced to a lesser degree.

(b) Tryptic digestion. The effect of tryptic digestion on immunological activity increases with the progress of bond cleavage. At values of alkali uptake equivalent to one bond broken per mole, no differences are observed in comparison with untreated HCS. At two equivalents, only 68% of the initial activity is retained, the hydrolysis of the second bond representing a critical point. The successive cleavage of three and four linkages reduces the activity to 61 and 44%, respectively. If these losses are referred to the number of bonds cleaved, constant values are obtained. This indicates that, within this range, each cleavage of a bond has a comparable effect to the previous cleavage.

3. *Effects of chemical and tryptic modifications on prolactin-like activity (Table III)*

(a) Chemical modifications. With the exception of RCAM-HCS, the activity of which was not significantly different from that of HCS at the dose level assayed, all the modifica-

TABLE III

Effect of various modifications on prolactin-like activity of HCS as assayed by the pigeon crop sac test

Experimental groups	Total dose (μg)	Increase in mucosal dry weight (mg)*	P value vs. HCS**	P value vs. control**
Prolactin	4	8.3 ± 0.8	>0.25 (94)	0.00 –0.005 (82)
HGH	80	8.9 ± 1.0	>0.25 (80)	0.00 –0.005 (68)
HCS	80	8.5 ± 0.4	–	0.00 –0.005 (128)
a. Chemically modified HCS:				
Mal – HCS	80	2.1 ± 0.6	0.00 –0.005 (83)	0.10 –0.25 (71)
Guan – HCS	80	3.5 ± 0.6	0.00 –0.005 (86)	0.00 –0.005 (74)
CHD – HCS	80	0.5 ± 0.3	0.00 –0.005 (80)	>0.25 (68)
CMC – HCS	80	2.0 ± 0.4	0.00 –0.005 (81)	0.005–0.025 (69)
NPS – HCS	80	0.00	0.00 –0.005 (79)	0.10 –0.25 (67)
HNB – HCS	80	3.6 ± 0.3	0.005–0.025 (78)	0.00 –0.005 (66)
TNM – HCS	80	1.2 ± 0.5	0.00 –0.005 (81)	>0.25 (69)
RCAM – HCS	80	7.4 ± 0.8	>0.25 (86)	0.00 –0.005 (74)
RCOM – HCS	80	0.4 ± 0.3	0.00 –0.005 (83)	>0.25 (71)
b. Tryptic digestion; alkali uptake:				
1 eq./mole	80	7.0 ± 1.3	>0.25 (79)	0.00 –0.005 (67)
2 eq./mole	80	6.5 ± 0.8	>0.25 (77)	0.00 –0.005 (65)
3 eq./mole	80	4.2 ± 0.1	0.025–0.10 (78)	0.00 –0.005 (65)
4 eq./mole	80	2.9 ± 0.6	0.00 –0.005 (78)	0.005–0.025 (66)

See Table I for abbreviations of chemically modified HCS.
* Mean ± standard error.
** Degrees of freedom are reported in brackets.

tions tested impaired the prolactin-like activity to a certain extent. Only one disulphide bridge was reduced and carbamidomethylated, while complete modification was achieved in RCOM-HCS. Since in this latter case, impaired activity was detected, it can be inferred that at least one disulphide bridge is inessential for lactogenic activity.

(b) Tryptic digestion. The chief purpose of these experiments was the demonstration of parallel behaviour between the prolactin-like and immunological activities. The lactogenic potency was, in fact, gradually lost as the digestion of HCS proceeded. Only 50% of the initial activity was retained at values of alkali uptake equivalent to an average cleavage of four bonds per molecule. This finding suggests that the sites of the two activities are closely connected.

4. *Effects of chemical and tryptic modifications on growth promoting activity (Table IV)*

(a) Chemical modification. The main differences between the effects on prolactin-like and growth promoting activity are seen after the modification of tryptophan and disulphide bonds. It was found that, while the lactogenic activity was lost when HCS was allowed to react with NPS, this treatment had no effect on the growth promoting activity. This suggests that tryptophan is not implicated in the latter activity. On the other hand, reduced and carbamido-methylated HCS shows impaired potency in comparison with untreated HCS when assayed by the tibia test, while the same derivative was fully active in the crop-sac assay.

TABLE IV

Effect of various modifications on growth promoting activity of HCS as assayed by the tibia test

Experimental groups	Total dose (μg)	Width of tibia cartilage* (μ)	P value vs. HCS**	P value vs. control**
Control	Saline	139 ± 1	0.00 –0.005 (56)	–
HGH	20	200 ± 8	0.00 –0.005 (22)	0.00 –0.005 (38)
HCS	200	174 ± 1	–	0.00 –0.005 (56)
a. Chemically modified HCS:				
Mal – HCS	200	157 ± 3	0.00 –0.005 (31)	0.00 –0.005 (47)
Guan – HCS	200	165 ± 3	0.005–0.025 (39)	0.00 –0.005 (55)
CHD – HCS	200	157 ± 3	0.00 –0.005 (32)	0.00 – 0.005 (48)
CMC – HCS	200	151 ± 5	0.00 –0.005 (31)	0.00 –0.005 (47)
NPS – HCS	200	181 ± 4	0.025–0.10 (23)	0.00 –0.005 (39)
HNB – HCS	200	152 ± 2	0.00 –0.005 (30)	0.00 –0.005 (46)
TNM – HCS	200	146 ± 2	0.00 –0.005 (27)	0.005–0.025 (43)
RCAM – HCS	200	163 ± 5	0.005–0.025 (25)	0.00 –0.005 (41)
RCOM – HCS	200	151 ± 4	0.00 –0.005 (26)	0.00 –0.005 (40)
b. Tryptic digestion; alkali uptake:				
1 eq./mole	200	169 ± 3	0.10 –0.25 (22)	0.00 –0.005 (38)
2 eq./mole	200	170 ± 2	0.10 –0.25 (26)	0.00 –0.005 (42)
3 eq./mole	200	169 ± 3	0.025–0.10 (27)	0.00 –0.005 (43)
4 eq./mole	200	181 ± 5	0.025–0.10 (25)	0.00 –0.005 (41)

See Table I for abbreviations of chemically modified HCS.
* Mean ± standard error.
** Degrees of freedom are reported in brackets.

(b) Tryptic digestion. The limited proteolysis does not affect the growth promoting activity of HCS. Full potency was retained up to the cleavage of four bonds per molecule, at which point the immunological activity was already significantly decreased. This suggests different structural requirements for the integrity of the antigenic determinants and the maintenance of the somatotropic activity in the HCS molecule.

DISCUSSION

In Table V we have tried to collect the effects of the various modifications we have described, on the activities of HCS. We have also reported the effects of the same modifications on HGH as recorded in the literature or as detected in experiments which we have carried out but not yet published. By comparing the activities of variously modified HCS as well as those of the corresponding HGH derivatives, the following considerations can be made:

1. The regions of the HCS molecule concerned with the lactogenic and somatotropic activity do not appear to overlap, as these activities are differentially affected in the experiments involving partial proteolysis. Moreover, tryptophan and disulphide bonds play different roles in maintaining these activities, as lactogenic activity is not affected by the modification of one disulphide bond, and somatotropic potency is not impaired when tryptophan is arylated with NPS.

TABLE V

Comparison between the biological and immunological activities of HCS and HGH after chemical modifications or tryptic digestion

Activities	Hormone	Treatment or derivatives (modified groups)									
		Trypsin (4 bonds)	Mal (free NH_2)	Guan (ε-NH_2)	CHD (arg)	CMC (-COOH)	HNB (trp)	NPS (trp)	TNM (tyr)	RCAM (-S-S-)	RCOM (-S-S-)
Prolactin-like	HCS	↓	—	—	—	—	±	—	—	+	—
	HGH	(+)	↓		↓	↓	(↓)	(↓)	(+)		(+)
Growth promoting	HCS	+	—	±	—	—	—	+	—	±	—
	HGH	(+)	↓	(+)	↓	↓	(↓)	(+)	(+)		(↓)
Immunological	HCS	↓	3	53	0	35	25	7	4	35	15

See Table I for abbreviations of chemically modified HCS.
'Trypsin' refers to digestion with trypsin.

Symbols for biological activities:
+ unaffected
↓ decreased
± decreased ($0.01 < P < 0.05$)
— decreased ($0.00 < P < 0.01$)
* symbols in brackets refer to activities of HGH derivatives reported by other authors, *i.e.* trypsin-digested (Li and Samuelsson, 1965); guanidinated (Geschwind and Li, 1957); HNB and NPS (Brovetto-Cruz and Li, 1969); RCOM (Bewley *et al.*, 1969).

Immunological activity:
numbers refer to % of residual activity.

2. Antigenic determinants and the site of the growth promoting activity similarly do not overlap, since the ability to bind the antibodies is more rapidly lost than the growth promoting potency, when HCS is digested with trypsin. Moreover, the chemical modifications have unequal effects. Consequently, it would be interesting to investigate whether this happens because the somatotropic activity is less sensitive to the conformational changes induced in the molecule by the aforesaid modifications, or because the two activities are independently located in the molecule. In this case, one might speculate that the core of somatotropic activity is buried and/or does not participate in the structure of the antigenic determinants.

3. This study provides additional, even if indirect, evidence that HCS and HGH share structural similarities. In fact, the modifications of the same residues in the two molecules have nearly equivalent effects. However, two exceptions are observed. Firstly, lactogenic activity is not so easily impaired when HGH is submitted to partial proteolysis, as it is in the case of HCS. Secondly, the nitration of tyrosine residues impairs the growth promoting activity of HCS, but leaves unaffected that of HGH. Further investigations are obviously necessary, and we expect to shed more light by comparison of the primary structures of HGH and HCS, as soon as these have been completely determined.

ACKNOWLEDGEMENTS

The authors are indebted to R. Botti, G. Barbarulli, M. Bartalini, G. Gazzei, G. Fantozzi, A. Ruspetti and F. Zappalorto for their skilful technical assistance. They also wish to thank Mrs. Susan Taddeo for her help in the preparation of the manuscript.

REFERENCES

ARMSTRONG, M. D. and LEWIS, J. D. (1951): Thioether derivatives of cysteine and homocysteine. *J. org. Chem.*, *16*, 749.

ARMSTRONG, J. McD. and McKENZIE, H. A. (1967): A method for modification of carboxyl groups in proteins: its application to the association of bovine β-lactoglobulin A. *Biochim. biophys. Acta (Amst.)*, *147*, 93.

AUBERT, M. L., NERI, P., GENAZZANI, A. R., COCOLA, F. and FELBERT, J.-P. (1970): Methodological aspects of the radioimmunoassay of human chorionic somatomammotropin (HCS or HPL). *Ann. Sclavo*, *12*, 689.

BEWLEY, T. A., BROVETTO-CRUZ, J. and LI, C. H. (1969): Human pituitary growth hormone. Physicochemical investigations of the native and reduced-alkylated protein. *Biochemistry*, *8*, 4701.

BEWLEY, T. A., DIXON, J. S. and LI, C. H. (1968): Human pituitary growth hormone. XVI. Reduction with dithiothreitol in the absence of urea. *Biochim. biophys. Acta (Amst.)*, *154*, 420.

BROVETTO-CRUZ, J. and LI, C. H. (1969): Human pituitary growth hormone. Studies of the tryptophan residue. *Biochemistry*, *8*, 4695.

DAVIS, B. J. (1964): Disc electrophoresis. II. Method and application to human serum proteins. *Ann. N.Y. Acad. Sci.*, *121*, 404.

FALCONI, G. and ROSSI, G. L. (1964): Transauricular hypophysectomy in rats and mice. *Endocrinology*, *74*, 301.

FLORINI, J. R., TONELLI, G., BREUER, C. B., COPPOLA, J., RINGLER, I. and BELL, P. H. (1966): Characterization and biological effects of purified placental protein (human). *Endocrinology*, *79*, 692.

FRIESEN, H. (1965): Purification of a placental factor with immunological and chemical similarity to human growth hormone. *Endocrinology*, *76*, 369.

GESCHWIND, I. I. and LI, C. H. (1957): The guanidination of some biologically active proteins. *Biochim. biophys. Acta (Amst.)*, *25*, 171.

GREENSPAN, F. S., LI, C. H., SIMPSON, M. E. and EVANS, H. H. (1949): Bioassay of hypophyseal growth hormone: the tibia test. *Endocrinology*, *45*, 455.

HIRS, C. H. W. (Ed.) (1967): Reduction and S-carboxymethylation of proteins. In: *Methods in Enzymology, Vol. XI*, Article 20, pp. 199–209. Academic Press, New York.

JOSIMOVICH, J. B. and MACLAREN, J. A. (1962): Presence in the human placenta and term serum of a highly lactogenic substance immunologically related to pituitary growth hormone. *Endocrinology*, *71*, 209.

KIMMEL, J. R. (1967): Analysis for homoarginine. In: *Methods in Enzymology*, *Vol. XI*, Article 71, pp. 589–590. Editor: C. H. W. Hirs. Academic Press, New York.

KOSHLAND, D. E., KARKHANIS, Y. D. and LATHAM, H. G. (1964): An environmentally sensitive reagent with selectivity for the tryptophan residue in proteins. *J. Amer. chem. Soc.*, *86*, 1448.

LEACH, S. J. and PARKHILL, E. M. J. (1955): *Proceedings International Wool Textiles Research Conference Australia C*, 92. (Quoted by Wilcox.)

LI, C. H. (1968): The chemistry of human pituitary growth hormone. In: *Growth Hormone*, p. 3. Editors: A. Pecile and E. Müller. ICS 158, Excerpta Medica, Amsterdam.

LI, C. H. (1970): On the characterization of human chorionic somatomammotropin. *Ann. Sclavo*, *12*, 651.

LI, C. H. and SAMUELSSON, G. (1965): Human pituitary growth hormone. XI. Rate of hydrolysis by trypsin, chymotrypsin and pepsin: effect of trypsin on the biological activity. *Molec. Pharmacol.*, *1*, 47.

MOKRASCH, L. C. (1967): Use of 2,4,6-trinitrobenzenesulfonic acid for the coestimation of amines, amino acids and proteins in mixtures. *Analyt. Biochem.*, *18*, 64.

MOORE, S. and STEIN, W. H. (1963): In: *Methods in Enzymology*, *Vol. VI*, Article 117, pp. 819–831. Editors: S. P. Colowick and N. O. Kaplan. Academic Press, New York.

NERI, P., TARLI, P., AREZZINI, C., COCOLA, F., PALLINI, V. and RICCI, C. (1970*a*): Effects of pH on conformation and biological activity of human chorionic somatomammotropin (HCS). *Ann. Sclavo*, *12*, 663.

NERI, P., TARLI, P. and COCOLA, F. (1970*b*): Effects of pH and protein concentration on the molecular weight of human chorionic somatomammotropin. *Ital. J. Biochem.*, *19*, 111.

NICOLL, C. S. (1967): Bio-assay of prolactin. Analysis of the pigeon crop-sac response to local prolactin injection by an objective and quantitative method. *Endocrinology*, *80*, 641.

RABEN, M. S. (1957): Preparation of growth hormone from pituitaries of man and monkey. *Science*, *125*, 883.

SCOFFONE, E., FONTANA, A. and ROCCHI, R. (1968): Sulfenyl halides as modifying reagents for polypeptides and proteins. I. Modification of tryptophan residues. *Biochemistry*, *7*, 971.

SIA, C. L. and HORECKER, B. L. (1968): Dissociation of protein subunits by maleylation. *Biochem. biophys. Res. Commun.*, *31*, 731.

SOKOLOVSKY, M., RIORDAN, J. F. and VALLEE, B. L. (1966): Tetranitromethane. A reagent for the nitration of tyrosil residues in proteins. *Biochemistry*, *5*, 3582.

SPIES, J. R. and CHAMBERS, D. C. (1949): Chemical determination of tryptophan in proteins. *Analyt. Chem.*, *21*, 1249.

TERAO, P. and UKITA, T. (1969): Modification of tryptophan residue in ribonuclease T_1 with 2-hydroxy-5-nitrobenzyl bromide. *Biochim. biophys. Acta (Amst.)*, *181*, 347.

TOI, K., BYNUM, E., NORRIS, E. and ITANO, H. A. (1967): Studies on the chemical modification of arginine. I. The reaction of 1,2-cyclohexanedione with arginine and arginyl residues of proteins. *J. biol. Chem.*, *242*, 1036.

WILCOX, P. E. In: *Methods in Enzymology*, *Vol. XI*, Article 7, pp. 63–65. Editor: C. H. W. Hirs. Academic Press, New York.

YAMASAKI, N., KIKUTANI, M. and SONENBERG, M. (1970): Peptides of a biologically active tryptic digest of bovine growth hormone. *Biochemistry*, *9*, 1107.

COMPARISON OF THE STRUCTURE AND FUNCTION OF HUMAN PLACENTAL LACTOGEN AND HUMAN GROWTH HORMONE*

LOUIS M. SHERWOOD, STUART HANDWERGER, WILLIAM D. McLAURIN and ELAINE C. PANG

Endocrine Unit, Department of Medicine, Beth Israel Hospital and Harvard Medical School, Boston, Mass., U.S.A.

Although they are secreted by different organs, human placental lactogen (HPL) and human growth hormone (HGH) have many chemical, biologic and immunologic features in common. The placental hormone, originally described by Josimovich and MacLaren (1962), is secreted in large quantities during pregnancy and has important effects on the development and differentiation of the mammary gland as well as on maternal metabolism (Grumbach *et al.*, 1968). Like growth hormone (Chadwick *et al.*, 1961; Ferguson and Wall, 1961), HPL has intrinsic lactogenic activity (Josimovich and MacLaren, 1962), but its somatotrophic potency on a weight basis is much less than that of the pituitary peptide (Friesen, 1965). The physicochemical properties of the two hormones are very similar (Sherwood and Handwerger, 1969), and antibodies directed against HGH cross-react well with HPL (Josimovich and MacLaren, 1962).

Because these two molecules have such great similarities in molecular weight, amino acid composition, and biologic and immunologic properties, we initiated studies of the chemistry of HPL and a detailed comparison of its properties with those of HGH (Li *et al.*, 1966). Based on the amino acid composition of the tryptic peptides of HPL and its carboxyl-terminal amino acid sequence, we predicted in earlier reports (Sherwood, 1967, 1969) that there would be marked homologies in the structure of the two hormones. Now that the complete amino acid sequence of HPL has been determined in our laboratory (Sherwood *et al.*, 1971), this prediction has proven correct. The two hormones are strikingly similar, with each molecule containing 190 amino acid residues and two intrachain disulfide bonds in the same location. A total of 86% or 163 out of 190 residues in the two hormones are identically placed, and most of the substitutions present in HPL are highly favored on the basis of genetic analysis (Dayhoff, 1969).

Since the hormones are so similar in structure, it is reasonable to suggest that their differences in growth-promoting activity are due to relatively minor amino acid substitutions in a portion of the polypeptide backbone. The chemical features essential for the prolactin-like activity, on the other hand, probably lie in identical or nearly identical regions of the molecule, since parallel increases or decreases in lactogenic activity result from chemical or enzymic modifications of the two hormones (Handwerger *et al.*, 1970, 1971). The studies that led to the structure of the hormone and a correlation of structure with function are described herein.

* This work was supported by grants HD 03388, CA 10736, and T01 AM 05116 from the U.S.P.H.S. and a grant from the John A. Hartford Foundation, Inc. Dr. Sherwood is the recipient of a Research Career Development Award #KO4 AM 46, 426 and Dr. Handwerger of a Special Fellowship F03 GM40345 from the U.S.P.H.S.

MATERIALS AND METHODS

Partially purified HPL (Florini *et al.*, 1966) was subjected to gel filtration on Sephadex G-100 as previously described (Sherwood, 1967). The final product was homogeneous by amino-terminal end group analysis, by sedimentation in the ultracentrifuge and gel filtration, and in a variety of immunologic tests. Minor heterogeneity detected on polyacrylamide gel electrophoresis was due to deamidation (Sherwood, 1967), a common problem encountered with acidic peptides such as HGH and ovine prolactin (Lewis and Cheever, 1965). The purified product stimulated lactation *in vivo* in midpregnant rabbits (Friesen, 1966) and *in vitro* in explants of mouse mammary tissue in organ culture (Turkington and Topper, 1966).

Chemical studies

Following purification and amino acid analysis (Hirs *et al.*, 1954) the native molecule was cleaved with a variety of enzymes and chemical reagents in order to determine its amino acid sequence. The composition of the tryptic peptides was previously reported (Sherwood, 1967). Peptides obtained from cleavage of the molecule with chymotrypsin, pepsin (Canfield, 1963; Canfield and Anfinsen, 1963*a*), and cyanogen bromide (Gross and Witkop, 1962) were purified by ion exchange or paper chromatography, gel filtration, and high voltage electrophoresis. Chymotrypsin (Worthington) was incubated with performic oxidized HPL (Hirs *et al.*, 1956; Hirs, 1967) in which the two disulfide bonds were disrupted and the cysteine residues converted to cysteic acid. Pepsin (Worthington) cleavage was performed on the native molecule. One large peptide in the pepsin digest consisting of two peptides joined by a disulfide bond was oxidized with performic acid and the fragments separated by high voltage electrophoresis (Katz *et al.*, 1959). Cleavage at the six methionine residues of HPL with cyanogen bromide also produced a similar fragment (C5) and five smaller pieces. The disulfide bond was disrupted by performic oxidation and the two peptides (C5A and C5B) separated by gel filtration on Sephadex G-50. Analysis of the cysteine-containing peptides in the peptic and cyanogen bromide digests identified the location of the two disulfide bonds in the molecule.

The amino acid sequence of individual peptides was determined by sequential Edman degradation (Edman, 1950) using the subtractive method as modified by Elzinga (1970). Relatively small peptides were degraded for an average of 6 to 8 cycles. In some instances, the remaining amino-terminal group was identified by thin layer chromatography (Woods and Wang, 1967) of the dansyl derivative (Gray, 1967). The tryptophan-containing peptide was identified by Ehrlich stain (Canfield and Anfinsen, 1963*b*) and the single tryptophan residue in peptic peptide 9B cleaved with n-bromosuccinimide (Ramachandran and Witkop, 1967) to identify its location.

Formation of derivatives

Purified HPL and HGH (kindly supplied by the National Pituitary Agency and National Institute of Arthritis and Metabolic Diseases) were oxidized with performic acid by the method of Hirs (1967). The proteins were incubated with performic acid (90% formic acid—10% hydrogen peroxide) for two hours at -10°C and the reaction stopped by dilution with water and lyophilization. Complete conversion was indicated by amino acid analysis and confirmed by radioimmunoassay.

Reduction and alkylation were performed by the method of Crestfield *et al.* (1963) using recrystallized iodoacetamide and iodoacetic acid. Complete conversion to the derivatives was confirmed by amino acid analysis.

Partial deamidation was accomplished by incubating HPL at increasing pH from 8.4 to 12.7 for 16 hours in solutions of ammonium bicarbonate, ammonium carbonate and sodium hydroxide respectively. Ammonia release was documented by Conway (1957) diffusion and direct measurement of ammonia on the amino acid analyzer.

Biologic assays

The prolactin activity of various hormone preparations was assayed *in vitro* by a modification of the method of Turkington and Topper (1966). Midpregnant mouse mammary glands were incubated in organ culture and the effects of hormones and their derivatives on histologic differentiation of the tissue and on casein synthesis tested. Mammary glands from midpregnant CD1 mice (Charles River) were incubated as one cubic mm explants on stainless steel grids in Medium 199 (Microbiological Associates) containing 5 μg/ml insulin, 5 μg/ml hydrocortisone and the hormone to be tested. After 44 hours of incubation in 5% CO_2 in air, the explants were pulsed for 4 hours with ^{14}C amino acids (New England Nuclear), following which the glands were homogenized and aliquots removed for protein determination. After centrifugation at 12000 x *g* for 60 min, casein was precipitated by calcium chloride and rennin and the pellet washed three times with alcohol-ether. The pellet was solubilized in 5 N acetic acid and an aliquot counted in Bray's solution. Histologic confirmation of glandular proliferation and differentiation was obtained in each experiment.

Prolactin activity was tested *in vivo* by injection of 10 mg native HPL or 10 mg performic oxidized HPL (PER-HPL) into midpregnant rabbits daily for 3 days. The animals were then sacrificed and the mammary glands examined grossly and microscopically.

Immunologic studies

The derivatives of the hormones were tested by radioimmunoassay. Growth hormone assays were performed by a double-antibody modification of the method of Glick *et al.* (1963) and HPL by methods previously described (Sciarra *et al.*, 1968). The hormones were iodinated by the method of Greenwood *et al.* (1963) and the immunoreactivity of the derivatives was tested in comparison with native hormone over a wide range of concentrations.

TABLE I

Tryptic peptides of human placental lactogen

Peptide	Sequence
T18C	val,gln,thr,val,pro,leu,ser,*arg*
T26	leu,phe,asp,his,ala,met,leu,gln,ala,his,*arg*
T18	ala,his,gln,leu,ala,ile,asp,thr,tyr,gln,glu,phe,glu,glu,thr,tyr,ile.pro,*lys*
T7B	asp,glx,*lys*
T12	tyr,ser,phe,leu,his,asp,ser,glx,thr,ser,phe,cys,phe,ser,asx,ser,ile,pro,thr,pro,ser,asx,met, glx,glx,thr,glx,*lys*
T17A	ser,asx,leu,glx,leu,leu,*arg*
T24	ile,ser,leu,leu,leu,ile,glx,ser,trp,leu,glx,pro,val,*arg*
T14	phe,leu,*arg*
T13B	ser,met,phe,ala,asx,asx,leu,val,tyr,asx,thr,ser,asx,asx,asx,ser,tyr,his,leu,leu,*lys*
T8A	asx,leu,glx,glx,gly,ile,glx,thr,leu,met,gly,*arg*
T9B	leu,glx,asx,gly,ser,*arg*,*arg*
T16C	thr,gly,glx,ile,leu,*lys*
T7A	glx,thr,tyr,ser,*lys*
T20D	phe,asx,thr,asx,ser,his,asx,his,asx,ala,leu,leu,*lys*
T16A	asx,tyr,gly,leu,leu,tyr,cys,phe,*arg*
T12D	*lys*
T14A	asx,met,asx,*lys*,val,glx,thr,phe,leu,*arg*
T13C	met,val,gln,cys,*arg*
T5A	ser,val,glu,gly,ser,cys,gly,phe

TABLE II

Chymotryptic peptides of human placental lactogen

Peptide	Structure
CT17D	val,gln,thr,val,pro,leu
CT31D	asp,his,ala,met,leu
CT29A	ala,ile,asp,thr
CT49B	ile,pro,lys,asp,glx,lys,tyr
CT37C	ser,phe,leu,his,asp,ser,glx,thr,ser,phe,cys,phe
CT13D	cys,phe,ser,asx,ser,ile,pro,thr,pro,ser,asx,met,glx,glx,thr,glx,lys,ser,asx,leu,glx,leu,leu
CT47C	arg,ile,ser,leu
CT37E	glx,pro,val,arg,phe,leu
CT52B	phe,leu,arg,ser
CT47B	arg,ser,met,phe
CT51C	ala,asx,asx,leu,val
CT37F	asx,thr,ser,asx,asx,asx,ser,tyr,his,leu,leu,lys
CT30D	lys,asx,leu,glx,glx,gly,ile,glx,thr,leu,met,gly
CT24B	leu,glx,glx
CT32C	asx,thr,asx,ser,his,asx,his,asx,ala,leu,leu
CT45C	lys,asx,tyr
CT37B	gly,leu,leu
CT37D	cys,phe,arg,lys,asx,met,asx,lys,val,glx,thr,phe
CT32A	arg,met,val,gln,cys,arg,ser,val,glu,gly,ser,cys,gly,phe

TABLE III

Peptic peptides of placental lactogen

Peptide	Structure
P5C	val,gln,thr,val,pro,leu,ser,arg,leu
P10C	asp,his,ala,met,leu,gln
P4C	tyr,gln,glu,phe,glu
P11B	glu,glu,thr,tyr,ile,pro,lys,asp,glx,lys,tyr,ser,phe,leu
P14-1	cys,phe,ser,asx,ser,ile,pro,thr,pro,ser,asx,met
3P1H	ser,asx,met,glx,glx,thr,glx,lys,ser,asx,leu
P3B	glx,glx,thr,glx,lys,ser,asx,leu,glx,leu
P10D	ile,ser,leu,leu
P11C	arg,ile,ser,leu,leu
P3C	leu,leu,ile,glx,ser
P9B	leu,ile,glx,ser,trp,leu,glx,pro,val
P5A	leu,val,tyr,asx,thr,ser,asx,asx,asx,ser,tyr,his,leu,leu,lys,asx,leu,glx,glx,gly
3P1E	leu,glx,glx,gly,ile,glx,thr,leu
3P1K	leu,met,gly,arg,leu,glx,asx,gly,ser,arg,arg,thr,gly
P5B	gly,arg,leu,glx,asx,gly,ser
P11A	leu,glx,asx,gly,ser
3P1M	ile,leu,lys,glx,thr,tyr,ser,lys,phe
P11B	ala,leu,leu,lys,asx,tyr,gly
P14-2	cys,phe,arg
P11D	leu,arg,met
P6D	gln,cys,arg,ser,val,glu,gly,ser,cys,gly,phe

RESULTS

Structure

Cleavage of reduced S-carboxymethylated HPL with trypsin produced the twenty-one tryptic peptides theoretically anticipated from complete cleavage at twenty lysine and arginine

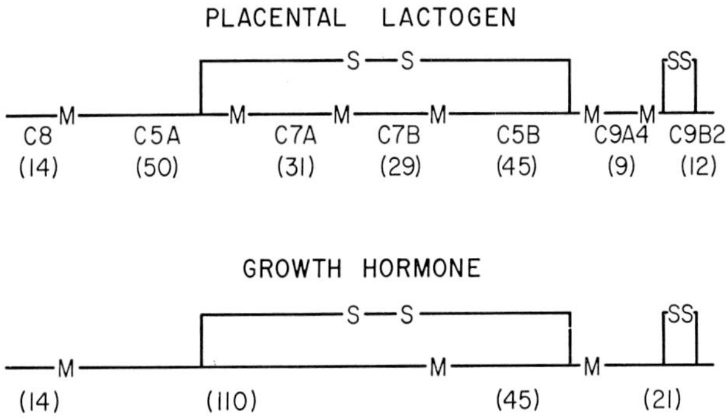

Fig. 1. Diagrammatic model indicating the sites of cyanogen bromide cleavage at methionine residues (M) in HPL and HGH. Cleavage of HPL produced six fragments; cleavage of HGH three fragments. The number of amino acid residues in each fragment is indicated.

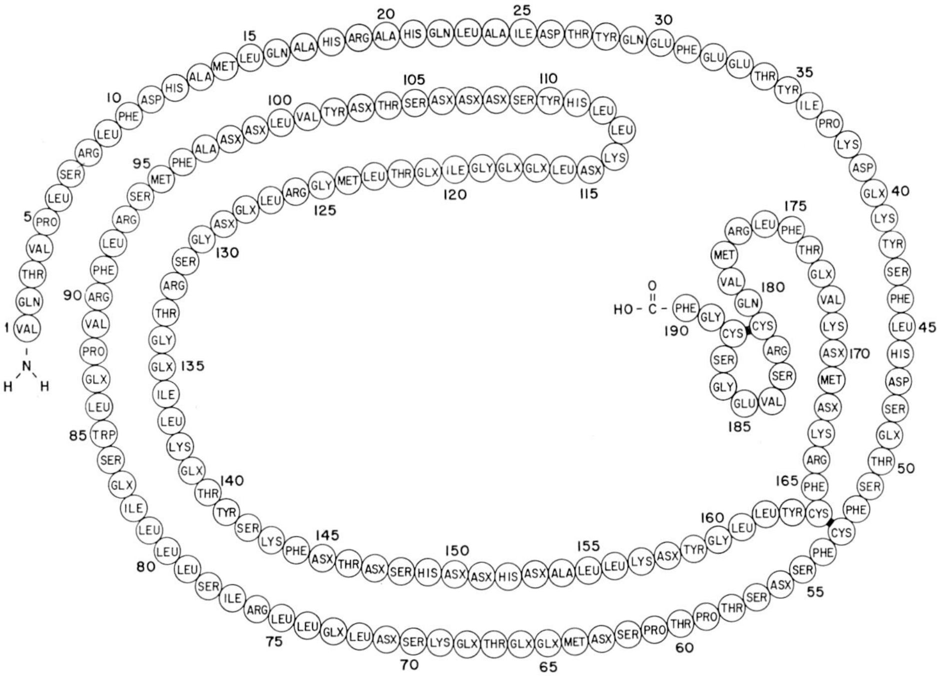

Fig. 2. The amino acid sequence of HPL based on the overlapping peptides from tryptic, chymotryptic, peptic and cyanogen bromide cleavage and Edman degradation at each position. Amide groups not yet determined finally are indicated as Glx and Asx respectively.

residues (Table I). Homogeneous peptides were eluted from paper for sequential degradation by the Edman method (1950). The large tryptic peptides were cleaved with chymotrypsin in order to produce smaller fragments which could then be sequenced. Small fragments were generally purified by high voltage electrophoresis and/or paper chromatography and eluted from paper with acetic acid and ammonia.

A total of nineteen chymotryptic (Table II) and twenty-one peptic peptides (Table III) were obtained from cleavage of the whole molecule with enzymes. These peptides permitted alignment of all but two of the tryptic peptides, the latter being aligned by the results of cyanogen bromide cleavage. These overlaps provided a unique sequence of peptides which was confirmed by the seven fragments produced by cleavage with cyanogen bromide (Fig. 1).

Individual peptides of relatively small size were sequenced by Edman degradation. On the basis of the overlapping peptides and sequential degradation of amino acids at each of the 190 positions, the complete amino acid sequence of the molecule was determined (Fig. 2). The disulfide bonds were located between half-cystine residues 53 and 164 and 181 and 188 respectively, and the six methionine residues placed at positions 14, 64, 95, 124, 169 and 178. The single tryptophan residue was at position 85. Amide content (where completed) was determined by leucine aminopeptidase digestion and modified amino acid analysis (Sherwood, 1969); most of the amide groups in HPL and HGH are identical (McLaurin, Handwerger and Sherwood, unpublished).

Function

In order to correlate chemical structure with biologic function, several chemical modifications of HPL and HGH were made and the biologic and immunologic properties of the derivatives compared with the native hormones. Three of the derivatives in which the two disulfide bonds of each molecule were disrupted are shown in Figure 3. Reduction and alkylation with ioacetamide selectively cleaved the disulfide bonds and converted cysteine to the S-carbamidomethyl derivative. Reduction and alkylation with iodoacetic acid, in addition to cleaving the disulfide bonds, introduced four additional negative charges into the

DISULFIDE BONDS IN HPL AND HGH

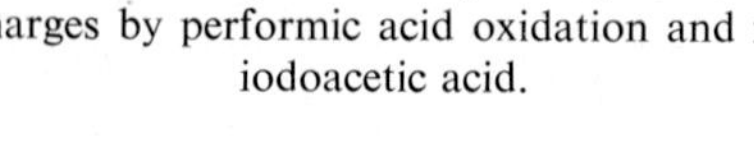

MODIFICATIONS:

Performic acid oxidized	— SO_3^-	[PER-HPL, PER-HGH]
Alkylated with iodoacetic acid	— SCH_2COO^-	[RA-SCMC-HPL, RA(COOH)HPL]
Alkylated with iodoacetamide	— SCH_2CONH_2	[RA-HPL, RA-HGH]

Fig. 3. Modifications of HPL and HGH produced by disruption of the disulfide bonds. Note the introduction of 4 negative charges by performic acid oxidation and reduction and alkylation with iodoacetic acid.

molecule. Oxidation with performic acid more drastically altered each hormone by (1) cleaving the disulfide bonds and introducing four new negative charges as cysteic acid, (2) oxidizing methionine to methionine sulfone and (3) destroying tryptophan. In the hormone treated with cyanogen bromide, selective cleavage occurred at the methionine residues, producing six fragments of HPL and three of HGH. One of the fragments of each hormone was comprised of two smaller peptides joined by a single disulfide bond (Fig. 1).

When tested in the mouse mammary system, native HPL and HGH stimulated casein synthesis and histologic differentiation at concentrations of 1 to 10 ng/ml, with casein

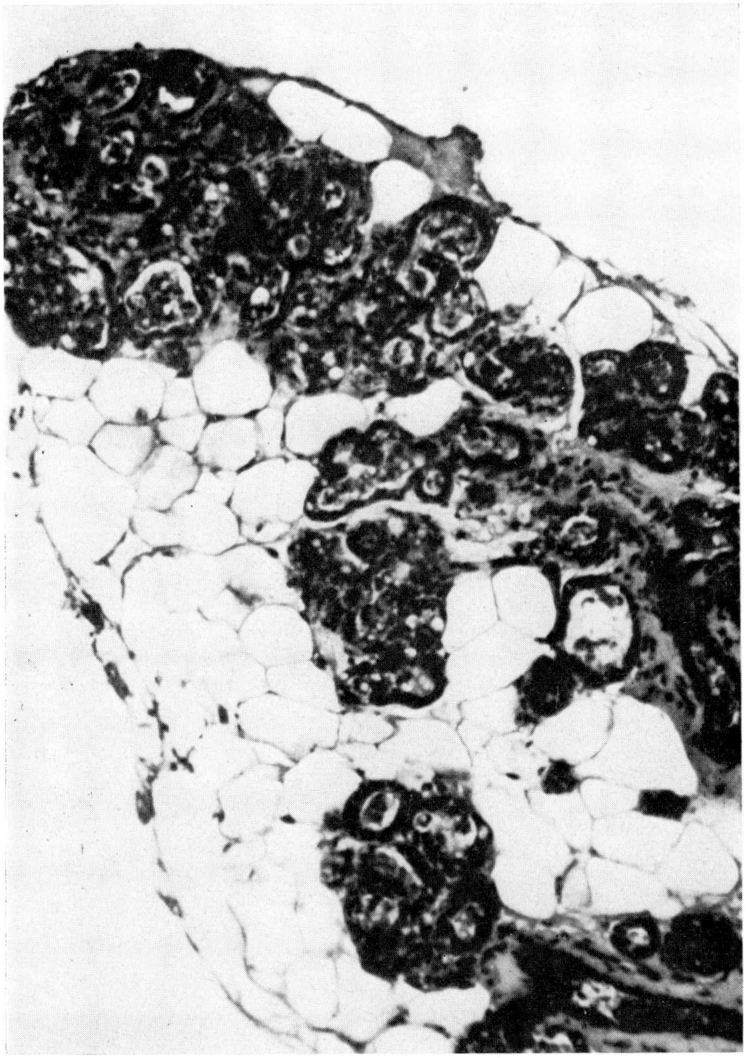

Fig. 4. Midpregnant mouse mammary tissue incubated for 48 hours with 10 ng/ml HPL in presence of 5 μg/ml of hydrocortisone and 5 μg/ml insulin. Note ductal proliferation and secretory activity.

synthesis in stimulated tissue averaging 1.5 to 3.2 times the control value (Figs. 4 and 5). At higher concentrations of hormone, no further increase in casein synthesis was noted. Despite some variation in the magnitude of stimulation from assay to assay, consistent results were obtained when different hormone preparations were compared with each other. In five separate experiments in which native HPL and HGH were compared, identical lactogenic responses were noted, providing firm evidence that the lactogenic potency of the two hormones is the same (Fig. 5).

Each of the derivatives of HPL in which the disulfide bonds were disrupted (RA-HPL, RA-SCMC-HPL, and PER-HPL) retained full lactogenic activity when compared with the native hormone (Fig. 6). Likewise, RA-HGH and PER-HGH were fully active in similar

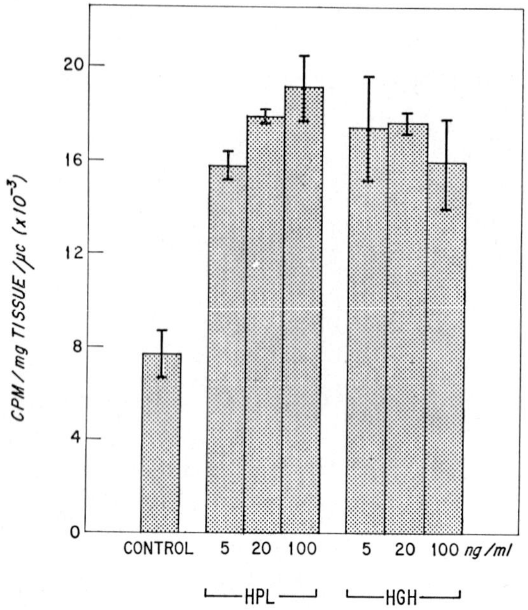

Fig. 5. The effects of HPL and HGH on casein synthesis by mouse mammary tissue *in vitro*. Stimulation 1.5 to 3.2 times control was observed at 1 to 10 ng/ml; higher concentrations produced no greater effect. HPL and HGH were equally active in all experiments. Results expressed as CPM ± SEM.

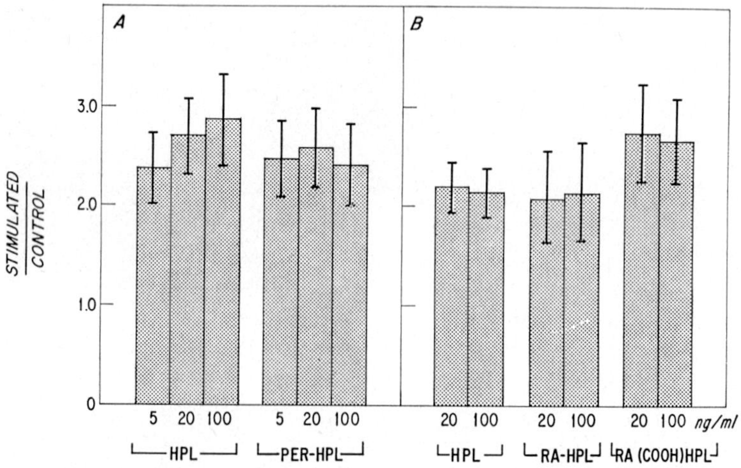

Fig. 6. The effects of performic acid oxidation (A) and reduction and alkylation (B) on the lactogenic activity of HPL. Results expressed as ratio of treated to control ± SEM.

tests (Fig. 7). These observations clearly indicated that the integrity of the disulfide bonds of both HPL and HGH, which are located in identical positions in each molecule, are not essential for prolactin-like activity. Furthermore, oxidation of methionine, modification of tryptophan and the addition of four new negative charges does not destroy lactogenic potency.

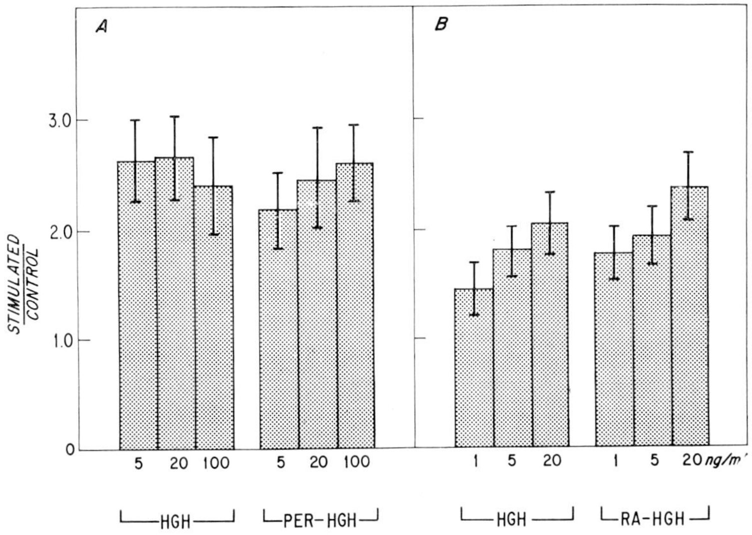

Fig. 7. The effects of performic acid oxidation (A) and reduction and alkylation with iodoacetamide (B) on the lactogenic activity of HGH. Results expressed as ratio of treated to control ± SEM.

In three separate experiments, in which the cyanogen bromide derivatives were compared with the native hormone, CNBR-HPL and CNBR-HGH were fully active, indicating that the integrity of the entire polypeptide backbone of each molecule was not essential for its prolactin effects (Fig. 8). Complete cleavage of HPL with trypsin, on the other hand, destroyed its lactogenic activity. In three experiments, deamidation of HPL produced a small but significant increase in prolactin activity which was linear in the pH range 8.4 to 11.7 (Fig. 9). Hormone incubated at pH 12.7 was fully active, even though there was disruption of the polyacrylamide gel pattern.

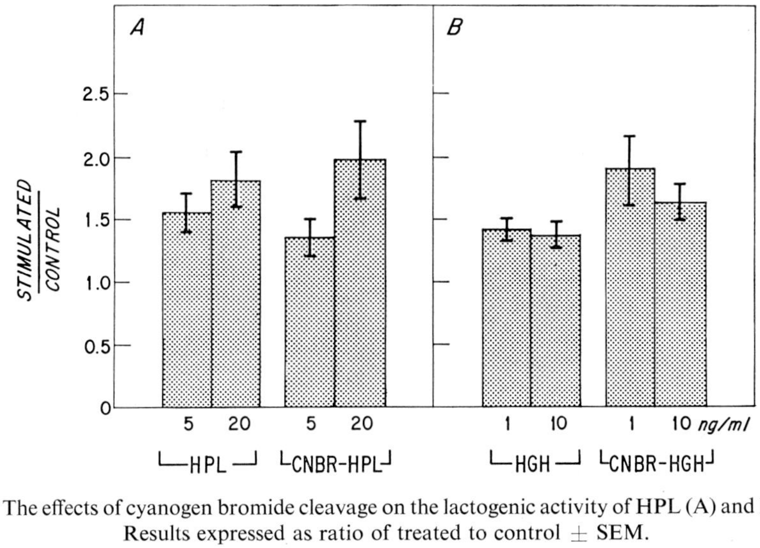

Fig. 8. The effects of cyanogen bromide cleavage on the lactogenic activity of HPL (A) and HGH (B). Results expressed as ratio of treated to control ± SEM.

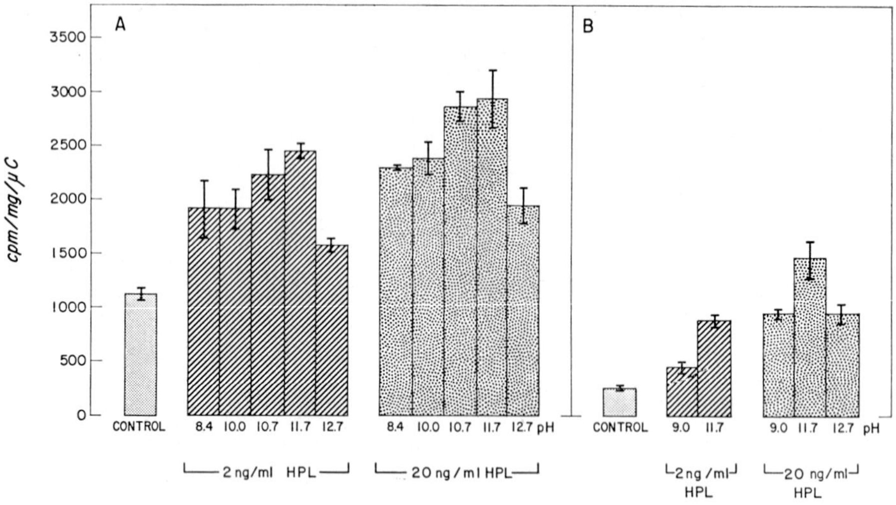

Fig. 9. Increase in lactogenic activity produced by partial deamidation of HPL. Maximal response was observed for the hormone incubated at pH 11.7 with the effect being significant at the 5% level. Results from two separate experiments are shown and are expressed as CPM/mg protein/μg tracer ± SEM.

In radioimmunoassay studies, the derivatives in which the disulfide bonds were disrupted (RA-HPL, RA-SCMC-HPL, PER-HPL; RA-HGH, PER-HGH) retained little or no immunologic activity when compared with the native molecule (Fig. 10). Likewise, both HPL and HGH cleaved completely with cyanogen bromide retained only a small fraction of its immunologic activity, and the slope of the response was markedly altered (Fig. 11). These findings indicated that the derivatives not only lost their ability to react with antibodies to native hormone (implying either a change in conformation or alteration of an immunodominant group in the primary sequence), but also that no native hormone was left in the

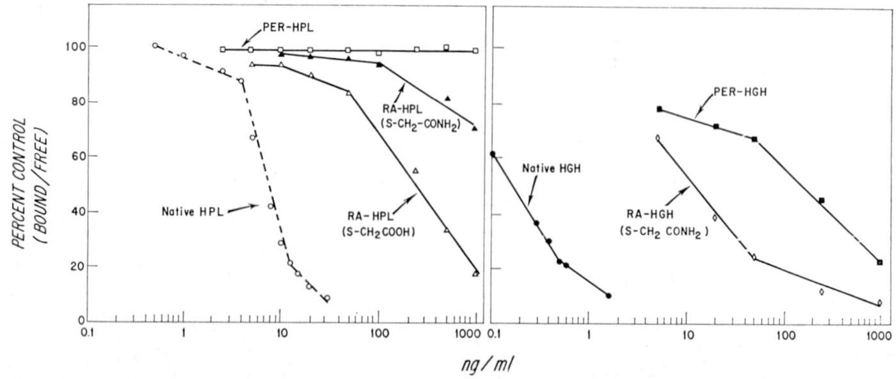

Fig. 10. Standard curves indicating the reactivity of modified HPL and HGH in the radioimmunoassay with native HPL (A) and HGH (B). PER-HPL and PER-HGH showed virtually no reactivity against antisera to the native hormones, while reduced and alkylated derivatives showed only minimal activity. Results expressed as percent of control bound/free ratios.

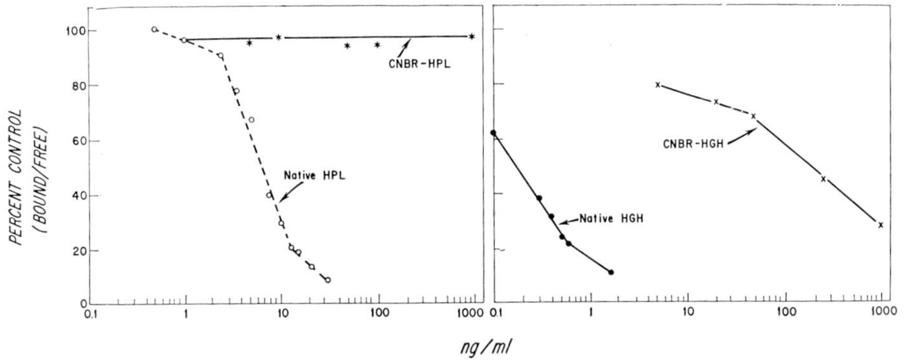

Fig. 11. Standard curves indicating the marked loss of immunologic activity associated with cyanogen bromide cleavage of HPL (A) and HGH (B). Results expressed as percent of control bound/free ratios.

modified preparation. On the other hand, extensive deamidation of HPL and even exposure to strong alkali caused no loss of immunologic activity.

DISCUSSION

As the sequence studies were performed in our laboratory, it soon became apparent that HPL and HGH were closely related chemically, particularly at the carboxyl-terminal end. Since potent lactogenic activity is an intrinsic feature of both HPL (Josimovich and Mac Laren, 1962; Friesen, 1966) and HGH (Chadwick *et al.*, 1961; Ferguson and Wall, 1961), it was reasonable to suggest that the chemical features responsible for this effect in both hormones were the same. A logical extension of this hypothesis would be that those features responsible for the somatotrophic activity of both molecules would be distinct or at least only partially overlapping, since HGH is 100 to 1000 times more potent than HPL as a growth hormone (Friesen, 1965). In the case of ACTH, parathyroid hormone, and other proteins, biologic activity has been defined in a portion of the polypeptide chain which is responsible for all of its physiologic effects. In HGH and HPL, on the other hand, there are two distinct biologic properties of different magnitude. Therefore, these hormones may contain two separate biologically active sites within the same molecule, making them unique in this respect. In order to test this hypothesis further, we produced a number of derivatives of the two hormones and tested their effects on biologic activity.

As described earlier, a series of derivatives of HPL and HGH showed parallel responses in lactogenic activity when tested for their ability to stimulate casein synthesis in mouse mammary tissue *in vitro*. Disruption of the two disulfide bonds, introduction of four negative charges and oxidation of methionine and tryptophan all failed to destroy the lactogenic potency of the hormone. Furthermore, these studies indicated that the entire polypeptide backbone of HPL and HGH was not essential for this effect since the cyanogen bromide-cleaved molecules were still lactogenic. The largest fragments in the cyanogen bromide digest of HPL and HGH was 95 and 155 residues in size respectively. More extensive digestion of these proteins with trypsin, which cleaves at all lysine and arginine residues, destroyed all biologic activity. These findings are of interest in view of the recent report of Nutting *et al.* (1970) that cyanogen bromide cleaved bovine growth hormone still retains somatotrophic activity when tested *in vitro*. In preliminary studies performed in collaboration with Drs. David Nutting and Jack Kostyo of Emory University, reduced and alkylated HGH remained fully active in growth-promoting assays, while performic oxidized HGH was inactive.

In earlier studies (Dixon and Li, 1966; Bewley *et al.*, 1969) it was shown that the disulfide bonds of HGH were not essential for its growth promoting or crop sac stimulating effects except when alkylation was performed with iodoacetic acid which destroyed somatotrophic activity. Mills and Wilhelmi (1966) also suggested that reduction of only the carboxyl-terminal disulfide bond of HGH caused no loss of lactogenic activity. Breuer (1969) suggested that reduction and alkylation inactivated the costal cartilage stimulating effects of HPL, but further studies need to be performed. Since the somatotrophic effects of HPL are very weak and require large doses (Friesen, 1965; Florini *et al.*, 1966), most of the information on the chemical features essential for somatotrophic activity must, by necessity, come from structure-function studies of HGH. Such studies are currently in progress in our laboratory.

TABLE IV

A comparison of the amino acid sequences of placental lactogen and growth hormone

```
HPL:  val-gln-thr-val-pro-leu-ser-arg-leu-phe-asp-his-ala-met-leu-gln-ala-his-arg-ala-his-gln-leu-ala-ile-
               X       X   X   X   X   X   X   X       X   X   X       X   X   X       X   X   X   X
HGH:  phe-pro-thr-ile-pro-leu-ser-arg-leu-phe-asp-asn-ala-met-leu-arg-ala-his-arg-leu-his-gln-leu-ala-phe-
       1                                                                                               25

HPL:  asp-thr-tyr-gln-glu-phe-glu-glu-thr-tyr-ile-pro-lys-asp-glx-lys-tyr-ser-phe-leu-his-asp-ser-glx-thr-
       X   X   X   X   X   X   X   X       X   X   X   X       X   X   X   X   X   X               X   X
HGH:  asp-thr-tyr-gln-glu-phe-glu-glu-ala-tyr-ile-pro-lys-glu-gln-lys-tyr-ser-phe-leu-gln-asn-pro-gln-thr-
       26                                                                                              50

HPL:  ser-phe-cys-phe-ser-asx-ser-ile-pro-thr-pro-ser-asx-met-glx-glx-thr-glx-lys-ser-asx-leu-glx-leu-leu-
       X       X   X   X       X   X   X   X   X   X   X       X   X   X   X   X   X   X   X   X   X   X
HGH:  ser-leu-cys-phe-ser-glu-ser-ile-pro-thr-pro-ser-asn-arg-glu-glu-thr-gln-lys-ser-asn-leu-gln-leu-leu-
       51                                                                                              75

HPL:  arg-ile-ser-leu-leu-leu-ile-glx-ser-trp-leu-glx-pro-val-arg-phe-leu-arg-ser-met-phe-ala-asx-asx-leu-
       X   X   X   X   X   X   X   X   X   X   X   X   X   X       X   X   X   X       X   X   X       X
HGH:  arg-ile-ser-leu-leu-leu-ile-gln-ser-trp-leu-glu-pro-val-gln-phe-leu-arg-ser-val-phe-ala-asn-ser-leu-
       76                                                                                             100

HPL:  val-tyr-asx-thr-ser-asx-asx-asx-ser-tyr-his-leu-leu-lys-asx-leu-glx-glx-gly-ile-glx-thr-leu-met-gly-
       X   X           X   X       X       X       X   X   X   X   X   X   X   X   X   X   X   X   X   X
HGH:  val-tyr-gly-ala-ser-asn-ser-asp-val-tyr-asp-leu-leu-lys-asp-leu-glu-glu-gly-ile-gln-thr-leu-met-gly-
      101

HPL:  arg-leu-glx-asx-gly-ser-arg-arg-thr-gly-glx-ile-leu-lys-glx-thr-tyr-ser-lys-phe-asx-thr-asx-ser-his-
       X   X   X   X   X   X       X   X   X   X   X       X   X   X   X   X   X   X   X   X   X   X   X
HGH:  arg-leu-glu-asp-gly-ser-pro-arg-thr-gly-gln-ile-phe-lys-gln-thr-tyr-ser-lys-phe-asp-thr-asn-ser-his-
      126                                                                                             150

HPL:  asx-his-asx-ala-leu-leu-lys-asx-tyr-gly-leu-leu-tyr-cys-phe-arg-lys-asx-met-asx-lys-val-glx-thr-phe-
       X       X   X   X   X   X   X   X   X   X   X   X   X   X   X   X   X   X   X   X   X   X   X   X
HGH:  asn-asp-asp-ala-leu-leu-lys-asn-tyr-gly-leu-leu-tyr-cys-phe-arg-lys-asp-met-asp-lys-val-glu-thr-phe-
      151                                                                                             175

HPL:  leu-arg-met-val-gln-cys-arg-ser-val-glu-gly-ser-cys-gly-phe
       X   X       X   X   X   X   X   X   X   X   X   X   X   X
HGH:  leu-arg-ile-val-gln-cys-arg-ser-val-glu-gly-ser-cys-gly-phe
      176                                                     190
```

The attempts to define the chemical features essential for both the lactogenic and somatotrophic effects were greatly facilitated by our determination of the complete amino acid sequence of HPL (Sherwood *et al.*, 1971) and a comparison with the structure of HGH (Li) as revised by Niall (1971). The two molecules are identical in length (190 amino acids each) and contain two disulfide bonds in identical locations. Their overall amino acid compositions are similar although they differ slightly in charge. A total of 163 (or 86%) of the amino acids in the two molecules are identically placed (Table IV) and 24 of 27 substitutions are highly favored or acceptable mutations due primarily to a single base change in the triplet codon (Dayhoff, 1969). Most of the substitutions are found near the amino terminus (15 in residues 1-75, 8 in residues 76-125, and 4 in residues 126-190), and two of the unfavored substitutions are present in the amino terminal half (residues 1, 20).

On the basis of structure-function studies to date, one is led to speculate that the lactogenic activity of HPL and HGH might be confined to the carboxyl-terminal half of each molecule where the amino acid sequences of the two hormones are nearly identical. Somatotrophic activity might reside in the amino-terminal end of HPL and HGH where the two hormones differ the most in sequence. Further support for this hypothesis comes from the chemical information available on ovine prolactin whose amino acid sequence has recently been determined (Li *et al.*, 1969). Ovine prolactin is more closely related to HPL and HGH at its carboxyl-terminal end than at its amino terminus which includes an extra disulfide bond not found in either HPL or HGH. Of great interest in the future will be the purification and chemical evaluation of human pituitary prolactin (Sherwood, 1971) which almost certainly will have marked chemical similarities at least in part with HPL and HGH. Although much work on structure-function needs to be done, the close chemical similarities between HPL and HGH imply that small differences in sequence may account for the wide differences in somatotrophic activity. It should be possible with the production of appropriate derivatives or synthetic fragments to identify a fragment of the total HGH molecule which has sufficient somatotrophic activity. Although Li and Yamashiro (1970) have recently synthesized an analogue of HGH which contained partial growth promoting and lactogenic activity, large scale synthesis of a 190 amino acid protein is not now feasible. Furthermore, attempts to synthesize HGH on a large scale in tissue culture of human pituitaries have not yet been possible. Since HPL and HGH undoubtedly are descended from a common ancestral polypeptide, which was probably smaller in size than 190 amino acids, it should be possible to resolve this problem (Sherwood, 1967; Niall *et al.*, 1971).

ACKNOWLEDGEMENTS

We are indebted to Mr. Michael Lanner for his expert technical assistance; to Dr. M. Elzinga for many helpful discussions in the course of this work: to Drs. Charles Breuer and Paul Bell of Lederle Laboratories for their generous gifts of HPL; to Dr. Alfred Wilhelmi and the National Pituitary Agency for generous gifts of HGH; to Dr. H. Hiatt for his support and encouragement, and to Mrs. Anita Heber for excellent secretarial assistance.

REFERENCES

Beck, C. and Catt, K. J. (1971): Effects of enzymatic and chemical digestion on the immunological reactivity of human chorionic somatomammotropin. *Endocrinology*, *88*, 777.

Bewley, T. A., Brovetto-Cruz, J. B. and Li, C. H. (1969): Human pituitary growth hormone. Physicochemical investigations of the native and reduced-alkylated protein. *Biochemistry*, *8*, 4701.

Breuer, C. B. (1969): Stimulation of DNA synthesis in cartilage of hypophysectomized rats by native and modified placental lactogen and anabolic hormones. *Endocrinology*, *85*, 989.

Brovetto-Cruz, J. and Li, C. H. (1969): Human pituitary growth hormone. Studies of the tryptophan residue. *Biochemistry*, *8*, 4695.

Canfield, R. E. (1963): Peptides derived from tryptic digestion of egg white lysozyme. *J. biol. Chem.*, *238*, 2691.

Canfield, R. E. and Anfinsen, C. B. (1963*a*): Chromatography of pepsin and chymotrypsin digests of egg white lysozyme on phosphocellulose. *J. biol. Chem.*, *238*, 2684.

Canfield, R. E. and Anfinsen, C. B. (1963*b*): Primary structure of proteins. In: *The Proteins*, *1st Ed.*, *Vol. I*, pp. 311–378. Editor: H. Neurath. Academic Press, New York.

CATT, K. J., MOFFAT, B. and NIALL, H. D. (1967): Human growth hormone and placental lactogen: structural similarity. *Science*, *157*, 321.
CHADWICK, A., FOLLEY, S. J. and GEMZELL, C. A. (1961): Lactogenic activity of human pituitary growth hormone. *Lancet*, *2*, 241.
CONWAY, E. J. (1957): *Microdiffusion Analysis and Volumetric Error*, *4th Ed.*, p. 90. Crosby-Lockwood, London.
CRESTFIELD, A. M., MOORE, S. and STEIN, W. H. (1963): The preparation and enzymatic hydrolysis of reduced and S-carboxymethylated proteins. *J. biol. Chem.*, *238*, 622.
DAYHOFF, M. O. (1969): The chemical meaning of amino acid mutations. In: *Atlas of Protein Sequence and Structure*, *Vol. IV*, pp. 85–87. National Biomedical Research Foundation, Silver Spring, Md.
DIXON, J. S. and LI, C. H. (1966): Retention of the biological potency of human pituitary growth hormone after reduction and carbamidomethylation. *Science*, *154*, 785.
EDMAN, P. (1950): Method for determination of the amino acid sequence in peptides. *Acta chim. scand.*, *4*, 283.
ELZINGA, M. (1970): Amino acid sequence studies on rabbit skeletal muscle actin. *Biochemistry*, *9*, 1365.
FERGUSON, K. A. and WALL, A. L. C. (1961): Prolactin activity of human growth hormone. *Nature (Lond.)*, *190*, 632.
FLORINI, J. R., TONELLI, G., BREUER, C. B., COPPOLA, J., RINGLER, I. and BELL, P. H. (1966): Characterization and biological effects of purified placental protein. *Endocrinology*, *79*, 692.
FRIESEN, H. G. (1965): Purification of a placental factor with immunologic and chemical similarity to human growth hormone. *Endocrinology*, *76*, 369.
FRIESEN, H. G. (1966): Lactation induced by human placental lactogen and cortisone acetate in rabbits. *Endocrinology*, *79*, 212.
GLICK, S. M., ROTH, J., YALOW, R. S. and BERSON, S. A. (1963): Immunoassay of human growth hormone in plasma. *Nature (Lond.)*, *199*, 784.
GRAY, W. R. (1967): Sequential degradation plus dansylation. In: *Methods in Enzymology*, *Vol XI*, pp. 469–475. Editor: C. H. W. Hirs. Academic Press, New York.
GREENWOOD, F. C., HUNTER, W. M. and GLOVER, J. S. (1963): The preparation of 1–131 labelled human growth hormone of high specific activity. *Biochem. J*, *89*, 114.
GROSS, F. and WITKOP, B. (1962): Nonenzymatic cleavage of peptide bonds: The methionine residues in bovine pancreatic ribonuclease. *J. biol. Chem.*, *237*, 1856.
GRUMBACH, M. M., KAPLAN, S. L., SCIARRA, J. J. and BURR, I. M. (1968): Chorionic growth hormone-prolactin: secretion, disposition, biologic activity in man and postulated function as the 'growth hormone' of the second half of pregnancy. *Ann. N. Y. Acad. Sci.*, *148*, 501.
HANDWERGER, S., CHAN, E. and SHERWOOD, L. M. (1970): The role of tertiary structure in the biologic and immunologic activity of human placental lactogen and human growth hormone. *Clin. Res.*, *18*, 455.
HANDWERGER, S., PANG, E. C., ALOJ, S. M. and SHERWOOD, L. M. (1971): Correlations in the structure and function of human placental lactogen and human growth hormone, I and II. Submitted.
HIRS, C. H. W. (1967): Performic acid oxidation. In: *Methods in Enzymology*, *Vol. XI*, pp. 197–199. Editor: C. H. W. Hirs. Academic Press, New York.
HIRS, C. H. W., MOORE, S. and STEIN, W. H. (1956): Peptides obtained by tryptic hydrolysis of performic acid-oxidized ribonuclease. *J. biol. Chem.*, *219*, 623.
HIRS, C. H. W., STEIN, W. H. and MOORE, S. (1954): The amino acid composition of ribonuclease. *J. biol. Chem.*, *211*, 941.
JOSIMOVICH, J. B. and MACLAREN, J. A. (1962): Presence in the human placenta and term serum of a highly lactogenic substance immunologically related to pituitary growth hormone. *Endocrinology*, *71*, 209.
KATZ, A. M., DREYER, W. J. and ANFINSEN, C. B. (1959): Peptide separation by two-dimensional chromatography and electrophoresis. *J. biol. Chem.*, *234*, 2897.
LEWIS, U. J. and CHEEVER, E. V. (1965): Evidence for two types of conversion reactions for prolactin and growth hormone. *J. biol. Chem.*, *240*, 247.
LI, C. H., DIXON, J. S., LO, T.-B., SCHMIDT, K. D. and PANKOV, Y. A. (1969): Amino acid sequence of ovine lactogenic hormone. *Nature (Lond.)*, *224*, 695.
LI, C. H., LIU, W. K. and DIXON, J. S. (1966): Human pituitary growth hormone XII. The amino acid sequence of the hormone. *J. Amer. chem. Soc.*, *88*, 2050.

LI, C. H. and YAMASHIRO, D. (1970): The synthesis of a protein possessing growth-promoting and lactogenic activities. *J. Amer. chem. Soc.*, *92*, 7608.

MILLS, J. B. and WILHELMI, A. E. (1968): Effects of treatment of bovine, porcine, and human growth hormones with sulfite. *Ann. N.Y. Acad. Sci.*, *148*, 343.

NIALL, H. D. (1971): Revised primary structure for human growth hormone. *Nature (Lond.)*, *230*, 90.

NIALL, H. D., HOGAN, M. L., SAVER, R., ROSENBLUM, I. Y. and GREENWOOD, F. C. (1971): Sequences of pituitary and placental lactogen and growth hormones: evolution from a primordial peptide by gene reduplication. *Proc. nat. Acad. Sci. USA*, *68*, 866.

NUTTING, D. F., KOSTYO, J. L., GOODMAN, H. M. and FELLOWS, R. E. (1970): Biologically active cyanogen bromide fragments of bovine growth hormone. *Endocrinology*, *86*, 416.

RAMACHANDRAN, L. K. and WITKOP, B. (1967): N-Bromosuccinimide cleavage of peptides. In: *Methods in Enzymology, Vol. XI*, pp. 283–298. Editor: C. H. W. Hirs. Academic Press, New York.

SCIARRA, J. J., SHERWOOD, L. M., VARMA, A. A. and LUNDBERG, W. B. (1968): Human placental lactogen and placental weight. *Amer. J. Obstet. Gynec.*, *101*, 413.

SHERWOOD, L. M. (1967): Similarities in the chemical structure of human placental lactogen and pituitary growth hormone. *Proc. nat. Acad. Sci. (Wash.)*, *58*, 2307.

SHERWOOD, L. M. (1969): Human placental lactogen: partial analysis of chemical structure and comparison with pituitary growth hormone. In: *Progress in Endocrinology*, pp. 394–401. Editors: C. Gual and F. J. G. Ebling. ICS 184, Excerpta Medica, Amsterdam.

SHERWOOD, L. M. (1971): Human prolactin. *New Engl. J. Med.*, *284*, 774.

SHERWOOD, L. M. and HANDWERGER, S. (1969): Correlations between structure and function of human placental lactogen and human growth hormone. In: *Proceedings of the Fifth Rochester Trophoblast Conference*, pp. 230–255. Editors: C. J. Lund and J. W. Choate, Jr. Rochester Press, Rochester, N.Y.

SHERWOOD, L. M., HANDWERGER, S., MAC LAURIN, W. D. and LANNER, M. (1971): The amino acid sequence of human placental lactogen. *New Biol.*, *233*, 59.

TURKINGTON, R. W. and TOPPER, Y. J. (1966): Stimulation of casein synthesis and histological development of mammary gland by human placental lactogen. *Endocrinology*, *79*, 175.

WILHELMI, A. E. (1961): Fractionation of human pituitary glands. *Canad. J. Biochem.*, *39*, 1659.

WILHELMI, A. E. (1968): Comment in discussion. *Recent Progr. Hormone Res.*, *24*, 434.

WOODS, K. and WANG, K. T. (1967): Separation of dansyl amino acids by polyamide layer chromatography. *Biochim. biophys. Acta (Amst.)*, *133*, 369.

YAMASAKI, N., KIKUTANI, M. and SONENBERG, M. (1970): Peptides of a biologically active tryptic digest of bovine growth hormone. *Biochemistry*, *9*, 1107.

THE SYNTHESIS AND SECRETION OF HUMAN AND MONKEY PLACENTAL LACTOGEN (HPL AND MPL) AND PITUITARY PROLACTIN (HPr AND MPr)*

H. FRIESEN, B. SHOME, C. BELANGER,
P. HWANG, H. GUYDA and R. MYERS

McGill University Clinic, Royal Victoria Hospital, Montreal, Que., Canada; and National Institutes of Health, Bethesda, Md., U.S.A.

A number of methods for the purification of HPL have been described (Cohen *et al.*, 1964; Friesen, 1965*b*; Josimovich and MacLaren, 1962), all of which yield preparations which are fairly similar in terms of their amino acid composition, molecular weight, immunochemical behavior and biological activity. Although several reports have shown that monkey placental lactogen (MPL) exists, and that it is immunologically related to monkey and human growth hormone (Grant *et al.*, 1970; Josimovich and Mintz, 1968; Kaplan and Grumbach, 1964), there is almost no information available on the circulating levels of MPL throughout pregnancy. In previous studies we have demonstrated the biosynthesis of monkey placental lactogen following the injection of labeled amino acids into the placenta *in situ* (Friesen, 1965*b*). Subsequently we observed that a partially purified MPL preparation inhibited the binding of HGH-^{131}I in a radioimmunoassay for HGH to a greater extent than did HPL; and in a bioassay for somatotropic activity, MPL was more potent than HPL (Friesen *et al.*, 1970). Because of these interesting differences we decided to purify MPL in order to compare its chemical composition and hopefully its amino acid sequence with that of HPL and HGH in order to gain a better understanding of the structure-function relationship of these three hormones. In addition we were interested in attempting to define the secretory pattern of MPL throughout pregnancy, in order to see if it were similar to that of HPL. Therefore a radioimmunoassay for MPL was developed which allowed us to measure MPL under normal circumstances in pregnant rhesus monkeys and also following a variety of experimental surgical procedures in monkeys (Friesen *et al.*, 1969*a*, *b*; Friesen and Guyda, 1971; Hwang *et al.*, 1971*a*). These studies were performed to try and shed some light on the regulatory mechanisms controlling the secretion of MPL and hopefully HPL. Indirectly these experiments led us to an investigation of protein biosynthesis in primate pituitaries (Friesen *et al.*, 1969, 1970*a*, *b*; Hwang *et al.*, 1971*a*) which ultimately allowed us to identify prolactin as a separate pituitary hormone in both man and monkeys. Finally as a result of these studies we were able to develop a specific radioimmunoassay for primate prolactin with which we have studied some of the factors which control prolactin synthesis and secretion *in vitro* and *in vivo* in man and monkeys.

Purification and chemistry of MPL

Full term rhesus monkey placentas were obtained through the courtesy of various regional primate centres as well as from our own laboratory. The placentas were maintained in a frozen state until the extraction of MPL was carried out. A flow sheet of the extraction

* The research was supported by Grants MRC- MA 1862 and USPHS – NIH-01727-06.

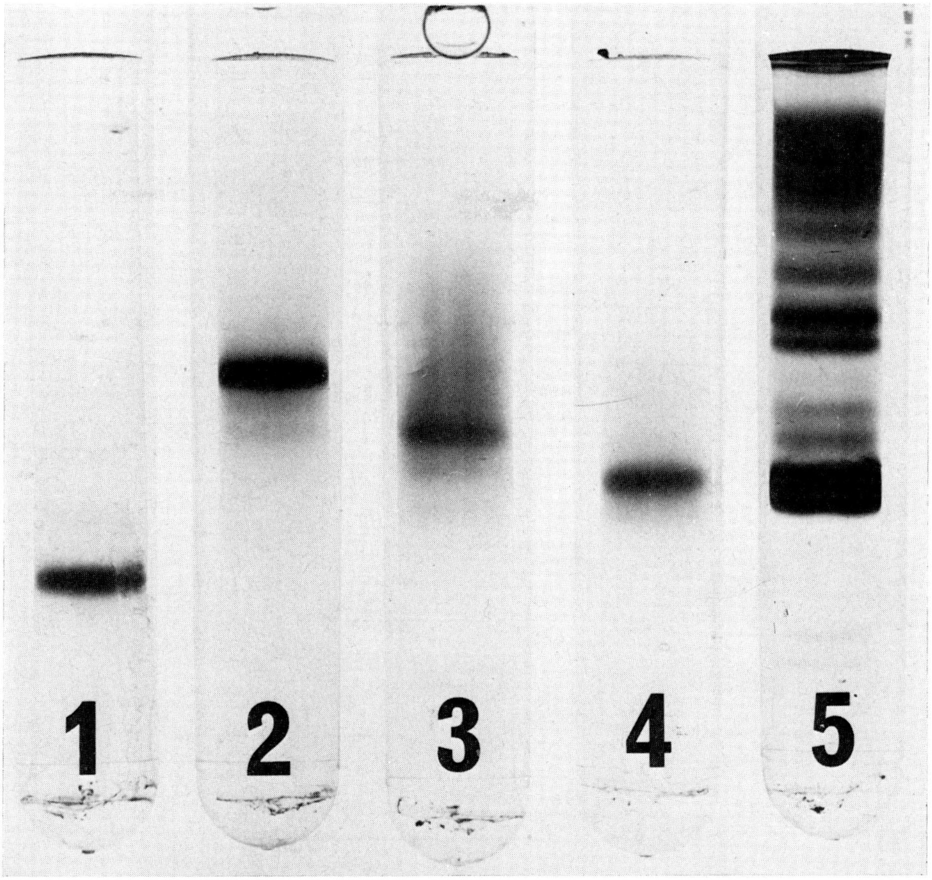

Fig. 1. The electrophoretic mobility upon acrylamide gel electrophoresis of MPL-2, MPL-1, HGH, HPL, and human serum proteins in channels 1 to 5, respectively.

procedure which was used is outlined in Table I and in greater detail by Shome and Friesen (1971). At each purification step the fractions containing MPL were identified by the formation of precipitin lines on Ouchterlony plates and by acrylamide gel electrophoresis, performed according to the method of Davis (1964). HPL was purified according to the method of Friesen (1965*b*), and HGH by Raben's method (1959). Amino acid analyses were carried out in a Beckman Model 120 C amino acid analyzer according to Spackman *et al.* (1958). Tryptophan was estimated by the spectrophotometric method of Edelhoch (1967) and also by titration with N-bromosuccinimide (NBS) as described by Spande *et al.* (1966). Trypsin digests were obtained after the protein had been reduced and carboxymethylated by the method of Crestfield *et al.* (1963), and the derivative was incubated with trypsin for 4 hours at 37° C. Two dimensional peptide maps were obtained by a modified method of Katz *et al.* (1959).

The total yield of MPL was between 25 and 30 mg/kg frozen placental tissue, which is considerably lower than the yield of HPL which is reported to be 150-300 mg/kg (Cohen *et al.*, 1964; Friesen *et al.*, 1969). The amount of MPL-1 was consistently only one half that of MPL-2. Figure 2 shows the relative mobility of the MPL preparations, HPL and HGH upon acrylamide gel electrophoresis. The relative mobilities of each hormone are as follows:

TABLE I

Flow sheet for purification of monkey placental lactogen (MPL)

Monkey placental tissue
↓ extracted at pH 8.5
↓ pH lowered to 5.8
↓
Isoelectric ppt'n
$(NH_4)_2SO_4$ – 50% saturat.
↓
Precipitate redissolved
↓
Gel filtration Sephadex G-100
↓
Fractions containing MPL applied
to carboxymethyl cellulose
↓
stepwise gradient
↓ NaCl in NH_4Ac
Fractions containing MPL applied to
DEAE with Tris buffer
↓ stepwise gradient NaCl

MPL-1 | MPL-2 (principal component)

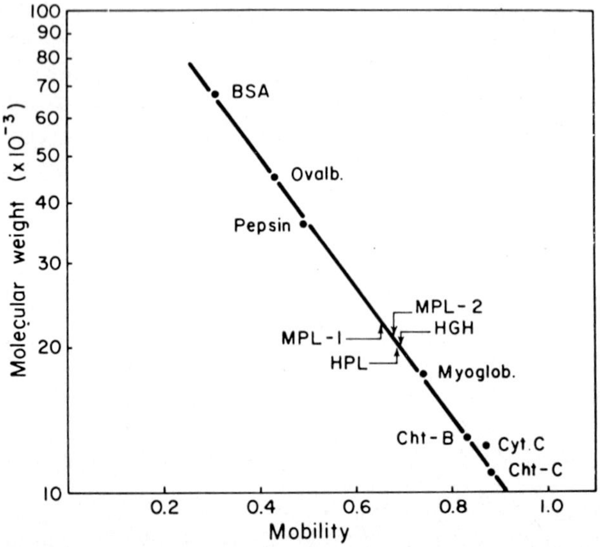

Fig. 2. The relative mobility of different proteins after acrylamide gel electrophoresis in gels containing sodium dodecyl sulfate, plotted against the molecular weight of proteins of known molecular weight. MPL-1 and MPL-2 have a molecular weight which is identical to that of HGH and HPL.

TABLE II

*Number of residues per mole**

Amino acid	MPL-1	MPL-2	HPL•	HGH∞	
				Amino acid analysis	From sequence
Lysine	10	10	9	9	9
Histidine	4	8	6	3	3
Arginine	8	8	10	10	10
Aspartic acid	19	19	22	20	20
Threonine	10	9	11–12	10	10
Serine	15	16	16	18	18
Glutamic acid	27	25	24	27	26
Proline	9	8	7	9	8
Glycine	11	9	8	9	8
Alanine	8	9	7	8	7
Half-cystine	4	4	4	4	4
Valine	9	9	8–9	7	7
Methionine	5	5	5	3	3
Isoleucine	6–7	7	7	6	8
Leucine	21	23	24	23	25
Tyrosine	6	4	7	7	8
Phenylalanine	9	10	10–11	12	13
Tryptophan	2	2	1	1	1
Total	183–185	185	185–188	185	188

* On the basis of a molecular weight of 21,500 except for HPL for which it was 19,000.
• From Sherwood (1967)
∞ From Li (*Arch. Biochem.*, 1969, *133*, 70).

MPL-2 0.80, HPL 0.72, HGH 0.58, and MPL-1 0.52. Estimates of molecular weight for MPL-1 and 2 are very similar to HGH and HPL on SDS acrylamide gel electrophoresis, namely 21,000-22,000. The amino acid composition of the different placental and pituitary preparations are summarized in Table II. Both MPL-1 and MPL-2 contain 4 half-cystine residues, most likely indicating the presence of two disulfide bonds, which is identical to the number found in HPL and HGH (Sherwood, 1967). Unlike HGH and HPL, both MPL fractions have two tryptophan residues. Apart from the difference in histidine content between MPL-1 and MPL-2 only minor differences in amino acid composition are present. Unfortunately the yield of MPL-1 was so small that we could not repeat the studies of the amino acid composition to verify these small differences. Comparison of tryptic peptides of MPL, HPL and HGH revealed a greater number of peptides which were common to HGH and HPL than which were common to HGH and MPL. The growth promoting activity of purified MPL preparations was tested by the tibia assay (Fig. 3) and the results obtained indicated that it was considerably more potent than HPL, but the growth promoting activity of MPL is much less than that of HGH.

In our previous studies on the biosynthesis of monkey placental proteins, we observed a large radioactive peak of high specific activity after gel filtration of the tissue supernatant. This peak had the same elution volume as MPL and after starch gel electrophoresis of aliquots of this fraction, two radioactive peaks of almost equal magnitude were observed consistently (Friesen, 1968) (Fig. 4). The more anodal radioactive peak was coincident with

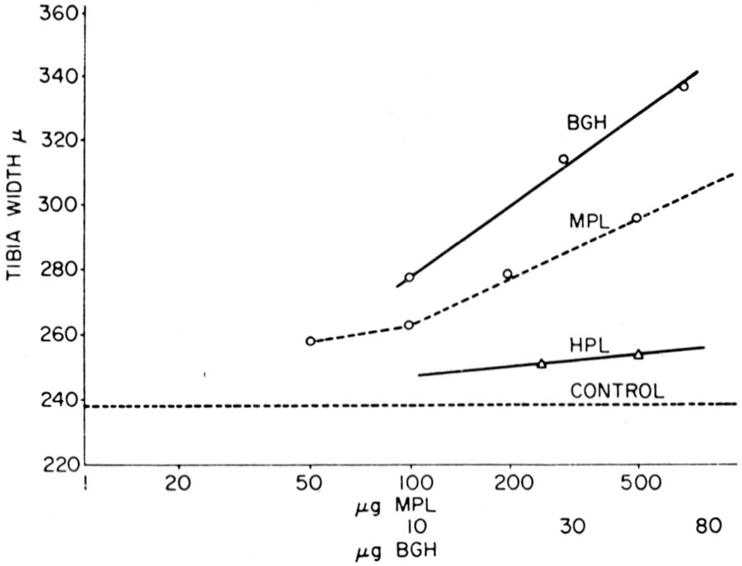

Fig. 3. Epiphyseal cartilage width in hypophysectomized rats after administration of BGH, MPL or HPL for 4 days. The daily dose administered is indicated in the abscissa which is a 2 cycle log scale. The units for MPL and HPL are 10 fold greater than for BGH. Each dose was tested in 8 rats. The horizontal interrupted line is the cartilage width of control rats receiving only saline.

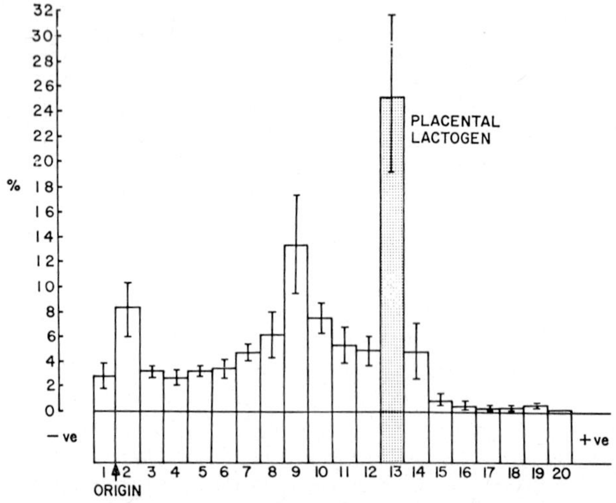

Fig. 4. The distribution of radioactivity in starch gel segments after electrophoresis of monkey placental tissue extracts. ^{3}H-leucine was injected directly into the placenta *in vivo*, and 20 to 30 min later, the placenta was removed, homogenized and the soluble proteins in the 60,000 x g supernatant were separated upon Sephadex G-100. The fractions containing MPL were then applied to starch gel electrophoresis and the distribution of radioactive proteins is shown. MPL-2 would be found in segment 13 and MPL-1 in segment 9.

MPL-2, but the nature of the second peak remained uncertain in those studies. It now appears that the Rf of this component is identical with that of MPL-1, so it is likely that both MPL components are synthesized separately and that MPL-2 is not simply a deamidated form of MPL-1 or an artifact of the isolation procedure.

Radioimmunoassay for MPL

Our initial attempts to develop a radioimmunoassay for MPL using an HGH or HPL radioimmunoassay or a hybrid assay with anti-HPL serum and HGH-^{131}I or vice versa, proved unsuccessful because serial dilutions of pregnant monkey serum samples or placental extracts did not give parallel inhibitions when either HPL or HGH were used as standards. Figure 5 shows the inhibition of binding of HGH-^{131}I to anti-HPL when increasing amounts of HGH or MPL were incubated in the assay. It is apparent, however, that MPL preparations cross-reacted to a lesser degree than HGH in this hybrid assay. Because of the difficulties encountered in attempting to use a hybrid assay for measuring MPL and with the availability of a highly purified MPL preparation, we immunized several guinea pigs with this fraction and developed a homologous double antibody radioimmunoassay for MPL.

Using the established procedures we determined the specificity, sensitivity and reproducibility of the assay and found it to be quite satisfactory. Figure 6 shows a typical radio-

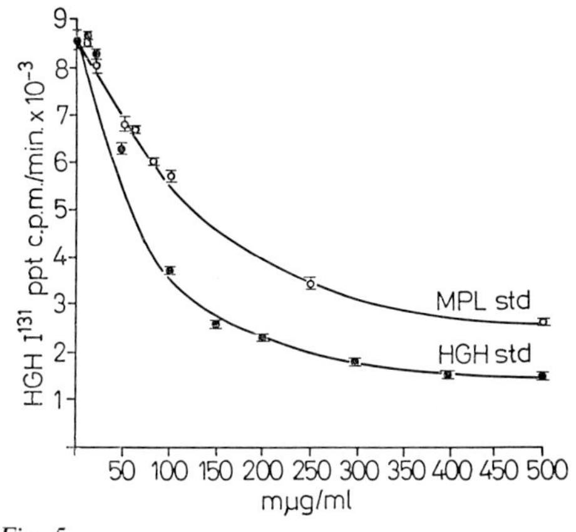

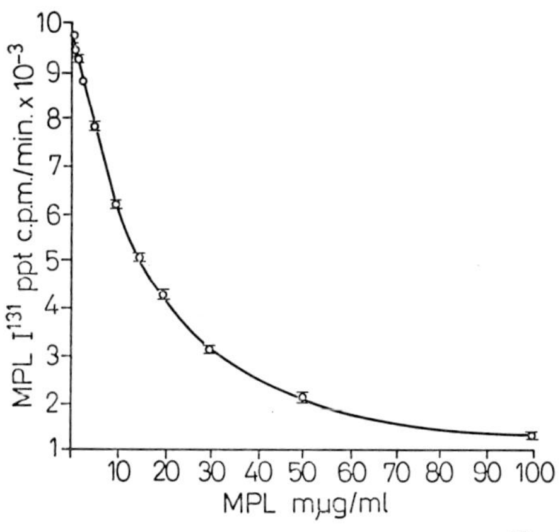

Fig. 5 *Fig. 6*

Fig. 5. A hybrid radioimmunoassay for MPL and HGH using guinea-pig antiserum to HPL and HGH-^{131}I. It is apparent that MPL and HGH do not inhibit the binding of HGH-^{131}I to anti-HPL in an identical manner, hence this assay is not suitable for measuring MPL.

Fig. 6. A radioimmunoassay for MPL. The sensitivity of the assay is less than 5 ng/ml HGH and HPL fail to inhibit the binding at concentrations less than 100 ng/ml.

immunoassay. HGH, HPL and MPL failed to cross-react to any major degree in the assay unless the concentrations of these hormones exceeded 100 ng/ml. With this assay serial dilutions of pregnant monkey sera inhibited MPL-^{131}I binding to the antibody in a manner parallel to the MPL standard. We used this assay to measure MPL concentrations throughout normal pregnancy and following a variety of experimental surgical procedures during pregnancy in monkeys in an attempt to define the factors which regulate MPL secretion. Figure 7 shows the concentrations of MPL, which were measured at various stages of pregnancy in the rhesus monkey. The earliest samples examined, day 40, already had a very high concentration of MPL and as pregnancy advances there is a progressive increase in MPL con-

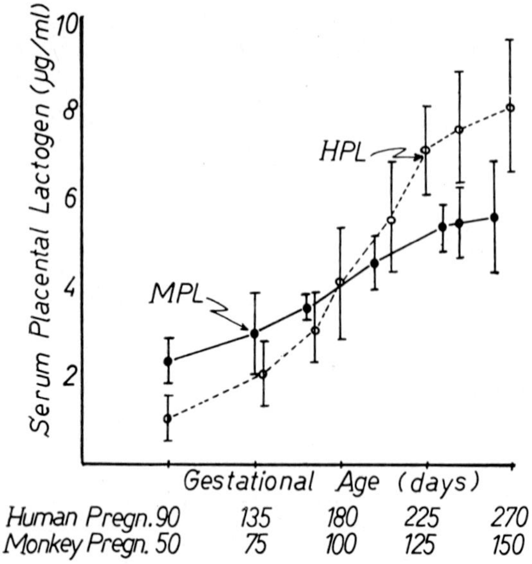

Fig. 7. HPL and MPL concentrations at different periods of gestation in pregnant women and monkeys. The increase in HPL and MPL is very similar in both groups.

centration until term, when the mean concentration is 8 μg/ml. This pattern is very similar to the mean concentration of HPL reported in a number of studies (Kaplan and Grumbach, 1965; Samaan *et al.*, 1966; Singer *et al.*, 1970). Fetal concentrations of MPL like those of HPL are almost undetectable, the average concentration is less than 1% of the maternal concentration, demonstrating that the placental barrier to both MPL and HPL is very effective. The half-time disappearance rate of both MPL and HPL is very similar in the initial phase, and somewhat longer in the second phase for MPL. Figure 8 shows the MPL concentrations as a percentage of the control level following removal of the placenta in 5 animals. Assuming that the volume of distribution of MPL and HPL is similar, one can calculate that the MPL production rate per day at term is 0.36 g/day in rhesus monkeys weighing 10 kg compared to estimates of 1.0 g/day in women at term (Kaplan *et al.*, 1968). Because the tissue content of MPL is lower than that of HPL, one can calculate that the

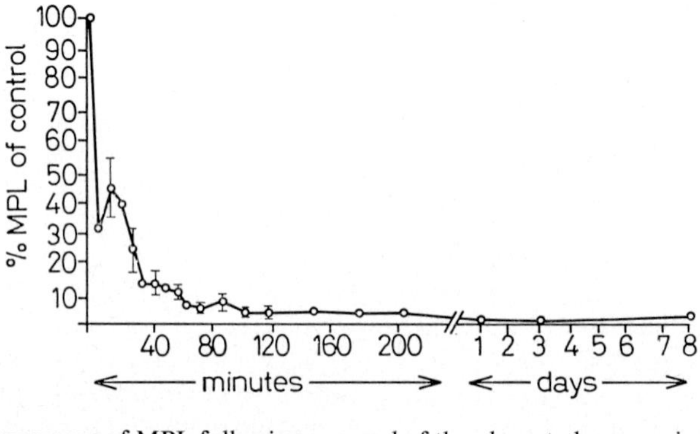

Fig. 8. The disappearance of MPL following removal of the placenta by caesarian section in 6 monkeys. The MPL concentrations are expressed as a percentage of the initial concentrations. From this data it is possible to calculate that the half-life of MPL is less than 20 min in the initial phase.

TABLE III

Comparison of HPL and MPL secretion

	HPL	MPL
1. Earliest period detected	30 days	40 days
2. Maternal conc. at term	3–10 μg/ml	3–7 μg/ml
3. Fetal conc. at term	$<$ 100 ng/ml	$<$ 100 ng/ml
4. T1/2 phase 1	12 min	20 min
phase 2	75 min	36 hr
5. Production rate/day	1 g	0.30 g
6. Placental content mg/g wet wt.	0.3	0.07
Average placental wt.	500 g	150 g
7. P.L. tissue pool turnover/day	6	20

tissue pool of MPL turns over more rapidly than that of HPL and may do so as many as 20 times per day. Table III summarizes the comparison of MPL and HPL concentrations in serum and placenta of each species.

In view of the close similarity of MPL and HPL secretory patterns, it appeared to us that one might explore a number of factors which might influence the secretion of MPL. A large number of reports have appeared in which attempts have been made to use HPL as an index of placental function or of fetal jeopardy. In the monkey it is possible to create experimental situations in which the fetus is removed completely or in which placental function is compromised deliberately. We have studied MPL levels following the removal of the fetus but leaving the placenta *in situ*. Figure 9 shows the MPL concentrations as a percentage of the control level in 9 monkeys in whom fetectomy was performed at various stages of pregnancy. In all animals the placenta remained firmly attached until the time at which a second operation was carried out from 1 to 8 weeks later. The figure shows that the decrease in MPL concentrations immediately after fetectomy was relatively small, but with time, MPL concentrations gradually declined instead of increasing as would be expected in a normal pregnancy. These studies suggest that the fetus does not exert an acute control over MPL secretion, but may do so indirectly because the fetus appears to influence the rate of placental growth. When an abruption of the secondary placental disc was produced manually we observed an immediate decrease in MPL concentrations to approximately 50% of the preoperative value demonstrating that both placental discs contribute about equally to maintaining the circulating MPL levels.

The concentration of MPL appears to be very closely related to placental perfusion of maternal blood. Figure 10 shows MPL levels measured in peripheral venous blood at various times following constriction of the aorta at a point immediately above the bifurcation. It is apparent that when uterine placental perfusion was restricted an immediate reduction in MPL levels was observed in all cases. The decreased levels persisted in 3 of 4 experiments where blood flow was reduced for an extended period. In the one experiment where flow was reduced for only 20 min, MPL levels promptly returned to the control values. In these studies placement of the clamp reduced blood flow to the uterus by at least 50%. The results observed may provide an explanation for the finding of Spellacy *et al.* (1971) that HPL determinations were low in patients with spastic toxemic hypertension and in chronic hypertension exacerbated by pregnancy, two situations where uterine and placental perfusion are reduced.

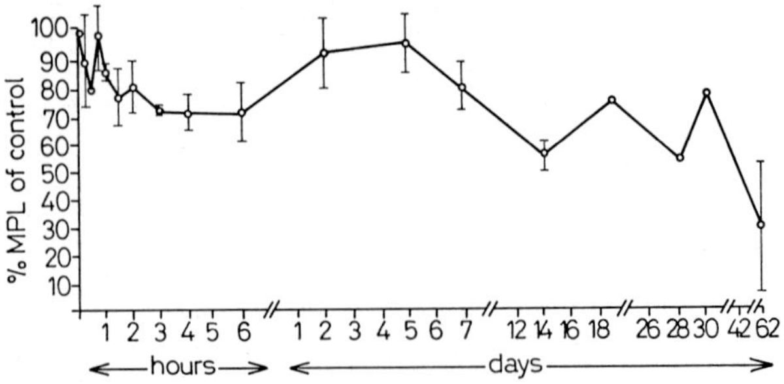

a

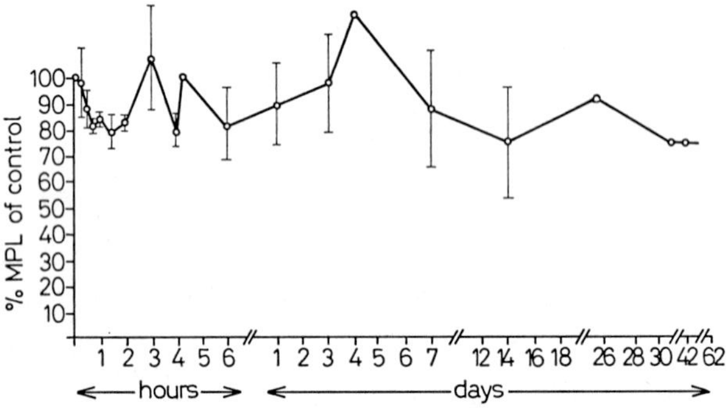

b

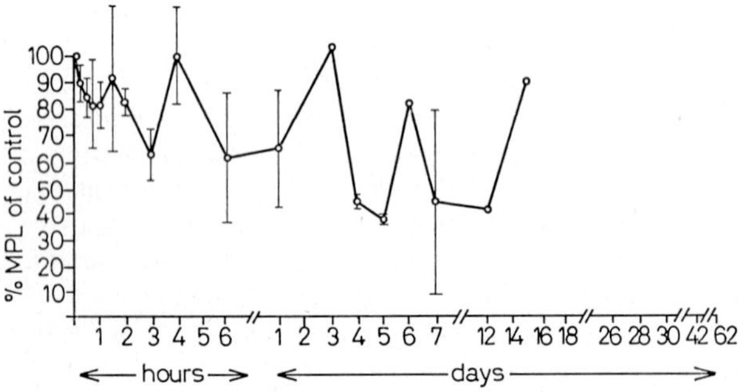

c

Fig. 9a,b,c. MPL levels following fetectomy performed in 9 monkeys, at 3 different periods of gestation, *a*: 120 days; *b*: 80 days; *c*: 150 days. There were 3 monkeys in each group. Despite the removal of the fetus, there was only a small decrease in MPL concentrations in all 3 groups. In the 3 monkeys operated on at 150 days, unexplained random fluctuations in MPL concentrations were observed after fetectomy. It is apparent that in all 3 groups MPL levels decreased gradually in the postoperative period, whereas normally MPL levels increase during pregnancy as shown in Figure 7.

Primate prolactin levels studied by radioimmunoassay

In some of the studies in which monkeys had to be sacrificed at the end of the experimental procedure, we removed the pituitary and incubated it in Krebs Ringer bicarbonate buffer with ^{3}H-l 4-5 leucine (50 μc/ml) in the same manner as we had incubated placental tissue previously (Suwa and Friesen, 1969*a*, *b*). After gel filtration of the incubation media we observed a single large radioactive protein peak which accounted for 60-70% of the TCA

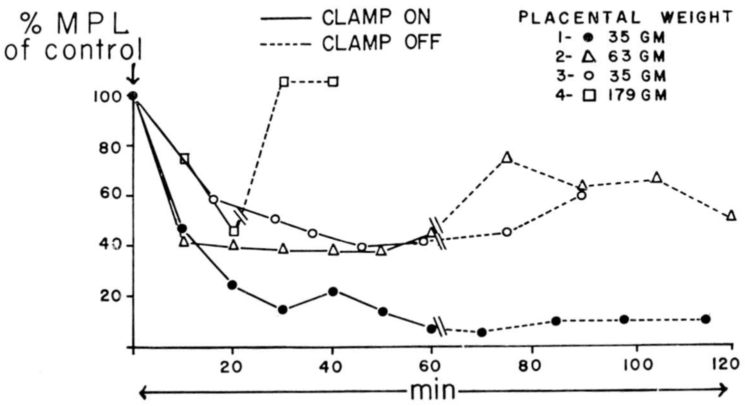

Fig. 10. MPL levels following aortic constriction. A clamp was placed just above the bifurcation of the aorta in each case. The MPL levels are expressed as a percent of the initial control samples. The continuous lines indicate the period during which the aorta was constricted, and the interrupted line indicates the MPL levels following the release of the aortic clamp. Only in the case of placenta no. 4 did MPL levels return to the control level.

precipitable proteins in the media and which had an elution volume similar to monkey growth hormone (Friesen *et al.*, 1970*b*; Friesen and Guyda, 1971). However, only a small fraction of the radioactive proteins in this pool was precipitated by antisera to HGH and a much greater proportion was precipitated with antisera to sheep prolactin. We interpreted this observation to mean that primate growth hormone could be separated from prolactin immunologically. We therefore employed affinity chromatography using antibodies to HPL to remove growth hormone completely from pituitary incubation media and obtained a prolactin rich fraction with a prolactin potency as high as 14 I.U./mg by pigeon crop sac assay and less than 1 μg/mg GH protein as measured by radioimmunoassay (Guyda and Friesen, 1971*a*). We immunized several rabbits with these fractions and obtained antisera to monkey prolactin which cross-reacted with sheep prolactin and to a lesser extent with HGH. With this antiserum and sheep prolactin-^{131}I, we developed a specific radioimmunoassay for primate prolactin which differentiated between primate growth hormone and prolactin, but the sensitivity of the hybrid assay did not allow us to measure circulating levels of prolactin (Guyda and Friesen, 1971*b*). However when we used a homologous monkey or human prolactin-^{131}I tracer which had been purified by affinity chromatography and either antisera to human or monkey prolactin, we found that the sensitivity of the assay was improved greatly (Hwang *et al.*, 1971*b*). Our radioimmunoassay was standardized by measuring the prolactin concentration in a postpartum serum sample kindly supplied to us by Dr. A. Frantz of Columbia University, New York, who had measured prolactin activity in the same sample by bioassay using a pregnant mouse mammary gland assay (Frantz and Kleinberg, 1970). Our results are expressed in ng/ml because we assumed that a 'pure' preparation of primate and ovine prolactin would be equipotent, that is 30 I.U./mg. The sensitivity of our assay for human and monkey prolactin at present is between 1 and 2 ng/ml (Fig. 11). HGH, MGH, HPL and MPL, do not cross-react in the assay unless the concentrations exceed 200-500 ng/ml for MGH and HGH and more than 10 μg/ml for placental lactogen. With this assay we have been able to measure prolactin concentrations in pituitary preparations as well as in serum of patients under a great variety of different circumstances. Figure 12 shows the HPr and MPr concentrations at various stages of pregnancy. It is obvious that the pattern between the two is very different; only occasionally is there an increase in MPr concentration. Finally

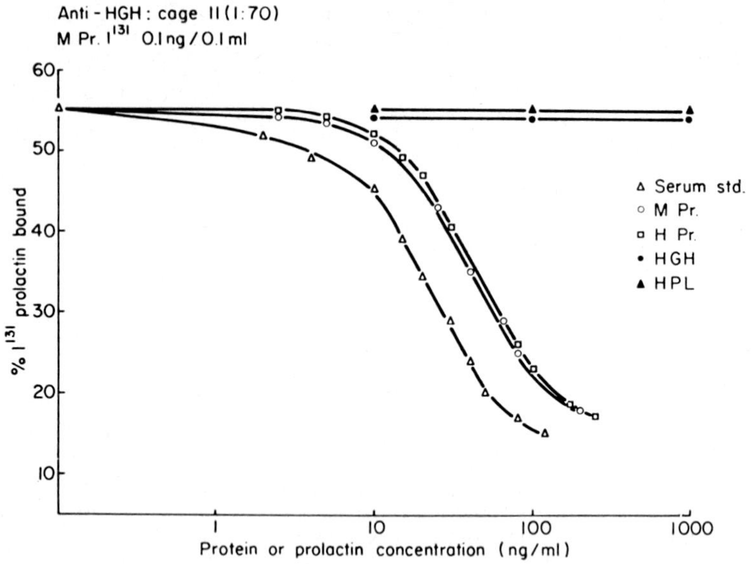

Fig. 11. A radioimmunoassay for primate prolactin. HGH and HPL do not cross-react at 1000 ng/ml and 10 μg/ml respectively. The MPr and HPr standards behave in an identical manner with each other and with serial dilutions of a postpartum serum sample.

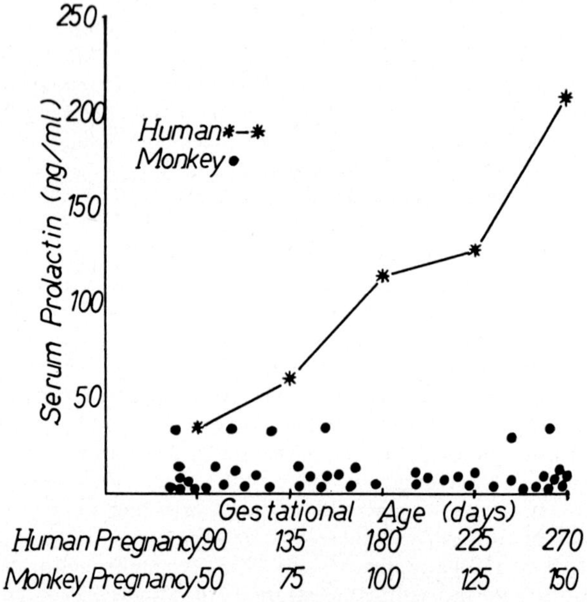

Fig. 12. The serum concentration of human and monkey prolactin (HPr and MPr) at different periods of pregnancy. The mean HPr concentrations were calculated from determination of 10 to 20 different serum samples at each point.

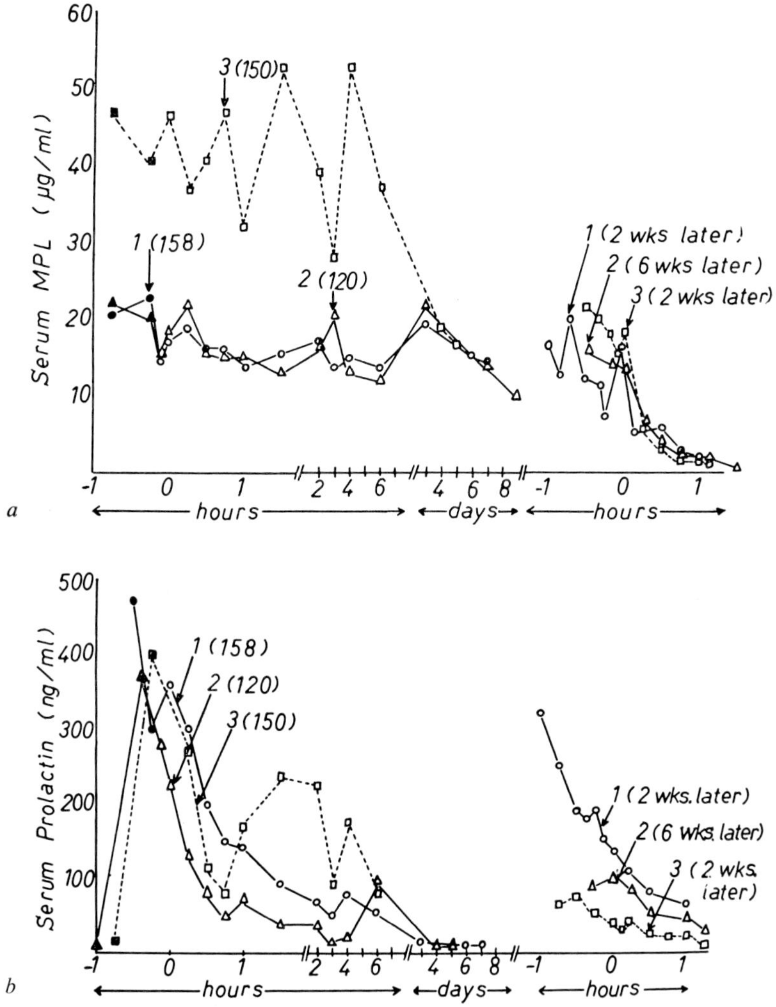

Fig. 13a and b. Serum MPL (*a*) and MPr (*b*) concentrations following surgery for removal of the fetus in three monkeys. During the fetectomy, the first operation, the initial sample was always taken before anesthesia, whereas the first sample at the time of the second operation, when the placenta was removed, was obtained after anesthesia. The solid symbols indicate the period before surgery began. The numbers in parentheses refer to the gestational age and those designated in weeks indicate the time which elapsed between the first and second operations. In general, there is an immediate and striking increase in MPr levels following anesthesia, while MPL concentrations usually fell. The highest serum prolactin level we have seen during surgery on pregnant monkeys was 1000 ng/ml.

in Figure 13 are shown the MPL and MPr concentrations in several rhesus monkeys in whom fetectomy was carried out. There is only a minor change in MPL levels, whereas there is a remarkable increase in MPr concentrations in the first two hours and a second increase following anesthesia and surgery at the time of the second operation two months later. In several other monkeys both pregnant and non-pregnant following the induction of anesthesia with nembutal we also have observed striking increases in MPr concentration. In humans the

TABLE IV

Pituitary prolactin levels in mother and fetus

No. exp't.	Gestational age (days)	MPr (ng/ml) Maternal	MPr (ng/ml) Fetal
1	80	220	14.0
2	120	225	7.4
3	150	270	44.0
4	158	200	56.5

stress of surgery also appears to increase prolactin levels, but the maximum concentrations are not as high as those observed in monkeys. Neill (1970) has also reported that stress results in the release of prolactin from rat pituitaries. The blood levels of prolactin in the monkey fetus are somewhat lower than in the mother (Table IV) and considerably lower than in the human (Hwang *et al.*, 1971*b*).

DISCUSSION

In this presentation we have outlined a method for the purification of MPL and have compared the chemistry of MPL, HPL and HGH. Immunologically and biologically MPL appears to resemble HGH more closely than HPL, because MPL cross-reacts to a greater degree in an HGH assay and also because it has greater somatotropic activity than HPL. There is, however, a very similar secretion pattern of HPL and MPL during pregnancy suggesting that a study of factors controlling MPL secretion will shed some light on the control of HPL secretion. This information should be helpful in obtaining a better understanding of the physiological role of both MPL and HPL in pregnancy.

During pregnancy there is a very consistent progressive increase in HPr levels beginning early in pregnancy, whereas in the pregnant rhesus monkey only occasionally are MPr levels elevated. In the fetus of both man and monkey we have observed markedly elevated prolactin levels. The role of prolactin in the mother during pregnancy is uncertain, but several possibilities might be considered. The first is that pituitary prolactin may be acting synergistically with other mammotropic factors in promoting breast development in preparation for lactation. The reason lactation fails to occur prior to delivery is uncertain, but may be a consequence of the high circulating levels of steroids, either estrogen or progesterone, or some unidentified steroid, which may block the action of prolactin on its target tissue. Prolactin may also have a role in maintaining sodium homeostasis during pregnancy and lactation, since sheep prolactin has been shown to increase sodium retention in rats (Lockett and Nail, 1965) and possibly in humans. Whether prolactin plays a role in modulating the changes which occur in intermediary metabolism during pregnancy is unknown. Certainly high levels of prolactin may in some species exhibit growth hormone-like effects (Apostalakis, 1968).

One wonders whether the difference in MPr and HPr levels in rhesus monkeys and humans respectively is the result of different circulating levels of estrogen during pregnancy in the two species. In humans at term, total urinary estrogen excretion is 32,000 μg/24 hr compared to 35 μg/24 hours in rhesus monkeys (Tullner, 1969). Even after allowances are made for differences in body weight, the excretion of estrogens in monkeys is relatively low. Chen and Meites (1970) have shown that in the rat there is a direct correlation between the amount of estrogen administered and the subsequent increase in prolactin levels. With the availability of a specific and sensitive radioimmunoassay for primate prolactin for the first time it is possible to explore many factors which might regulate prolactin secretion in man under different circumstances. Such studies hopefully will lead to a better understanding of the physiological role of prolactin in man and other primates.

ACKNOWLEDGEMENTS

We wish to acknowledge the technical assistance of Miss Jean Henderson, Mrs. Judy Halmagyi and Mrs. Klara Holmwood. We are also indebted to Miss Francine Dupuis for the figures, and to Mrs. Inara Leimanis for typing the manuscript.

REFERENCES

APOSTALAKIS, M. (1968): Prolactin. *Vitam. and Horm.*, *26*, 197.

BELANGER, C., SHOME, B., FRIESEN, H. and MYERS, R. E.: Studies of the secretion of monkey placental lactogen (MPL). *J. clin. Invest.*, submitted for publication.

CHEN, C. L. and MEITES, J. (1970): Effects of estrogen and progesterone on serum and pituitary prolactin levels in ovariectomized rats. *Endocrinology*, *86*, 503.

COHEN, H., GRUMBACH, M. M. and KAPLAN, S. L. (1964): Preparation of human chorionic 'growth hormone prolactin'. *Proc. Soc. exp. Biol. Med. (N.Y.)*, *117*, 438.

CRESTFIELD, A. M., MOORE, S. and STEIN, W. H. (1963): The preparation and enzymatic hydrolysis of reduced and S-carboxymethylated proteins. *J. biol. Chem.*, *238*, 622.

DAVIS, B. J. (1964): Disc electrophoresis. II. Method and application of human serum proteins. *Ann. N.Y. Acad. Sci.*, *121/2*, 404.

EDELHOCH, H. (1967): Spectroscopic determination of tryptophan and tyrosine in proteins. *Biochemistry*, *6*, 1948.

FRANTZ, A. G. and KLEINBERG, D. L. (1970): Prolactin: evidence that it is separate from growth hormone in human blood. *Science*, *170*, 745.

FRIESEN, H. (1965*a*): Purification of placental factor with immunological and chemical similarity to human growth hormone. *Endocrinology*, *76*, 369.

FRIESEN, H. (1965*b*): Further purification and characterization of a placental protein with immunological similarity to human growth hormone. *Nature (Lond.)*, *208*, 1214.

FRIESEN, H. (1968): Biosynthesis of placental proteins and placental lactogen. *Endocrinology*, *83*, 744.

FRIESEN, H. and GUYDA, H. (1971): The biosynthesis of monkey growth hormone and prolactin. *Endocrinology*, *88*, 1353.

FRIESEN, H., GUYDA, H. and HARDY, J. (1970*a*): Biosynthesis of human growth hormone and prolactin. *J. clin. Endocr.*, *31*, 611.

FRIESEN, H., GUYDA, H. and HWANG, P. (1971): Prolactin biosynthesis in primates. *Nature (Lond.)*, *232*, 19.

FRIESEN, H., SUWA, S. and PARE, P. (1969): Synthesis and secretion of human placental lactogen and other proteins by placenta. *Recent Progr. Hormone Res.*, *25*, 161.

GRANT, D. B., KAPLAN, S. L. and GRUMBACH, M. M. (1970): Studies on a monkey placental protein with immunochemical similarity of human growth hormone and human chorionic somatomammotropin. *Acta endocr. (Kbh.)*, *63*, 730.

GUYDA, H. and FRIESEN, H. (1971*a*): The separation of monkey prolactin from monkey growth hormone by affinity chromatography. *Biochem. biophys. Res. Commun.*, *42*, 1068.

GUYDA, H. and FRIESEN, H. (1971*b*): Immunological evidence for monkey and human prolactin (MPr and HPr). *J. clin. Endocr.*, *32*, 120.

HWANG, P., GUYDA, H., FRIESEN, H., HARDY, J. and WILANSKY, D. (1971*a*): Biosynthesis of human growth hormone and prolactin by normal pituitary glands and pituitary adenomas. *J. clin. Endocr.*, *31*, 1.

HWANG, P., GUYDA, H. and FRIESEN, H. (1971*b*): A radioimmunoassay for human prolactin. *Proc. Nat. Acad. Sci. USA*, *68*, 1902.

JOSIMOVICH, J. B. and MACLAREN, J. A. (1962): Presence in the human placenta and term serum of a highly lactogenic substance immunologically related to pituitary growth hormone. *Endocrinology*, *71*, 209.
JOSIMOVICH, J. B. and MINTZ, D. H. (1968): Biological and immunochemical studies on human placental lactogen. *Ann. N.Y. Acad. Sci.*, *148*, 488.
KAPLAN, S. L. and GRUMBACH, M. M. (1964): Studies of a human and simian hormone with growth hormone-like and prolactin-like activities. *J. clin. Endocr.*, *24*, 80.
KAPLAN, S. L. and GRUMBACH, M. M. (1965): Serum chorionic growth hormone prolactin and serum pituitary growth hormone in mother and fetus at term. *J. clin. Endocr.*, *25*, 1370.
KAPLAN, S. L., GURPIDE, E., SCIARRA, J. J. and GRUMBACH, M. M. (1968): Metabolic clearance rate and production rate of chorionic growth hormone prolactin in late pregnancy. *J. clin. Endocr.*, *28*, 1450.
KATZ, A. M., DREYER, W. J. and ANFINSEN, C. B. (1959): Peptide separation by two-dimensional chromatography and electrophoresis. *J. biol. Chem.*, *234*, 2897.
LOCKETT, M. and NAIL, B. (1965): A comparative study of the renal actions of growth and lactogenic hormones in rats. *J. Physiol. (Lond.)*, *180*, 147.
NEILL, J. D. (1970): Effect of 'stress' on serum prolactin and luteinizing hormone levels during the estrous cycle of the rat. *Endocrinology*, *87*, 1192.
RABEN, M. S. (1959): Human growth hormone. *Rec. Progr. Hormone Res.*, *15*, 71.
SAMAAN, N., YEN, S. C. C., FRIESEN, H. and PEARSON, O. H. (1966): Serum placental lactogen levels during pregnancy and in trophoblastic disease. *J. clin. Endocr.*, *26*, 1303.
SHERWOOD, L. M. (1967): Similarities of the chemical structure of human placental lactogen and pituitary growth hormone. *Proc. nat. Acad. Sci. (Wash.)*, *58*, 2307.
SHOME, B. and FRIESEN, H. G. (1971): Purification and characterization of monkey placental lactogen. *Endocrinology*, in press.
SINGER, W., DESJARDINS, P. and FRIESEN, H. (1970): Human placental lactogen: an index of placental function. *Obstet. and Gynec.*, *36*, 222.
SPACKMAN, D. H., STEIN, W. H. and MOORE, S. (1958): Automatic recording apparatus for use in the chromatography of amino acids. *Anal. Chem.*, *30*, 1190.
SPANDE, T. F., GREEN, N. M. and WITKOP, B. (1966): The reactivity toward N-bromosuccinimide of tryptophan in enzymes, zymogenes and inhibited enzymes. *Biochemistry*, *5*, 1926.
SPELLACY, W., TEOH, E., BUHI, W., BIRK, S. and MCCREARY, S. (1971): Value of human chorionic somatomammotropin in managing high-risk pregnancies. *Amer. J. Obstet. Gynec.*, *109*, 588.
SUWA, S. and FRIESEN, H. (1969*a*): Biosynthesis of human placental proteins and human placental lactogen (HPL) in vitro I: identification of ^{3}H-labeled HPL. *Endocrinology*, *85*, 1028.
SUWA, S. and FRIESEN, H. (1969*b*): Biosynthesis of human placental proteins and human placental lactogen (HPL) in vitro II: dynamic studies of normal term placentas. *Endocrinology*, *85*, 1037.
TULLNER, W. (1969): Comparative studies of chorionic gonadotropins and estrogens in urine and serum of subhuman primates during pregnancy. In: *Transcripts of the Fifth Rochester Trophoblast Conference*, pp. 363–377. Editors: C. J. Lund and J. C. Choate. Rochester University Press.

AN APPRAISAL OF THE ROLE OF SERUM HCS IN MONITORING PREGNANCY

RADIOIMMUNOASSAY OF HCG AND HCS IN PREGNANCIES AT RISK

P. G. CROSIGNANI and T. NENCIONI

Section of Obstetrical and Gynecological Endocrinology, Department of Obstetrics and Gynecology, University of Milan, and Centre of Gynecological Endocrinology C.N.R., Milan, Italy

In the course of the last ten years a new chapter in preventive medicine has been started: exploration of the foeto-placental function. The problem has come to affect, in turn, several sectors of research. In the area of endocrine exploration, after the contradictory results given by the biological method for the determination of human chorionic gonadotropin (HCG) (Loraine and Matthew, 1953), the first real advance was the evaluation of urinary oestriol. However, there are some situations (Rh-incompatibility, diabetes) where the evaluation of this steroid has no clinical value (Klopper, 1969). Recently, a protein hormone, the human chorionic somatomammotropin (HCS), has been suggested as an index of placental function (Josimovich, 1968). We thought, therefore, that it would be useful to study the two protein hormones of trophoblastic origin, HCG and HCS, with a view to assessing their value as different indices of the placenta's capacity to produce molecules of high biological significance and thus as measures of the functional integrity of the placenta itself. The serum profiles of the two hormones also differ during normal pregnancy (Fig. 1). The study therefore deals with the parallel radioimmunoassay of human chorionic gonadotropin (HCG) and of human chorionic somatomammotropin (HCS) in abnormal pregnancies.

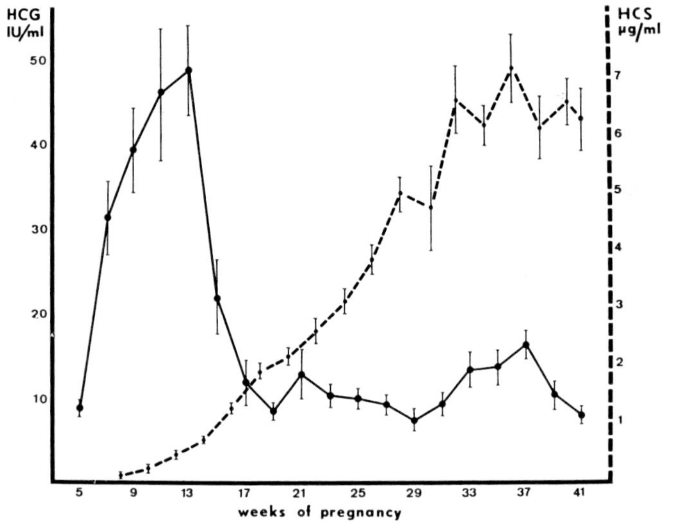

Fig. 1. Serum HCG and HCS (mean ± SE derived from 233 samples) in normal pregnancy.

MATERIALS AND METHOD

Method: Double antibody, and solid phase radioimmunoassay (Crosignani *et al.*, 1970).
Antisera: Anti-HCG = rabbit anti-HCG serum F.3 prepared by 4 successive injections of commercial HCG preparation; anti-HCS was kindly supplied by Dr. D. Mishell Jr.

Pure preparations: HCG = purified HCG was kindly donated by Dr. Donini, specific activity 12,000 I.U./mg. HCS = 95/98% pure Lederle preparation (Lot # Prep. 4508-C-75) was kindly supplied by Dr. D. Mishell Jr.
Standards: HCG = IInd international reference preparation. HCS = 95/98% pure Lederle preparation (Lot # Prep. 4508-C-75).

Abortion

Our cases comprised 37 patients hospitalised with the diagnosis of threatened abortion. Figure 2 shows the HCG and HCS values of these patients. Our subjects were divided into two groups depending on the course of their condition as observed during stay in the clinic: subjects whose pregnancy proceeded regularly and those who had a miscarriage within one week or on whom a curettage was performed after a verified incomplete abortion.

It can be seen that the plasma profiles of the first group fit into the area of normality both in respect of HCG and of HCS, while those of the second group showed rapidly falling levels up to the 18th week and were often already abnormally low at the time of the first determination.

After the 18th week, even in the cases of impending or incomplete abortion, the HCG values remained normal until the time of abortion labour. On the other hand, the HCS levels were normal in the 7 cases of regularly continued pregnancy. In the 6 cases that had a miscarriage within one week of the first determination, 5 already showed at the time of admission pathological levels that diminished subsequently. The case with a normal HCS profile was a subject with cervical incompetence, who had an abortion in the 25th week with a live-born foetus which only survived 3 hours, due to its severe prematurity.

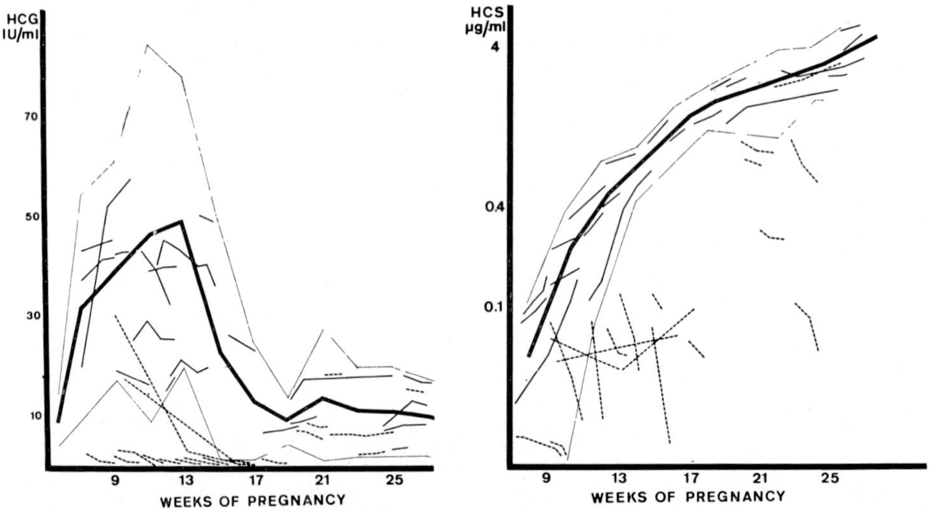

Fig. 2. Serum HCG and HCS levels in threatened abortion compared with mean and twice the standard deviation of the normal values.

Dotted line: cases that subsequently aborted; solid line: cases with good prognosis.

In conclusion, as previously reported (Crosignani *et al.*, 1971*a*) and as already observed by others (Brody and Carlstrom, 1965; Genazzani *et al.*, 1969; Singer *et al.*, 1970; Selenkov *et al.*, 1969) both the HCG and HCS levels proved useful as indices in monitoring the cases of threatened abortion prior to the 18th week.

After this period, while the HCG production of the placenta was unaffected by the death of the product of conception, the HCS production of the placenta was, on the contrary, considerably reduced. Therefore after the 18th week, only the HCS concentration provides a clinically useful index.

Toxaemia

Recently, Spellacy *et al.* (1970) concluded that, in pregnancies complicated by hypertension, if the maternal HCS levels drop below 4 μg/ml after the 30th week, the prognosis of impending foetal death is certain. Figure 3 shows our data for 20 pregnant women with toxaemia of varying severity. It can be seen that there are opposite trends in the blood concentrations of HCG and HCS. Seven subjects show pathologically high HCG levels, whereas 6 have definitely very low HCS levels. In the group with high HCG concentrations, 3 intrauterine foetal deaths were recorded, whereas in the low HCS group 5 foetal deaths were observed. Only 1 case with severe anaemia exhibited high concentrations of both hormones. On the basis of these data we tried to establish a severity score, using as indices the following: the age of pregnancy, arterial pressure, albuminuria, the presence of high HCG and low HCS levels. Table I shows the results of this trial.

From Table II it can be seen that: in the group with score 12 no foetus survived; in the

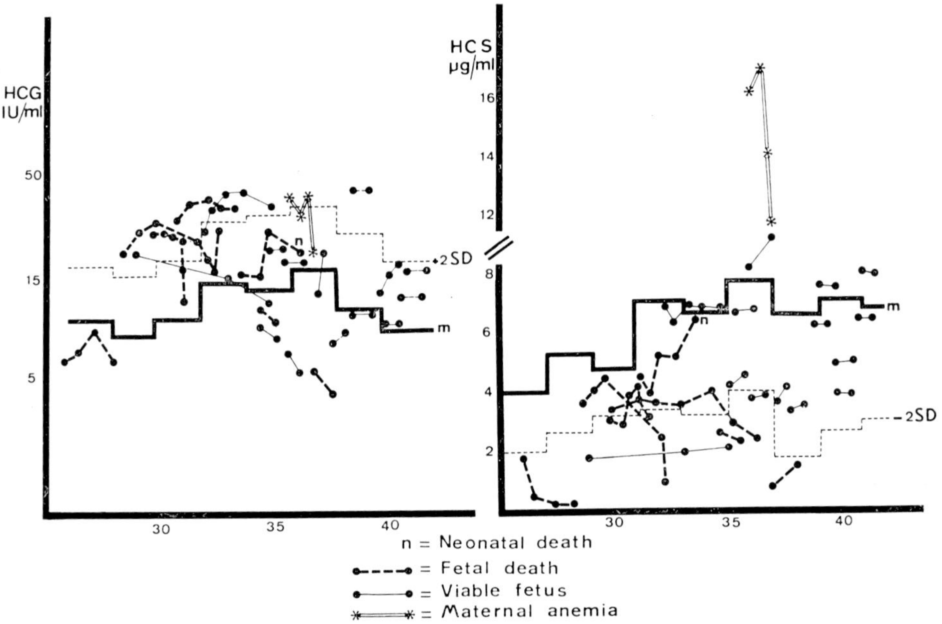

Fig. 3. Maternal serum CHG and HCS in toxemic patients (reprinted from Crosignani *et al.*, 1971*a*, by kind permission of the publishers).

TABLE I

Toxaemia: score table

Weeks gestation	Score	Blood pressure	Score	Proteinuria	Score	Serum HCG	Score	Serum HPL	Score
35–41	0	below $\frac{140}{90}$	0	—	0	below +1SD	0	above–1SD	0
30–35	2	$\frac{140}{90} \rightarrow \frac{160}{100}$	2	below $1^0/_{00}$	2	+1SD→ +2SD	2	–1SD→ –2SD	2
25–30	4	above $\frac{160}{100}$	4	above $1^0/_{00}$	4	above +2SD	4	below –S2D	4

group with score 10, there was 1 neonatal death and 1 severely immature foetus; in the group with score 8 to 0, all foetuses survived and their weight was normal. It seems, therefore, that opposite trends in the two protein hormones of the maternal serum are somehow correlated with the severity of the chronic placental insufficiency and the foetal distress characterising toxaemia.

TABLE II

Toxaemia: score and foetal outcome

Score	Length of gestation	Foetal outcome
16	31	intrauterine foetal death
16	32	intrauterine foetal death
14	35	intrauterine foetal death
12	38	intrauterine foetal death
12	37	intrauterine foetal death
12	33	small for date – neonatal death
10	35	intrauterine foetal death
10	37	small for date
10	40	small for date
8	36	normal weight
8	40	normal weight
8	39	normal weight
8	37	normal weight
8	39	normal weight
8	36	normal weight
6	39	normal weight
6	41	normal weight
4	38	normal weight
4	40	normal weight
2	41	normal weight

Post-term pregnancy

The identification of this condition is quite difficult as it requires a precise knowledge of the time of commencement of pregnancy. This may explain why consistent results are not available either for oestriol levels or for HCG and HCS concentrations. However Saxena *et al.* (1969) and Gusdon (1969) recently reported a drop in the maternal serum HCS level in cases of pregnancy beyond term. Our data concern 13 subjects with a history of normal cycles and reported to be in the 42nd week of pregnancy. The mean HCG value shown by these subjects did not differ from that obtained from 22 normal subjects in the 41st week of pregnancy. The average HCS concentration in the 42nd week (4.7 μg/ml) was found to be

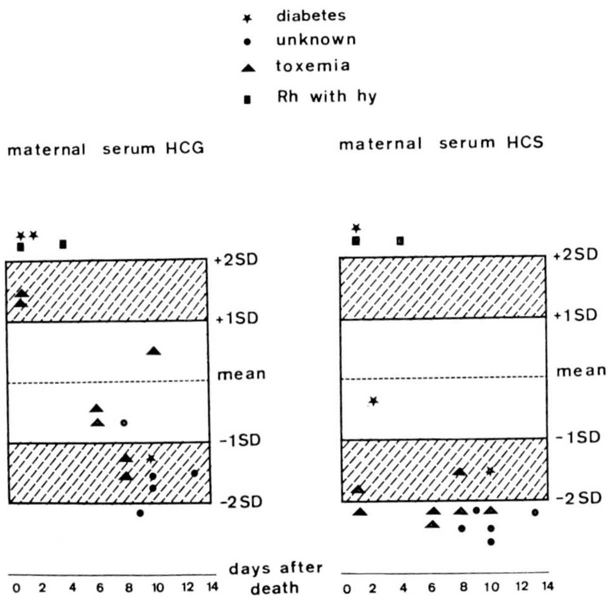

Fig. 4. Serum HCG and HCS levels after intrauterine fetal death plotted against mean, 1st and 2nd SD of the normal values.

significantly lower ($P < 0.01$) than that of the 41st week (6.13 μg/ml). In a group of 6 pregnant women we have been able to take 2-4 subsequent blood samples in the period between the beginning of the 41st and the end of the 42nd week.

The series of determinations made on these samples did not reveal any significant variations in somatomammotropin between the first and the last blood sample. These data can be interpreted in one of two ways: either the placenta reduced its HCS production before the 41st week in these subjects, or the drop will be more easily observed in post-term pregnancies among subjects with constitutionally low HCS plasma levels.

Diabetes

Nineteen diabetic women were investigated between the 18th and the 40th week of gestation. The patients, adequately treated with insulin, showed normal levels of both hormones.

Abnormal levels (above the 2nd S.D.) and pathological profiles (absence of the normal fall in the 16th-18th week) for HCG have been found in 2 cases of gestational diabetes that had not been treated with insulin and in 2 severe diabetics with poor metabolic control. In 1 of the 2 latter patients, whose macrosomic foetus died in utero, we also observed pathologically raised HCS levels.

In 3 diabetic women, who carried dead foetuses, both the HCG and HCS serum concentration appeared abnormally high even one week after foetal death.

The abnormal HCG concentration found in the 2 subjects with gestational diabetes seems to exclude the existence of a relationship between HCG serum levels and the severity of the syndrome. The excess of chorionic gonadotropin is probably related to failure of villus maturation with retention of the cytotrophoblastic layer (Fox, 1971).

Neither our limited series of observations nor the contradictory data obtained by other authors (Beck *et al.*, 1965; Gusdon, 1969; Samaan *et al.*, 1969; Saxena *et al.*, 1969; Spellacy *et al.*, 1970) seem to point to an unambiguous behaviour. The different diet or substitutive treatment used, the appearance of complications such as toxaemia, maternal ketoacidosis, retarded development of the foetus, foetal macrosomia, are all factors which make this type of investigation particularly difficult and any conclusions untimely. Probably these are the reasons why so many contradictory data are also observed in oestriol determinations (Klopper, 1969).

Rh-immunisation

Table III shows the two protein hormone concentrations in the maternal serum and in the amniotic fluid of 26 pregnant women with varying degrees of Rh-isoimmunisation. In the mildly or moderately affected cases, normal HCG and HCS values were observed in the maternal serum: in the amniotic fluid, although the HCG mean level did not change, there were often values outside the range. In the 6 severely affected cases, complicated by foeto-placental hydrops, the two biological fluids yielded constantly abnormal levels of HCG and HCS. In the severe cases there was a gradual pathological rise in the hormone concentrations with a parallel trend in the maternal serum and in the amniotic fluid, as the foeto-placental situation deteriorated and the hydropic condition set in.

TABLE III

Rh pregnancy: HCG and HCS levels in maternal serum and in amniotic fluid

HCG I.U./ml		HCS μg/ml	
Maternal serum	Amniotic fluid	Maternal serum	Amniotic fluid
Mildly-medially Rh-immunised women			
9.77 ± SE 1.94	0.40 ± SE 0.10	7.09 ± SE 0.70	0.41 ± SE 0.08
Rh women with hydrops (mean with range)			
87.41 (24.5–200)	13.4 (0.5–60)	12.8 (5.1–26)	1.88 (0.48–7)

The abnormal profile of HCG and HCS concentrations is sometimes observed for a period of several days before the intrauterine death of the foetus (Crosignani *et al.*, 1971*b*). It is also interesting to note that the abnormal rise persists (3 cases) after the verified death of the foetus in utero.

There is a certain amount of agreement that in this instance the urine and plasma oestriol determinations are ill-suited as a means for monitoring the foeto-placental unit (Klopper, 1969). With regard to HCG and HCS, the various authors have so far failed to find any clear correlation between the plasma and amniotic levels of the single hormones and the foetal prognosis (Samaan *et al.*, 1969; Josimovich *et al.*, 1970; Spellacy *et al.*, 1970).

Only Singer *et al.* (1970) have recently reported high HCS values in severely immunized pregnant women, whereas Josimovich (1971) describes a rise in amniotic HPL concentration corresponding to a deterioration of the condition.

On the basis of our data we feel that the indices reported here might open up new prospects for the investigation of pregnancy with Rh-immunisation.

Intrauterine foetal death

Selenkov *et al.* (1969) observed no variations in serum HCS 12-36 hours after having injected into the amniotic fluid a hypertonic saline solution in order to induce abortion. Saxena *et al.* (1969) observed in 2 out of 3 diabetic women, whose foetuses died in utero, a decrease in the level of this hormone upon foetal death. On the other hand, Singer *et al.* (1970) found no significant modifications in the HCS upon the death of the foetus in utero.

Our data showed normal HCG values in 16 out of 17 cases within two weeks from the intrauterine death of the foetus. In 2 cases of diabetes and Rh-isoimmunisation, the levels reached even beyond twice the standard deviation from the normal mean. As far as the HCS concentrations are concerned, 10 out of 17 cases exhibited values below twice the standard deviation from the mean and 3 cases (2 of toxaemia and 1 of diabetes) showed low values ranging between once and twice the standard deviation of a normal population. In 3 subjects with HCS levels abnormally raised beyond twice the standard deviation, intrauterine foetal death was associated, in 1 case, with maternal diabetes and, in 2 cases, with severe Rh-isoimmunisation.

So, excluding association with diabetes and Rh-isoimmunisation, only 1 case out of 16 displayed HCS values within the range of a single standard deviation.

Therefore the HCS level appears to provide, just as much as does oestriol, a useful index of intrauterine foetal death.

As to the possibility of forecasting the death of the foetus, the data recorded in patients with toxaemia are quite interesting. In 5 out of 6 cases of impending foetal death, low or decreasing HPL levels were found in the maternal serum. Obviously, from the data previously reported – and in agreement with Spellacy *et al.* (1970) – it emerges that the HCS assay by itself cannot forecast the impending death of the foetus in pregnancies with diabetes and Rh-isoimmunisation. From a general point of view it might be said that the abnormally high HCG and HCS production described in some cases of diabetes and Rh-isoimmunisation persists even after the death of the foetus. The presence of high levels of serum HCG and HCS after foetal death seems to provide a measure of the persisting abnormal trophoblastic activity in these conditions: the histological appearance of the placenta in diabetes and Rh-isoimmunisation is in some respects consistent with the biochemical data (Fox, 1971).

CONCLUSIONS

Placenta HCG and HCS production is often altered in abnormal pregnancies:

- modification of either of the two protein hormones is not always matched by an analogous modification of the other, and even opposite trends may be observed. For these reasons the two hormones can be potentially different indices;
- measurement of HCS seems to provide a useful index of intrauterine death of the foetus;
- in the diagnosis of abortion, the parallel assay of HCG and HCS has a diagnostic value, whereas in the study of Rh-immunisation and toxaemia conditions, the HCG and HCS data, if properly used, seem to be correlated with the severity of the disease;
- the phenomenon of foetal distress is not a well-known one and its pathogenesis may be different in various instances, depending on different aetiologies, so that it is not safe to draw conclusions on the basis of a single index.

The method of utilising several indices seems to provide a practical means for reducing errors and improving the reliability of the information obtained;
parallel radioimmunoassays of steroid and protein hormones in the maternal serum and in the amniotic fluid are likely to open up new prospects for monitoring pregnancy.

REFERENCES

BECK, P., PARKER, M. L. and DAUGHADAY, W. H. (1965): Radioimmunologic measurement of human placental lactogen in plasma by a double antibody method during normal and diabetic pregnancies. *J. clin. Endocr.*, *25*, 1457.

BRODY, S. and CARLSTROM, G. (1965): Human chorionic gonadotropin in abnormal pregnancy. Serum and urinary findings using various immunoassay techniques. *Acta obstet. gynec. scand.*, *44/1*, 32.

CROSIGNANI, P. G., NAKAMURA, R. M., HOVLAND, D. N. and MISHELL, Jr, D. R. (1970): A method of solid phase radioimmunoassay utilising polypropylene discs. *J. clin. Endocr.*, *30/2*, 153.

CROSIGNANI, P. G., NENCIONI, T. and BRAMBATI, B. (1971*a*): Parallel assay of HCG and HCS in pregnancies at risk. In: *Fetal Evaluation during Pregnancy and Labor*, pp. 94–109. Editors: P. G. Crosignani and G. Pardi. Academic Press, New York and London.

CROSIGNANI, P. G., NENCIONI, T., BRAMBATI, B. and POLVANI, F. (1971*b*): The prognostic value of HCG and HPL in Rh sensitized women. In: *Perinatal Medicine.* Editor: P. J. Huntingford. S. Karger A. G., Basel. In press.

FOX, H. (1971): Relationship between placental morphology and fetal distress. In: *Fetal Evaluation during Pregnancy and Labor*, pp. 5–10. Editors: P. G. Crosignani and G. Pardi. Academic Press, New York and London.

GENAZZANI, A. R., ALBERT, M. and CASOLI, M. (1969): Use of human placental lactogen radioimmunoassay to predict outcome in cases of threatened abortion. *Lancet*, *2*, 1385.

GUSDON JR, J. P. (1969): Improved hemagglutination-inhibition assay: Clinical application to measurement of human placental lactogen. *Obstet. and Gynec.*, *33/3*, 397.

KLOPPER, A. (1969): The assessment of placental function in clinical practice. In: *Foetus and Placenta, 1st ed.*, Chapter 4, pp. 471–555. Editors: A. Klopper and E. Diczfalusy. Blackwell Scientific Publications, Oxford.

JOSIMOVICH, J. B. (1968): The human placental lactogen. *Clin. Endocr.*, *2*, 658.

JOSIMOVICH, J. B. (1971): Human chorionic somatomammotropin (HCS) in high-risk pregnancies. In: *Fetal Evaluation during Pregnancy and Labor*, pp. 87–93. Editors: P. G. Crosignani and G. Pardi. Academic Press, New York and London.

JOSIMOVICH, J. B., KOSOR, B., BOCCELLA, L., MINTZ, D. H. and HUTCHINSON, D. L. (1970): Placental lactogen in maternal serum as an index of fetal health. *Obstet. and Gynec.*, *36*, 244.

LORAINE, J. A. and MATHEW, G. D. (1953): The placental concentration of chorionic gonadotropin in normal and abnormal pregnancy. *J. Obstet. Gynaec. Brit. Cwlth*, *60*, 640.

SAMAAN, N. A., BRADBURY, J. T. and GOPLERUD, C. P. (1969): Serial hormonal studies in normal and abnormal pregnancy. *Amer. J. Obstet. Gynec.*, *104*, 781.

SAXENA, B. N., EMERSON, K. and SELENKOV, H. A. (1969): Serum placental lactogen as index of placental function. *New Engl. J. Med.*, *281*, 225.

SELENKOV, H. A., SAXENA, B. N., DANA, C. L. and EMERSON, K. (1969): In: *Foeto-Placental Unit*, Chapter V, pp. 340–362. Editors: A. Pecile and C. Finzi. ICS 183, Excerpta Medica, Amsterdam

SINGER, W., DESJARDINS, P. and FRIESEN, H. G. (1970): Human placental lactogen, an index of placental function. *Obstet. and Gynec.*, *36/2*, 222.

SPELLACY, W., TEOH, E. S. and BUHI, W. C. (1970): Human chorionic somatomammotropin (HCS) levels prior to fetal death in high-risk pregnancies. *Obstet. and Gynec.*, *35/5*, 685.

VI. Regulation of secretion

GROWTH HORMONE-RELEASING HORMONE (GH-RH) OF THE HYPOTHALAMUS; ITS CHEMISTRY AND IN VIVO AND IN VITRO EFFECTS*

ANDREW V. SCHALLY and AKIRA ARIMURA

Endocrine and Polypeptide Laboratories, Veterans Administration Hospital, and Department of Medicine, Tulane University School of Medicine, New Orleans, La., U.S.A.

At least three comprehensive reviews of purification and of *in vitro* and *in vivo* studies with growth hormone releasing hormone (GH-RH) have been published in recent years (Schally *et al.*, 1968*a*, *c*; 1970). In the following paragraphs we have attempted to gather the significant data concerning the chemistry and biological properties of GH-RH collected since these reviews were written. Some earlier papers of others as well as ourselves will be quoted as essential key references.

Chemistry of GH-RH

Several groups attempted to purify GH-RH (Franz *et al.*, 1962; Dhariwal *et al.*, 1965), but because of the difficulty of assays for GH-RH and the complexity of isolation procedures only one group has succeeded in isolating it (Schally *et al.*, 1969, 1970). This work utilized porcine hypothalami and the details of isolation were reported previously (Schally *et al.*, 1969, 1970). Isolated GH-RH was homogeneous by electrophoresis and chromatography. No evidence was found for any active trace components or factor non-covalently bound to

TABLE I

Amino acid composition of porcine growth hormone releasing hormone (GH-RH)[1]

Amino acid	Molar ratio[3]	Integral residues[2]
Lysine	1.18	1
Histidine	0.94	1
Ammonia	2.07	2
Serine	0.95	1
Glutamic acid	3.10	3
Alanine	1.78	2
Valine	1.03	1
Leucine	1.0	1

[1] Amino acids account for 98% dry weight.
[2] To the nearest integer assuming 10 amino acid residues.
[3] Accepting leucine as 1.0.

* Supported in part by USPHS grant AM-07467 and AM-09094.

the GH-RH (Schally *et al.*, 1970). The amino acid composition of this decapeptide is illustrated in Table I. GH-RH shows a high glutamic acid content in agreement with its acidic isoelectric point (Schally *et al.*, 1969, 1970). The structural work (Schally *et al.*, 1971) was based on preliminary cleavage with trypsin and papain, enzymes which were previously shown to inactivate GH-RH (Schally *et al.*, 1970).

Trypsin split GH-RH into two fragments which were readily separable by paper electrophoresis (Schally *et al.*, 1971). Digestion with papain followed by electrophoresis showed that 8 fragments were formed from GH-RH. The fragments resulting from cleavage with these two endopeptidases were subjected to sequential degradation from the N-terminus by the method of Edman (Edman, 1950) followed by the dansyl procedure (Gray and Hartley, 1963; Gray, 1967). The intact GH-RH molecule was also subjected to the combined Edman-Dansyl procedure. In addition valuable information was provided by the digestion of GH-RH and its fragments thereof with aminopeptidase M, leucine aminopeptidase and carboxypeptidase A and B. On the basis of these data, the sequence was derived as Val-His-Leu-Ser-Ala-Glu-Glu-Lys-Glu-Ala. Porcine GH-RH is thus a straight chain acidic decapeptide with an N-terminal valine and C-terminal alanine (Schally *et al.*, 1971). The structure of GH-RH is shown in Figure 1. The structure of GH-RH was confirmed by synthesis

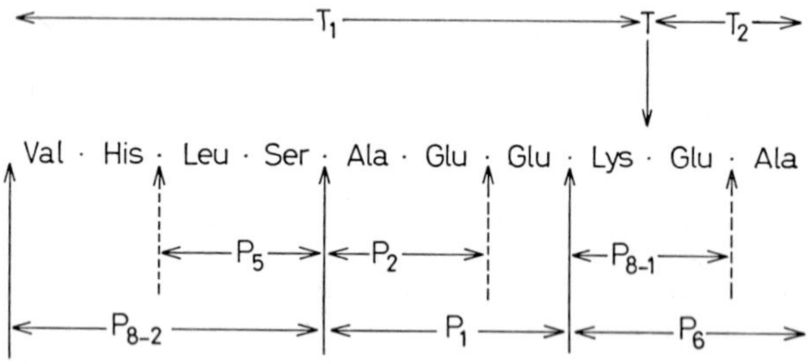

Fig. 1. Structure of porcine growth hormone-releasing hormone (GH-RH). (From Schally *et al.*, 1971.)

(Veber *et al.*, 1971). The synthetic decapeptide possesses biological and chemical properties similar to natural porcine GH-RH (Schally *et al.*, 1972; Arimura *et al.*, 1972). Thus, synthetic GH-RH stimulates the release of GH *in vitro* (Schally *et al.*, 1968*b*) and depletes pituitary GH-content *in vivo* (Pecile *et al.*, 1965), when the measurement of GH is performed by the 'tibia test' method of Greenspan *et al.* (1949). However, neither natural nor synthetic GH-RH was active in tests where the release of GH was measured by the radioimmunoassay for rat GH (Schalch and Reichlin, 1966). The suitability of radioimmunoassay for measuring GH in rat plasma has been questioned recently by Müller *et al.* (1971). Moreover, the possibility of the presence of two pools of GH, one of which is biologically active while the other is immunologically active, cannot be discounted at present. Nevertheless, the lack of effect of GH-RH on the release of radioimmunoassayable-GH (RIA-GH) introduces some element of uncertainty which at the moment remains unresolved. Whether this GH-RH molecule represents the true hypothalamic hormone which is responsible for the stimulation of GH release and synthesis under physiological conditions, will have to be determined by future studies.

Biological studies with porcine GH-RH

Several important studies utilizing purified porcine GH-RH were published recently. Sawano *et al.* (1968) previously determined in our laboratory that partially purified porcine

GH-RH induces, in addition to the depletion of pituitary GH content, a rise in GH-like activity in plasma as measured by the 'tibia test'. Müller *et al.* (1971), using homogeneous porcine GH-RH, confirmed that the depletion of pituitary GH content in rats is accompanied by a simultaneous rise in plasma GH-activity. This suggests, that depletion of pituitary GH content in rats caused by GH-RH reflects the release of the hormone.

Electron microscopy studies of rat pituitaries by Couch *et al.* (1969) clearly showed the increased extrusion of secretion granules from the somatotrophs into the perivascular space 1 min after intracarotid injection of porcine GH-RH. This indicates that GH-RH acts selectively on the somatotrophs. We have previously observed an effect of GH-RH on the synthesis as well as the release of GH during short term incubation of rat pituitary tissue (Schally *et al.*, 1968*b*).

These findings were confirmed and extended by the work of Mittler *et al.* (1970) in our laboratory. They investigated the effect of GH-RH on GH synthesis in 5 day tissue cultures of rat pituitaries. Addition of GH-RH caused a rise in bioassayable GH in both medium and tissue and increased the incorporation of radioactive amino acids into GH. These results indicate that GH-RH stimulates *de novo* synthesis of GH.

The search for a GH-RH able to stimulate the release of immunoreactive GH

We have recently made attempts, during the purification of pig hypothalamic extracts, to follow GH-RH activity by a rise in plasma GH-levels in rats as measured by radioimmunoassay for rat GH (Schalch and Reichlin, 1966). Rats anesthetized with urethane, injected with reserpine were used as suggested by Malacara and Reichlin (1971). Test materials were injected into the carotid artery and blood was collected 15 min thereafter. Plasma GH levels were determined by radioimmunoassay (Schalch and Reichlin, 1966). In our hands pretreatment with the sex steroids did not increase the responses to GH-RH fractions which stimulated the release of RIA-GH. The GH-RH active materials, as determined by RIA for rat GH, were located in exactly the same areas in effluents from Sephadex G-25 columns as well as in fractions from free-flow electrophoresis as the GH-RH previously reported to stimulate the release of bioassayable GH (Schally *et al.*, 1972*b*). Infusion into a hypophysial portal vessel of GH-RH purified by gel filtration on Sephadex also stimulated the release of RIA-GH in the rat (Sandow *et al.*, 1972).

Other laboratories also made attempts to find a GH-RH able to stimulate the release of RIA-GH in rats. Malacara and Reichlin (1971) claimed that GH-RH purified from pig hypothalamic extracts by gel filtration on Sephadex G-10 induced upon intravenous injection in very large doses, an elevation of RIA-GH in rats pretreated with estrogen and progesterone and anesthetized with ether or pentobarbital.

Frohman *et al.* (1971) reported that GH-RH purified from sheep hypothalamic extracts by gel filtration on Sephadex caused a rise in plasma RIA-GH levels in rats after intrapituitary infusion. The same fractions were said to stimulate the release of stored RIA-GH from isolated rat pituitaries *in vitro* (Stachura *et al.*, 1971; Peake *et al.*, 1971).

Wilbur *et al.* (1971) reported partial purification of growth hormone releasing material from pig hypothalami which stimulated release of RIA-GH from rat pituitaries *in vitro*. This material appeared to be much more basic than the GH-RH of Schally *et al.* (1969, 1970, 1971). The significance of findings on hypothalamic GH-RH active materials, able to stimulate the release of RIA-GH *in vivo* and *in vitro*, remains to be evaluated.

REFERENCES

Arimura, A., Wakabayashi, I., Sandow, J. and Schally, A. V. (1972): In preparation.

Couch, E. F., Arimura, A., Schally, A. V., Saito, M. and Sawano, S. (1969): Electron microscope studies of somatotrophs of rat pituitary after injection of purified growth hormone releasing factor (GRF). *Endocrinology*, *85*, 1084.

DHARIWAL, A. P. S., KRULICH, L., KATZ, S. H. and MCCANN, S. M. (1965): Purification of growth hormone-releasing factor. *Endocrinology*, *77*, 932.

EDMAN, P. (1950): Methods for determination of the amino acid sequence in peptides. *Acta chem. scand.*, *4*, 283.

FRANZ, J. C., HASELBACH, C. H. and LIBERT, O. (1962): Studies of the effect of hypothalamic extracts on somatotrophic pituitary function. *Acta endocr. (Kbh.)*, *41*, 336.

FROHMAN, L. A., MARAN, J. W., YATES, F. E. and DHARIWAL, A. P. S. (1971): Growth hormone responses to intrapituitary injection of growth hormone releasing factor in the rat as measured by radioimmunoassay. *Fed. Proc.*, *30*, 198.

GRAY, W. R. (1967): Sequential degradation plus dansylation. In: *Methods in Enzymology*, *Vol. XI*, pp. 469–475. Editor: C. H. W. Hirs. Academic Press, New York.

GRAY, W. R. and HARTLEY, B. S. (1963): The structure of chymotryptic peptide from pseudomonas cytochrome C-551. *Biochem. J.*, *89*, 379.

GREENSPAN, F. S., LI, C. H., SIMPSON, M. E., and EVANS, H. M. (1949): Bioassay of hypophysial growth hormone: the tibia test. *Endocrinology*, *45*, 455.

MALACARA, J. M. and REICHLIN, S. (1971): Elevation of radioimmunoassayable plasma growth hormone (RIA-GH) in the rat induced by porcine hypothalamic extracts. *Fed. Proc.*, *30*, 198.

MITTLER, J. C., SAWANO, S., WAKABAYASHI, I., REDDING, T. W. and SCHALLY, A. V. (1970): Stimulation of release and synthesis of growth hormone (GH) in tissue cultures of anterior pituitaries in response to GH-releasing hormone (GH-RH). *Proc. Soc. exp. Biol. Med. (N.Y.)*, *133*, 890.

MÜLLER, E. E., SCHALLY, A. V. and COCCHI, D. (1971): Increase in plasma growth hormone (GH)-like activity after administration of porcine GH-releasing hormone. *Proc. Soc. exp. Biol. Med. (N.Y.)*, in press.

PECILE, A., MÜLLER, E. E., FALCONI, G. and MARTINI, L. (1965): Growth hormone releasing activity of hypothalamic extracts at different ages. *Endocrinology*, *77*, 241.

PEAKE, G. T., DAUGHADAY, W. H. and DHARIWAL, A. P. S. (1971): Growth hormone releasing factor (GRF): effect on *in vitro* growth hormone (GH) release and the pituitary adenyl cyclose system. *Program 53rd Meeting of the Endocrine Society*, *88*, A-81.

SANDOW, J., ARIMURA, A. and SCHALLY, A. V. (1972): *Endocrinology*, in press.

SAWANO, S., ARIMURA, A., BOWERS, C. Y., REDDING, T. W. and SCHALLY, A. V. (1968): Pituitary and plasma growth hormone-like activity after administration of highly purified pig growth hormone-releasing factor. *Proc. Soc. exp. Biol. Med. (N.Y.)*, *127*, 1010.

SCHALCH, D. S. and REICHLIN, S. (1966): Plasma growth hormone concentration in the rat determined by radioimmunoassay: Influence of sex, pregnancy, lactation, anesthesia, hypophysectomy and extrasellar pituitary transplants. *Endocrinology*, *79*, 275.

SCHALLY, A. V., ARIMURA, A., BOWERS, C. Y., KASTIN, A. J., SAWANO, S. and REDDING, T. W. (1968*a*): Hypothalamic neurohormones regulating anterior pituitary function. *Recent Progr. Hormone Res.*, *24*, 497.

SCHALLY, A. V., MÜLLER, E. E. and SAWANO, S. (1968*b*): Effect of porcine growth hormone-releasing factor on the release and synthesis of growth hormone *in vitro*. *Endocrinology*, *82*, 271.

SCHALLY, A. V., SAWANO, S., MÜLLER, E. E., ARIMURA, A., BOWERS, C. Y., REDDING, T. W. and STEELMAN, S. L. (1968*c*): Hypothalamic growth hormone-releasing hormone (GRH). Purification and *in vivo* and *in vitro* studies. In: *Growth Hormone*, pp. 185–203. Editors: A. Pecile and E. E. Müller. ICS 158, Excerpta Medica, Amsterdam.

SCHALLY, A. V., SAWANO, S., ARIMURA, A., BARRETT, J. F., WAKABAYASHI, I. and BOWERS, C. Y. (1969): Isolation of growth hormone-releasing hormone (GRH) from porcine hypothalami. *Endocrinology*, *84*, 1493.

SCHALLY, A. V., ARIMURA, A., WAKABAYASHI, I., SAWANO, S., BARRETT, J. F., BOWERS, C. Y., REDDING, T. W., MITTLER, J. C. and SAITO, M. (1970): The chemistry of hypothalamic growth hormone-releasing hormone (GRH). In: *Hypophysiotropic Hormones of the Hypothalamus: Assay and Chemistry*, pp. 208–222. Editor: J. Meites. The Williams and Wilkins Co., Baltimore, Md.

SCHALLY, A. V., BABA, Y., NAIR, R. M. G. and BENNETT, C. (1971): The amino acid sequence of porcine growth hormone-releasing hormone. *J. biol. Chem.*, *246*, 6647.

SCHALLY, A. V., ARIMURA, A. and WAKABAYASHI, I. (1972*a*): In preparation.

SCHALLY, A. V., ARIMURA, A. and WAKABAYASHI, I. (1972*b*): In preparation.

STACHURA, M. E., FROHMAN, L. A. and DHARIWAL, A. P. S. (1971): Effect of purified hypothalamic extract (HTE) and acromegalic plasma on growth hormone (GH) synthesis and release *in vitro*. *Program 53rd Meeting of the Endocrine Society*, *88*, A-81.

VEBER, D. F., BENNETT, C. D., MILKOWSKI, J. D., GAL, G., DENKEWALTER, R. G. and HIRSCHMANN, R. (1971): Synthesis of a proposed growth hormone releasing factor. *Biochem. biophys. Res. Commun.*, *45*, 235.

WILBER, J., NAGEL, T. and WHITE, W. F. (1971): Hypothalamic growth hormone releasing activity (GRA): Characterization by the *in vitro* pituitary and radioimmunoassay. *Abstracts 2nd International Symposium on Growth Hormone.* ICS 236, Excerpta Medica, Amsterdam.

INFLUENCE OF AGE, SEX AND ESTROUS CYCLE ON PITUITARY AND PLASMA GH LEVELS IN RATS*

ELIAS DICKERMAN, SAMUEL DICKERMAN** and JOSEPH MEITES

Department of Physiology, Michigan State University, East Lansing, Mich., U.S.A.

The relationship between age and pituitary growth hormone (GH) levels in the rat has been studied with biological and radioimmunological assay methods. Contopoulos and Simpson (1957) detected GH in the fetal rat pituitary at day 19 of gestation by bioassay. Solomon and Greep (1958) and Bowman (1961) showed that the total amount of GH present in the rat pituitary increased with age, but that the concentration of the hormone per mg of tissue remained constant in rats between 10 and 630 days of age. By contrast, radioimmunological assays by Birge *et al.* (1967*a*) and Garcia and Geschwind (1968) showed increases in pituitary GH concentration with age in both male and female rats. To our knowledge, only Pecile *et al.* (1965) measured the content of hypothalamic growth hormone-releasing factor (GH-RF) as it relates to age in the rat. These authors observed that hypothalamic content of GH-RF was greater in 30-day-old than in 2-year-old rats.

None of the above reports considered the estrous cycle in relation to GH and age in female rats, nor measured the levels of GH in plasma or serum as related to age. It was the purpose of this study, therefore, to measure pituitary and plasma GH in male and female rats of different ages, and in females during different stages of the estrous cycle.

MATERIALS AND METHODS

Male and female Sprague-Dawley rats of different ages were obtained from Spartan Research Animals (Haslett, Michigan) and housed in a temperature controlled room (75±1° F) with automatically controlled lighting (lights on from 7 a.m. to 9 p.m. daily). A group of female rats 180 days old was maintained under constant illumination (C.L.) for 3 weeks.

Individual blood samples were taken under ether anesthesia via heart puncture into syringes containing 0.1 ml of a 100 mg% solution of Na-heparin/ml of blood withdrawn. The blood samples were kept in an ice bath during collection and centrifuged immediately thereafter at 2200 r.p.m. for 20 min. The plasma was separated by pipette and stored at —20° C until assayed. Within an hour after the blood was collected the animals were killed by guillotine and their pituitaries were removed, weighed individually and homogenized in 0.01 M phosphate buffer in 0.14 M NaCl (phospho-saline buffer, PSB), pH 7.2, with a Sonifier cell disruptor. The individual homogenates were also stored at —20° C until assayed. Blood samples from male rats were collected between 10:00 and 12:00 a.m. Vaginal smears were taken daily between 8:00 and 10:00 a.m., and the female rats were bled between 12:00 and 2:00 p.m. of the same day.

---

* This work was supported in part by NIH grants AM 04784 and CA 10771.

** Postdoctoral fellow, University of Honduras Medical School, Tegucigalpa, Honduras, C.A.

Plasma and pituitary GH were measured by a double antibody radioimmunoassay for rat GH (Dickerman and Mack, 1970; Dickerman, 1971), with antiserum to rat GH produced in monkeys (NIAMD-A-Rat GHS-1) and antiserum to monkey gamma-globulin produced in goats. Two highly purified GH preparations were used for iodination: Dr. A. Parlow's NIAMD-RGH-I-1 and Dr. S. Ellis' HVII-38-C. Under the conditions described (Dickerman, 1971), a preparation of RGH-^{125}I with specific activity of 36.5 μc/μg was obtained. Figure 1 shows typical standard curves obtained in the rat GH radioimmunoassay when adding cold hormone of iodination purity (NIAMD-RGH-I-1) or standard reference preparation (NIAMD-RGH-RP-1). It may be noted that NIAMD-RGH-I-1, on the average, is 3.2 times more potent than NIAMD-RGH-RP-1. A 1 : 50,000 dilution of the monkey antiserum to rat GH (NIAMD-A-Rat GHS-1) was required to achieve this range.

The radioimmunoassay used in this study was not affected by plasma proteins or other

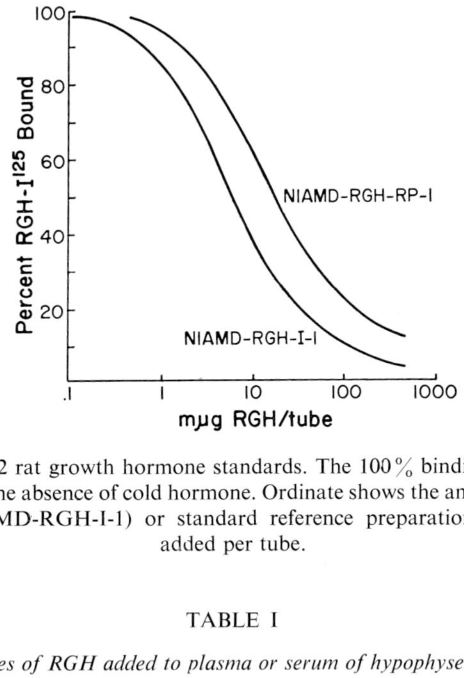

Fig. 1. Comparison of 2 rat growth hormone standards. The 100% binding in abscissa represents binding of RGH-^{125}I in the absence of cold hormone. Ordinate shows the amount of cold hormone of iodination purity (NIAMD-RGH-I-1) or standard reference preparation (NIAMD-RGH-RP-1) added per tube.

TABLE I

Recovery rates of RGH added to plasma or serum of hypophysectomized rats

mμg RGH added	mμg RHG recovered					
	1 day incubation		3 day incubation		5 day incubation	
	Serum*	Plasma*	Serum*	Plasma*	Serum*	Plasma*
125.0	130.0	125.0	128.0	126.5	125.0	120.0
62.5	68.0	60.0	64.3	61.8	63.2	62.3
31.0	33.0	35.0	32.5	33.0	31.0	33.0
16.0	15.0	17.0	15.7	16.0	15.8	17.0
8.0	7.6	9.3	7.7	8.3	8.2	8.5
4.0	4.1	3.3	3.9	3.9	4.1	3.8
Mean % recovered	101.7	102.3	100.3	102.9	100.8	101.6

* Average of 2 determinations

TABLE II

Biological and immunological estimates of potency for male rat pituitary homogenates

Exp. No.	Treatment and No. of rats	GH by RIA[1,2] (μg/mg)	GH by tibia[1,3] bioassay (μg/mg)	RIA-GH[4] / Tibia-GH	λ[5]
I	Intact (20)	26.8 ± 1.3	42.26 (23.42–76.19)	0.63	0.188
	Gonadectomized (20)	17.7 ± 1.5	23.88 (14.14–39.44)	0.74	
II	Intact (20)	29.4 ± 2.1	39.90 (23.55–65.59)	0.74	0.160
	Thyroidectomized (20)	1.9 ± 0.3	2.82 (1.61– 4.93)	0.67	

[1] Expressed as μg equivalents of NIAMD-RGH-RP-1.
[2] Mean ± standard error.
[3] Mean and 95% confidence limits.
[4] Index of discrimination: GH concentration by radioimmunoassay as μg/mg ÷ GH concentration by tibia test as μg/mg.
[5] Index of precision of bioassay.

purified rat anterior pituitary hormones. Table I shows the recovery rate when exogenous rat GH was added to serum or plasma from hypophysectomized male rats. In all cases the first antibody was incubated for 3 days. The second antibody was incubated for 1, 3 and 5 days. It can be seen that the recovery of exogenous RGH in plasma or serum from hypophysectomized male rats was about 100%.

The correlation between radioimmunological and biological activity of GH was determined in three different pituitary preparations using both methods of assay. Pituitary homogenates of control and experimental animals were bioassayed at two dose levels (1 and 4 mg/assay rat/4 days, or 2 and 8 mg/assay rat/4 days) by the tibia test of Greenspan *et al.* (1949). Radioimmunoassays of the pituitary homogenates were carried out on 4 dilutions of a solution containing 0.0125 mg AP/ml (intact and gonadectomized rats) or 0.25 mg AP/ml (thyroidectomized rats). Five replications of the four dilutions were used for each homogenate. The results, shown in Table II, were analyzed by the statistical methods of Bliss (1952). In all cases the radioimmunoassay values were below those obtained by bioassay, with indices of discrimination ranging from 0.63 to 0.74. However, over the range of concentrations tested, the potency estimates by bioassay and radioimmunoassay were not significantly different since the radioimmunoassay values were within the 95% confidence limits of the bioassay determinations.

Plasma and pituitary GH were measured by radioimmunoassay at 3 different concentrations for each sample. Ten animals were used per age group or stage of the estrous cycle within a particular age group. The results were analyzed by one way analysis of variance followed by the new multiple range test of Duncan (Bliss, 1967). All results are expressed in terms of the NIAMD-RGH-RP-1 reference standard preparation.

RESULTS

Pituitary and plasma GH were measured in male rats at 23, 33, 43, 64, 84, 104 and 120 days of age. Plasma GH was also measured in a group of male rats at about 240 days of age. Pituitary GH as a function of age in the male rat is shown in Table III. Figure 2 shows the levels of plasma GH in male rats at different ages.

The concentration and content of pituitary GH in the male rat increased significantly with age up to about 84 days. No further increase in pituitary GH concentration was ob-

TABLE III

Pituitary GH as a function of age in the male rat

Age (days)	Body wt. (g)	AP wt. (mg)	μg RGH/AP	μg RGH/mg AP
23	62.9 ± 1.4	1.87 ± 0.05	37.6 ± 3.1	20.1 ± 2.1
33	113.6 ± 0.7	3.28 ± 0.15	84.9 ± 2.9	25.9 ± 1.7
43	169.3 ± 1.7	4.63 ± 0.18	155.6 ± 12.1	33.6 ± 2.4
64	315.5 ± 6.1	7.58 ± 0.31	372.9 ± 19.2	49.2 ± 3.3
84	394.7 ± 4.6	9.07 ± 0.28	542.8 ± 13.4	59.8 ± 4.0
104	434.8 ± 9.0	8.81 ± 0.21	569.1 ± 20.2	64.6 ± 3.9
120	501.2 ± 22.4	10.18 ± 0.38	637.3 ± 27.4	62.6 ± 6.3

All values expressed as mean ± S.E.

served with advancing age, although a significant increase in pituitary GH content was observed at 120 days. Significant increases were found in plasma GH as male rats advanced from 23 to 64 days of age. No significant difference was found in plasma GH in rats between 64, 84, 104 and 120 days of age. At 240 days the concentration of plasma GH was significantly reduced to levels similar to those found in rats of about 33 to 43 days of age.

Pituitary and plasma GH was also determined in female rats ranging in age from 21 to 560 days old. The rats were divided into the following age groups: 21, 28, 34 days when the

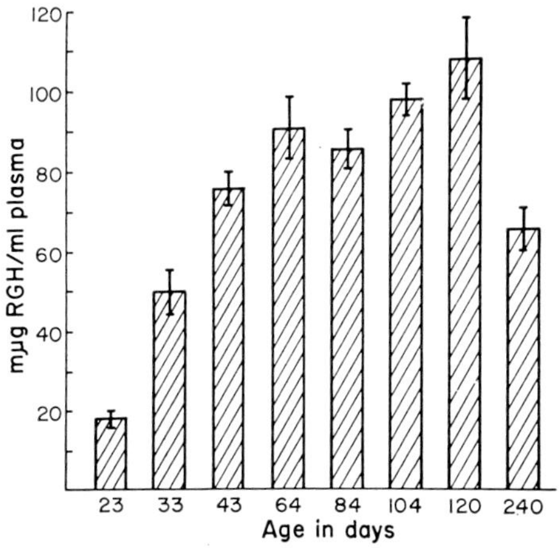

Fig. 2. Plasma GH as a function of age in the male rat.

vaginal canal was closed (vg. canal cld.), and 36 and 43 days when the vaginal canals were open. Rats 60 and 120 days old were used for determining the levels of GH in the different stages of the estrous cycle. A 180 day old group was subjected to constant illumination to elicit constant estrus (c. estrus, C.L.), and 560 day old rats found to be in constant estrus were also used. The results of this experiment are shown in Table IV and Fig. 3.

A significant increase in pituitary content and concentration of GH occurred from 21 to 60 days of age in female rats. The concentration of GH did not change significantly from 60 to 180 days, while a significant decrease was observed in the 560 day old group. Although

TABLE IV

Pituitary GH as a function of age in the female rat: the estrous cycle

Age (days)	Stage of cycle	Body wt. (g)	AP wt (mg)	μg RGH/AP	μg RGH/mg AP
21	——	60.0 ± 0.6	2.31 ± 0.07	53.6 ± 4.7	23.2 ± 1.7
28	——	85.9 ± 0.6	2.93 ± 0.52	85.8 ± 7.4	29.3 ± 2.3
34	vg. canal closed	108.3 ± 0.8	3.21 ± 0.12	89.2 ± 9.3	27.8 ± 2.1
36	vg. canal open	122.1 ± 1.8	6.83 ± 0.27	206.3 ± 14.5	30.2 ± 2.9
43	vg. canal open	146.2 ± 2.9	6.19 ± 0.35	237.7 ± 17.0	38.4 ± 3.3
60	Proestrus	238.7 ± 7.2	10.66 ± 0.40	625.7 ± 29.3	58.7 ± 5.2
	Estrus	224.4 ± 5.6	10.91 ± 0.47	558.6 ± 35.2	51.2 ± 4.8
	Metestrus	230.6 ± 3.8	10.53 ± 0.28	563.4 ± 32.1	53.5 ± 4.7
	Diestrus	224.4 ± 4.3	10.65 ± 0.52	535.7 ± 27.3	50.3 ± 5.0
120	Proestrus	264.0 ± 4.0	11.03 ± 0.25	662.9 ± 35.3	60.1 ± 5.4
	Estrus	259.0 ± 2.7	12.68 ± 0.37	680.9 ± 31.0	53.7 ± 4.8
	Metestrus	260.0 ± 5.3	12.25 ± 0.42	674.9 ± 37.1	55.1 ± 3.9
	Diestrus	255.0 ± 6.1	11.83 ± 0.35	618.7 ± 30.2	52.3 ± 4.9
180	C. Estrus (C.L.)	274.5 ± 9.0	16.27 ± 0.85	863.9 ± 42.3	53.1 ± 6.3
560	C. Estrus	340.5 ± 13.3	18.2 ± 0.87	697.1 ± 45.2	38.3 ± 9.3

All values expressed as mean ± S.E.

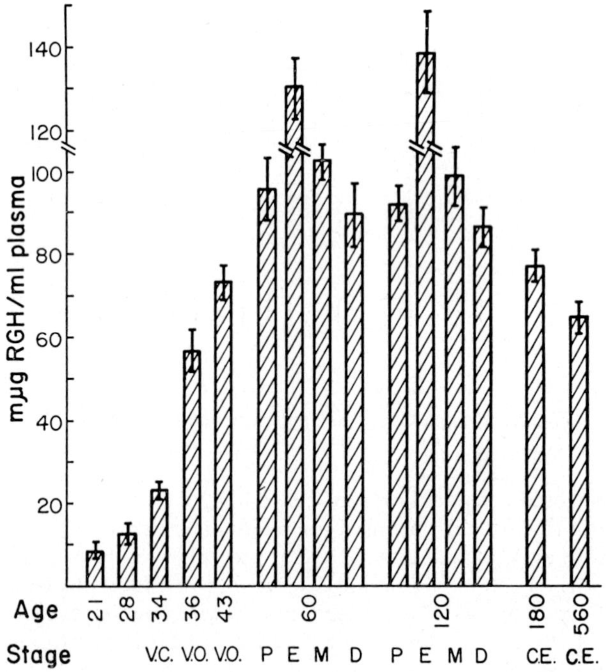

Fig. 3. Plasma GH as a function of age and the estrous cycle in the female rat.

a small increase in concentration was observed during proestrus, the levels of pituitary GH did not differ significantly from those found in estrus, metestrus, or diestrus. In contrast, the pituitary content of GH increased at about 180 days, with a significant decrease occurring at 560 days of age. No change in pituitary GH content was observed during the different stages of the estrous cycle.

Plasma concentration of GH rose significantly from 21 to 60 days of age, with the sharpest increase at 34 days in rats with a closed vagina and at 36 days in rats with an open vagina. In normally cycling female rats the mean plasma GH concentration during estrus was significantly higher than in proestrus, metestrus or diestrus, when no differences were observed. The plasma levels of GH decreased significantly in constant estrous rats (as a result of continuous light) at 180 days of age and decreased even more in old constant estrous rats at 560 days of age.

A direct comparison of GH levels in the pituitary and plasma between male and female rats of approximately the same ages failed to reveal any significant differences between the two sexes, except for the peak in plasma GH observed during estrus.

In view of the changes in female rats during the estrous cycle and at vaginal canalization, an experiment was designed to test the possible influence of estrogen on plasma GH. Female rats 180-200 g were divided into the following groups: *(a)* intact controls, *(b)* unilateral ovariectomy, *(c)* bilateral ovariectomy, *(d)* bilateral ovariectomy and 0.2 ml corn oil daily, and *(e)* bilateral ovariectomy and 5 μg of estradiol benzoate (E.B.) in 0.2 ml corn oil daily. The animals were treated for 2 weeks, at the end of which time plasma and pituitaries were collected and assayed for GH as described previously.

It can be seen in Figure 4 that unilateral ovariectomy did not alter anterior pituitary GH concentration. On the other hand, bilateral ovariectomy with or without corn oil injections resulted in higher pituitary GH concentration and in a decrease in plasma GH. Administration of estradiol benzoate to bilateral ovariectomized rats produced a decrease in pituitary GH with a concomitant increase in plasma GH levels.

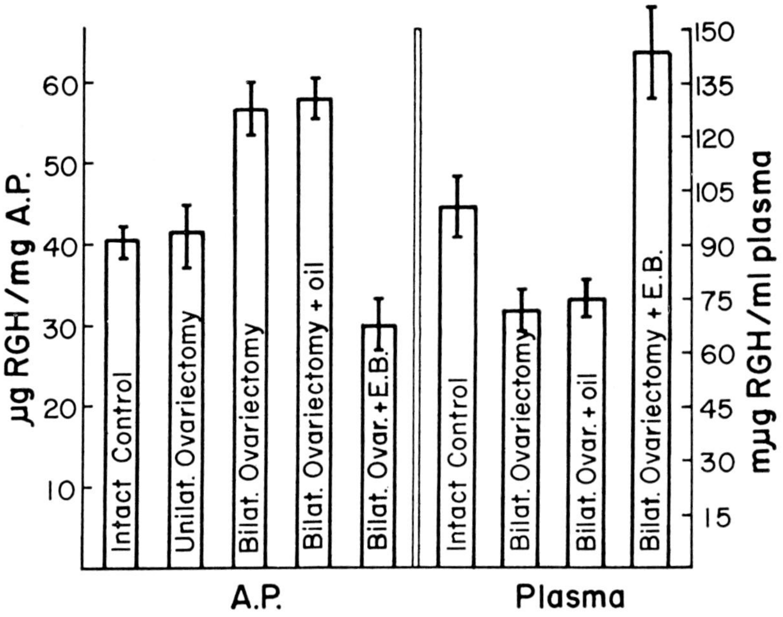

Fig. 4. Pituitary and plasma RGH after ovariectomy and estradiol benzoate injections.

DISCUSSION

The results presented here on pituitary GH levels are in basic agreement with those of Garcia and Geschwind (1968) who measured pituitary GH concentration in male and female rats of 5 to 75 days of age, and with Burek and Frohman (1970) who used rats weighing from 61 to 485 grams each. Our results appear to differ from those of Birge *et al.* (1967*a*) who reported that pituitary GH concentration in male rats continued to increase into old age, whereas female pituitary concentration plateaued at maturity. No difference in pituitary GH content and concentration between males and females was found up to eight weeks of age, after which pituitary content and concentration in males were significantly higher than in females. It is of interest to note that Birge *et al.* (1967*a*) found little or no increase in weight of the pituitary gland in male rats after 49 days of age. Upon substituting their reported pituitary weights with those reported by others and us for animals of the same age or weight, little or no difference in pituitary GH concentration was found between male and female rats. Furthermore, such a substitution also indicates that male pituitary GH concentration also shows a plateau between 63 and 77 days of age, in agreement with our data.

By biological assay methods, Solomon and Greep (1958) and Bowman (1961) reported increases in pituitary GH content but not in concentration in female and male rats 10 to 630 days old. Close inspection of their data, however, reveals that increases in pituitary GH concentration were observed up to about 6 to 9 weeks of age, but their significance was discounted.

Our results show a steady increase in plasma GH from 21 to about 64 days of age in both male and female rats. In addition, female rats show a significant elevation during estrus. If one assumes that GH is required for body growth generally, it is peculiar that the most rapid rate of body growth in young rats occurs during a time when both pituitary and plasma GH are lowest. There is a possibility that during the most rapid growth phase of life in rats other factors may be more important for body growth than GH. It need only be mentioned that removal of fetal pituitaries (Jost, 1947) in rabbits or rats does not reduce birth weight of the young. Also, rats hypophysectomized after birth continue to show some growth up to about 30 days of age (Walker *et al.*, 1952).

The ability of young rats to synthesize and release GH and the rate at which the body utilizes GH must also be considered. Burek and Frohman (1970) recently reported that pituitaries from adult male rats were able to synthesize more GH than pituitaries from young male adult rats, and that pituitaries from the latter synthesized more GH than pituitaries from weanling rats *in vitro*. If one assumes that rate of body growth is related to the utilization rate of GH, then these results could be interpreted as reflecting a low synthesis rate with almost all of the GH released utilized by the body. This could account for the relatively low levels of GH in plasma and pituitary observed during this period of rapid growth. It should be pointed out that in the human, Greenwood *et al.* (1964*a*, *b*) and Cornblath *et al.* (1965) found the highest plasma HGH levels in the fetus and at parturition, with a subsequent decline in children in whom GH levels were higher than in adults. Purchas *et al.* (1970) similarly reported that plasma GH levels were higher at birth than at any other age in bulls. Gershberg (1957) found no difference in pituitary HGH concentrations among fetal, adolescent and mature male human subjects, indicating that pituitary GH levels do not necessarily reflect GH values in the blood.

Our observation that ovariectomy increased while estradiol benzoate decreased pituitary GH concentration confirms the previous reports of Jones *et al.* (1965) and Birge *et al.* (1967*a*). Our experiments also show that ovariectomy decreases whereas estradiol benzoate increases plasma levels of GH. Similar increases in blood GH levels have been reported after administration of oral contraceptives containing estrogen in human subjects (Garcia *et al.*, 1967). It appears therefore, that estrogen is responsible for the elevation of GH after canalization of the vagina, and at estrus during each cycle. During the estrous cycle in the rat, GH appears to rise shortly after the peak in serum prolactin, LH and FSH on the late afternoon of proestrus.

Birge *et al.* (1967*b*) reported that diethylstilbestrol produced suppression of GH release from the rat pituitary *in vitro*. It remains to be demonstrated that this is a physiological and not a pharmacological effect. MacLeod *et al.* (1969) reported that estrogen administration to male rats had no consistent effect on *in vitro* incorporation of leucine-4,5-^{3}H into GH. On the other hand, the same authors reported that estrogen administration into female rats significantly decreased synthesis of GH by per mg of pituitary gland incubated *in vitro*, but not when calculated on a whole gland basis. In view of the peripheral antagonism of estrogen on GH actions on some body tissues (Josimovich *et al.*, 1967; Roth *et al.*, 1968), it could be argued that the increase in plasma GH is the result of peripheral inhibition by estrogen which reduces the amount of GH utilized by the body per unit time.

The data presented here indicate the need for caution in interpretation of results based on content or concentration of hormones in plasma or pituitary tissue. The need for such caution assumes particular importance in relation to GH since no one tissue can be called a 'target organ' for GH and provide an indirect parameter of utilization rate. There may be differences in rates of secretion and utilization of GH among very young, adult and old animals. The physiological significance of the increased plasma GH at estrus or after estrogen administration may be clarified by future studies on metabolic clearance and secretion rates of GH under these conditions, and by studies on the effects of estrogen on hypothalamic GH-RF and directly on pituitary GH release.

REFERENCES

Birge, C. A., Peake, G. T., Mariz, I. K. and Daughaday, W. H. (1967*a*): Radioimmunoassayable growth hormone in the rat pituitary gland: effects of age, sex and hormonal state. *Endocrinology*, *81*, 195.

Birge, C. A., Peake, G. T., Mariz, I. K. and Daughaday, W. H. (1967*b*): Effects of cortisol and diethylstilbestrol on growth hormone release by rat pituitary *in vitro*. *Proc. Soc. exp. Biol. (N.Y.)*, *126*, 342.

Bliss, C. I. (1952): *The Statistics of Bioassay*. Academic Press, New York.

Bliss, C. I. (1967): *Statistics in Biology, Vol. I.* McGraw-Hill, New York.

Bowman, R. H. (1961): Growth hormone activity of the anterior pituitary lobe of the male rat at various ages. *Nature (Lond.)*, *192*, 976.

Burek, C. L. and Frohman, L. A. (1970): Growth hormone synthesis by rat pituitary *in vitro:* effect of age and sex. *Endocrinology*, *86*, 1361.

Contopoulos, A. N. and Simpson, N. E. (1957): Presence of trophic hormones in fetal rat pituitary. *Fed. Proc.*, *16*, 24.

Cornblath, M., Parker, M. L., Reisner, S. H., Forbes, A. E. and Daughaday, W. H. (1965): Secretion and metabolism of growth hormone in pre-mature and full-term infants. *J. clin. Endocr.*, *25*, 209.

Dickerman, E. (1971): *Radioimmunoassay for rat growth hormone; further studies on the control of growth hormone secretion in the rat.* Ph. D. Thesis, Michigan State University, East Lansing, Mich.

Dickerman, E. and Mack, W. N. (1970): Radioimmunoassay of rat growth hormone (GH). *Fed. Proc.*, *29*, 509.

Garcia, J. F., Linfoot, J. A., Manougian, E., Born, J. L. and Lawrence, J. H. (1967): Plasma growth hormone studies in normal individuals and acromegalic patients. *J. clin. Endocr.*, *27*, 1395.

Garcia, J. F. and Geschwind, I. I. (1968): Investigations of growth hormone secretion in selected mammalian species. In: *Proceedings of the First International Symposium on Growth Hormone*, pp. 267–291. Editors: A. Pecile and E. E. Müller. ICS 158, Excerpta Medica, Amsterdam.

Gershberg, H. (1957): Growth hormone content and metabolic actions of human pituitary glands. *Endocrinology*, *61*, 160.

Greenspan, F., Li, C. H., Simpson, M. E. and Evans, H. M. (1949): Bioassay of hypophyseal growth hormone: the tibia test. *Endocrinology*, *45*, 455.

GREENWOOD, F. C., HUNTER, W. M. and KLOPPER, A. (1964*a*): Assay of human growth hormone in pregnancy, at parturition and in lactation. Dectection of a growth hormone-like substance from the placenta. *Brit. med. J.*, *1*, 22.

GREENWOOD, F. C., HUNTER, W. M. and MARRIAN, V. J. (1964*b*): Growth hormone levels in children and adolescents. *Brit. med. J.*, *1*, 25.

JONES, A. E., FISCHER, J. N., LEWIS, U. J. and VANDERLAAN, W. F. (1965): Electrophoretic comparison of pituitary glands from male and female rats. *Endocrinology*, *76*, 578.

JOSIMOVICH, J. B., MINTZ, D. H. and FINSTER, J. L. (1967): Estrogenic inhibition of growth hormone induced tibial epiphyseal growth in hypophysectomized rats. *Endocrinology*, *81*, 1428.

JOST, A. (1947): Experiences de decapitation de l'embryon du lapin. *C. R. Soc. Biol. (Paris)*, *225*, 322.

MACLEOD, R. M., ABAD, A. and EIDSON, L. L. (1969): *In vivo* effect of sex hormones on the *in vitro* synthesis of prolactin and growth hormone in normal and pituitary tumor-bearing rats. *Endocrinology*, *84*, 1475.

PECILE, A., MÜLLER, E. E., FALCONI, G. and MARTINI, L. (1965): Growth hormone releasing activity of hypothalamic extracts at different ages. *Endocrinology*, *77*, 241.

PURCHAS, R. W., MACMILLAN, K. L. and HAFS, H. D. (1970): Pituitary and plasma growth hormone levels in bulls from birth to one year of age. *J. Anim. Sci.*, *31*, 358.

ROTH, J., GORDEN, P. and BATES, R. W. (1968): Studies of growth hormone and prolactin in acromegaly. In: *Proceedings of the First International Symposium on Growth Hormone*, pp. 124–128. Editors: A. Pecile and E. E. Müller. ICS 158, Excerpta Medica, Amsterdam.

SOLOMON, J. and GREEP, R. O. (1958): Relationship between pituitary growth hormone content and age in rats. *Proc. Soc. exp. Biol. Med. (N.Y.)*, *99*, 725.

WALKER, D. G., ASLING, C. W., SIMPSON, M. E., LI, C. H. and EVANS, H. M. (1952): Structural alterations in rats hypophysectomized at six days of age and their correlation with growth hormone. *Anat. Rec.*, *114*, 19.

NERVOUS SYSTEM PARTICIPATION IN GROWTH HORMONE RELEASE FROM ANTERIOR PITUITARY GLAND*

A. PECILE, E. E. MÜLLER, M. FELICI and C. NETTI

Department of Pharmacology, School of Medicine, University of Milan, Milan, Italy

The role of the hypothalamus in regulating the growth hormone secretion of the pituitary gland has received much attention in recent years (Müller and Pecile, 1968). Various factors that modulate the secretion of the hormone have been systematically considered. There is growing evidence for the involvement of the central nervous system in all instances in which growth hormone (GH) release from the pituitary is evoked. Hypoglycemia appears to trigger GH secretion through a central mechanism. Lesions of the hypothalamus or pituitary stalk in monkeys impair GH release after hypoglycemia; variations of growth hormone-releasing factor (GHRF) in the median eminence have been noticed after insulin-induced hypoglycemia (Katz *et al.*, 1967). Clinical findings of Roth *et al.* (1963) demonstrated that in patients with surgical transections of the pituitary stalk GH response to insulin hypoglycemia is reduced

TABLE I

Effects of intravenous infusion of saline or amino acid solutions on pituitary growth hormone (GH) activity in the rat

Group	Treatment[1] (7 animals per group)	GH evaluation (width of tibial cartilage μm)	Blood glucose (mg/100 ml)	Significance vs appropriate control (P)	
				Width of tibial cartilage	Blood glucose
A	Saline (5 ml/100 g)	276 ± 5.9	96 ± 2.6	—	—
B	L-lysine (10 mg/100 g in 5 ml)	226 ± 8.8	120 ± 3.6	<0.01	NS
C	Saline (5 ml/100 g)	265 ± 4.7	82 ± 4.1	—	—
D	L-arginine (10 mg/100 g in 5 ml)	221 ± 3.8	93 ± 4.4	<0.001	NS

GH in pituitary measured by 'tibia test'. 6–8 hypox. rats in each assay group.
NS = not significant.
[1] Intravenous infusion time 30 min.
(From Pecile *et al.*, 1970, *Rivista di Farmacologia e Terapia*, *1*, 471, by kind permission of the editors).

* This work was supported by PHS Research Grant HD 01109-04 and 05 from the National Institute of Child Health and Human Development, Public Health Service, U.S.A.

or abolished. The possibility of central nervous system involvement should be considered in all instances in which GH release from the pituitary is elicited.

Since amino acids have been recognized as potent releasers of GH in the human (Merimee *et al.*, 1965, 1967; Knopf *et al.*, 1965, 1966; Parker *et al.*, 1967) it appeared of interest to examine whether or not their stimulating action on GH release was exerted through a mobilizing effect on GHRF in the hypothalamus (Pecile *et al.*, 1970).

Thus we have first of all verified that in the rat GH is effectively released from the pituitary as a consequence of intravenous infusion of various amino acids (Table I). Subsequently the GHRF activity of hypothalamic stalk median eminence (SME) extracts has been studied in animals submitted to amino acid infusions following the *in vivo* bioassay method (Pecile *et al.*, 1965). It has been noticed that, concomitantly to GH depletion of the pituitary there is a decrease of GHRF activity of SME extracts (Table II).

TABLE II

Growth hormone releasing activity (GHRF) of hypothalamic stalk median eminence (SME) of rats after intravenous infusion of saline or amino acid solutions

Group	Treatment of donors of SME[1] (7 animals per group)	Treatment of recipients of SME[2] (5 animals per group)	Measurement of pituitary GH of recipients of SME extract (width of tibial cartilage μm)	Significance (P)
A	——	Saline (0.25 ml)	262 ± 4.6	—
B	——	1 SME (in 0.25 ml)	205 ± 7.6	B vs A <0.001
C	Saline (5 ml/100 g)	1 SME (in 0.25 ml)	203 ± 5.6	C vs A <0.001
D	L-lysine (10 mg/100 g in 5 ml)	1 SME (in 0.25 ml)	265 ± 5.5	D vs A NS
E	L-arginine (10 mg/100 g in 5 ml)	1 SME (in 0.25 ml)	254 ± 8.8	E vs A NS

GH in pituitary measured by 'tibia test'. 6–8 hypox. rats in each assay group.
NS = not significant.
[1] Intravenous infusion time 30 min.
[2] All injections of SME were intracarotid.
(From Pecile *et al.*, 1970, *Rivista di Farmacologia e Terapia*, *1*, 471, by kind permission of the editors).

This fact appeared to be of significance and the investigations were extended to include injections of some amino acids in decreasing doses into the lateral brain ventricles of the rat. Our hope was not only to confirm the action of amino acids both on GH release and GHRF mobilization but to discriminate, if possible, among amino acids in terms of their relative potencies in influencing pituitary GH and hypothalamic GHRF. Although we have not completed the study of all amino acids the information obtained to date is presented.

From Table III it appears that lysine and arginine given in doses of 100 μg into the lateral brain ventricles produce a clear-cut mobilization of hypothalamic GHRF which is considerably reduced. When the dose of amino acids used is brought down to 10 μg the effect of intraventricular injection is still evident but of a lower magnitude. If the dosage of amino acids is further reduced (5 μg) the effect seen with higher doses is no longer evident.

On the basis of these data it is clear that amino acids, in eliciting GH release from the

TABLE III

Growth hormone releasing activity (GHRF) of hypothalamic stalk median eminence (SME) of rats after injection of saline or amino acid solutions into the brain lateral ventricles

Group	Treatment of donors of SME (7 animals per group)	Treatment of recipients of SME (6 animals per group)	Measurement of pituitary GH of recipients of SME extracts (width of tibial cartilage μm)	Significance (P)
A	——	Saline (0.25 ml)	259 ± 2.6	—
B	Saline (0.02 ml)	1 SME (in 0.25 ml)	209 ± 4.8	B vs A <0.001
C	L-lysine (100 μg in 0.02 ml)	1 SME (in 0.25 ml)	254 ± 2.9	C vs A NS
D	L-arginine (100 μg in 0.02 ml)	1 SME (in 0.25 ml)	249 ± 6.2	D vs A NS
E	——	Saline (0.25 ml)	264 ± 4.8	—
F	Saline (0.02 ml)	1 SME (in 0.25 ml)	204 ± 5.1	F vs E <0.001
G	L-lysine (10 μg in 0.02 ml)	1 SME (in 0.25 ml)	231 ± 6.9	G vs E <0.01
H	L-arginine (10 μg in 0.02 ml)	1 SME (in 0.25 ml)	229 ± 11.2	H vs E <0.01
I	——	Saline (0.25 ml)	251 ± 3.8	—
L	Saline (0.02 ml)	1 SME (in 0.25 ml)	205 ± 5.2	L vs I <0.001
M	L-lysine (5 μg in 0.02 ml)	1 SME (in 0.25 ml)	213 ± 4.7	M vs I <0.01
N	L-arginine (5 μg in 0.02 ml)	1 SME (in 0.25 ml)	208 ± 5.3	N vs I <0.01

GH in pituitary measured by 'tibia test'. 6–8 hypox. rats in each assay group.
NS = not significant.

(From Pecile *et al.*, 1970, *Rivista di Farmacologia e Terapia*, *1*, 471, by kind permission of the editors).

pituitary, may provoke a remarkable mobilization of GHRF from the hypothalamus. This effect seems to be dose dependent.

The action of amino acids on GHRF activity in hypothalamus coincident with GH release from the pituitary strongly suggests a possible involvement of CNS in the mechanism of GH release induced by amino acids. A direct action of amino acids on hypothalamic receptors which may sense the level of amino acids or the ratio among certain amino acids seems a reasonable hypothesis. Although the amino acid stimulation of GH release may well be different from that of other important metabolic signals as, for instance, hypoglycemia (the release of GH by amino acids is only partially prevented by a parallel glucose infusion (Burday *et al.*, 1968)), the participation of the CNS is strongly suggested by variations of GHRF in the median eminence after minute injections of amino acids into the lateral brain ventricles.

A close correlation between pituitary GH depletion and decrease of GHRF activity of SME extracts has been repeatedly reported when stress was the stimulus of GH release from the pituitary. In animals submitted to cold exposure, 4°C for 1 hr (a stressful stimulus which elicits a significant pituitary GH depletion), GHRF activity of SME extracts was reduced suggesting that also in this case GH release is triggered by hypothalamic GHRF (Müller *et al.*, 1967).

Even in neonatal rats, in which the hypothalamic neurohumoral control of GH secretion may not be fully developed (Pecile *et al.*, 1969), an adequate CNS control mechanism was demonstrated when following intermittent electric shocks, a close correlation existed between pituitary GH release and a decrease in hypothalamic GHRF activity (Tables IV and V).

A conclusion from these studies might be that the CNS participates in every instance of GH release from the pituitary and acts via hypothalamic GHRF.

A further series of experiments on the nervous control of growth hormone release concerned the role of glucose receptors within the hypothalamus in eliciting GH responses to known stimuli.

Insulin, although not necessary for penetration of glucose into neuronal cells, can enhance glucose uptake by neurons *in vitro* or *in vivo* (Rafaelsen, 1961). When injected in minute

TABLE IV

Effect of electric shock treatment on growth hormone (GH) pituitary activity of neonatal rats

Group	Age of animals (days) (10 animals per group)	Treatment (10 animals per group)	Evaluation of GH activity of pituitaries from treated animals (width of tibial cartilage μm)	Significance (P)
A	1	—	242 ± 6.5	A vs B <0.001
B	1	Shock	155 ± 4.7	
C	5	—	269 ± 9.9	C vs D <0.001
D	5	Shock	178 ± 5.5	
E	10	—	278 ± 5.4	E vs F <0.001
F	10	Shock	176 ± 4.6	

GH in pituitary measured by 'tibia test'. 6–8 hypox. rats in each assay group. NS = not significant. Parameters of shock treatment: voltage, 75 V; single shock duration, 0.2 sec; shock frequency, 4 per min; number of shocks, 10; treatment time, 2 min 30 sec.

(From Pecile *et al.*, 1969, *Proceedings of the Society for Experimental Biology and Medicine*, *130*, 425, by kind permission of the editors).

TABLE V

Growth hormone releasing activity (GHRF) of hypothalamic extracts of neonatal rats submitted or not to electric shock treatment

Group	Material injected into the carotid artery of recipient animals (8 animals per group)	Evaluation of GH activity of pituitaries from treated animals (width of tibial cartilage μm)	Significance (P)
A	Saline	256 ± 3.4	
B	2 SME of 5-day-old rats untreated	193 ± 4.2	B vs A <0.001
C	2 SME of 5-day-old rats shocked	229 ± 9.2	C vs A <0.001 C vs B <0.01

GH in pituitary measured by 'tibia test'. 6–8 hypox. rats in each assay group. NS = not significant. Parameters of shock treatment as in Table IV.

(From Pecile *et al.*, 1969, *Proceedings of the Society for Experimental Biology and Medicine*, *130*, 425, by kind permission of the editors).

amounts into the lateral brain ventricles, insulin could probably affect GH release in response to known stimuli (Pecile *et al.*, 1971*b*). First of all we established that insulin, injected in small amounts into the lateral ventricles of the brain, did not modify the GH content of the pituitary. Then we submitted animals intraventricularly injected with saline or insulin to a hypoglycemic stimulus by systemic administration of insulin (2 U./kg).

As it appears from the recently reported results (Pecile *et al.*, 1971*b*) insulin injected into the lateral ventricles does not modify pituitary content and does not induce any change of blood glucose level. When systemic hypoglycemia was induced in intraventricularly saline injected animals a clear-cut depletion of pituitary GH content was noted.

When the same hypoglycemic values were obtained in the blood of animals pretreated intraventricularly with insulin (0.2 or 0.02 I.U.) there was a block of GH release and pituitary GH content remained at normal levels (Table VI).

TABLE VI

Effect of insulin (0.02 or 0.01 I.U.) injected into the lateral ventricles of the rat brain on the release of growth hormone (GH) induced by hypoglycemia produced by systemic (intraperitoneal) insulin administration

Group	Treatment (7 animals per group)	Experiment I: Blood glucose (mg/100 ml)	Experiment I: GH activity (width of tibial cartilage μm)	Experiment II: Blood glucose (mg/100 ml)	Experiment II: GH activity (width of tibal cartilage μm)
A	Saline (intraventricular)	80 ± 3.1	242 ± 2.8	—	—
B	Saline (intraventricular) +insulin (2 I.U./kg, i.p.)	27 ± 2.8	208 ± 4.1	—	—
C	Insulin (0.02 I.U., intraventricular)	71 ± 2.3	242 ± 2.3	—	—
D	Insulin (0.02 I.U., intraventricular) +insulin (2 I.U./kg, i.p.)	12 ± 1.1	247 ± 1.7	—	—
E	Saline (intraventricular)	70 ± 1.9	245 ± 0.7	79 ± 2.3	250 ± 2.8
F	Saline (intraventricular) +insulin (2 I.U./kg, i.p.)	26 ± 1.2	200 ± 16.3	29 ± 2.3	179 ± 10.5
G	Insulin (0.01 I.U., intraventricular)	—	—	80 ± 3.1	243 ± 2.1
H	Insulin (0.01 I.U., intraventricular) +insulin (2 I.U./kg i.p.)	19 ± 1.0	188 ± 9.6	26 ± 2.6	234 ± 0.6

Significance (P)	Experiment I: Blood glucose	Experiment I: Width of tibial cartilage	Experiment II: Blood glucose	Experiment II: Width of tibial cartilage
B vs A	<0.001	<0.001	—	—
C vs A	NS	—	—	—
D vs A	<0.001	—	—	—
F vs E	<0.001	NS	<0.001	<0.001
G vs E	—	—	NS	NS
H vs E	<0.001	<0.005	<0.001	<0.05

GH in pituitary measured by 'tibia test'. 6–8 hypox. rats in each assay group. NS = not significant.
(From Pecile *et al.*, 1971, *Journal of Endocrinology*, *50*, 51, by kind permission of the editors).

A similar block of pituitary GH depletion is present also when animals pretreated intraventricularly with insulin are submitted to the metabolic stress of cold exposure, a known inducer of GH release (Table VII).

The growth hormone releasing factor (GHRF) in SME was measured in rats after the intraventricular injection of saline or insulin following or not hypoglycemia induced by intraperitoneal insulin. Table VIII shows that saline injections into the lateral brain ventricles did not modify the GHRF content of the hypothalamic extracts which, injected into the carotid of recipient rats, evoked an unequivocal depletion of the GH content of the

TABLE VII

Effect of insulin (0.2 or 0.02 I.U.) injected into the lateral ventricle of the rat brain on the release of growth hormone (GH) induced by cold exposure

Group	Treatment (7 animals per group)	Blood glucose (mg/100 ml)	GH evaluation: Width of tibial cartilage (μm)	GH evaluation: Pituitary GH content (μg GH/mg wet pit.)	Significance vs appropriate control (P): Blood glucose	Significance vs appropriate control (P): Width of tibial cartilage
A	Saline (intraventricular)	77 ± 4.2	255 ± 1.7	53.0	—	—
B	Saline (intraventricular) + cold (4°C, 1 hr)	83 ± 5.0	219 ± 5.3	10.2	NS	<0.001
C	Insulin (0.2 I.U., intraventricular)	71 ± 4.3	256 ± 2.3	54.5	NS	NS
D	Insulin (0.2 I.U., intraventricular) + cold (4°C, 1 hr)	78 ± 0.1	251 ± 3.2	47.5	NS	NS
E	Insulin (0.02 I.U., intraventricular)	84 ± 8.3	254 ± 3.3	51.5	NS	NS
F	Insulin (0.02 I.U., intraventricular) + cold (4°C, 1 hr)	85 ± 4.4	259 ± 5.8	43.5	NS	NS

GH in pituitary measured by 'tibia test'. 6–8 hypox. rats in each assay group. NS = not significant. (From Pecile *et al.*, 1971, *Journal of Endocrinology*, *50*, 51, by kind permission of the editors).

pituitary. Insulin-induced hypoglycemia was followed by a drop in the GH content of the pituitary and by a decrease of GHRF activity. When insulin was injected into the ventricle in sufficient doses (0.02–0.2 I.U.) depletion of the pituitary gland of GH consequent to hypoglycemia was prevented and the GHRF activity of the SME extracts remained unchanged.

The reported data again suggest that the CNS really mediates between metabolic signals and GH release. Not only does the GH release evoked by metabolic stimuli induce changes in hypothalamic GHRF but variations provoked in the CNS can also clearly interfere with the known mechanism(s) for the release of GH in response to known stimuli.

The importance of the glucose-sensitive regulatory system in the hypothalamus in the neural control of GH secretion is further emphasized by the block of release obtained after intraventricular insulin; not only is the insulin induced GH secretion, with peripheral hypoglycemia prevented, but the GH secretion induced by cold exposure is also blocked.

The last series of experiments on CNS participation in the control of GH secretion employed hypothalamic lesioning.

TABLE VIII

Growth hormone releasing activity (GHRF) of hypothalamic stalk median eminence (SME) of rats after injection of saline or insulin into the brain lateral ventricle and subjection to hypoglycemia produced by systemic (intraperitoneal) insulin administration

Group	Treatment of donors of SME (7 animals per group)	Treatment of recipients of SME[1] (6 animals per group)	Measurement of pituitary GH of recipients of SME extract		Significance (P)
			Width of tibial cartilage (μm)	Pituitary GH content (μg GH/mg wet pit.)	
A	——	Saline	251 ± 2.1	59.0	—
B	Saline (intraventricular)	1 SME	215 ± 8.2	11.5	B vs A <0.005
C	Saline (intraventricular) + insulin (2 I.U./kg, i.p.)	1 SME	249 ± 3.1	56.5	C vs A NS
D	Insulin (0.02 I.U., intraventricular) + insulin (2 I.U./kg, i.p.)	1 SME	217 ± 5.2	15.0	D vs A <0.001
E	——	Saline	248 ± 1.8	55.0	—
F	Saline (intraventricular)	1 SME	226 ± 4.7	26.0	F vs E <0.001
G	Saline (intraventricular) + insulin (2 I.U./kg, i.p.)	1 SME	256 ± 2.4	66.0	G vs E NS
H	Insulin (0.01 I.U., intraventricular)	1 SME	209 ± 6.0	5.0	H vs E <0.001
I	Insulin (0.01 I.U., intraventricular) + insulin (2 I.U./kg, i.p.)	1 SME	228 ± 5.8	28.5	I vs E <0.01

GH in pituitary measured by 'tibia test'. 6–8 hypox. rats in each assay group. NS = not significant.
[1] All injections were intracarotid.
(From Pecile *et al.*, 1971, *Journal of Endocrinology*, *50*, 51, by kind permission of the editors).

TABLE IX

Effect of lesions of hypothalamic ventromedial nuclei (VMN) of the rat on the release of growth hormone (GH) induced by hypoglycemia produced by intraperitoneal insulin administration

Group	Hypothalamic lesion (12 rats/group)	Treatment	GH evaluation (width of tibial cartilage μm)	Blood glucose	Significance vs appropriate control (P)	
					Width of tibial cartilage	Blood glucose
A	—	—	251 ± 2.62	83 ± 2.8	—	—
B	—	Insulin (2 U./kg)	219 ± 3.61	25 ± 7.2	<0.001	<0.001
C	Sham-operated	—	253 ± 2.80	74 ± 2.0	NS	NS
D	Sham-operated	Insulin (2 U./kg)	202 ± 5.67	22 ± 1.2	<0.001	<0.001
E	VMN lesions	—	267 ± 2.76	71 ± 3.5	NS	NS
F	VMN lesions	Insulin (2 U./kg)	246 ± 4.57	13 ± 1.3	NS	<0.001

GH in pituitary measured by 'tibia test'. 6–8 hypox. rats in each assay group. NS = not significant.

We followed the brilliant observations of Frohman and Bernardis (1968, 1970) who demonstrated the outstanding relevance of the hypothalamic ventromedial nuclei (VMN) for the control of GH secretion. Small to medium sized lesions were placed in VMN or in posterior nuclei (PN) or in the paraventricular nuclei (PVN) of the hypothalamus (Pecile *et al.*, 1971*a*).

Pituitary GH and SME growth hormone releasing activity was evaluated 15 days after the operation in lesioned animals after insulin hypoglycemia. It appears (Table IX) that while intact and sham-operated rats show a depletion of GH content those submitted to lesions of VMN do not respond to the hypoglycemic stimulus.

On the other side the animals lesioned in the posterior nuclei of the hypothalamus behave like control ones demonstrating that the hypothalamic locus corresponding to VMN is of particular relevance for the control of GH secretion.

In VMN lesioned animals the impairment of growth hormone release after an appropriate stimulus was associated with changes in growth hormone releasing activity of the hypothalamic median eminence. While the growth hormone releasing activity of median eminence present in sham-operated as well as in animals lesioned in the posterior hypothalamic nuclei, is similar to that found in intact controls (Table X) in VMN lesioned rats GHRF activity of SME is sharply reduced. This fact again indicates that the integrity of VMN is necessary for the production of GHRF and/or its storage in the SME.

TABLE X

Effect of lesions of ventromedial nuclei (VMN) or posterior nuclei (PN) of the rat hypothalamus on the growth hormone releasing activity (GHRF) of the stalk median eminence (SME)

Group	Hypothalamic lesions of donors of SME (7 animals per group)	Treatment of recipients of SME[1] (6 animals per group)	Measurement of pituitary GH of recipients of SME extracts (width of tibial cartilage, μm)		
			Exp. I	Exp. II	Exp. III
A	—	Saline (0.25 ml)	260 ± 3.7	240 ± 3.6	251 ± 2.1
B	Sham-operated	1 SME	191 ± 1.6	218 ± 3.1	209 ± 6.1
C	VMN lesions	1 SME	241 ± 7.0	252 ± 3.1	242 ± 2.9
D	PH lesions	1 SME	203 ± 6.4	210 ± 4.2	227 ± 1.8

Significance (P)	Exp. I	Exp. II	Exp. III
B vs A	<0.001	<0.01	<0.001
C vs A	NS	NS	NS
D vs A	<0.001	<0.01	<0.01

GH in pituitary measured by 'tibia test'. 6–8 hypox. rats in each assay group. NS = not significant.
[1] All injections were intracarotid.

The effect of PVN lesions on GHRF activity of pituitary stalk median eminence has been also studied but at the moment we cannot make firm conclusions. It appears from the available data that PVN lesions have an influence although of different degree on GHRF availability at hypothalamic level. GHRF activity in PVN lesioned animals is significantly higher than that of VMN lesioned animals but an impairment of GHRF availability also seems to be part of the response to PVN lesions (significant difference of GHRF activity of SME of sham-operated versus PVN lesioned rats).

Before drawing definitive conclusions more work is necessary. The reported data are substantially in agreement with those reported by Frohman *et al.* (1968) and Bernardis and Frohman (1971) obtained with electrical stimulation and also with those of Frohman and Bernardis (1968, 1970) obtained with electrolytic lesions and using RIA-GH determinations. Together, the data point to the existence within the CNS, of areas of prominent importance in the control of GH secretion.

In conclusion, it may be again pointed out that all the stimuli capable of evoking GH release, although through different mechanism(s), apparently always involve the CNS. At least the common pathway within the CNS for triggering the release of the hormone, the hypothalamic GHRF discharge, always plays a prominent role.

ACKNOWLEDGEMENTS

The standard bovine growth hormone used in these studies was a gift of the Endocrinology Study Section of National Institutes of Health through the courtesy of Prof. A. Wilhelmi. The skilful technical assistance of Mr. Vincenzo Olgiati is gratefully acknowledged.

REFERENCES

BERNARDIS, L. L. and FROHMAN, L. A. (1971): Plasma growth hormone responses to electrical stimulation of the hypothalamus in the rat. *Neuroendocrinology*, *7*, 193.

BURDAY, S. Z., FINE, P. H. and SCHALCH, D. S. (1968): Growth hormone secretion in response to arginine infusion in normal and diabetic subjects: relationship to blood glucose levels. *J. Lab. clin. Med.*, *71*, 897.

FROHMAN, L. A. and BERNARDIS, L. L. (1968): Growth hormone and insulin levels in weanling rats with ventromedial hypothalamic lesions. *Endocrinology*, *82*, 1125.

FROHMAN, L. A. and BERNARDIS, L. L. (1970): Effect of lesion size in the ventromedial hypothalamus on growth hormone and insulin levels in weanling rats. *Neuroendocrinology*, *6*, 319.

FROHMAN, L. A., BERNARDIS, L. L. and KANT, K. J. (1968): Hypothalamic stimulation of growth hormone secretion. *Science*, *162*, 580.

KATZ, S. H., DHARIWAL, A. P. S. and MC CANN, S. M. (1967): Effect of hypoglycemia on the content of pituitary GH and hypothalamic GHRF in the rat. *Endocrinology*, *81*, 333.

KNOPF, R. F., CONN, J. W., FAJANS, S. S., FLOYD JR, J. C., GUNTSCHE, E. M. and RULL, J. A. (1965): Plasma growth hormone response to intravenous administration of aminoacids. *J. clin. Endocr.*, *25*, 1140.

KNOPF, R. F., CONN, J. W., FLOYD Jr, J. C., FAJANS, S. S., RULL, J. A., GUNTSCHE, E. M. and THIFFAULT, C. A. (1966): The normal endocrine response to ingestion of protein and infusions of amino acids: sequential secretion of insulin and growth hormone. *Trans. Ass. Amer. Phycns*, *79*, 312.

MERIMEE, T. J., LILLICRAP, D. A. and RABINOWITZ, D. (1965): Effect of arginine on serum levels of human growth hormone. *Lancet*, *2*, 668.

MERIMEE, T. J., RABINOWITZ, D., RIGGS, L., BURGESS, J. A., RIMOIN, D. L. and MCKUSICK, V. A. (1967): Plasma growth hormone after arginine infusion: clinical experiences. *New Engl. J. Med.*, *276*, 434.

MÜLLER, E. E., ARIMURA, A., SAWANO, S., SAITO, T. and SCHALLY, A. V. (1967): Growth hormone releasing activity in the hypothalamus and plasma of rats subjected to stress. *Proc. Soc. exp. Biol. (N.Y.)*, *125*, 874.

MÜLLER, E. E. and PECILE, A. (1968): Studies on the neural control of growth hormone secretion. In: *Growth Hormone*, pp. 253–266. Editors: A. Pecile and E. E. Müller. ICS 158, Excerpta Medica, Amsterdam.

PARKER, M. L., HAMMOND, J. M. and DAUGHADAY, W. H. (1967): The arginine provocation test: an aid in the diagnosis of hyposomatotropism. *J. clin. Endocr.*, *27*, 1129.

PECILE, A., FELICI, M. and MÜLLER, E. E. (1971*a*): Growth hormone releasing (GRF) activity in rats submitted to hypothalamic electrolytic lesions. *Program 53rd Meeting, Endocrine Society*, p. 244.

PECILE, A., FERRARIO, G., FALCONI, G. and MÜLLER, E. E. (1969): Pituitary growth hormone content and hypothalamic growth hormone releasing activity in neonatal rats after stress. *Proc. Soc. exp. Biol. (N.Y.)*, *130*, 425.

PECILE, A., MÜLLER, E. E., FALCONI, H. and MARTINI, L. (1965): Growth hormone-releasing activity of hypothalamic extracts at different ages. *Endocrinology*, *77*, 241.

PECILE, A., MÜLLER, E. E., FELICI, M. and MASARONE, M. (1970): Partecipazione del sistema nervoso nella liberazione di ormone somatotropo dall'ipofisi indotta da aminoacidi. *Riv. Farm. Terap.*, *1*, 471.

PECILE, A., MÜLLER, E. E., FELICI, M., NETTI, C. and COCCHI, D. (1971*b*): Influence of insulin injected into the lateral ventricle on pituitary growth-hormone release in the rat. *J. Endocr.*, *50*, 51.

RAFAELSEN, O. J. (1961): Studies on a direct effect of insulin on the central nervous system: a review. *Metabolism*, *10*, 99.

ROTH, J., GLICK, S. M., YALOW, R. S. and BERSON, S. (1963): Secretion of human growth hormone: physiologic and experimental modification. *Metabolism*, *12*, 577.

HYPOTHALAMIC CONTROL OF GROWTH HORMONE SECRETION IN THE RAT*

LAWRENCE A. FROHMAN, LEE L. BERNARDIS, LYNNE BUREK,
JANICE W. MARAN and ANAND P. S. DHARIWAL

Departments of Medicine and Pathology, State University of New York at Buffalo, Buffalo, N.Y., and Department of Physiology, Stanford University, Stanford, Calif., U.S.A.

During the past several years our investigations have been concerned with various aspects of the hypothalamic control of growth hormone secretion. The studies to be described will be divided into three separate parts which include: (1) the effects of hypothalamic destruction on growth hormone secretion *in vivo*, and on growth hormone synthesis and release *in vitro*; (2) effects of hypothalamic stimulation on growth hormone secretion; and (3) effects of purified growth hormone releasing factor *in vivo*.

I. Effects of hypothalamic destruction on growth hormone secretion

The hypothalamic control of growth was first recognized by Hetherington and Ranson (1940) 30 years ago, but it was only in 1961 that Reichlin (1961) reported decreases in pituitary growth hormone levels after destructive lesions of the ventromedial hypothalamus that also included the median eminence. Our initial studies were directed at separating the effects

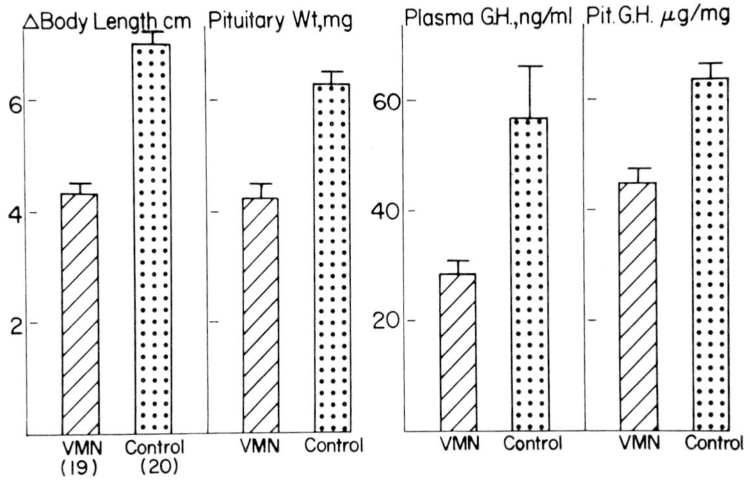

Fig. 1. Changes in linear growth, pituitary weight, and immunoreactive growth hormone levels (mean ± SE) in pituitary and plasma after bilateral destruction of the ventromedial hypothalamic nuclei in weanling rats.

* This work was supported in part by USPHS Grants AM 11456, HD 03331 and AM 04612.

of destruction of the median eminence from those of other hypothalamic structures. When small bilateral electrolytic lesions limited primarily to the ventromedial nucleus and sparing the median eminence were produced in weanling rats (Frohman and Bernardis, 1968) we were able to show a decrease in linear growth and in pituitary weight associated with reduced levels of immunoreactive growth hormone in both the plasma and the pituitary (Fig. 1). These studies suggested that the ventromedial hypothalamic nuclei are directly involved in the regulation of growth hormone secretion either as the site of synthesis of growth hormone releasing factor or as a locus required for its secretion. Subsequent studies in which lesions were produced more dorsally or in varying anterior-posterior directions from the plane of the ventromedial nucleus have not resulted in any decrease in pituitary growth hormone concentrations and have supported the concept that the hypothalamic locus involved in growth hormone regulation is related to the ventromedial nucleus. Because the lesions also impinged on structures beyond the ventromedial nucleus, however, the specificity of this locus as the critical hypothalamic structure remained open to question.

Further evidence linking the ventromedial nucleus to growth hormone regulation has been obtained in an experiment in which the effects of different sized lesions were evaluated (Bernardis and Frohman, 1970). By varying the duration of the current applied to the tissue, three different sized electrolytic lesions were produced (Fig. 2). The smallest (2.5 mC) was limited entirely to the ventromedial nucleus but was estimated to have destroyed only 25-50% of this locus; the intermediate (7.5 mC) resulted in a 50-90% destruction of the ventromedial nucleus but also impinged on the arcuate nucleus and the acellular area surrounding the ventromedial nucleus; while the largest (17.5 mC) destroyed virtually all of the ventromedial nucleus but also produced extensive damage to the remainder of the ventromedial hypothalamus and some portions of the dorsomedial hypothalamus. In all

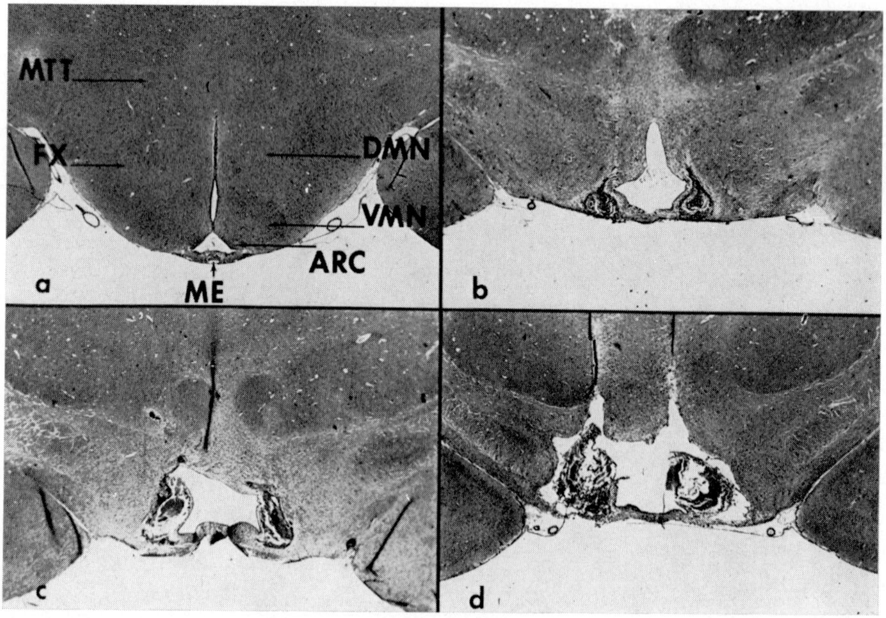

Fig. 2. Microphotographs of coronal sections through the hypothalami of (a) a control rat, (b) a rat with small (2.5 mC) lesions, (c) a rat with medium-sized (7.5 mC) lesions, and (d) a rat with large (17.5 mC) lesions. MTT: mammillothalamic tract; FX: fornix; DMN: dorsomedial nucleus; VMN: ventromedial nucleus; ARC: arcuate nucleus; ME: median eminence. (From Bernardis and Frohman, 1970, *Neuroendocrinology*, *6*, 319; by kind permission of the editors).

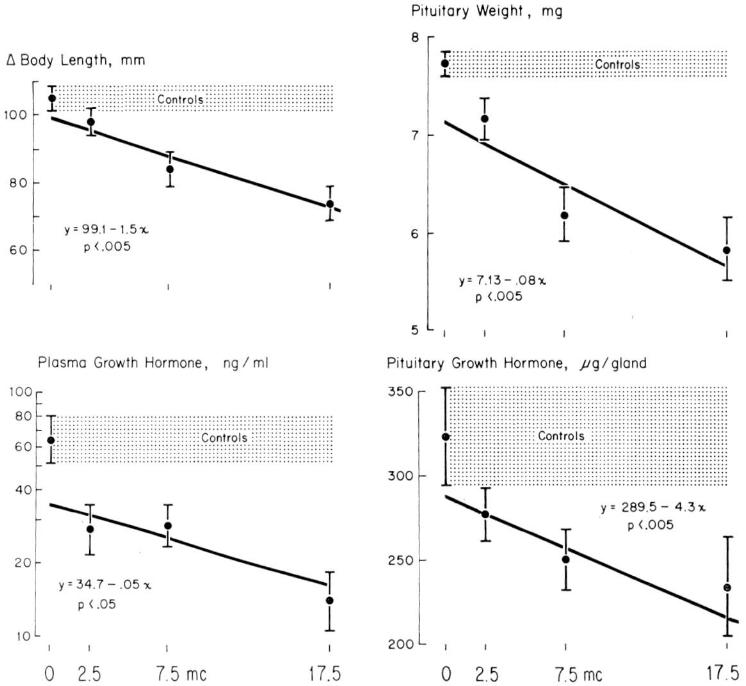

Fig. 3. Relationship between lesion size (mC) and representative parameters of growth (body length change and pituitary weight) and growth hormone secretion (plasma and pituitary growth hormone levels) in weanling rats with ventromedial hypothalamic lesions of varying sizes. Shown are the mean ± SE and the combined linear regression equations. (From: Bernardis and Frohman, 1970, *Neuroendocrinology*, *6*, 319; by kind permission of the editors).

animals, however, the median eminence was intact. A progressive impairment in linear growth and pituitary weight could be demonstrated with increasing hypothalamic destruction though the changes in the group of animals bearing the smallest lesions were of borderline significance (Fig. 3). Similarly, progressive decreases were observed in the levels of growth hormone in both pituitary and plasma with increasing lesion size. These results have suggested to us that the critical locus is actually the ventromedial nucleus inasmuch as changes could be shown with the smallest lesions which were limited to this structure. These studies have not, however, excluded the possibility that surrounding loci may also be involved in the control of growth hormone secretion.

In order to determine the sequence of events following the destruction of the ventromedial nucleus with respect to the observed alterations in pituitary growth hormone stores, we have studied the sequential effect of ventromedial nucleus lesions in weanling rats on pituitary growth hormone levels and on the synthesis and release of pituitary growth hormone *in vitro* during the first two postoperative weeks. Rats were killed at varying times after hypothalamic operation and anterior pituitaries from individual animals incubated in 1 ml of Media 199 containing 5 μC of ^{3}H-leucine for 3 hours. Following this period the glands were transferred to nonradioactive media for a 4th hour. Incorporation of ^{3}H-leucine into growth hormone in both tissue and media was measured by an immunoprecipitation method using an antibody specific for growth hormone (Burek and Frohman, 1970). Sham-operated animals served as controls. In this system, growth hormone synthesis has been shown to be linear for at least 6 hours and the release of newly synthesized growth hormone easily detectable between the

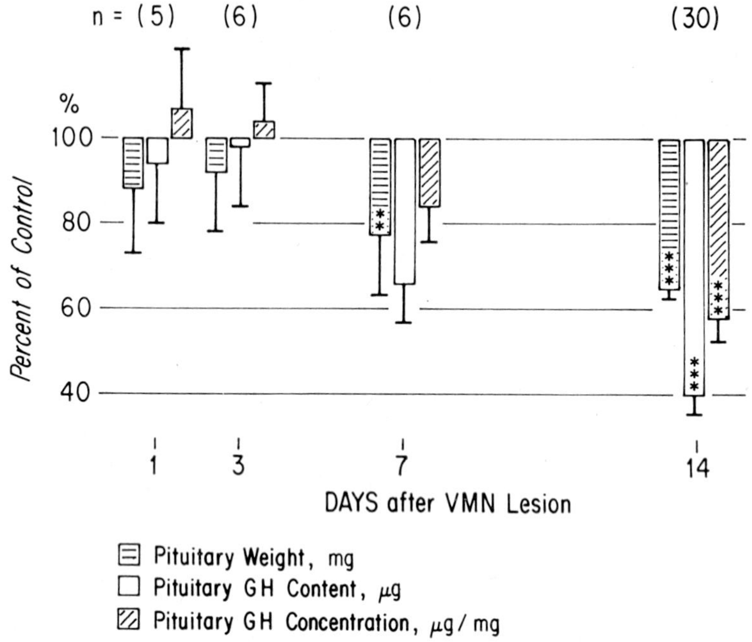

Fig. 4. Sequential changes in pituitary weight and growth hormone levels in weanling rats with ventromedial nucleus (VMN) lesions. Mean ± SE of the changes in VMN rats are compared to sham operated controls. Two and three stars represent p values of <.01 and <.001 respectively when compared to appropriate controls.

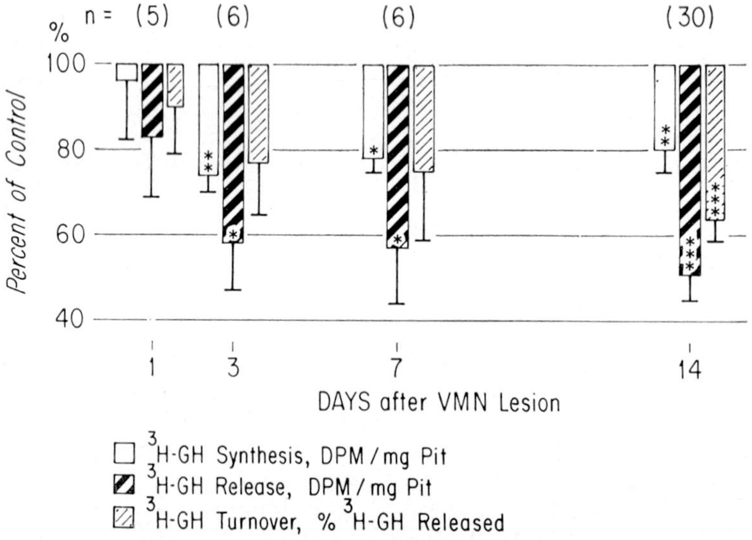

Fig. 5. Sequential changes in synthesis and release of growth hormone *in vitro* by pituitaries of weanling rats with VMN lesions. Mean ± SE of the changes in VMN rats are compared to sham operated controls. One, two and three stars represent p values of <.05, <.01 and <.001 respectively when compared to appropriate controls.

2nd and 3rd hours. As shown in Figure 4, no significant differences were noted in pituitary weight or growth hormone on either the first or third postoperative day. If anything, pituitary growth hormone concentration was slightly increased. By the seventh day, pituitary weight was significantly reduced and an even greater reduction was present in pituitary growth hormone content, though variability among control animals precluded statistical significance. By two weeks, significant reductions were present in all measuremets. The effects on growth hormone synthesis and release are shown in Figure 5. 'Synthesis' refers to the total ^{3}H-leucine incorporation into both pituitary and media growth hormone, 'release' refers to the ^{3}H-growth hormone present in the media during the 4th hour, while 'turnover' is the ratio of these two measurements, or the percentage of newly synthesized growth hormone which was released into the media during the 4th hour.

Although all three parameters were somewhat decreased on the first postoperative day, none of the changes were significant. However, by the third day significant decreases were present in both synthesis and release and these changes persisted throughout the two week period. Decreases in pituitary growth hormone turnover were present throughout this period but were not significant until 14 days because of greater individual variability. It can be noted that the decrease in growth hormone release was persistently greater than that of growth hormone synthesis and by the 14th day these differences were significant. These results, together with thc slight increase in pituitary growth hormone concentration during the first 3 days after ventromedial nucleus destruction (Fig. 4), suggest that the primary effect of the destruction is an inhibition of growth hormone release and that changes in growth hormone synthesis may be secondary.

II. Effects of hypothalamic stimulation on growth hormone secretion

Because of the limitation in the precision of hypothalamic lesion production and the possibility of functional, rather than histologically visible derangement of other hypothalamic loci, we next studied the hypothalamic control mechanism by means of electrical stimulation experiments (Frohman *et al.*, 1968). Using adult female rats anesthetized with

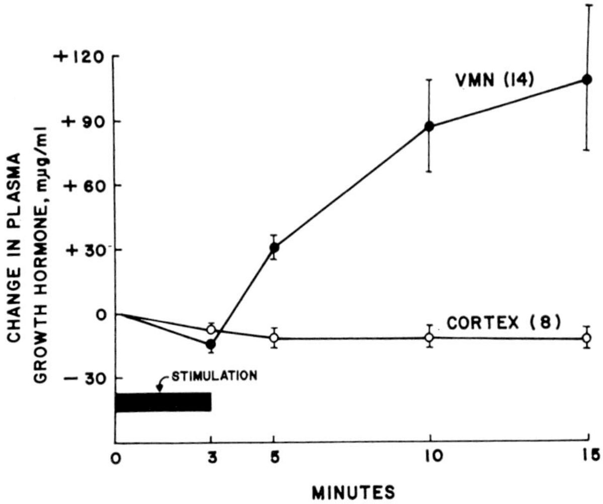

Fig. 6. Effect of electrical stimulation in the ventromedial hypothalamic nucleus and the cerebral cortex on plasma immunoreactive growth hormone levels (mean ± SE) in the rat. (From: Frohman *et al.*, 1968, *Science*, *162*, 580; by kind permission of the editors).

either pentobarbital or hexobarbital in which a catheter had been inserted into the carotid artery, stereotaxic electrical stimulation was performed using biphasic square wave pulses of 1 mA intensity, 5 msec duration, 50 cycles/sec for 3 min. Stimulation of the ventromedial nucleus was followed in approximately 80% of rats by rises in plasma growth hormone within 5 min which peaked at 10 to 15 min (Fig. 6). In experiments involving several hundred animals we have never observed plasma growth hormone rises prior to 5 min after the onset of stimulation. The responses have been observed using both monopolar and bipolar electrodes and in animals of both sexes. Decreasing the stimulus duration or the current flow

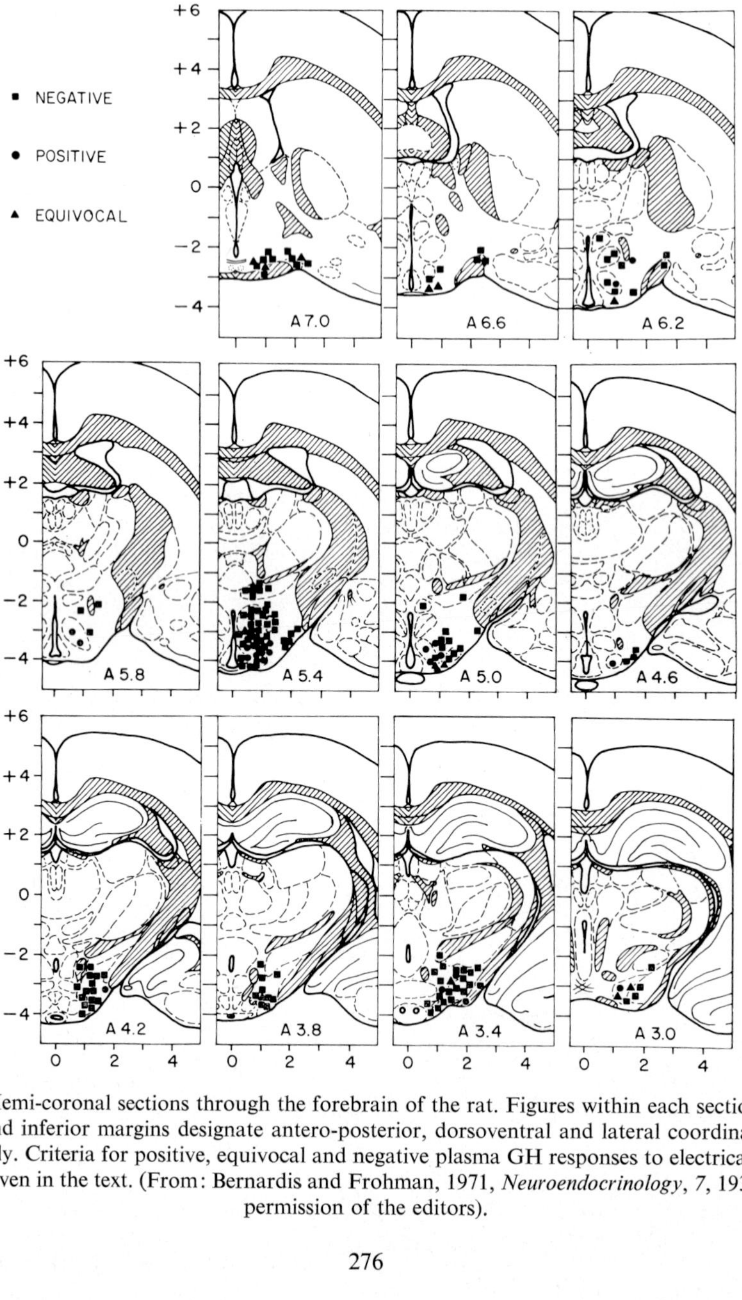

Fig. 7. Hemi-coronal sections through the forebrain of the rat. Figures within each section and on the left and inferior margins designate antero-posterior, dorsoventral and lateral coordinates (mm) respectively. Criteria for positive, equivocal and negative plasma GH responses to electrical stimulation are given in the text. (From: Bernardis and Frohman, 1971, *Neuroendocrinology*, 7, 193; by kind permission of the editors).

intensity resulted in a smaller frequency of positive responses. Using the stimulation parameters which have just been described, we stimulated sites other than the ventromedial nucleus throughout the hypothalamus and the limbic system in an attempt to trace possible neuronal circuits involved in growth hormone secretion (Bernardis and Frohman, 1971). Figure 7 shows a series of hemi-coronal diagrams through the rat hypothalamus indicating the stimulation sites. 'Positive' responses were defined as a rise in plasma growth hormone of 10 ng/ml or greater at 10 or 15 min after the onset of stimulation, whereas 'equivocal' responses were either smaller in magnitude, or were detected at an earlier time period but did not persist. Positive responses were seen only when the tip of the stimulating electrode was located within the ventromedial nucleus, in the acellular area immediately lateral to it, in the arcuate nucleus or in the median eminence. No differences in response parameters were noted between stimulation of the ventromedial nucleus and the median eminence, although the number of animals stimulated in the latter location was small. Equivocal responses were noted in a few animals stimulated just lateral to the ventromedial nucleus and also in a few animals stimulated in the substantia nigra. A series of 50 rats was also stimulated in extrahypothalamic limbic system sites including the piriform cortex, amygdala, median forebrain bundle, hippocampus, preoptic area, and the entorhinal cortex. All of these stimulations have resulted in negative results.

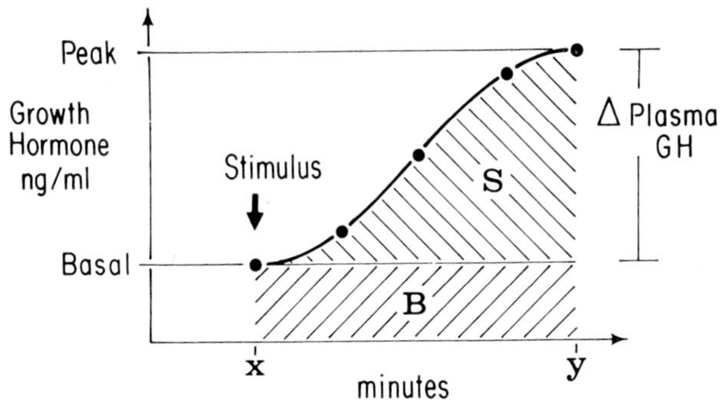

Basal Secretion = M.C.R. x Integrated Area **B**

Stimulatory Secretion = M.C.R. x Integrated Area **S** + Δ GH x Plasma Volume

Fig. 8. Method used for calculating the GH secretory response to a stimulus based on changes in plasma levels, metabolic clearance rate (MCR) of the hormone and plasma volume.

Thus, the ability to induce a growth hormone release by electrical stimulation appears to involve a rather circumscribed area including the ventromedial nucleus, arcuate nucleus and the median eminence and is identical to that area which, when destroyed, has resulted in decreases in plasma and pituitary growth hormone levels.

The growth hormone secretory response to hypothalamic stimulation was next investigated. Whether this response can be considered a prototype of other anterior pituitary hormone responses to hypothalamic stimulation is not known. However, a similar pattern of response has recently been reported for TSH secretion by Martin and Reichlin (1970). With the knowledge of the metabolic clearance rate of growth hormone from rat plasma which we determined by both single injection and continuous infusion techniques (Frohman and Bernardis, 1970) it was possible to quantitate the growth hormone secreted in response to a specific stimulus.

The total growth hormone secretion during this period is comprised of the basal secretion and that specifically due to the stimulus (Fig. 8). The former is equal to the metabolic clearance rate multiplied by the area 'B'. This secretion would have occurred even in the absence of the stimulus. The stimulus related secretion is the sum of 2 components: (1) the additional growth hormone cleared from plasma as a result of the stimulus, shown as the area 'S' multiplied by the metabolic clearance rate, and (2) the quantity of growth hormone added to the plasma pool to elevate the growth hormone level from preinjection to post-injection value, shown as the peak minus the basal level multiplied by an estimate of the plasma volume.

TABLE I

Parameters of growth hormone secretion following VMN stimulation

Stimulation site	Plasma GH rise at 10 min (ng/ml)	Secretory response in 10 min (μg)	Pit. GH conc. at 10 min (μg/mg)
VMN (15)	206 ± 28	2.34 ± 0.31	43.3 ± 2.5
Cortex (8)	−8 ± 3	−0.02 ± 0.11	39.3 ± 4.7
P	<0.001	<0.001	NS

In a series of 15 animals stimulated in the ventromedial nucleus, (Table I) the secretion of growth hormone in the first 10 min ranged from 0.51 to 4.13 μg with a mean ± SE of 2.34 ± 0.31. No significant secretion of growth hormone occurred in 8 control animals which were stimulated in the cerebral cortex. This quantity of growth hormone represents less than 1% of the pituitary growth hormone content and would not be expected to result in any measurable depletion of pituitary growth hormone. This was confirmed by direct measurement as shown in Table I. Pituitary growth hormone concentration in the ventro-medial nucleus stimulated rats was 43.3 ± 2.5 μg/mg as compared to 39.3 ± 4.7 in cortex stimulated controls.

We have calculated the growth hormone secretory response to insulin hypoglycemia in humans in a similar manner using the metabolic clearance rate estimations obtained by continuous infusion studies (MacGillivray *et al.*, 1970). Our results indicate that no more than 1-2% of the pituitary growth hormone content was released during the first 60 min and are quantitatively similar to those in the rat after hypothalamic stimulation. Both are in considerable contrast to the reports of 30 to 60% depletion of pituitary growth hormone as measured by the tibia test bioassay following stimuli expected to cause the release of endogenous growth hormone releasing factor such as insulin hypoglycemia and cold (Müller *et al.*, 1967).

III. *Effects of purified growth hormone releasing factor in vivo*

Having established that acute rises in plasma growth hormone could be demonstrated by the endogenous release of growth hormone releasing factor (GRF), we attempted to use this model for the demonstration of GRF in both crude and purified hypothalamic extracts. For several years, we were uniformly unsuccessful in demonstrating any rise in plasma immunoreactive growth hormone after the intravenous or intracarotid injection of any type of hypothalamic extract either prepared in our laboratory or provided by others. Recently, however, we have injected purified hypothalamic extracts by a direct intrapituitary injection method and have obtained positive results (Frohman *et al.*, 1971).

A polyethylene catheter was inserted into the femoral vein of mature male rats, brought subcutaneously to the back and the coiled end attached to a small midscapular incision. On the following day, and under pentobarbital anesthesia, the catheter was uncoiled and attached to a heparinized syringe and the animal placed in a stereotaxic apparatus. After obtaining a preinjection blood sample, the test substance was injected through a small glass cannula in a volume of 0.5 µl into the pituitary followed by a similar volume of fast green dye to mark the injection site. The animal's head was flexed so that the cannula tract passed posterior to the hypothalamus. Repeated blood samples were obtained after injection for up to a 20 min period. The pituitary was examined *in situ* after the last sample was collected and adequacy of the injection determined by the localization of the dye marker.

GRF was prepared from an acetone powder of pooled ovine hypothalami by glacial acetic acid extraction followed by gel filtration on Sephadex G-25 in 0.1 M ammonium acetate buffer, pH 5. The fractions used in the present studies were chosen on the basis of their position relative to the location of other releasing factors as determined by previous bioassay experience (Dhariwal *et al.*, 1965). Thus, the fractions used were free of other hypothalamic releasing factors, of vasopressin and oxytocin, and were those expected to give positive results using the pituitary depletion bioassay. Aliquots were concentrated 100-fold by drying under nitrogen or by desiccation over NaOH and H_3PO_4 to remove the ammonium acetate. In both cases the dried sample was then redissolved in acid saline, pH 3.5. The half µl injected was estimated to contain the GRF present in one hypothalamus. Column buffer, similarly treated and redissolved in acid-saline served as a control solution. Since the concentration technique did not affect the responses to either GRF or the column buffer used as a control, the results were pooled for subsequent analysis. As an additional control, a group of animals were injected with 100 µg bovine serum albumin in a similar volume.

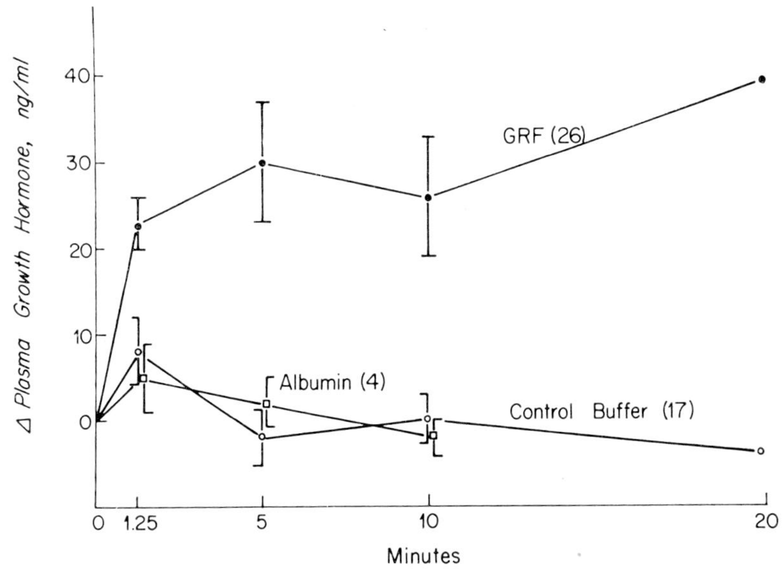

Fig. 9. Effect of intrapituitary injection of partially purified growth hormone releasing factor (GRF), column control buffer and bovine serum albumin on plasma immunoreactive GH levels (mean ± SE) in the rat. (From: Frohman *et al.*, 1971, *Endocrinology*, *88*, 1483; by kind permission of the editors).

Mean preinjection plasma growth hormone values were similar in the three groups of rats, ranging from 15 to 20 ng/ml. Technically, satisfactory injections of GRF were achieved in 26 animals. A decreasing number of samples were available at the later periods because animals were killed at varying times after injection for electron microscopic study of the pituitary gland.

Following the injection of GRF (Fig. 9) a prompt rise in plasma growth hormone occurred which peaked in individual animals at 5 or 10 min and was still evident at 20 min. When examined individually, 24 of 26 GRF injected rats demonstrated a positive response, defined as a rise of 5 ng/ml or greater.

Only 6 of 17 acid saline vehicle injected rats demonstrated a positive plasma growth hormone response and as a group, the slight rise at 1.25 min was not significant. No rises were observed at subsequent sampling periods, in contrast to the GRF responses which remained positive throughout the 20 min period.

In 4 animals injected with bovine serum albumin the responses were similar to those in the acid saline injected controls. In 3 additional rats, the data for which are not shown, GRF was inadvertently injected into the intermediate lobe. Positive but delayed responses were seen in all 3 animals.

The secretion of growth hormone during the first 10 min in response to intrapituitary GRF was determined as after hypothalamic stimulation (Fig. 10). The secretory response in 10 animals ranged from 0.21 to 2.60 μg with a mean of 0.78 μg. No significant secretion occurred in response to either of the control solutions. This response was about one-third

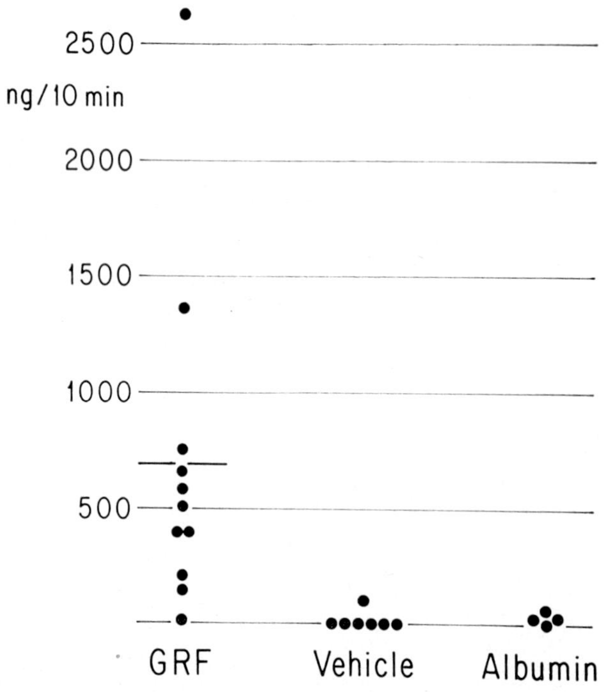

Fig. 10. Growth hormone secretory response in the rat during the first 10 min after intrapituitary injection of partially purified GRF, column control buffer (vehicle), and albumin.

that after hypothalamic stimulation and equivalent to about 0.2% of the pituitary growth hormone content of these animals. Consequently, no measurement of pituitary growth hormone was performed in these animals.

These results have added GRF to the list of hypothalamic releasing factors which include LRF and TRF whose action has been demonstrated by producing elevations of specific pituitary hormones in plasma measured by immunoassay. The results, however, also raise a number of questions which still require resolution. The foremost of these is again the discrepancy between the small quantity of growth hormone secreted in response to GRF as judged from our plasma changes and values several hundred fold greater, based on the pituitary depletion bioassay (Ishida *et al.*, 1965) which have not been confirmed by concomitant radioimmunoassay (Reichlin and Schalch, 1969). It would be tempting to conclude that two different and unrelated hypothalamic preparations were involved. However, our GRF preparation was selected on the basis of previous bioassay experience. The results of both the hypothalamic stimulation and GRF injection experiments, therefore, must raise the question of the interpretation of the pituitary depletion bioassay particularly as to its specificity for growth hormone.

A second question relates to the consistently negative responses observed by both us and by others (Reichlin and Schalch, 1969) after systemic injections of purified GRF preparations. The possibility of plasma peptidase destruction, of inhibition by a growth hormone inhibiting factor, or a problem of insufficient quantity reaching the pituitary must all be raised. Evidence favoring the latter possibility has recently been presented by Malacara and Reichlin (1971) who reported rises in plasma growth hormone levels after the intravenous injection of 10-25 porcine hypothalamic equivalents in rats pretreated with estrogen and progesterone.

Finally, it is now apparent that all future purification studies involving GRF will require validation by the demonstration of rises in plasma immunoreactive growth hormone.

REFERENCES

BERNARDIS, L. L. and FROHMAN, L. A. (1970): Effect of lesion size in the ventromedial hypothalamus on growth hormone and insulin levels in weanling rats. *Neuroendocrinology*, *6*, 319.

BERNARDIS, L. L. and FROHMAN, L. A. (1971): Plasma growth hormone responses to electrical stimulation of the hypothalamus in the rat. *Neuroendocrinology*, *7*, 193.

BUREK, C. L. and FROHMAN, L. A. (1970): Growth hormone synthesis by rat pituitaries in vitro: Effect of age and sex. *Endocrinology*, *86*, 1361.

DHARIWAL, A. P. S., KRULICH, L., KATZ, S. H. and MCCANN, S. M. (1965): Purification of growth hormone-releasing factor. *Endocrinology*, *77*, 932.

FROHMAN, L. A. and BERNARDIS, L. L. (1968): Growth hormone and insulin levels in weanling rats with ventromedial hypothalamic lesions. *Endocrinology*, *82*, 1125.

FROHMAN, L. A. and BERNARDIS, L. L. (1970): Growth hormone secretion in the rat: Metabolic clearance and secretion rates. *Endocrinology*, *86*, 305.

FROHMAN, L. A., BERNARDIS, L. L. and KANT, K. J. (1968): Hypothalamic stimulation of growth hormone secretion. *Science*, *162*, 580.

FROHMAN, L. A., MARAN, J. W. and DHARIWAL, A. P. S. (1971): Plasma growth hormone responses to intrapituitary injections of growth hormone releasing factor (GRF) in the rat. *Endocrinology*, *88*, 1483.

HETHERINGTON, A. W. and RANSON, S. W. (1940): Hypothalamic lesions and adiposity in the rat. *Anat. Rec.*, *78*, 149.

ISHIDA, Y., KUROSHIMA, A., BOWERS, C. Y. and SCHALLY, A. V. (1965): In vivo depletion of pituitary growth hormone by hypothalamic extracts. *Endocrinology*, *77*, 759.

MACGILLIVRAY, M. H., FROHMAN, L. A. and DOE, J. (1970): Metabolic clearance and production rates of human growth hormone in subjects with normal and abnormal growth. *J. clin. Endocr.*, *30*, 632.

MALACARA, J. M. and REICHLIN, S. (1971): Elevation of radioimmunoassayable plasma growth hormone (RIA-GH) in the rat induced by porcine hypothalamic extracts. *Fed. Proc.*, *30*, 198.
MARTIN, J. B. and REICHLIN, S. (1970): Thyrotropin secretion in rats after hypothalamic electrical stimulation or injection of synthetic TSH-releasing factor. *Science*, *168*, 1366.
MÜLLER, E. E., SAITO, T., ARIMURA, A., and SCHALLY, A. V. (1967): Hypoglycemia, stress and growth hormone release: blockade of growth hormone release by drugs acting on the central nervous system. *Endocrinology*, *80*, 109.
REICHLIN, S. (1961): Growth hormone content of pituitaries from rats with hypothalamic lesions. *Endocrinology*, *69*, 225.
REICHLIN, S. and SCHALCH, D. S. (1969): Growth hormone releasing factor. In: *Progress in Endocrinology*, pp. 584–594. Editor: C. Gual. ICS 184, Excerpta Medica, Amsterdam.

ANALOGOUS PATTERN OF BIOASSAYABLE AND RADIOIMMUNOASSAYABLE GROWTH HORMONE IN SOME EXPERIMENTAL CONDITIONS OF RAT AND MOUSE

E. E. MÜLLER[1], G. GIUSTINA[2], D. MIEDICO[2], D. COCCHI[1] and A. PECILE[1]

[1]2nd Chair Department of Pharmacology and [2]Department of Medicine, School of Medicine, University of Milan, Milan, Italy

Seven years have elapsed since Deuben and Meites (1964) demonstrated for the first time that rat hypothalamic extracts promoted the release of growth hormone (GH) from 6-day cultured rat pituitary glands.

The results of Deuben and Meites were confirmed by Schally *et al.* (1965) who showed that release of GH could be elicited in rat pituitaries incubated with porcine or bovine extracts. Pecile *et al.* (1965) obtained the first *in vivo* evidence for growth hormone releasing activity in the rat and their findings were subsequently confirmed by Ishida *et al.* (1965) and by Krulich *et al.* (1965). These *in vivo* and *in vitro* experiments seemed to show clearly that hypothalamic extracts possess a GH releasing (GRF) activity. In the rat acute administration of GRF evoked a decline of pituitary GH that was associated with an increase in GH activity in the plasma (Sawano *et al.*, 1968). GRF has been purified and separated from other releasing factors and the same fractions that are active in releasing GH *in vivo* enhance GH release *in vitro* (Schally *et al.*, 1969). A variety of dissimilar stimuli such as exposure to cold, insulin hypoglycemia, high doses of vasopressin, etc., have been shown to considerably reduce the levels of pituitary GH in the rat (Müller *et al.*, 1967*b*; Krulich and McCann, 1966), mobilizing the hypothalamic stores of GRF (Katz *et al.*, 1967; Müller *et al.*, 1967*a*). Alterations of stored GRF have been reported in situations thought to be associated with altered GH release (Pecile *et al.*, 1965; Meites and Fiel, 1965). All these results were based on the determination of pituitary bioassayable (BA) growth hormone. In primates the introduction of a specific and sensitive radioimmunoassay for measurements of GH in plasma, allowed the demonstration of increased levels of circulating hormone following the administration of crude hypothalamic extracts (Garcia and Geschwind, 1966; Knobil, 1966). Similar effects were reported in sheep and pigs infused with hypothalamic extracts (Machlin *et al.*, 1968).

Results on GH control obtained in rodents prompted further studies which were extended to include experiments on GH control in primates. Thus, for instance, the participation of the adrenergic system in the neurohormonal mechanism(s) of GH release, described firstly in the rat (Müller *et al.*, 1967*c*) was similarly found to be present in the human (Blackard and Heidingsfelder, 1968; Imura *et al.*, 1968).

When in 1966 a radioimmunoassay (RIA) for measurement of GH in the plasma and pituitary gland of the rat was developed (Schalch and Reichlin, 1966) one might have predicted confirmation of the findings on GH control found in rodents by bioassay; but, surprisingly and disappointingly this was not the case. Cold exposure, insulin hypoglycemia, prolonged fasting, neither affected the pituitary GH concentration nor circulating GH levels; if anything, both insulin and cold produced a decrease of plasma RIA-GH (Schalch and Reichlin, 1968; Garcia and Geschwind, 1968).

Preparations of rat stalk median eminence extracts, previously found to be capable of producing marked depletion of pituitary GH when measured by bioassay, failed to elevate RIA plasma GH levels or to consistently diminish pituitary GH content in the rat (Garcia and Geschwind, 1968; Frohman, 1970).

Since then, these disquieting reports on the lack of correlation between BA- and RIA-GH data in rodents have been extended and the significance of the bioassay of pituitary GH in *in vivo* experiments has been questioned (Rodger *et al.*, 1969). We will report here some studies on GH control in rodents (rat and mouse), in which, whenever possible, both bioassay and immunoassay have been used to determine GH levels. It was hoped that this approach would give some insight into the divergent behavior of BA- and RIA-GH in rodents.

For reasons of economy, details on materials and methods have been incorporated in the section on Results. The reader is also referred to the papers by Müller *et al.* (1969, 1970, 1971*a*, *b*). In all of our studies a double antibody radioimmunoassay for rat GH was used to measure pituitary or plasma GH in rats and mice (Müller *et al.*, 1969). It is known in fact that mouse pituitary reacts with antibody to rat GH with an affinity equal to that shown by the rat growth hormone standard (Garcia and Geschwind, 1968). Dilution curves of mouse pituitary extracts were parallel to those of the purified rat growth hormone. Pituitary bioassayable GH was measured by the 'tibia test' method of Greenspan *et al.* (1949).

RESULTS

Effect of acute stimuli on pituitary and plasma GH

While in the rat many negative results have been reported on the determinations of RIA-GH in plasma, fewer reports are available on GH levels in the pituitary and, to our knowledge in only one study pituitary GH measurements have been performed using both types of assay on the same pituitary extracts (Daughaday *et al.*, 1968). Thus we thought it was worthwhile determining BA- and RIA-GH levels in the pituitary of animals subjected

TABLE I

Simultaneous immunological and biological assays of GH in rat under different experimental conditions

Exp.	Treatment	Pituitary growth hormone		Blood glucose (mg/100 ml)
		RIA-GH (μg GH/mg AP)*	BA-GH (expressed as epiphyseal width μ**; mean ± S.E.)	
1	Saline	46.32 ± 6.14	257 ± 2.8	69 ± 2.4
	Insulin (0.2 U./100 g b.w. ip)	64.59 ± 14.6	202 ± 9.5††	42 ± 9.9†
2	Saline	22.00 ± 4.2	267 ± 4.6	74 ± 5.6
	Insulin (0.2 U./100 g b.w. ip)	34.57 ± 5.2	213 ± 4.8††	40 ± 4.2††
	Cold exposure (4°C – 1 hr)	31.91 ± 3.1	213 ± 5.2††	82 ± 6.6
	Fasting (60 hr)	22.80 ± 1.4	224 ± 13.9††	52 ± 4.3†

* Standard rat GH, Ellis (1968), potency 2.7 I.U./mg
** 6–8 hypophysectomized assay animals per group were used
† $p < 0.01$ *vs* saline
†† $p < 0.001$ *vs* saline

(From Müller *et al.*, 1969, *Experientia (Basel)*, *25*, 1146, by kind permission of the editors).

TABLE II

Simultaneous immunological and biological assays of GH in rat pituitary under different experimental conditions

Treatment	Pituitary growth hormone		Blood glucose (mg/100ml)
	RIA-GH* (μg GH/mg AP)	BA-GH expressed as epiphyseal width μ**; mean ± S.E.)	
Saline	25.81 ± 4.4	234 ± 6.0 (70.0)°	48 ± 1.4
Insulin (0.3 U./100 g b.w. ip)	38.82 ± 5.2	194 ± 5.0††(14.2)°	25 ± 2.2††
Cold exposure (4°C – 1 hr)	48.52 ± 1.7††	180 ± 8.7†† (4.1)°	51 ± 1.2††
Fasting (60 hr)	37.24 ± 6.2	211 ± 4.6† (37.5)°	26 ± 2.3

* Standard rat GH, Ellis (1968), potency 2.7 I.U./mg
** 6–8 hypophysectomized assay animals per group were used
† $p < 0.01$ *vs* saline
†† $p < 0.001$ *vs* saline
° Estimates of GH content (μg GH/mg AP) according to a bracketed 3-point assay

(From Müller *et al.*, 1969, *Experientia (Basel)*, *25*, 1146, by kind permission of the editors).

to insulin hypoglycemia, cold exposure or starvation; these conditions are known to induce release of GH from the pituitary in primates (Glick *et al.*, 1965; Glick, 1968).

From the results reported in Tables I and II it appears that none of the three experimental conditions investigated induced significant changes in pituitary GH as measured by RIA, with the exception of cold exposure which in one experiment (Table II) induced a clearcut increase of pituitary GH levels. In contrast with these negative results are the profound decreases of pituitary BA-GH measured in the same pituitary extracts. Insulin hypoglycemia not only failed to modify pituitary RIA-GH, but failed also to produce consistent changes

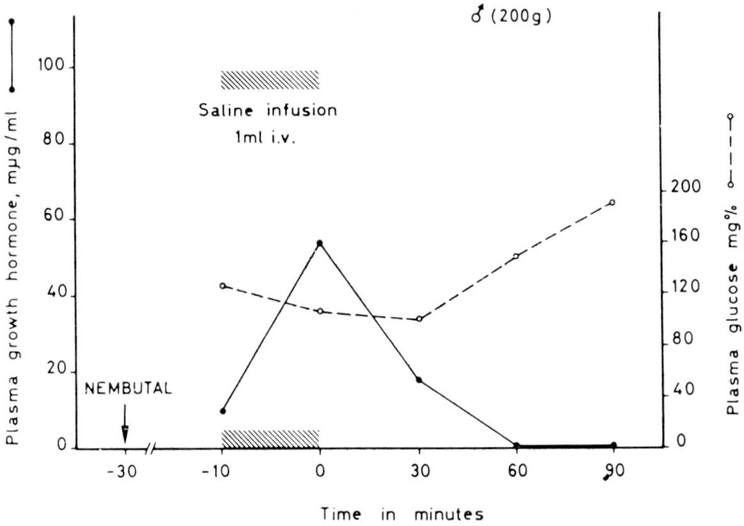

Fig. 1. Time course of plasma GH levels after the intravenous infusion of saline through an in-dwelling catheter in intact rats.

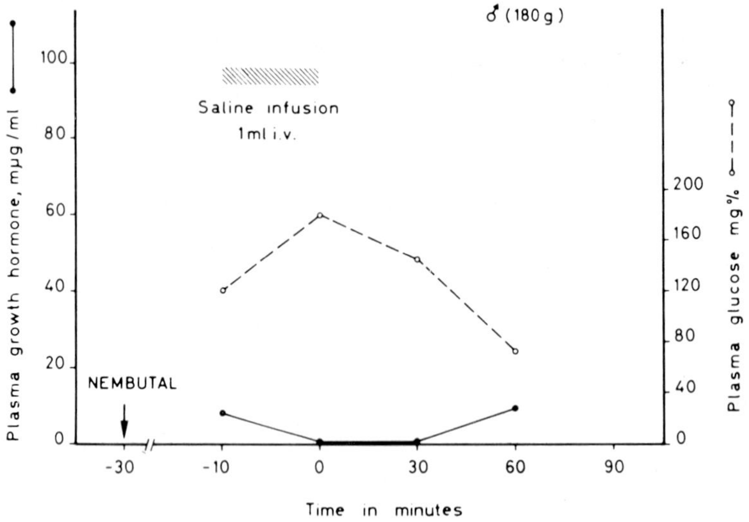

Fig. 2. Time course of plasma GH levels after the intravenous infusion of saline through an indwelling catheter in intact rats.

in the plasma concentration of RIA-GH. Figures 1 and 2 are representative of the time-course of GH in the plasma of rats in which an indwelling catheter was placed in the carotid artery for the withdrawal of serial blood samples. It would appear that prolonged cannulation tends to diminish or to maintain low values of plasma GH; in this experimental condition blood glucose values were rather elevated.

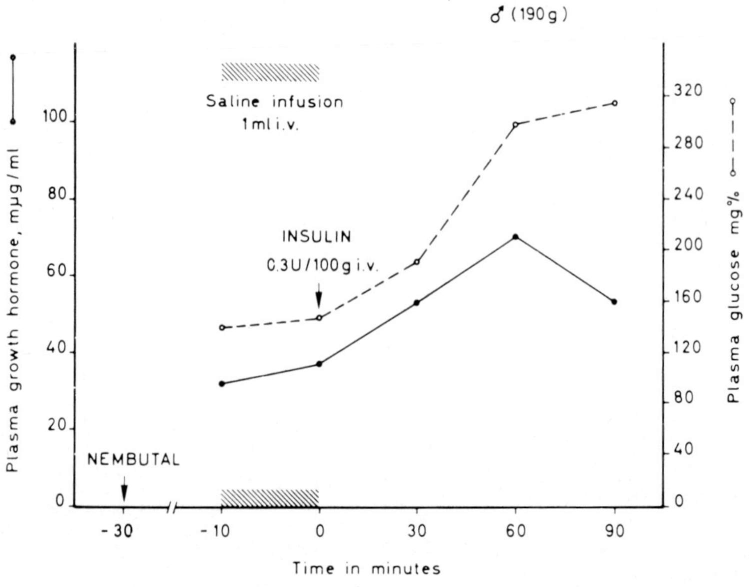

Fig. 3. Effect of intravenous insulin administration on plasma GH and glucose levels in intact rats.

Administration of insulin through the saphenous vein (0.3 U./100 g body weight) at a dose which is highly effective in decreasing blood glucose levels in unmanipulated rats (Müller *et al.*, 1967*c*), did not modify the elevated blood glucose (Fig. 3). Insulin given at a higher dose (0.5 U./100 g body weight) induced a sustained decrease of blood glucose levels but showed only an erratic effect on the levels of plasma GH (Fig. 4). A similar negative result was obtained in another animal although in this instance a more prompt decrease of blood glucose was evoked by the injected insulin (Fig. 5).

To rule out the possibility that adrenal cortical stimulation was affecting our results, the

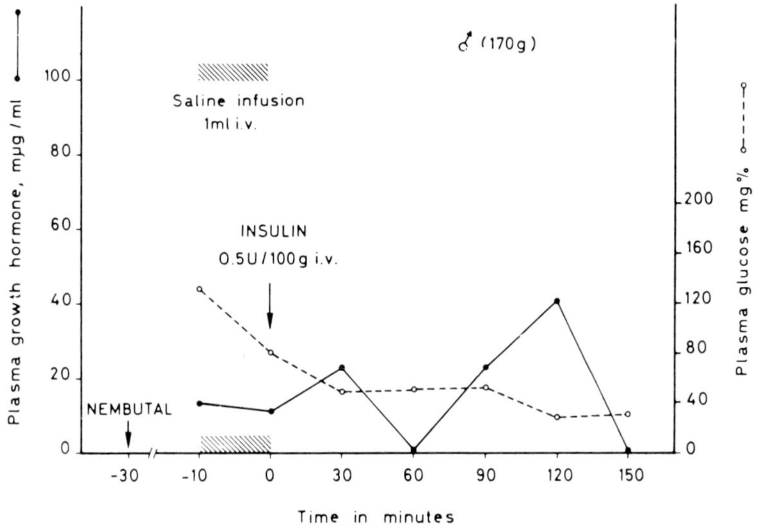

Fig. 4. Effect of intravenous insulin administration on plasma GH levels in intact rats.

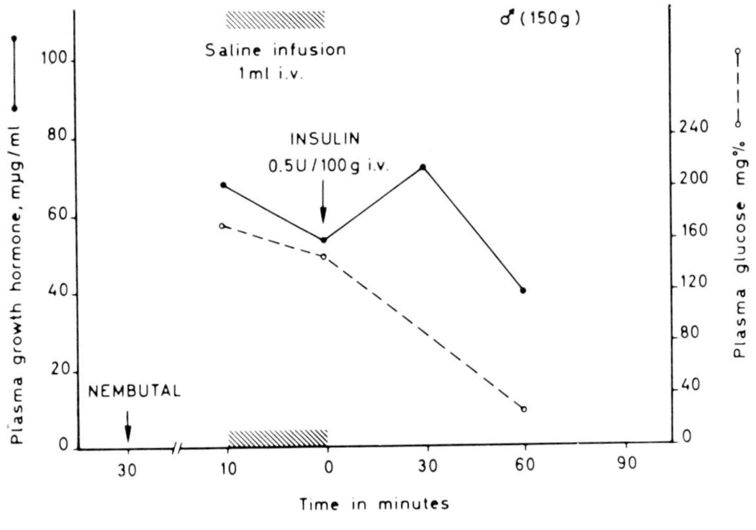

Fig. 5. Effect of intravenous insulin administration on plasma GH levels in intact rats.

insulin tolerance test was performed on adrenalectomized rats. Again no significant increase in plasma RIA-GH occurred following induction of insulin hypoglycemia.

Analogous pattern of radioimmunoassayable and bioassayable GH in some experimental conditions

The bulk of the data so far reported confirm the original observations of other authors that a variety of stressful stimuli fail to elicit significant changes in pituitary or plasma levels of RIA-GH, and in addition confirm the effectiveness of the same stimuli in modifying pituitary BA-GH. In some experimental conditions, however, we were able to observe changes in the titers of pituitary and plasma RIA-GH and to show that they are paralleled by analogous changes in pituitary BA-GH.

Studies in mice

Goldthioglucose induced hypothalamic obesity in mice is correlated with bilateral destruction of the hypothalamic ventromedial nucleus (VMN), the satiety area (Liebelt and Perry, 1957; Brecher *et al.*, 1965), thus eliminating one of the inhibitory mechanisms to feeding. There is now convincing evidence showing that VMN exerts a significant control over growth hormone secretion in the rat (Frohman and Bernardis, 1968; Frohman *et al.*, 1968).

In the light of these observations the study of GH secretion in obese mice seemed to be of particular interest. A single injection of goldthioglucose (GTG) was given intraperitoneally at a dose corresponding approximately to the LD_{50}.

Two separate and similar experiments were carried out. Figure 6 shows the growth curves of mice in the two experiments. Animals were killed by decapitation sixteen (exp. 1) or nine (exp. 2) weeks after GTG-injection.

Table III shows that pituitary weight in contrast to body weight was considerably reduced in GTG-mice in comparison with controls and the effect was even more evident in the heavier GTG-mice of exp. 2 than in exp. 1.

The results summarized in Table IV show that the pituitary BA-GH was markedly

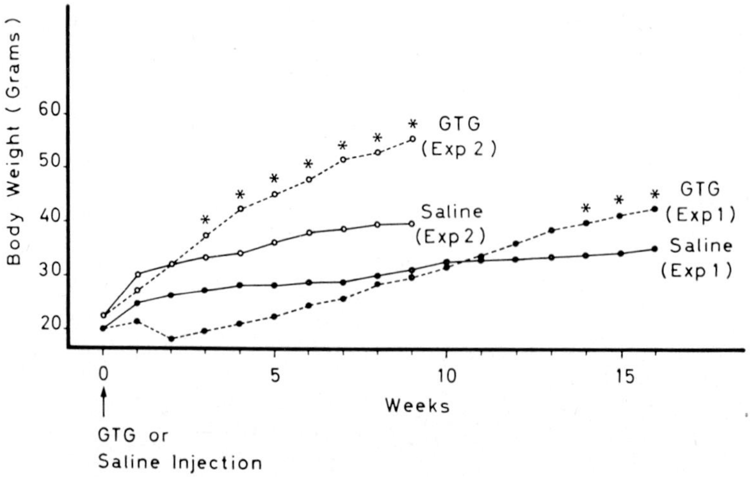

Fig. 6. Body weight curves of control and goldthioglucose (GTG)-obese mice. (From Müller *et al.*, 1971, *Endocrinology*, *89*, 56, by kind permission of the editors).

TABLE III

Body and pituitary weights of control or goldthioglucose (GTG)-obese mice

Exp.	Treatment (dose in parentheses)	No. mice	Body weight (g ± S.E.)		Increment (g ± S.E.)	Pituitary weight (mg ± S.E.)	
			Initial	Final*		Absolute	Relative (mg/100 g body weight)
1	Saline	29	21 ± 0.24	36 ± 0.92	15 ± 0.80	2.3 ± 0.02	6.28 ± 0.25
	GTG (0.8 mg/g ip)	29	20 ± 0.37	43 ± 1.08	23 ± 1.10††	2.1 ± 0.02†	4.83 ± 0.15††
2	Saline	30	22 ± 0.80	39 ± 0.30	17 ± 0.50	2.3 ± 0.10	5.70 ± 0.22
	GTG (0.42 mg/g ip)	29	22 ± 0.20	55 ± 1.00	32 ± 1.10††	1.7 ± 0.07††	3.00 ± 0.14††

* Body weight recorded 112 days (exp. 1) or 63 days (exp. 2) after GTG administration.
† $p < 0.05$ *vs* saline
†† $p < 0.001$ *vs* saline
(From Müller *et al.*, 1971, *Endocrinology*, *89*, 56, by kind permission of the editors).

decreased in the obese mice (exp. 1). In the lower section of the same Table the simultaneous values of pituitary BA- and RIA-GH are presented (exp. 2). It can be seen that both BA- and RIA-GH values were considerably reduced in the obese mice as was the plasma RIA-GH (Fig. 7).

Thus, it is apparent that both synthesis and release of GH are highly impaired in the obese mice and that a concomitant decrease of pituitary BA- and RIA-GH took place under these experimental conditions.

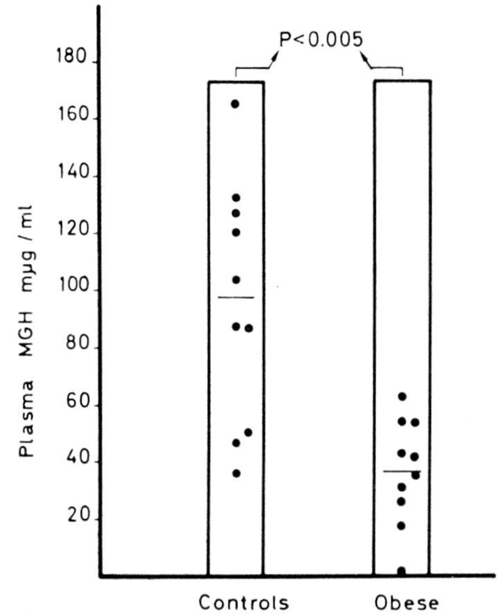

Fig. 7. Radioimmunoassayable plasma growth hormone in controls and GTG-obese mice. (From Müller *et al.*, 1971, *Endocrinology*, *89*, 56, by kind permission of the editors).

TABLE IV

Bioassayable (BA) and radioimmunoassayable (RIA) growth hormone (GH) in the pituitary of control or gold-thioglucose (GTG)-obese mice

Exp.	Treatment (dose and No. of animals in parentheses)	BA-GH				RIA-GH†† μg GH/mg pit. (mean ± S.E.)
		Pit. equivalents assay rat total dose /4 days (mg)	Width of tibia cartilage μ † (mean ± S.E.)	μg GH/ mg pit.	95% fiducial limits	
1	(29) Saline	5.0	255 ± 3.1			
		2.5	228 ± 2.4	46.1	34.2 –62.0	—
		1.25	206 ± 11.4			
	(29) GTG	2.5	203 ± 6.9			
	(0.80 mg/g ip)	1.25	187 ± 6.1	22.5*	17.3 –29.2	—
	GH	200 μg	250 ± 8.6			
		50 μg	198 ± 4.7			
		12.5 μg	163 ± 6.1			
2	(30) Saline	5.0	266 ± 1.6			
		2.5	252 ± 1.3	82.02	70.22–95.79	135.06 ± 20.49
		1.25	236 ± 1.7			
	(29) GTG	5.0	251 ± 3.7			
	(0.42 mg/g ip)	2.5	233 ± 4.2	37.17*	29.35–47.06	51.71 ± 10.96**
		1.25	209 ± 5.2			
	GH	200 μg	257 ± 3.3			
		50 μg	206 ± 2.3			
		12.5 μg	173 ± 2.3			

Pit. = Pituitary
* $p < 0.01$ *vs* saline by factorial analysis
** $p < 0.005$ *vs* saline by Student's t test
† 6–8 hypophysectomized assay rats per group were used. Mean value of epiphyseal width of hypophysectomized rats was 151 ± 3.7 (exp. 1) and 155 ± 2.8 (exp. 2).
†† Mean values of 10 single determinations for each group

(From Müller *et al.*, 1971, *Endocrinology*, *89*, 56, by kind permission of the editors).

Studies in rats

Resting levels of BA- and RIA-pituitary GH and RIA-plasma GH were determined at intervals during a controlled 24-hour light-dark cycle in intact prepuberal female rats. On the day of the experiments at each designated time, unanesthetized rats were decapitated with a guillotine. A significant daily periodicity in the content of both pituitary BA- and RIA-GH was found in these animals (Fig. 8).

A gradual parallel increase in GH values was present from late morning through afternoon, with peak levels at 18.00-21.00 hr. Reattainment of values close to the morning levels took place during the night. Plasma RIA-GH exhibited wide fluctuations throughout the day, which are difficult to reconcile with the pituitary values (Fig. 9).

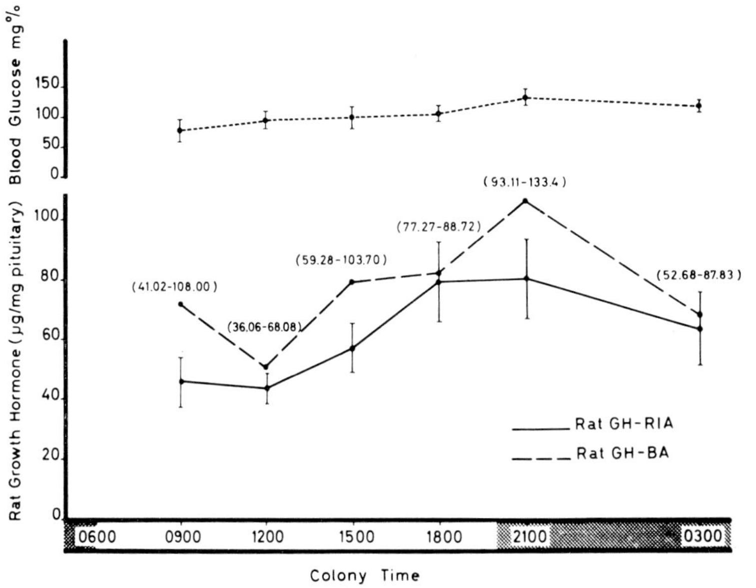

Fig. 8. Radioimmunoassayable (RIA) and bioassayable (BA) pituitary growth hormone in intact female rats during a 24-hour light-dark cycle. (From Müller *et al.*, 1970, *Proc. Soc. exp. Biol. (N.Y.)*, *135*, 934, by kind permission of the editors).

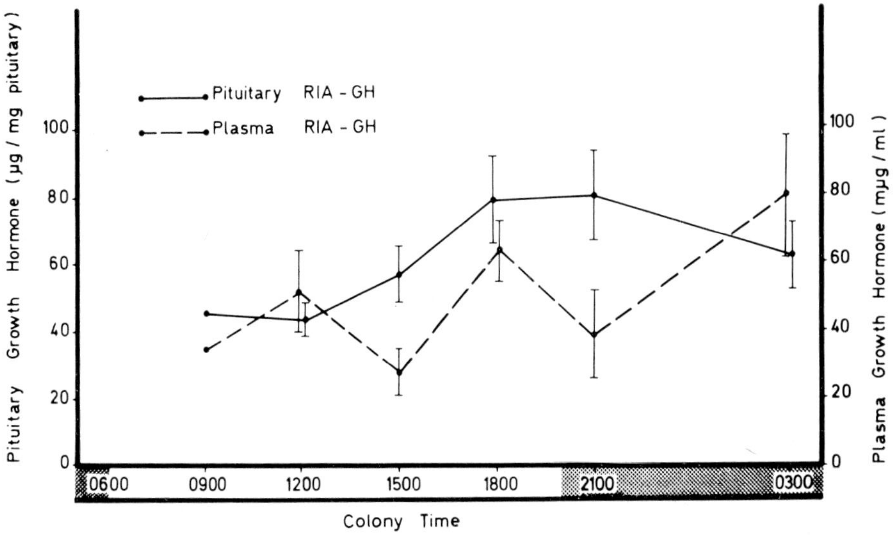

Fig. 9. Radioimmunoassayable (RIA) pituitary and plasma growth hormone in intact female rats during a 24-hr light-dark cycle. (From Müller *et al.*, 1970, *Proc. Soc. exp. Biol. (N.Y.)*, *135*, 934, by kind permission of the editors.)

Growth hormone-releasing activity of rat SME extracts evaluated by increase in plasma RIA-GH

Efforts to demonstrate a GRF activity in hypothalamic extracts from many animal species, by measuring their effects on pituitary or plasma RIA-GH in the rat, have, to date, been unsuccessful (Garcia and Geschwind, 1968; Rodger *et al.*, 1969; Frohman, 1970). Studies were performed to determine whether intrapituitary administration of rat SME extracts would result in changes of pituitary BA- and RIA-GH and plasma RIA-GH in the rat.

Figure 10 shows that in rats anesthetized with nembutal (35 mg/kg body weight), a 1 minute infusion of rat SME extracts (1/33 SME-1/66 SME/lobe; 2 lobes – 1.5 μl/min/lobe) significantly increased plasma RIA-GH, above the values present in saline-infused rat

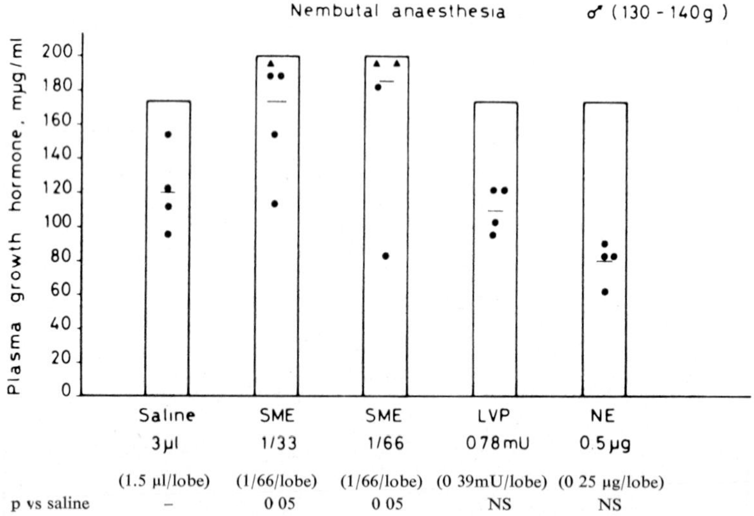

Fig. 10. Effect of intrapituitary infusions of saline, rat stalk median eminence (SME) extracts, lysine vasopressin (LVP) and norepinephrine (NE) on plasma radioimmunoassayable (RIA) GH in the rat.

(172.0 ± 8.1 *vs* 120.5 ± 12.2 mμg GH/ml) 10 minutes after the infusion had stopped. Infusion into one lobe (1/66 SME) resulted in an increase of plasma RIA-GH which was of even greater magnitude (192.0 ± 29.0 mμg GH/ml) even if not statistically different from that of rats infused into two lobes. Lysine vasopressin (0.78 mU.) infused in an amount corresponding to that present in 1/50 SME equivalent (McCann and Haberland, 1959), and NE (0.5 μg), regarded as another control material, did not increase plasma RIA-GH above the values of saline injected rats. Simultaneous measurements of pituitary GH (Table V) showed that SME extracts at the higher dose infused (1/33 SME) induced a small insignificant decrease of pituitary RIA-GH and a significant decrease of pituitary BA-GH. The decrease of RIA-GH induced by SME extracts at the lower dose (1/66 SME) was not statistically significant. LVP and NE did not, in the doses used, modify pituitary BA- and RIA-GH content.

Since there was the possibility that infusion of saline *per se* elicited an aspecific release of the hormone, in a further experiment values of saline-infused rats were compared with those

TABLE V

Effect of intrapituitary infusions of saline, rat stalk median eminence (SME) extracts, lysine vasopressin (LVP) and norepinephrine on pituitary bioassayable (BA) and radioimmunoassayable (RIA) growth hormone (GH) in the rat

No. of animals infused	Treatment	Pituitary growth hormone	
		RIA-GH (μg GH/mg AP)*	BA-GH (expressed as epiphyseal width μ†, mean ± S.E.)
4	Saline (3 μl) 1.5 μl/lobe	18.1 ± 1.62	217 ± 8.5
5	SME extracts (1/33) 1/66/lobe	15.5 ± 0.67	194 ± 3.9††
4	SME extracts (1/66) 1/66/lobe	16.8 ± 0.62	—
4	LVP (0.78 mU) 0.39 mU/lobe	19.4 ± 0.72	205 ± 2.8
4	NE (0.5 μg) 0.25 μg/lobe	18.3 ± 0.50	205 ± 3.7

† 2 mg AP given in four days to hypophysectomized assay animals (Sprague-Dawley)
* AP = anterior pituitary
†† p <0.02 *vs* saline

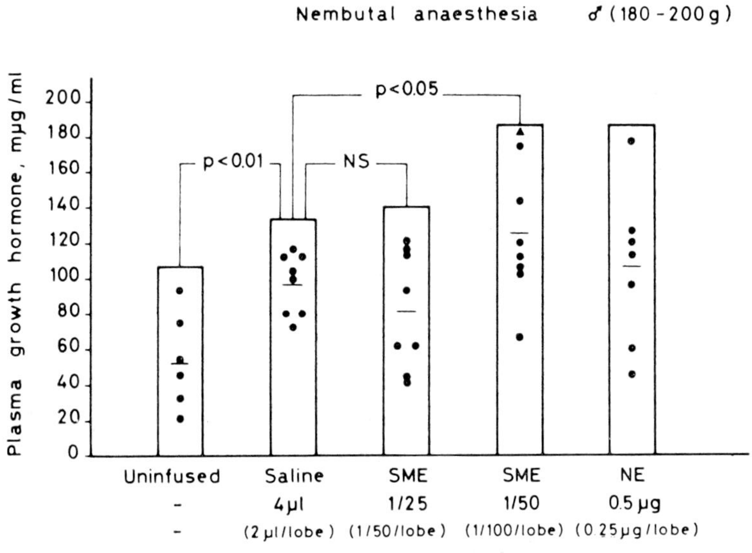

Fig. 11. Effect of intrapituitary infusions of saline, rat stalk median eminence (SME) extracts and norepinephrine (NE) on plasma RIA-GH in the rat.

of rats uninfused or infused with SME extracts. From Figure 11 it appears that in our experimental conditions saline infusion (4 μl) induced *per se* higher levels of plasma RIA-GH (54.0 ± 11.0 *vs* 98.3 ± 5.76 mμg GH/ml); infusion of 1/25 SME did not significantly modify plasma RIA-GH (81.9 ± 11.7 mμg GH/ml), while surprisingly enough, halving the dose (1/50 SME) resulted in a significant increase of GH over the values of saline-infused rats (126.8 ± 11.2 mμg GH/ml). Again no effect was noticed with the infusion of NE (0.5 μg).

Determinations of pituitary GH (Table VI) showed firstly that RIA-GH was unchanged

TABLE VI

Effect of intrapituitary infusions of saline or rat stalk median eminence (SME) extracts on pituitary bioassayable (BA) and radioimmunoassayable (RIA) growth hormone (GH) in the rat

No. of animals	Treatment	Pituitary growth hormone	
		RIA-GH (μg GH/mg AP)†	BA-GH (expressed as epiphyseal width μ*, mean ± S.E.)
6	Uninfused	11.68 ± 0.74	213 ± 5.1
8	Saline (4 μl) 2 μl/lobe	12.07 ± 1.89	194 ± 6.0††
8	SME extracts (1/25) 1/50/lobe	12.68 ± 0.76	188 ± 5.6**
8	SME extracts (1/50) 1/100/lobe	11.36 ± 1.15	215 ± 4.8

† 2.5 mg AP administered in four days to hypophysectomized assay animals (Wistar)
†† p <0.02 *vs* uninfused
** p <0.005 *vs* uninfused
* Mean value of epiphyseal width of hypophysectomized rats was 125 ± 1.5 μ

following all treatments; infusion of saline slightly decreased pituitary BA-GH while a more marked decrease followed infusion of SME extracts (1/25 SME).

Administration of 1/50 SME extracts, the dose which was the most effective in increasing plasma levels of immunoreactive GH, did not modify pituitary BA-GH.

Plasma radioimmunoassayable GH in hypophysectomized-transplanted rats

To assess how titers of plasma RIA-GH might truly reflect the events connected with body growth, the hormone was determined in hypophysectomized Wistar rats, transplanted

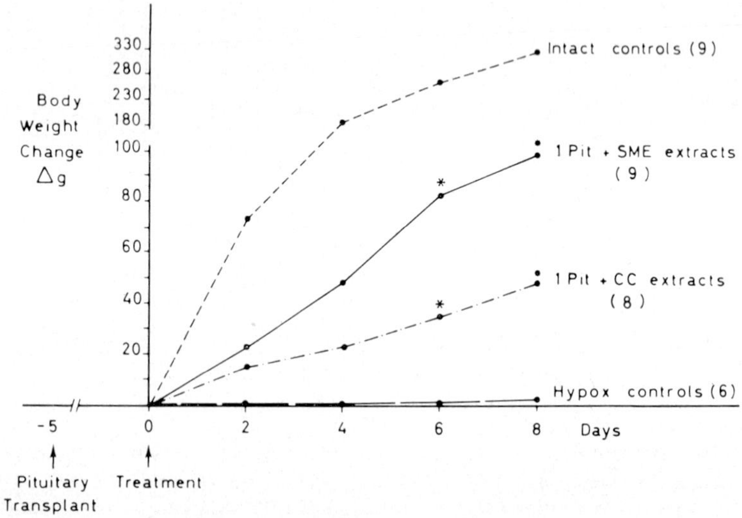

Fig. 12. Pooled daily changes in body weight of hypophysectomized (Hypox) rats with a pituitary transplanted beneath the kidney capsule treated with rat stalk median eminence (SME) extracts or rat cerebral cortex (CC) extracts. Asterisks indicate significant differences between SME- and CC-treated rats.

(Muller et al, 1972)

TABLE VII

ect of a pituitary transplanted beneath the kidney capsule of hypophysectomized rats (hypox) treated with rat stalk median eminence (SME) extract or rat cerebral cortex (CC) extract on body weight, body length and tibia cartilage width

oup treatment (= transplant)	No. of rats	Body weight (g) (mean ± S.E.)		Body length (cm) (mean ± S.E.)		Tibia cartilage width (μ) (mean ± S.E.)
		Initial	Final	Initial	Final	
hypox controls	6	85.66 ± 4.02	86.20 ± 6.15	27.90 ± 0.30	27.76 ± 0.50	133.36 ± 15.70
hypox + T + CC**	8	96.77 ± 1.82	102.50 ± 2.86†	28.95 ± 0.25	29.72 ± 0.09†	189 .93 ± 7.83^{+}
hypox + T + SME*	9	97.85 ± 1.11	108.85 ± 1.62††°	28.93 ± 0.22	30.15 ± 0.03†††	198.46 ± 16.42^{++}
intact controls	9	91.50 ± 1.20	134.12 ± 3.09	27.37 ± 0.43	31.47 ± 0.46	267.57 ± 4.14

1.5 rat SME equivalents twice daily ip
Amounts in mg corresponding to 1.5 SME equivalents (twice daily)
II *vs* I $p < 0.05$
III *vs* I $p < 0.001$
III *vs* II $p < 0.05$
III *vs* II $p = 0.05$
II *vs* I $p < 0.01$
III *vs* I $p < 0.02$

with a pituitary gland beneath the kidney capsule and treated with stalk median eminence (SME) extracts or cerebral cortex (CC) extracts. Hypophysectomized-untransplanted rats or intact rats served as controls.

Figure 12 and Table VII report data on body and linear growth of these animals.

It is apparent that pituitary transplantation resulted in a prompt resumption of body growth which was more evident in the rats treated with SME extracts (Fig. 12, Table VII).

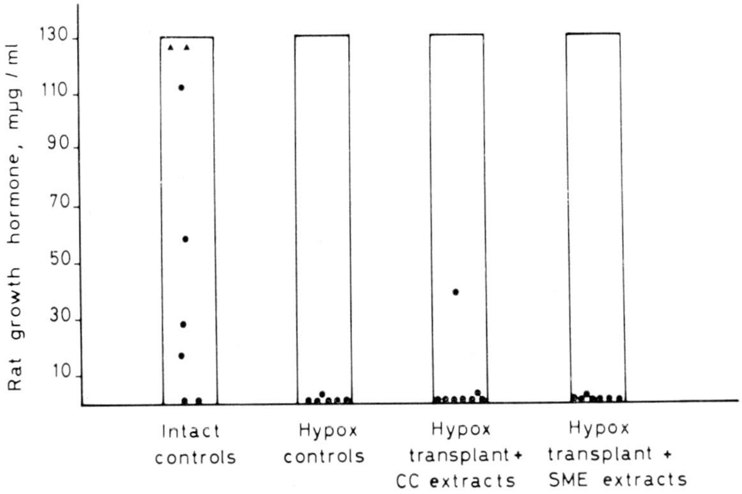

Fig. 13. Plasma GH levels in blood samples obtained by decapitation from intact rats, hypophysectomized (Hypox) rats or hypox rats with a pituitary beneath the kidney capsule treated with rat stalk median eminence (SME) extracts or rat cerebral cortex (CC) extracts.

Irrespective of the type of treatment used, plasma RIA-GH was undetectable in hypophysectomized-transplanted rats, indistinguishable from that present in hypophysectomized-untransplanted rats (Fig. 13).

DISCUSSION

Simultaneous measurements of bioassayable and radioimmunoassayable growth hormone gave, in some of the experimental conditions examined, a good agreement between the results for BA- and those for RIA-GH. Thus in mice made obese by a single administration of goldthioglucose, in agreement with the BA data, the pituitary RIA-GH was found markedly decreased and in addition plasma RIA-GH was also diminished. In partial agreement with these results Frohman and Bernardis (1968) have observed a tendency towards decreased pituitary BA-GH content, in addition to decreased amounts of immunoassayable GH in both plasma and pituitary of rats with bilateral stereotactic electrolytic lesions limited to the VMN. A significant daily periodicity in the content of pituitary growth hormone both by using bioassay and radioimmunoassay was found in nonstressed fed female rats during a controlled 24-hour light-dark cycle.

These findings might be surprising in view of the marked discrepancies between BA and RIA results reported previously for GH. If one recalls, however, that BA and RIA data are well correlated in measurement of growth hormone content in the unstimulated pituitary (Daughaday *et al.*, 1968), the present data obtained in animals which were untreated or manipulated only once are more readily explicable. Thus from these and previous results a pattern is emerging which indicates that both bioassay and radioimmunoassay give approximately the same values for the concentration of GH in the pituitary of animals which have not been acutely manipulated. Acute manipulation of the animal (by exposure to cold, insulin, or hypothalamic extracts) induces a profound discrepancy between BA and RIA data.

The observation that an acute stimulus results in a dramatic decrease of BA-GH in the pituitary, with no or barely detectable pituitary RIA-GH change, suggests the possibility that as a consequence of the stimulus, even if of maximal intensity, only a limited amount of RIA-GH enters the circulation. Concomitantly, however, a reversible conformational change of GH structure may take place in the pituitary, without modifying the immunological characteristic of the molecule. This may change the biological activity (being responsible for the results of the 'depletion assay') and impair further release of the hormone. It has been reported by Li (1968) that oxidation of human growth hormone with performic acid resulted in loss of biological activity but retention of the immunological activity, suggesting that the antigenic site of human GH is not related to that responsible for GH activity.

Growth hormone might be released from the pituitary and appears in plasma as RIA-GH under the effect of a given stimulus within a certain range of the 'strength' of the stimulus, beyond which the release of the hormone, far from being further increased, on the contrary is diminished.

The results of the intrapituitary infusion study which showed that paradoxically the lower doses of infused SME extracts were the ones which were the most effective in increasing RIA-GH in the plasma would be in keeping with such an explanation. This hypothesis, although appealing, does not at present account for the elevated values of plasma BA-GH which have been reported after administration of purified pig GRF (Sawano *et al.*, 1968) or pure pig GRF (Müller *et al.*, 1971*c*), unless one admits a preferential release of BA-GH into plasma following an acute stimulus to the adenohypophysis. In this context the observation is particularly interesting that in the plasma of rats bearing a GH-secreting tumor the GH-like activity determined by bioassay is significantly higher than that detected by radioimmunoassay (S. Ellis, personal communication).

Marked variability of plasma RIA-GH under normal physiological conditions often precludes any valid interpretation of this parameter; in addition, at least sometimes, it apparently does not reflect the events connected with body growth. It cannot be excluded *a priori* that pituitary hormones besides GH, *e.g.* prolactin, might be responsible for the effect on growth resumption observed in the transplantation experiments. This hypothesis, however, seems to be rather untenable when it is considered that the most pronounced growth resumption was present in the animals stimulated with SME extracts, in a condition in which prolactin secretion, far from being stimulated, should actually be inhibited (Meites and Nicoll, 1966).

To conclude, in the light of these and previous findings the hope of reconciling BA and RIA data is far from lost; on the contrary, the discrepancy may be explained by putting into a proper perspective the contribution of both methods in each instance.

ACKNOWLEDGEMENTS

We are grateful to Drs. S. Ellis, A. Parlow and D. S. Schalch for the supply of rat growth hormone or anti-rat growth hormone serum and to WHO for bovine growth hormone. The technical assistance of Miss Patrizia Filisetti and Miss Carmela Moraschini is also acknowledged.

REFERENCES

BLACKARD, W. G. and HEIDINGSFELDER, S. A. (1968): Adrenergic receptor control mechanism for growth hormone secretion. *J. clin. Invest.*, *47*, 1407.

BRECHER, G., LAQUEUR, G. L., CRONKITE, E. P., EDELMAN, P. M. and SCHWARTZ, I. L. (1965): The brain lesion of goldthioglucose obesity. *J. exp. Med.*, *121*, 395.

DAUGHADAY, W. H., PEAKE, G. T., BIRGE, C. A. and MARIZ, I. K. (1968): The influence of endocrine factors on the concentration of growth hormone in rat pituitary. In: *Growth Hormone*, p. 238. Editors: A. Pecile and E. E. Müller. ICS 158, Excerpta Medica, Amsterdam.

DEUBEN, R. and MEITES, J. (1964): Stimulation of pituitary growth hormone release by a hypothalamic extract 'in vitro'. *Endocrinology*, *74*, 408.

FROHMAN, L. A. (1970): In: *Hypophysiotropic Hormones of the Hypothalamus: Assay and Chemistry*, p. 164. Editor: J. Meites. The Williams and Wilkins Co., Baltimore, Md.

FROHMAN, L. A. and BERNARDIS, L. L. (1968): Growth hormone and insulin levels in weanling rats with ventromedial hypothalamic lesions. *Endocrinology*, *82*, 1125.

FROHMAN, L. A., BERNARDIS, L. L. and KANT, K. J. (1968): Hypothalamic stimulation of growth hormone secretion. *Science*, *162*, 580.

GARCIA, J. F. and GESCHWIND, I. I. (1966): Increase in plasma growth hormone level in the monkey following the administration of sheep hypothalamic extracts. *Nature (Lond.)*, *211*, 372.

GARCIA, J. F. and GESCHWIND, I. I. (1968): Investigation of growth hormone secretion in selected mammalian species. In: *Growth Hormone*, p. 267. Editors: A. Pecile and E. E. Müller. ICS 158, Excerpta Medica, Amsterdam.

GLICK, S. M. (1968): Normal and abnormal secretion of growth hormone. *Ann. N.Y. Acad. Sci.*, *148*, 471.

GLICK, S. M., ROTH, J., YALOW, R. S. and BERSON, S. A. (1965): The regulation of growth hormone secretion. *Recent Progr. Hormone Res.*, *21*, 241.

GREENSPAN, F. S., LI, C. H., SIMPSON, M. E. and EVANS, H. M. (1949): Bioassay of hypophysial growth hormone: the tibia test. *Endocrinology*, *54*, 455.

IMURA, H., KATO, Y., IKEDA, M., MORIMOTO, M., YAWATA, M. and FUKASE, M. (1968): Increased plasma levels of growth hormone during infusion of propranolol. *J. clin. Endocr.*, *28*, 1079.

ISHIDA, Y., KUROSHIMA, A., BOWERS, C. Y. and SCHALLY, A. V. (1965): In vivo depletion of pituitary growth hormone by hypothalamic extracts. *Endocrinology*, *77*, 759.

KATZ, S. H., DHARIWAL, A. P. S. and MCCANN, S. M. (1967): Effect of hypoglycemia on the content of pituitary GH and hypothalamic GH-RF in the rat. *Endocrinology*, *81*, 333.

KNOBIL, E. (1966): Tenth Bowditch Lecture. The pituitary growth hormone: an adventure in physiology. *Physiologist*, *9*, 25.

KRULICH, L., DHARIWAL, A. P. S. and MCCANN, S. M. (1965): Growth-hormone releasing activity of crude ovine hypothalamic extracts. *Proc. Soc. exp. Biol. (N.Y.)*, *120*, 180.

KRULICH, L. and MCCANN, S. M. (1966): Effect of alterations in blood sugar on pituitary growth hormone content in the rat. *Endocrinology, 78*, 759.

LI, C. H. (1968): The chemistry of human pituitary growth hormone. In: *Growth Hormone*, p. 3. Editors: A. Pecile and E. E. Müller. ICS 158, Excerpta Medica, Amsterdam.

LIEBELT, R. A. and PERRY, J. H. (1957): Hypothalamic lesions associated with goldthioglucose induced obesity. *Proc. Soc. exp. Biol. (N.Y.), 95*, 774.

MACHLIN, L. J., TAKAHASHI, Y., HORINO, M., HERTELENDY, F., GORDON, R. S. and KIPNIS, D. (1968): Regulation of growth hormone secretion in non-primate species. In: *Growth Hormone*, p. 292. Editors: A. Pecile and E. E. Müller. ICS 158, Excerpta Medica, Amsterdam.

MCCANN, S. M. and HABERLAND, P. (1959): Vasopressin versus corticotrophin-releasing factor. *Proc. Soc. exp. Biol. (N.Y.), 102*, 319.

MEITES, J. and FIEL, N. J. (1965): Effect of starvation on hypothalamic content of 'somatotrophin releasing factor' and pituitary growth hormone content. *Endocrinology, 77*, 455.

MEITES, J. and NICOLL, C. S. (1966): Adenohypophysis: Prolactin. *Ann. Rev. Physiol., 28*, 57.

MÜLLER, E. E., ARIMURA, A., SAWANO, S., SAITO, T. and SCHALLY, A. V. (1967*a*): Growth hormone-releasing activity in the hypothalamus and plasma of rats subjected to stress. *Proc. Soc. exp. Biol. (N.Y.), 125*, 874.

MÜLLER, E. E., GIUSTINA, G., MIEDICO, D., PECILE, A., COCCHI, D. and WANG KING F. (1970): Circadian pattern of bioassayable and radioimmunoassayable growth hormone in the pituitary of female rats. *Proc. Soc. exp. Biol. (N.Y.), 135*, 934.

MÜLLER, E. E., MIEDICO, D., GIUSTINA, G. and COCCHI, D. (1971*a*): Ineffectiveness of hypoglycemia, cold exposure and fasting in stimulating GH secretion in the mouse. *Endocrinology, 88*, 345.

MÜLLER, E. E., MIEDICO, D., GIUSTINA, G. and PECILE A. (1969): Growth hormone immunological and biological assays in pituitary of rat under different experimental conditions. *Experientia (Basel), 25*, 1146.

MÜLLER, E. E., MIEDICO, D., GIUSTINA, G., PECILE, A., COCCHI, D. and MANDELLI, V. (1971*b*): Impaired secretion of growth hormone in goldthioglucose-obese mice. *Endocrinology, 89*, 56.

MÜLLER, E. E., SAITO, T., ARIMURA, A. and SCHALLY, A. V. (1967*b*): Hypoglycemia, stress and growth hormone release: blockade of growth hormone release by drugs acting on the central nervous system. *Endocrinology, 80*, 109.

MÜLLER, E. E., SAWANO, S., ARIMURA, A. and SCHALLY, A. V. (1967*c*): Blockade of release of growth hormone by brain norepinephrine depletors. *Endocrinology, 80*, 471.

MÜLLER, E. E., SCHALLY, A. V. and COCCHI, D. (1971*c*): Increase in plasma growth hormone (GH)-like activity after administration of porcine GH-releasing hormone. *Proc. Soc. exp. Biol. (N.Y.), 137*, 489.

PECILE, A., MÜLLER, E. E., FALCONI, G. and MARTINI, L. (1965): Growth hormone-releasing activity of hypothalamic extracts at different ages. *Endocrinology, 77*, 241.

RODGER, N. W., BECK, J. C., BURGUS, R. and GUILLEMIN, R. (1969): Variability of response in the bioassay for a hypothalamic somatotrophin releasing factor based on rat pituitary growth hormone content. *Endocrinology, 84*, 1373.

ROSSI, G. L. and FALCONI, G. (1965): Transauricular perfusion of the pituitary gland in rats and mice. *Nature (Lond.), 207*, 1200.

SAWANO, S., ARIMURA, A., BOWERS, C. Y., REDDING, T. W. and SCHALLY, A. V. (1968): Pituitary and plasma growth-hormone-like activity after administration of highly purified pig growth hormone-releasing factor. *Proc. Soc. exp. Biol. (N.Y.), 127*, 1010.

SCHALCH, D. S. and REICHLIN, S. (1966): Plasma growth hormone concentration in the rat determined by radioimmunoassay: influence of sex, pregnancy, lactation, anesthesia, hypophysectomy and extrasellar pituitary transplants. *Endocrinology, 79*, 275.

SCHALCH, D. S. and REICHLIN, S. (1968): Stress and growth hormone release. In: *Growth Hormone*, p. 211. Editors: A. Pecile and E. E. Müller. ICS 158, Excerpta Medica, Amsterdam.

SCHALLY, A. V., SAWANO, S., ARIMURA, A., BARRETT, J. F., WAKABAYASHI, I. and BOWERS, C. Y. (1969): Isolation of growth hormone-releasing hormone (GRH) from porcine hypothalami. *Endocrinology, 84*, 1493.

SCHALLY, A. V., STEELMAN, S. L. and BOWERS, C. Y. (1965): Effects of hypothalamic extracts on release of growth hormone 'in vitro'. *Proc. Soc. exp. Biol. (N.Y.), 119*, 208.

ELEVATION OF PLASMA RADIOIMMUNOASSAYABLE GROWTH HORMONE IN THE RAT INDUCED BY PORCINE HYPOTHALAMIC EXTRACTS*

JUAN M. MALACARA** and SEYMOUR REICHLIN

Department of Medical and Pediatric Specialties, University of Connecticut School of Medicine, Farmington, Conn., U.S.A.

In 1967, at the time of the First International Symposium on Growth Hormone in Milan, the role of the hypothalamus in the regulation of Growth Hormone secretion was accepted with certainty on the basis of extensive physiological study. Evidence for neural control of GH secretion has since been strengthened by further experiments, the most important of which have been those dealing with cyclic GH secretion related to EEG patterns (Parker *et al.*, 1969), the blockade of GH secretion by neuropharmacologic means (Blackard and Heidingsfelder, 1968), and the demonstration by Frohman *et al.* (1968) that electrical stimulation of the basal median eminence and ventromedial nucleus of the rat bring about a prompt elevation of plasma GH. In support of the portal vessel chemotransmitter hypothesis of GH regulation, excellent evidence has been adduced to indicate that hypothalamic extracts (Krulich *et al.*, 1968) or plasma from the cut end of the pituitary stalk (Wilber and Porter, 1970), added *in vitro* to pituitary incubates caused release of GH into the media, and that crude extracts of hypothalamic origin injected into either monkeys (Knobil *et al.*, 1968) or sheep (Machlin *et al.*, 1968) were capable of inducing GH hypersecretion.

Despite this overwhelming evidence for neural control of GH secretion through the mediation of a hypothalamic growth hormone releasing factor (SRF), workers in the field of GH secretion control have been plagued by seeming inconsistencies between the methods used for the demonstration of SRF in the rat. These methods are primarily those of the pituitary depletion assay, utilizing tibial plate bioassay, and those utilizing plasma radioimmunoassayable GH. This problem has been of major practical importance in that purification of and proof of the activity of authentic SRF is based on bioassays still not universally accepted, and the same uncertainties affect conclusions in physiological studies of GH regulation as, for example, in stress (Müller *et al.*, 1967*a*) and in neuropharmacological analysis (Müller *et al.*, 1967*b*). Reviews by Reichlin and Schalch (1968) and by Daughaday *et al.* (1970) have summarized this problem, and a detailed criticism of the pituitary depletion assay for SRF has been published (Rodger *et al.*, 1969). Since levels of plasma radioimmunoassayable GH can be elevated in the rat by electrical stimulation of the hypothalamus and in the monkey or man by emotional and physical stress, it has seemed most surprising that authentic releasing factor would not similarly increase plasma radioimmunoassayable GH as well. In this paper, similar to one presented in April 1971 at the Spring Meeting of the American Physiological Society (Malacara and Reichlin, 1971), we demonstrate that under appropriate conditions hypothalamic extract is capable of raising plasma radioimmunoassayable GH in the rat.

---

* Supported in part by U.S.P.H.S. Grant No. AM 13695.

** Recipient of a Population Council Traveling Fellowship. Present address: University of Guanajuato, School of Medicine, Leon, Gto, Mexico.

Successful demonstration of SRF after intravenous injection in the rat has depended upon the pretreatment of the assay animals with estradiol benzoate 50 μg/day and progesterone 25 mg/day for 3 days, a finding discovered incidently in the course of experiments on LRF using the Ramirez-McCann assay (1963). Without steroid sensitization, neither 10 nor 25 porcine hypothalamic equivalents raised plasma GH (Table I); in earlier work from our laboratory, 5 porcine equivalents were without effect (Reichlin and Schalch, 1968) as was 1 equivalent in the work of Garcia and Geschwind (1968). We have not as yet critically determined the relative importance of progesterone and estradiol in sensitizing the GH response, but it seems most likely that the important agent is estradiol since estrogens in the human are known to increase GH responsiveness to exercise (Frantz and Rabkin, 1965) and to arginine infusion (Merimee *et al.*, 1966). Assays were conducted under pentobarbital anesthesia, allowing 30 min for stabilization of values, but this is not a completely satisfactory procedure, because initial values are high, and if above 50 ng/ml, the assay is quite insensitive. Ether anesthesia does ensure an initially low plasma GH, but also apparently inhibits the responsiveness of the preparation to hypothalamic extracts. Methanol (90%) has been used for extraction of the freeze-dried porcine hypothalamic fragments (obtained from Oscar Meyer) because of our previous success in using this solvent for extraction of TRH and LRF as first used by Burgus *et al.* (1967).

The effect of intravenous injection of porcine hypothalamic extract on plasma GH level is shown in Figure 1. GH was determined by radioimmunoassay using NIH Pituitary Distri-

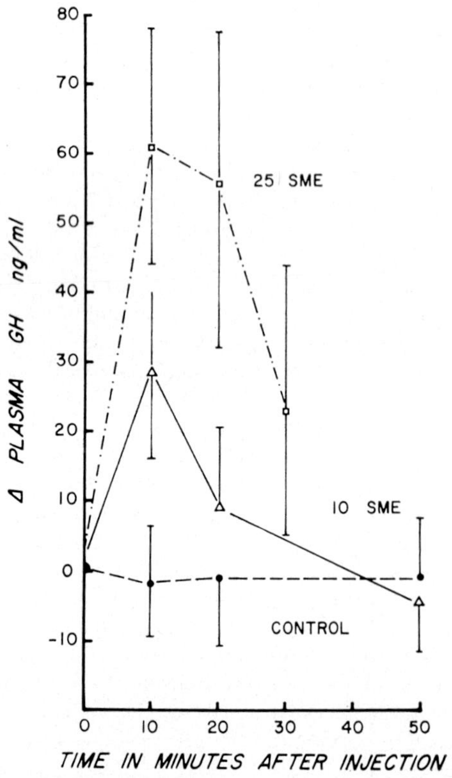

Fig. 1. Change in plasma radioimmunoassayable growth hormone concentration in estrogen-progesterone-pretreated rats following injection of methanol extract of porcine hypothalamic fragments (6 rats per group, 11 in control group).

bution Agency RGH, rhesus anti-porcine GH antibody prepared in our laboratory and the coated charcoal separation method. Plasma levels rise promptly, being maximal within 10 min of injection and then fall rapidly. The shape of the response curve is similar to that seen after electrical stimulation of the hypothalamus by Frohman *et al.* (1968) and confirmed by Martin and Reichlin (1971). Larger amounts of extract cause larger increases in plasma GH. The injected material itself is free of GH activity, an important control observation, since porcine GH is immunologically cross-reactive with RGH. Among the known constituents of hypothalamic extract are oxytocin, vasopressin and TRH, and it was important to determine whether the SRF effects observed were due to these substances. Oxytocin has not previously been reported to alter GH secretion but vasopressin is known to increase plasma GH in man (Brostoff *et al.*, 1968) and in monkeys (Knobil *et al.*, 1968), and has been

TABLE I

Plasma radioimmunoassayable GH levels ten minutes after the intravenous injection of extracts of porcine stalk median eminence tissue, oxytocin, vasopressin and TRH in rats

				Plasma GH (ng/ml)						
Anes- thesia	Steroid	Test material	No. of animals	Initial (mean ± S.E.)		Final (mean ± S.E.)		Increment (mean ± S.E.)		p value
Pento- barbital	E + P	Saline	11	26.8	3.7	26.2	7.9	–1.6	7.9	n.s.
		5 SME	5	23.5	6.7	19.9	4.6	–3.6	3.4	n.s.
		10 SME	6	21.3	2.9	49.5	10.5	28.5	12.3	<0.05
		25 SME	6	24.8	5.6	84.1	14.5	61.9	17.7	<0.005
	0	10 SME	7	17.5	5.6	20.9	8.1	–13.3	6.9	n.s.
		25 SME	7	27.9	6.9	28.0	7.9	0.2	2.5	n.s.
	E + P	Vasopressin 5 U.	5	41.2	3.9	126.9	20.7	85.7	19.5	<0.005
		Oxytocin 2 U.	6	24.2	3.9	57.5	12.8	33.1	11.1	<0.02
		TRH 100 ng	3	36.3	13.5	21.7	6.8	–14.7	12.7	n.s.

E = estradiol benzoate 50 μg daily for 3 days.
P = progesterone 25 mg daily for 3 days.
n.s. = not significant.

linked to GH regulation in the rat by several workers (*cf.* Pecile and Müller, 1966). In our laboratory even 5 U. had previously been shown to have no effect on plasma GH in the rat (Reichlin and Schalch, 1968), but the experiment had not been carried out in steroid 'sensitized' animals. TRH has been reported to have inconsistent and delayed effects on GH secretion in man (Anderson *et al.*, 1971), but no data were available on TRH effects on GH secretion in the rat. We found that oxytocin 2 U. and vasopressin 5 U. were indeed capable of releasing GH in the steroid sensitized rat (Table I), but that TRH 100 ng was without effect. Critical dose response studies of the neurohypophysial hormones have not as yet been carried out but it is known from immunoassay studies that a single porcine hypothalamic fragment contains approximately 3.5 mU. of lysine vasopressin (courtesy of Dr. Myron Miller), indicating that 10 hypothalamic fragments contain 35 mU.

Since vasopressin and oxytocin (at least in large doses) can cause the release of GH, we next separated extracts by Sephadex G-10 chromatography using 0.1 N acetic acid in 5 ml

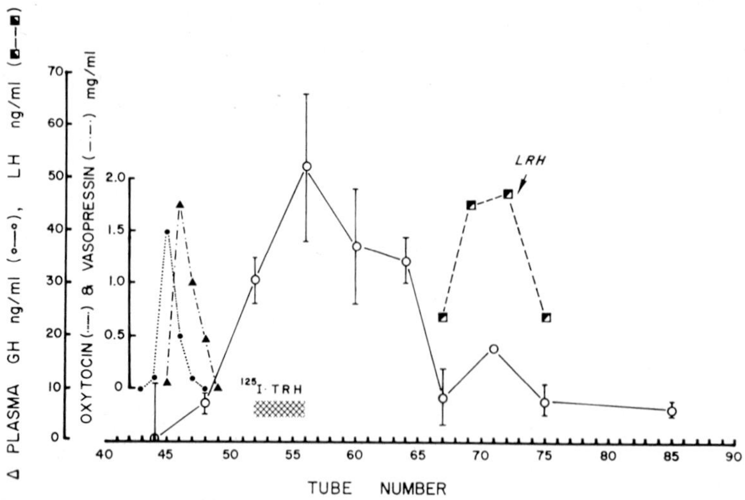

Fig. 2. Illustrated here is the distribution on a Sephadex G-10 column of biological SRF activity, oxytocin and vasopressin (the latter two estimated by Lowrey protein measurements), radioactive TRH, and biologically active LRF. Column was 2.5 × 100 cm, eluent was 0.1 N acetic acid, 5 ml samples were used. Samples were prepared for injection by lyophilization and solution in 4 ml of normal saline to obtain 4 doses of 1 ml for each fraction tested. Alternate fractions were tested. Vasopressin and oxytocin positions were obtained in separately run columns after addition of 0.49 mg and 6 mg respectively.

fractions on a column 2.5 by 100 cm. The SRF activity was separate from oxytocin, vasopressin and LRF, but overlapped with TRF (Fig. 2). Since TRF does not release GH, it can be inferred that the SRF activity is distinct from the associated TRF activity.

SRF activity can also be separated from vasopressin on Silica Gel thin layer chromatography using chloroform-methanol-ammonia (60:40:20). In work done in collaboration with Dr. Carlos Valverde-R., Dr. Angela Cabeza and Miss M. Mitnick, it has been found that the SRF activity travels in a zone with an rF value between 0.3 and 0.6, whereas vasopressin has an rF of 0.17. Oxytocin has an rF value of 0.55.

Much early work on SRF effect has utilized incubation preparations. To determine whether the material active in releasing GH *in vivo* was effective *in vitro*, the contents of fraction 59 (from the column illustrated in Fig. 2 which represented the separation of 1000 hypothalamic fragments) was lyophilized, dissolved in Krebs-Ringer bicarbonate buffer solution and incubated with halves of normal male rat pituitary tissue, paired with control incubates after a one half hour preincubation in buffer alone. The extract was used with 4 half pituitaries, and was incubated at 37° with shaking under 95% O_2 5% CO_2. After one half hour, control media contained 3.0 ± 0.2 μg/ml of GH and the extract treated 7.3 ± 1.1 μg/ml. The stimulating effect was significant statistically ($p < .005$).

We next examined the effect of extracts active in raising plasma radioimmunoassayable GH on pituitary GH concentration, an experiment of interest because of the extensive work on pituitary depletion assay of SRF. In two series of experiments (Table II), one done under pentobarbital anesthesia and the other under ether anesthesia, injections of hypothalamic extract which were effective by plasma assay did not cause a depletion of pituitary GH concentration. In fact, in each experiment, the mean GH concentration of extract treated animals was slightly higher than control, but the differences were not significant by statistical test. This observation accords well with the theoretical calculations of Frohman and Bernardis (1970), who, on the basis of GH secretory turnover studies predicted that

TABLE II

Plasma and pituitary radioimmunoassayable GH concentration after intravenous injection of methanol extract of stalk median eminence tissue

Anes-thesia	Steroid	Test material	No. of animals	Plasma GH (ng/ml) Initial (mean ± S.E.)		Final (mean ± S.E.)		Increment (mean ± S.E.)		Pituitary GH μg/mg (mean ± S.E.)	
Pento-barbital	E — P	Control	4	30.1	11.2	33.1	7.2	3.0	14.3	19.5	1.5
		25 SME	6	24.8	5.6	84.1	14.5	*61.9*	17.7	22.6	2.7
Ether	E — P	Control	8	18.8	6.9	17.9	7.9	−0.9	9.4	21.9	3.0
		10 SME	4	18.0	7.5	34.9	14.3	16.8	12.2	21.1	2.3
		20 SME	3	13.8	3.9	44.2	18.1	*30.3*	9.9	28.8	4.3

Italicized values significantly different from control ($p < 0.05$).

significant elevations of plasma GH would occur in association with pituitary depletions of less than 2%, and with the demonstration by Schalch and Reichlin (1968) that the intravenous injection of one fourth of a rat pituitary (equivalent to approximately 30 μg of GH) led to an increase in plasma GH concentration of 980 ng/ml at 10 min. On the basis of the latter observation, it would be anticipated that a rise in plasma GH similar to that reported here (61.9 ng/ml) would lead to a depletion of less than 2% of gland GH content (based on the assumption of a gland weight of 6 mg and concentration of 20 μg/mg wet weight). These findings have considerable bearing on the interpretation of the pituitary GH depletion assay as it has been used for the evaluation of SRF in the rat, and the evaluation of GH secretory responses to various physiological and pharmacological procedures. In collaborative studies with Dr. Andrew Schally and colleagues, our group was unable to demonstrate a rise in plasma radioimmunoassayable GH concentration, or a decrease in pituitary radioimmunoassayable GH concentration in animals in which tibial epiphysial plate bioassay had indicated a depletion of GH-like activity (Reichlin and Schalch, 1968). Similar lack of correlation was observed in collaborative studies with Rodger *et al.* (1969) who in addition point out problems in the statistical evaluation of the depletion method. In an attempt to prove that hypothalamic extracts do indeed discharge GH into the blood, Sawano *et al.* (1968) have measured by bioassay plasma growth stimulating activity and show effects equivalent to 2.5 μg/ml. The blood containing growth stimulating activity does not have increased immunoassayable GH (Reichlin and Schalch, 1968), despite the fact that the method used would permit the detection of as little as 5 ng/ml which is but 1/500 of the concentration claimed.

Our studies do not explain the discrepancy between bioimmunoassayable and radioimmunoassayable GH concentration, but are entirely compatible with all previously reported studies on electrical stimulation and stress, which have used GH immunoassay. Consideration must be given to the possibility that prolactin and not growth hormone is the growth stimulating material which accounts for the observed discrepancies between bio- and radioimmunoassayable GH. In the rat, ether stress increases prolactin release (Wuttke and Meites, 1970), while inhibiting GH release (Schalch and Reichlin, 1968) and hypothalamic extracts discharge prolactin into the blood as inferred from changes in plasma prolactin (Valverde-R. and Chieffo, 1971).

We would interpret the results of the present experiment to mean that when an authentic growth hormone releasing factor is identified, it will cause a discharge of immunoassayable

GH into the blood, with or without an associated depletion of immunoassayable GH content of the pituitary.

ACKNOWLEDGEMENT

We are grateful to Mrs. Marcia Van Camp for help with the bioassay and to Miss Judy Bollinger for the immunoassays.

REFERENCES

ANDERSON, M. S., BOWERS, C. Y., KASTIN, A. J., SCHALCH, D. S., SCHALLY, A. V., UTIGER, R. D., SNYDER, P. J., WILBER, J. F. and WISE, A. J. (1971): Synthetic thyrotropin releasing hormone (TRH): A specific and potent stimulator of thyrotropin (TRH) release in man. *Clin. Res., 19*, 366 (Abstract).

BLACKARD, W. G. and HEIDINGSFELDER, S. A. (1968): Adrenergic receptor control mechanism for growth secretion. *J. clin. Invest., 47*, 1407.

BROSTOFF, J., JAMES, V. H. T. and LANDON, J. (1968): Plasma corticosteroid and growth hormone response to lysine vasopressin in man. *J. clin. Endocr., 28*, 511.

BURGUS, R., AMOSS, M. S. and GUILLEMIN, R. (1967): Solubility of the hypothalamic hormones TSH-releasing factor (TRF) and LH-releasing factor (LRF) in organic and alcoholic solvents. *Experientia (Basel), 23*, 417.

DAUGHADAY, W. H., PEAKE, G. T. and MACHLIN, L. J. (1970): Assay of the growth hormone releasing factor. In: *Hypophysiotropic Hormones of the Hypothalamus: Assay and Chemistry*, pp. 151–170. Editor: J. Meites. Williams and Wilkins, Baltimore.

FRANTZ, A. G. and RABKIN, M. T. (1965): Effects of estrogen and sex differences on secretion of human growth hormone. *J. clin. Endocr., 25*, 1470.

FROHMAN, L. A. and BERNARDIS, L. L. (1970): Growth hormone secretion in the rat: Metabolic clearance and secretion rates. *Endocrinology, 86*, 305.

FROHMAN, L. A., BERNARDIS, L. L. and KANT, K. J. (1968): Effect of hypothalamic stimulation on pituitary and plasma growth hormone levels in rats. *Science, 162*, 580.

GARCIA, J. F. and GESCHWIND, I. I. (1968): Investigation of growth hormone secretion in selected mammalian species. In: *Growth Hormone*, pp. 267–291. Editors: A. Pecile and E. E. Müller. ICS 158, Excerpta Medica, Amsterdam.

KNOBIL, E., MEYER, V. and SCHALLY, A. V. (1968): Hypothalamic extracts and the secretion of growth hormone in the Rhesus monkey. In: *Growth Hormone*, pp. 226–237. Editors: A. Pecile and E. E. Müller. ICS 158, Excerpta Medica, Amsterdam.

KRULICH, L., DHARIWAL, A. P. S. and MCCANN, S. M. (1968): Stimulatory and inhibitory effects of purified hypothalamic extracts on growth hormone release from rat pituitary *in vitro*. *Endocrinology, 83*, 783.

MACHLIN, L. J., TAKAHASHI, Y., HORINO, M., HERTELENDY, F., GORDON, R. S. and KIPNIS, D. (1968): Regulation of growth hormone secretion in non-primate species. In: *Growth Hormone*, pp. 292–305. Editors: A. Pecile and E. E. Müller. ICS 158, Excerpta Medica, Amsterdam.

MALACARA, J. M. and REICHLIN, S. (1971): Elevation of plasma radioimmunoassayable growth hormone in the rat induced by porcine hypothalamic extract. *Fed. Proc., 30*, 198 (Abstract).

MARTIN, J. and REICHLIN, S. (1971): Anatomical specificity of hypothalamic sites for the release of thyrotropin and growth hormone. *Program 23rd Meeting of the American Academy of Neurology*, p. 112 (Abstract).

MERIMEE, T. J., BURGESS, J. A. and RABINOWITZ, D. (1966): Sex-determined variation in serum insulin and growth hormone response to amino acid stimulation. *J. clin. Endocr., 25*, 1470.

MÜLLER, E. E., ARIMURA, A., SAWANO, S., SAITO, T. and SCHALLY, A. V. (1967*a*): Growth hormone-releasing activity in the hypothalamus and plasma of rats subjected to stress. *Proc. Soc. exp. Biol. Med. (N.Y.), 125*, 874.

MÜLLER, E. E., SAITO, T., ARIMURA, A. and SCHALLY, A. V. (1967*b*): Hypoglycemia, stress and growth hormone release: blockade of growth hormone release by drugs acting on the central nervous system. *Endocrinology, 80*, 109.

PARKER, D. C., SASSIN, J. F., MACE, J. W., GOTLIN, R. W. and ROSSMAN, L. G. (1969): Human growth hormone release during sleep: electroencephalographic correlation. *J. clin. Endocr.*, *29*, 871.

PECILE, A. and MÜLLER, E. E. (1966): Control of growth hormone secretion. In: *Neuroendocrinology, Vol. I,* pp. 537–564. Editors: L. Martini and W. F. Ganong. Academic Press, New York.

RAMIREZ, V. D. and MCCANN, S. M. (1963): A new sensitive test for LH-releasing activity: the ovariectomized, estrogen progesterone-blocked rat. *Endocrinology*, *73*, 193.

REICHLIN, S. and SCHALCH, D. S. (1968): Growth hormone releasing factor. In: *Progress in Endocrinology*, pp. 584–594. Editor: C. Gual. ICS 184, Excerpta Medica, Amsterdam.

RODGER, N. W., BECK, J. C., BURGUS, R. and GUILLEMIN, R. (1969): Variability of response in the bioassay for a hypothalamic somatotrophin releasing factor based on rat pituitary growth hormone content. *Endocrinology*, *84*, 1373.

SAWANO, S., ARIMURA, A., BOWERS, C. Y., REDDING, T. W. and SCHALLY, A. V. (1968): Pituitary and plasma growth hormone-like activity after administration of highly purified pig growth hormone-releasing factor. *Proc. Soc. exp. Biol. (N.Y.)*, *127*, 1010.

SCHALCH, D. S. and REICHLIN, S. (1968): Stress and growth hormone release. In: *Growth Hormone*, pp. 211–225. Editors: A. Pecile and E. E. Müller. ICS 158, Excerpta Medica, Amsterdam.

WILBER, J. F. and PORTER, J. C. (1970): Thyrotropin and growth hormone releasing activity in hypophysial portal blood. *Endocrinology*, *87*, 807.

WUTTKE, W. and MEITES, J. (1970): Effect of ether and pentobarbital in serum prolactin and LH levels in proestrus. *Proc. Soc. exp. Biol. (N.Y.)*, *135*, 648.

VALVERDE-R., C. and CHIEFFO, V. (1971): Prolactin releasing factor(s) in porcine hypothalamic extracts. *Program 53rd Meeting of the Endocrine Society*, p. 84.

DUAL HYPOTHALAMIC REGULATION OF GROWTH HORMONE SECRETION*

L. KRULICH, P. ILLNER, C. P. FAWCETT, M. QUIJADA and S. M. McCANN

Department of Physiology, University of Texas Southwestern Medical School, Dallas, Tex., U.S.A.

The history of the hypothalamic growth hormone releasing factor (GRF) and growth hormone inhibiting factor (GIF) is not long but already voluminous and full of controversy (for recent reviews see Burgus and Guillemin, 1970; Bala, 1970; McCann and Porter, 1969; McCann *et al.*, 1968).

The first tentative evidence for the presence of GRF in hypothalamic extracts was reported by Franz *et al.* (1962). Later and almost simultaneously, workers from several laboratories published results documenting the presence of GRF in crude hypothalamic extracts by demonstrating either augmentation of GH release and production in pituitary tissue cultures (Deuben and Meites, 1964) or depletion of GH from pituitaries *in vivo* (Krulich *et al.*, 1965; Müller and Pecile, 1965; Ishida *et al.*, 1965), or stimulation of release of GH from glands incubated *in vitro* (Schally *et al.*, 1966). Soon thereafter purification of GRF was achieved. Dhariwal *et al.* (1965) obtained a highly active preparation by gel filtration of the crude extract followed by chromatography on carboxymethylcellulose. Schally *et al.* (1968, 1969) carried the purification further arriving at approximately the same estimation of the molecular weight of GRF. Dhariwal *et al.* (1965) suggested that GRF might be a basic peptide on the basis of its retention on CMC, but Schally *et al.* (1968) concluded that it was acidic after electrophoresis.

GRF was detected in blood of hypophysectomized rats during insulin-induced hypoglycemia (Krulich and McCann, 1966*c*) and in blood of long-term hypophysectomized rats (Müller *et al.*, 1967*a*); Schally *et al.* (1968) even claimed an increase in GH in plasma of rats injected with hypothalamic extracts.

In pursuing the work with purified hypothalamic extracts, Krulich *et al.* (1968) presented evidence for the existence of a growth hormone inhibiting factor (GIF), which has been shown to decrease the release of GH from pituitaries *in vitro*. This factor was then further purified (Dhariwal *et al.*, 1969) and was shown to counteract the action of GRF on both synthesis and release of GH *in vitro* (Krulich and McCann, 1969).

In all the work thus far mentioned the evaluation of GRF or GIF activities was based on changes in GH release from pituitaries *in vitro* or depletion of GH from pituitaries *in vivo* as criteria of the respective hypothalamic activities and bioassay for GH (tibia test) to measure the GH changes in the incubation media or in the pituitaries of the test animals.

This technique did not find unanimous approval and the validity of the above cited results was questioned on several grounds. In the hands of Rodger *et al.* (1969) the pituitary depletion method as a test for GRF activity proved to be so variable as to preclude its use for this purpose; however, it should be noted that these authors did not attempt to repeat

* Supported by NIH Grant AM 10073 and Robert A. Welch Foundation.

the earlier work but instead introduced alterations in the procedures at almost every step. Daughaday *et al.* (1968, 1969) and Schalch and Reichlin (1968) using radioimmunoassay for detection of GH were unable to find any depletion of rat pituitary GH following an injection of hypothalamic extract or during insulin-induced hypoglycemia or exposure to cold, both of which were shown to deplete pituitary GH by bioassay (Krulich and McCann, 1966*b*; Müller *et al.*, 1967). Schalch and Reichlin (1968) as well as Garcia and Geschwind (1968) also failed to observe any changes in the radioimmunoassayable GH in plasma of rats injected with hypothalamic extracts, although similar extracts were shown to increase plasma GH concentrations effectively in monkeys (Smith *et al.*, 1968) and sheep (Machlin *et al.*, 1967).

The failure to increase the concentration of circulating GH in the rat by injection of hypothalamic extracts was probably due to insufficient dosage, because recently Malacara and Reichlin (1971) observed a significant increase of plasma GH after large doses of hypothalamic extracts in spayed rats sensitized by pretreatment with estrogen and progesterone, and Frohman *et al.* (1971) noted similar effects in normal animals if hypothalamic extract purified according to the technique of Dhariwal *et al.* (1965) was introduced directly into the pituitary gland. The discrepancy between the findings based on bioassay and immunoassay of GH in measuring the depletion of pituitary GH was confirmed in a direct comparative study by Müller *et al.* (1969). This question still awaits its final solution.

In view of all these problems, we resumed the studies on GRF and GIF in purified hypothalamic fractions when radioimmunoassay for rat GH became available to us as well as a more sophisticated fractionation procedure. At the same time an attempt was made to localize both activities in the hypothalamus of the rat. The results of these studies are given below.

Purification of GRF and GIF

We have been searching for the factors which control GH secretion in the fractions obtained during the procedure which has been established for the isolation of the gonadotropin-releasing factors.

The early stages of this procedure are shown in Table I. Sheep ME tissue was first lyophilized. Then, the lyophilized material was extracted with acetone followed by a me-

TABLE I

1. Sheep ME tissue	*a.* lyophilized
	b. extracted with acetone
	c. extracted with methanol-chloroform (2/1)
2. Residue homogenized in 2 N acetic acid, heated 8 min at 100°C	
3. Ultrafiltration	
	a. through Diaflo UM-10 membrane
	b. through Diaflo UM-2 membrane
	c. each residue washed exhaustively with 0.5 N acetic acid

```
          |    ↓    |
UM-10     |---------|
          |    ↓    |→ UR₂
UM-2      |---------|
          |    ↓    |
              UF₂
```

4. Gel-filtration, on Sephadex G-15 in 0.5 N acetic acid, of samples UF_2 and UR_2 (separately)

thanol/chloroform mixture (2/1). The residue was extracted with 2 N acetic acid, heated briefly, and the supernate was filtered through two Diaflo membranes, first through grade UM-10 and then through UM-2. These membranes permit the filtration of solutes of sizes corresponding to molecular weights of approximately 10,000 and 1,000 respectively.

The fraction which passed through the UM-10 membrane but not the finer UM-2 membrane after exhaustive washing is known as UR_2, while the fraction which has passed through both membranes is known as UF_2.

Each of these (UR_2 and UF_2) was further purified by gel-filtration on Sephadex G-15 (in 0.5 N acetic acid). UR_2 was treated directly while UF_2 was first lyophilized and only the material which was soluble in 90% ethanol/water was placed on the gel-filtration column.

The GH content of media from incubations of the ultrafiltration samples and their respective gel-filtration fractions are shown in Table II. These were obtained during routine screening for LRF etc.; that is, the incubations were not performed exclusively for GH measurement. Four pituitary halves incubated for 3 hours were used.

TABLE II

	Ultrafiltration sample (SME equivalents)	GH content of medium (μg/ml)
UR_2	6	4.4
	12	4.6
	Control	15.4
	—	—
UF_2	10	115
	20	120
	Control	72
	—	
	12	52
	Control	30
	—	—

The material (UR_2) which is able to pass through the first membrane (UM-10) causes a marked lowering of the basal release of GH, while that which filters through both membranes (UF_2) produces a stimulation of GH release. In neither case was a satisfactory dose-response relationship observed.

Figure 1 shows the control of GH release by the column fractions obtained by gel-filtration of UR_2 and UF_2. As anticipated from the results in Table II, regions of inhibition and stimulation were clearly observed and in agreement with the membrane filtration properties, the inhibitory activity appears to be associated with a larger molecule than the stimulatory substance, since the former (GIF) is eluted from the column soon after the void volume while the latter (GRF) has an elution position similar to TRF and LRF. In fact, we are currently undertaking the further purification of GRF by the methods developed for LRF. It would appear that alternative methods will have to be applied to the isolation of GIF, since GIF may be a substance having a molecular weight in the range from 2,000-4,000.

Assays conducted on the pooled active fractions from each of these columns gave the results shown in Table III. The values are means from four separate incubations of paired hemipituitaries in which the test sample was added to only one half of each pair. The pooled GIF sample caused up to 50% inhibition while the pooled GRF showed a 2.5 fold stimulation.

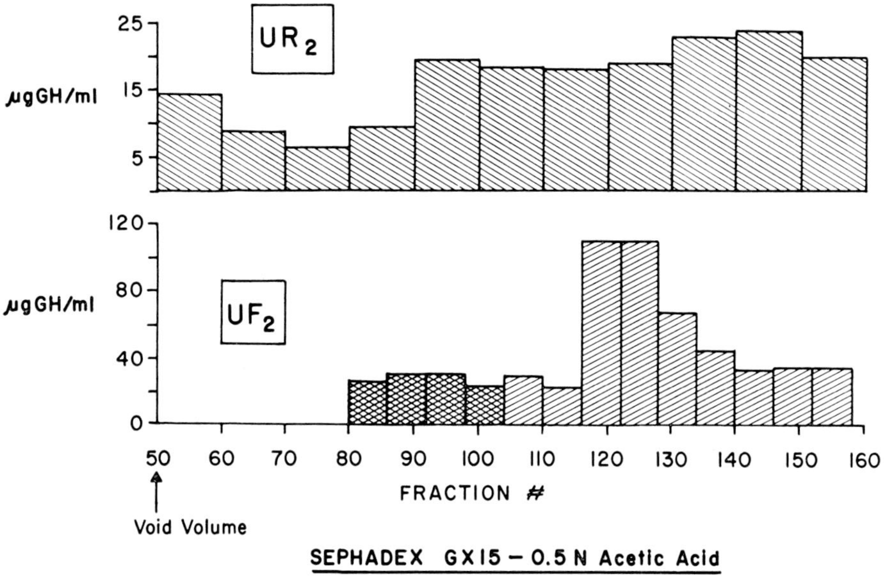

Fig. 1. Control of GH release by fractions obtained from gel-filtration of ultrafiltration samples.

TABLE III

	UR_2 (tubes 60–80)			
Dose (ME/flask)	Control	6	24	120
GH (µg/ml)	9.36 ± 0.8	7.61 ± 1.1	5.62 ± 0.7	5.26 ± 1.0
	UF_2 (tubes 115–124)			
Dose (ME/flask)	Control	7	17.5	35
GH (µg/ml)	6.74 ± 0.45	13.1 ± 1.8	16.53 ± 1.9	12.54 ± 0.5

In summary, we feel confident not only of the existence of the two factors controlling GH secretion since both the assays and isolation procedures use methods quite different from those used in the earlier work, but also of their eventual isolation and characterization.

Localization of GRF and GIF in the hypothalamus

The localization studies were performed as follows: Normal adult male rats were killed by decapitation and the brains were quickly removed and frozen on dry ice. The hypothalami were then sectioned at 400 µ in a cryostat along the frontal, horizontal and sagittal planes. Appropriate histological controls were taken at the same time. In some experiments, sections 600 µ in thickness were obtained, and also in some experiments sagittal and horizontal sections were prepared from the isolated anterior portion of the hypothalamus containing anterior hypothalamus and preoptic area or from the isolated posterior part containing middle and posterior hypothalamus. The individual sections were extracted with 0.1 N HCl. The homogenates were immersed for 10 min in a boiling water bath, centrifuged and the supernates assayed in an *in vitro* system for their respective growth hormone releasing or inhibiting activities. The assay system consisted of six pituitary halves incubated at 37° C in a Dubnoff shaker for 3 hr in 2 ml of tissue culture medium 199. The period of incubation was preceded by 30 min of preincubation. Cortical extracts prepared in the same manner served as controls.

The change in GH release into the media in comparison to the control was taken as the criterion of the respective GH releasing and inhibiting activities in the original sections.

The results obtained with frontal and horizontal sections are shown in Figure 2. The curve below the parasagittal diagram of hypothalamic structures corresponds to the results from frontal sections and the curve to the left to the results from horizontal sections. The changes are expressed as % of the controls, the vertical bars are standard errors of the mean. The millimeter scale gives the stereotaxic coordinates according to Konig and Klippel's rat brain atlas (1963). The average number of sections assayed was 15 and 10 for each frontal level and for each horizontal level, respectively.

In the frontal sections there appears to be a small but significant decrease in GH release in a narrow region corresponding to the stalk and posterior part of the median eminence. The inhibition is again apparent in the anterior hypothalamus. Between these zones of inhibitory activity, a broader zone of increased release is located, with the peak at the level of the ventromedial nucleus.

On the horizontal sections only the releasing activity is apparent again corresponding to the position of the ventromedial nucleus.

Figure 3 shows analogous results obtained with the sagittal sections. In this case the

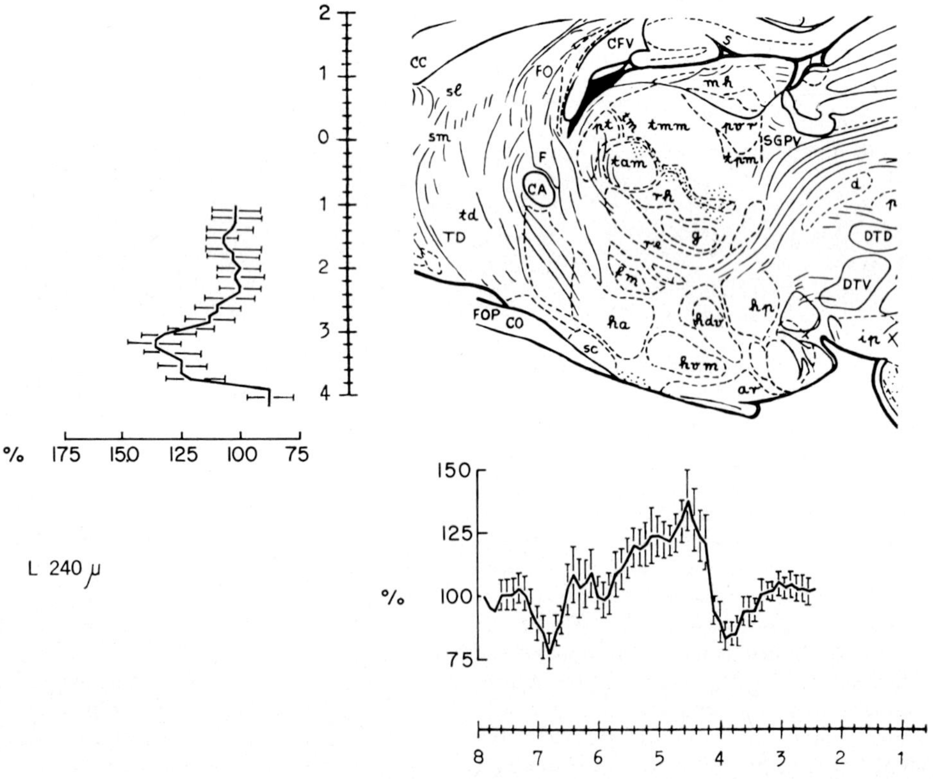

Fig. 2. GRF-GIF distribution.

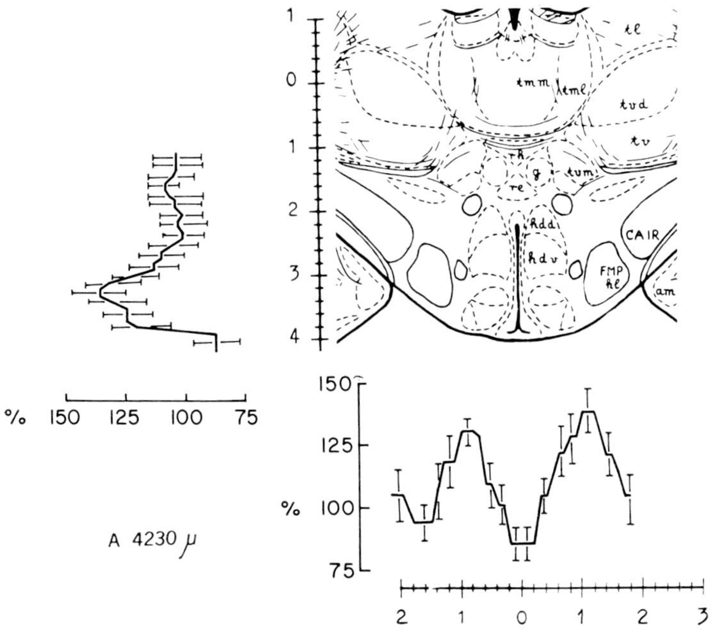

Fig. 3. GRF-GIF distribution.

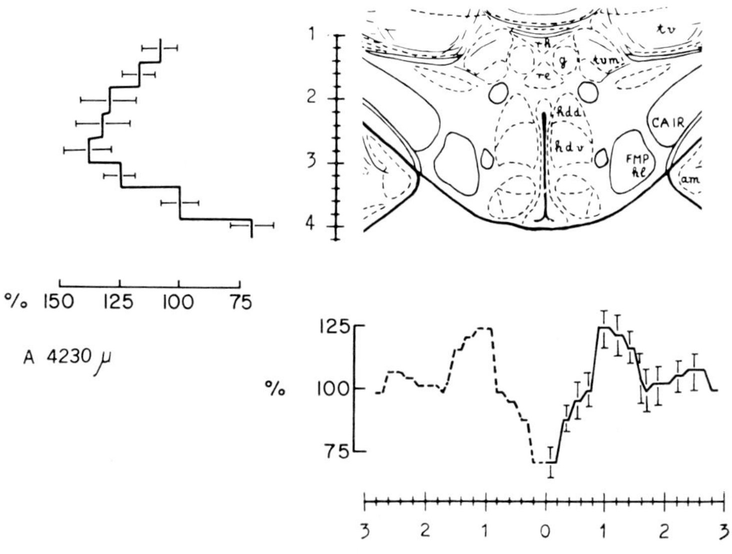

Fig. 4. GRF-GIF caudal hypothalamus.

hypothalami were sectioned from side to side so that both corresponding halves were assayed simultaneously. A fairly symmetrical figure was obtained, showing two small but significant loci of releasing activity just touching the lateral border of the ventromedial nucleus. The decreased release of GH from sections corresponding to the median eminence was not significant.

In order to obtain a better spatial resolution in this obviously complex system, sagittal and horizontal sections were obtained separately from the isolated posterior or anterior portions of the hypothalamus. Figure 4 presents data obtained from the posterior portion. Horizontal sections were in this case 600 μ thick (curve to the left), the sagittal ones were cut at 400 μ as in the previous experiments. Sagittal sections were cut laterally starting from the midline, so that only the right part of the curve indicates the changes actually obtained. The dotted curve is a mirror image. In this set of experiments a significant inhibitory activity appeared in both the horizontal as well as sagittal sections which corresponded to the region of the median eminence. The releasing activity in the sagittal sections is located in the same plane as in the previous experiment. In the horizontal sections the releasing activity starts at the same level as in the first experiment but seemed to extend further dorsally in this case.

Analogous sections, that is sagittal and horizontal through the anterior portion of the hypothalamus, failed to reveal any GH releasing or inhibiting activity.

For comparison, results on TSH release measured simultaneously with GH release are shown in Figure 5. As can be seen, very high TRF activity is apparent not only in the median

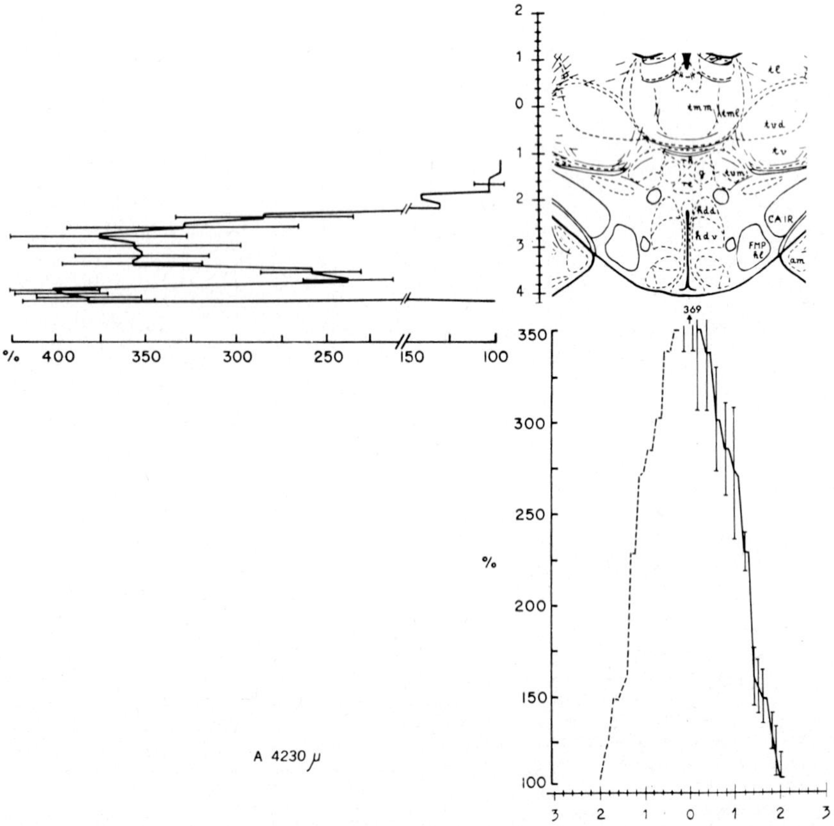

Fig. 5. TRF caudal hypothalamus.

eminence region but also more dorsally in an area roughly corresponding to the dorsomedial nucleus. These results show not only that the TRF distribution in the middle hypothalamus differs from that for GRF, but also that the small changes in GH release were not due to a faulty assay system but to weak GRF and GIF activity present in the hypothalamic extracts or to low sensitivity of the test pituitaries or to both.

To study further the apparent presence of inhibitory activity in the median eminence, two additional experiments were performed. In the first one extracts were prepared either from the whole hypothalamus, in which the block included median eminence and parts of the medial, anterior and posterior hypothalamus, or from an isolated median eminence (the most ventral 400 μ removed in a horizontal plane from a frozen hypothalamus with a microtome). The results of 10 simultaneous assays are shown in Figure 6. The extracts from the

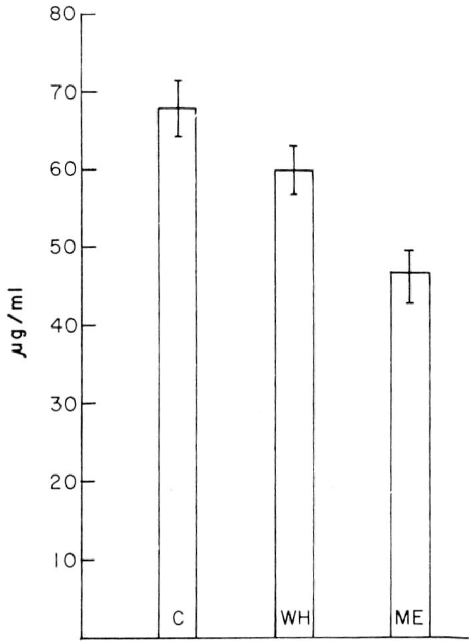

Fig. 6. Effect of crude hypothalamic extract on GH release *in vitro*. C = cortex, WH = whole hypothalamus, ME = median eminence.

isolated median eminence had a clear inhibitory effect on GH release in comparison with both the cortical extract and whole hypothalamic extract, which in itself showed a slight inhibition.

In the second experiment the thickness of the horizontal slice which included the median eminence was increased to 600 μ, and the block included not only the whole hypothalamus but the preoptic hypothalamic area as well. The results of four experiments are shown in Figure 7. In a dose corresponding to one hypothalamus both extracts inhibited the release of GH, the extract from isolated median eminence significantly more than the whole hypothalamic extract. At a lower dose of one-fourth of a hypothalamic equivalent only the median eminence extract was inhibitory.

Although the changes in GH release obtained with the hypothalamic sections were by no means dramatic, we think that they show the presence of both GRF and GIF activities clearly enough. Not only were the observed changes statistically significant, but they appeared in corresponding loci in all three planes of sectioning.

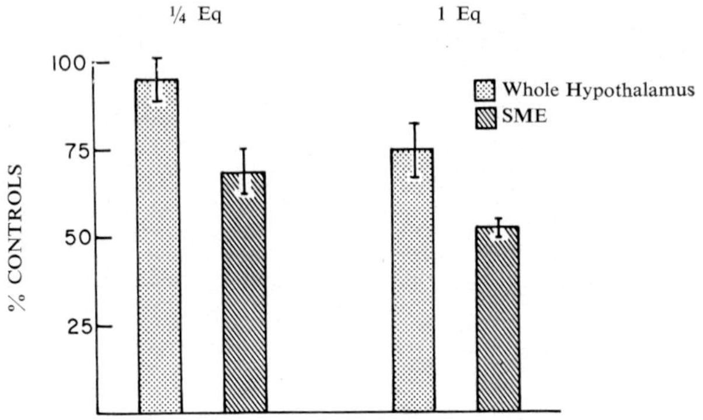

Fig. 7. Effect of crude hypothalamic extract on GH release *in vitro.*

The GRF activity judging from the horizontal and frontal sections appears to be associated with the ventromedial nucleus. On the sagittal sections it touches the lateral border of the ventromedial nucleus and extends then slightly laterally from it. However, the true extent of the GRF activity toward the midline may be obscured by the presence of GIF in the median eminence, because both activities may partly overlap in the sagittal plane.

In general this localization of GRF is in good agreement with other evidence indicative of the importance of the ventromedial nucleus for GRF release. Abrams *et al.* (1966) were able to block the hypoglycemia-induced GH discharge in monkeys by destructive lesions in this area, whereas Frohman and Bernardis (1968) observed an increase in the circulating GH in rats following electrical stimulation of this region. Recently, Toivola and Gale (1971) were able to stimulate GH release in baboons by microinjection of norepinephrine into the region of the ventromedial nucleus.

As far as the localization of GIF is concerned, we obtained evidence for its presence only in the median eminence. But we are nevertheless inclined to believe that the median eminence is not the only area of the hypothalamus which contains GIF. In our own experiments there seemed to be another area with inhibitory activity on the border between the anterior hypothalamus and preoptic area as determined by frontal sections. Our failure to detect this GIF activity in sagittal and horizontal sections through this area may be due to the relative insensitivity of our assay system. This type of distribution would be in accord with the distribution of LRF and FRF (Watanabe and McCann, 1968; Crighton *et al.*, 1969; Quijada *et al.*, 1971) and TRF (Quijada *et al.*, 1971) in which activity is found not only in the median eminence but also in sites distant from it in the hypothalamus and preoptic area.

SUMMARY

Using ultrafiltration on Diaflo membranes followed by gel-filtration on Sephadex, it was possible to separate and purify the growth hormone releasing factor (GRF) and the growth hormone inhibiting factor (GIF) from sheep hypothalamic extracts. GIF appears to be a larger molecular moiety than GRF. Both activities were also found in crude hypothalamic extracts prepared from hypothalamic sections. GRF activity seems to be located in the ventromedial nucleus or its immediate vicinity. GIF activity was found in the median eminence. The presence of GRF or GIF in the hypothalamic fractions and extracts was assessed from their effects on the release of GH (determined by specific radioimmunoassay) from rat pituitaries *in vitro*.

REFERENCES

ABRAMS, R. L., PARKER, M. L., BLANCO, S., REICHLIN, S. and DAUGHADAY, W. H. (1966): Hypothalamic regulation of growth hormone secretion. *Endocrinology*, *78*, 605.

BALA, R. M., BURGUS, R., FERGUSON, K. A., GUILLEMIN, R., KUDO, C. F., OLIVIER, G. C., RODGER, N. W. and BECK, J. C. (1970): Control of growth hormone secretion. In: *The Hypothalamus*, pp. 401–408. Editors: L. Martini, M. Motta and F. Fraschini. Academic Press, New York.

BURGUS, R. and GUILLEMIN, R. (1970): Hypothalamic releasing factors. *Ann. Rev. Biochem.*, *39*, 499.

CRIGHTON, D. B., SCHNEIDER, H. P. G. and MCCANN, S. M. (1969): Localization of LH-releasing factor in the hypothalamus and neurohypophysis as determined by an *in vitro* assay. *Endocrinology*, *87*, 323.

DAUGHADAY, W. H., PEAKE, G. T., BIRGE, C. A. and MARIZ, I. K. (1968): The influence of endocrine factors on the concentration of growth hormone in rat pituitary. In: *Growth Hormone*, pp. 238-252. Editors: A. Pecile and E. E. Müller. ICS 158, Excerpta Medica, Amsterdam.

DAUGHADAY, W. H. PEAKE, G. T. and MACHLIN, L. J. (1969): Assay of the growth hormone releasing factor. In: *Hypophysiotropic Hormones of the Hypothalamus: Assay and Chemistry*, pp. 157–170. Editor: J. Meites. Williams and Wilkins, Baltimore.

DEUBEN, R. R. and MEITES, J. (1964): Stimulation of pituitary growth hormone release by a hypothalamic extract *in vitro*. *Endocrinology*, *74*, 408.

DHARIWAL, A. P. S., KRULICH, L., KATZ, S. H. and MCCANN, S. M. (1965): Purification of growth-hormone releasing factor. *Endocrinology*, *77*, 939.

DHARIWAL, A. P. S., KRULICH, L. and MCCANN, S. M. (1969): Purification of a growth hormone-inhibiting factor (GIF) from sheep hypothalamus. *Neuroendocrinology*, *4*, 282.

FRANZ, J., HASELBACH, C. H. and LIBERT, O. (1962): Studies on the effect of hypothalamic extracts on somatotropic pituitary function. *Acta endocr. (Kbh.)*, *41*, 336.

FROHMAN, L. A. and BERNARDIS, L. L. (1968): Growth hormone secretion in the rat: metabolic clearance and secretion rates. *Endocrinology*, *86*, 305.

FROHMAN, L. A., MORAN, J. W., YATES, F. E. and DHARIWAL, A. P. S. (1971): Growth hormone response to intrapituitary injection of growth hormone releasing factor in the rat as measured by radioimmunoassay. *Fed. Proc.*, *30*, 198 (Abstract).

GARCIA, J. F. and GESCHWIND, I. I. (1968): Investigation of growth hormone secretion in selected mammalian species. In: *Growth Hormone*, pp. 267–291. Editors: A. Pecile and E. E. Müller. ICS 158, Excerpta Medica, Amsterdam.

ISHIDA, Y., KUROSHIMA, A., BOWERS, C. Y. and SCHALLY, A. V. (1965): *In vivo* depletion of pituitary growth hormone by hypothalamic extracts. *Endocrinology*, *77*, 759.

KONIG, T. F. R. and KLIPPEL, R. A. (1963): *The Rat Brain. A Stereotaxic Atlas*. Williams and Wilkins, Baltimore.

KRULICH, L., DHARIWAL, A. P. S. and MCCANN, S. M. (1965): Growth hormone releasing activity of crude ovine hypothalamic extracts. *Proc. Soc. exp. Biol. (N.Y.)*, *120*, 180.

KRULICH, L., DHARIWAL, A. P. S. and MCCANN, S. M. (1968): Stimulatory and inhibitory effects of purified hypothalamic extracts on growth hormone release from the pituitary *in vitro*. *Endocrinology*, *83*, 783.

KRULICH, L. and MCCANN, S. M. (1966*a*): Effect of alterations in blood sugar on pituitary growth hormone content in the rat. *Endocrinology*, *78*, 759.

KRULICH, L. and MCCANN, S. M. (1966*b*): Influence of stress on the growth hormone GH content in the pituitary of the rat. *Proc. Soc. exp. Biol. (N.Y.)*, *122*, 612.

KRULICH, L. and MCCANN, S. M. (1966*c*): Evidence for the presence of growth hormone-releasing factor in blood of hypoglycemic, hypophysectomized rats. *Proc. Soc. exp. Biol. (N.Y.)*, *122*, 668.

KRULICH, L. and MCCANN, S. M. (1969): Effects of GH-releasing factor and GH-inhibiting factor on the release and concentration of GH in pituitaries incubated *in vitro*. *Endocrinology*, *85*, 319.

MACHLIN, L. J., HORINO, M., KIPNIS, D. M., PHILIPS, S. L. and GORDON, R. S. (1967): Stimulation of growth hormone secretion by median eminence extracts in the sheep. *Endocrinology*, *80*, 205.

MALACARA, J. M. and REICHLIN, S. (1971): Elevation of radioimmunoassayable plasma growth hormone (RIA-GH) in the rat by porcine hypothalamic extract. *Fed. Proc.*, *30*, 198 (Abstract).

MCCANN, S. M., DHARIWAL, A. P. S. and PORTER, J. C. (1968): Regulation of the adenohypophysis. *Ann. Rev. Physiol.*, *30*, 589.

MCCANN, S. M. and PORTER, J. C. (1969): Hypothalamic pituitary stimulating and inhibiting hormones. *Physiol. Rev.*, *49*, 240.

MÜLLER, E. E., ARIMURA, A., SAITO, T. and SCHALLY, A. V. (1967*a*): Growth hormone releasing activity in plasma of normal and hypophysectomized rats. *Endocrinology*, *80*, 77.

MÜLLER, E. E., MIEDICO, D., GIUSTINA, G. and PECILE, A. (1969): Growth hormone immunological and biological assay in pituitary of rat under different experimental conditions. *Experientia (Basel)*, *25*, 1146.

MÜLLER, E. E. and PECILE, A. (1965): Growth hormone releasing factor of a guinea pig hypothalamic extract: its activity in guinea pig and rat. *Proc. Soc. exp. Biol. (N.Y.)*, *119*, 1191.

MÜLLER, E. E., PECILE, A. and SMIRNE, S. (1965): Substances present at the hypothalamic level and growth hormone releasing activity. *Endocrinology*, *77*, 390.

MÜLLER, E. E., SAITO, T., ARIMURA, A. and SCHALLY, A.V. (1967*b*): Hypoglycemia, stress and growth hormone release: blockade of growth hormone release by drugs acting on the central nervous system. *Endocrinology*, *80*, 109.

QUIJADA, M., KRULICH, L., FAWCETT, C. P., SUNDBERG, D. K. and MCCANN, S. M. (1971): Localization of thyroid stimulating hormone-releasing factor, luteinizing hormone-releasing factor and follicle-stimulating hormone-releasing factor in rat hypothalamus. *Fed. Proc.*, *30*, 197 (Abstract).

RODGER, N. W., BECK, J. C., BURGUS, R. and GUILLEMIN, R. (1969): Variability of response in the bioassay for a hypothalamic somatotrophin releasing factor based on rat pituitary growth hormone content. *Endocrinology*, *89*, 1373.

SCHALCH, D. S. and REICHLIN, S. (1968): Stress and growth hormone release. In: *Growth Hormone*, pp. 211–225. Editors: A. Pecile and E. E. Müller. ICS 158, Excerpta Medica, Amsterdam.

SCHALLY, A. V., KUROSHIMA, A., ISHIDA, Y., ARIMURA, A., SAITO, T., BOWERS, C. Y. and STEELMAN, S. L. (1966): Purification GH-RF from beef hypothalamus. *Proc. Soc. exp. Biol. (N.Y.)*, *122*, 821.

SCHALLY, A. V., SAWANO, S., ARIMURA, A., BARRET, S. F., WAKABAYASHI, I. and BOWERS, C. Y. (1969): Isolation of growth hormone-releasing hormone (GRH) from porcine hypothalami. *Endocrinology*, *84*, 1493.

SCHALLY, A. V., SAWANO, S., MÜLLER, E. E., ARIMURA, A., BOWERS, C. Y., REDDING, T. W. and STEELMAN, S. E. (1968): Hypothalamic growth hormone-releasing hormone (GRH). Purification and *in vivo* and *in vitro* studies. In: *Growth Hormone*, pp. 185–203. Editors: A. Pecile and E. E. Müller. ICS 158, Excerpta Medica, Amsterdam.

SMITH, G. P., KATZ, S. H., ROOT, A., DHARIWAL, A. P. S., BONGIOVAUNNI, A., EBERLEIN, W. and MCCANN, S. M. (1968): Growth hormone releasing activity in rhesus monkeys of crude ovine stalk median eminence extracts. *Endocrinology*, *83*, 25.

TOIVOLA, P. T. K. and GALE, C. C. (1971): Growth hormone release by microinjections of norepinephrine into hypothalamus of conscious baboons. *Fed. Proc.*, *30*, 197.

WATANABE, S. and MCCANN, S. M. (1968): Localization of FSH-releasing factor in the hypothalamus and neurohypophysis as determined by *in vitro* assay. *Endocrinology*, *82*, 664.

EFFECT OF PROSTAGLANDINS AND CYCLIC 3′5′-ADENOSINE MONOPHOSPHATE (AMP) ON THE SYNTHESIS AND RELEASE OF GROWTH HORMONE*

ROBERT M. MACLEOD** and JOYCE E. LEHMEYER

Department of Internal Medicine, University of Virginia School of Medicine, Charlottesville, Va., U.S.A.

Hormonal secretions of the anterior pituitary gland are thought to be regulated by agents released by the hypothalamus into the hypophyseal portal blood vessels leading to the gland (McCann and Porter, 1969). Very little is known about the mechanisms by which these agents, or neurohormones, effect the release of pituitary hormones, or about the factors that stimulate synthesis of these hormones, including growth hormone. The rapidity with which the neurohormones stimulate growth hormone release suggests that initially, stored hormone is discharged from the cell, and that renewed biosynthesis of the hormone is a secondary event.

When hormone-secreting pituitary tumors are transplanted into rats, the growth hormone or prolactin secreted by the tumor causes atrophy of the host's pituitary gland and decreases the concentration of the hypophyseal hormones within the gland, presumably through a short-loop feedback mechanism (MacLeod *et al.*, 1966). Since many hormonal mechanisms involve cyclic AMP, work was initiated to determine if growth hormone synthesis and release are influenced by the cyclic nucleotide and to determine if the tumor hormones affect the cyclic AMP-adenyl cyclase system in the host's pituitary gland.

RESULTS

The *in vitro* incorporation of radioactive amino acids into pituitary hormones and the separation of the hormones by polyacrylamide gel electrophoresis have been previously described (MacLeod and Abad, 1968). Pituitary glands were incubated with radioactive leucine in Tissue Culture Medium 199 for various periods of time, at 37°C and gassed with 95% O_2–5% CO_2. At the termination of the incubation, aliquots of the incubation media and of homogenates of the pituitary glands are subjected to gel electrophoresis. This method achieves excellent separation of prolactin and growth hormone and produced very distinct bands. After the amount of hormone in the band has been measured microdensitometrically, the gel is sectioned and the radioactivity in the bands determined by conventional techniques.

When pituitary glands of female rats were incubated with labeled leucine, prolactin and growth hormone were synthesized at comparable rates but the cells of the glands regulated the release of the two hormones quite differently. While prolactin was released quickly into the incubation medium, little being retained by the gland, the newly-synthesized growth hormone accumulated within the tissue and little of the hormone was released.

* This investigation was supported in part by USPHS Grant CA-07535 from the National Cancer Institute.

** USPHS Research Career Development Awardee.

That the incorporation of the radioactive amino acid into the pituitary hormones was truly the result of *de novo* protein synthesis is indicated by the fact that in the presence of puromycin the synthesis of prolactin and growth hormone was almost completely inhibited. Actinomycin D, however, did not effect hormone synthesis, a result which suggests minimal nuclear involvement during the biosynthesis of the hormones.

We have recently published evidence that the *in vitro* addition of prostaglandins to normal rat pituitary glands increased the release of labeled growth hormone from the glands (MacLeod *et al.*, 1970). The data presented in Table I show that 10^{-6} and 10^{-7} M PGE_1 significantly increased the release of labeled growth hormone into the incubation medium. At a concentration of 10^{-8} M, PGE_1 did not stimulate release of the hormone but caused the labeled protein to accumulate within the gland. It is evident from these results that

TABLE I

Effect of prostaglandin PGE_1 on the incorporation of leucine-4,5-³H into growth hormone

	cpm / mg pituitary		
	Pituitary gland	Incubation medium	Total
Control	1381 ± 36	307 ± 47	1688
10^{-6} M PGE_1	1369 ± 97	820 ± 108	2190
10^{-7} M PGE_1	1756 ± 168	623 ± 57	2379
10^{-8} M PGE_1	2199 ± 171	152 ± 33	2352

Pituitary glands were incubated with 10 μC radioactive leucine for 7 hours.

TABLE II

Effect of various prostaglandins on the incorporation of leucine-4,5-³H into growth hormone

Type	Number of determinations	cpm / mg pituitary		Absorbancy units / 10 mg pituitary
		Pituitary gland	Incubation medium	Incubation medium
Control	6	1241 ± 63	385 ± 36	3.1
PGE_1	4	1103 ± 56	556 ± 37[2]	7.9
PGE_2	4	1091 ± 22[1]	557 ± 48[1]	8.0
PGA_1	4	1562 ± 48[2]	370 ± 30	3.5
$PGF_{2\alpha}$	4	1054 ± 49	282 ± 40	3.7

Values are mean ± S.E.M.
[1]$p < 0.05$ [2]$p < 0.01$
Pituitary glands were incubated with radioactive leucine. Prostaglandins were added at a concentration of 10^{-6} M.

PGE_1 initially increases synthesis of growth hormone and, subsequently, increases its release from the pituitary gland.

The effects of other prostaglandins on growth hormone synthesis and release are shown in Table II. While 10^{-6}M PGE_1 or PGE_2 significantly increased labeled growth hormone release, PGA_1 and $PGF_{2\alpha}$ did not effect release. It should be noted, however, that PGA_1 did increase the amount of labeled growth hormone contained within the gland. The absorbancy data confirm the fact that only PGE_1 and PGE_2 stimulated release of growth hormone into the incubation medium.

Since prostaglandins have been shown to influence the cyclic AMP concentration in certain tissues, it was decided to study their effect on adenyl cyclase activity in the pituitary gland. Despite the fact that only prostaglandins of the E series increase growth hormone release, all prostaglandins tested demonstrated an ability to stimulate adenyl cyclase activity in the gland (Table III).

TABLE III

Effect of prostaglandins on adenyl cyclase activity in anterior pituitary gland

	cAMP-^{32}P formed/gland/10 min (% of control)
Control	100
PGE_1	178 ± 12
$PGF_{2\alpha}$	177 ± 29
PGA_1	156 ± 14

The conversion of α-ATP-^{32}P to cyclic AMP-^{32}P by pituitary glands during a 10 minute incubation was studied according to the method of Zor *et al.* (1969).

TABLE IV

Effect of dibutyryl 3′5′-AMP on growth hormone synthesis and release

	cpm / mg pituitary: Pituitary gland	cpm / mg pituitary: Incubation medium
Experiment 1		
Control	2330 ± 146	570 ± 10
5 mM BcAMP	1290 ± 18	2221 ± 110[1]
Experiment 2		
Control	2850 ± 96	780 ± 80
0.1 mM BcAMP	2400 ± 124	1880 ± 200[1]

cpm leucine-4,5-^{3}H incorporated into pituitary gland after 7 hours incubation.

[1] $p < 0.01$

The product of adenyl cyclase activity, cyclic AMP, is important in regulating the release of growth hormone. The *in vitro* addition of 0.1 or 5 mM dibutyryl cyclic AMP to pituitary glands significantly increased the release of labeled hormone into the incubation medium (Table IV). The fact that radioactive growth hormone did not accumulate within the gland is taken as evidence that the nucleotide was acting primarily on release rather than on synthesis of the hormone.

The *in vitro* levels of cyclic AMP in pituitary glands can be increased by both the phosphodiesterase inhibitor theophylline and sodium fluoride which stimulates adenyl cyclase activity. The data presented in Table V show that theophylline significantly increased release of growth hormone into the incubation medium. Although sodium fluoride completely inhibited the incorporation of the labeled amino acid into the hormone, the absorbancy data show that the salt also stimulated release of the protein. These results strongly suggest that the release of growth hormone from the gland is independent of the mechanisms which govern synthesis of the hormone.

TABLE V

Effect of theophylline and NaF on the incorporation of leucine-4,5-^{3}H into growth hormone

	Pituitary gland	Incubation medium	Medium (Absorbancy units/mg pituitary)
Control	3196 ± 265	1362 ± 132	17.6 ± 2.3
Theophylline 10mM	2847 ± 239	2109 ± 103[1]	38.0 ± 0.9[1]
NaF 10mM	76 ± 5[1]	65 ± 6[1]	37.1 ± 1.5[1]

Pituitary glands were incubated with agents for 7 hours.
[1] $p < 0.01$

We have previously reported that the transplantation of hormone-secreting pituitary tumors into rats causes atrophy of the host's pituitary gland and decreases synthesis of pituitary hormones by the gland (MacLeod and Abad, 1968). When the glands of rats bearing the growth hormone-secreting tumors MtTW5 or StW5 were incubated with radioactive leucine it was found that incorporation of the labeled amino acid into growth hormone was significantly reduced. There is good evidence of specificity for this inhibition because tumor 7315a, which secretes prolactin and ACTH but not growth hormone, does not effect growth hormone synthesis in the host's pituitary gland.

Investigations were initiated to determine if the inhibitory action of the tumor hormones on synthesis of pituitary hormones by the gland was regulated by mechanisms involving cyclic AMP. The experiment in Table VI compares adenyl cyclase activity in the pituitary glands of normal animals with that in glands of animals bearing pituitary tumors. One observes that only very small decreases in enzyme activity were produced by the presence of the growth hormone secreting tumors MtTW5 and StW5. The prolactin and ACTH secreting tumor 7315a, however, significantly decreased adenyl cyclase activity in the host's pituitary gland. None of the tumors had a significant effect on cyclic AMP phosphodiesterase activity in the pituitary gland (Table VII). It would appear, then, that the growth hormone secreting tumors do not alter the flux of cyclic AMP in the pituitary gland of the host animal.

When adenyl cyclase activity was measured in the presence of 10 mM sodium fluoride, it was found that enzyme activity in glands of both normal and tumor-bearing animals was

TABLE VI

Effects of pituitary hormone-secreting tumors on adenyl cyclase activity in anterior pituitary gland

Group	Pituitary gland weight (mg)	pmoles cAMP-^{32}P/10 min					
		per mg tissue	% change	per mg protein	% change	per gland	% change
Experiment 1							
Control	7.88	15.6		143.1		123.6	
MtTW5	6.82	15.5	−1	139.8	−2	107.3[1]	−13
Experiment 2							
Control	6.33	13.0		101.1		82.6	
StW5	5.70	13.3	+2	98.1	−3	75.9	−10
7315a	7.87	7.4[2]	−43	60.6[2]	−40	57.5[2]	−30

[1] $p < 0.05$ [2] $p < 0.005$

TABLE VII

cAMP phosphodiesterase activity in anterior pituitary gland

Group	Pituitary gland weight (mg)	PO_4^{-3} formed per mg tissue (nmoles)	% change	PO_4^{-3} formed per gland (nmoles)	% change
Experiment 1					
Control	9.15	49.9 ± 1.5		455 ± 11	
MtTW5	7.43	53.2 ± 2.0	+7	394 ± 25	−13
Experiment 2					
Control	9.35	93.8 ± 2.9		872 ± 16	
StW5	7.11	107.9 ± 1.5[1]	+15	756 ± 50	−13
Experiment 3					
Control	10.74	67.2 ± 0.1		721 ± 4	
7315a	6.98	111.7 ± 1.3[1]	+66	779 ± 0	+ 8

Cyclic phosphodiesterase activity was measured according to the method of Butcher and Sutherland (1962). Pituitary homogenates were incubated for 20 min with cyclic AMP substrate.
[1] $p < .005$

increased 500% (Table VIII). These results indicate that the tumor hormones do not affect either the normal levels of adenyl cyclase activity nor the potential enzyme activity in the host's pituitary gland.

We have observed that the *in vitro* addition of sodium fluoride or theophylline to glands of normal animals stimulated growth hormone release. When glands from tumor-bearing rats were incubated in the presence of theophylline or sodium fluoride, they also released increased amounts of growth hormone into the medium (Table IX).

TABLE VIII

Adenyl cyclase activity in anterior pituitary gland

Group[1]	Pituitary gland weight (mg)	Mean body weight (g)	cAMP-^{32}P accumulated in 10 min					
			per mg tissue		per mg protein		per gland	
			(cpm)	(pmoles)	(cpm)	(pmoles)	(cpm)	(pmoles)
Control	4.99	176	2,850 ± 480	13.0	19,800 ± 3400	92.5	14,000 ± 2100	65.5
Control + NaF[2]			18,200 ± 1250	84.7	126,000 ± 8000	590.2	90,000 ± 2200	421.7
MtTW5	6.04	220	2,330 ± 150	10.4	15,400 ± 300	71.7	14,000 ± 700	65.5
MtTW5 + NaF[2]			15,930 ± 1120	74.3	105,800 ± 7200	495.5	95,600 ± 3800	447.7

[1] Six control rats and six MtTW5-bearing rats were used in this experiment.
[2] 10mM NaF.

TABLE IX

Effect of theophylline and NaF on the in vitro *release of growth hormone by pituitary glands from tumor-bearing rats*

	Absorbancy units/mg pituitary
Control	27.9 ± 4.6
StW5	9.9 ± 2.5
StW5 + 10^{-2} M theophylline	22.7 ± 2.6
StW5 + 10^{-2} M NaF	16.9 ± 1.2

TABLE X

Effect of prostaglandin PGE_1 on the incorporation of leucine-4,5-^{3}H into growth hormone by pituitary glands from tumor-bearing rats

Group	Number of determinations	cpm / mg pituitary	
		Pituitary gland	Incubation medium
MtTW5 control	4	1062 ± 22	151 ± 10
MtTW5 + 10^{-6} M PGE_1	4	1036 ± 40	554 ± 24*
StW5 control	4	511 ± 106	78 ± 16
StW5 + 10^{-6} M PGE_1	4	677 ± 57	153 ± 12*
Non-tumor control	7	1250 ± 83	293 ± 32
7315a control	7	1116 ± 228	262 ± 121
7315a + 10^{-6} M PGE_1	4	878 ± 123	540 ± 132*
7315a + 10^{-6} M PGE_1	4	1095 ± 89	393 ± 28*

* $p < 0.05$.

It has also been shown that prostaglandins increase the release of labeled growth hormone from the pituitary glands of normal rats. The experiments presented in Table X confirm this result with glands of animals bearing pituitary tumors. One observes that the *in vitro* addition of 10^{-6} M PGE_1 to the glands of animals with the MtTW5 or StW5 tumor increased release of labeled hormone. The prostaglandin also stimulated growth hormone release from glands of rats with the non-growth hormone-secreting tumor 7315a.

It was noted earlier that the *in vitro* addition of dibutyryl cyclic AMP to pituitary glands of normal rats increased labeled growth hormone release. When the glands of animals with the pituitary tumors were incubated in the presence of the nucleotide, the release of labeled growth hormone from these glands was also stimulated (Table XI).

TABLE XI

Effect of dibutyryl cAMP on the in vitro *synthesis and release of growth hormone by pituitary glands from tumor-bearing rats*

	cpm / mg pituitary	
	Pituitary	Incubation medium
Control	1168 ± 311	245 ± 13
Control + 10^{-3} M DBcAMP	844 ± 172	491 ± 68
MtTW5	584 ± 72	209 ± 25
MtTW5 + 10^{-3} M DBcAMP	620 ± 97	349 ± 45
Control	578 ± 55	249 ± 18
Control + 5×10^{-3} M DBcAMP + 1×10^{-3} M caffeine	651 ± 47	381 ± 66
MtTW5	383 ± 51	105 ± 7
MtTW5 + 5×10^{-3} M DBcAMP + 1×10^{-3} M caffeine	345 ± 35	170 ± 20

From these results it would appear that the decreased *in vitro* release of growth hormone by glands of animals bearing growth hormone secreting pituitary tumors is not mediated by changes in adenyl cyclase of cyclic AMP phosphodiesterase activities. Furthermore, pituitary glands from normal animals and glands of rats bearing the tumors responded similarly to prostaglandins and dibutyryl cAMP; both compounds stimulated growth hormone release. Hence it is concluded that the tumor hormones effect primarily synthesis rather than release of pituitary gland growth hormone by a mechanism which is independent of cyclic AMP production.

These investigations were extended to include a study of the effects of pituitary tumor hormones on RNA metabolism in the host's pituitary gland. Pituitary glands were incubated *in vitro* in Tissue Culture Medium 199 containing radioactive uridine for 4 hours. The nuclear and microsomal RNA fractions were prepared by the phenol-SDS method described by DiGirolamo *et al.* (1964) and after repetitive precipitation with ethanol, the RNA species were separated by sucrose gradient centrifugation. The data in Figures 1 and 2 show that less RNA is present in the nuclear and microsomal fractions of pituitary glands from rats bearing the pituitary tumors. It is important to note, however, that more radioactivity was incorporated into the RNA of the tumor-bearing rats.

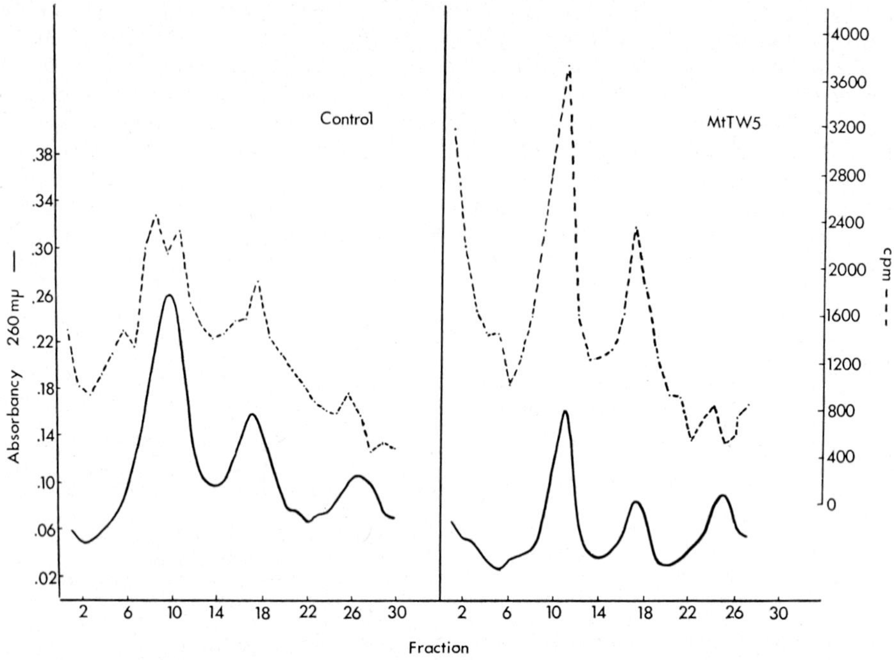

Fig. 1. Sucrose density gradient profiles of pituitary gland nuclear RNA.

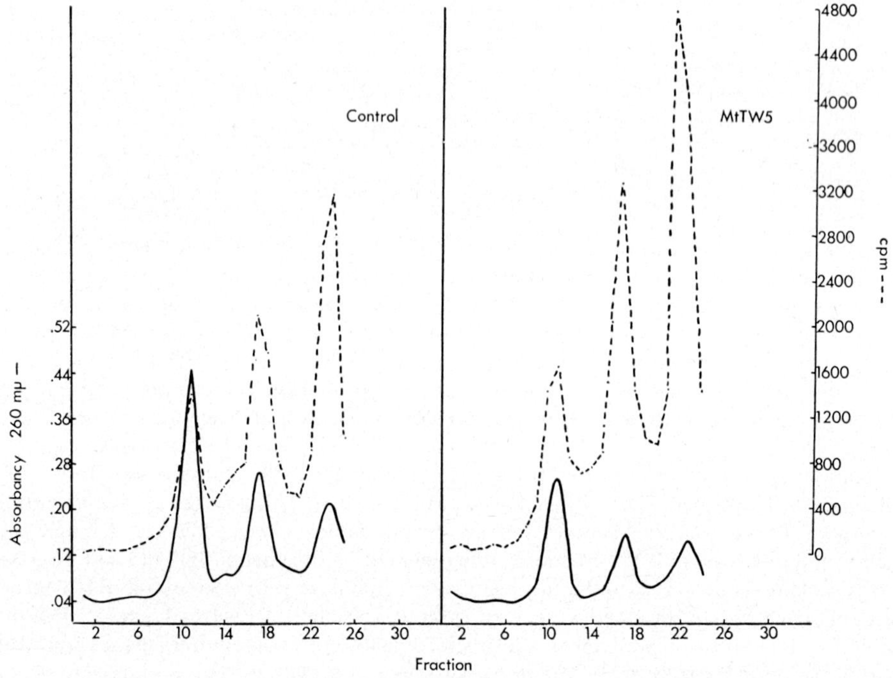

Fig. 2. Sucrose density gradient profiles of pituitary gland microsomal RNA.

TABLE XII

Effect of MtTW5 on pituitary gland RNA

	No. glands	mg	28 S μg RNA	28 S μg RNA / 100 mg pituitary	18 S μg RNA	18 S μg RNA / 100 mg pituitary
Nuclear RNA						
Control	10	106	53	50	30	28
MtTW5	10	83	24	29	16	19
		−22%	−55%	−42%	−47%	−32%
Microsomal RNA						
Control			97	91	61	57
MtTW5			43	52	31	37
			−56%	−43%	−49%	−35%

The tumor hormones, which cause atrophy of the host's pituitary gland and decreased hormone synthesis by the gland, also produced a 50% decrease in the amount of 28 S and 18 S RNA in the nuclear and microsomal fractions of the gland (Table XII). The data in Table XIII show that the specific activity of the labeled pituitary gland RNA was greatly increased by the tumor hormones, in both cellular fractions. The total incorporation of labeled uridine into RNA was more modestly increased. When radioactive guanosine was substituted for uridine, it was again found that RNA of glands of tumor animals incorporated more radioactivity than did controls (Table XIV). Since both purine and pyrimidine precursors were incorporated more rapidly, it is unlikely that the observed effect was due to changes in precursor pool size.

The decreased RNA content of the glands of tumor-bearing animals, coupled with the increased incorporation of RNA precursors, suggests that RNA degradation was also increased. An experiment was designed to study RNA turnover by incubating the pituitary glands as usual with radioactive uridine and then, after rinsing the glands, re-incubating

TABLE XIII

Incorporation of uridine-5-^{3}H into pituitary gland RNA of tumor-bearing rats

	Nuclear RNA 28S	Nuclear RNA 18S	Microsomal RNA 28S	Microsomal RNA 18S
cpm/μg RNA				
Control	94	65%	64	88
MtTW5	163	107	118	162
	+73%	+65%	+85%	+84%
cpm/mg pituitary				
Control	47.0	18.4	58.5	50.6
MtTW5	47.3	20.5	61.1	60.5
	NC	+11%	+ 4%	+20%

TABLE XIV

In vitro incorporation of guanosine-8-^{3}H into pituitary gland RNA of tumor-bearing rats

	Nuclear RNA		Microsomal RNA	
	28S	18S	28S	18S
cpm/μg RNA				
Control	15	18	10	13
MtTW15	32	29	18	20
	+131%	+56%	+80%	+54%
cpm/mg pituitary				
Control	77	45	51	42
MtTW15	76	44	74	71
	NC	NC	+45%	+69%

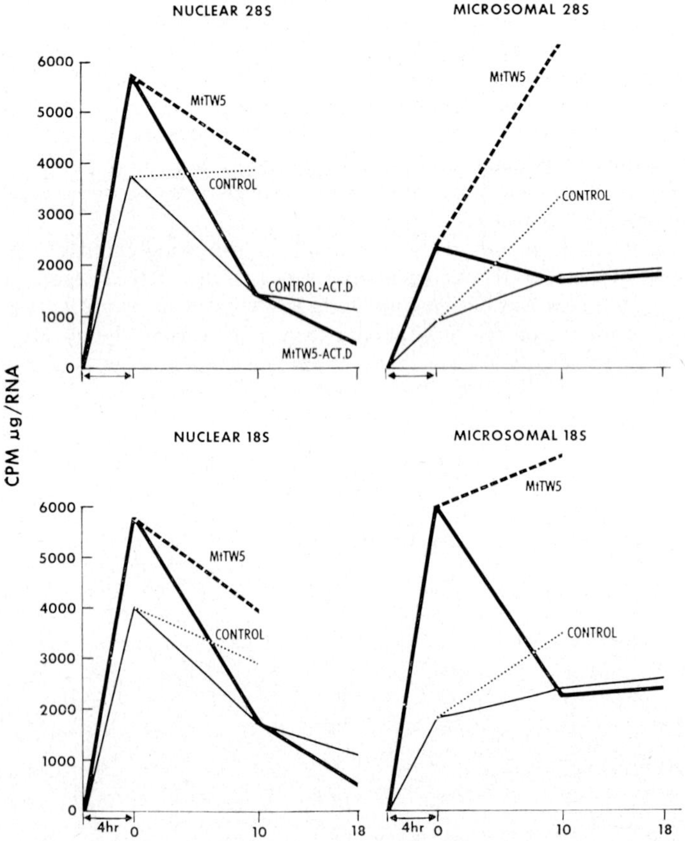

Fig. 3. Effect of MtTW5 on the turnover of pituitary gland RNA.

in non-radioactive medium in the presence and absence of actinomycin D. The glands of tumor-bearing rats synthesized more RNA during the 4-hour incubation with labeled precursor than did glands from control animals (Fig. 3). In the presence of actinomycin D, the specific activity of the nuclear RNA of glands of tumor-bearing rats decreased more rapidly than the controls. Microsomal RNA continued to accumulate in control glands even in the presence of the antibiotic, but the specific activity of microsomal RNA decreased in the glands of tumor-bearing animals. No difference in the rate of degradation of total RNA was observed, however, between the two groups of animals, suggesting the presence of two different pools of RNA.

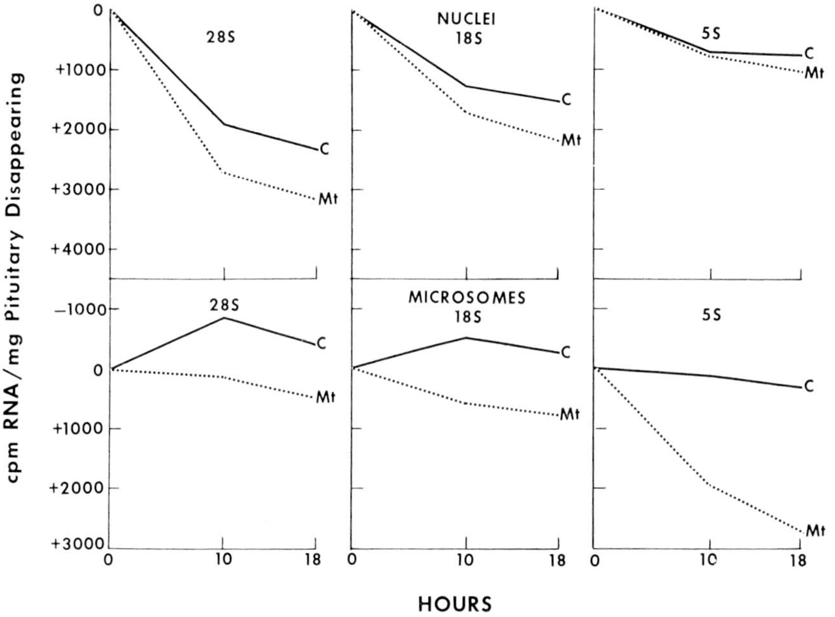

Fig. 4. Rate of RNA disappearance during actinomycin D chase.

Figure 4 shows the rates of RNA disappearance in the presence of actinomycin D. In all nuclear species studied, the RNA disappeared more rapidly from the glands of tumor-bearing animals. The 28 S and 18 S microsomal RNA's of control glands tended to accumulate while the RNA of glands exposed to tumor hormones degraded steadily. We conclude that the synthesis and degradation of pituitary gland RNA is increased as a result of the presence of the growth hormone secreting pituitary tumors.

DISCUSSION

The present findings confirm the work of Zor *et al.* (1970) in showing that prostaglandins increase pituitary gland adenyl cyclase activity and further demonstrate that prostaglandins increase the *in vitro* release of growth hormone. The fact that dibutyryl cyclic AMP also increased growth hormone release suggests that prostaglandin may stimulate hormone release by activating the adenyl cyclase-cyclic AMP systems. Prostaglandins apparently have some degree of specificity with regard to hormone release since they do not effect release of prolactin or luteinizing hormone. Other agents which influence adenyl cyclase or phosphodiesterase activity also stimulate growth hormone release. Hormone release was increased by the *in vitro* addition of theophylline and sodium fluoride, the latter exerting its effect even in the absence of protein synthesis. Since puromycin completely inhibited the *de novo* synthesis of growth hormone but had no effect on the increase in hormone release induced by prostaglandin, theophylline or dibutyryl cyclic AMP (MacLeod *et al.*, 1970), it is concluded that the mechanism governing hormone release is independent of the regulation of synthesis. Geschwind (1970) has recently reported results which agree with the hypothesis that synthesis and release of pituitary hormones are controlled independently.

Although the adenyl cyclase system is responsive to norepinephrine in many tissues, the pituitary gland enzyme is not activated by the catecholamine (Zor *et al.*, 1970; MacLeod and Lehmeyer, 1970). Catecholamine-depleting agents (MacLeod *et al.*, 1970) and norepinephrine (MacLeod, 1969) do not effect growth hormone synthesis and release.

The mechanisms are unknown whereby the hormones secreted by pituitary tumors cause atrophy of the host's pituitary gland and decrease its ability to synthesize growth hormone *in vitro*. The current results demonstrate that the pituitary glands of pituitary tumor-bearing rats retain their responsiveness to the stimulatory effects that prostaglandin and dibutyryl cyclic AMP exert on growth hormone release. It would appear, then, that the mechanisms which govern the release of growth hormone are still functional in glands of tumor-bearing rats and that the primary effect of the tumor hormones is an inhibition of synthesis of the pituitary gland growth hormone.

These results are consistent with the hypothesis that the hormones secreted by the pituitary tumors cause atrophy of the host's pituitary gland and decrease growth hormone synthesis by accelerating RNA turnover and decreasing the amount of RNA in the hypophysis.

It is doubtful whether the day-to-day fluctuation in growth hormone secretion is controlled at the transcriptional level of protein synthesis, although there is no experimental evidence either to substantiate or refute this. It is more likely that the physiological control of growth hormone release is mediated by mechanisms governing cyclic AMP production.

SUMMARY

The effects of prostaglandins and 3′5′-AMP on growth hormone synthesis and release were studied. Pituitary glands were incubated in Tissue Culture Medium 199 containing leucine-4,5-^{3}H in an atmosphere of 95% O_2–5% CO_2. Aliquots of the pituitary gland homogenates and incubation medium were subjected to polyacrylamide gel electrophoresis. This method achieves an excellent separation of growth hormone and prolactin from the other pituitary hormones. The amount of hormone present in the gels was determined by microdensitometry prior to sectioning the gels and measuring the amount of isotope incorporated into growth hormone and prolactin. Addition of PGE_1 or PGE_2 (10^{-6} M) increased growth hormone release 100-300% while 10^{-7} and 10^{-8} M increased growth hormone synthesis but not release. PGA_1 and $PGF_{2\alpha}$ were ineffective at these concentrations. No reproducible effects of prostaglandins on prolactin synthesis or release were observed. Adenyl cyclase activity in pituitary glands was measured and the addition of 10^{-4} M PGE_1, PGA_1 or $PGF_{2\alpha}$ increased the activity of the enzyme 100-200%. 10 mM NaF stimulated the enzyme 500% or more. Addition of 10^{-4} M and 5×10^{-3} M dibutyryl 3′5′-AMP to pituitary glands increased the release of newly-synthesized growth hormone 200-300% but had only a very minor effect on prolactin. Theophylline caused a significant increase in the release of growth hormone but not of prolactin. Implantation of growth hormone secreting pituitary tumors into rats caused a decrease in incorporation of leucine-4,5-^{3}H into the growth hormone retained by the gland

and that released into the incubation medium. When pituitary glands from rats with pituitary tumors were incubated with 10^{-6} M PGE_1 a significant increase in labeled growth hormone release was observed, but the stimulation was less than observed in control tissue. Similarly, dibutyryl 3′5′-AMP added to glands from tumor-bearing rats caused an increase in growth hormone release. Comparison of adenyl cyclase activity in pituitary glands from normal and tumor animals revealed no difference in enzyme activity and both enzymes were stimulated equally by fluoride. No difference in phosphodiesterase activity was found. Although these studies demonstrate that prostaglandins stimulate pituitary adenyl cyclase activity and growth hormone release and that dibutyryl 3′5′-AMP also stimulates release of the hormone, we do not yet have definitive evidence that growth hormone synthesis and release are regulated by prostaglandins through the mediation of intracellular 3′5′-AMP. It was also found that the RNA concentration of pituitary glands from tumor-bearing rats was decreased but that the *in vitro* incorporation of RNA precursors was increased. Studies conducted in the presence of actinomycin D showed that RNA degradation in glands from rats with the pituitary tumors was also increased, thus indicating that RNA turnover was increased.

REFERENCES

BUTCHER, R. W. and SUTHERLAND, E. W. (1962): Adenosine 3′,5′-phosphate in biological materials. I. Purification and properties of cyclic 3′,5′-nucleotide phosphodiesterase and use of this enzyme to characterize adenosine 3′,5′-phosphate in human urine. *J. biol. Chem., 237/4*, 1244.

DIGIROLAMO, A., HENSHAW, E. C. and HIATT, H. H. (1964): Messenger ribonucleic acid in rat liver nuclei and cytoplasm. *J. molec. Biol., 8/4*, 479.

GESCHWIND, I. (1970): Mechanism of action of hypothalamic adenohypophysiotropic factors. In: *Hypophysiotropic Hormones of the Hypothalamus; Assay and Chemistry*, pp. 298–319. Editor: J. Meites. Williams and Wilkins, Baltimore.

MACLEOD, R. M. (1969): Influence of norepinephrine and catecholamine-depleting agents on the synthesis and release of prolactin and growth hormone. *Endocrinology, 85/5*, 916.

MACLEOD, R. M. and ABAD, A. (1968): On the control of prolactin and growth hormone synthesis in rat pituitary glands. *Endocrinology, 83/4*, 799.

MACLEOD, R. M., FONTHAM, E. H. and LEHMEYER, J. E. (1970): Prolactin and growth hormone production as influenced by catecholamines and agents that affect brain catecholamines. *Neuroendocrinology, 6/5-6*, 283.

MACLEOD, R. M. and LEHMEYER, J. E. (1970): Release of pituitary growth hormone by prostaglandins and dibutyryl adenosine cyclic 3′:5′-monophosphate in the absence of protein synthesis. *Proc. nat. Acad. Sci. (Wash.), 67/3*, 1172.

MACLEOD, R. M., SMITH, M. C. and DEWITT, G. W. (1966): Hormonal properties of transplanted pituitary tumors and their relation to the pituitary gland. *Endocrinology, 79/6*, 1149.

MCCANN, S. M. and PORTER, J. C. (1969): Hypothalamic pituitary stimulating and inhibiting hormones. *Physiol. Rev., 49/2*, 240.

ZOR, U., KANEKO, T., SCHNEIDER, H. P. G., MCCANN, S. M. and FIELD, J. B. (1970): Further studies of stimulation of anterior pituitary cyclic adenosine 3′,5′-monophosphate formation by hypothalamic extract and prostaglandins. *J. biol. Chem., 245/11*, 2883.

ZOR, U., KANEKO, T., SCHNEIDER, H. P. G., MCCANN, S. M., LOWE, I. P., BLOOM, G., BORLAN, B. and FIELD, J. B. (1969): Stimulation of anterior pituitary adenyl cyclase activity and adenosine 3′:5′-cyclic phosphate by hypothalamic extract and prostaglandin E_1. *Proc. nat. Acad. Sci. (Wash.), 63/3*, 918.

ENVIRONMENTAL CONTROL OF GROWTH HORMONE AND GROWTH*

SANDY SORRENTINO Jr,** DON S. SCHALCH and RUSSEL J. REITER†

Departments of Anatomy and Medicine, University of Rochester School of Medicine and Dentistry, Rochester, N.Y., U.S.A.

The neural control and physiologic effects of growth hormone (GH) are perhaps the most poorly understood and complicated of all the hormones of the anterior pituitary gland. There are many factors which contribute to the confusion, three of which will be considered here. First, the technique used to measure GH must be considered. Workers using the radioimmunoassay have frequently obtained results that are contradictory to those of others using the standard tibial assay (*cf.* Garcia and Geschwind, 1968 and Müller and Pecile, 1968). Secondly, experimental species used must be taken into consideration. A well known dichotomy exists between the response of plasma immunoassayable GH in humans (Glick *et al.*, 1965) compared to rats (Schalch and Reichlin, 1968) after the stress of exercise. Thirdly, changes in GH secretion may be a secondary result of alterations in other hormones which might accompany experimental manipulation. For example, Daughaday *et al.* (1968) have clearly shown that GH decreases in the pituitary glands of hypothyroid or castrate rats, while Frantz and Rabkin (1965) have demonstrated that estrogen-treated human males have increased GH secretion similar to females after ambulation.

With these considerations in mind we have undertaken a number of experiments in order to determine the importance of environmental cues in regulating GH and growth (body weight and tibial length). In order to assess the importance of each environmental stimulus in determining normal growth, the effects of various experimental manipulations were studied. Thus, the effects on growth of light and olfactory deprivation, complete medial-basal hypothalamic isolation, exposure to reduced ambient temperature, and decreased daily food intake were examined. The role of the pineal gland in mediating the abnormal growth and GH response in blind and blind-anosmic rats was also studied, since it has been established that the pineal is an indispensable mediator of reproductive organ hypotrophy following blinding and anosmia (Reiter *et al.*, 1970).

METHODS

All animals used in these studies were albino rats of the Sprague Dawley strain, housed either 2-3 per stainless steel cage or 4-5 per clear plastic cage in a temperature- (75°F) and

* This investigation was supported by Postdoctoral Fellowship IF024042856-01, General Research Grant RR-05403 (from General Research Grant Division of Research Facilities, National Institutes of Health), and U.S.P.H.S. Grants HD-02937 and AM-08943.

** Postdoctoral Fellow.

† Career Development Awardee.

light- (lights on 6:00 a.m. to 8:00 p.m.) regulated room with free access to water and standard laboratory rat chow (except where food was restricted).

The various operations were performed using sodium pentobarbital anesthesia, the only exception being in the case of complete medial-basal hypothalamic isolation where ether was used. Blinding was done by bilateral enucleation, olfactory bulb removal by bilateral aspiration of bulbs, pinealectomy according to the technique described by Hoffman and Reiter (1965), and medial-basal hypothalamic isolation with a modified Halász knife similar to that of Palka *et al.* (1969).

Body weights were usually recorded at various intervals during the course of a given study. To minimize any effects of stress, rats were taken to the autopsy room one or two days prior to the termination of the experiments. Rats were sacrificed by rapid decapitation and blood was collected, except in the cold-induced stress study when blood was collected by venipuncture of the external jugular vein while the rats were anesthetized with ether. Anterior pituitary glands were weighed, homogenized in 0.05 M phosphate buffer (pH 7.5), frozen, and assayed after dilution with physiological saline. Plasma and pituitary glands were assayed essentially according to the radioimmunoassay technique of Schalch and Reichlin (1966) except that anti-porcine GH antibody was used instead of anti-rat GH. A highly purified rat GH (R331A: 1.91 U./mg, Dr. A. E. Wilhelmi) standard was used. Data were analyzed using a one way analysis of variance and a 't' test between several means or a student 't' test.

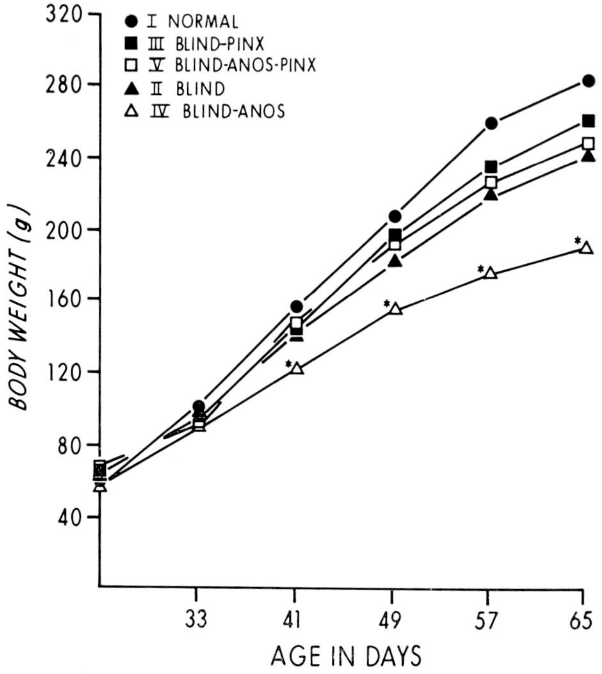

Fig. 1. Body weight gain is significantly depressed after blinding and blinding and anosmia compared to normal rats. Blind-pinealectomized and blind-anosmic-pinealectomized grew better than sensory deprived rats with intact pineals, but not as well as normals. Asterisks indicate a significant difference ($p < 0.01$) from respective pinealectomized rats. Number of rats per group: I (7), II (10), III (12), IV (10), V (11). (From S. Sorrentino Jr, R. J. Reiter and D. S. Schalch, 1971, *Neuroendocrinology*, *7*, 210, by kind permission of the editors).

RESULTS AND DISCUSSION

Blinding and anosmia in young male rats

In the first experiment rats were subjected to sensory deprivation at 25 ± 2 days of age and sacrificed at 65 ± 2 days. As had been previously described by other workers (Luce-Clausen and Brown, 1939; Browman, 1940; Eayrs and Ireland, 1949; Reiter, 1967), we observed a slight depression in body weight gain of light deprived rats relative to control rats (Fig. 1). Rats that were both blinded and pinealectomized had heavier body weights than blinded rats, but they were not as heavy as unoperated control animals. Blinding and anosmia severely retarded body weight gain, unless the rats were pinealectomized, in which case their body weights were nearly normal. Table I shows that commensurate with an inhibition of body weight there was also a retardation in tibial length and accessory organ weight in blind-anosmic rats which was not seen in blind-anosmic-pinealectomized rats.

TABLE I

Tibial length and accessory organ weight after sensory deprivation

Treatment (No. of rats)	Tibial length (cm)	Accessory organ weight (mg)
I Normal (7)	3.90 ± 0.05[1] III*, II, IV, V***[2]	398 ± 42 III, V*, II, IV***
II Blind (10)	3.74 ± 0.05 III*, I, IV***	191 ± 62 I, III-V***
III Blind-Pinx (12)	3.83 ± 0.02 I, II, V*, IV***	336 ± 41 I*, II, IV***
IV Blind-Anos (10)	3.53 ± 0.03 I-III, V***	72 ± 74 I-III, V***
V Blind-Anos-Pinx (11)	3.70 ± 0.03 I, III, IV***	334 ± 63 I*, II, IV***

[1] Mean ± SE

[2] Numerals indicate groups whose means are different and the asterisk designates the degree, *e.g.*, that 3.90 (I) is the same as 3.74 (II), 3.53 (IV) and 3.70 (V), is less than 0.01.

* $p < 0.05$

*** $p < 0.01$

(From S. Sorrentino Jr., R. J. Reiter and D. S. Schalch, 1971, *Neuroendocrinology*, 7, 210, by kind permission of the editors).

It is now well established that the pineal gland is the neuroendocrine mediator of the atrophic response of the gonads and accessory organs of rats after blinding and anosmia (Reiter *et al.*, 1970). Studies on the effect of blinding and anosmia on pituitary GH content (Fig. 2) revealed a marked decrease in this constituent which was not observed in blind-anosmic-pinealectomized rats. However, the concentration of GH was unchanged because dual sensory deprivation also reduced the size of the anterior pituitary gland considerably. Because of the marked variability found in plasma GH concentration within all groups no valid interpretation of these data could be made. This experiment demonstrated that specific sensory deprivations (blinding and anosmia) could modify growth and GH with the pineal gland assuming an important part in this response.

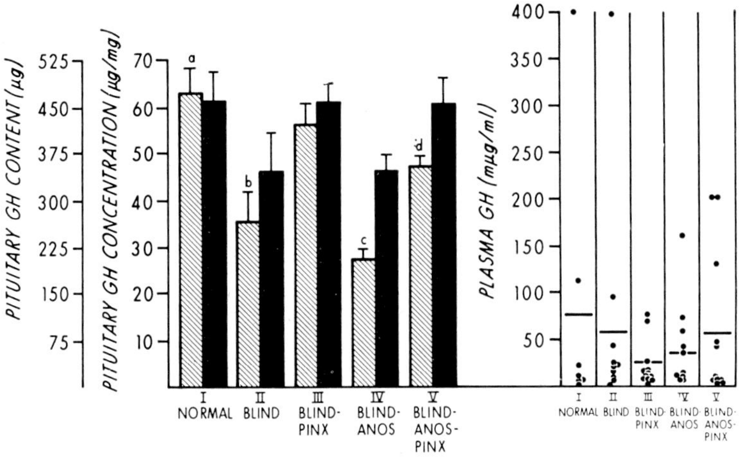

Fig. 2. Pituitary and plasma GH after blinding and anosmia. Cross hatched bars refer to pituitary GH content and solid bars to pituitary GH concentration. Horizontal lines in plasma GH graph indicate means. [a]±SE; [b]different from I and III ($p < 0.01$); [c]different from I, III and V ($p < 001$); [d]different from I ($p < 0.02$). (From S. Sorrentino Jr, R. J. Reiter and D. S. Schalch, 1971, *Neuroendocrinology*, *7*, 210, by kind permission of the editors).

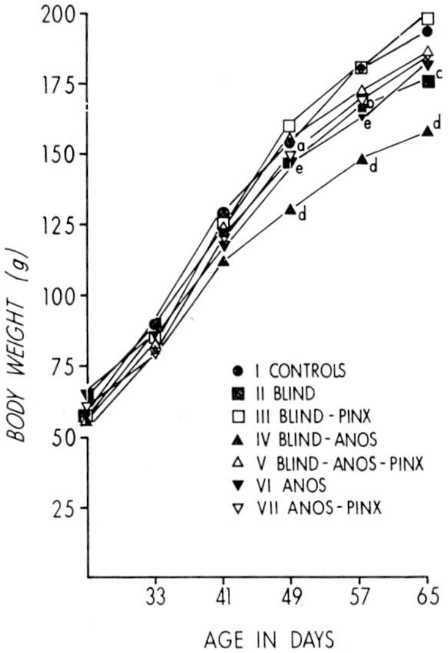

Fig. 3. Body weight gain is significantly depressed after blinding, anosmia, and blinding and anosmia compared to controls. [a]different from I and III ($p < 0.05$) and IV ($p < 0.01$); [b]different from I, III, and IV ($p < 0.01$); [c]different from I and IV ($p < 0.02$) and III ($p < 0.01$); [d]different from I-VII ($p < 0.01$); [e]different from I, III and IV ($p < 0.01$). Number of rats per group: I (9), II (9), III (8), IV (8), V (10), VI (9), VII (7).

Blinding and anosmia in young female rats

If young (25-day-old) female rats are rendered blind and anosmic, the growth response is similar to that in males (Fig. 3). Blinded females weighed less than sham operated (controls) and blind-pinealectomized rats at 49, 57 and 65 days of age. Blinding, anosmia, or anosmia and pinealectomy appeared to be equally as effective in depressing body weight. The combination of sensory deficits (*i.e.* sight and smell) had the most marked effect on subsequent

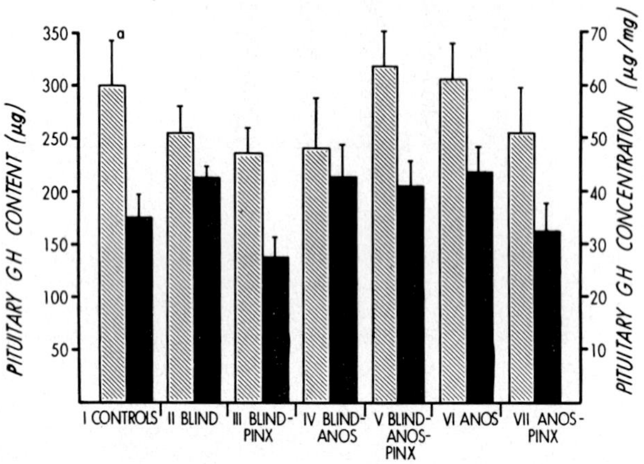

Fig. 4. Pituitary GH after blinding and anosmia. Cross hatched bars refer to pituitary GH content and solid bars to pituitary GH concentration. [a]±SE; no significant differences exist between any groups.

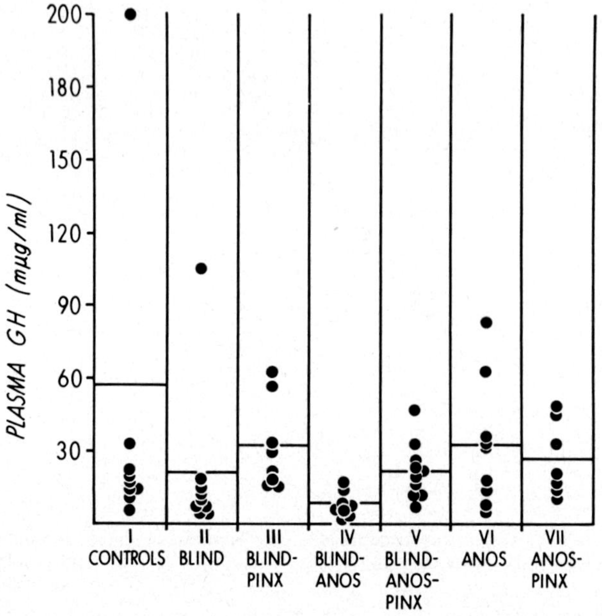

Fig. 5. Plasma GH after blinding and anosmia. No significant differences exist between any groups.

body weight gain. Absence of the pineal gland negated the detrimental effect of blinding but apparently not that due to anosmia. Anosmic-pinealectomized rats did not grow as well as controls, and blinded-anosmic-pinealectomized rats grew at a similar rate. Blind-anosmic rats possessed significantly shorter tibiae than blind-anosmic-pinealectomized rats (3.47 cm *vs* 3.64 cm, respectively).

Pituitary content and concentration of GH did not reflect changes which were observed in body weights or tibial length (Fig. 4). Furthermore, plasma GH levels were similar in all groups (Fig. 5). However, an important point to consider here is that the GH values in blind-anosmic rats resembled values observed in stressed rats as measured by immunoassay.

These results of sensory deprivation on GH in young females are in contrast to those seen in young males. A feasible explanation for these findings is that secretion of steroids from the ovaries of blind-anosmic females was inhibited since these rats had atrophic uteri which may have in turn caused increased amounts of pituitary GH. According to Birge *et al.* (1967) castrated female rats have increased amounts of pituitary GH compared to intact females and estrogen treatment of intact male rats leads to a reduction in pituitary GH content. Blinding and anosmia in young female rats induces a demonstrable inhibition in growth probably by reducing GH secretion. This combination of sensory deficits inhibited growth more severely than blinding, anosmia, or anosmia and pinealectomy which appeared to be equally as effective in growth retardation. Absence of the pineal gland negated the detrimental effects of blinding, but apparently not that of anosmia.

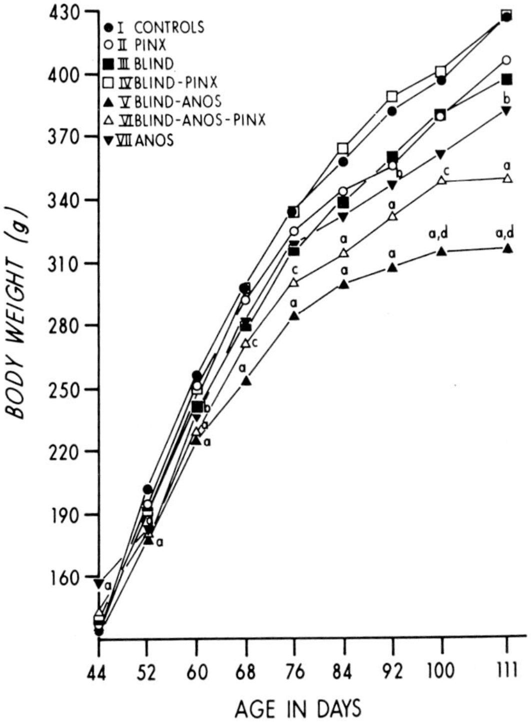

Fig. 6. Body weight gain is significantly depressed by anosmia, blinding and anosmia, and blinding, anosmia and pinealectomy compared to controls. [a] different from I ($p < 0.01$); [b] different from I ($p < 0.05$); [c] different from I ($p < 0.02$); [d] different from VI ($p < 0.05$). Number of rats per group: I (9), II (10), III (8), IV (9), V (9), VI (7), and VII (5). (From S. Sorrentino Jr, R. J. Reiter and D. S. Schalch, 1971, *Neuroendocrinology*, *8*, 116, by kind permission of the editors).

Blinding and anosmia in adult male rats

It is now established that the antigonadotropic properties of the pineal are more apparent in immature than in mature rats (Reiter, 1967; Sorrentino, 1969). It was of interest to examine the pineal's potential in inhibiting growth and GH of adult male rats after sensory deprivation. Male rats that were 44 days of age were subjected to various combinations of operations including blinding, anosmia, and pinealectomy. In this experiment all rats that were not blinded were unilaterally enucleated and those that were not pinealectomized were subjected to a sham pinealectomy in order to minimize growth effects due to the trauma of the operation. Figure 6 illustrates the growth curves of rats in this experiment. It is obvious that these rats are in a more static phase of growth compared to young male rats. As in the immature animals, dual sensory deprivation significantly retarded body weight gain. The presence of the pineal is partially, but not solely, responsible for this inhibition, as evidenced by the subnormal growth in blind-anosmic-pinealectomized rats. This group's growth is approximately equal to that in anosmic rats, which indicates that the pineal probably mediates the growth response after blinding while anosmia inhibits growth via another neuroendocrine mechanism. Blinding and pinealectomy alone were relatively ineffective in inhibiting growth. As can be seen in Table II accessory organ weight and tibial alterations accompanied the growth changes. Removal of the pineal gland was effective in reversing the alterations in reproductive organs and the inhibitory effects of blinding on tibial lengths.

Blind-anosmic rats had the lowest mean concentration of plasma GH (Fig. 7), but due to variability, no statistical differences were achieved. Blind and blind-anosmic rats had reduced pituitary GH stores relative to pinealectomized controls (Fig. 8).

TABLE II

Tibial length and accessory organ weight after sensory deprivation

Treatment (No. of rats)	Tibial length (cm)	Accessory organ weight (mg)
I Controls (9)	4.60 ± 0.03[1] V***, VI*	639 ± 27 III, V, VII***, VI**
II Pinx (10)	4.57 ± 0.03 V***	653 ± 38 III, V-VII***
III Blind (8)	4.55 ± 0.03 V***, IV*	336 ± 45 I, II, IV-VII***
IV Blind-Pinx (9)	4.67 ± 0.07 V***, III, VI, VII*	553 ± 31 II, V***
V Blind-Anos (9)	4.42 ± 0.03 I-IV***	118 ± 28 I-VII***
VI Blind-Anos-Pinx (7)	4.46 ± 0.07 I, IV*	496 ± 74 II, III, V***, I**
VII Anos (5)	4.50 ± 0.04 IV*	479 ± 42 I-III, V***

[1] Mean ± SE
* $p < 0.05$
** $p < 0.02$
*** $p < 0.01$

(From S. Sorrentino Jr, R. J. Reiter and D. S. Schalch, 1971, *Neuroendocrinology*, *18*, 116, by kind permission of the editors).

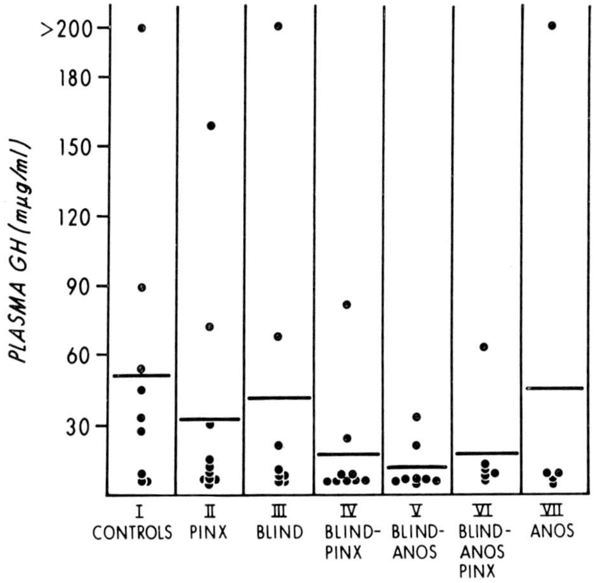

Fig. 7. Plasma GH after blinding and anosmia. No significant differences exist between any groups. (From S. Sorrentino Jr, R. J. Reiter and D. S. Schalch, 1971, *Neuroendocrinology*, *8*, 116, by kind permission of the editors).

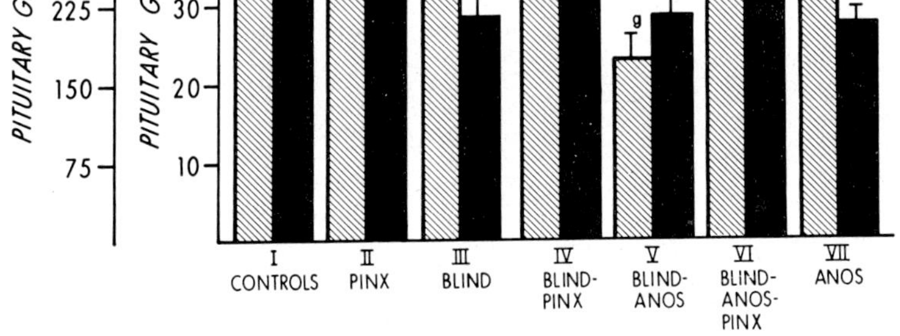

Fig. 8. Pituitary GH after blinding and anosmia. Cross hatched bars refer to pituitary GH content and solid bars to pituitary GH concentration. [a]±SE; [b]different from II ($p < 0.05$), III, VI, and VII ($p < 0.01$); [c] different from III, V, VI, and VII ($p < 0.01$); [d] different from I ($p < 0.05$), V and VII ($p < 0.01$); [e] different from I ($p < 0.01$) and IV ($p < 0.02$; [f] dfferend fıom V, VII ($p < 0.01$) and VI ($p < 0.05$); [g] different from VI (p 0.05). (From S. Sorrentino Jr, R. J. Reiter and D. S. Schalch, 1971, *Neuroendocrinology*, *8*, 116, by kind permission of the editors).

This experiment demonstrated that just as the pineal's capability to regulate reproductive organ size in the rat is dependent on age, so is its growth regulating capability. Blinding the adult rat merely altered pituitary GH; whereas both growth and GH were altered in the prepuberally blind rat.

The previous three experiments provided insight into the regulation of growth and GH by light and smell in rats. They revealed that the pineal gland is intimately linked in the chain of events initiated by purely nervous stimuli, or their absence, and consummated by endocrine changes. They did not tell us whether the pineal influenced growth directly by affecting the activity of the hypothalamic or extra-hypothalamic neurons responsible for the production of growth hormone releasing factor (GRF) or by determining GH secretion of the somatotrophs of the anterior pituitary gland. Its effectiveness in retarding growth may be of a non-specific nature by first diminishing secretion of other hormones (testosterone or thyroxine) which in turn inhibit growth. Possibly the pineal's effect on growth and GH may be as indirect as a reduction in food intake. The pineal substance, melatonin, has been shown to depress food intake in female rats (Narang *et al.*, 1967) and a reduction in food consumption in rats leads to retarded growth (Mulinos and Pomerantz, 1940).

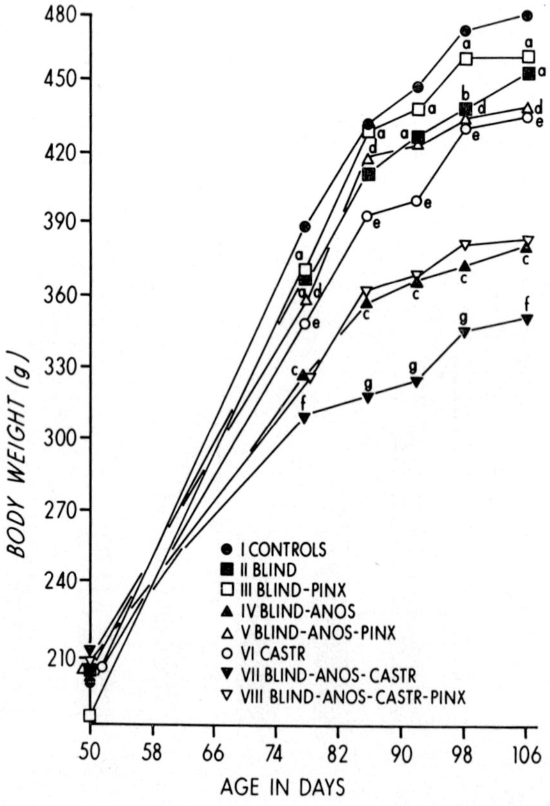

Fig. 9. Body weight gain is significantly depressed after blinding, anosmia and castration compared to controls. [a] different from IV, VII and VIII (p <0.01); [b] different from I, III (p <0.02), IV, VII and VIII (p <0.01); [c] different from I and V (p <0.01); [d] different from VII and VIII (p <0.01); [e] different from I, VII and VIII (p <0.02); [f] different from I, II, III, V and VI (p <0.01); [g] different from I-III and V-VIII (p <0.01). Number of rats per group: I (8), II (8), III (7), IV (12), V (8), VI (5), VII (10), VIII (8).

The possible role of food intake in determining growth in a similar experiment was assessed by measuring the food intake in sensory deprived rats. Blind-anosmic adult male rats consumed 16.0 g/rat/day of rat chow compared to 19.7 g/rat/day in blind-anosmic-pinealectomized rats. Normal rats ate 20.3 g/rat/day. It does not seem feasible that this decrease in food consumption in blind-anosmic and blind-anosmic-pinealectomized rats could be totally responsible for their poor growth, because sensory deprived rats ate only 3.7 g/rat/day of food less than sensory deprived pinealectomized rats and yet weighed approximately 40 g less at sacrifice (Reiter *et al.*, 1971).

Blinding and anosmia in adult male castrated rats

This experiment was designed to determine if the pineal gland is capable of growth retardation in the absence of testicular steroids. In addition to the usual sensory deprivation at 50 days of age a group of castrates, blind-anosmic-castrates, and blind-anosmic-pinealectomized-castrates were also included.

Figure 9 shows the effects of these surgical manipulations on body weights. It is apparent that castration itself retarded body weight gain. Dual sensory deprivations in rats with intact pineal glands seemed to be equally as effective in inhibiting growth regardless of status of the gonadal steroids. Plasma GH concentrations (Fig. 10) were depressed in blind rats, blind-anosmic rats, blind-anosmic-castrates, and blind-anosmic-pinealectomized castrates. In fact, group VIII and especially group VII contained many animals in which plasma GH was undetectable. Blinding failed to reduce pituitary GH in this experiment (Fig. 11). However, the combination of blinding and anosmia depressed pituitary GH levels, an effect which was not seen in similarly treated pinealectomized rats. GH levels of rats in group VII did not differ significantly from those of rats in group VIII. Castration (groups VI-VIII) did reduce

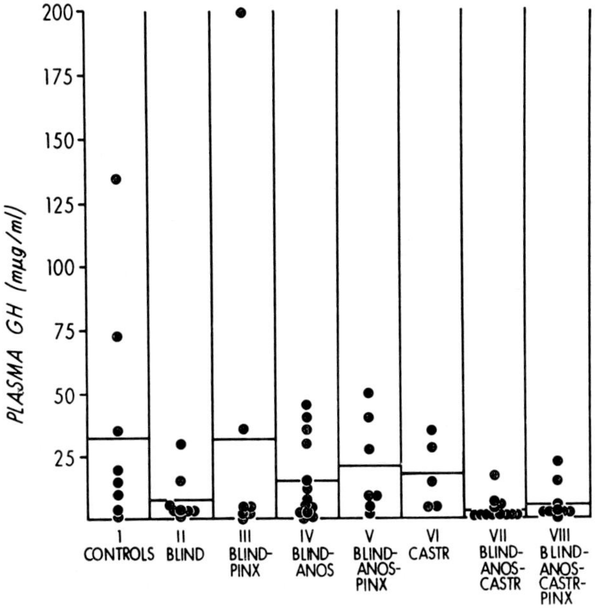

Fig. 10. Plasma GH after blinding, anosmia and castration. No significant differences exist between any groups.

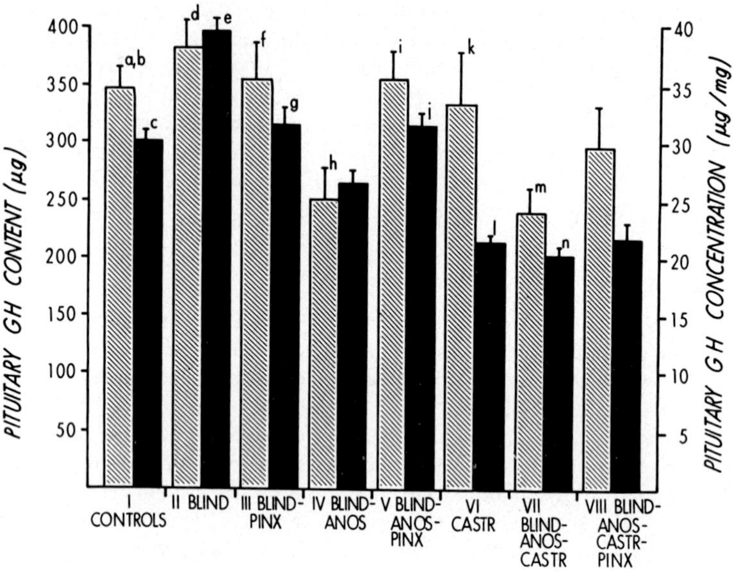

Fig. 11. Pituitary GH after blinding, anosmia and castration. Cross hatched bars refer to pituitary GH content and solid bars to pituitary GH concentration. [a] ±SE; [b] different from IV ($p < 0.05$) and VII ($p < 0.02$); [c] different from II ($p < 0.02$) and VII ($p < 0.05$); [d] different from IV and VII ($p < 0.01$); [e] different from VI-VIII ($p < 0.01$); [f] different from IV ($p < 0.02$) and VII ($p < 0.01$); [g] different from VIII ($p < 0.02$), VI and VII ($p < 0.01$); [h] different from I, VI ($p < 0.05$), II, III, and V ($p < 0.01$); [i] different from IV and VII ($p < 0.01$); [j] different from VI, VIII (0 < 0.02) and VII ($p < 0.01$); [k] different from IV ($p < 0.05$) and VII ($p < 0.02$); [l] different from II, III ($p < 0.01$) and V ($p < 0.02$); [m] different from I ($p < 0.02$), II, III, V, VI ($p < 0.01$); [n] different from I ($p < 0.02$), II, III, V ($p < 0.01$).

GH concentrations compared to non-castrates, but this was probably a mathematical artifact in that pituitary glands of castrated animals were significantly heavier than those of the non-castrates.

The discrepancy between body weight gain and total pituitary GH in this experiment is rather disturbing. This experiment did show, however, that the pineal is capable of inhibiting body weight increase regardless of the presence or absence of the testes. In other words, the pineal seemed capable of inhibiting growth by a means which does not involve the testicular steroids. A modification of this experiment was done for two reasons. First, because young rats seem to be more sensitive to the activated pineal, similar experiments were carried out using immature animals. Second, since castration itself adversely affected growth, it was decided to use castration accompanied with androgen replacement therapy to study this response.

Blinding and anosmia in young male castrated rats with androgen supplementation

Rats were operated on at 27 days of age. One group was castrated along with blinding and anosmia and received 0.75 mg/3 days of testosterone propionate for the first 9 days and 1.5 mg/3 days subcutaneously in sesame oil throughout the remainder of the experiment.

The body weight graph (Fig. 12) shows that sensory deprived rats grew poorly when androgen levels were high due to androgen replacement (*cf.* groups II and IV). As a matter of fact the parallelism in growth between these two groups is remarkable. The pineal gland can inhibit growth in these two groups equally as well. Table III shows that tibial length was also inhibited in these animals. Furthermore, total thyroxine measurements, assayed by a competitive protein blinding technique, in rats of groups IV and V indicated that hypothyroidism is probably not the cause of growth abnormalities.

Plasma GH levels (Fig. 13) were the lowest in rats of group I and IV. Levels in rats of group I are unusually low and similar to those of stressed rats. This observation remains unexplained. Once again, pituitary levels of GH (Fig. 14) were somewhat confusing. As has

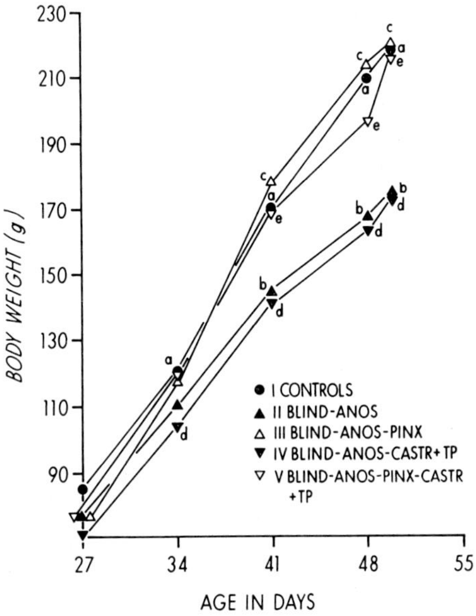

Fig. 12. Body weight gain is significantly depressed after blinding and anosmia and blinding, anosmia, castration and testosterone propionate replacement compared to controls. [a] different from II and IV ($p < 0.01$); [b] different from I, III, and V ($p < 0.01$); [c] different from II and IV ($p < 0.01$); [d]different from I, III, and V ($p < 0.01$); [e]different from II and IV ($p < 0.01$). Number of rats per group: I (8), II (11), III (11), IV (10), V (12).

TABLE III

Tibial length and total thyroxine after sensory deprivation

Treatment (No. of rats)	Tibial length (cm)	Total thyroxine (μg/100 ml)
I Controls (8)	3.50 ± 0.03[1] II, IV***	
II Blind-Anos (11)	3.28 ± 0.05 I, III, V***	
III Blind-Anos-Pinx (11)	3.53 ± 0.07 II, IV***	
IV Blind-Anos-Castr + TP (10)	3.33 ± 0.04 I, III, V***	3.4 ± 0.3[2]
V Blind-Anos-Pinx-Castr + TP (12)	3.50 ± 0.03 II, IV***	4.0 ± 0.2[2]

[1] Mean ± SE
[2] Represents mean from 5 rats
*** $p < 0.01$

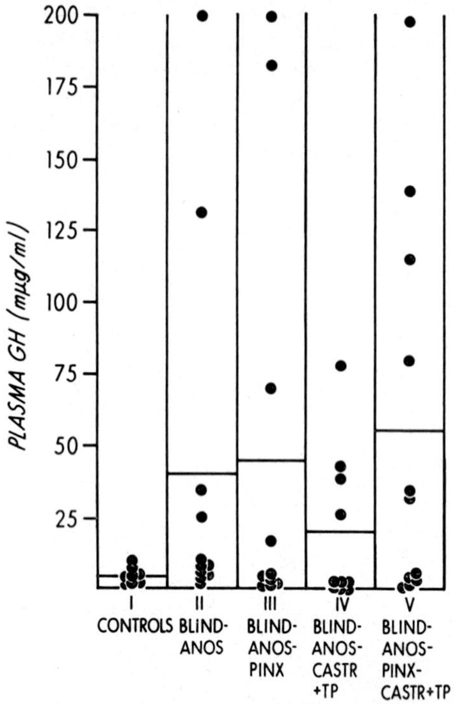

Fig. 13. Plasma GH after blinding and anosmia, and blinding, anosmia, castration and testosterone propionate replacement. No significant differences exist between any groups.

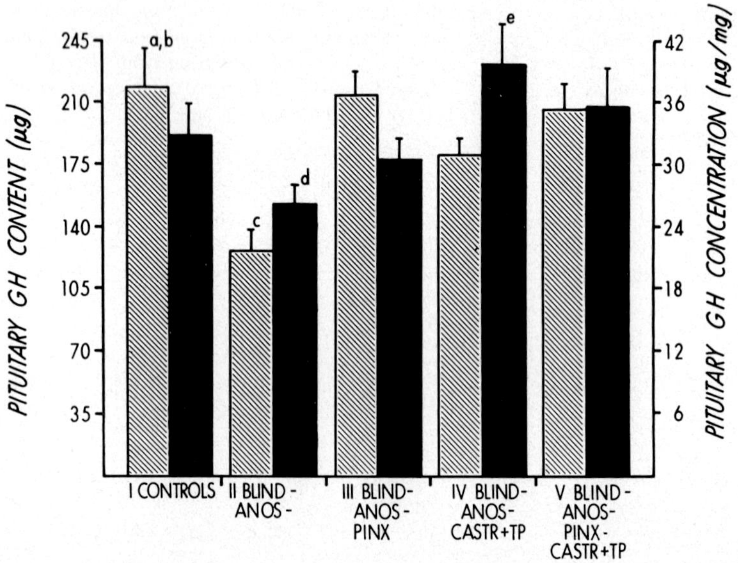

Fig. 14. Pituitary GH after blinding and anosmia, and blinding, anosmia, castration and testosterone propionate replacement. Cross hatched bars refer to pituitary GH content and solid bars to pituitary GH concentration. [a] ±SE; [b] different from II ($p < 0.01$) and IV ($p < 0.05$); [c] different from I and III-V ($p < 0.01$); [d] different from IV and V ($p < 0.01$); [e] different from II ($p < 0.01$) and III ($p < 0.02$).

been repeatedly observed, blind-anosmic rats have exceedingly low total pituitary GH levels and yet blind-anosmic-castrate-androgen treated rats which grew just as poorly have higher levels of GH. These levels do not differ from levels in controls or in similarly treated pinealectomized rats. A point worthy of consideration is that pituitary GH concentration in rats of group IV was increased compared to groups II and III suggesting that secretion of GH in these rats may have been inhibited.

These experiments conclusively demonstrated that the influence of the pineal on growth is not mediated by diminished gonadal or thyroidal secretions. The apparent discrepancy between growth and pituitary GH is rather disturbing and remains unexplained.

Effects of medial-basal-hypothalamic isolation and cold exposure in adult male rats

Rats were subjected to complete medial-basal-hypothalamic isolation (MBHI) at 50 days of age. This lesion consistently isolated a block of neural tissue which had a radius of 0.75 mm and a depth of 1.5 mm. The isolated tissue characteristically included the majority of the median eminence (Fig. 15). The block also contained the ventromedial and arcuate nuclei and separated medial hypothalamus from the lateral hypothalamus. Most brains examined at the termination of the experiment did not display the cut extending to the base of the brain. This was thought to be due to healing rather than an incomplete cut, since all rats killed immediately after the lesion exhibited complete MBHI. The 'healing' phenomenon has been seen by others (Palka *et al.*, 1969). Cold-induced stress was accomplished when the rats were 85 days of age by placing them in a cold room (3°C with 4 rats per cage).

Figure 16 clearly shows that rats subjected to MBHI were significantly heavier than sham operated controls as early as a week postoperatively and remained heavier throughout the experiment. All rats with MBHI became overtly obese. This result contradicts that of Halász (1968) who has observed that rats with complete deafferentation of the hypophysiotrophic area have retarded body weights. The cuts in the present report were considerably more medial than were those in Halász's work (1969). It is presumed that our knife isolated the medial hypothalamus from the lateral and this accounted for the hyperphagia and obesity which resulted. A lesion of this type has been reported to produce obesity in rats

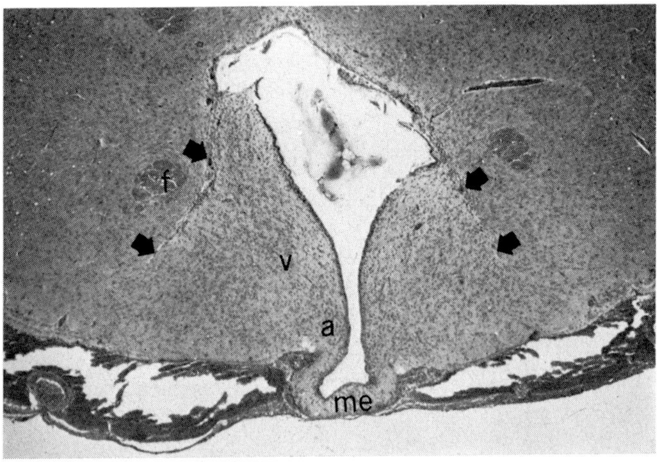

Fig. 15. Frontal section of rat hypothalamus after medial-basal-hypothalamic isolation (× 76). Note ventromedial nucleus (v), arcuate nucleus (a), fornix (f), and median eminence (me). Arrows indicate cut and nervous scar.

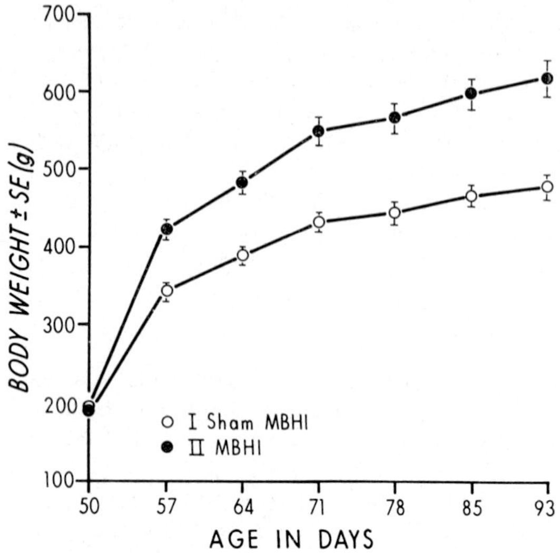

Fig. 16. Body weight gain is significantly elevated after medial-basal-hypothalamic isolation (MBHI) compared to sham operated rats. Mean of I differs from II ($p < 0.01$) at all ages except at 50 days. Number of rats per group: I (8) and II (12).

(Palka *et al.*, 1969). When bone growth was considered, as judged by tibial length (Table IV), rats with MBHI grew less than did sham operated controls. This observation confirms Halász's report (1968) but it is inconsistent with that of Critchlow *et al.* (1970) who showed increased nose-anal length in obese rats with complete MBHI. A comparison between lesions in terms of size and location, would be necessary to determine if the results were due to differences in the extent of the cuts.

After two hours of cold exposure plasma GH fell significantly in both groups (Fig. 17). This confirms earlier reports by Schalch and Reichlin (1968) but is diametrically opposite to the results of Müller and Pecile (1968) who used the standard tibial assay and to the results of Machlin *et al.* (1968) who demonstrated an increase in immunoassayable plasma GH after cold exposure in sheep. Plasma levels of GH in rats with MBHI tended to be lower than sham operated at 85 days (Fig. 17) but this was not seen at sacrifice (Fig. 18). Total pituitary GH (Fig. 18) was decreased in rats with MBHI compared to controls, whereas pituitary GH concentration was also reduced in rats of Halász's report (1968). This may be due to interruption of neural afferents or secondarily due to diminished androgen secretion (both our and Halász's rats had small accessory organs).

TABLE IV

Tibial length after medial-basal-hypothalamic isolation

Treatment (No. of rats)	Tibial length (cm)
I Sham (8)	4.70 ± 0.03[1] II***
II MBHI (12)	4.49 ± 0.06 I***

[1] Mean ± SE
*** $p < 0.01$

It is of interest that rats with MBHI can respond to sensory stimulation such as cold. This has also been shown by other workers investigating plasma GH response to ether stress in rats (Critchlow *et al.*, 1970). These results indicate either that axonal regeneration across the lesion may have occurred, or that the isolated neural tissue can respond to stressful stimuli after being informed chemically or physically of the stress. Stress may cause a curtailment of growth hormone releasing factor or increased release of growth hormone inhibitory factor.

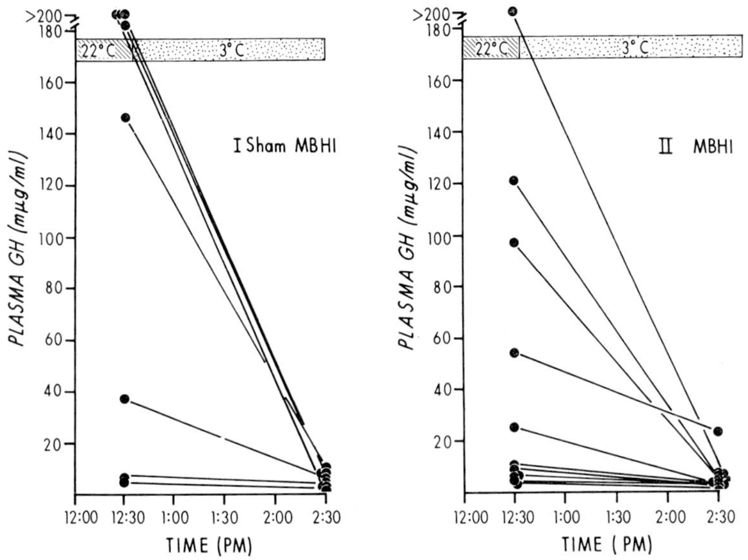

Fig. 17. Plasma GH levels in nonstressed (12:30 P.M.) and cold-induced stressed rats (2:30 P.M.). Nonstressed value (mean ± SE) of rats in I is 123 ± 32 mμg/ml and that of rats in II is 49 ± 2 mμg/ml (not significantly different). Stressed values are 6 ± 1 mμg/ml and 6 ± 2 mμg/ml for rats in I and II, respectively (not significantly different from each other but significantly different from nonstressed value in respective groups – $p < 0.01$).

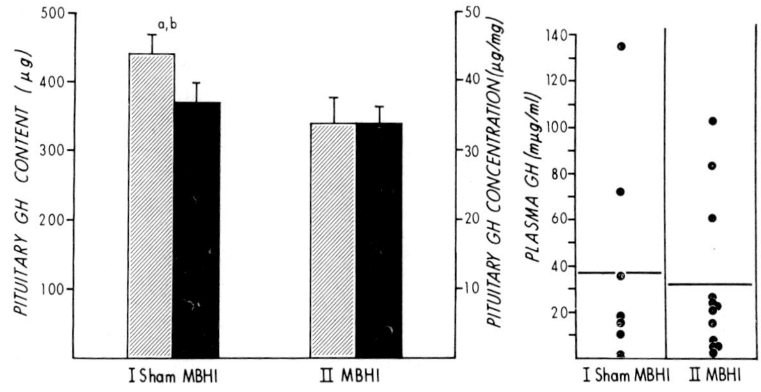

Fig. 18. Pituitary GH and plasma GH at sacrifice after medial-basal-hypothalamic isolation (MBHI). Cross hatched bars refer to pituitary GH content and solid bars to pituitary GH concentration. [a]±SE; [b] different from II ($p < 0.05$). No significant differences exist between plasma GH levels.

Underfeeding in young male rats

Another important environmental input which was examined relative to growth and GH was food intake. Beginning when the rats were 24-25 days old, they were fed daily one-half the food intake of similarly aged rats and sacrificed 35 days later. In addition to underfeeding they were unilaterally enucleated since they were part of an experiment dealing with blind rats. Table V shows that in addition to a retardation in body weight, tibial length, and pituitary GH content and concentration, plasma GH was also severely reduced compared to *ad libitum* controls. These data support the conclusion of Trenkle (1970) that both pituitary and plasma GH levels are reduced in acutely starved rats. Underfeeding probably interferes with normal GH synthesis and release via a neuroendocrine mechanism or simply by causing a lack of anabolic substrates necessary for constructing the GH or growth hormone releasing factor molecule.

TABLE V

Body weight, tibial length, pituitary GH content and concentration, and plasma GH after underfeeding

			Anterior pituitary GH		
Treatment (No. of rats)	Body weight (g)	Tibial length (cm)	Total (μg)	Concentration (μg/mg)	Plasma GH (mμg/ml)
I Fed (7)	277 ± 6[1] II***	3.80 ± 0.04 II***	516 ± 55 II***	65 ± 7 II***	64 ± 22 II***
II Underfed (7)	132 ± 10 I***	2.47 ± 0.03 I***	119 ± 13 I***	34 ± 3 I***	6 ± 1 I***

[1] Mean ± SE
*** $p < 0.01$

(From S. Sorrentino Jr, R. J. Reiter and D. S. Schalch, 1971, *Neuroendocrinology*, *7*, 105, by kind permission of the editors).

SUMMARY

By selectively deleting or diminishing several environmental inputs we have successfully modified growth and GH synthesis and/or release. The importance of light in determining normal growth patterns was realized in the poor growth of rats deprived of photic stimuli by blinding. The modification of GH synthesis and/or release after light deprivation is not a totally new concept. Browman (1940) first observed retarded growth in blind rats. Recently Müller *et al.* (1970) have observed a diurnal fluctuation in immunoassayable and bioassayable rat GH.

Normal olfactory stimuli in rats also seem to play an integral role in determining normal growth. Giammanco *et al.* (1968) have clearly shown that anosmia in young rats can significantly inhibit growth rates.

Dual sensory deprivation (light and smell) has a devastating effect on growth. This was first observed by Reiter in 1967. The regulation of growth and GH by the pineal gland in the dual sensory deprived rats is a most interesting observation. Prior to this observation the pineal's depressant effect on reproductive size of the blind-anosmic rat was well established (Reiter *et al.*, 1970). This presumably occurred by inhibition of gonadotrophin release. Our data show that in the blind-anosmic rat another endocrine malady occurs – deficient GH production and/or release resulting in poor growth. Considering the data available at present it seems that the pineal's influence on growth and GH is not mediated by other endocrine abnormalities that may exist in the blind-anosmic rat.

The first workers to suggest a relationship between the pineal gland and growth were Oestreich and Slawyk (1899). These workers observed a young boy who at the age of four was a giant and experienced precocious puberty. He also possessed a pineal tumor. If the tumor was stromal it may have destroyed the parenchymal cells of the pineal or the cells responsible for production of the growth inhibiting factor. This idea has been proposed by Wurtman to explain precocious puberty following a pineal tumor (Cohen *et al.*, 1964).

More recent observations by Malm *et al.* (1959) and Kincl and Benagiano (1967) that pinealectomy accelerates the increase in body weight of rats support the contention that the pineal restrains body weight gain. In a rather prophetic monograph Wiener suggests that the human pineal secretes an anti-growth hormone which effectively inhibits the action of GH (1968).

We have also shown that both temperature and food intake are important determinants of production and fate of GH. Whether hypothermia or malnutrition affect GH because of their stressful nature is not known. The significance of this work is that GH production and/or release can obviously be changed by alterations in afferent stimuli and that these changes are probably accomplished by the releasing and inhibiting factors with dual natures (the neural hormones).

ACKNOWLEDGEMENT

The authors wish to thank Louyse M. Lee and Anna Chornobil for their excellent technical assistance.

REFERENCES

BIRGE, C. A., PEAKE, G. T., MARIZ, I. K. and DAUGHADAY, W. H. (1967): Radioimmunoassayable growth hormone in the rat pituitary gland: effects of age, sex and hormonal state. *Endocrinology*, *81*, 195.

BROWMAN, L. G. (1940): The effect of optic enucleation on the male albino rat. *Anat. Rec.*, *78*, 59.

COHEN, R. A., WURTMAN, R. J., AXELROD, S. and SNYDER, S. (1964): Some clinical, biochemical and physiological actions of the pineal gland. *Ann. intern. Med.*, *61*, 1144.

CRITCHLOW, V., SMYRL, R., PALKA, Y., BORDELON, C., HUTCHINS, M., LIEBELT, R. and SCHINDLER, W. (1970): Obesity and increased growth following isolation of medial-basal hypothalamus. *Fed. Proc.*, *29*, 377.

DAUGHADAY, W. H., PEAKE, G. T., BIRGE, C. A. and MARIZ, I. K. (1968): The influence of endocrine factors on the concentration of growth hormone in rat pituitary. In: *Growth Hormone*, pp. 238–252. Editors: A. Pecile and E. E. Müller. ICS 158, Excerpta Medica, Amsterdam.

EAYRS, S. T. and IRELAND, K. F. (1949): The effect of total darkness on the growth of the newborn albino rat. *J. Endocr.*, *6*, 386.

FRANTZ, A. G. and RABKIN, M. T. (1965): Effects of estrogen and sex difference on secretion of human growth hormone. *J. clin. Endocr.*, *25*, 1470.

GARCIA, J. F. and GESCHWIND, I. I. (1968): Investigation of growth hormone secretion in selected mammalian species. In: *Growth Hormone*, pp. 267–291. Editors: A. Pecile and E. E. Müller. ICS 158, Excerpta Medica, Amsterdam.

GIAMMANCO, S., TESSITORE, V., LAGRUTTA, V. and DIBERNARDO, C. (1968): Ricerche sulle modificazioni del sistema neuroendocrino del ratto sottoposto ad asportazione dei bulbi olfattivi in epoca prepuberale. *Biol. lat. (Milano)*, *21*, 121.

GLICK, S. M., ROTH, J., YALOW, R. S. and BERSON, S. A. (1965): The regulation of growth hormone secretion. *Recent Progr. Hormone Res.*, *21*, 241.

HALÁSZ, B. (1968): The role of the hypothalamic hypophysiotrophic area in the control of growth hormone secretion. In: *Growth Hormone*, pp. 204–210. Editors: A. Pecile and E. E. Müller. ICS 158, Excerpta Medica, Amsterdam.

HALÁSZ, B. (1969): The endocrine effects of isolation of the hypothalamus from the rest of the brain. In: *Frontiers in Neuroendocrinology*, pp. 307–342. Editors: W. F. Ganong and L. Martini. Oxford University Press, New York.

HOFFMAN, R. A. and REITER, R. J. (1965): Rapid pinealectomy in hamsters and other small rodents. *Anat. Rec.*, *153*, 19.

KINCL, F. A. and BENAGIANO, G. (1967): The failure of the pineal gland removal in neonatal animals to influence reproduction. *Acta endocr. (Kbh.)*, *54*, 189.

LUCE-CLAUSEN, E. M. and BROWN, E. F. (1939): The use of isolated radiation in experiments with the rat: II. Effects of darkness, visible, and infrared radiation on three succeeding generations of rats. (a) Growth and storage of vitamin A. *J. Nutr.*, *18*, 537.

MACHLIN, L. J., TAKAHASHI, Y., HORINO, M., HERTELENDY, F., GORDON, R. S. and KIPNIS, D. (1968): Regulation of growth hormone secretion in non-primate species. In: *Growth Hormone*, pp. 292–305. Editors: A. Pecile and E. E. Müller. ICS 158, Excerpta Medica, Amsterdam.

MALM, O. J., SKAUG, O. E. and LINGJOERDE, P. (1959): The effect of pinealectomy on bodily growth, survival rate and P^{32} uptake in the rat. *Acta endocr. (Kbh.)*, *30*, 22.

MULINOS, M. G. and POMERANTZ, L. (1940): Pseudo-hypophysectomy. A condition resembling hypophysectomy produced by malnutrition. *J. Nutr.*, *19*, 493.

MÜLLER, E. E., GIUSTINA, G., MIEDICO, D., PECILE, A., COCCHI, D. and KING, F. W. (1970): Circadian pattern of bioassayable and radioimmunoassayable growth hormone in the pituitary of female rats. *Proc. Soc. exp. Biol. (N.Y.)*, *135*, 934.

MÜLLER, E. E. and PECILE, A. (1968): Studies on the neural control of growth hormone secretion. In: *Growth Hormone*, pp. 253–266. Editors: A. Pecile and E. E. Müller. ICS 158, Excerpta Medica, Amsterdam.

NARANG, G. D., SINGH, D. V. and TURNER, C. W. (1967): Effect of melatonin on thyroid hormone secretion rate and feed consumption of female rats. *Proc. Soc. exp. Biol. (N.Y.)*, *125*, 184.

OSTREICH, R. and SLAWYK, (1899): Riesenwuchs und Zirbeldrüssen-Geschwulst. *Virchows Arch. path. Anat.*, *157*, 475.

PALKA, Y., COYER, D. and CRITCHLOW, V. (1969): Effects of isolation of medial basal hypothalamus on pituitary-adrenal and pituitary-ovarian functions. *Neuroendocrinology*, *5*, 333.

REITER, R. J. (1967): The pineal gland: A report of some recent physiological studies. U.S. Army Edgewood Arsenal Technical Report 4110. U.S. Army Edgewood Arsenal, Maryland.

REITER, R. J., SORRENTINO JR, S. and ELLISON, N. M. (1970): Interaction of photic and olfactory stimuli in mediating pineal – induced gonadal regression in adult female rats. *Gen. comp. Endocr.*, *15*, 326.

REITER, R. J., SORRENTINO JR, S., RALPH, C. L., LYNCH, H. J., MULL, D. and JARROW, E. (1971): Some endocrine effects of blinding and anosmia in adult male rats with observations on pineal melatonin. *Endocrinology*, *81*, 895.

SCHALCH, D. S. and REICHLIN, S. (1966): Plasma growth hormone concentration in the rat determined by radioimmunoassay. Influence of sex, pregnancy, lactation, anesthesia, hypophysectomy and extrasellar pituitary transplants. *Endocrinology*, *79*, 275.

SCHALCH, D. S. and REICHLIN, S. (1968): Stress and growth hormone release. In: *Growth Hormone*, pp. 211–225. Editors: A. Pecile and E. E. Müller. ICS 158, Excerpta Medica, Amsterdam.

SORRENTINO JR, S. (1969): Pineal-mediated alterations of reproductive organs in blinded hamsters and rats. Ph.D. Thesis University of Tennessee Medical Units, Memphis, Tenn.

TRENKLE, A. (1970): Effect of starvation on pituitary and plasma growth hormone in rats. *Proc. Soc. exp. Biol. (N.Y.)*, *135*, 77.

WIENER, H. (1968): External chemical messengers. IV. Pineal gland and V. More functions of the pineal gland. *N.Y. St. J. Med.*, *68*, 912 and 1019.

CONTINUING SECRETION OF MAMMOTROPHIN BUT NOT SOMATOTROPHIN BY INTRAMAMMARY PITUITARY GRAFTS IN RATS*

W. R. LYONS, J. ASTRIN, C. AMSTERLAW and P. E. PETROPOULOS

Department of Anatomy, School of Medicine, University of California, San Francisco, Calif., U.S.A.

One finds ample reason to regret that the case for organ transplantation has been tried in the amphitheatres of cardiac and renal surgeons. One reason for the partial success of the latter in contrast to the failure of the former lies in the duality of kidneys; and the great advantage this gave the kidney transplanter in terms of more acceptable donors, during the early days of imperfect technic and knowledge. Why has there been such a paucity of grafting of other, less complex organs? Probably because of the unsuccessful, earlier attempts at grafting in general before newer knowledge had accrued with respect to immuno-suppressive methods. Then too, having or not having certain organs is not a matter of life or death. This is especially true of most endocrine organs for which a specific hormone or an unexpensive synthetic pharmaceutical agent can be substituted. The pituitary gland, however, offers a particular challenge in this respect because it synthesizes at least 6 anterior lobe hormones, at least one intermediate lobe hormone, and is the receptor of numerous hypothalamic neuro-endocrines inhibitory or stimulatory to its own hormones or to distant organs. With the synthesis of some of the relatively simple hypothalamic inhibitors and activators of the anterior pituitary cells the grafting of acceptable pituitaries in a region accessible to long-acting preparations (pellets, etc.) of these compounds makes the challenge more interesting.

Although the pituitary gland is considered unnecessary in the maintenance of life of an individual, one would have to admit that it is necessary for the survival of all species in which it has evolved. It is also necessary for the normal growth, metabolism and reproduction in all individuals endowed with it. Its hormones can be used as partial substitutes for the gland as shown in experimental animals and in some human beings to greater or lesser degrees. When some of its hormones are undersecreting or not secreting it is possible to substitute the target hormone instead of the deficient trophic hormone of the pituitary (*e.g.*, thyroxin instead of thyrotrophin; one or two of the adrenocortical steroids instead of ACTH; estrogen or androgen instead of the FSH-ICSH gonadotrophic pair; progestogen instead of the luteotrophic hormone). The direct action of somatotrophin (STH) on general body growth and certain aspects of metabolism; and of mammotrophin (MH) on the mammary gland have not, as yet, any substitutes. Of these two hormones, MH has been shown to be secreted by pituitary grafts or *in vitro* cultures independently of a hypothalamic activator and uninhibited by the hypothalamic inhibitor (PIF) (see Jacobsohn, 1966) (Everett, 1966). A large amount of evidence supports the contention that somatotroph cells require a hypothalamic releasor for synthesis and secretion of STH. Some experimenters have claimed that pituitary grafts independently secrete subnormal amounts of STH and others disagree (see Reichlin, 1966). This report represents an attempt to resolve this problem in

* Supported by Grant HD-02731 from the National Institutes of Health.

an experimental animal (the Long-Evans rat), and is a continuation of the one made at the First International Symposium on Growth Hormone in 1967 (Lyons *et al.*, 1968).

In the earlier paper we introduced new data on a procedure in which 1-month-old hypophysectomized rats may be used for the simultaneous assays of MH (mammotrophin, prolactin, lactogen) and STH (somatotrophin, growth hormone). At this time data based on that same procedure are presented to support the contention that a pituitary graft in the Long-Evans rat continues to secrete MH but not STH, the latter being released from the graft only during the first few days post implantation. Apparently the somatotroph cells are without the necessary hypothalamic neurohormonal releasor whereas the mammotrophs do not need (under these conditions) a hypothalamic releasor, and are indeed more effective without its inhibitor (PIF – prolactin inhibitor).

METHODS

Immediately after hypophysectomy ($\overline{\text{H}}$) 1-month-old Long-Evans female rats received one pituitary (or in some cases 1/8, 1/4 or 1/2 of a pituitary) as a graft in the subcutaneous fat embedding the No. 4 right mammary gland. Some of these rats received 25 μg of a prednisolone acetate (PA) suspension daily for 4 days in the graft area. This was necessary in synergism with the graft's MH to induce milk secretion; although it was also known to be inhibitory to STH's chondrogenic activity. Four control groups were also studied: (A) normal, untreated; (B) $\overline{\text{H}}$ and given a graft, but no PA; (C) $\overline{\text{H}}$ and untreated; and (E) $\overline{\text{H}}$ and treated only with PA in the same dosage as the main experimental group D. Rats in each of these 5 groups were necropsied after 4, 10, 15, 20 and 25 days of treatment (or no treatment) respectively. In all but the 4-day series the dosage of PA when given was 100 μg every 7 days. Thus although the daily amount of PA to which a rat was exposed varied slightly between series this was controlled by having 4 control groups for each experimental group. The main experimental group (D) was the one in which MH and STH may be assayed simultaneously in a $\overline{\text{H}}$ rat receiving one or both of these hormones plus PA. Mammary glands and tibias were processed for histologic examination as previously reported (Lyons *et al.*, 1968). Ovaries, adrenals and thyroids were also studied until the evidence became convincing that ACTH, TSH, FSH and ICSH may be released from the graft only during its adjustment and partially necrotizing period of 4 days (or slightly more). Thereafter it seemed unnecessary to process these organs for evaluation of their obviously regressive tendency due to a lack of the hypothalamic releasors.

The secretory response of each of 4 quadrants of the main body of the sectioned mammary glands was rated on a 0 to 4+ scale; and an average value was derived as the lactogenic index. The proximal epiphysial tibial cartilage was measured in median longitudinal sections with an oculomicrometer calibrated with an object micrometer. This proximo-distal measurement is reported in micra. It includes the zone of provisional calcification of the hypertrophic, degenerating chondrocytes; and is thus usually about 25% greater than the dimension measured by the cruder silver nitrate technic which is limited to the zones of non-calcified matrix.

This work will be reported in two parts: (1) the graft's excellent continuing potency in lactogenic activity; and (2) its lack of somatotrophic secretory activity except for that present at the moment of explantation and released with the aid of residual 'releasor' or merely 'leaked out' during the necrotizing process that is usually detectable in the central core of the explant.

In this experiment and in all earlier ones in which attempts were made to titrate a graft by implanting it (even in highly favorable recipients such as in auto or sibling grafting), the data derived were of little use because in the case of MH one not only detected the momentary content, but also the continuing secretion of grafts that are readily acceptable in the mammary area as well as under the renal capsule. For the momentary content of any given pituitary we have had to use an extractive procedure which is quite feasible. The grinding of a single gland to a fine emulsion is only useful if no living MH cells remain to function and divide – a process we have observed.

In an attempt to determine the smallest fraction of the anterior pituitary capable of releasing MH and STH a procedure was used that could not be considered quantitative; but it gave results that seemed worth reporting. The anterior lobe with posterior lobe removed was sectioned with a sharp razor blade along the mid-line furrow. These halves were then halved through the transverse axis to make 1/4's and these were halved to make 1/8's. Our intention in studying graft sections was not so much to quantitate activities but rather to test this procedure in determining the effect of hypothalamic releasing factors (or PIF) infiltrated in and around a pituitary, the secretions of which could be detected by their effects on the specific targets (including cartilage). The work on pituitary fractions is reported to show that although MH may be detected in 1/8th of an anterior lobe, a whole one is necessary to show the small, initial release of STH.

RESULTS AND DISCUSSION

The graft's mammotrophic activity

Intramammary auto- or homografts of rat pituitaries are easily transplanted and well-accepted by the 1-month-old host. There are enough secretory units in the mammary gland of these immature recipients to respond quickly to the lactogenic activity of the graft if an adequate dose of an adrenocorticoid is infiltrated in the implant area. Since control series of grafts without additional prednisolone showed equal if not better acceptance there was no evidence that prednisolone provided any benefit other than through its necessary adjunct role in lactogenesis. The assessments given to the grafts upon gross and histologic examination were arbitrary and comparative. Planimetry studies on serial sections and counts of specific cell types would present formidable difficulties in any attempt at quantitation. It would be possible to approach more acceptable, objective quantitation of the graft's lactogenic activity by chemical analyses of the milk products or enzymes. However, in these preliminary experiments, arbitrary lactogenic indices were assigned to each gland after averaging the degree of secretory response in 4 different sectors of the greatest mammary area including the graft.

In Table I it may be noted that the values assigned to the grafts varied greatly in the whole pituitary series within a given time group, and regardless of whether or not prednisolone was also administered. The lactogenic indices were essentially equal after all 5 intervals, but the graft values deteriorated after 15 days. Two explanations can be offered for this apparent discrepancy: (1) most of these small glands respond maximally in 4 days to either a whole graft or about 50 μg of locally injected human MH, and evidence of this degree of stimulation remains in the form of over-distended alveoli and ducts containing inspissated milk (since it cannot escape via the nipples); and (2) this pathologic milieu becomes increasingly unfavorable toward a previously healthy and functioning graft (see Figs. 5, 10, 11, 12). Some of the mammary glands especially in the 20 and 25 day D Groups showed barely recognizable graft parenchyma and yet these grafts had induced good lactogenesis before deteriorating.

Since all of the 24 whole pituitary implants were accepted, and all induced milk secretion, fractions of 1/2, 1/4 and 1/8th were also graded for lactogenesis. The results proved to be interesting in showing that cutting a pituitary into fractions as small as an estimated 1/8th does not always prevent its acceptability as a viable and functional graft. Obviously, this procedure should lead to methods of evaluating stimulatory or inhibitory agents of the

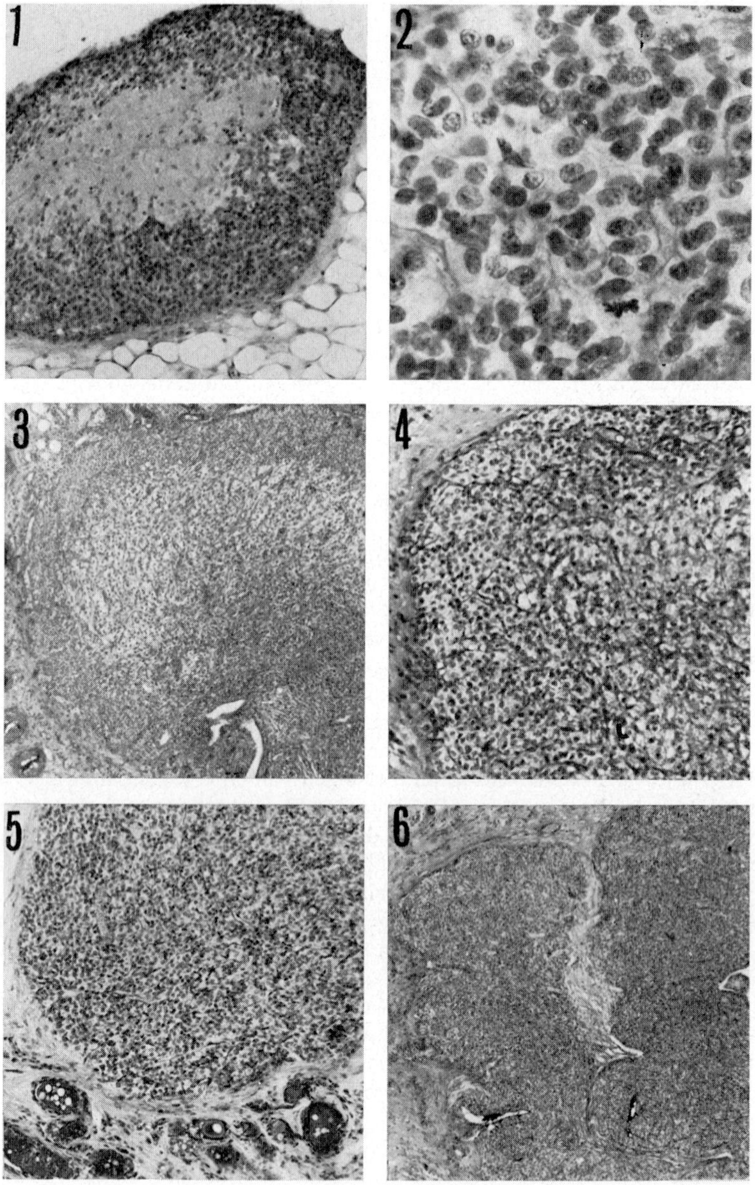

Fig. 1. A 100 × magnification of a secretion of 1/8th of a male sibling anterior pituitary 4 days after implantation in the mammary area of a hypophysectomized ($\overline{H}$), 1-month-old female. Most grafts show the central necrosis seen here, and a well-accepted peripheral area which becomes highly vascularized.

Fig. 2. A higher magnification (600 ×) of a section of the same pituitary graft shown in Fig. 1. The cells with abundant cytoplasm showed Orange G positive granules (presumably mammotrophs). One mitotic figure is present.

Fig. 3. A 30 × magnification of a section of a whole male non-sibling pituitary 4 days after intra-mammary grafting in a 1-month-old $\overline{H}$ female. No prednisolone was given with this graft and no milk secretion followed. Centrally there are degenerative changes in the parenchymal cells (pycnosis and atrophy) and lymphocytic infiltration.

TABLE I

Average whole graft assessments and lactogenic indices

Groups*	No. of rats	Days	Average graft assessment					Lactogenic index***
			-**	1+	2+	3+	4+	
B	16	4	0	2	11	1	2	0
B	10	10	0	1	3	4	2	0
B	9	15	0	1	0	4	4	0
B	9	20	0	0	3	1	5	0
B	11	25	0	3	5	3	0	0
D	24	4	0	12	9	3	0	3.1
D	22	10	1	9	4	7	1	2.9
D	32	15	2	12	11	5	2	3.1
D	32	20	8	15	6	0	3	3.8
D	33	25	13	13	4	2	1	3.2

* As in Figs. 14 to 18, Groups D received PA and Groups B did not.
** A negative assessment indicates that too little of the graft remained even though the remnant was found in a gland well filled with condensed milk in D groups.
*** Based on assignment of arbitrary values to each quadrant of the largest gland area.

pituitary's trophic hormones. Fourteen $\overline{H}$ rats implanted with 1/2 of a pituitary from non-sibling females and injected daily for 4 days with 25 μg PA all accepted the grafts in varying degrees and all showed milk secretion with an average index of 1.4. Thirteen rats similarly treated with 1/4 of a pituitary from non-sibling females showed functional graft acceptance in 10 and the average lactogenic index in these was 1.2. Seventeen rats similarly treated with 1/8th of a pituitary from non-sibling females showed functional graft acceptance in 9 and the average lactogenic index was 1.4. In 4 of the 8 non-lactating rats fragments of what appeared to be viable pituitary were seen, indicating that in terms of lactogenic activity whether from hormone release due to damage or lysis or from a continuing synthesis and secretion in viable cells, one approximates in 1/8th of a pituitary the minimal size of fragment that could be used in this 4-day test. It should be recalled that (Nikitovitch-Winer and Everett, 1959) illustrated the central necrosis that occurs after whole gland transplantation. That it also may occur after grafting 1/8th of a pituitary is shown in Fig. 1. In spite of this considerable degeneration in an area not so accessible to ingrowing blood vessels, this graft induced milk secretion and judging by mitoses in parenchymal cells was capable of proliferating. Other grafts with some areas showing degeneration (nuclear pycnosis, lymphocytic

←

Fig. 4. A 100 × magnification of a section of a whole male non-sibling pituitary, 20 days after intramammary grafting in a 1-month-old $\overline{H}$ female. A fibrous capsule has formed around the graft. The peripheral areas are well vascularized; and the Orange G cells appeared as in a normal gland. Centrally there is necrosis, fibrosis and lymphocytic and macrophagic infiltration.

Fig. 5. A 60 × magnification of a male non-sibling pituitary, 25 days post implantation. Prednisolone (100 μg) was infiltrated in the area of the graft every week. Mammary alveoli contain inspissated milk.

Fig. 6. 30 × magnification of a section of a male non-sibling pituitary. Almost the entire graft is shown and was considered to be well vascularized and in good condition 25 days after implantation. No prednisolone was injected and the mammary gland did not show lactogenesis.

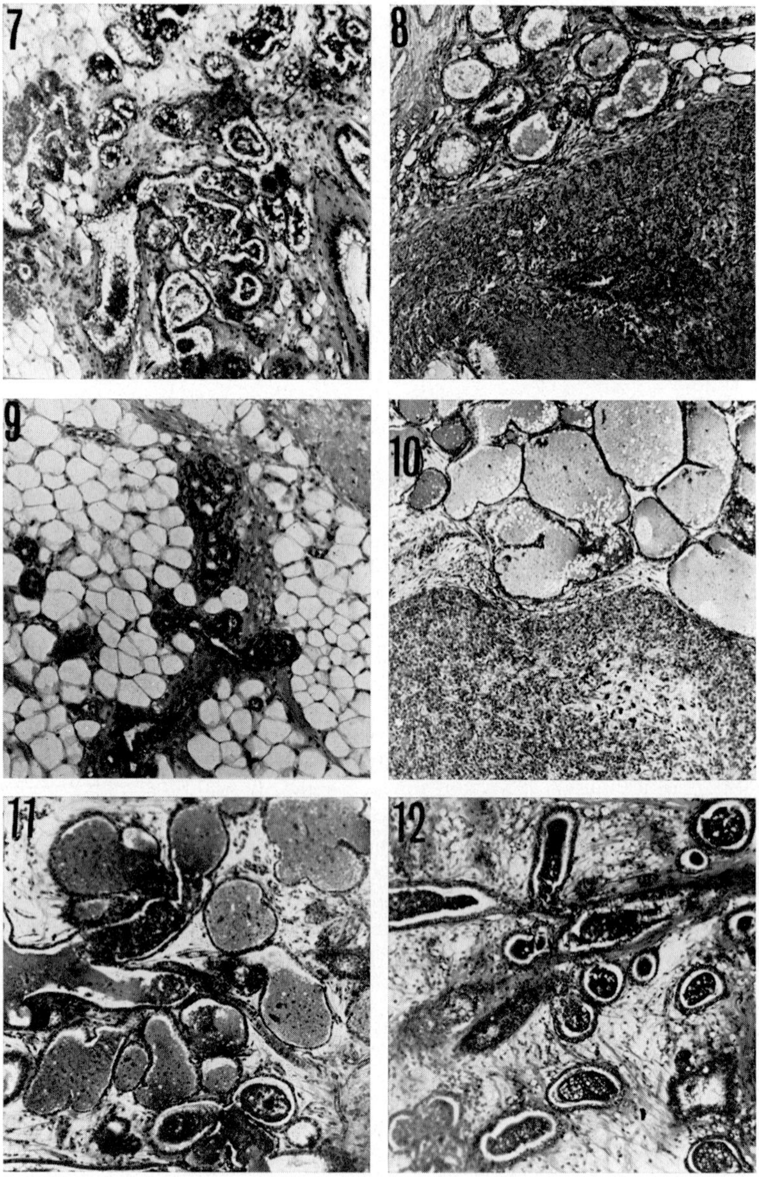

Figs. 7–12. Sections of mammary glands (× 60) from rats hypophysectomized at 1 month of age grafted (as indicated) with a pituitary and necropsied after stated intervals.

Fig. 7. Milk secretion (2+) in a gland 4 days after receiving 1/8th of a male sibling pituitary.

Fig. 8. Milk secretion (4+ in most areas) in a gland 4 days after receiving one male non-sibling pituitary plus 25 μg of prednisolone daily. The graft shown here was in good condition except for the region of central necrosis (bottom of section).

Fig. 9. A control left No. 4 mammary section from a rat that had responded to a graft in the opposite gland 4 days post implantation. This section showing only duct and non-secreting alveolar buds is typical of all mammary glands not grafted and not receiving prednisolone.

or macrophagic infiltration and fibrous replacement and encapsulation) or a 'healthy' normal appearance may be seen in Figs. 3, 4, 5, 6, 8 and 10. That 1/8th of a pituitary from a 1-month-old non-sibling female may indeed survive and function in the usual test rat was shown in a 10-day experiment with 7 rats. Six of these showed a 1.8 average lactogenic index and grafts with a 'healthy' appearance were detected in all but the one non-lactating gland.

As shown in Table I none of the 55 $\overline{H}$ rats with accepted grafts but uninjected with PA showed milk secretion nor did any of the 74 $\overline{H}$ rats receiving only PA nor any of the 67 $\overline{H}$ untreated rats nor any of the 41 normal untreated controls.

An accompanying feature in the mammary glands that received pituitary grafts was the regression of the large fat cells (Fig. 9) to an essentially lipid-free mesenchymal cell. This was particularly true in milk-secreting glands (Figs. 10, 11, 12) in which the developing parenchyma usurps the rich capillary plexus around the fat cells. This lipolysis results in a direct transfer of fatty acids to be used in milk fat synthesis, and in the absorption into blood and lymph vessels for general distribution. It is known (Lyons, 1966*a*) that hormones may compel an animal to deplete its tissue stores in order to synthesize milk and to support the energy involded in this process. Grafts without prednisolone, that did not induce milk secretion, did however cause some lipolysis and this was ascribed tentatively to the pars intermedia cells that survived in excellent condition in the grafts and would therefore be a good source of MSH which is a potent lipolytic agent.

The graft's lack of continuing somatotrophic activity

The results of our attempts to determine a somatotrophic influence of the grafts on the tibial epiphysial cartilage are illustrated in Figs. 13-18, which also serve as tabulations. As noted in the legends for each of these figures, the bar graphs represent the mean disk dimension and except in Fig. 13 are labelled as follows: A = normal terminal control for each time interval; B, C, D and E = hypophysectomized rats treated thus: B = only grafted with 1 pituitary; C = no treatment; D = grafted with 1 pituitary and injected in the graft area with PA; E = PA only. The lettering serves in the same way for the underlying, corresponding photomicrographs of the epiphysial disks.

In Fig. 13 the lettering is used on the bar graphs representing epiphysial disk averages and designates the 4-day treatment regimens in the usual $\overline{H}$ 1-month-old female, all of which received 25 μg of PA in the graft area plus the following amounts of anterior lobe graft: A, 1/8; B, 1/4; C, 1/2; D, 1; E, 0. The figure for the animals that received one anterior lobe, D = 257 ± 5.7, was found to be significantly different from those of the other 4 groups (with a p value < 0.01 in all cases). The latter groups were not significantly different from each other ($p = 0.1$ or greater). It would seem therefore, that a whole anterior lobe grafted in the mammary locus, infiltrated with PA releases a detectable amount of chondrotrophic substance (presumably STH). This average increase of 37 μ over the PA controls may also be obtained with a daily dose of approximately 20 μg of the bovine International Standard (B-10) of STH containing 1 I.U. per mg.

←

Fig. 10. Over-distension of alveoli with milk (some had ruptured and milk was seen in the interalveolar areas) in a gland that had received one male sibling pituitary plus 100 μg of prednisolone on days 1 and 8. Necropsy 10 days after grafting. An area of necrosis and fibrosis is seen; but the graft was in excellent condition elsewhere.

Fig. 11. Essentially the same picture as shown in Fig. 10. Treatment was the same except for day of necropsy which was 15 days post grafting. After 4 to 10 days and a 4+ lactogenic response, the milk becomes inspissated and is slowly resorbed as in a normally lactating adult rat when its litter is removed. The alveolar and ductal epithelium regresses and cells are cast off into the lumina.

Fig. 12. Section of a gland that had also received similar treatment except that necropsy was performed 20 days after grafting. Further condensation of the milk and parenchymal regression is shown.

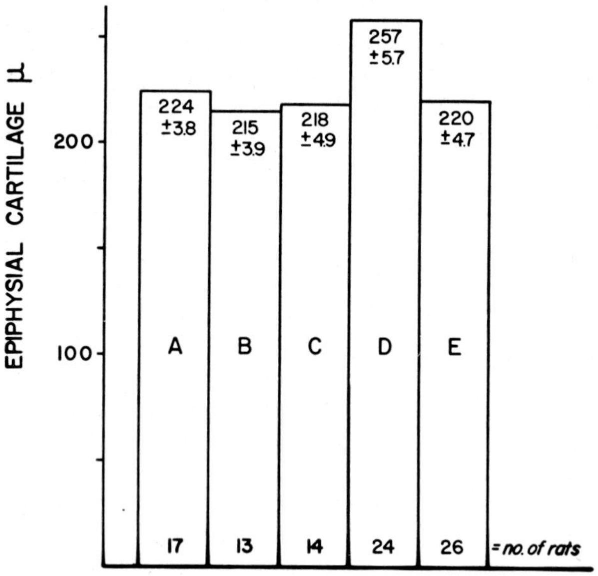

Fig. 13. Bar graphs showing average measurements in μ (± S.E.) of the proximal tibial epiphysial cartilages from female rats hypophysectomized ($\overline{\text{H}}$) at 1 month of age, and implanted in the right No. 4 mammary area immediately after $\overline{\text{H}}$ as follows: (A) received 1/8th of an anterior lobe only on day of $\overline{\text{H}}$ plus 25 μg of PA daily for 4 days; (B) received 1/4 of an anterior lobe plus PA as in (A); (C) received 1/2 of an anterior lobe plus PA as in (A); (D) received 1 anterior lobe plus PA as in (A); (E) received only 25 μg of PA daily for 4 days. All of the fractional grafts (A, B, C,) were from female non-siblings. The 24 whole grafts (D) were of 5 types, *viz.* autografts and grafts from male or female siblings and from male or female non-siblings. No significant difference was found in the average response to these 5 types of whole grafts (all p values >0.1) nor was there any significant difference between the average cartilage measurements of groups A, B, and C receiving less than 1 pituitary and the control group E receiving no grafts (all p values >0.1). Group D receiving one pituitary showed a significant difference between its disk average and that of all other 4 groups (p = <0.001).

In Fig. 14 the difference between bars A (467, normal) and C (305, $\overline{\text{H}}$ only) or 162 μ is representative of the disk's regression due to $\overline{\text{H}}$ 4 days previously. One pituitary graft without PA prevents some of this regression and offers further evidence of some STH becoming available from the graft. That PA causes considerably more disk regression in $\overline{\text{H}}$ rats than $\overline{\text{H}}$ alone is clearly shown in the difference between bars C and E and their corresponding photomicrographs. Bars D and E repeat what has already been shown in Fig. 13 and the corresponding disk sections show that one graft prevents some of the chondrolytic action of PA. It should be re-emphasized that the PA is a necessary synergist in MH's lactogenic effect, in contrast to its antagonism toward STH's chondrotrophic activity.

In noting the gradual decrement in the proximo-distal dimension of the normal disk (467 to 310 μ in 25 days), it should be remembered that the whole tibia is growing rapidly in length and girth during this period, and the true disk volume could only be determined by including the circumferential dimension which is increased.

The figures on the normal rats are presented only to re-emphasize how important STH or an intact pituitary are in the maintenance of the disk and to serve as guide marks in future experiments with multiple grafts or single grafts plus an STH-releasing factor. As a corollary, the average disk figures for the untreated $\overline{\text{H}}$ rat are equally useful in that they give some indication of the rate of regression of the disk due to the absence of STH. The average

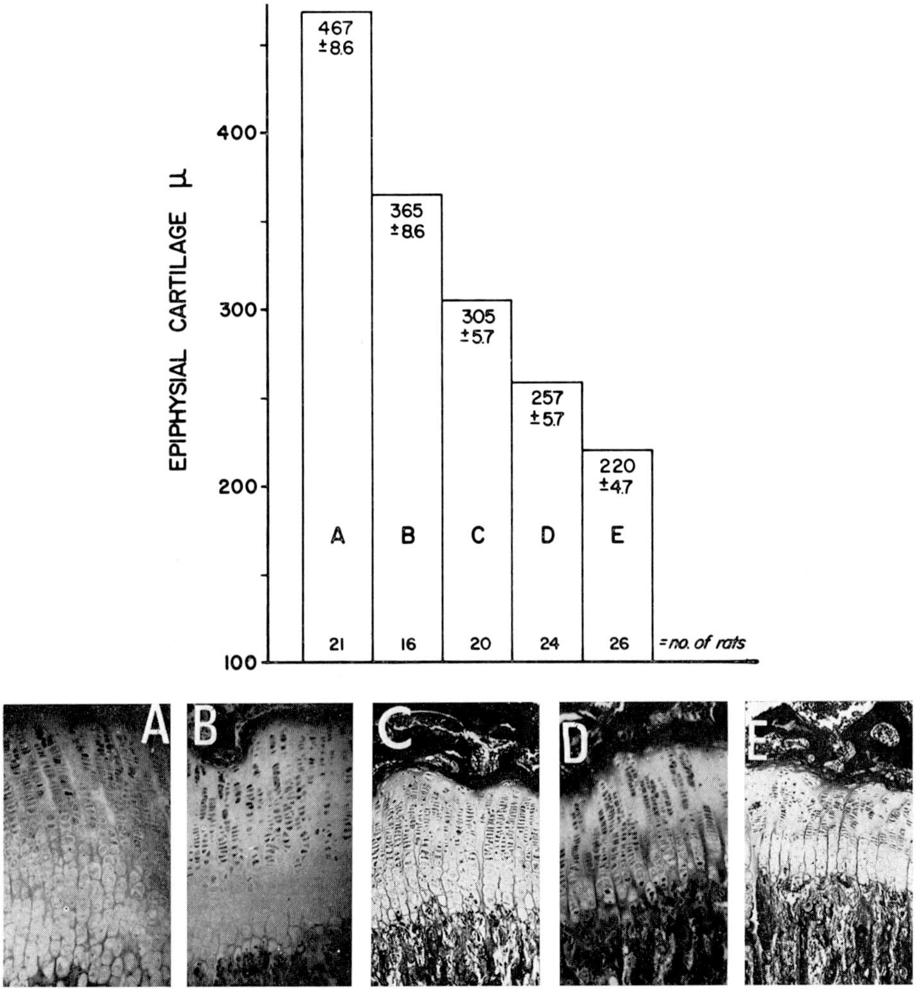

Fig. 14. Upper: bar graphs showing average measurements in μ of the proximal tibial epiphysial cartilage in the 4-day tests. A = normal terminal controls; B = hypophysectomized ($\overline{H}$) plus 1 pituitary graft in the right No. 4 mammary area; C = $\overline{H}$ only; D = $\overline{H}$ + 1 pituitary graft plus 25 μg PA daily; E = $\overline{H}$ + 25 μg PA daily for 4 days.
Lower: Photomicrographs of typical disks of the above 5 groups.

dimension of 305 ± 5.7 μ after 4 days is significantly different from 234 ± 7.9 μ after 10 days ($p < 0.001$) and these 2 values are significantly different from 201 ± 4.5 at 15 days, 187 ± 7.3 at 20 days, and 194 ± 5.6 at 25 days. It is interesting that the latter 3 values do not differ significantly from each other, implying not necessarily a complete growth stasis in the disk, but rather an equilibrium between slight chondrogenesis and chondrolysis. In order to determine whether or not the graft is secreting or releasing a chondrogenic hormone these figures for the 5 different intervals in $\overline{H}$ untreated rats (Group C), and $\overline{H}$ rats treated only with a whole pituitary graft (Group B) should be compared. They are as follows: 4 days, 60 μ; 10 days, 33 μ; 15 days, 43 μ; 20 days, 29 μ and 25 days, 26 μ. These differences are all significant after each time interval ($p < 0.01$). However, one cannot conclude from

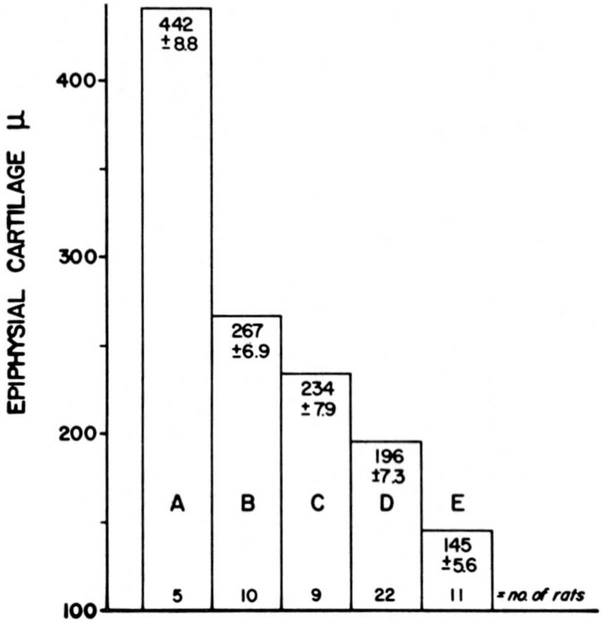

Fig. 15. As Fig. 14 except that these are bar graphs only showing the results of the 10-day tests. The PA was injected in 100 μg doses every 7 days.

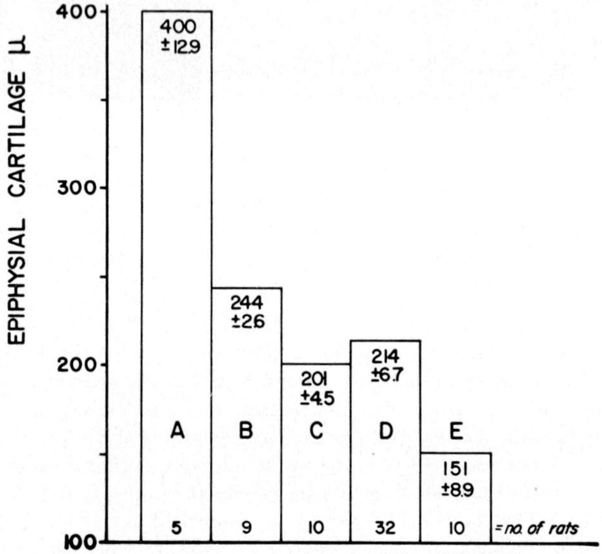

Fig. 16. As Fig. 15 except that these are bar graphs only showing the results of the 15-day tests.

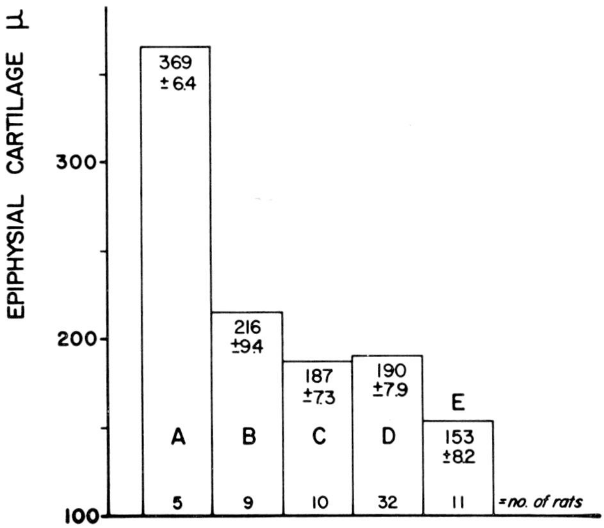

Fig. 17. As Fig. 15 except that these are bar graphs only showing the results of the 20-day tests.

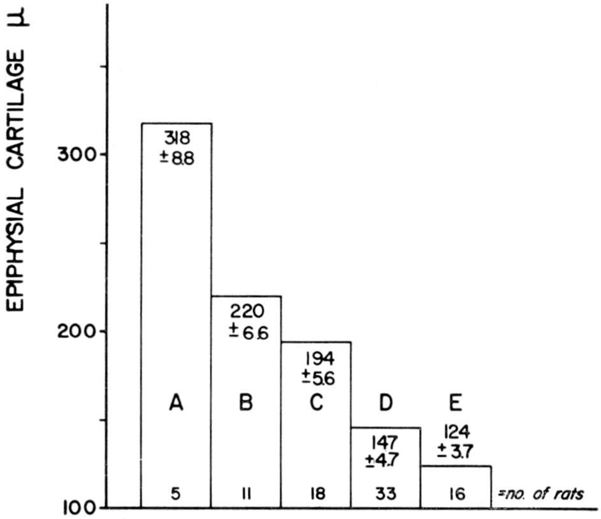

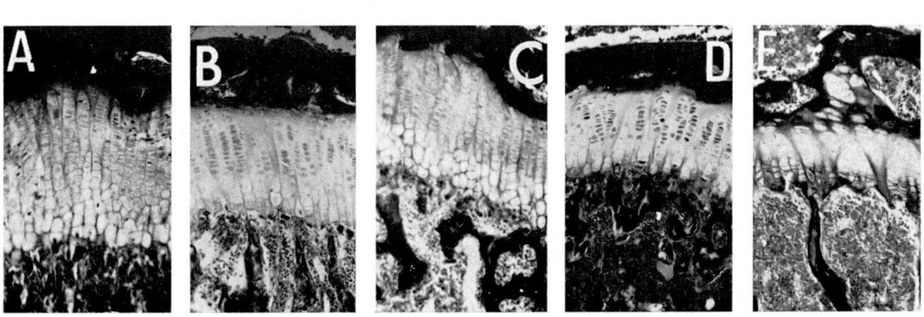

Fig. 18. As in Fig. 14 except that these graphs and photos illustrate the results of the 25-day tests, and 100 µg of PA was injected in the graft area every 7 days.

these data that STH continues to be released from the graft for any of the periods after 4 days. An alternative explanation would be that the slight, but highly significant activity of the small amount of STH in the gland at the time of implantation 'leaked out' or was somehow released and did prevent the greater degree of disk regression noted in the untreated $\overline{H}$ rats and that this accrued benefit had been dissipated by only about 50% at the end of 25 days. An earlier experiment from this laboratory (Ugrob *et al.*, 1957) supports this interpretation. Ten normal rats gained an average of 160 $\pm$ 7 g between the ages of 40 and 75 days. They were then $\overline{H}$ and in the ensuing 28 days lost an average of 59 $\pm$ 3 g. Ten 40-day-old rats, $\overline{H}$ and treated daily with STH, gained an average of 148 $\pm$ 5 g in 35 days and when treatment was stopped lost an average of 52 $\pm$ 5 g during the following 28 days. This same capacity to retain much of the gain accomplished through STH treatment was confirmed with pellets of STH which may be heated at 100° C for sterilization and implanted subcutaneously for gradual release of the hormone and eventual recovery and weighing.

In an experiment designed to test both MH and STH simultaneously in the same $\overline{H}$ rat the use of PA is mandatory in order to establish a lactogenic index and, even though it is detrimental to cartilage growth, this is readily controlled with $\overline{H}$ rats receiving only the same regimen of PA. Thus in groups D and E there was essentially the same opportunity to estimate STH activity as was found in Groups B and C. The disk differences between Group D receiving one pituitary plus PA and Group E receiving only PA for the 5 different intervals were as follows: 4 days, 37 μ; 10 days, 51 μ; 15 days, 63 μ; 20 days, 37 μ; and 25 days, 23 μ. In each instance the difference between the D and E groups was found to be significant ($p < 0.01$). This is obvious from the corresponding photomicrographs in Figures 14 and 18, which may be considered typical of each group.

GENERAL DISCUSSION

The mammotrophic and somatotrophic activities of the pituitary grafts have been discussed with the data presented in the above two sections. None of these data support the contention that intramammary grafts without benefit of a 'releasing' agent secrete STH beyond the 4-day post implantation interval. The experiments seeming to support subnormal STH secretion of ectopic pituitaries, and those clearly refuting this are well presented by Reichlin (1966), and it is not our intention to review the controversy. Judging by effective early release of the graft's content of a chondrotrophic substance, one might expect, by multiple and particularly sequential grafting, to restore a $\overline{H}$ rat's skeletal growth to normal. As shown by Peng *et al.* (1969) thyroxin increased the effectiveness of their renal grafts. This group of workers using sibling grafts in Long-Evans rats over a 15-week experimental period observed subnormal body weight gains and nose-tail measurements among many other parameters studied. They did not use epiphysial disk measurements which we find more accurate than body weight changes. They indicated their awareness of the fact that body weight increase does not necessarily only reflect an STH effect and presumably they used the nose-tail measurement instead of the epiphysial disk as adequate evidence of skeletal growth. This paper is mentioned here because it was published after the Reichlin (1966) review and because, after suggesting other possible interpretations of their results, *viz.* (1) an STH-like effect of MH, (2) a body weight increase due to fat deposits, and (3) the availability in the systemic circulation of an STH-releasing factor, they conclude that the renal grafts 'can synthesize and secrete growth hormone as well as the graft placed under the median eminence if thyroxine is given daily.' Their third interpretation is derived from the work of Müller *et al.* (1967) who found that a neural STH-releasing factor (GRF) is detectable in the blood of rats $\overline{H}$ for periods of over 1 week (3 months). Considering the short half-life of GRF and its dilution by systemic blood before reaching an intramammary graft, it is not surprising that our disk measurements within the 10-25 day post hypophysectomy period showed no indication of its activity. However, our experimental model of a $\overline{H}$ rat with a graft easily available to infiltration by GRF, preferably with a resorption-retarding agent, should prove useful in detecting this or other releasing and inhibiting neuro-secretions especially since the graft may be excised and assayed for its momentary content of the different trophic hormones.

Another relevant reference is that of Rubenstein and Ahren (1965). They concluded that there was a slight but significant body growth in Sprague-Dawley male rats $\overline{H}$ at 8 weeks of age and implanted with 1 pituitary autograft or 1 autograft plus 3 isografts over a 7-week period. The average body weight increase in the 10 rats with 1 autograft was 23 g; and in the 10 rats with 1 autograft plus 3 isografts was 54 g. Their 7 normal controls gained an average of 248 g during the same period. Of their 2 explanations: (1) a residual secretion of small amounts of STH, and (2) a possible effect of considerable quantities of MH, we find ourselves in accord with the first.

There would be no need to discuss the inability of a MH-secreting graft to stimulate milk secretion without its co-lactogen ACTH (indirect) or an adrenocorticoid (direct) if one could disregard the contribution of Dao and Gawlak (1963). These authors, in spite of considerable evidence to the contrary (consult Cowie, 1966; Lyons, 1966*b*) have claimed that intramammary pituitary grafts were capable of inducing milk secretion without the aid of any steroid. It seems unlikely that strain differences would account for this discrepancy (Sprague-Dawley *vs* Long-Evans rats) or that food and drink would provide the difference. That a small residue of a pituitary would be sufficient to activate the adrenal cortex to supply the minute amount of either a gluco- or mineralo-corticoid necessary for synergism with MH should not be disregarded.

SUMMARY

Long-Evans female rats hypophysectomized at the age of 28-30 days were immediately implanted with whole pituitaries (or fractions thereof) from themselves or siblings or non-siblings of either sex. In 4-day tests 25 μg of prednisolone acetate was injected daily around the implant and in longer experiments, up to 25 days, it was given in 100 μg doses every week. At necropsy, the mammary gland containing the graft as well as its contralateral gland were inspected for milk. They were then fixed in Bouin's fluid, as were the ovaries, adrenals, thyroids and tibias, and all were processed for histologic study and measurement of the proximal epiphysial cartilage.

Grafts of from 1/8th to 1 pituitary secreted mammotrophin (MH) and induced milk secretion in the 4-day tests. There was histologic evidence that small amounts of ACTH, FSH, ICSH, STH and TSH were released only in the 4-day whole graft tests from their respective secretory cells which became atrophic and inactive due presumably to the absence of hypothalamic stimulation.

In the longer experiments in which the grafts appeared viable only MH and possibly MSH (from the pars intermedia) were secreted. In control series, injected with prednisolone and not grafted, no milk secretion was observed and this was also true in cases that retained good grafts but did not receive prednisolone. Some of the beneficial effect of the early release of a chondrogenic hormone (? STH) was still noticeable in the tibial epiphysial disks even after the 25-day post grafting period.

ADDENDA

Two short experiments in support of suggestions made in the 'Discussion' were performed since this manuscript was submitted: (1) one homograft from 1 month old donors implanted weekly for 4 weeks induced a disk measurement in 8 rats $\overline{H}$ at 1 month of age averaging 300 ± 10.9 μ after 30 days. This is within the normal range for 2 month old rats; (2) three homografts implanted in mammary areas infiltrated with prednisolone for 4 weeks had induced good milk secretion; and were then transplanted to a second set of 3 $\overline{H}$ rats also given prednisolone. They also induced milk secretion in these rats.

REFERENCES

COWIE, A. T. (1966): Anterior pituitary function in lactation. In: *The Pituitary Gland, Vol. II*, Chapter 13, pp. 412–443. Editors: G. W. Harris and B. T. Donovan. Butterworths, London.

DAO, T. L. and GAWLAK, D. (1963): The direct mammotrophic effect of a pituitary homograft in rats. *Endocrinology, 72/6*, 884.

EVERETT, J. W. (1966): The control of the secretion of prolactin. In: *The Pituitary Gland, Vol. II*, Chapter 5, pp. 166–194. Editors: G. W. Harris and B. T. Donovan. Butterworths, London.

JACOBSOHN, D. (1966): The technics and effects of hypophysectomy, pituitary stalk section and pituitary transplantation in experimental animals. In: *The Pituitary Gland. Vol. II*, Chapter 1, pp. 1–21. Editors: G. W. Harris and B. T. Donovan. Butterworths, London.

LYONS, W. R. (1966*a*): Hormonal treatment of lactating rats on a protein-free diet. *Endocrinology, 78/3*, 575.

LYONS, W. R. (1966*b*): The physiology and chemistry of the mammotrophic hormone. In: *The Pituitary Gland, Vol. I*, pp. 527–581. Editors: G. W. Harris and B. T. Donovan. Butterworths, London.

LYONS, W. R., LI, C. H., AHMAD, N. and RICE-WRAY, E. (1968): Mammotrophic effects of human hypophysial growth hormone preparations in animals and man. In: *Growth Hormone*, pp. 349–363. Editors: A. Pecile and E. E. Müller. ICS 158, Excerpta Medica, Amsterdam.

MÜLLER, E. E., ARIMURA, A., SAITO, T. and SCHALLY, A. P. (1967): Growth-hormone releasing activity in plasma of normal and hypophysectomized rats. *Endocrinology, 80/1*, 77.

NIKITOVITCH-WINER, M. and EVERETT, J. W. (1959): Histo-cytologic changes in grafts of rat pituitary on the kidney and upon re-transplantation under the diencephalon. *Endocrinology, 65/3*, 357.

PENG, M. T., PI, W. P. and WU, C.-I. (1969): Growth hormone secretion by pituitary grafts under the median eminence or renal capsule. *Endocrinology, 85/2*, 360.

REICHLIN, S. (1966): Regulation of somatotrophic hormone secretion. In: *The Pituitary Gland, Vol. II*, Chapter 8, pp. 270–298. Editors: G. W. Harris and B. T. Donovan. Butterworths, London.

RUBENSTEIN, L. and AHREN, K. (1965): Growth hormone secretion in hypophysectomized rats with multiple pituitary transplants. *J. Endocr., 32*, 99.

UGROB, J. M., LYONS, W. R., JORDAN, C. W. and LI, C. H. (1957): Body weight changes in hypophysectomized rats following implantation of somatotrophin pellets. *Endocrinology, 61/4*, 477.

VII. Clinical investigations

SOME METABOLIC CHANGES INDUCED BY ACUTE ADMINISTRATION OF HGH AND ITS REDUCED-ALKYLATED DERIVATIVE IN MAN*

EROL CERASI, CHOH HAO LI and ROLF LUFT

Department of Endocrinology and Metabolism, Karolinska Hospital, Stockholm, and the Hormone Research Laboratory, University of California, San Francisco, Calif., U.S.A.

Human growth hormone (HGH) is a polypeptide comprising 190 amino acid residues with two disulfide bonds (Li *et al.*, 1966). It is a general belief that disulfide bonds serve the purpose of stabilizing the conformation of a molecule and may be involved in the active sites of the hormone. Thus, the reduction of the disulfide bonds of the insulin molecule causes complete loss of the biological and immunological activities of the hormone (Dixon and Wardlaw, 1960; Yalow and Berson, 1961). It is therefore surprising that complete reduction and carbamidomethylation of the cystine residues of HGH did not alter the lactogenic and growth promoting properties of the protein (Bewley *et al.*, 1969).

The present study was undertaken in order to evaluate some of the acute actions of native HGH and of reduced-tetra-S-carbamidomethylated HGH (RA-HGH) in man. In addition, the immunologic potencies of these preparations were compared with that of the local HGH standard.

MATERIAL AND METHODS

The studies were performed on five young healthy non-obese volunteers. Each subject served as its own control. The time interval between the studies was at least 48, usually 72 hours.

The studies were started at 8 a.m. after an overnight fast. In the first experiment, a slow intravenous saline infusion of 30 min duration (resting period) was followed by the rapid intravenous injection of 0.5 g of glucose per kg body weight. Blood samples were drawn into heparinized syringes at 0, 15, 30, 40, 45, 50, 55, 60, 70, 80 and 90 min.

In the second experiment, 4 mg of HGH in saline were infused intravenously over a period of 60 min after the initial resting period. The HGH infusion was followed by a 90 min saline infusion at the end of which glucose was injected rapidly as described above. Blood samples were drawn at 0, 15, 30, 40, 50, 60, 75, 90, 120, 150, 180, 185, 190, 195, 200, 205, 210, 220, 230 and 240 min.

In a third experiment, an identical protocol was followed except that RA-HGH was administered instead of native HGH.

In subjects 3 and 4, the second and third experiments were repeated after several weeks with a smaller hormone dose, 0.9 mg over 60 min.

In one additional subject, a woman of 27 with panhypopituitarism of unknown origin,

* These studies were supported by generous grants from the Swedish Medical Research Council (B70-19x-34-06B and B71-19x-34-07C), the Knut and Alice Wallenberg Foundation, and partly from the American Cancer Society and the Geffen Foundation.

native HGH and RA-HGH were each administered intramuscularly for 4 days in a daily dose of 10 mg. The intravenous glucose tolerance was determined before and on the third day of the experiments. A glucose infusion test was performed before and on the fourth day of the study: 500 mg of glucose were given as a priming injection followed by continuous infusion of 20 mg/kg/min during 1 hour. Blood samples for determination of glucose and insulin were taken at 0, 10, 20, 30, 40, 50, 60, 80, 100 and 120 min.

Native HGH was prepared according to Li *et al.* (1962). RA-HGH was prepared as described by Bewley *et al.* (1968). The local HGH standard was prepared according to Roos *et al.* (1963).

Blood glucose was determined enzymatically with a commercial glucose oxidase preparation (Kabi, Stockholm), glycerol in plasma according to Chernick (1969), insulin in plasma by the double-antibody immunoassay of Hales and Randle (1963), plasma HGH by the double-antibody technique of Cerasi *et al.* (1966) and alanine in plasma with an enzymatical technique (Karl and Kipnis, to be published). The disappearance rate of glucose in blood (k-value) was calculated as described by Ikkos and Luft (1957).

The insulin response to glucose was expressed as the insulinogenic index: the area below the plasma insulin curve with the base at the fasting insulin level divided by the area below the curve of the absolute blood glucose values.

RESULTS

The affinity of native HGH and RA-HGH to antibodies, prepared by immunization with HGH-Roos, was tested by using two dilutions of the hormones, 6.25 and 25 mμg/ml. Figure 1 shows the displacement of ^{131}I-labelled HGH-Roos from the antibodies induced by the three HGH preparations. There was no significant difference in this respect between native HGH and HGH-Roos. On the other hand, RA-HGH could only partially compete for the binding sites of the antibodies, the immunological activity being about 50% of that of native HGH.

Table I demonstrates the effect of intravenous administration of native HGH and RA-

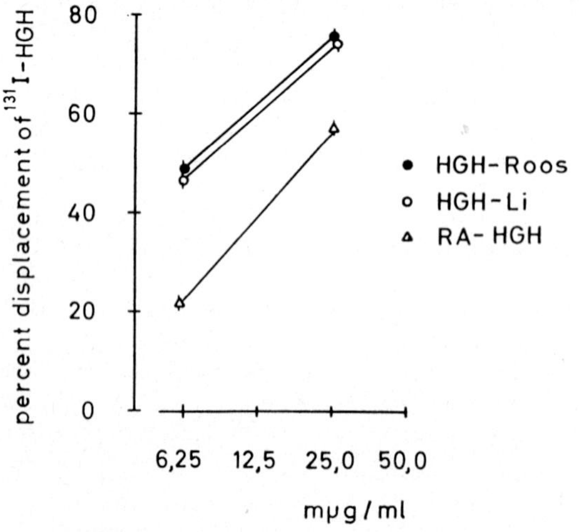

Fig. 1. Displacement of ^{131}I-HGH-Roos from antibodies by unlabelled HGH-Roos, HGH-Li and RA-HGH. Vertical bars denote S.E.M.

TABLE I

Effect of native HGH and RA-HGH on glucose tolerance (k-value), insulin response to glucose (insulin area and insulinogenic index) and on plasma levels of glycerol and alanine

	Subject nr	k-value (per cent $\times$ min^{-1})	Insulin response		Maximal glycerol release (μmol/l above control)	Maximal decrease in alanine (μmol/l below control)
			Insulin area (μU $\times$ ml^{-1} $\times$ min)	Insulinogenic index ($\times$ 100)		
Control	1	3.47	1413	17.2	—	—
	2	3.55	2160	30.3	—	—
	3	5.54	1363	20.8	—	—
	4	3.75	1635	20.7	—	—
	5	1.59	1598	15.5	—	—
Mean $\pm$ SEM		3.58 $\pm$ 0.63	1634 $\pm$ 142	20.9 $\pm$ 2.6	—	—
Native HGH	1	0.50	808	6.4	51	24
	2	0.95	2430	22.6	88	75
	3	1.44 (3.85)	1145 (630)	12.6 (7.1)	17 (38)	61
	4	1.27 (1.39)	1660 (1755)	14.4 (15.2)	34 (35)	—
	5	1.90	1913	18.5	44	—
Mean $\pm$ SEM		1.21 $\pm$ 0.24	1591 $\pm$ 285	14.9 $\pm$ 2.7	47 $\pm$ 10	53 $\pm$ 15.2
Significance of paired differences *vs* control		$p < 0.025$	NS	$p < 0.05$	NS	NS
RA-HGH	1	0.62	1205	9.9	52	161
	2	0.61	1935	19.6	57	61
	3	1.16 (0.82)	853 (915)	8.9 (8.0)	59 (42)	72
	4	1.69 (1.08)	2385 (1145)	22.6 (9.3)	55 (36)	—
	5	1.12	1634	13.5	31	—
Mean $\pm$ SEM		1.04 $\pm$ 0.20	1604 $\pm$ 269	14.9 $\pm$ 2.7	51 $\pm$ 5	98 $\pm$ 31.6
Significance of paired differences *vs* control		$p < 0.01$	NS	$p < 0.05$	$p < 0.01$	NS

HGH on the k-value, on the insulin response to glucose, on the release of glycerol and on alanine in plasma. The mean k-value decreased from 3.58 to 1.21 with native HGH and to 1.04 with RA-HGH. Significant paired differences were obtained between the k-values of the control experiments and those obtained with native HGH and RA-HGH, whereas there were no differences between the k-values obtained with the two hormone preparations.

The plasma insulin response to glucose injection, when expressed as the area below the insulin curve, was not altered by the two hormone preparations. Taking into consideration the prolonged hyperglycemic stimulus to insulin secretion, corresponding to the decreased k-values, the insulin response expressed as the insulinogenic index was moderately but significantly decreased both with native HGH and RA-HGH; from 20.9 for the control experiments to 14.9 for the HGH experiments ($p < 0.05$).

The lipolytic effects of the hormones are expressed as the maximal increase in plasma glycerol concentration above the fasting level of 53 $\pm$ 6 μmol per liter in the experiments with native HGH and above 47 $\pm$ 3 in those with RA-HGH. Both HGH preparations had similar effects, increasing the glycerol levels with 47 and 51 μmol per liter, respectively. Only the latter change was significant ($p < 0.01$), however.

Plasma alanine was measured in only three of the subjects. Both hormones decreased the alanine concentration with 53 and 98 μmol per liter from the fasting values of 231 $\pm$ 5.0 μmol per liter for the experiments with native HGH and 272 $\pm$ 8.0 for RA-HGH. The small number of experiments does not permit any evaluation of this effect, however.

The results obtained with the low doses of the HGH preparation in subjects 3 and 4 are presented within brackets in Table I. The small number of experiments and the great variability in the results obtained do not allow a detailed comparison of the effects of different doses of the hormones. It can only be stated that 0.9 mg of the two preparations seemed to induce approximately the same metabolic changes as 4 mg.

Figure 2 demonstrates the mean concentrations of glucose in blood and insulin, HGH and glycerol in plasma during the control experiments and those with native HGH and RA-HGH. The basal levels of glucose and insulin were not influenced by the hormone infusions. Glycerol in plasma increased gradually during the infusions and decreased again after the administration of glucose. The marked difference in plasma HGH levels reached with the two hormones is obviously due to the poor immunological reactivity of RA-HGH in the growth hormone assay system.

The disappearance rates of the HGH preparations were measured by plotting on a

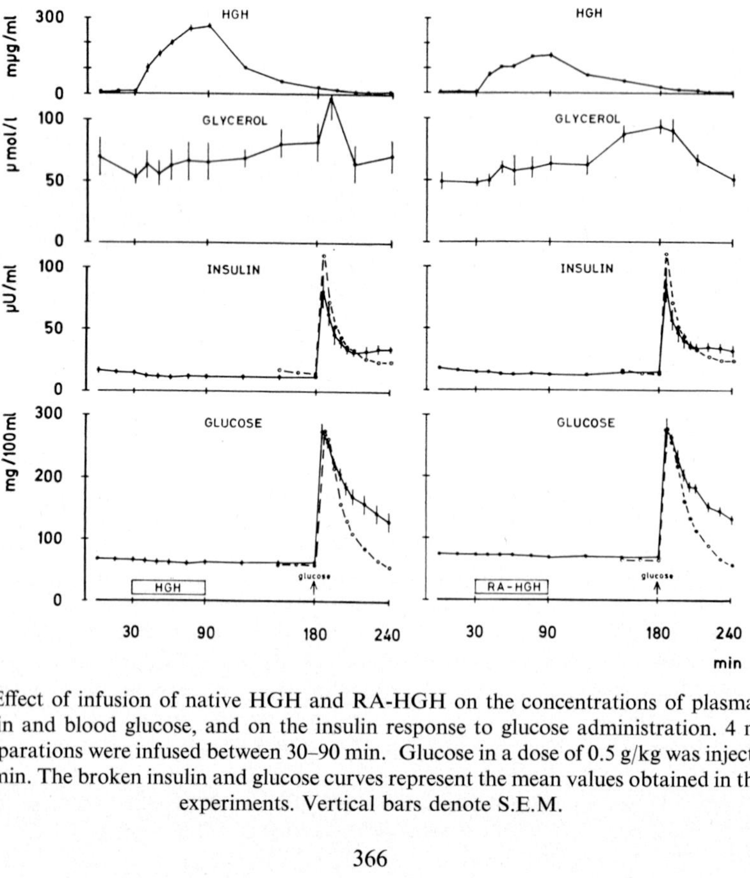

Fig. 2. Effect of infusion of native HGH and RA-HGH on the concentrations of plasma glycerol and insulin and blood glucose, and on the insulin response to glucose administration. 4 mg of the HGH preparations were infused between 30–90 min. Glucose in a dose of 0.5 g/kg was injected rapidly at 180 min. The broken insulin and glucose curves represent the mean values obtained in the control experiments. Vertical bars denote S.E.M.

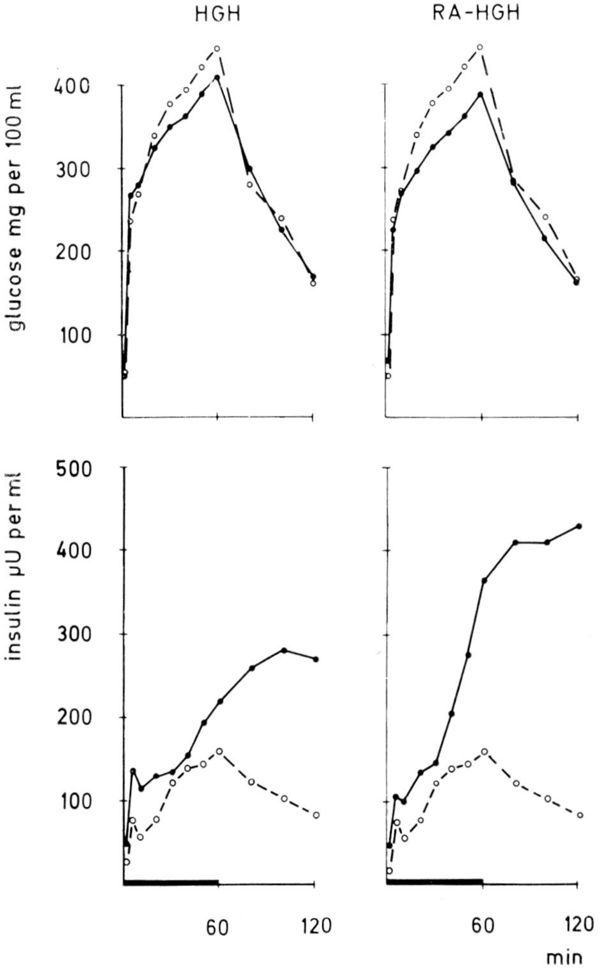

Fig. 3. Effect of four days administration of native HGH and RA-HGH on the insulin response to glucose infusion in a patient with panhypopituitarism. Glucose was infused between 0–60 min. Open circles and broken lines represent the control experiments, filled circles and solid lines the experiments with the HGH preparations.

semilogarithmic scale the plasma HGH values between 90 (end of the infusion of HGH) and 180 min. The half-life of the hormone in plasma was identical for both preparations, 31 ± 2.4 min for native HGH and 31 ± 2.7 min for RA-HGH.

Figure 3 illustrates the effects of four days' administration of native HGH and RA-HGH. The k-value decreased from 1.63 to 0.81 with native HGH and to 1.19 with RA-HGH. The insulin response to glucose infusion was markedly enhanced with both preparations, more so when RA-HGH was given.

DISCUSSION

The acute biological actions of native HGH-Li and its reduced and alkylated derivative (RA-HGH) were qualitatively and quantitatively similar in the present experiments, where glycerol release, glucose tolerance, insulin secretion and concentration of plasma alanine were measured. Their effects on glucose tolerance and insulin release were qualitatively similar also when the hormones were given for 4 days. This last statement is based on one single experiment however, due to the limited amount of RA-HGH available. The finding of a less decreased k-value and more enhanced insulin secretion with RA-HGH may be of some interest but must await confirmation.

It is surprising that such a radical change in the structure of the hormone as breaking up the disulfide bridges did not alter its biological activity. Our findings are consistent with the studies of Dixon and Li (1966) and Bewley *et al.* (1969), who demonstrated similarity in the actions of HGH and RA-HGH in the tibia test and in the pigeon crop-sac assay. These findings indicate that the biologically active site of the hormone is not dependent on the presence of intact sulfhydryl bridges. This does not mean that an intact conformation of the molecule is unnecessary for its biological effect, since the physico-chemical studies of Bewley *et al.* (1969) demonstrated that the disulfide bonds were not required for the formation of the secondary and tertiary structure of HGH.

On the other hand, our studies indicate clearly that reduction and alkylation of HGH diminished its affinity to antibodies by about 50%. It is thus obvious that the immunologically active site of the molecule is located in the vicinity of the sulfhydryl bridges or depends to some extent on their presence. Antibodies against polypeptides are usually heterogenous molecules with varying affinity to the different antigenic determinants of the polypeptide. A 50% reduction of the immunological action of RA-HGH indicates, therefore, that roughly half of the antigenic determinants may be located in the vicinity of the disulfide bonds. The studies of Bewley *et al.* (1969) seem to exclude heterogeneity of RA-HGH or secondary or tertiary structure changes as a possible explanation for the reduction in immunological activity.

The difference in immunological activity of native HGH and RA-HGH in spite of their identical biological effects is of considerable interest. This might indicate a dissociation of these two properties of HGH induced by the alterations of the molecule. On the other hand, it must be taken into account that complete dose-response studies with the hormones have not yet been performed. It cannot still be excluded that we have been using supramaximal doses of the two preparations, and that an eventual 50% reduction of the biological activity of RA-HGH therefore could not be demonstrated. The finding in one instance of a higher k-value with a low dose of native HGH than with a low dose of RA-HGH is not enough evidence against this reservation. Reduction and alkylation does not seem to alter markedly the metabolism of HGH since the half-life in plasma of the native hormone and its derivative were identical. The half-life found in this study, 31 min, is in agreement with that reported by Glick *et al.* (1964) and Parker *et al.* (1962).

It is well established that HGH, besides suppressing plasma amino acid concentration (Snipes, 1968), increasing lipolysis (Goodman, 1968) and decreasing glucose tolerance (Ikkos *et al.*, 1962), also stimulates insulin response to glucose (Luft and Cerasi, 1964; Bouman and Bosboom, 1965; Martin and Gagliardino 1967; Luft *et al.*, 1969). Figure 3 in the present study illustrates this. It was surprising, therefore, that in acute experiments with HGH and RA-HGH insulin response to glucose was depressed while lipolysis and glucose tolerance were altered as expected. This finding suggests that one of the initial events in the diabetogenic action of HGH is inhibition of insulin secretion. The enhancement of insulin response into prolonged administration of HGH then might either be due to the stimulatory action of chronic elevation in blood glucose or reflect a later direct action of HGH on the pancreatic β-cells.

Several authors have investigated the acute effects of HGH on glucose tolerance and insulin release, reporting unaltered or somewhat enhanced insulin secretion (Daughaday and Kipnis, 1966; Frohman *et al.*, 1967; Doar *et al.*, 1969). However, their results have not been expressed as insulin secretion per magnitude of stimulation (insulinogenic index). The higher blood glucose levels achieved after HGH administration may therefore have obscured a moderate inhibition of insulin response to glucose. Differences in the experimental procedure (rapid injection of HGH, timing of the experiment etc.) may also have contributed to the differences in the results.

The acute effects of HGH – inhibition of insulin release together with increased lipolysis and decreased glucose tolerance – are qualitatively similar to the actions of catecholamines, and emphasize the role of HGH as a stress hormone.

SUMMARY

Reduction and carbamidomethylation of the disulfide bridges of HGH were followed by a 50% decrease of the immunological activity of the hormone. This indicates that the immunological active site of the HGH molecule is located in the vicinity of the disulfide bridges.

The effect of HGH in man on glucose in blood and glycerol, alanine and insulin in plasma, as well as on glucose tolerance and insulin response to glucose, were not altered by the reduction of the molecule. Thus, there seems to exist a dissociation of the sites of the molecule which are responsible for the biological and immunological properties of the hormone.

Both native HGH and its reduced derivative, when infused for one hour, inhibited partially the insulin response to glucose administration. On prolonged administration, both preparations had an enhancing effect on glucose-induced insulin release. It is suggested that HGH initially suppresses insulin release, thereby acting as a stress hormone in a similar fashion to the catecholamines. The later enhancement of insulin secretion may be due either to prolonged hyperglycemia or to long-term effects of HGH on the pancreatic islets.

ACKNOWLEDGEMENTS

The authors are grateful to Drs. Irene E. Karl and D. M. Kipnis, St. Louis, Missouri, for the determinations of alanine in plasma, and to Dr. J. Östman, Stockholm, for the glycerol determinations. Dr. Sigbritt Werner and the nurse staff of the Department of Endocrinology and Metabolism, Karolinska Hospital, Stockholm, kindly assisted in the metabolic studies.

REFERENCES

Bewley, T. A., Brovetto-Cruz, J. and Li, C. H. (1969): Human pituitary growth hormone. Physico-chemical investigations of the native and reduced-alkylated protein. *Biochemistry*, *8*, 4701.

Bewley, T. A., Dixon, J. S. and Li, C. H. (1968): Human pituitary growth hormone. XVI. Reduction with dithiothreitol in the absence of urea. *Biochim. biophys. Acta (Amst.)*, *154*, 420.

Bouman, P. R. and Bosboom, R. S. (1965): Effects of growth hormone and of hypophysectomy on the release of insulin from rat pancreas *in vitro*. *Acta endocr. (Kbh.)*, *50*, 202.

Cerasi, E., Della Casa, L., Luft, R. and Roovete, A. (1966): Determination of human growth hormone in plasma by a double antibody radioimmunoassay. *Acta endocr. (Kbh.)*, *53*, 101.

Chernick, S. S. (1969): Determination of glycerol in acyl glycerols. In: *Methods in Enzymology*, *Vol. XIV*, pp. 627–630. Editor: J. M. Lowenstein. Academic Press, New York.

Daughaday, W. H. and Kipnis, D. M. (1966): The growth-promoting and anti-insulin actions of somatotropin. *Recent Progr. Hormone Res.*, *22*, 49.

Dixon, J. S. and Li, C. H. (1966): Retention of the biological potency of human pituitary growth hormone after reduction and carbamidomethylation. *Science*, *154*, 785.

Dixon, G. H. and Wardlaw, A. C. (1960): Regeneration of insulin activity from separated and inactive A and B chains. *Nature (Lond.)*, *188*, 721.

Doar, J. W. H., Maw, D. S. J., Simpson, R. D., Audhya, T. K. and Wynn, V. (1969): The effects of growth hormone on plasma glucose, NEFA, insulin and blood pyruvate during intravenous glucose tolerance tests. *J. Endocr.*, *45*, 137.

FROHMAN, L. A., MACGILLIVRAY, M. H. and ACETO, T. (1967): Acute effects of human growth hormone in insulin secretion and glucose utilization in normal and growth hormone deficient subjects. *J. clin. Endocr.*, *27*, 561.

GLICK, S. M., ROTH, J. and LONERGAN, E. T. (1964): Survival of endogenous human growth hormone in plasma. *J. clin. Endocr.*, *24*, 501.

GOODMAN, H. M. (1968): Growth hormone and the metabolism of carbohydrate and lipid in adipose tissue. *Ann. N.Y. Acad. Sci.*, *148*, 419.

HALES, C. N. and RANDLE, P. J. (1963): Immunoassay of insulin with insulin–antibody precipitate. *Biochem. J.*, *88*, 137.

IKKOS, D. and LUFT, R. (1957): On the intravenous glucose tolerance test. *Acta endocr. (Kbh.)*, *25*, 312.

IKKOS, D., LUFT, R., GEMZELL, C. A. and ALMQVIST, S. (1962): Effect of human growth hormone on glucose tolerance and some intermediary metabolites in man. *Acta endocr. (Kbh.)*, *39*, 547.

LI, C. H., LIU, W. K. and DIXON, J. S. (1962): Human pituitary growth hormone. VI. Modified procedure of isolation acid NH^2 terminal animo acid sequence. *Arch. Biochem.*, *Suppl. 1*, 327.

LI, C. H., LIU, W. K. and DIXON, J. S. (1966): Human pituitary growth hormone. XII. The amino acid sequence of the hormone. *J. Amer. chem. Soc.*, *88*, 2050.

LUFT, R. and CERASI, E. (1964): Effect of human growth hormone on insulin production in panhypopituitarism. *Lancet*, *II*, 124.

LUFT, R., CERASI, E. and WERNER, S. (1969): The effect of moderate and high doses of human growth hormone on the insulin response to glucose infusion in prediabetic subjects. *Hormone metab. Res.*, *1*, 111.

MARTIN, J. M. and GAGLIARDINO, J. J. (1967): Effect of growth hormone on the isolated pancreatic islets of rat *in vitro*. *Nature (Lond.)*, *213*, 630.

PARKER, M. L., UTIGER, R. D. and DAUGHADAY, W. H. (1962): Studies on human growth hormone. II. The physiological disposition and metabolic fate of human growth hormone in man. *J. clin. Invest.*, *41*, 262.

ROOS, P., FEVOLD, H. R. and GEMZELL, C. A. (1963): Preparation of human growth hormone by gelfiltration. *Biochim. biophys. Acta (Amst.)*, *74*, 525.

SNIPES, C. A. (1968): Effects of growth hormone and insulin on amino acid and protein metabolism. *Quart. Rev. Biol.*, *43*, 127.

YALOW, R. S. and BERSON, S. A. (1961): Immunology aspects of insulin. *Amer. J. Med.*, *31*, 882.

GROWTH HORMONE RELEASE IN MAN REVISITED: SPONTANEOUS VS STIMULUS-INITIATED TIDES

I. SPITZ, B. GONEN and D. RABINOWITZ*

Department of Chemical Endocrinology, Hadassah University Hospital, Jerusalem, Israel

It has been known for some considerable time that growth hormone (HGH) release in man can be evoked by a number of stimuli, amongst the commonest being insulin hypoglycemia (Roth *et al.*, 1963), and arginine infusion (Knopf *et al.*, 1965; Merimee *et al.*, 1965). However, all observations regarding HGH release are complicated by the fact that spontaneous elevations in HGH secretion occur (Glick *et al.*, 1965). These elevations are apparently not clearly related to an initiating event ('agnogenic' or 'spontaneous' tides of HGH). Often there has been failure to appreciate this phenomenon by many workers, including our own group. The facts were pointed out clearly at the First International HGH Symposium (Glick and Goldsmith, 1968; Hunter *et al.*, 1968; Knobil *et al.*, 1968). Best *et al.* (1968) stated that the rise in HGH after arginine infusion is no more frequent than after control saline administration and consequently doubted the specificity of arginine in provoking HGH release.

Because of the obvious complexities of the problem, we have re-evaluated patterns of HGH release in a group of healthy human male subjects. In this report, we show the results of tests in which subjects were challenged either with repeated pulses of normal saline or with one or more pulses of arginine monochloride.

MATERIALS AND METHODS

The control subjects investigated were university students. With the exception of one subject (aged 36), their ages ranged from 20 to 30 and all were within 10% of their ideal body weight (Metropolitan Life Insurance Tables). All were paid volunteers and the nature of the proposed investigation was clearly explained prior to the procedure. All had consumed an adequate carbohydrate diet for at least one week prior to study.

The tests were performed at the Hadassah University Hospital. The subjects were studied as outpatients and were taken to the hospital by motor car on the morning of the procedure. All tests were commenced between 7:30 a.m. and 8 a.m. after a 10-hour overnight fast, during which time only plain water was allowed. Smoking was not permitted for a number of hours before and during the procedure.

After arriving at the hospital, the subjects were placed at bed rest in a quiet room. The experimental protocol is outlined in Fig. 1. Initially, needles were placed in both antecubital

* Supported in part by grants from the Joint Research Fund of Hebrew University-Hadassah Medical School and the Israel Cancer Society. During part of this work Dr. Spitz was supported by a special fellowship of the Israeli Cancer Society.

fossae. Throughout the whole experiment, the needles were kept patent by the slow administration of normal saline. Blood samples were taken from one vein via a three-way stop cock, the first two samples being taken during placement of the needles.

Following a basal control period of at least 60 minutes during which time intermittent blood samples were taken, repeated pulses of normal saline or L-arginine monochloride were given through the contralateral vein. All pulses were administered via a three-way stop cock connected to a second intravenous administration set which was modified to contain a graduated reservoir which filled from above. It was thus possible to administer the pulses without any stress.

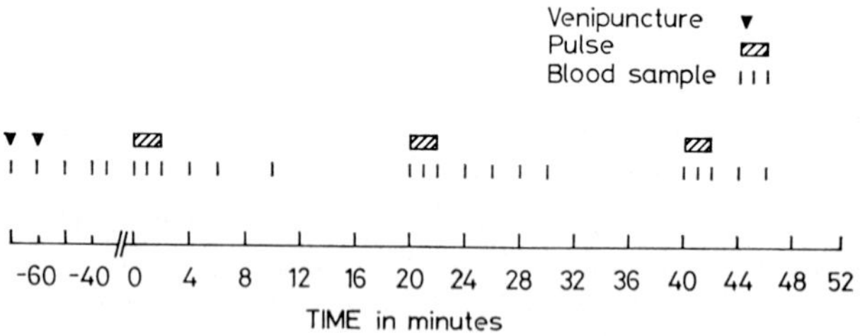

Fig. 1. The general experimental protocol used in the study.

Pulses were usually given at 20 minute intervals up to a maximum of five pulses. In the arginine pulse experiments, 5 g of L-arginine monochloride in a volume of 80 ml was administered over 2 minutes starting at time zero. Blood samples were taken at zero time and thereafter at 1, 2, 4, 6, 10 and 20 minute intervals. The 20 minute sample was withdrawn immediately prior to the administration of the following pulse, and represents the zero time for this pulse. In the saline experiments, 80 ml of saline was administered over 2 minutes, blood samples being taken at 2, 4, 6, 10 and 20 minute intervals.

After taking the blood, it was allowed to clot at 4°C, then separated, and kept frozen at −20°C, until the various parameters were measured. Serum HGH was estimated by an immunoassay procedure using a method of charcoal separation (Herbert *et al.*, 1965); serum glucose was estimated by a glucose oxidase method on the Autoanalyser; cortisol by the fluorometric method of Mattingly (1962); and α-amino nitrogen by the ninhydrin method of Yemm and Cocking (1955).

RESULTS

Multiple saline pulses

From the results of the 12 control male subjects given saline, two basic patterns of HGH response have emerged. In five subjects HGH levels remained low throughout the experiments ('non-responders'). In seven subjects, however, spontaneous elevations of HGH were observed ('responders'). We believe that we can provide a reasonable framework by which it will be possible to differentiate this spontaneous or agnogenic tide from stimulus-related elevations of HGH.

The hormonal profile displayed in Fig. 2 is representative of that seen among non-responders. Needles were inserted at −80 and −70 minutes. Five saline pulses were then administered at 20 minute intervals starting at zero time. Levels of HGH, insulin, glucose and α-amino nitrogen were all acceptably stable in the post-placement control period, and following the administration of saline pulses.

Fig. 3 displays the hormonal profile observed in one subject from the 'responder' group.

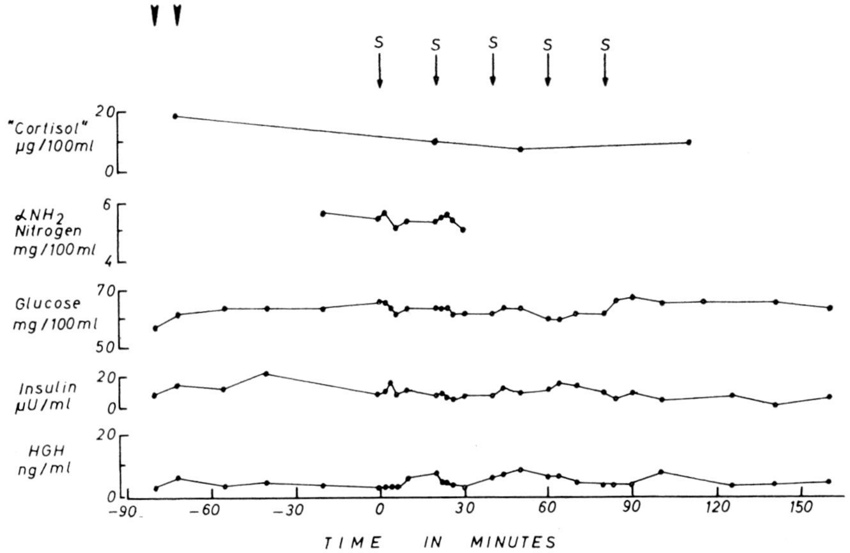

Fig. 2. The metabolic profile following the administration of repeated saline pulses in experiment No. 9. A = arginine, S = saline, in this figure and those following.

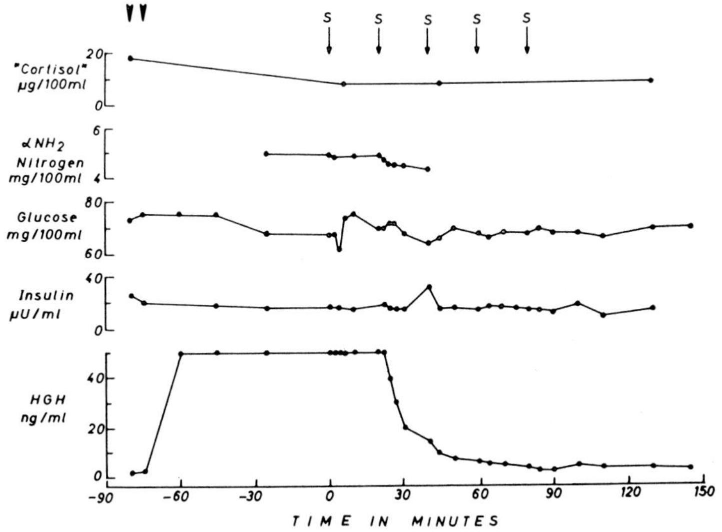

Fig. 3. The metabolic profile following the administration of repeated saline pulses in experiment No. 1.

Cannulations were at −80 and −75 minutes respectively. Again, throughout the whole experiment there were no significant changes in the concentration of α-amino nitrogen, glucose and insulin. Serum cortisol levels fell from fasting control values of 18 µg/100 ml to 8 µg/100 ml at +130 minutes. HGH levels during the needle placement were low (2.6 ng/ml). However, HGH had risen markedly and abruptly to over 50 ng/ml at the time of the third blood sample, taken at −60 minutes, *i.e.* 15 minutes after the second cannulation. These high levels continued for 82 minutes and then decreased abruptly to less than 10 ng/ml within 22 minutes. It should be noted that there was no response to the 5 saline pulses – in fact the HGH rise anticipated the first saline pulse.

The needle related responses (or 'early tide') which occurred in six of the twelve control male subjects are summarized in Fig. 4. In the first four subjects, five saline pulses were administered, whilst the last two subjects received two pulses. With one exception, HGH levels taken during needle placement were low, values ranging between zero and 4.5 ng/ml.

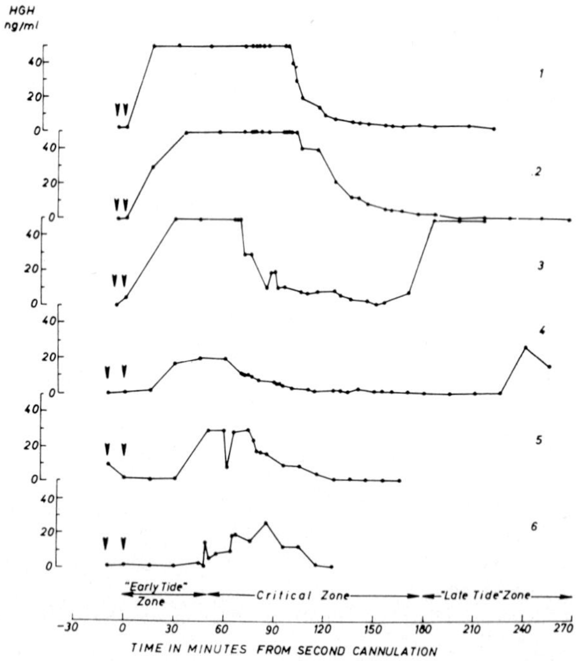

Fig. 4. The HGH response of control subjects demonstrating spontaneous tides.

In Study No. 5, the initial value of 10 ng/ml had decreased to 1 ng/ml by the time of the second needle placement and two post-placement values were also low. In three subjects significant elevations of HGH were observed in the first post-placement sample; in one subject (Study No. 4) in the second, and in two subjects (Numbers 5 and 6) in the third and fifth post-placement sample respectively. The rise in HGH secretion was not momentary, but persisted from 24 to 100 minutes. Saline pulses did not appear to play a role in the HGH release. However, were one to ignore the presence of an early HGH tide, one could well have attributed the rise in HGH to the administration of the saline pulse in Study No. 6.

In Study Numbers 3 and 4 late rises in HGH secretion occurred 190 and 250 minutes after commencing the experiment. The rise was not associated with any fall in blood glucose. This phenomenon we refer to as the 'late tide' of HGH. These results indicate that a needle related surge of HGH secretion, or early tide, can occur up to an hour after the last venipuncture. The late tide of HGH secretion is generally not seen until 3 hours after venipuncture.

One control 'responder' subject demonstrated an atypical response. He was cannulated at −70 and −65 minutes. HGH remained low until +30 minutes (*i.e.* 95 minutes following the second cannulation) and then the hormone concentration abruptly increased. There were no changes in cortisol, glucose or insulin levels. Subsequent to performing the experiment, it was discovered that the subject was taking a serotonin inhibitor. The precise significance of this is unknown but we do not feel justified in excluding this subject from the data. On the contrary, it indicates that exceptions to the general pattern of early and late tides are likely to occur. Nevertheless our studies may provide a reasonable framework for differentiating agnogenic from stimulus-related elevations of HGH.

Since HGH levels usually were low at the time of needle insertion, it seems reasonable that the rise in HGH, often apparent in the first post-cannulation period sample, is triggered by the venipunctures. Copinschi *et al.* (1967) observed spontaneous peaks of HGH in eight out of nine male subjects following arterial cannulations. Possibly the needle placement could be regarded as a form of stress. This is not, however, reflected by any similar elevation in serum cortisol levels. Yalow *et al.* (1969) also observed a non-parallel responsiveness of HGH and cortisol to stressful procedures. They noted dissociation of ACTH and HGH responses to various stimuli including electroconvulsive therapy and surgical procedures. The mechanism underlying the late tide of HGH secretion occurring after 180 minutes is not clear to us.

It is evident that rises in HGH levels occurring within 50 minutes or more than 3 hours after venipuncture, cannot with confidence be attributed to any coincidentally given stimulus. Rises of HGH occurring 1 hour after needle placement can more reasonably be interpreted for a finite time interval, the 'critical zone' (Fig. 4). Furthermore, when two cannulations are performed, needle related surges of HGH secretion appear to be common, and can occur in 50% of control subjects.

Our data in male subjects lead us to suggest the following model of HGH release (Fig. 5).

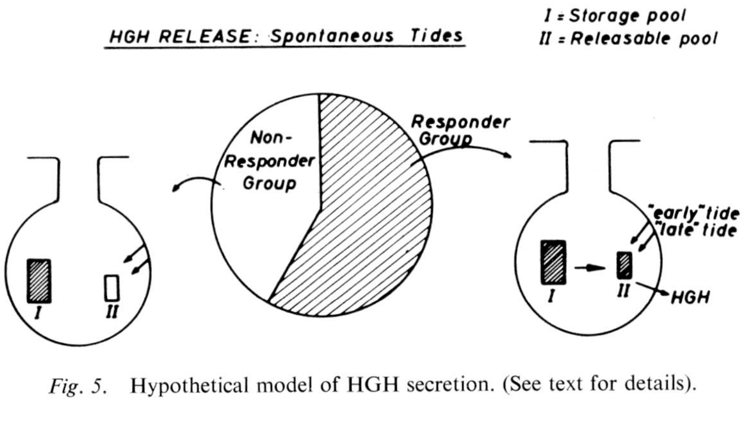

Fig. 5. Hypothetical model of HGH secretion. (See text for details).

If we look at any group of male subjects, we know that in the course of a morning, following two venipunctures, about 50% will release HGH. This group we call 'responders' and the other group 'non-responders.' We propose that there are two pools of HGH – a storage pool and a releasable pool. Subjects with a releasable pool of HGH constitute those subjects in whom HGH may be discharged by a number of stimuli, two of which are 'early' and 'late' spontaneous tides. On the other hand, the 'non-responder' group lacking a releasable pool, do not demonstrate 'spontaneous' tides and possibly also fail to respond to other forms of stimuli.

From our limited series, we would suggest that the responder group is round 50%, but this is a minimum figure. We are not suggesting that every patient who has a releasable pool will always discharge spontaneously. Indeed, as we will see, we suggest that it is this responder group who can release HGH in response to arginine and possibly to other stimuli. Unfortunately, at the moment we are unable to differentiate between the two groups on the basis of the other parameters which were determined.

Fig. 6 shows the suggested mode of action of insulin hypoglycemia in stimulating HGH secretion within the framework of our working model. Insulin hypoglycemia universally

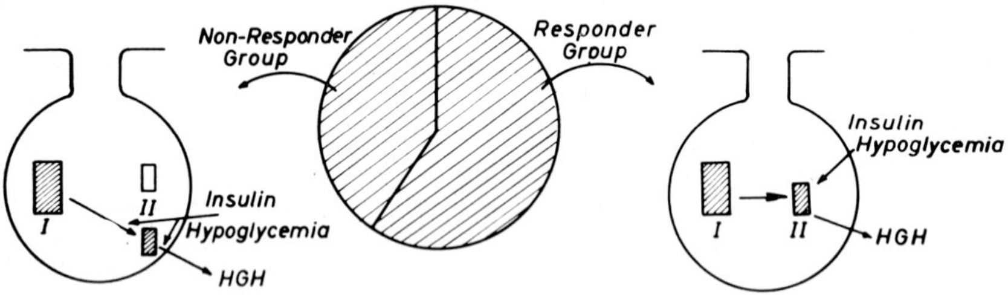

Fig. 6. HGH release following insulin hypoglycemia based on the hypothetical model. (See text for details).

elevates HGH levels. In those subjects with a readily releasable HGH pool this could, of course, be discharged as a 'spontaneous' tide, although insulin could readily effect HGH release from this same pool. In the non-responder group, we suggest that insulin hypoglycemia has the ability to induce filling of the readily releasable pool and then to discharge its contents.

From our earlier data, we were impressed that arginine infusion was reproducibly followed by a rise in HGH secretion in normal females (Rabinowitz *et al.*, 1968) but the response occurred in less than half of the males studied. Since half of our normal males are 'responders', it is probable that an arginine infusion in males acts in the responder group, *i.e.* those with a releasable pool who similarly can discharge HGH as an early or late spontaneous tide.

We have asked whether we may be able to effect HGH release among non-responders by changing the mode of administration of arginine. In other words, will arginine, given as one or multiple pulses, be able to effect HGH release, not only from the responder but also the non-responder group? In the execution of the study, two basic considerations have been entertained.

The first is a temporal one. Our experiments with arginine pulses have been specifically constructed around the framework of spontaneously occurring 'early' and 'late' tides. We have invariably administered arginine 60 minutes following venipuncture. This then falls in the 'critical zone' (Fig. 4) and with this proviso we believe that our data can be more confidently interpreted, *i.e.* that in a particular subject the arginine pulse(s) was the trigger for HGH. However, a second quantitative consideration is also demanded. More than 50% of our subjects are responders in the absence of an arginine stimulus. Hence we require close to 100% responsiveness in our experimental group with arginine pulses to establish that our pulses also effect release from non-responders.

Our results show that, despite the administration of one or more pulses of arginine to 24 subjects, there still exists a definite group of non-responders. They presumably do not have an available releasable pool of HGH, and this cannot be harnessed by pulse administration of arginine.

Single arginine pulse

Single arginine pulses were administered to eight subjects. All the HGH data are summarized in Fig. 7. In experiments 16 and 17, three and four venipunctures respectively were performed because the initial attempts at cannulation were unsuccessful. In each instance 60 minutes elapsed between the last venipuncture and the arginine pulse, which was administered at zero time. The arginine pulse was followed by four saline pulses given at 20 minute intervals in experiments 15, 16, 17 and 19.

HGH levels at the time of needle placement were low in all eight subjects. Figure 7 shows that in the first two panels a needle related surge was evident, appreciable HGH elevation already being present in the first post-placement sample. In the following two panels late tides were evident, being observed 190 and 220 minutes from the commencement of the experiment. These results are typical of the 'early' and 'late' tides which we have described. In the fifth and sixth panels the HGH release was temporally related to the pulse, appreciable elevations being noticed eight minutes after completion of the two minute pulse. In the final two subjects, challenged with a single arginine pulse, no elevation of HGH occurred during the experiment.

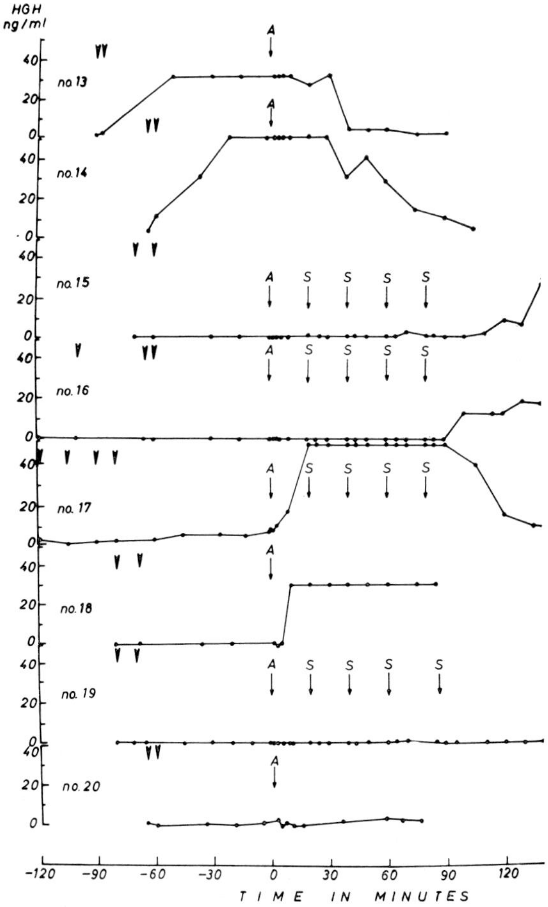

Fig. 7. The HGH response to a single arginine pulse in 8 subjects.

We suggest that in this group of eight subjects, six were responders and two were non-responders. Among the six, the releasable pool was discharged as a needle related tide in two; as a late tide in two; and in two other subjects this pool was probably discharged by the pulse of arginine. The two non-responders we suggest lacked a dischargable pool and this was not generated by the arginine pulse.

Multiple arginine pulses

Multiple arginine pulses were administered to 16 subjects. Again, the HGH responses were heterogeneous; that is, a 'responder' and a 'non-responder' group were observed.

Those HGH responses which were not temporally related to the arginine pulses are depicted in Fig. 8. In experiments 21 to 24, two arginine pulses were administered and in experiment 25, four pulses. In the remaining subject the two arginine pulses were followed by the administration of saline pulses. In all instances HGH values were low in samples

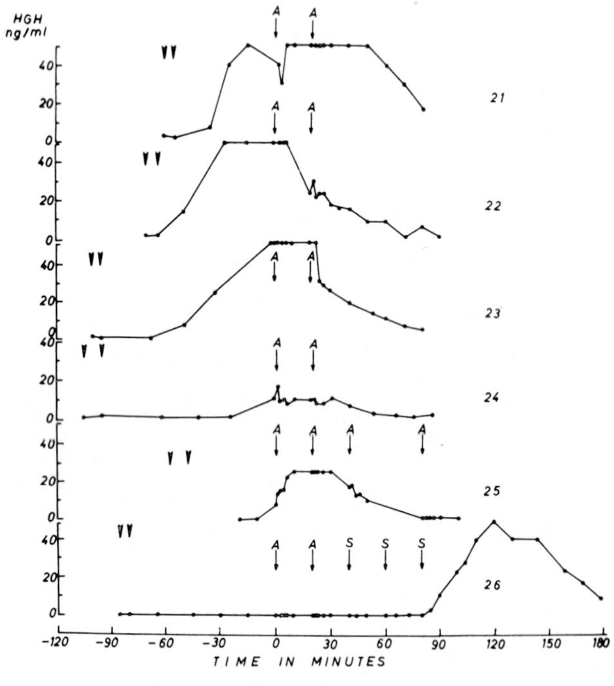

Fig. 8. The HGH responses to multiple arginine pulses. In these 6 subjects there is no temporal relation between the HGH rise and the pulse administration.

obtained at placement of the i.v. needles. We attribute the HGH responses of the first five subjects to an early or needle related tide, and, with the exception of experiment No. 24, HGH elevation was already apparent within 50 minutes of the second needle cannulation. The response in experiment 26 is consistent with a 'late' spontaneous tide.

Fig. 9 depicts those subjects who demonstrated a pulse related response, as well as those who failed to elevate HGH with arginine pulses. Double pulses followed by three saline pulses were given in experiments 27, 28 and 33. Three pulses were administered in experiment 29, and four in 30 and 31. In experiment No. 32, a total of 5 arginine pulses were given,

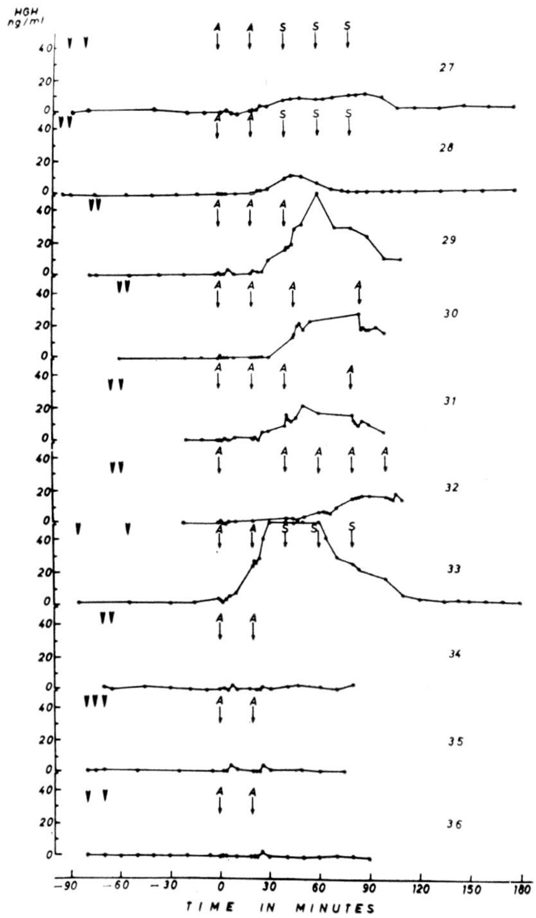

Fig. 9. The HGH responses to multiple arginine pulses. In the first 7 subjects the HGH response is temporally related to one or more arginine pulses. The last three subjects have no HGH response.

the first two being separated by an interval of 40 minutes. The remaining three subjects all received two pulses. Initial levels of HGH were all low. In the first six subjects, a burst of HGH secretion occurred in relation to the second arginine pulse. HGH levels were significantly elevated from 6 to 25 minutes following the commencement of the second pulse. Elevation of HGH was temporally related to the second pulse in experiment 32 where an interval of 40 minutes separated the two pulses. In experiment 33, the HGH rise was temporally related to the first arginine pulse. In the final three subjects there was no response to the arginine pulses, levels remaining low throughout.

Thus out of a total of sixteen subjects challenged with two pulses of arginine, six showed spontaneous tides, seven responded to either one or two pulses of arginine, and three were non-responders. Thus a non-responder group persists despite the administration of two pulses of arginine. In terms of our model, the repeated arginine pulses did not induce filling of a releasable pool.

Summarizing the data, of the 24 subjects who were challenged with one or more arginine pulses:

7 demonstrated a spontaneous 'early' needle related tide,
3 showed a late tide,
9 showed a response to 1 or 2 arginine pulses,
5 were unresponsive.

That is, there were a significant number who did not respond to arginine pulses, neither did they show spontaneous elevations.

In terms of our model, this latter group lacks a readily releasable pool of HGH. Arginine pulses do not harness such a pool. Conversely the responders, having such a pool, can release HGH either as an early, or as a late tide, or following one or two pulses of arginine.

These data agree well with the findings of Best *et al.* (1968). They found that all subjects responding to arginine also showed HGH elevation during control study or following saline infusion.

Further evidence supporting the concept of a responder and non-responder group is shown in Fig. 10. During the project, several subjects were investigated twice, and three such examples are shown. The first case was studied as experiment 8 with saline pulses and two arginine pulses were given in experiment 36. On neither occasion was there any HGH response. The second subject was studied as experiment 1 with repeated saline pulses and as

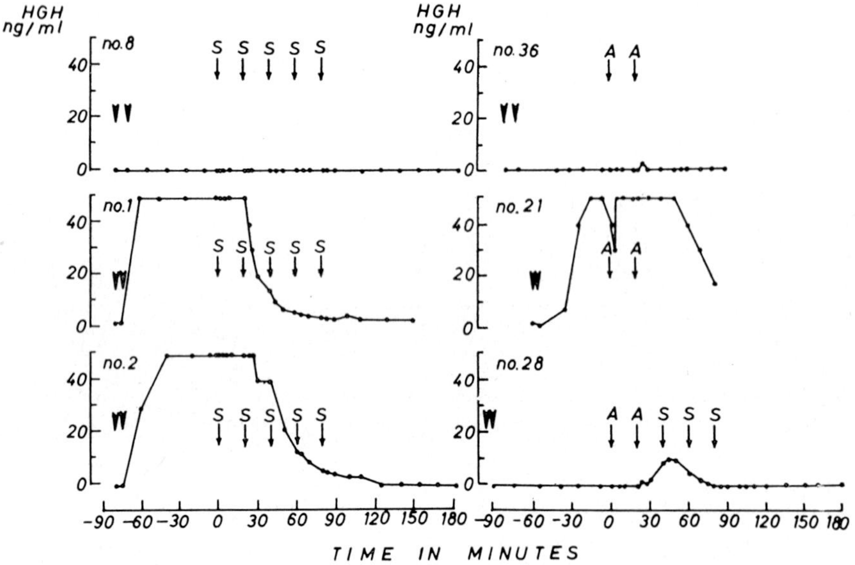

Fig. 10. The HGH response of three subjects when challenged on separate occasions with multiple saline pulses (left-hand panel) and two arginine pulses (right-hand panel).

experiment 21 with a double arginine pulse. In both instances a clear needle related rise in HGH was evident. In the final subject, to whom saline pulses were given in experiment 2, a clear needle related surge was apparent. In experiment 28, the subject was given two arginine pulses and clear elevation of HGH occurred following the second pulse. The first subject thus falls into the non-responder group, the final two into the responder group. The second subject discharged his releasable HGH pool on 2 occasions following needle placement. In the final case, the HGH pool was released as an early tide on one occasion and as an arginine related rise on the second occasion.

SUMMARY

We have re-evaluated patterns of HGH release in 36 studies on normal male subjects. Repeated saline pulses were administered to twelve controls and two response patterns have emerged. In the first, HGH levels remained low throughout the duration of the experiment. In seven subjects, notable elevations in HGH occurred usually within 60 minutes of placing a needle (needle related HGH tide). Two of these subjects demonstrated additional HGH elevations after the third hour of the test (late HGH tides).

Twenty-four subjects were challenged with one or repeated pulses of arginine, and of these: 7 demonstrated a spontaneous 'early' needle related tide; 3 showed a late tide; 9 showed a response to one or two arginine pulses; 5 were unresponsive.

Our working model divides male subjects into 'responder' and 'non-responder' groups. In the former, there is a readily available pool of HGH, which can be discharged spontaneously either consequent to needle placement (the 'early' tide) or as a 'late' tide. Should the HGH not be released spontaneously, then one or two arginine pulses have the capacity to effect its release. The non-responder group do not manifest spontaneous tides and are unresponsive to arginine.

REFERENCES

Best, J., Catt, K. J. and Burger, H. G. (1968): Non-specificity of arginine as a test for growth hormone secretion. *Lancet*, *2*, 124.

Copinschi, G., Hartog, M., Earll, J. M. and Havel, R. J. (1967): Effect of various blood sampling procedures on serum levels of immunoreactive human growth hormone. *Metabolism*, *16*, 402.

Glick, S. M. and Goldsmith, S. (1968): The physiology of growth hormone secretion. In: *Growth Hormone*, pp. 84–88. Editors: A. Pecile and E. E. Müller. ICS 158, Excerpta Medica, Amsterdam.

Glick, S. M., Roth, J., Yalow, R. S. and Berson, S. A. (1965): The regulation of growth hormone secretion. *Recent Progr. Hormone Res.*, *21*, 241.

Herbert, W., Lau, K. S., Gottlieb, C. W. and Bleicher, S. J. (1965): Coated charcoal immunoassay of insulin. *J. clin. Endocr.*, *25*, 1375.

Hunter, W. M., Rigal, W. M. and Sukkar, M. Y. (1968): Plasma growth hormone during fasting. In: *Growth Hormone*, pp. 408–417. Editors: A. Pecile and E. E. Müller. ICS 158, Excerpta Medica, Amsterdam.

Knobil, E., Meyer, V. and Schally, A. V. (1968): Hypothalamic extracts and the secretion of growth hormone in the rhesus monkey. In: *Growth Hormone*, pp. 226–237. Editors: A. Pecile and E. E. Müller. ICS 158, Excerpta Medica, Amsterdam.

Knopf, R. F., Conn, J. W., Fajans, S. S., Floyd, J. C., Guntsche, E. M. and Rull, J. A. (1965): Plasma growth hormone response to intravenous administration of amino acids. *J. clin. Endocr.*, *25*, 1140.

Mattingly, D. (1962): A simple fluorimetric method for the estimation of free 11-hydroxycorticoids in human plasma. *J. clin. Path.*, *15*, 374.

Merimee, T. J., Lillicrap, D. A. and Rabinowitz, D. (1965): Effect of arginine on serum-levels of human growth-hormone. *Lancet*, *2*, 668.

Rabinowitz, D., Merimee, T. J., Nelson, J. K., Schultz, R. B. and Burgess, J. A. (1968): The influence of proteins and amino acids on growth hormone release in man. In: *Growth Hormone*, pp. 105–115. Editors: A. Pecile and E. E. Müller. ICS 158, Excerpta Medica, Amsterdam.

Roth, J., Glick, S. M., Yalow, R. S. and Berson, S. A. (1963): Hypoglycemia: A potent stimulus to secretion of growth hormone. *Science*, *140*, 987.

Yalow, R. S., Varsano-Aharon, N., Echemendia, E. and Berson, S. A. (1969): HGH and ACTH secretory responses to stress. *Hormone metab. Res.*, *1*, 3.

Yemm, E. W. and Cocking, E. C. (1955): The determination of amino acids with ninhydrin. *Analyst*, *80*, 209.

THE ONTOGENESIS OF HYPOTHALAMIC-HYPOPHYSIOTROPIC RELEASING FACTOR REGULATION OF HGH SECRETION*

SELNA L. KAPLAN** and MELVIN M. GRUMBACH

Department of Pediatrics, University of California at San Francisco, San Francisco, Calif., U.S.A.

The capacity of the pituitary gland to synthesize and secrete hormones is evident during the early gestational period. Integrated regulation by hypothalamic and higher CNS centers of this function of the fetal pituitary gland, however, may not be fully operative until the postnatal period. Storage and release of HGH are demonstrable at an early stage in development using immunologic methods (Kaplan and Grumbach, 1962), immunofluorescent techniques (Ellis *et al.*, 1966), or radioimmunoassay measurements of GH in serum and in pituitary homogenates (Kaplan and Grumbach, 1967; Kaplan *et al.*, 1971).

The anlage of the anterior hypophysis appears by the 7th to 8th week of gestation with demonstration of acidophiles by the 9th to 10th week (Falin, 1961). Cytologic differentiation proceeds concomitantly with changes in pituitary weight so that a marked increase in acidophile representation is evident by the 24th week of gestation. Although there is a gradual change in the weight of the pituitary gland during the first few months of gestation, a 4-fold increment occurs by mid-gestation (20th to 28th week). Continual enlargement of the pituitary gland results in a 10 to 15 times greater weight at term than at 20 weeks of gestation (Daikoku, 1958).

Development of the hypothalamus and portal system proceeds simultaneously with these observed changes in the pituitary gland. The first hypothalamic nuclei and the fibers of the supraoptic tract appear by 55 days of gestation (Weill and Bernfeld, 1954; Kuhlenbeck, 1954). By 16 weeks of gestation, differentiation of the pars tuberalis, median eminence and the remainder of the hypothalamic nuclei occurs. Neurosecretory material is present in the supraoptic and paraventricular nuclei soon thereafter (Raiha and Hjelt, 1957). The hypothalamic nuclei continue to increase in size until the early postnatal period when structural development is completed.

The primary plexus of the portal vascular system is initiated by 100 days at which time capillaries are in abundance in the pituitary gland. Establishment of the continuity of the primary and secondary plexus of the portal system has been demonstrated by 130 to 150 days of gestation (Espinasse, 1933; Niemineva, 1949).

These anatomic transformations can be correlated with physiologic changes in the fetal pituitary gland. Thus, the growth hormone content of the pituitary parallels the increase in weight of the fetal pituitary gland. Growth hormone rises from a content of 0.13 μg at 9 weeks of gestation to 400 μg at 26 weeks and 1.5 mg at term (Kaplan *et al.*, 1971) (Fig. 1).

Not only synthesis but secretion of growth hormone occurs in the human fetus. As early

* This work was supported in part by grants from the National Institutes of Child Health and Human Development and the National Institute of Arthritis and Metabolic Diseases, NIH, USPHS.

** Dr. Kaplan is a recipient of a Research Career Development Award, National Institute of Child Health and Human Development, NIH.

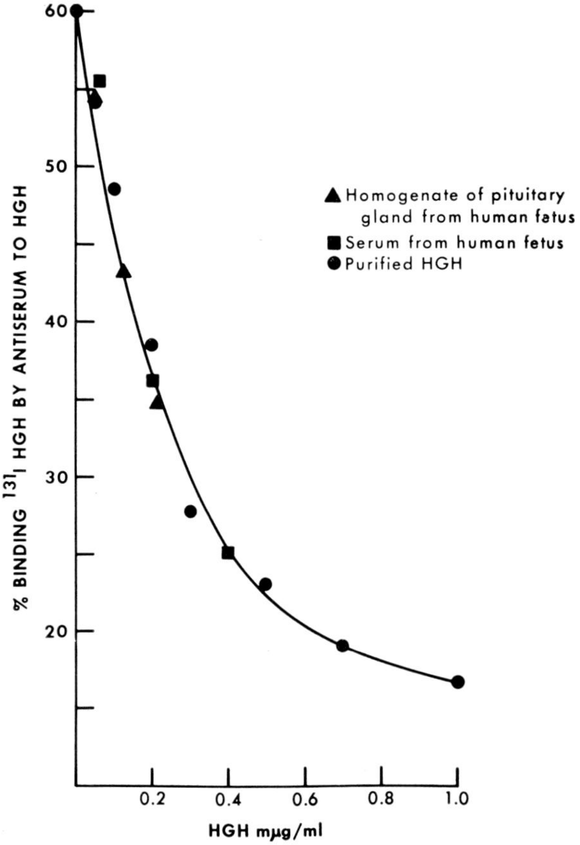

Fig. 1. Standard curve for HGH in radioimmunoassay double-antibody method. The percent binding of ^{131}I-HGH to anti-HGH serum is indicated on the ordinate and HGH in increasing concentration on the abscissa. Note the displacement of the binding of ^{131}I-HGH by serial dilutions of pituitary homogenate and serum from a human fetus is parallel to that of the standard curve for HGH.

as 70 days of gestation, GH was detectable in the circulation at a concentration of 14.5 ng/ml in the youngest fetus studied. Between 124 to 162 days of gestation, the concentration varied from 100 to 320 ng/ml, which is within the acromegalic range (Kaplan *et al.*, 1971).

The immunologic reactivity of the growth hormone in serum and pituitary gland of the human fetus was similar to that of purified pituitary human growth hormone. Serial dilutions of assayed pituitary homogenates and sera of fetuses show a linear relationship to the standard curve for purified human growth hormone (Fig. 2). Further evidence for the similarity of the growth hormone in the pituitary gland of the fetus and adult are the comparable physicochemical and immunochemical properties observed following disc electrophoresis and immunoelectrophoresis.

The concentration of serum growth hormone is 2 to 5 times higher at mid-gestation than at term. This suggests a marked decrease in circulating HGH levels during the late gestational period. The mean concentration of serum growth hormone in umbilical cord specimens, as determined in our laboratory, was 33.5 ng-ml (Kaplan and Grumbach, 1965), which is comparable to that reported by Cornblath *et al.* (1965). In the anencephalic fetuses at term, a mean concentration of 10.5 ng/ml of HGH was detected which is at the lowest range of

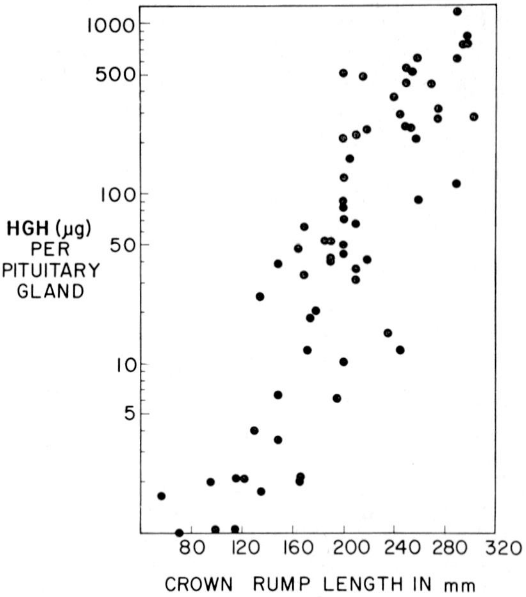

Fig. 2. The content of HGH in micrograms (μg) per fetal pituitary gland on the ordinate is plotted against crown-rump length in millimeters (mm) on the abscissa. Note the continual use throughout gestation.

normal values for infants at term. A gradual decrease in the serum concentration of growth hormone thus occurs during late gestation and the first few days of life, but the levels remain elevated for a period of 14 to 16 days during the neonatal period (Laron *et al.*, 1966). In the premature infants, the increased growth hormone concentration is sustained for a more prolonged period (Cornblath *et al.*, 1965; Milner and Wright, 1966).

Increased levels of circulating growth hormone in the fetus and neonate suggest that the CNS mechanisms regulating release of hypothalamic hypophysiotropic factors, in particular regulation of growth hormone releasing factor secretion, may not be operative until the postnatal period. In the monkey fetus, Mintz *et al.* (1969) were unable to elicit an increased GH response to arginine or tolbutamide nor suppression of the elevated fetal GH by administration of glucose. Similar findings were observed in sheep by Bassett *et al.* (1970). This lack of suppression of GH secretion by hyperglycemia has been reported in a limited number of subjects by Cornblath and associates in the human neonate and by Mintz and associates in the monkey, and apparently persists for the first 2 to 4 weeks of life. In contrast, an increase in serum GH following insulin-induced hypoglycemia or arginine infusion can be demonstrated in the newborn human or monkey (Chez *et al.*, 1970).

This pattern of hormonal secretion in the fetus may reflect the immaturity of the neurophysiologic function of the brain. Bergstrom (1968) has suggested that the electrical activity of the brain progresses from a simplistic primitive stage during the first 70 to 120 days of gestation to a phasic asynchronous stage, and finally in late gestation and early postnatal period to synchronous activity with inhibitory restraint. The EEG changes and motor activity in the human fetus support this interpretation (Ellingson, 1964; Humphrey, 1964). No EEG or motor activity is apparent until the second month of gestation. Dysrhythmic low voltage EEG activity occurs from 2 to 4 months, at which time tonic slow irregular muscular movements become apparent. By 5 months of fetal life, increased brain wave activity is noted with onset of activity in the diencephalic area. There is no change with sleep-wakefulness cycle. Increased motor activity is noted but is not fully coordinated. GH secretion is at peak concentrations during this period. By the 8th month of fetal life, a distinction between sleep and wakefulness stages can be demonstrated on EEG, but asynchrony between hemispheres persists throughout the immediate neonatal period. Motor activity is restrained coincident with the appearance of inhibitory circuits in the higher brain centers. The secretion of GH is drastically reduced at this time postnatally. Synchronous hemispheric activity with EEG changes in response to sleep becomes apparent by 2 months of age (Ellingson, 1964). Finkelstein *et al.* (1971) have demonstrated the absence of sleep-induced GH release in the human neonate during the first 2 months of life. By one year of age, the magnitude of the EEG electrical activity increases and alpha rhythm appears. These changes correlate well with observed coordinative and volitional motor activity in infants.

The chronology of events in the development of hypothalamic-hypophyseal function may be interpreted as follows: onset of GH secretion is coincident with appearance and increase in acidophiles in the fetal pituitary; by mid-gestation with appearance of hypothalamic nuclei and electrical activity of the diencephalon, secretion of growth hormone releasing factor (GRF) may occur with resultant increased unrestrained release of GH by the fetal pituitary; in the late gestational period, the operative inhibitory influences could lead to decreased GRF and GH secretion; regulatory mechanisms for control of GH secretion may not become fully functional until infancy at which time myelination, cortical development and synchronous EEG activity become evident. A similar developmental cycle may be operative in the control of secretion of other fetal pituitary hormones, such as fetal FSH and LH based on our studies (Kaplan *et al.*, 1969), for fetal TSH from studies of Fisher and Odell (1969), and Greenberg *et al.* (1969).

Elevated fasting levels observed during the neonatal period persist during the first year of life. Thereafter, the fasting concentration is similar in prepubertal and pubertal children (Kaplan *et al.*, 1968).

The GH response to provocative stimuli is also affected by maturation. In the child, the

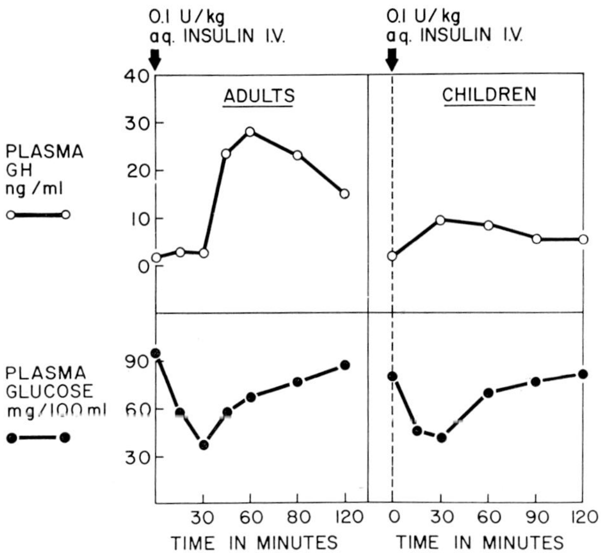

Fig. 3. The response of growth hormone (ng/ml) and glucose (mg/100 ml) to insulin-induced hypoglycemia in children and adults. Note that the maximal GH rise in children (on right panel) is significantly less than for adults (on left panel) despite an equivalent decrease in blood glucose.

mean maximum elevation in GH following insulin-induced hypoglycemia is 11.5 ng/ml (Youlton *et al.*, 1969), less than that observed in the adult which is 32.5 ng/ml (Abrams *et al.*, 1971) (Fig. 3). Sex steroid administration can convert the maximum GH response in a child to that observed in an adult. There is no sex difference in responsiveness during the prepubertal period. At puberty, the GH response to insulin-induced hypoglycemia is greater than in the prepubertal child (Kaplan *et al.*, 1968; Frasier *et al.*, 1970).

Circulating levels of 'estrogen' in the adult influence GH responsiveness to provocative stimuli and to ambulation. The concentration of plasma GH in the fasting state at rest is 1.0 ng/ml and is comparable in the adult male and female. Ambulation induces an increase in fasting GH levels in females which is less pronounced or not observed in males. Oral estrogen administration to males converts their fasting post-ambulation concentration of plasma GH to that observed in females (Frantz and Rabkin, 1965). Further evidence in women of the relationship of estrogen to GH secretion is the variation in GH response to provocative stimuli at different stages of the menstrual cycle (Merimee *et al.*, 1969*a*; Fiedler *et al.*, 1969). At mid-cycle, when peak estradiol secretion occurs, the GH response to insulin-induced hypoglycemia or to arginine infusion is of a greater magnitude than during the late phases of the menstrual cycle. In the male adult, arginine infusion does not consistently elicit a significant GH response without prior administration of oral estrogens (Merimee *et al.*, 1969*b*).

Administration of oral estrogen also affects GH release in the prepubertal child. Diethylstilbestrol in doses of 10 mg daily for 1 to 2 days to normal short children of either sex, results in increased concentrations of serum GH following insulin-induced hypoglycemia or arginine infusion, when compared to non-primed states (Youlton *et al.*, 1969; Root *et al.*, 1969). Similarly, conjugated estrogen (Premarin) in a dose of 2 to 10 mg daily has a similar enhancing effect on GH release in girls with gonadal dysgenesis or tall stature (Sadeghi-Nejad *et al.*, 1971). Circulating 'estrogen' in the prepubertal child converts the GH response to provocative stimuli to that observed in the pubertal individual or adult.

Administration of testosterone enhances the GH response to provocative stimuli in children as reported by Deller *et al.* (1966), by Martin *et al.* (1968), and by Illig and Prader (1970).

The alteration in GH responsiveness with advancement in sexual development may reflect the facilitating effect of circulating sex steroids on GH release rather than maturational changes in the neurohypothalamic control mechanisms. A possible synergistic action of GH and sex steroids in the induction of growth acceleration during pubertal development and the development of secondary sexual characteristics, has been suggested by Zachmann and Prader (1970) and supported by our own observations.

There is only limited information concerning GH release in those of advanced age. In 2 patients, ages 8 and 2-11/12 years, with progeria (premature aging syndrome) (Youlton *et al.*, 1969; Rosenbloom *et al.*, 1970), the GH levels following provocative stimuli were within normal limits for chronologic age. Normal GH responses have been demonstrated in adults 60 to 102 years of age (Laron *et al.*, 1970; Vinik *et al.*, personal communication).

The ontogeny of GH secretion progresses from hypersecretory phase induced by unrestrained GRF stimulation during fetal development to a stage of partial controlled release in infancy, and finally to the regulated release of GRF secretion as affected by circulating levels of GH and sex steroids, adrenergic receptors, and higher CNS center responsiveness.

REFERENCES

ABRAMS, R. L., GRUMBACH, M. M. and KAPLAN, S. L. (1971): The effect of administration of human growth hormone on the plasma growth hormone, cortisol, glucose, and free fatty acid response to insulin: Evidence for growth hormone autoregulation in man. *J. clin. Invest.*, *50*, 940.

BASSETT, J. M., THORBURN, G. D. and WALLACE, A. L. C. (1970): The plasma growth hormone concentration of the foetal lamb. *J. Endocr.*, *48*, 251.

BERGSTROM, R. M. (1968): Development of EEG and unit electrical activity of the brain during ontogeny. In: *Ontogenesis of the Brain*, p. 61. Editors: L. Jilek and S. Trojan. Charles University Press, Prague.

CHEZ, R. A., MINTZ, D. H., HORGER, E. O. III and HUTCHINSON, D. L. (1970): Factors affecting the response to insulin in the normal subhuman pregnant primate. *J. clin. Invest.*, *49*, 1517.

CORNBLATH, M., PARKER, M. L., REISNER, S. H., FORBES, A. E. and DAUGHADAY, W. H. (1965): Secretion and metabolism of growth hormone in premature and full-term infants. *J. clin. Endocr.*, *25*, 209.

DAIKOKU, S. (1958): Studies on the human foetal pituitary. 1. Quantitative observations. 2. On the form and histological development, especially that of the anterior pituitary. *Tokushima J. exp. Med.*, *5*, 200, 213.

DELLER, J. J., PLUNKET, D. C. and FORSHAM, P. H. (1966): Growth hormone studies in growth retardation – therapeutic response to administration of androgen. *Calif. Med.*, *104*, 359.

ELLINGSON, R. J. (1964): Studies of the electrical activity of the developing human brain. In: *Progress in Brain Research, Vol. IX*, p. 26. Editors: W. A. Himwich and H. E. Himwich. Elsevier Publishing Company, Amsterdam and New York.

ELLIS, S. T., BECK, J. S. and CURRIE, A. R. (1966): The cellular localization of growth hormone in the human fetal adenohypophysis. *J. Path. Bact.*, *92*, 179.

ESPINASSE, P. G. (1933): The development of the hypophysial-portal system in man. *J. Anat.*, *68*, 11.

FALIN, L. I. (1961): The development of human hypophysis and differentiation of cells of its anterior lobe during embryonic life. *Acta anat. (Basel)*, *44*, 188.

FIEDLER, A. J., TYSON, J. E. and MERIMEE, T. J. (1969): Arginine induced growth hormone release after clomiphene treatment. *J. clin. Endocr.*, *29*, 1110.

FINKELSTEIN, J. W., ANDERS, T. R., SACHAR, E. J., ROFFWARG, H. P. and HELLMAN, L. D. (1971): Behavioral state, sleep stage and growth hormone levels in human infants. *J. clin. Endocr.*, *32*, 368.

FISHER, D. A. and ODELL, W. D. (1969): Acute release of thyrotropin in the newborn. *J. clin. Invest.*, *48*, 1670.

FRANTZ, A. G. and RABKIN, M. T. (1965): Effects of estrogen and sex difference on secretion of human growth hormone. *J. clin. Endocr.*, *25*, 1470.

FRASIER, S. D., HILLBURN, J. M. and SMITH, F. G. (1970): Effect of adolescence on the serum growth hormone response to hypoglycemia. *J. Pediat.*, *77*, 465.

GREENBERG, A. H., CZERNICHOW, P., REBA, R. C., TYSON, J. and BLIZZARD, R. M. (1970): Observations on the maturation of thyroid function in early fetal life. *J. clin. Invest.*, *49*, 1790.

HUMPHREY, T. (1964): Some correlations between the appearance of human fetal reflexes and the development of the nervous system. In: *Progress in Brain Research, Vol. IV*, p. 93. Editors: D. P. Purpura and J. P. Schade. Elsevier Publishing Company, Amsterdam.

ILLIG, R. and PRADER, A. (1970): Effect of testosterone on growth hormone secretion in patients with anorchia and delayed puberty. *J. clin. Endocr.*, *30*, 615.

KAPLAN, S. L. and GRUMBACH, M. M. (1962): Immunologic assay and characteristics of growth hormone in the pituitary gland of the human fetus. *Amer. J. Dis. Child.*, *104*, 528.

KAPLAN, S. L. and GRUMBACH, M. M. (1965): Serum chorionic 'growth hormone-prolactin' and serum pituitary growth hormone in mother and fetus at term. *J. clin. Endocr.*, *25*, 1370.

KAPLAN, S. L. and GRUMBACH, M. M. (1967): Growth hormone secretion in the human fetus and in anencephaly. *Pediat. Res.*, *1*, 308.

KAPLAN, S. L., ABRAMS, C. A. L., BELL, J. J., CONTE, F. A. and GRUMBACH, M. M. (1968): Growth and growth hormone. I. Changes in serum level of growth hormone following hypoglycemia in 134 children with growth retardation. *Pediat. Res.*, *2*, 43.

KAPLAN, S. L., GRUMBACH, M. M. and SHEPARD, T. H. (1969): Gonadotropins in serum and pituitary of human fetuses and infants. *Pediat. Res.*, *3*, 512.

KAPLAN, S. L., GRUMBACH, M. M., and SHEPARD, T. H. (1971): Immunoreactive growth hormone in serum and pituitary of the human fetus. In preparation.

KUHLENBECK, H. (1954): *The Human Diencephalon*, p. 17. S. Karger, Basel.

LARON, Z., DORON, M. and AMIKAM, B. (1970): Plasma growth hormone in men and women over 70 years of age. *Med. Sport*, *4*, 126.

LARON, Z., MANNHEIMER, S., PERTZELAN, A. and NITZAN, M. (1966): Serum growth hormone concentration in full term infants. *Israel J. med. Sci.*, *2*, 770.

MARTIN, L. G., CLARK, J. W. and CONNOR, T. B. (1968): Growth hormone secretion by androgens. *J. clin. Endocr.*, *28*, 425.

MERIMEE, T. J., FINEBERG, S. E. and TYSON, J. E. (1969*a*): Fluctuations of human growth hormone secretion during menstrual cycle: response to arginine. *Metabolism*, *18*, 606.

MERIMEE, T. J., RABINOWITZ, P. and FINEBERG, S. E. (1969*b*): Arginine-initiated release of human growth hormone. *New Engl. J. Med.*, *280*, 1434.

MILNER, R. D. G. and WRIGHT, A. D. (1966): Blood glucose, plasma insulin, and growth hormone response to hyperglycemia in the newborn. *Clin. Sci.*, *31*, 309.

MINTZ, D. H., CHEZ, R. A. and HOGER, E. O. (1969): Fetal insulin and growth hormone metabolism in the subhuman primate. *J. clin. Invest.*, *48*, 176.

NIEMINEVA, K. (1949): Observations on the development of the hypophysial-portal system. *Acta paediat. scand.*, *38*, 366.

RAIHA, N. and HJELT, L. (1957): The correlation between the development of the hypophysial portal system and onset of neurosecretory activity in the human fetus and infant. *Acta paediat. scand.*, *46*, 610.

ROOT, A. W., SAENZ-RODRIGUEZ, C., BONGIOVANNI, A. M. and EBERLEIN, W. R. (1969): The effect of arginine infusion on plasma growth hormone and insulin in children. *J. Pediat.*, *74*, 187.

ROSENBLOOM, A. L., KARACAN, I. J. and DEBUSK, F. L. (1970): Sleep characteristics and endocrine response in progeria. *J. Pediat.*, *77*, 692.

WEILL, J. and BERNFELD, J. (1954): Le syndrome hypothalamique. In: *Library of Academy of Medicine*, p. 11. Masson et Cie., Paris.

YOULTON, R., KAPLAN, S. L. and GRUMBACH, M. M. (1969): Growth and growth hormone: IV. Limitations of the growth hormone response to insulin and arginine and of the immunoreactive insulin response to arginine in the assessment of growth hormone deficiency in children. *Pediatrics*, *43*, 989.

ZACHMAN, M. and PRADER, A. (1970): Anabolic and androgenic effect of testosterone in sexually immature boys and its dependency on growth hormone. *J. clin. Endocr.*, *30*, 85.

HUMAN GROWTH HORMONE IN PROTEIN-CALORIE MALNUTRITION*

B. L. PIMSTONE, D. J. BECKER and J. D. L. HANSEN

Departments of Medicine and Child Health, University of Cape Town Medical School, Observatory, Cape Province, South Africa

Endocrine function has for many years been considered to be depressed in protein-calorie malnutrition; indeed the term 'pseudohypophysectomy' was coined in this condition (Mulinos and Pomerantz, 1940) to emphasize the apparent clinical and biochemical hypopituitarism. Recently, with the increasing availability of more sophisticated tests of endocrine function, this observation has been seriously questioned. The studies to be reviewed attempt to define the status of human growth hormone (HGH) in protein-calorie malnutrition as it has been reported that the secretion of this hormone is low (Srebnik and Nelson, 1962) and that children with marasmus do not grow adequately on refeeding unless growth hormone is administered (Mönckeberg *et al.*, 1963). Of additional relevance in considering HGH homeostasis in malnutrition is its known effect on nitrogen retention (Committee of MRC, 1959) and protein synthesis (Korner, 1959), actions highly desirable in a disorder characterised by very low protein intake.

The review below outlines published studies on fasting serum HGH levels in protein-calorie malnutrition and the response to induced hyperglycaemia as well as to milk and carbohydrate feeding (Pimstone *et al.*, 1966, 1967, 1968, 1969). Further data on the relationships with serum albumin and plasma amino acids are presented and discussed.

PATIENTS AND METHODS

106 children suffering from protein-calorie malnutrition (PCM) were studied. All had the clinical and biochemical evidence either of kwashiorkor (underweight, oedema, hypoalbuminaemia and frequent dermatosis) or marasmus (less than 60% expected weight for age and no oedema). The children were admitted to the Metabolic Ward of the Red Cross War Memorial Children's Hospital for study and treatment. Full consent was obtained from the parents in every case.

All patients were treated with antibodies, potassium chloride, vitamins and, where indicated, intravenous fluid therapy. Studies were commenced prior to the initiation of protein feeding. These studies were repeated at the times after treatment discussed in the individual sections to follow. In all cases blood was withdrawn by venipuncture starting at 9 a.m. after a 9 hour overnight fast with patients at rest in bed. Serum was separated and frozen for 2-10 weeks before assay for HGH by the method of Glick *et al.* (1963) and in later studies by a modification of the method of Morgan and Lazarow (1963). Blood sugar was

* This study was supported by grants from South African Medical Research Council, South African Atomic Energy Board, Wellcome Trust and U.S.P.H.S. Grant AM03995.

estimated by the method of Somogyi (1952) and serum albumin by the biuret method (Wolfson *et al.*, 1948). Plasma amino acids were estimated after salicylsulphonic acid deproteinization on an automated ion exchange column by the method of Hamilton (1963).

DETAILS OF INDIVIDUAL STUDIES AND RESULTS

Fasting HGH in kwashiorkor and marasmus

In 28 children with kwashiorkor and 20 children with marasmus, fasting HGH was assayed on admission and again after 2-3 weeks feeding at which time signs of malnutrition had disappeared and serum albumin had risen considerably (for details see Pimstone *et al.*, 1966, 1967). The results are shown in Figures 1 and 2. There is a substantial elevation of fasting HGH in both kwashiorkor and marasmus which drops to control values after 2-3 weeks of feeding.

HGH during induced hyperglycaemia

Twenty-eight children with protein-calorie malnutrition were given glucose (2 g/kg) orally and samples of venous blood were taken fasting and at 30, 60, 90, 120 and 150 min after the glucose load. A typical result is shown in Fig. 3. There is grossly impaired and

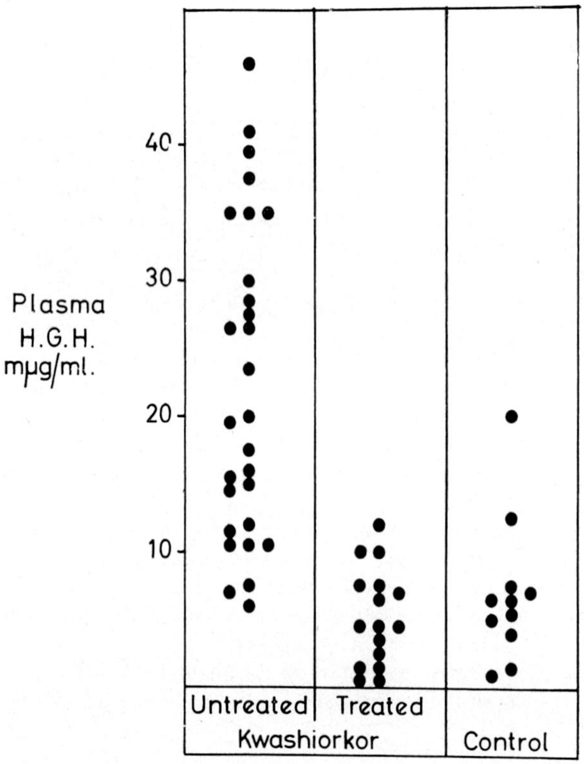

Fig. 1. Fasting HGH in kwashiorkor before and after treatment and in controls. (From Pimstone *et al.*, 1966, *Lancet*, *2*, 779, by kind permission from the editors).

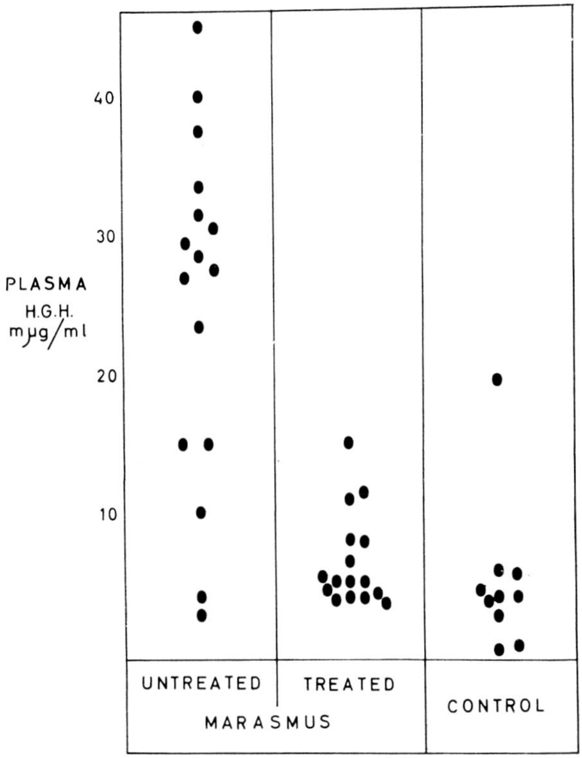

Fig. 2. Fasting HGH in marasmus before and after treatment and in controls. (From Pimstone *et al.*, 1968, *Amer. J. clin. Nutr.*, *21*, 482, by kind permission from the editors).

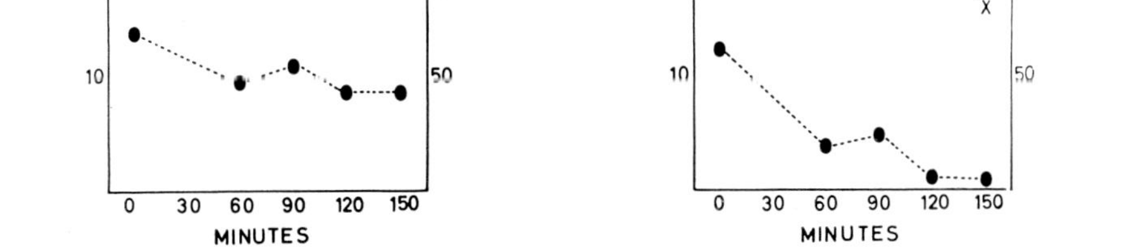

Fig. 3. Impaired suppression of HGH ● --- ● by hyperglycaemia associated with an abnormal glucose tolerance curve X ----- X in a patient with kwashiorkor. HGH suppression returns to normal and glucose tolerance improves after treatment.

delayed suppression of HGH by hyperglycaemia, suppressibility returning to normal after 3 weeks of feeding. On occasions, paradoxical rises of growth hormone after glucose were encountered.

HGH response to protein or carbohydrate feeding

Thirteen children with kwashiorkor had fasting blood sampled on admission, after 3 days of a protein-free diet containing 80-100 calories/kg/day and finally after 3 days of an isocaloric diet containing protein. A further 9 patients had fasting blood sampled on the day of admission after which protein was immediately introduced into the diet and further sampling done 3 and 6 days later. A final blood specimen was taken 21-28 days after admission at which stage the children were clinically and biochemically greatly improved. The results are shown in Figures 4 and 5.

The elevated HGH does not drop when adequate carbohydrate alone is given but is lowered promptly after the administration of oral protein.

The relationship between fasting HGH and glucose in PCM

Fasting HGH has been plotted against fasting blood glucose in untreated PCM. These results are shown in Fig. 6. No correlation between these two measurements is apparent.

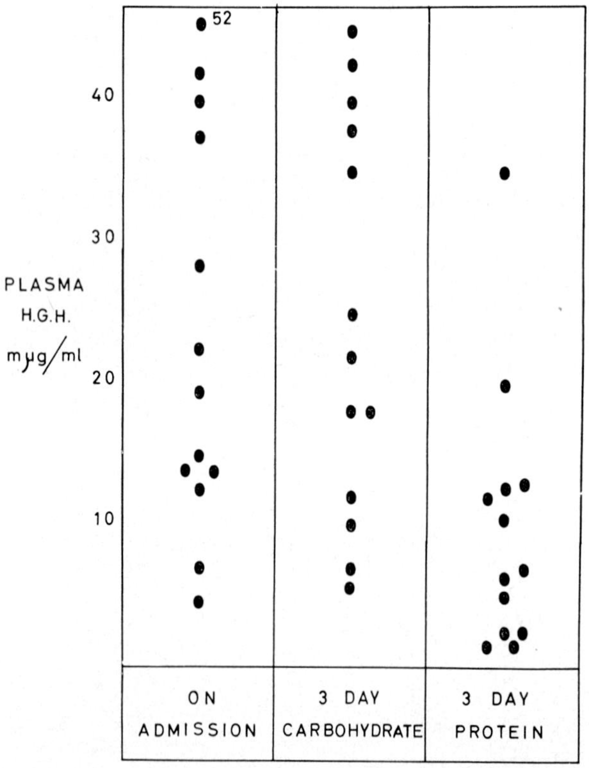

Fig. 4. HGH in kwashiorkor after protein-free and protein-containing diets. HGH levels after protein feeding differ significantly from the other 2 groups ($p < 0.005$). (From Pimstone *et al.*, 1968, *Amer. J. clin. Nutr.*, *21*, 482, by kind permission from the editors).

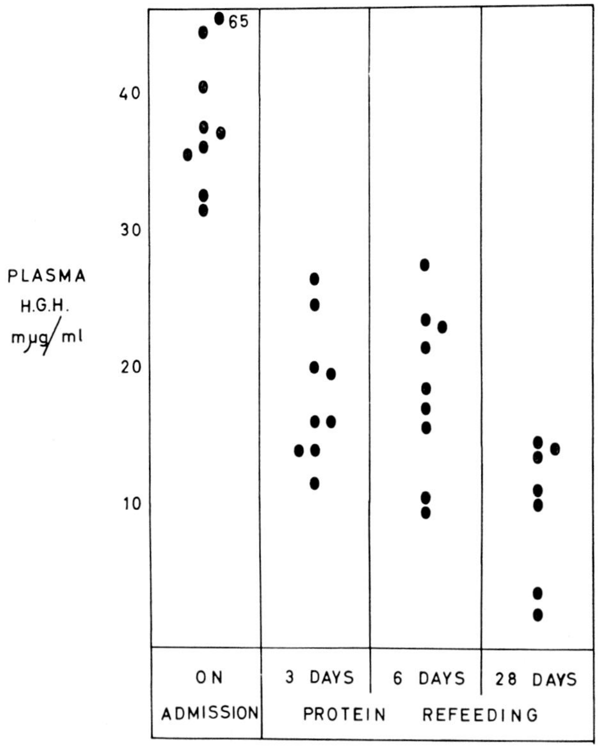

Fig. 5. HGH in kwashiorkor after immediate protein feeding. HGH levels drop significantly ($p < 0.001$) after 3 days of protein feeding. (From Pimstone *et al.*, 1968, *Amer. J. clin. Nutr.*, *21*, 482, by kind permission from the editors).

The relationship between albumin and growth hormone in PCM

Fasting HGH has been plotted against serum albumin in 30 patients with kwashiorkor and marasmus. As shown in Fig. 7, a significant inverse correlation is apparent ($p < 0.01$). When this relationship is re-examined after an albumin or amino acid infusion or 1 to 2 days of milk feeding, the correlation is less significant ($p < 0.05$). In some cases there was a substantial drop in HGH while serum albumin remained low (after milk feeding). In others HGH was unchanged despite increased serum albumin levels (after albumin infusion). After 3-6 weeks of treatment the inverse correlation returns once more ($p < 0.001$) as at this stage the albumin has risen and HGH has dropped to normal values.

In 10 patients with kwashiorkor, albumin was acutely infused intravenously over 24 or 48 hours in order to elevate rapidly the serum albumin levels. Oral or intravenous glucose tolerance tests were performed before and on the first or second day following the albumin infusion and then finally repeated after 3-6 weeks of protein feeding. A representative pattern is shown in Fig. 8. Although albumin is raised by the acute infusion, both the glucose tolerance and high non-suppressible growth hormone remain unchanged in the great majority of cases. When serum albumin levels rise to normal after 3-6 weeks of protein feeding however, both glucose tolerance and HGH more closely approximate the expected normal.

In 7 patients intravenous glucose tolerance tests were performed on admission and 36-72 hours after initiating milk feeding. A representative pattern is shown in Fig. 9. Glucose tolerance shows some improvement but growth hormone levels drop strikingly and sup-

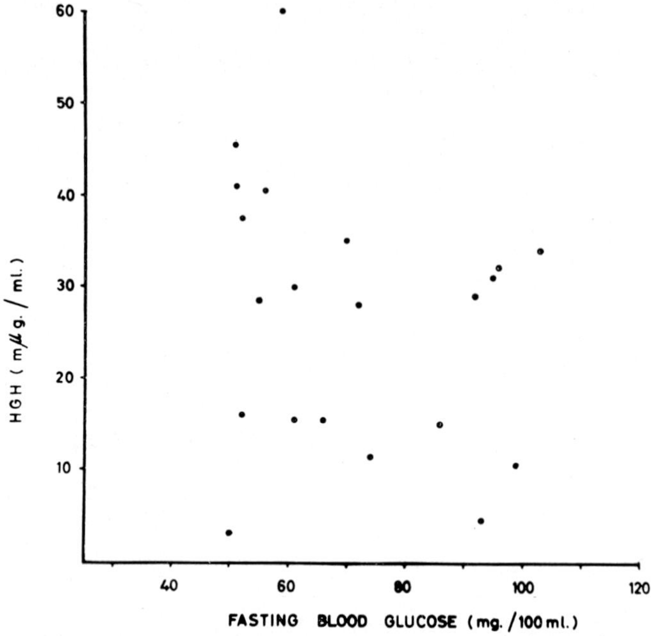

Fig. 6. Lack of relationship between HGH and fasting blood glucose levels in kwashiorkor ($p < 0.1$) before treatment.

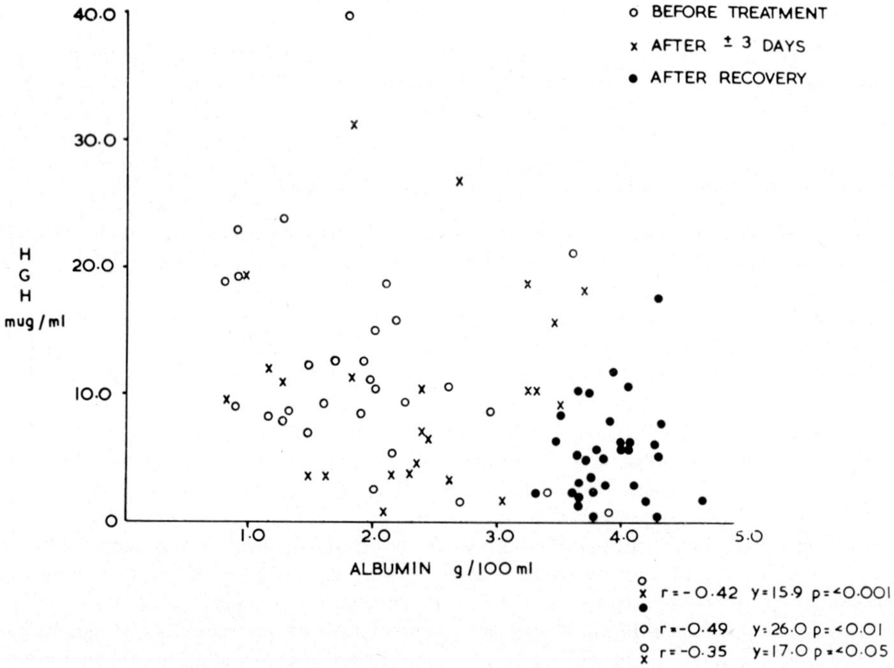

Fig. 7. Relationship between HGH and serum albumin levels in kwashiorkor and marasmus before treatment, after 1–3 days of albumin or amino acid infusion or milk feeding, and after 3–6 weeks of therapy.

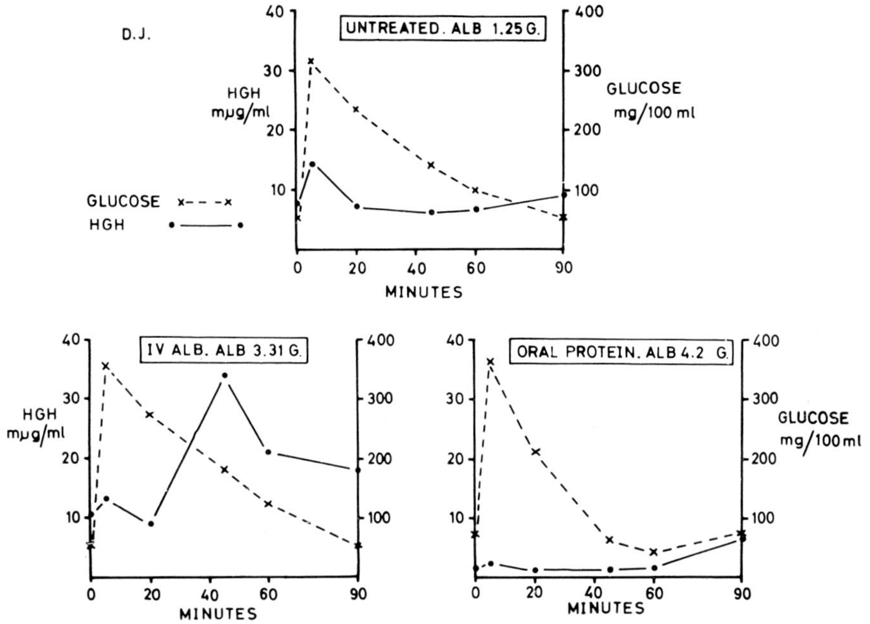

Fig. 8. Non-suppressible HGH after glucose loading in a patient with kwashiorkor before and after an albumin infusion. Suppression of HGH returns to normal after 3 weeks of protein feeding and glucose tolerance improves.

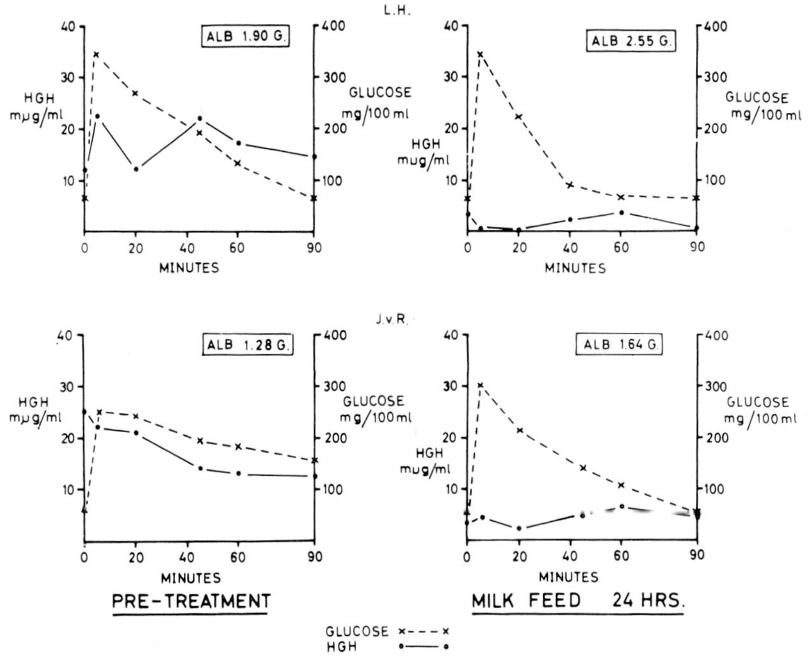

Fig. 9. Raised non-suppressible HGH and impaired glucose tolerance in 2 patients with kwashiorkor. After 36 and 72 hours of protein feeding respectively HGH levels drop with normal suppressibility and glucose tolerance improves despite minimal changes in serum albumin.

pressibility returns. These improvements occur at a time when serum albumin values have risen only minimally, in contrast to the results after albumin infusion.

Oral glucose tolerance tests (as outlined above) were performed in 8 patients suffering from a nephrotic syndrome. Typical responses in 4 patients are shown in Fig. 10. Notice that growth hormone levels are low throughout the study in spite of gross hypoalbuminaemia.

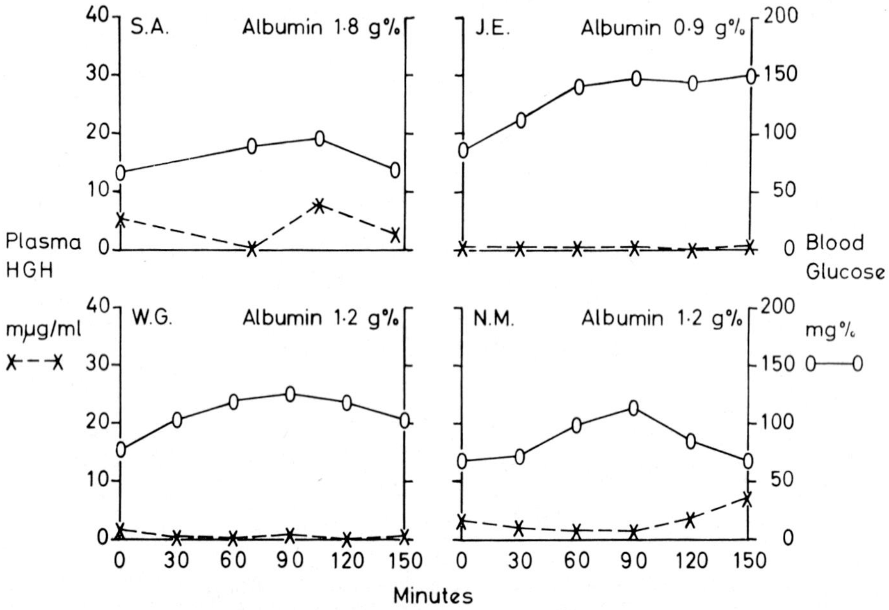

Fig. 10. Normal basal and glucose-induced suppression of HGH in 4 patients with a nephrotic syndrome associated with low serum albumin levels.

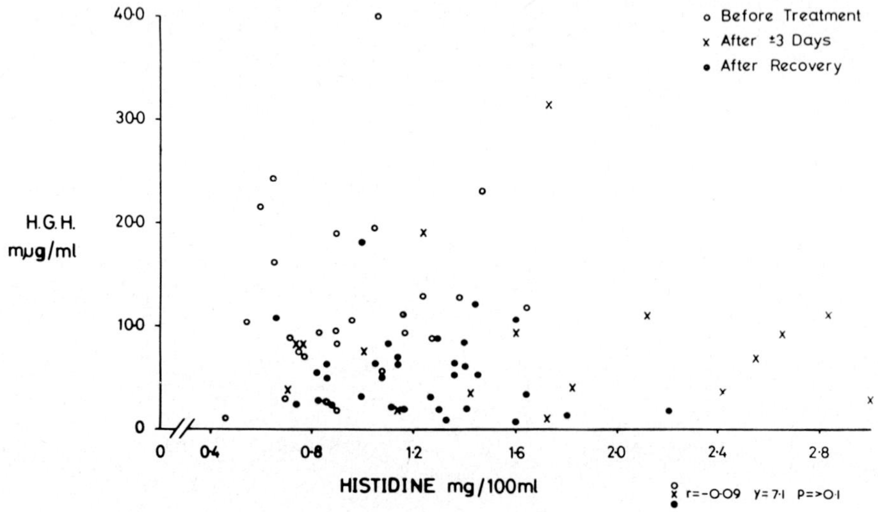

Fig. 11. Lack of correlation between HGH and plasma histidine in kwashiorkor and marasmus before and after treatment.

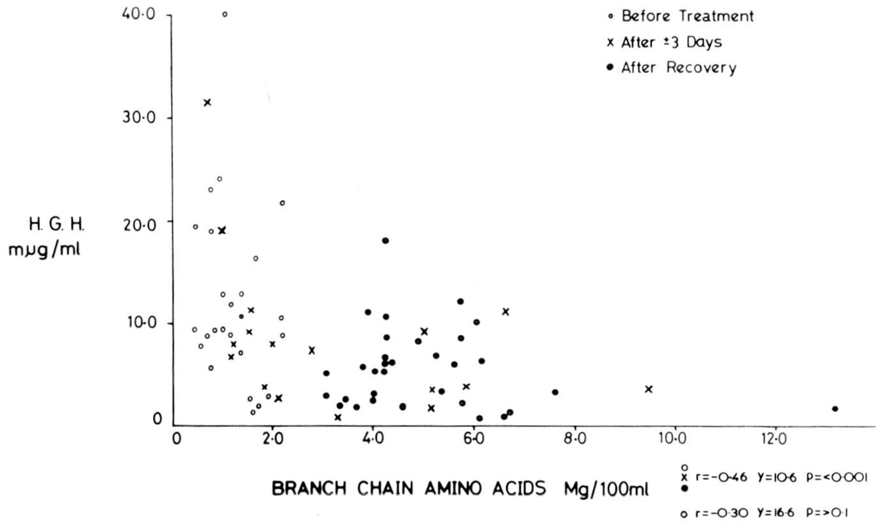

Fig. 12. Inverse correlation between HGH and the plasma branch chain amino acids (leucine, isoleucine and valine) in kwashiorkor and marasmus before treatment, after 1–3 days of milk feeding or albumin and amino acid infusion, and after 3–6 weeks of treatment.

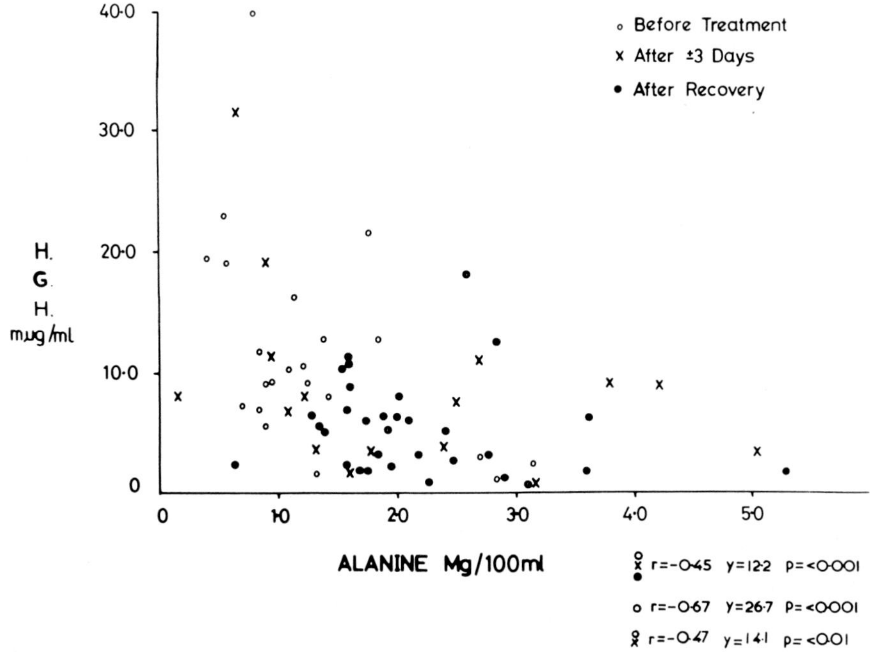

Fig. 13. Inverse correlation between HGH and plasma alanine levels in kwashiorkor and marasmus before treatment, after 1–3 days of milk feeding or albumin or amino acid infusion, and after 3–6 weeks of treatment.

The relationship between amino acids and growth hormone in PCM

The relationship between fasting serum HGH and various amino acids has been compared in untreated PCM, after 1-3 days of milk feeding or amino acid or albumin infusions, and again after 3-6 weeks at the initiation of clinical cure. No correlation is noted between HGH and histidine ($p > 0.1$) (Fig. 11), glycine and arginine, although early data (Pimstone *et al.*, 1969) suggested that one may exist in the former. A strong inverse correlation exists however between HGH and the branch chain amino acids leucine, iso-leucine and valine ($p < 0.001$) although there is none prior to therapy ($p > 0.1$) (Fig. 12). A highly significant correlation between HGH and alanine is found in the untreated patient ($p < 0.001$) as well as after 3 days ($p < 0.01$) and 3 weeks of treatment ($p < 0.001$) (Fig. 13).

DISCUSSION

It is quite evident that a profound disturbance in growth hormone homeostasis is the rule in protein-calorie malnutrition. Contrary to previous reports (Srebnik and Nelson, 1962; Mönckeberg *et al.*, 1963), fasting HGH levels are high in both kwashiorkor and marasmus. Furthermore, these elevated levels are not suppressed normally by induced hyperglycaemia, do not correlate with fasting blood glucose or drop when an adequate carbohydrate but protein-free diet is given. The return of fasting HGH and glucose induced suppressibility to normal follows the administration of oral protein, requiring as little as 36-72 hours of milk feeding. This occurs without significant change in serum albumin levels. An inverse correlation is noted between the serum albumin levels and the elevated growth hormone in untreated PCM as well as after adequate and prolonged therapy. However, this correlation is not evident after 2-3 days of milk feeding or artificial elevation of the serum albumin levels. After albumin infusion, growth hormone usually remains high and is poorly suppressed by glucose. Finally, an inverse correlation is clear between the fasting HGH levels and certain plasma amino acids. In patients before and after treatment, HGH levels correlate well with plasma branch chain amino acids, (leucine, iso-leucine and valine) as well as alanine. The correlation with branch chain amino acids is less significant prior to therapy but that between HGH and alanine is maintained at all stages of treatment. Other amino acids such as histidine, arginine and glycine correlate poorly with HGH.

A number of conclusions can be drawn. 'Pseudohypophysectomy' as envisaged by Mulinos and Pomerantz (1940), Mönckeberg *et al.* (1963) and others does not occur in protein-calorie malnutrition. Indeed the secretion of growth hormone, regarded by many as being the most sensitive test of anterior pituitary function, appears normal or possibly supranormal in these children. This is consistent with the report of Alleyne and Young (1967) in respect of adrenocortical function and PCM where similar elevations of plasma cortisol were described. Since the initial observations of high growth hormone were made in our laboratories (Pimstone *et al.*, 1966, 1967, 1968) the work has been confirmed by others (Hadden, 1967; Graham *et al.*, 1969; Milner, 1971). It is clear therefore that protein malnourished children fail to grow because of a deficiency in the building blocks, *i.e.* amino acids, rather than a deficiency of the appropriate hormonal stimulus.

The finding of high but poorly suppressible growth hormone levels does not necessarily imply hypersecretion of this hormone. Some pathologists have described eosinophil hyperplasia (Tejada and Russfield, 1957) or increased numbers of secretory eosinophils (Golden *et al.*, 1961) in the anterior pituitary glands of protein depleted individuals, suggesting increased secretion. However, the possibility must be entertained that the high serum HGH levels may result from impaired degradation of the hormone, especially as other hormones, for example adrenocorticosteroids (Alleyne and Young, 1967), have been shown to exhibit a prolonged half disappearance time in PCM. Studies are underway at present to elucidate this problem.

Irrespective of whether secretion or impaired degradation of HGH is dominant in PCM, the net result, *i.e.* high serum levels, is profound. It remains to consider whether these levels are simply non-specific or whether they represent a real adaptive consequence of protein depletion. The relationship of HGH to protein as opposed to calorie deprivation is apparent from the results outlined above. Neither glucose infusion nor a high carbohydrate diet altered HGH levels. On the other hand in the untreated case, inverse correlation was clearly noted between the low levels of albumin, and alanine and high HGH. A drop in albumin and alanine are late effects of protein depletion and reflect a substantial degree of protein deprivation.

It has recently been shown (Adibi, 1968; Abidi and Drash, 1970) that *short term* protein depletion is associated with normal HGH levels and a rise in plasma alanine although the branch chain amino acids drop early. Our observations have been similar (Pimstone and Kelman, unpublished data). Holt and Snyderman (1964) have reported that in mild kwashiorkor plasma alanine is elevated and only drops in more severe cases, a feature also evident in kwashiorkor in our population (Truswell and Hansen, personal communication). This would suggest therefore that a rise in serum HGH is not an immediate consequence of dietary protein restriction and that severe protein deprivation is necessary (as evidenced by a drop in plasma alanine levels and serum albumin) before this occurs. It is unlikely therefore that the responsiveness of growth hormone is critically geared to minor dietary protein deprivation in spite of its known effects of nitrogen retention (Committee of MRC, 1959) and protein synthesis (Korner, 1959), but its hypersecretion may be a very important compensatory factor in severe depletion. A further possibility exists that intracellular glucose may be relevant in the pathogenesis of the raised growth hormone levels. In modest protein depletion, insulin secretion is relatively unimpaired (Abidi and Drash, 1970). With a simultaneous good dietary carbohydrate intake the need for gluconeogenesis falls away (Cahill, 1970) and plasma alanine (a major precursor in hepatic gluconeogenesis) (Felig *et al.*, 1970*a*) rises because of its reduced hepatic utilization. With prolonged protein deprivation, insulin secretion is grossly impaired (Becker *et al.*, 1971) and insulin resistance may be important (Becker *et al.*, 1971; Heàrd, 1966). In spite of adequate carbohydrate intake therefore, glucose may be poorly available as a substrate for intracellular oxidation because of impaired transport or utilization in the cell. This could be reflected by a greater need for gluconeogenesis and a subsequent late drop in plasma alanine levels as has been demonstrated in diabetes (Felig *et al.*, 1970*b*). One could postulate therefore that the high HGH found in advanced PCM in association with low plasma alanine might reflect an intracellular glucose deficit as a consequence of advanced protein deprivation. Impaired cellular glucose utilization, even in the presence of hyperglycaemia, is a known stimulus of HGH secretion as has been shown by acute experiments using 2-deoxy-D-glucose (Wegienka *et al.*, 1967). However, as chronic hypoglycaemia is unassociated with serum HGH elevation (Marks *et al.*, 1967), the role of a prolonged intracellular glucose deficit in PCM must remain highly speculative.

Although it is tempting to apply a teleological explanation to the raised HGH levels of PCM, it must be conceded that without certain knowledge of the fate of growth hormone, one cannot exclude the effects of protein depletion on the degradative enzymes of HGH and consequently its non-specific elevation.

SUMMARY

Growth hormone levels are elevated and poorly suppressed by glucose in advanced protein-calorie malnutrition. Dramatic response to oral protein as opposed to carbohydrate occurs. This together with the marked inverse correlation between HGH and albumin and amino acids suggests a real relationship between protein deprivation and growth hormone levels. In the present state of knowledge it is impossible to be certain whether this is a result of hypersecretion of growth hormone, impaired degradation, or both processes. While the elevation of growth hormone might be an important compensatory factor for retaining protein in a person whose protein intake is meagre, it cannot be excluded that this might be a non-specific response to protein deprivation either as a result of impaired synthesis of those enzymes responsible for HGH degradation or as a consequence of impaired intracellular glucose utilization.

REFERENCES

ADIBI, S. A. (1968): Influence of dietary deprivations on plasma concentrations of free amino acids of man. *J. appl. Physiol.*, *25*, 52.

ADIBI, S. A. and DRASH, A. L. (1970): Hormone and amino acid levels in altered nutritional states. *J. Lab. clin. Med.*, *76*, 722.

ALLEYNE, G. A. O. and YOUNG, V. H. (1967): Adrenocortical function in children with severe protein calorie malnutrition. *Clin. Sci.*, *33*, 189.

BECKER, D. J., PIMSTONE, B. L., HANSEN, J. D. L. and HENDRICKS, S. (1971): Insulin secretion in protein-calorie malnutrition: I. Quantitative abnormalities and response to treatment. *Diabetes*, *20*, 542.

CAHILL Jr, G. F. (1970): Starvation in man. *New Engl. J. Med.*, *282*, 668.

Committee of Medical Research Council (1959): The effectiveness in man of human growth hormone. *Lancet*, *1*, 7.

FELIG, P., POZEFSKY, T., MARLISS, E. and CAHILL Jr, G. F. (1970*a*): Alanine. Key role in gluconeogenesis. *Science*, *167*, 1003.

FELIG, P., MARLISS, E., OHMAN, J. L. and CAHILL, Jr., G. F. (1970*b*): Plasma amino acid levels in diabetic ketoacidosis. *Diabetes*, *19*, 727.

GLICK, S. M., ROTH, J., YALOW, R. S. and BERSON, S. A. (1963): Immunoassay of human growth hormone in plasma. *Nature (Lond.)*, *199*, 784.

GOLDEN, A., BONDY, P. and CHAMBERS, R. (1961): Adenohypophyseal changes in patients dying of neoplastic disease. *Yale J. Biol. Med.*, *34*, 299.

GRAHAM, G. G., CORDANO, A., BLIZZARD, R. M. and CHEEK, D. B. (1969): Infantile malnutrition. Changes in body composition during rehabilitation. *Paediat. Res.*, *3*, 579.

HADDEN, D. R. (1967): Glucose, free fatty acid and insulin interrelations in kwashiorkor and marasmus. *Lancet*, *2*, 589.

HAMILTON, P. B. (1963): Ion exchange chromatography of amino acids. *Analyt. Chem.*, *35*, 2055.

HEARD, C. R. C. (1966): Effects of severe protein-calorie deficiency on the endocrine control of carbohydrate metabolism. *Diabetes*, *15*, 78.

HOLT, L. E. and SNYDERMAN, S. E. (1964): Anomalies of amino acid metabolism. In: *Mammalian Protein Metabolism*, *Vol. II*, Chapter 18, p. 321–372. Editors: H. N. Munro and J. B. Allison. Academic Press, New York.

KORNER, A. (1959): The effect of hypophysectomy of the rat and of treatment with growth hormone on the incorporation of amino acids into liver protein in a cell-free system. *Biochem. J.*, *73*, 61.

MARKS, V., GREENWOOD, F. C., HOWORTH, P. J. N. and SAMOLS, E. (1967): Plasma growth hormone levels in chronic hypoglycaemia. *J. clin. Endocr.*, *27*, 523.

MILNER, R. D. G. (1971): Metabolic and hormonal responses to glucose and glucagon in patients with infantile malnutrition. *Paediat. Res.*, *5*, 33.

MÖNCKEBERG, F., DONOSO, G., OXMAN, S., PAK, N. and MENEGHELLO, J. (1963): Human growth hormone in infant malnutrition. *Paediatrics*, *31*, 58.

MORGAN, C. R. and LAZAROW, A. (1963): Immunoassay of insulin. Two antibody systems. *Diabetes*, *12*, 115.

MULINOS, M. C. and POMERANTZ, L. (1940): Pseudohypophysectomy: a condition resembling hypophysectomy produced by malnutrition. *J. Nutr.*, *19*, 493.

PIMSTONE, B., BARBEZAT, G., HANSEN, J. D. L. and MURRAY, P. (1967): Growth hormone and protein-calorie malnutrition. Impaired suppression during induced hyperglycaemia. *Lancet*, *2*, 1333.

PIMSTONE, B. L., BARBEZAT, G., HANSEN, J. D. L. and MURRAY, P. (1968): Studies on growth hormone secretion in protein-calorie malnutrition. *Amer. J. clin. Nutr.*, *21*, 482.

PIMSTONE, B., SAUNDERS, S. J., HANSEN, J. D. L. and BUCHANAN-LEE, B. (1969): Observations on the serum albumin and plasma amino acids in association with the elevated growth hormone levels found in protein deficiency states. In: *Protein and Polypeptide Hormones. Proceedings, First International Symposium, Liège, 1969*, pp. 906–907. Editor: M. Margoulies. ICS 161, Excerpta Medica, Amsterdam.

PIMSTONE, B. L., WITTMANN, W., HANSEN, J. D. L. and MURRAY, P. (1966): Growth hormone and kwashiorkor. Role of protein in growth hormone homeostasis. *Lancet*, *2*, 779.

SREBNIK, H. H. and NELSON, M. M. (1962): Anterior pituitary function in male rats deprived of dietary protein. *Endocrinology*, *70*, 723.

SOMOGYI, M. (1952): Notes on sugar determination. *J. biol. Chem.*, *195*, 19.

TEJADA, C. and RUSSFIELD, A. B. (1957): A preliminary report in the pathology of the pituitary gland in children with malnutrition. *Arch. Dis. Childh.*, *33*, 343.

WEGIENKA, L. C., GRODSKY, G. M., KARAM, J. H., GRASSO, S. G. and FORSHAM, P. H. (1967): Comparison of insulin and 2-deoxy-D-glucose-induced glycopenia as stimulators of growth hormone secretion. *Metabolism*, *16*, 245.

WOLFSON, W. Q., COHN, G., CALVARY, E. and ICHABA, F. (1948): Studies in serum proteins. *Amer. J. clin. Path.*, *18*, 723.

EFFECT OF CHRONIC POTASSIUM DEPLETION ON GROWTH HORMONE RELEASE IN MAN*

S. PODOLSKY, B. A. BURROWS, H. J. ZIMMERMAN and C. PATTAVINA

Departments of Nuclear Medicine, Internal Medicine and Intermediate Care Service, Veterans Administration Hospital and Boston University School of Medicine, Boston, Mass., U.S.A.

Potassium is quantitatively the most prominent cation in intracellular water and more than 99% of the total body potassium is intracellular. Chronic potassium depletion is associated with reversibly impaired glucose tolerance and reduced insulin response in primary aldosteronism and other conditions (Conn, 1965). We have confirmed these findings with serial studies of total body potassium content before and after potassium repletion (Podolsky *et al.*, 1970*a*, *b*, 1971). In studies of the isolated, perfused rat pancreas Grodsky and Bennett (1966) established that both potassium and calcium directly stimulated insulin release, and insulin production is impaired in the absence of these cations. The present study provides evidence that patients with potassium depletion have reduced growth hormone response to stimulation, which is reversible by potassium replacement.

MATERIALS AND METHODS

Twelve hospitalized male patients with stable chronic hepatic disease were serially studied with insulin induced hypoglycemia and arginine tolerance tests before and after a period of KCl loading. Serum glucose levels were measured by the *o*-toluidine technique on a Technicon Auto-Analyzer (Frings *et al.*, 1970), and plasma growth hormone levels determined by a modification of the dextran coated charcoal radioimmunoassay (Lau *et al.*, 1966). Total body potassium content was assessed by serial measurement of the naturally occurring radioisotope of potassium, ^{40}K, with the patient lying supine for 40 min in a low background whole body counter (Podolsky *et al.*, 1970*a*; Tyson *et al.*, 1970). Seven patients had no change in total body ^{40}K after KCl treatment, and constituted the control group. The other five patients demonstrated total body ^{40}K increases from 20% to 40% higher after identical oral KCl supplementation (180-200 mEq K plus diet daily for a period of 14 days or longer). Intracellular potassium depletion in this group, due to a variety of causes, including diuretics, secondary aldosteronism, vomiting, cirrhosis, etc., might ordinarily have been overlooked since only 20% (1/5) showed one or more serum potassium levels below 3.5 mEq/l.

RESULTS

Arginine infusion over a 30-min period produces subnormal insulin output in potassium depleted patients (Podolsky *et al.*, 1970*a*, *b*; 1971). Figure 1 shows before and after arginine tolerance tests in a potassium deficient patient who increased total body potassium signifi-

* Supported by Veterans Administration Central Office Special Research Funds (Clinical Investigator Program of S.P.).

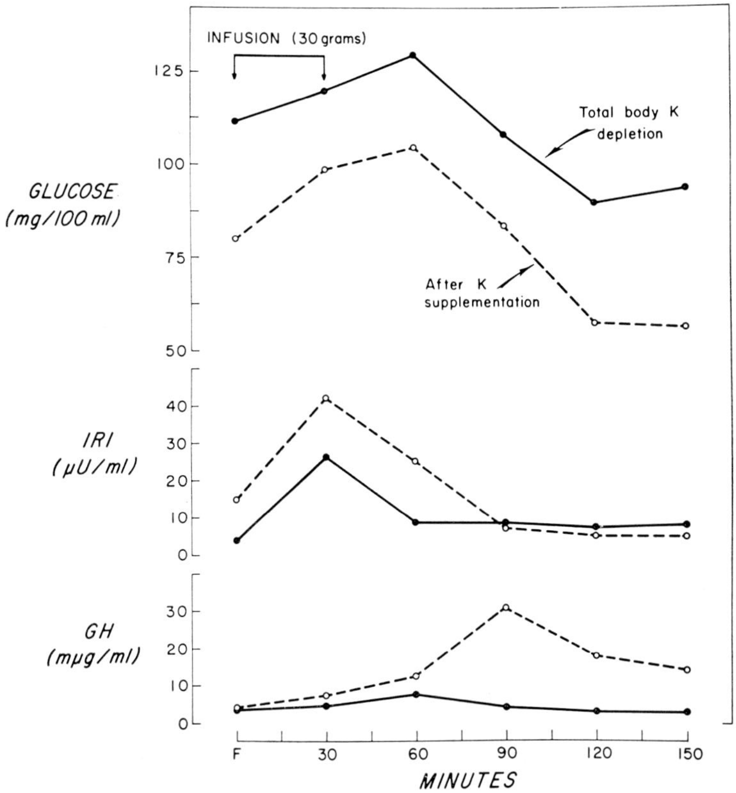

Fig. 1. Effect of potassium repletion on arginine tolerance test. Original hypokalemia (3.0–3.5 mEq/l) corrected by KCl therapy, with total body potassium increase of 40%.

cantly after KCl therapy. During the baseline test there was modest insulin response and negligible growth hormone response. When the test was repeated after KCl supplementation, however, there was a markedly increased growth hormone response paralleling the replenishment of potassium stores. Estrogen priming did not precede either arginine tolerance test, and the patient's clinical status was little changed except for potassium repletion.

Figure 2 demonstrates that the glucose values on the seven non-potassium depleted control patients almost overlap on repeat insulin tolerance tests. In each test porcine insulin in a dose of 0.1 U./kg was injected intravenously. There was also no change in endogenous ^{40}K count rate after KCl loading. Furthermore, there was very little difference in growth hormone response after KCl supplementation in this control or non-potassium repletable group (Figure 3). Mean fasting glucose level was 92.1 ± 6.4 mg/100 ml (S.E.M.) on the first test and 92.3 ± 2.7 on the repeat test after KCl. Mean fasting growth hormone level on Test 1 was 5.6 ± 1.7 mμg/ml, rising to a peak of 43.0 ± 7.7 at 60 min. Test 2 in this control group revealed a mean fasting growth hormone level of 6.6 ± 1.6 with a peak response of 39.4 ± 5.5 at 60 min. Fasting growth hormone levels were somewhat higher than normal, an observation related to their chronic liver disease (Hernandez *et al.*, 1969).

In contrast to the studies on control patients, the five potassium depleted patients demonstrated a rise in total body potassium with KCl treatment, and concomitant changes in insulin tolerance tests occurred. In Figure 4 note the lowering of the mean fasting glucose level after potassium replenishment. Glucose level on Test 1 was 96.8 ± 7.1 mg/100 ml

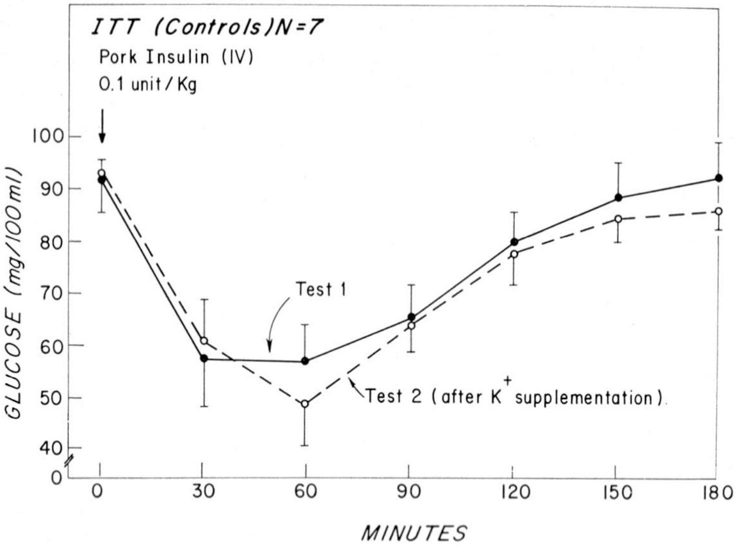

Fig. 2. Mean ± SEM serum glucose values during repeat insulin tolerance tests before and after KCl supplementation (control patients).

compared to 77.6 ± 4.4 on Test 2. KCl therapy also led to significantly increased growth hormone response to hypoglycemia as well as increased total body potassium content (Figure 5). Lower than control growth hormone levels rose with KCl therapy. Mean fasting growth hormone on Test 1 was 3.5 ± 0.8 mμg/ml with a maximal level of only 21.5 ± 5.0 at 60 minutes. Test 2, after potassium repletion, demonstrated a mean fasting growth hormone of 5.2 ± 1.6 with a peak response of 40.8 ± 10.9 at 60 minutes. Analogous results were noted during serial arginine tolerance tests.

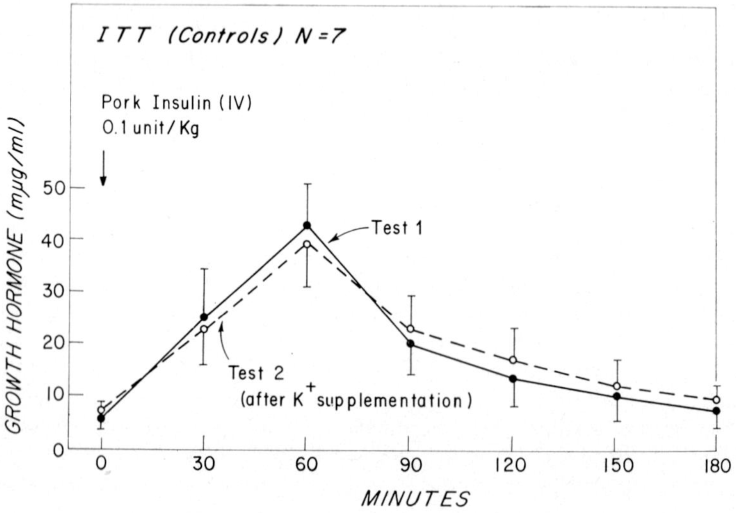

Fig. 3. Mean ± SEM plasma growth hormone values during repeat insulin tolerance tests before and after KCl supplementation (control patients).

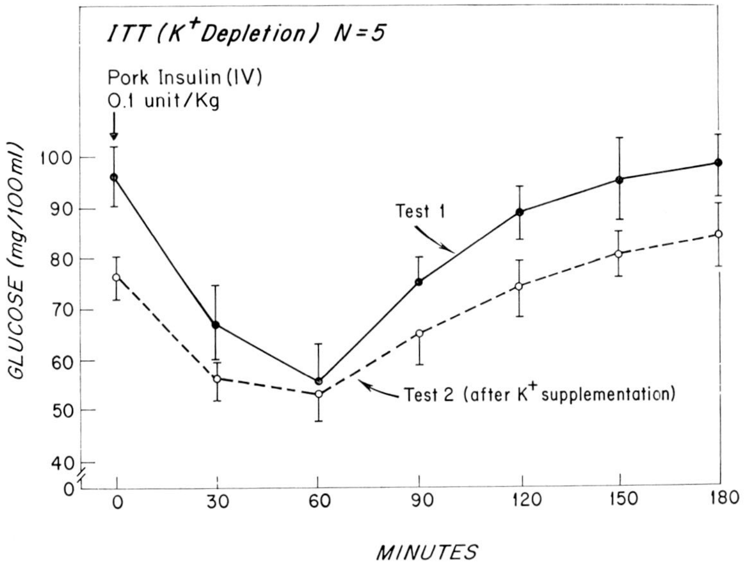

Fig. 4. Mean ± SEM serum glucose values during repeat insulin tolerance tests before and after KCl supplementation (potassium depleted patients).

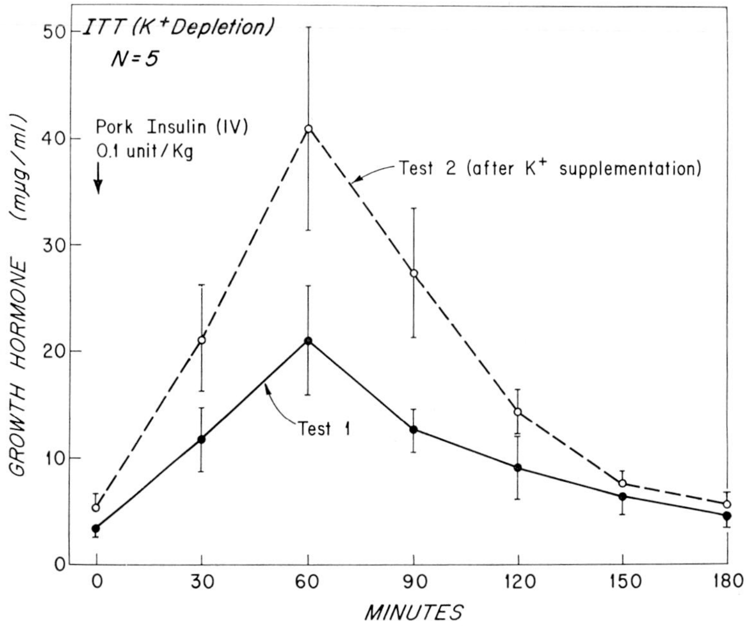

Fig. 5. Mean ± SEM plasma growth hormone values during repeat insulin tolerance tests before and after KCl supplementation (potassium depleted patients).

DISCUSSION

Data using earlier techniques suggest that a majority of patients with cirrhosis may have large potassium deficits (Burrows *et al.*, 1953). Although these patients can become hypokalemic for a variety of reasons, it is noteworthy that four of our five potassium depleted patients had multiple normal serum potassium levels. Even in the absence of hypokalemia, positive potassium balance during KCl therapy indicates the presence of intracellular potassium deficiency. ^{40}K balance studies during KCl administration have shown little or no change in total body potassium in patients without potassium deficiency who had normal renal function. The concept that tissue potassium stores may be significantly reduced in the absence of hypokalemia has been greatly strengthened by radioisotope assessment of intracellular potassium concentration (Tyson *et al.*, 1970).

In our laboratory reliability of repeat insulin tolerance tests done under basal conditions has been such that conversion from a nonresponder or hypo-responder (growth hormone release) to a clear-cut responder has been unlikely unless there has been some change in clinical condition. Only the five individuals who were potassium depleted showed a definite enhancement of growth hormone response to insulin induced hypoglycemia after potassium replenishment. Fasting growth hormone also increased after KCl therapy, further indicating that the initial decreased level and response were related to potassium depletion. No such changes in baseline or maximal growth hormone level occurred after KCl therapy in the control patients, none of whom changed total body ^{40}K.

It has been noted that impaired growth hormone response to a specific stimulus should not be considered absolute evidence of pituitary insufficiency until the lack of responsiveness is confirmed by a different type of test (Martin *et al.*, 1968). For this reason growth hormone reserve was also tested by response to a standard 30-min arginine infusion (0.5 g/kg). Similar results were noted in the potassium depleted patients whose hyporesponsiveness was reversed by repair of total body potassium deficit. Control patients had good responses with insignificant changes after KCl therapy.

These data are in keeping with *in vitro* evidence that growth hormone release from rat pituitary glands is increased where there is increased potassium concentration in the incubation medium (Wakabayashi, 1970). Similar potassium mediated enhancement of hormone release has been demonstrated with luteinizing hormone (Samli and Geschwin, 1968) and thyrotropin (Wilber, 1970) output from rat pituitary glands.

SUMMARY

Chronic potassium depletion is associated with reversibly impaired glucose tolerance and reduced insulin response in primary aldosteronism and other conditions. The authors have confirmed these findings with serial studies of total body potassium content (endogenous ^{40}K balance studies measured in whole body counter). Five of 12 serially studied patients showed increased total body ^{40}K (20% to 40% higher) after KCl supplementation (up to 200 mEq of K plus diet daily for at least 14 days). Fasting growth hormone levels were lower in this potassium depleted group compared to the seven control patients who had no change in total body ^{40}K after identical KCl therapy. Peak growth hormone responses to insulin induced hypoglycemia or arginine infusion were also lower in the potassium depleted group. KCl therapy led to restoration of normal growth hormone response. Of further interest is the fact that only one of the five patients with potassium deficiency was hypokalemic before therapy.

The data presented demonstrate that chronically potassium depleted patients, even when normokalemic, have lower fasting growth hormone levels and decreased growth hormone response to standard stimuli. These abnormalities are reversible with potassium supplementation.

ACKNOWLEDGEMENTS

We are indebted to Dr. Alfred E. Wilhelmi, Atlanta, Ga., for his kind supply of purified human pituitary growth hormone (HS 1147) used as our immunoassay standard. Highly specific rabbit antiserum to human growth hormone was generously provided by Dr. Thomas J. Merimee, Boston, Mass., during the initial stages of this study. Miss Bernadine Miller, R. N., Head Nurse, Intermediate Care Service, and her staff provided highly professional nursing care.

REFERENCES

Burrows, B. A., Denton, J., Ferguson, B. and Ross, J. F. (1953): Changes in body potassium in hepatic decompensation. *Clin. Res. Proc.*, *1*, 111.

Conn, J. W. (1965): Hypertension, the potassium ion and impaired carbohydrate tolerance. *New Engl. J. Med.*, *273*, 1135.

Frings, C. S., Ratliff, C. R. and Dunn, R. T. (1970): Automated determination of glucose in plasma and urine by a direct o-toluidine procedure. *Clin. Chem.*, *16*, 282.

Grodsky, G. M. and Bennett, L. L. (1966): Cation requirements for insulin secretion in the isolated perfused pancreas. *Diabetes*, *15*, 910.

Hernandez, A., Zorilla, E. and Gershberg, H. (1969): Decreased insulin production, elevated growth hormone levels, and glucose intolerance in liver disease. *J. Lab. clin. Med.*, *73*, 25.

Lau, K., Gottlieb, C. W. and Herbert, V. (1966): Preliminary report on coated charcoal immunoassay of human chorionic 'growth-hormone-prolactin' and growth hormone. *Proc. Soc. exp. Biol. (N.Y.)*, *123*, 126.

Martin, M. M., Gaboardi, F., Podolsky, S., Raiti, S. and Calcagno, P. B. (1968): Intermittent steroid therapy. Its effect on hypothalamic-pituitary-adrenal function and the response of plasma growth hormone and insulin to stimulation. *New Engl. J. Med.*, *279*, 273.

Podolsky, S. (1971): Trace metals, cations and diabetes. In: *Joslin's Diabetes Mellitus, 11th Edition*, Chapter 28, pp. 722–766. Editors: A. Marble, P. White, R. F. Bradley and L. P. Krall. Lea and Febiger, Philadelphia, Pa.

Podolsky, S., Gutman, R. A., Zimmerman, H. J. and Burrows, B. A. (1971): Effects of potassium on insulin, proinsulin and growth hormone release in cirrhotics with abnormal carbohydrate tolerance. *Diabetes*, *20*, 372.

Podolsky, S., Zimmerman, H. J. and Burrows, B. A. (1970*a*): Relation between total body potassium and growth hormone and insulin response in cirrhosis. *J. clin. Invest.*, *49*, 75a.

Podolsky, S., Zimmerman, H. J. and Burrows, B. A. (1970*b*): Decreased growth hormone and insulin response in normokalemic cirrhosis with potassium depletion. *Clin. Res.*, *18*, 694.

Samli, M. H. and Geschwind, I. I. (1968): Some effects of energy-transfer inhibitors and of Ca^{++} – free or K^{+}-enhanced media on the release of luteinizing hormone (LH) from the rat pituitary gland in vitro. *Endocrinology*, *82*, 325.

Tyson, I., Genna, S., Jones, R. L., Bikerman, V. and Burrows, B. A. (1970): Studies of potassium depletion using direct measurements of total-body potassium. *J. nucl. Med.*, *11*, 126.

Wakabayashi, L. (1970): Effect of potassium on the release of growth hormone and prolactin by rat pituitary in vitro. *Program, 52nd Meeting, Endocrine Society, St. Louis, Mo.*

Wilber, J. F. (1970): Stimulation of thyrotropin (TSH) synthesis by thyrotropin releasing factor (TRF) and 59 mM K^{+}. *Program, 52nd Meeting, Endocrine Society, St. Louis, Mo.*

ON THE AETIOLOGY OF HYPOPITUITARY DWARFISM*

J. R. BIERICH

Universitätskinderklinik Tübingen, Federal Republic of Germany

The different forms of pituitary dwarfism may be appropriately divided into three main types: (1) manifestations of organic lesions of the hypothalamus or the hypophysis; (2) so-called idiopathic hypopituitary dwarfism; (3) genetically determined forms.

Table I is a compilation of several series of pituitary dwarfs, described in the last few years. I have taken into consideration only series including all kinds of the syndrome. The table contains a total of 325 cases: 21.9% fall into group 1 which is caused by organic lesions; 69.5%, by far the majority, are idiopathic cases; 8.6% belong to the different genetic forms.

Group 1

For better understanding of the organic defects and lesions the ontogenetic development of the hypophysis should be briefly recalled. The gland is formed by the junction of two ectodermal processes. Caudally Rathke's pouch is turned out from the epithelium of the upper pharynx; from this part the anterior lobe and the pars tuberalis of the hypophysis are developed. Cranially the floor of the third ventricle forms another pouch which gives rise to the posterior lobe. The functional contact between both of these structures continues within the pituitary stalk. Here neurosecretory granules pass over from the tractus tubero-hypophyseus to the portal system.

Three congenital malformations of this junction leading to pituitary dwarfism are known:

(1) Dystrophy of the neurohypophysis leads eventually to the isolation of the anterior from the posterior lobe and from the infundibulum. The first case was reported by Priesel (1920). Eighteen further cases have since been published (see Bahner, 1965).

(2) Combination of cheilognathopalatoschisis with pituitary dwarfism has been described by Prüsener (1933), Francés *et al.* (1966) and Laron *et al.* (1969). The development of Rathke's pouch and its junction with the infundibulum presumably does not take place in the normal way.

(3) Numerically the craniopharyngiomas are the most important of the organic lesions. The synonyms 'Rathke's pouch tumour' and 'suprasellar cysts' indicate their pathogenesis. The tumours arise from cysts representing remnants of Rathke's pouch which are localised mostly above the sella; eventually the hypophysis itself is entirely intact. Symptoms of diencephalic obesity and diabetes insipidus point to the hypothalamus as the primary site of the disease.

Other tumours, for instance gliomas, are rare, compared with craniopharyngiomas. Organic lesions, chronic infections, histiocytosis X and traumas of the skull also play a minor role.

---

* Supported by the Deutsche Forschungsgemeinschaft, Bad Godesberg.

TABLE I

Author	Tumor	Dysonto-genesis	Trauma	Organic lesion	Idiopathic	Hereditary
Kirchhoff *et al.* (1954)					11	
De Gennes and Royer (1962)	(7)			7	18	
Knorr and Butenandt (1962)	(1)		(2)	3	6	
Van der Werff ten Bosch (1962)	(3)		(1)	4	18	
Kogut *et al.* (1963)	(2)			2	13	
Bierich (1965)	(9)			9	21	9
Prader (1960)	(7)	(5)	(3)	15	25	4
Brasel *et al.* (1965)	(19)	(2)		21	54	
Hubble (1967)	(1)	(2 men.)	(3)	6	13	2
Trygstadt *et al.* (1969)				2	12	8
Lässker (1970)					8	2
Bierich (1971)	(2)			2	27	3
Total: 325				71 21.9%	226 69.5%	28 8.6%

Group 2

The term 'idiopathic hypopituitary dwarfism' indicates that the aetiology of this form of the disease which concerns the majority of the patients has been obscure for a long time. Information from the pathology is scanty because autopsies have only rarely been carried out. From the nine publications traced the following is apparent: the hypophyses are small; the anterior lobe is particularly involved. The chromophils, especially the acidophil cells, are considerably diminished numerically. Cysts and important increments of the connective tissue were found in several cases. In 1919 the pathologist Simmonds (who also gave an explanation for the postpartal pituitary necrosis), from such findings suspected that these alterations represent residues of birth injuries with replacement by connective tissue. Prader (1960) stressed the fact that the relation of boys to girls in pituitary dwarfism, as well as in birth trauma, is clearly biased to the male sex, supporting a causal connection.

Eighteen out of 25 patients observed by Prader had an abnormal birth history, 10 of them with fair certainty of a perinatal cerebral lesion. Bierich (1962) and Van der Werff ten Bosch (1962) reported on the birth histories of some pituitary dwarfs. Criteria for birth traumas were found in numerous cases. Moreover, the authors stated that nearly two-thirds of their patients were born in breech-, foot- or transverse position. Severe asphyxias, cyanoses and convulsions were frequently observed in the newborns.

Up to the early sixties the diagnosis of pituitary dwarfism could only be established by indirect parameters such as increased insulin sensitivity, secondary insufficiency of the thyroid, the adrenals and the gonads. In our second series of 27 children we have meanwhile obtained diagnosis by the radioimmunological determination of growth hormone. Twelve of these children had, at birth, distinct symptoms of perinatal damage. Fourteen were born in breech-, foot- or transverse position.

Table II gives the birth histories of a total of 45 patients. Twenty-eight children were born in breech-, foot- or transverse position. In half of them obstetrical interventions such as turning, extraction, Veit-Smellie and forceps were necessary. Seventeen children were born in vertex position; in 3 cases the delivery was finished by forceps or vacuum extraction.

Table III demonstrates the findings in the newborns. There were 35 boys and 10 girls. Eleven were prematures, some with very low birth weight, and 3 were postmatures. Severe

TABLE II

45 'Idiopathic' hypopituitary dwarfs
Position during birth

Position	Number	%		Normal %	
Breech and foot	22	49	62	3–4	4
Transverse	6	13		0.6	
Vertex	17	38		96	

Obstetrical operations:

In breech-, foot-, transverse-positions (28):	14
In vertex-position (17):	3

TABLE III

45 'Idiopathic' hypopituitary dwarfs
Findings in the newborn

Boys : girls	35 : 10
Prematures and twins < 2500 g	11
Postmatures > 12 days	3
Symptoms of cerebral lesion (severe asphyxia, convulsions)	20
Hints at difficulties during birth	35 = 78%

asphyxias, cyanoses and convulsions were observed in 20 babies. Summing up, criteria pointing to a perinatal lesion were present in 78%. Thus, the assumption that so-called 'idiopathic' pituitary dwarfism is due to intracranial lesions caused by birth is confirmed by much material.

The fact that parturitions proceeding from breech and transverse positions are connected with a high frequency of cerebral trauma and haemorrhage, with high mortality and with increased frequency of cerebral palsy and epilepsy, is well established. The majority of obstetricians, therefore, nowadays demand termination of such deliveries by caesarian section (Goethels, 1956; Hall, 1956; Wright, 1959; Trolle, 1960). Which factor in breech and transverse parturition is responsible for the pituitary lesion is not clear as yet. Van der Werff ten Bosch supposes that ruptures of the pituitary stalk are a major factor. In addition ischaemic and congestional infarctions of the pituitary gland should be considered as possible causes. The circulation of the umbilical cord is temporarily discontinued in these births by pressure of the skull against the pelvic bones (Fig. 1). Ischaemia due to the constricted umbilical cord was mentioned six times as a complication in our birth histories. Moreover early placental abruption is also frequently observed. This again leads to anoxia.

Group 3

Early reports on hereditary dwarfism which retrospectively has to be regarded as hypophysial in origin have been published by Rischbieth and Barrington (1912) and by Hanhart

(1925). Principally we differentiate between panhypopituitary dwarfs with multi-hormonal defects on the one hand and mono-hormonal deficiencies on the other in whom only growth hormone is lacking. Gilford's and Merimee's asexual ateliotic dwarfism belongs to the former group; their sexual ateliotics belong to the latter.

In recent years numerous reports on hereditary multi-hormonal deficiencies have been published (Grebe, 1959; Schönberg and Bierich, 1970; Butenandt and Knorr, 1970). As presumably more than one cell type of the anterior lobe is deficient an anatomical dysgenesis or hypoplasia of the adenohypophysis appears likely.

In fact abnormally small sellae turcicae have been found radiologically by Dzierzynski (1938), Rochlin and Simonson (1928) and Laron (1971) – although this finding is certainly not consistent. Autoptical reports on aplasias and hypoplasias of the hypophysis have been repeatedly published (Blizzard and Alberts, 1956; Ehrlich, 1957; Mosier, 1956; Reid, 1960). Mono-hormonal deficiencies have been described during the past few years in an even greater number. Merimee *et al.* (1969) have proposed a classification into 4 types which will now be discussed.

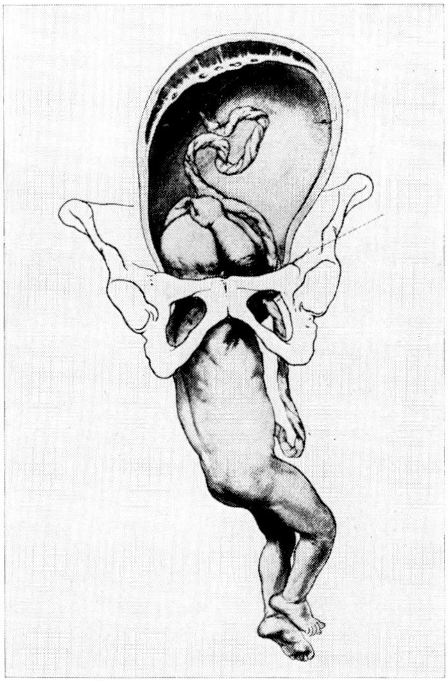

Fig. 1

Most often *type 1* is encountered in which only growth hormone is lacking. Insulin sensitivity is increased. Insulin secretion is diminished. Fetal or prenatal growth hormone deficiency may lead to a lack of immunotolerance against HGH and thus to early production of antibodies when these children are treated with HGH, as Illig *et al.* (1970) and Butenandt and Knorr (1970) have shown.

In *type 2* immunologically reactive HGH is also lacking. However, after glucose and arginine infusions plasma insulin reaches considerably high values. Recently we examined two patients, a 17-year-old boy and his 15-year-old sister, with this form of dwarfism (Figs.

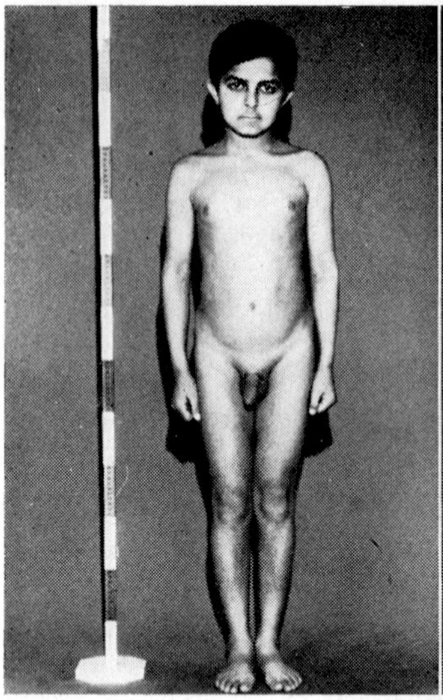

Fig. 2

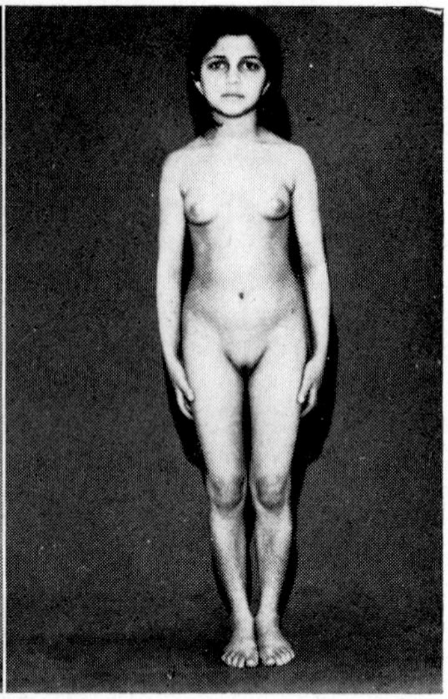

Fig. 3

2 and 3). These patients belong to the sexual ateliotics – in both of them puberty had begun. HGH was not detectable by radioimmunoassay. Insulin tolerance was not impaired. An intravenous glucose tolerance test yielded low K values: 1.2 and 1.4 respectively. At the same time, plasma insulin rose to 70 μU./ml in the girl and to 300 μU./ml in the boy. The latter value is clearly elevated. Merimee *et al.* suggest an abnormal growth hormone which does not react in the radioimmunoassay but still retains activity in carbohydrate metabolism. This hormone or residue of a hormone has not as yet been proven.

Nevertheless, Sonenberg *et al.* (1965, 1968) have furnished evidence for the possible existence of growth hormone fragments with dissociated biological activities. In addition, Cerasi, Li and Luft (This Volume, pp. 363–370) have shown a good model for such a hormonal fragment. Their 'reduced and alkylated HGH' gave only a 50% reaction in the radioimmunoassay compared with genuine HGH, but was strongly insulinogenic and led to a prediabetic state in the patients treated.

Type 3 is the kind of dwarfism first described by Laron *et al.* (1966). Growth hormone levels are elevated when measured by the radioimmunoassay, but the clinical picture and the laboratory tests resemble those in ordinary pituitary dwarfism. Dr. Laron discusses this disturbance himself, including the important question of whether it is in fact a pituitary disease or a defect of sulfation factor production (This Volume, pp. 458–482).

The last group, *type 4* according to Merimee *et al.*, is represented by the Babinga pygmees, an African dwarf race. Growth hormone is measurable in normal amounts (it increases normally after insulin and arginine), but it is obviously ineffective in promoting growth. The same holds true regarding exogenous growth hormone which neither causes nitrogen retention, nor elevation of free fatty acids, nor growth. Apparently this syndrome is neither a pituitary disorder nor a disease of the endocrine glands at all but represents an example of nonresponsiveness of the target organs which is genetically determined.

At the end of this necessarily incomplete brief survey I have to admit that there are certainly still more forms of dwarfism in which the pituitary is involved. We ourselves have data suggesting that short stature in Moon-Bardet-Biedl's syndrome and in Bloom's disease may be connected with a deficiency of growth hormone.

In addition, one of the abstracts of this symposium points to a new dwarfism syndrome in which the hypophysis participates (Kaplan *et al.*, 1971).

However, of even greater significance is the following. The diagnosis of pituitary dwarfism is defined by the result of the radioimmunoassay of HGH. Increases after insulin which amount to less than 5 ng/ml are considered as pathological and as proof for this diagnosis. However, the mean increase of healthy children is 15-20 ng/ml in our laboratory. Thus it is most probable that increments up to 10-12 ng/ml have to be regarded as incomplete or mild pituitary insufficiencies. These minor changes are in our experience clinically connected with a more or less *atypical* symptomatology. Consequently, in addition to the classical picture of hypopituitary dwarfism, we have to expect 'formes frustes' or minor forms of hypopituitary dwarfism. According to statistical probability, these incomplete cases will be even more frequent than the classical ones.

REFERENCES

Bahner, F. (1965): Zur Pathogenese des endokrinen Zwergwuchses. In: *Verh. 11. Symposion, Deutschen Gesellschaft für Endokrinologie, Düsseldorf 1964.* Editor: E. Klein. Springer Verlag, Berlin.

Bierich, J. R. (1962): The function of the adenohypophysis in pituitary dwarfism. *Acta endocr. (Kbh.), Suppl. 67*, 76.

Bierich, J. R. (1965): Ätiopathogenese und klinisches Bild hypothalamischer und hypophysärer Wachstumsstörungen. *Mschr. Kinderheilk., 113*, 269.

Blizzard, R. M. and Alberts, M. (1956): Hypopituitarism, hypoadrenalism and hypogonadism in the newborn infant. *J. Pediat., 48*, 782.

Brasel, J. A., Wright, J. C., Wilkins, L. and Blizzard, R. M. (1965): An evaluation of 75 patients with hypopituitarism beginning in childhood. *Amer. J. Med., 38*, 484.

Butenandt, O. and Knorr, D. (1970): Familiärer Hypopituitarismus. *Mschr. Kinderheilk., 118*, 470.

Cerasi, E., Li, C. H. and Luft, R. (1971): Short-term metabolic effects of HGH and a HGH-derivative in man. *This Volume*, pp. 363-370.

De Gennes, J. L. and Royer P., (1962): Reévaluation, à l'occasion de la révision analytique de 26 cas, des aspects cliniques et biologiques des nanismes hypophysaires. *Ann. Pédiat. (Paris), 38*, 240, 380.

Dzierzynski, W. (1938): Nanosomia pituitaria hypoplastica hereditaria. *Z. Neurol., 162*, 411.

Ehrlich, R. M. (1957): Ectopic and hypoplastic pituitary with adrenal hypoplasia. Case report. *J. Pediat., 51*, 377.

Goethels, T. R. (1956): Caesarean section as the method of choice in the management of breech delivery. *Amer. J. Obstet., 72*, 977.

Grebe, H. (1959): Erblicher Zwergwuchs. *Ergebn. inn. Med. Kinderheilk., 12*, 343.

Hall, J. E. and Kohl, S. (1956): Breech presentation. A study of 1456 cases. *Amer. J. Obstet. Gynec., 72*, 977.

Hanhart, E. (1925): Über heredodegenerativen Zwergwuchs mit Dystrophia adiposo-genitalis. *Arch. Klaus-Stift. Vererb. Forsch., 1*, 181.

Hubble, D. (1967): Growth hormone deficiencies in childhood. *Canad. med. Ass. J., 97*, 1144.

Illig, R., Prader, A., Ferrandez, A. and Zachmann, M. (1971): Hereditary prenatal growth hormone deficiency with increased tendency to growth hormone antibody formation. IXth Annual Meeting, European Society for Paediatric Endocrinology, Lyon 1970 (Abstracts). *Acta paediat. scand., 60*, 603.

Kaplan, S. L., Grumbach, M. M. and Hoyt, W. F. (1971): A syndrome of hypopituitary dwarfism, hypoplasia of optic nerves, and malformation of prosencephalon: Report on 6 patients. *Abstracts, II International Symposium on Growth Hormone, Milan*, p. 17. Editors: A. Pecile and E. E. Müller. ICS 236. Excerpta Medica, Amsterdam.

Kirchhoff, H. W., Lehmann, W. and Schäfer, U. (1954): Klinische, erbbiologische und körperliche Untersuchungen bei primordialen Zwergen. *Z. Kinderheilk., 75*, 243.

Knorr, D. and Butenandt, O. (1962): Zur Behandlung des hypophysären Zwergwuchses mit anabolen Steroiden. *Z. Kinderheilk., 86*, 489.

KOGUT, M. D., KAPLAN, S. A. and SHIMIZU, C. S. N. (1963): Growth retardation: Use of sulfation factor as a bioassay for growth hormone. *Pediatrics*, *31*, 539.

LÄSSKER, G. (1970): Ergebnisse der Langzeitbehandlung von Kindern mit hypophysär-hypothalamischem Minderwuchs mit homologem Wachstumshormon. *Dtsch. Gesundh.-Wes.*, *25*, 2328.

LARON, Z. (1971): Adenohypophyse und Hypothalamus. In: *Handbuch der Kinderheilkunde, 2nd Ed., Vol. I*, pp. 180–206. Editors: J. R. Bierich, R. Grüttner and K. H. Schäfer. Springer Verlag, Berlin.

LARON, Z., KARP, M., PERTZELAN, A., KAULI, R., KERET, R. and DORON, M. (1971): The syndrome of familial dwarfism and high plasma immunoreactive human growth hormone (IR-HGH). *This Volume*, pp. 458–482.

LARON, Z., PERTZELAN, A. and MANNHEIMER, S. (1966): Genetic pituitary dwarfism with high serum concentration of growth hormone. A new inborn error of metabolism? *Israel J. med. Sci.*, *2*, 152.

LARON, Z., TAUBE, E. and KAPLAN, I. (1969): Pituitary growth hormone deficiency associated with cleft lip and palate. An embryonal developmental defect. *Helv. paediat. Acta*, *24*, 576.

MERIMEE, T. J., RIMOIN, D. L., HALL, J. D. and MCKUSICK, V. A. (1969): A metabolic and hormonal basis for classifying ateliotic dwarfs. *Lancet*, *2*, 963.

MOSIER, H. D. (1956): Hypoplasia of the pituitary and adrenal cortex. *J. Pediat.*, *48*, 633.

PRADER, A. (1960): Hypothalamo-pituitary dwarfism. In: *Transactions, Club for Paediatric Research, Meeting, Groningen 1960.*

PRIESEL, A. (1920): Ein Beitrag zur Kenntnis des hypophysären Zwergwuchses. *Beitr. path. Anat.*, *67*, 220.

PRÜSENER, L. (1933): Hypophysäre Wachstumshemmung mit Kachexie beim Kinde. *Z. menschl. Vererb.-u. Konstit. Lehre*, *17*, 215.

REID, J. D. (1960): Congenital absence of the pituitary gland. *J. Pediat.*, *56*, 658.

RISCHBIETH, H. and BARRINGTON, A. (1912): *Treasure of Human Inheritance, Vol. I*, Sect. 15a, p. 355. Dulau and Co., London.

ROCHLIN, D. G. and SIMONSON, S. G. (1928): Über den Klein- und Zwergwuchs. *Fortschr. Röntgenstr.*, *37*, 467.

SCHÖNBERG, D. and BIERICH, J. R. (1970): Clinical evaluation and pituitary function in familial dwarfism. VIIIth Annual Meeting, European Society for Paediatric Endocrinology, Malmö (Sweden) 1969 (Abstracts). *Acta paediat. scand.*, *58*, 660.

SIMMONDS, M. (1919): Zwergwuchs bei Atrophie des Hypophysenvorderlappens. *Dtsch. med. Wschr.*, 487.

SONENBERG, M., FREE, C. A., DELLACHA, J. M., BONADONNA, G., HAYMOWITZ, A. and NADLER, A. C. (1965): The metabolic effects in man of bovine growth hormone digested with trypsin. *Metabolism*, *14*, 1189.

SONENBERG, M., KIKUTANI, N., FREE, C. A., NADLER, A. C. and DELLACHA, J. M. (1968): Chemical and biological characterization of clinically active tryptic digests of bovine growth hormone. *Ann. N.Y. Acad. Sci.*, *148*, 533.

TROLLE, D. (1960): Consideration on breech presentation as an indication for Caesarean section. *Dan. med. Bull.*, *7*, 117.

TRYGSTADT, O. (1969): Human growth hormone and hypopituitary growth retardation. *Acta paediat. scand.*, *58*, 407.

VAN DER WERFF TEN BOSCH, J. J. (1962): Hypofysaire dwerggroei. *Ned. T. Geneesk.*, *106*, 1282.

WRIGHT, R. C. (1959): Reduction of perinatal mortality and morbidity in breech delivery through routine use of Caesarean section. *Obstet. Gynec.*, *14*, 758.

PROVOCATION TESTS FOR GROWTH HORMONE DEFICIENCY

FREDERIC M. KENNY

Children's Hospital of Pittsburgh, Pittsburgh, Pa., U.S.A.

Our medical students, so overly concerned these days with the 'relevance' of the activities of faculty, have not questioned my trip to Milan. Everyone knows that a visit to this city will always be relevant. Had they questioned the pertinence of a discussion of growth hormone provocative tests I should have answered thus:

Children and adolescents, who accept each others' differences in terms of race, religion and moral values – dress, hair length and other such visible and invisible differences – go out of their way to make life difficult – or even Hell – for the playmate who is too fat or too skinny . . . too tall or too short . . . or too late in going through adolescence. It has been virtually impossible for the victim to alter these characteristics. But in 1971, the parent and child expect the physician to make the child 'normal' in these respects.

For 2,000 years we had accepted the fact that 'man cannot by taking thought add one cubit to his stature'. Within the past ten years this has changed, at least with regard to hypopituitarism.

Previously the diagnosis of hypopituitarism had served only to categorize the patient and no therapeutic benefit was derived from attempts at treatment with bovine growth hormone. In 1959 Knobil and Greep found simian and human growth hormones mutually effective. Between 1957 (Beck *et al.*) and 1960, HGH was shown to enhance the linear growth rate, with attendant metabolic balance parameters, in hypopituitary patients. Spurred by the new therapeutic possibilities, but in keeping with the limited supply of HGH, investigators now needed precise diagnosis. Radioimmunoassay (Hunter and Greenwood, 1962) proved superior to other methods of quantification of circulating HGH, but unlike the level of circulating thyroid hormone which in itself is usually diagnostic, basal levels of HGH were found to overlap in normal subjects and hypopituitary patients.

The first standardized provocative test capable of discriminating normal subjects from hypopituitary patients was insulin induced hypoglycemia (Frantz and Rabkin, 1964). Originally reported in 1963 by Roth *et al.*, within two years it was widely used by internists and pediatricians; and other tests, based on the same principle of intracellular hypoglycemia were adopted.

That intracellular hypoglycemia is the mechanism for GH release, rather than a specific effect of insulin on the GH secretory mechanism has been shown in two ways. If glucose is given concomitant with insulin, GH is not released. Furthermore, inhibition of glucose utilization in the GH regulatory centers by the analog 2-deoxyglucose, stimulates GH release.

A rapid fall in blood sugar to 50% of the initial level, or below 50 mg% is usually sufficient to stimulate GH release. The sensitivity of the sensor is capable of detecting a 10 mg% decrease in blood sugar, as shown by Luft *et al.* (1968).

The teleologic reason for growth hormone requirement during hypoglycemia can be considered the antagonism of insulin mediated transport of glucose into organs other than brain, thereby permitting the central nervous system which is not insulin dependent, to continue functioning when circulating glucose is low. Furthermore, the elevated growth hormone provides fuel through release of non-esterified fatty acids.

Less risky in terms of production of hypoglycemia in the GH or ACTH deficient patient is stimulation of GH release by levorotatory amino acid infusions, of which arginine is the most potent and most widely used. Proposed in 1965 by Knopf *et al.*, the mechanism of the stimulus is unrelated to any change in insulin or blood sugar. Arginine stimulates insulin release prior to GH release, and the GH response is elicited in diabetics who are without insulin.

Arginine mediated GH release would appear to provide the body with a mechanism for stimulating protein synthesis when amino-acid precursor is available.

On a given day, some normal individuals will 'fail' one or the other test. Therefore, in clinical assessment of patients, advantage is taken of the separate hypoglycemic and amino acid stimulatory mechanisms by linking the tests over a two hour period (Penny *et al.*, 1969).

Venipuncture itself stimulates GH release in 50% of children (Helge *et al.*, 1969). Therefore, as shown in Figure 1 two basal values are taken 30 min apart in order to permit the

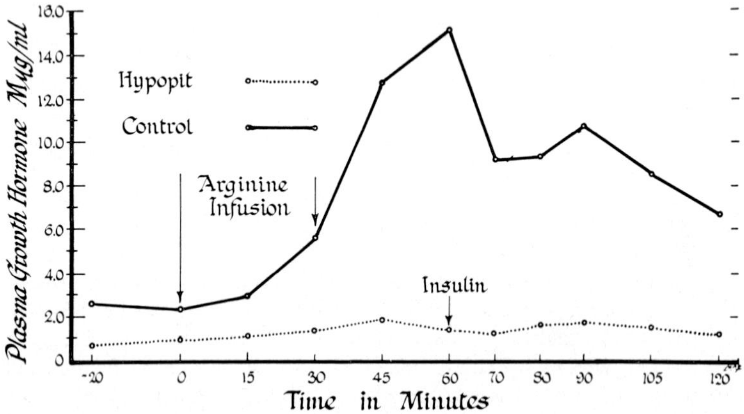

Fig. 1. Growth hormone response to arginine-insulin infusion.

patient to reach a relatively stable baseline. Arginine, 0.5 g/kg of body weight is administered over 30 min. Peak GH values are within 1 1/2 hours after the start of infusion and exceed basal levels by 5 ng. Following insulin, 0.1 U./kg, normal children attain a 5 ng increment. Hypopituitary children fall below this minimal response on both tests.

In prepubertal children, the magnitude of GH response is unrelated to age or sex (Sperling *et al.*, 1970). However, the enhancing effect of estrogen on growth hormone release is seen in menstruating girls whose GH peak after arginine exceeds that of pubertal males (Sperling *et al.*, 1970). The growth hormone response of women to exercise and hypoglycemia also exceeds that of men, and the male's response can be improved by brief administration of a potent estrogen (Frantz and Rabkin, 1965).

Originally it was supposed that the majority of patients with hypopituitarism lacked all functions of the anterior portion since in general, only the most blatant forms of diseases are the first to be recognized. Monotropic deficiencies were next noted, although only three of a large series of 75 hypopituitary patients collected by Wilkins and Blizzard and reviewed by Brasel *et al.* (1965) were considered to have monotropic growth hormone defects. However, that series was studied prior to the widespread use of the insulin and arginine provocative tests. Utilizing those procedures, Kaplan *et al.* (1968) found that 30% of their hypopituitary patients had isolated growth hormone deficiency. Since somatotrophin therapy is becoming more generally available, the benefit to those additional patients is evident.

Doctors Sperling, Drash and I (1971) have been interested to apply these procedures to patients with cortisone treated adrenogenital syndrome of the 21-hydroxylase deficiency variety. I had employed the therapeutic regimen of Wilkins for many years, that is cortisone acetate 25 mg intramuscularly every third day from diagnosis to age two years; then oral cortisone acetate 25 mg/day in divided dosage for the next few years. Figure 2 shows the poor

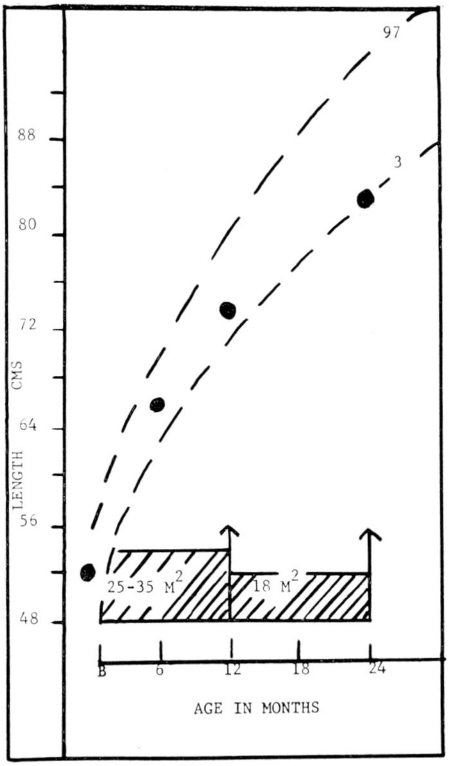

Fig. 2. CAH-salt losers in cortisone therapy.

growth which the patients experienced. Although the infants grew well until about age six months, when their cortisone dose averaged 30 mg/square meter of body surface area, the growth rate decelerated between one and two years when the mean length was at the third percentile for age, even though the cortisone dose was relatively lower.

By use of the *in vivo* isotope dilution method for cortisol production rate, we had learned that normal subjects of this age produce about 12 mg/m² of cortisol per day (Kenny *et al.*, 1970). Therefore, we suspected that the previously recommended, supraphysiologic therapy with cortisone, was either suppressing GH release, interfering with the peripheral action, or both. Since it had been shown that in spontaneous or iatrogenic Cushing's syndrome of long duration, the response of growth hormone to hypoglycemic provocation could be blunted, we assessed our patients with insulin tolerance tests and failed to detect diminished GH release. Arginine provocation was also assessed, and no diminution in GH secretion was found.

Therefore we attribute the growth failure of our patients to peripheral antagonism between cortisol and GH. This explanation is consistent with the work of Morris *et al.* (1968) who noted no diminution of GH response to insulin-hypoglycemia in long-term steroid treated asthmatics. Her patients, while on steroid, demonstrated metabolic unresponsiveness to exogenous HGH during nitrogen and mineral balance studies. Soyka and Crawford (1965) have shown that this type of antagonism can be overcome by progressively increasing the dose of GH while holding the steroid dose constant, suggesting a competitive inhibition.

For some time it was considered that GH responsiveness to provocative testing was obligatory for normal growth rate. However, in 1967 Zimmerman *et al.* reported two males with hypopituitarism and normal or increased height, in the presence of failure of elevation of serum growth hormone to hypoglycemic provocation. We (Kenny *et al.*, 1968), as well as Holmes *et al.* (1968), have described catch-up growth – that is a temporary acceleration over several years, in order to reach a higher growth rate channel – in patients following operations for craniopharyngioma. Failure of growth hormone release after insulin and arginine provocation was shown. We suspected that, since the pituitary glands of our patients were left in place and only the craniopharyngioma was removed, prolactin, now freed from inhibition could be released and could account for the catch-up growth. The finding of elevated prolactin during the rapid growth phase in the only patient whom we so tested, is consistent with this speculation.

Every few months one is aware of new provocative tests for GH release. Bovril®, a commercial beverage containing several amino acids, has gained acceptance and has the advantage over arginine infusion of ease of oral administration (Jackson *et al.*, 1968).

Bacterial pyrogen (Pyromen®) stimulates both ACTH and GH release (Kohler *et al.*, 1967). Hypopituitary patients may release ACTH following pyrogen, even though they may fail to respond to the eleven hydroxylase inhibitor SU 4885 (metyrapone). However, I am unaware of hypopituitary patients who respond to Pyromen and not to hypoglycemia or arginine. Furthermore, the discomfort to the patient of pyrogen argues against its use in diagnosis.

Metyrapone has been shown to stimulate GH as well as ACTH release in normal subjects (Kunita *et al.*, 1970). The data suggest that the mechanism is independent of a negative feedback of plasma cortisol, and could depend on the direct effects of metyrapone on the pituitary and/or hypothalamus as a nonspecific stimulus.

Exogenous pitressin consistently stimulates ACTH with inconsistent GH release (Yalow *et al.*, 1969). Similarly, histamine (Histalog®) evokes both ACTH and GH response. The HGH release occurs as a response to falling levels of free fatty acids indicating a response to decreasing energy supply to the hypothalamus (Tsushima *et al.*, 1970).

Glucagon was shown by Drash *et al.* (1968) to act as a GH stimulus in juvenile diabetic, but not in normal children although no explanation for the disparate response was evident. Recently it was shown that glucagon evokes GH release in normal adults at a time during the test when blood sugar levels are dropping (Cain *et al.*, 1970).

An exciting area of current investigation is the effect of sympathetic amines on GH release. Blockade of alpha transmitters with phentolamine abolishes insulin induced GH release, whereas beta blockade with propranolol leaves the response intact (Parra *et al.*, 1970). Hypopituitary children do not respond to these maneuvers. Furthermore, L-dopa administration to patients with Parkinson's disease is responsible for GH secretion (Boyd *et al.*, 1970). L-Dopa is a precursor of dopamine and catecholamines, and (unlike them) it is capable of crossing the blood-brain barrier. It is speculated that exogenous L-dopa enhances the already elevated levels of dopamine in the median eminence which is a site for growth-hormone releasing factor. Alternately, it could increase the norepinephrine levels in the hypothalamus or limbic system, thereby mediating GH release.

The ultimate provocative test for GH release awaits the isolation, characterization, and synthesis of GH releasing factor. One might hope that this may even provide an ideal therapy for GH deficiency.

REFERENCES

Beck, J. C., McGarry, E. E., Dyrenfurth, I. and Venning, E. H. (1957): Metabolic effects of human and monkey growth hormone in man. *Science, 125*, 884.

Boyd, A. E., Lebovitz, H. E. and Pfeiffer, J. B. (1970): Stimulation of human growth hormone secretion by L-dopa. *New Engl. J. Med., 283*, 1425.

Brasel, A. J., Wright, J. C., Wilkins, L. and Blizzard, R. M. (1965): An evaluation of seventy-five patients with hypopituitarism beginning in childhood. *Amer. J. Med., 38*, 484.

Cain, J. R., Williams, G. H. and Dluhy, R. G. (1970): Glucagon stimulation of growth hormone. *J. clin. Endocr., 31*, 222.

Drash, A. L., Field, J. B., Garces, L. Y., Kenny, F. M., Mintz, D. and Vasquez, A. (1968): Endogenous insulin and growth hormone response in children with newly diagnosed diabetes. *Pediat. Res., 2*, 94.

Frantz, A. G. and Rabkin, M. T. (1964): Human growth hormone. Clinical measurement, response to hypoglycemia and suppression by corticosteroids. *New Engl. J. Med., 271*, 1375.

Frantz, A. G. and Rabkin, M. T. (1965): Effects of estrogen on sex hormone difference on secretion of human growth hormone. *J. clin. Endocr., 25*, 1470.

Helge, H., Weber, B. and Quabbe, H. J. (1969): Growth-hormone release and venipuncture. *Lancet, 1*, 204.

Holmes, L. B., Franz, A. G., Rabkin, M. T., Soeldner, J. S. and Crawford, J. D. (1968): Normal growth with subnormal growth hormone levels. *New Engl. J. Med., 279*, 559.

Hunter, M. W. and Greenwood, F. C. (1962): Radio-immuno-electrophoretic assay for human growth hormone. *Acta endocr. (Kbh.), Suppl. 67*, 59.

Jackson, D., Grant, B. D. and Clayton, B. E. (1968): A simple oral test of growth hormone secretion in children. *Lancet, 2*, 373.

Kaplan, S. L., Abrams, C. A. L., Bell, J. J., Conte, F. A. and Grumbach, M. M. (1968): Growth and growth hormone. I. Changes in serum level of growth hormone following hypoglycemia in 134 children with growth retardation. *Pediat. Res., 2*, 43.

Kenny, F. M., Iturzacta, N. F., Mintz, D., Drash, A., Garces, L. Y., Susen, A. and Askari, H. A. (1968): Iatrogenic hypopituitarism in craniopharyngioma: Unexplained catch-up growth in three children. *J. Pediat., 72*, 766.

Kenny, F. M., Richards, C. and Taylor, F. H. (1970): Reference standards for cortisol production and 17-hydroxy-corticosteroid excretion during growth: Variation in pattern of excretion of radiolabeled cortisol metabolites. *Metabolism, 19*, 280.

Knobil, E. and Greep, R. O. (1959): The physiology of growth hormone with particular reference to its action in the rhesus monkey and the 'species specificity' problem. *Recent Progr. Hormone Res., 15*, 1.

Knopf, R. F., Conn, J. W., Fajans, S. S., Floyd, J. C., Guntsche, E. M. and Rull, J. A. (1965): Plasma growth hormone response to intravenous administration of amino acids. *J. clin. Endocr., 25*, 1140.

Kohler, P. O., O'Malley, B. W., Rayford, P. L., Lipsett, M. B. and Odell, W. D. (1967): Effect of pyrogen on blood levels of pituitary trophic hormones. Observations on the usefulness of the growth hormone response in the detection of pituitary disease. *J. clin. Endocr., 27*, 219.

Kunita, H., Takabe, K., Nakagawa, K., Sawano, S. and Horiuchi, Y. (1970): Effect of metyrapone on secretion of growth hormone in man. *J. clin. Endocr., 31*, 301.

Luft *et al.* (1968): Quoted in Williams, R. H.: *Textbook of Endocrinology*, p. 48. W. B. Saunders Cy., Philadelphia, Pa.

Morris, H. G., Jorgensen, J. R. and Elrick, H. (1968): Metabolic effects of human growth hormone in corticosteroid treated children. *J. clin. Invest., 47*, 435.

PARRA, A., SCHULTZ, R. B., FOLEY, T. P. and BLIZZARD, R. M. (1970): Influence of epinephrine-propranolol infusions on growth hormone release in normal and hypopituitary subjects. *J. clin. Endocr.*, *30*, 134.

PENNY, R., BLIZZARD, R. M. and DAVIS, W. T. (1969): Sequential arginine and insulin tolerance tests on the same day. *J. clin. Endocr.*, *29*, 1499.

ROTH, J., GLICK, S. M., YALOW, R. S. and BERSON, S. A. (1963): Hypoglycemia: potent stimulus to secretion of growth hormone. *Science*, *140*, 987.

SOYKA, L. F. and CRAWFORD, J. D. (1965): Antagonism by cortisone of the linear growth induced in hypopituitary patients and hypophysectomized rats by human growth hormone. *J. clin. Endocr.*, *25*, 469.

SPERLING, M. A., KENNY, F. M. and DRASH, A. L. (1970): Arginine-induced growth hormone responses in children: effect of age and puberty. *J. Pediat.*, *77*, 462.

SPERLING, M. A., KENNY, F. M., SCHUTT-AINE, J. and DRASH, A. L. (1971): Linear growth and growth hormone responsiveness in treated congenital adrenal hyperplasia. *Amer. J. Dis. Child.*, *122*, 408.

TSUSHIMA, T., MATSUZAKI, F. and IRIE, M. (1970): Effect of heparin administration on plasma growth hormone concentrations. *Proc. Soc. exp. Biol. (N.Y.)*, *133*, 1084.

YALOW, R. S., VARSANO-AHARON, N., ECHEMENDIA, E. and BERSON, S. A. (1969): HGH and ACTH secretory response to stress. *Hormones Metab. Res.*, *1*, 3.

ZIMMERMAN, T. S., WHITE, M. G., DAUGHADAY, W. H. and GOETZ, F. C. (1967): Hypopituitarism with increased height. Report of two cases with measurement of plasma growth hormone levels. *Amer. J. Med.*, *42*, 146.

EVALUATION OF GROWTH HORMONE DEFICIENCY BY METABOLIC TESTS

M. ZACHMANN, A. PRADER, A. FERRANDEZ and R. ILLIG

Department of Pediatrics, University of Zürich, Zürich, Switzerland

In recent years, the diagnosis of possible growth hormone (GH) deficiency has been considerably improved by widely used radioimmunoassay techniques. However, as experience with stimulation tests increased (insulin tolerance, arginine infusion, protein ingestion etc.), it became evident that these tests do not allow a clear diagnosis in all cases. In addition, tests of maximal secretory capacity do not necessarily reflect the physiological state of GH secretion. It is generally accepted that GH-deficiency may be excluded, if there is a definite response of plasma GH to one of the tests. Absence of response, however, is known to be possible in a number of conditions other than GH deficiency, including primary hypothyroidism, Cushing's syndrome and obesity. More physiological tests such as GH determinations during sleep or exercise, as well as integrated plasma GH levels over a 24-hour period, are difficult to perform and may also not provide clear-cut results.

In cases with inconclusive results in the tests mentioned, it is often desirable to have additional diagnostic tools, based on principles other than plasma GH determinations. This is especially true in the increasing number of cases without defects of other pituitary hormones, where a decision between isolated GH deficiency and dwarfism of non-endocrine origin has to be reached.

From general experience in endocrinology, one would expect that exogenous GH has a more marked metabolic effect in GH deficient than in normal subjects. In earlier studies, we and others have shown that such a difference between the two groups exists with respect to the influence of HGH on a number of metabolic parameters, including urinary nitrogen, calcium, creatinine and hydroxyproline, and blood urea nitrogen and free fatty acids.

Most of the significant differences, however, show too much overlap between the two groups to be diagnostically useful in an individual case. The influences of HGH on the carbohydrate metabolism (*e.g.* recovery index of blood glucose after insulin) are also not sufficiently different in the two groups.

In our experience, two tests of metabolic response to HGH are exceptions in that they allow separation of the two populations with reasonable accuracy: *(a) The nitrogen retention test* has been used for several years by our group (Prader *et al.*, 1968*a*, *b*) and by others; the results in an extended series of subjects are presented here. *(b)* The other test is new and concerns the short-term response of some *amino acids in plasma* to a single intravenous dose of HGH. Experience with this test is still limited and some preliminary results are presented.

The nitrogen retention test was carried out as previously described (Prader *et al.*, 1968*a*). The children were given a self-chosen constant diet. A 2-3 day adaptation period was followed by a 5-day control and a 5-day HGH period, during which 2 mg of HGH per square meter of body surface area were administered intramuscularly every day. In the earlier tests, HGH-Raben, and more recently, HGH-Roos was used. The potency of both preparations was identical (approximately 1.2 I.U./mg). The mean urinary nitrogen excretions of the 2nd

TABLE I

Nitrogen retention test: Groups of patients

Diagnosis	Criteria	n
1. Proven GH deficiency		
(a) idiopathic or organic	successful HGH treatment	18
(b) isolated = 'A-type'	initial success of HGH, then failure and antibodies	5
2. Possible GH deficiency	no HGH treatment or too short period	10
3. Normal GH secretion	normal stature and/or normal GH response to stimuli and/or dwarfism of other origin	25
	Total	58

to the 5th HGH-day were compared with the mean values of the control period and the nitrogen retention was expressed as per cent of the pretreatment excretion. Up to now, this test has been performed on 58 children (Table I). Eighteen patients had GH deficiency indicated by successful long-term treatment with HGH. This group includes patients with idiopathic hypopituitary dwarfism with and without defects of other pituitary hormones as well as patients with operated craniopharyngeoma. Five patients had a special type of isolated GH deficiency, termed 'A-type' characterized by a tendency for the formation of antibodies against HGH (Illig *et al.*, 1970). A short initial period of successful HGH treatment was followed in these patients by failure of continued therapy and by development of antibodies in high concentrations.

Another 10 patients had possible GH deficiency suggested by an absence of GH response to insulin and arginine and/or by the nitrogen retention test; they have not yet been treated with HGH or the treatment period is too short for an evaluation of success. In 25 additional patients, GH deficiency was excluded. They had normal stature and/or a normal GH response to insulin or arginine and/or dwarfism known not to be associated with GH deficiency, such as Silver or Turner syndromes.

To establish criteria for the nitrogen retention test, we have compared the results in the patients with proven GH deficiency with those in the subjects with normal GH secretion, and the patients of group 2 with possible GH deficiency were excluded.

The individual results, mean values and standard error of the mean in the patients with certain diagnosis (GH deficiency and normal GH secretion) are shown in Figure 1. The GH deficient patients have a mean nitrogen retention of 35.2% (SD 4.7%). The mean value minus two standard deviations gives a lower limit of 25.8%.

The patients with 'A-type'-deficiency show a mean retention of 45.8%. In spite of the small number of cases in this group and a standard deviation of 11.4%, this mean is significantly ($P < 0.01$) higher than that of the other GH deficient patients.

In contrast, the mean nitrogen retention in the subjects with normal GH secretion is only 14.3% (SD 6.0%). This mean value plus 2 standard deviations indicates an upper limit of 26.4% in this group. Only one case has a nitrogen retention higher than this limit (33.5%). With this unexplained exception (girl with Turner's syndrome, in whom treatment with HGH was a failure), there is no overlap between GH deficient and non-deficient subjects. The difference between the two groups is highly significant ($P < 0.001$).

The results may be summarized as follows: The nitrogen retention test is diagnostically

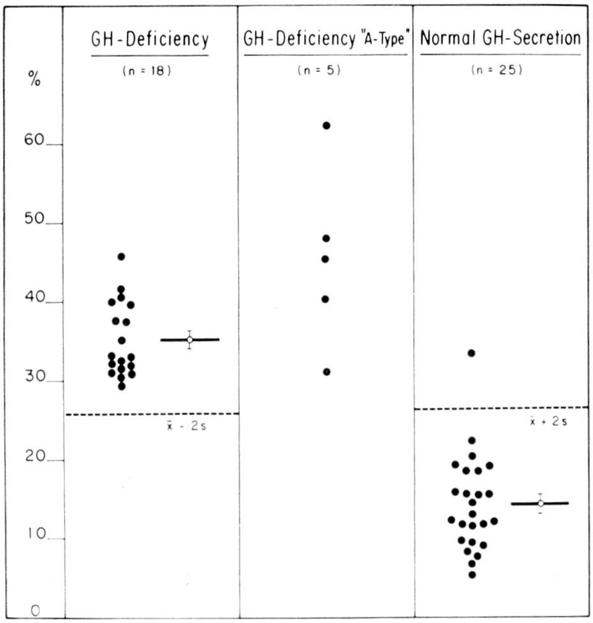

Fig. 1 Nitrogen retention (per cent of pre-treatment excretion) in patients with and without GH deficiency (individual values, mean and SEM).

useful, when the results of less time-consuming tests are inconclusive. Applying these criteria, a nitrogen retention higher than 26% of the pretreatment excretion indicates GH deficiency. With a figure higher than this, successful long-term HGH treatment may be expected. There is no evidence that deficiency of other pituitary hormones (untreated or treated) influences the test result, although further analysis is required. Patients with the 'A-type' of isolated GH deficiency (Illig *et al.*, 1970) retain more nitrogen than other GH deficient subjects. With the exception of the case with Turner's syndrome, all patients with a high nitrogen retention, and subsequently treated with HGH, responded well and had a growth velocity above the 97th percentile during the first year of treatment. The test has therefore a prognostic value with respect to the success of long-term HGH treatment. The major drawback of the test is its long duration and inconvenience.

We are currently investigating the possibility of a shorter similar test using the stable isotope N^{15}.

The second test concerns the *response of plasma amino acids to HGH.* In earlier studies (Zachmann, 1969), we have shown that some amino acids in plasma, notably threonine, serine, glycine and methionine, increase in GH deficient subjects after 3 to 5 days of intramuscular HGH administration. This effect is apparently different from that of testosterone on amino acids (Zachmann *et al.*, 1966). From other studies in the human and in animals (Daughaday, 1968; Rabinowitz *et al.*, 1968), it is known that shortly after HGH, especially if given intravenously, serum α-amino nitrogen falls, reflecting a shift of amino acids into the cells. In an extension of previous studies, the effects of intravenous HGH in GH deficient and control subjects have been investigated under the following test conditions. A basal blood sample was drawn in the fasting state at 8 a.m. HGH was then injected at a dosage of 2 mg/sq.m of body surface area and additional blood samples were drawn 1 and 2 hours after injection. Since complete amino acid analysis in a few cases has shown that threonine and

TABLE II

Influence of HGH on plasma amino acids: Groups of patients

Diagnosis	n
1. Proven GH deficiency	7
2. Possible GH deficiency	3
3. Normal GH secretion	10
4. Untreated hypothyroidism, primary or secondary	5
Total	25

serine change most markedly, only these two amino acids have been determined in the remaining cases using an automatic amino acid analyzer.

So far, we have carried out this test in 25 patients (Table II). Among the 7 GH deficient cases are 4 with idiopathic and 1 with familial isolated GH deficiency and 2 with panhypopituitarism after radical operation of a craniopharyngioma, who were receiving adequate replacement therapy with thyroxine and cortisol. The patients of the second group were excluded for evaluation since their diagnosis is not yet certain. The third group consists of 3 normal adult males and 1 female and of 6 children with a diagnosis other than GH deficiency, including Addison's disease, Silver syndrome, neurofibromatosis and delayed puberty (control group). In the fourth group, there are 5 patients with untreated hypothyroidism, 1 primary and 4 secondary, associated with GH deficiency.

The mean absolute plasma threonine values of the control subjects (group 3) and of the GH deficient patients are shown in Figure 2. In the controls there is a slight, but insignificant drop from 9.6 before to 8.2 μmol/100 ml, 2 hours after HGH. The GH deficient subjects

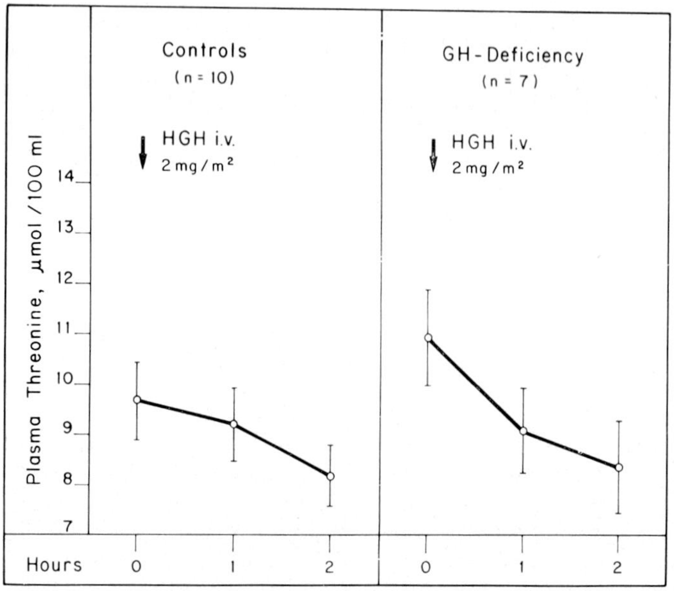

Fig. 2. Plasma threonine before and after i.v. administration of HGH in patients with and without GH deficiency (absolute values, mean and SEM).

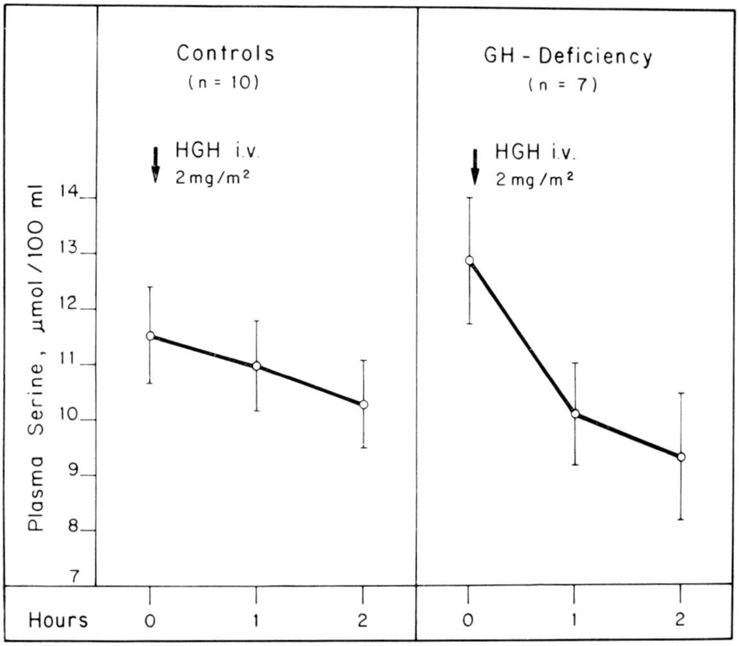

Fig 3. Plasma serine before and after i.v. administration of HGH in patients with and without GH deficiency (absolute values, mean and SEM).

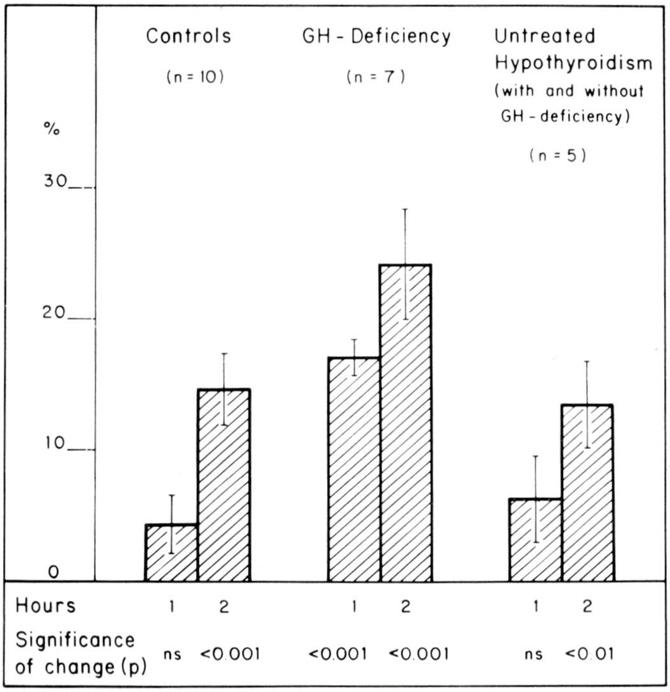

Fig. 4. Mean per cent reduction of plasma threonine (± SEM) 1 and 2 hours after i.v. administration of HGH (2 mg/sq.m) in patients with and without GH deficiency.

show a highly significant ($P < 0.001$) decrease from a higher basal value of 10.9 to 8.4 μmol/100 ml after 2 hours.

The same is true for plasma serine, as shown in Figure 3. In the controls, there is an insignificant change from 11.5 to 10.3 μmol/100 ml, while in the GH deficient subjects, a highly significant ($P < 0.001$) reduction from 12.9 to 9.3 μmol/100 ml was found. In the untreated hypothyroid patients, neither amino acid changed significantly, even in the presence of simultaneous GH deficiency.

The mean per cent reduction of plasma threonine is shown in Figure 4. While there is no significant change in the control and untreated hypothyroid patients after 1 hour, the reduction in the GH deficient patients is 17% after 1 and 24% after 2 hours. Similar results were obtained with respect to plasma serine (Fig. 5). The difference between the patients

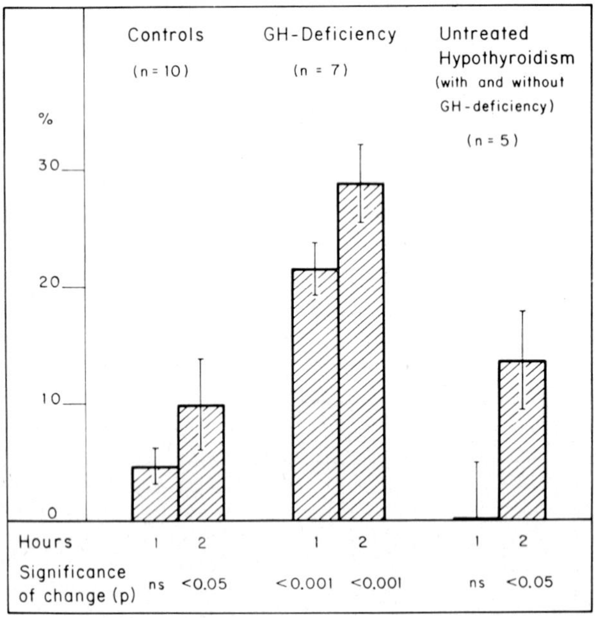

Fig. 5. Mean per cent reduction of plasma serine (± SEM) 1 and 2 hours after i.v. administration of HGH (2 mg/sq.m) in patients with and without GH deficiency.

with and without GH deficiency was found to be most significant ($P < 0.001$) for plasma serine 1 hour after HGH.

Figure 6 shows the individual per cent reduction of plasma serine 1 hour after HGH. If this amino acid and time are chosen for evaluation, the GH deficient subjects without hypothyroidism and the non-deficient subjects may be separated without overlap.

The reproducibility of the test was evaluated in 1 GH deficient patient. The decrease of plasma serine 1 hour after HGH was 32.6% in the first and 27.2% in the second test.

There is no evident theoretical explanation for the particular influence of HGH on plasma serine and threonine. In the case of serine, the change is possibly related to the influence of HGH on carbohydrate metabolism. Intravenous administration of HGH is known to induce transient hypoglycemia. Since serine may be formed from glucose through 3-phosphoglyceric acid (Ichihara and Greenberg, 1957), its reduction by HGH may be due to this mechanism.

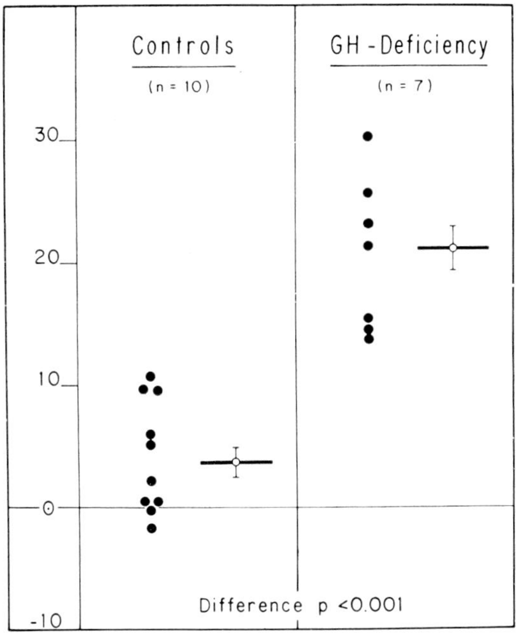

Fig. 6. Per cent reduction of plasma serine 1 hour after i.v. administration of HGH (2 mg/sq.m) in patients with and without GH deficiency (individual values, mean and SEM).

These preliminary results may be summarized as follows: intravenously administered HGH induces a drop of threonine and serine in plasma after 1 and 2 hours. This reduction is not significant in normal subjects, but very marked in patients with GH deficiency. A diagnostically useful separation of GH deficient from non-deficient subjects appears to be possible on this basis, provided that the patients do not have concomitant untreated primary or secondary hypothyroidism. If the value of this test is confirmed by further studies, it offers a simple and fast additional test to diagnose GH deficiency.

REFERENCES

DAUGHADAY, W. H. (1968): In: *Textbook of Endocrinology, 4th ed.*, Chapter 2, pp. 27–84. Editor: R. H. Williams. Saunders Co., Philadelphia.

ICHIHARA, A. and GREENBERG, D. M. (1957): Further studies on the pathway of serine formation from carbohydrates. *J. biol. Chem.*, *224*, 331.

ILLIG, R., PRADER, A., FERRANDEZ, A. and ZACHMANN, M. (1970): Familiärer pränataler Wachstumshormon-Mangel mit erhöhter Bereitschaft zur Bildung von Wachstumshormon-Antikörpern. In: *Endokrinologie der Entwicklung und Reifung, 16. Symposium der Deutschen Gesellschaft für Endokrinologie in Ulm, Februar 1970*, pp. 246–247. Editor: J. Kracht. Springer-Verlag, Berlin-Heidelberg-New York.

PRADER, A., ZACHMANN, M., POLEY, J. R. and ILLIG ,R. (1968*a*): The metabolic effect of a small uniform dose of human growth hormone in hypopituitary dwarfs and in control children. I. Nitrogen, α-amino-N, creatine-creatinine and calcium excretion and serum urea-N-α-amino-N, inorganic phosphorus and alkaline phosphatase. *Acta endocr. (Kbh.)*, *57*, 115.

PRADER, A., ZACHMANN, M., POLEY, J. R., ILLIG, R. and SZÉKY, J. (1968*b*): Studies with human growth hormone in hypopituitary dwarfism. In: *Growth Hormone. Proceedings, First International Symposium, Milan 1967*, pp. 388–397. Editors: A. Pecile and E. E. Müller. ICS 158, Excerpta Medica, Amsterdam.

RABINOWITZ, D., MERIMEE, T. J., RIMOIN, D. L., HALL, J. G. and MCKUSICK, V. A. (1968): Peripheral subresponsiveness to human growth hormone in a proportionate dwarf. *J. clin. Invest.*, *47*, 82a.
ZACHMANN, M. (1969): Influence of human growth hormone (HGH) on plasma and urine amino acid concentrations in hypopituitary dwarfs. *Acta endocr. (Kbh.)*, *62*, 513.
ZACHMANN, M., CLEVELAND, W. W., SANDBERG, D. H. and NYHAN, W. L. (1966): Concentrations of amino acids in plasma and muscle. *Amer. J. Dis. Child.*, *112*, 28.

THE PATTERN OF GROWTH IN CHILDREN WITH GROWTH HORMONE DEFICIENCY BEFORE, DURING AND AFTER TREATMENT

J. M. TANNER and R. H. WHITEHOUSE

Department of Growth and Development, Institute of Child Health, University of London, London, United Kingdom

One hundred patients (the roundness of the figure is quite fortuitous) have so far been measured at set intervals in the Department of Growth and Development and treated with human growth hormone (HGH) for a year or longer. All were participants in the national clinical trial of HGH being carried out by the Medical Research Council Sub-Committee on Human Pituitary Hormones. The numbers of patients treated in each diagnostic category are shown in Table I. A fuller account of these patients is given in Tanner *et al.* (1971).

Anthropometric measurements

Height, weight, four skinfolds, sitting height, upper arm, thigh and calf circumferences, biacromial and bi-iliac diameters and bicondylar diameters of humerus and femur were measured, always by the same anthropometrist.

Height was measured using the apparatus and technique described in Tanner (1964, 1972). The technique minimises variation in posture due to tiredness or boredom, which may otherwise amount to 1 or even 2 cm. A good anthropometrist can repeat his reading to 0.3 cm or

TABLE I

Numbers of patients treated in each diagnostic category

Diagnosis	Number treated
(1) Hyposomatotrophic (HS) or 'isolated' GH deficiency	36
(2) Craniopharyngiomas	15
(3) Other CNS tumours	3
(4) Panhypopituitarism	3
(5) Low birth weight (small-for-dates)	18
(6) Small/delay	4
(7) 'Sulphation-factor' lack	1
(8) Psychosocial short stature	4
(9) Uncertain diagnosis	4
(10) Turner's syndrome	6
(11) Steroid-induced short stature	2
(12) Coeliac disease	1
(13) Other metabolic disorders	3
	100

less on 95% of occasions, and two well-trained observers should seldom differ by more than 0.5 cm.

The accuracy of measurement is particularly important where calculation of increments or velocities of growth are concerned since there are two errors involved, which may summate. Casual observers without special training may differ in their measurements by 1.5 cm, and two summating errors of this magnitude introduce large errors into a yearly velocity and make a three-monthly velocity meaningless.

Skinfolds were measured as described in Tanner and Whitehouse (1962, 1967). The frequency distributions of these measurements are skewed, and throughout logarithmic transformations were used.

Sitting height was taken using the Harpenden anthropometer (Tanner, 1972).

Ratings of sexual development were made on the scale of 1 to 5 for genitalia or breast development, and pubic hair development, separately, using the Tanner (1969) scales. Size of testes was measured using the Prader orchidometer. In patients who showed some degree of sexual development the absolute velocities or SDS (see below) over the period in question have been omitted when means were calculated.

Skeletal maturity (bone age). This was estimated from the bones of the left hand and wrist using the method of Tanner *et al.* (1962) which rates each of 20 bones separately on a defined scale (1–8) of maturity. We think this has both theoretical and practical advantages over the Greulich-Pyle atlas method; greater accuracy and more information is obtainable, and standards for British children exist.

Radiographic measurements. At certain times in selected patients radiographic measurements were made of the bone, muscle and fat diameters midway down the left upper arm and at the maximum overall diameter of the left calf. The technique was that described by Tanner (1964; see also 1962, Appendix). The upper arm was taken in the lateral view so that the two epicondyles appeared superimposed on the film: the central vertical plane of the arm was exactly 5 cm in front of the film, and the anode of the X-ray machine was exactly 1.5 m from the film. The calf film was taken with the foot pointing directly towards the anode, the weight equally on both feet, the central plane of the calf 10 cm in front of the film and the anode-film distance also 1.5 m.

A lead marker was placed on the upper arm at the level at which the triceps skinfold was measured; a line was drawn on the radiograph passing down the long axis of the upper arm as nearly as possible parallel to the skin borders and a line perpendicular to this was drawn across the arm at the marked level. The widths of the whole arm and the anterior and posterior fat areas were measured along this second line. The humerus was measured at right angles to its own axis, as near as possible along this line. Fat width was the sum of the two measurements; muscle width was obtained by subtraction.

In the calf radiograph a line was drawn perpendicular to the long axis of the tibia, at the maximum overall width of the calf. The widths of total calf, lateral and medial fat, tibia and fibula were measured along this line. 'Bone' represents tibia only and 'muscle' was obtained by subtracting fat and tibia breadths from total width. The fibula width is included in muscle, therefore.

Normal standards for age for these limb tissue widths were available from unpublished data on the Harpenden Growth Study (see Tanner, 1962). SDS (see below) were calculated for arm and calf separately and the average of the two limb scores was used.

Treatment procedure

There was a *pre-treatment year* during which measurements, bone age assessment and photogrammetric photographs were made at three-monthly intervals. An uninterrupted year of treatment followed *(first treatment year)*, with the same schedule of measurements. The patient then entered a second control year of no HGH treatment *(coasting year)*, during which measurements were made at six-monthly intervals. The stature velocity during the first treatment year was compared with the velocity during the pre-treatment year, and if a significant and clinically useful increase, or acceleration, had occurred during treatment, the patient proceeded to a second year of HGH treatment. Treatment thereafter is planned to be continuous, until such time as an adult height within normal limits is achieved or can be brought about by inducing an adolescent growth spurt.

A small number of patients, started on HGH before the above procedure was established have been treated continuously, without the 'coasting' year. Some of these have had a year off treatment after three or four years on, and have subsequently been restarted on treatment. These long-term patients are discussed separately in the Results section.

Hormone and dose

Currently about 60,000 pituitaries a year are being collected and preserved in acetone by pathologists throughout the United Kingdom. The GH preparation was made by Dr. A. S. Hartree, using her modification of Raben's original method (Hartree, 1966). The potency of the batches varied between 0.8 and 1.2 I.U./mg. The preparation was free of measurable thyroid-stimulating or gonadotrophic activity.

The HGH was administered by intramuscular injection in saline, with two injections per week. The standard dose has been a nominal 20 I.U./week, but due to batch potency differences, this has varied from 18 to 24 I.U./week. The small difference in dose due to batch variation had no detectable influence on growth response. Doses of half and double this (a nominal 10 I.U./week and 40 I.U./week) have been used in a few patients.

Assessment of growth status and response

The calculations for assessing growth status and response are computerised. After each visit the patient's measurements were transferred to tape and the velocities etc. described below printed out.

All measurements (including bone age) are presented in terms of Standard Deviation Scores (SDS). If X is the measurement, $\bar{x}$ the mean at the relevant age or bone age, and s_x the SD at that age, then

$$\mathrm{SDS} = \frac{X - \bar{x}}{s_x}.$$

All SDS have a standard deviation approximating 1.0 if the population they come from is normally distributed.

The standards used for height, weight, height velocity and weight velocity were those of Tanner *et al.* (1966). We have taken the individual-type (longitudinal) standards (Tables V, VI, IX and X of that article) for ages over 8 in girls and 10 in boys. This enables meaningful SDS to be calculated up to the age of 10 in girls and 12 in boys; but above this the adolescent growth spurt of the standard makes their application to prepubescent children erroneous. Over the ages of 18 in boys and 16 in girls SDS are again valid. Since the bone ages of many of our patients are retarded, we have calculated the SDS for height and other measurements in relation to bone age (BA) as well as chronological age (CA).

Sitting height standards are cross-sectional (Tanner, 1972). For skinfolds we have used the log transformations of triceps and subscapular folds and then taken the average of the two SDS. The skinfold standards are also cross-sectional. For bone age, standard deviations diminish rapidly as full maturity is attained, and no SDS for ages over 16 in boys and 14 in girls have been calculated.

Height SDS, with allowance for parents' heights was assessed in children aged 2.0 to 9.0 using the standards of Tanner *et al.* (1970).

Height velocity during the pre-treatment year was estimated by fitting a straight line to the five quarterly measurements and taking its slope. Velocity during the treatment year was estimated in the same way, giving the average velocity over the year, even though the growth response on treatment is not really linear, being greater at the beginning of the year and subsequently less (see Results). In the coasting year only three measurements are available for fitting the line; in later treatment years five measurements are again available. The velocity standards apply to velocity over whole year periods only and would give erroneous SDS if applied over periods of less than a year (see Tanner *et al.*, 1966, p. 614). Seasonal differences in growth rates (Tanner, 1962) may be very great in normal children; indeed, on average, children grow about three times as rapidly in their fastest quarter-year as in their slowest (Marshall, 1971). Thus a change of velocity of 5 cm/yr from one three-monthly period to another (especially if from winter to spring) is to be expected and a change of over 10 cm/yr is not exceptional. Reports of growth responses to treatment based on periods of less than a full year should be viewed very critically.

The *height response to treatment* was measured by the amount the velocity in the treatment year exceeded that expected in the absence of treatment, that is, by the treatment acceleration. Since a normal child is decelerating from birth to adolescence, the expected velocity diminishes from year to year by approximately 1 cm/yr each year from 2 to 4, 0.5 cm/yr from 4 to 6, and 0.2 cm/yr each year from 6 till puberty (Tanner *et al.*, 1966). One method of compensating for the diminution would be to use the average of the pre-treatment and post-treatment control years as the baseline for comparison with the treatment velocity, but we would have to be sure that the post-treatment velocity had been unaffected by the treatment year. There is evidence (see Discussion) that in the post-treatment year a 'regulatory deceleration' occurred, at least if treatment was successful. Thus the average of pre- and post-treatment years would over-estimate the treatment acceleration, just as comparison with the pre-treatment year alone would under-estimate it. On balance we prefer the latter alternative and have used the treatment velocity less pre-treatment velocity figure as the treatment acceleration, uncorrected.

Since velocity standards are not available for skinfolds or sitting height the treatment response has been estimated simply by the change of SDS from beginning to end of treatment.

One important problem is to see whether on treatment the height has advanced relatively more or less rapidly than the bone age since the final predicted adult height depends on this. Methods for testing this to date are rather unsatisfactory. We have used the ratio:

$$\frac{\text{height velocity/expected height velocity for CA}}{\text{bone age velocity/expected bone age velocity}}.$$

RESULTS

The results will be given in two sections (1) Characteristics at diagnosis and (2) Response to HGH treatment.

CHARACTERISTICS AT DIAGNOSIS

Columns 1 to 15 of Table II give information obtained at the initial diagnostic examination of the patients. (This Table, not reproduced here, is available from the authors.)

Hyposomatotrophic, or isolated GH deficiency patients (Group 1)

Sex incidence and genital development. Out of 36 patients, 30 were boys, and of these 13 had an abnormally small penis and ill-developed scrotum. Only 3 of these had bilaterally

TABLE III

	HS				CR				LBW			
	No.	Mean	S.D.	S.E.	No.	Mean	S.D.	S.E.	No.	Mean	S.D.	S.E.
Bone age SDS	30	–3.21	1.30	0.24	17	–2.99	1.71	0.42	17	–1.78	1.35	0.33
Chronological age less bone age (yr)	32	3.29	1.50	0.27	17	3.21	1.48	0.35	17	1.46	1.14	0.28
Height SDS for chronological age	32	–4.66	1.07	0.19	18	–3.75	1.55	0.37	17	–3.65	0.83	0.20
Height SDS, parents allowed for	20	–5.40	1.23	0.27	—	—	—	—	12	–4.56	1.01	0.29
Height SDS for bone age	31	–1.58	1.65	0.30	18	–1.61	1.33	0.31	15	–1.78	1.69	0.44
Sitting height SDS	29	–4.56	1.06	0.20	14	–3.53	1.40	0.37	16	–3.61	0.81	0.20
Skinfold transform SDS (av. triceps + subscapula)	30	0.91	0.83	0.15	14	1.27	0.79	0.21	17	–0.88	1.03	0.25
Birth weight (kg)	30	3.38	0.51	0.09	5	3.20	—	—	17	1.86	0.38	0.09
Birth weight SDS, gestation and mother's height allowed for	30	–0.21	0.97	0.18	5	–0.46	—	—	17	–3.08	0.67	0.16
Mid-parent height SDS	32	–0.17	1.16	0.20	13	0.64	0.84	0.23	16	34.44	0.83	0.21
GH peak after insulin (μU./ml)	25	4.72	3.35	0.67	7	3.14	1.86	0.70	9	33.08	3.71	1.24
GH peak after bovril (μU./ml)	21	5.70	3.05	0.67	5	4.00	—	—	13	–2.82	11.50	3.19
Limb bone width SDS	18	–3.02	0.92	0.22	11	–2.26	1.03	0.31	12	–2.53	0.44	0.13
Limb muscle width SDS	18	–2.92	1.20	0.28	11	–2.52	0.88	0.27	12	–0.77	1.03	0.30
Limb fat width SDS	18	2.15	2.90	0.68	11	3.12	2.53	0.76	12	5.26	0.88	0.25
Height velocity pre-treatment year (cm/yr)	30	3.15	1.18	0.21	12	2.67	0.38	0.40	16	6.70	1.47	0.37
Height velocity 1st treatment year	28	9.10	2.43	0.46	12	6.13	1.35	0.39	12	0.89	1.58	0.46
Height acceleration 1st treatment year	28	6.00	2.82	0.53	12	3.48	1.37	0.39	12	–0.97	1.24	0.36
Ht. vel. SDS (chron. age) pre-treatment year	20	–2.76	1.08	0.24	2	–2.70	—	—	14	–1.38	1.21	0.32
Ht. vel. SDS (bone age) pre-treatment year	27	–3.28	1.36	0.26	7	–3.17	1.94	0.73	14	0.47	1.23	0.33
Ht. vel. SDS (chron. age) 1st treatment year	19	4.26	2.22	0.51	2	–3.80	—	—	11	–0.79	1.12	0.34
Bone age velocity pre-treatment year	31	0.86	0.58	0.10	12	0.59	0.48	0.14	15	0.97	0.44	0.12
Bone age velocity 1st treatment year	28	1.18	0.49	0.09	12	0.65	0.55	0.16	12	1.11	0.28	0.08
Ht. vel. %/bone age vel. pre-treatment year	19	0.85	0.34	0.08	2	0.45	—	—	13	0.88	0.30	0.08
Ht. vel. %/bone age vel. 1st treatment year	20	1.61	0.74	0.17	2	0.45	—	—	12	1.05	0.38	0.11
Av. skinfold trans. SDS begin treatment year	28	0.96	0.80	0.15	8	1.28	0.73	0.26	16	–0.71	0.80	0.20
Av. skinfold trans. SDS end 1st treatment year	25	0.32	0.70	0.14	6	0.77	0.61	0.25	12	–0.63	0.97	0.28
Bone age velocity pre-treatment year	16	0.86	4.08	1.02	9	1.07	3.42	1.14	8	1.65	1.71	0.60
Bone age velocity 1st treatment year	24	2.19	0.94	0.19	11	2.01	0.99	0.30	15	1.78	1.25	0.32
Limb muscle velocity pre-treatment year	15	0.43	12.03	3.11	9	3.51	19.88	6.63	8	–0.04	7.31	2.58
Limb muscle velocity 1st treatment year	24	7.75	5.53	1.13	11	4.85	5.78	1.74	15	4.79	4.14	1.07
Limb fat velocity pre-treatment year	15	1.81	5.57	1.44	9	5.86	8.81	2.94	8	–0.08	5.11	1.81
Limb fat velocity 1st treatment year	24	–3.81	4.51	0.92	11	–7.11	6.15	1.85	15	–0.30	2.31	0.60

undescended testes however, and a further 3 unilateral undescended testes. The more extreme examples of this maldevelopment quite resembled fused labia and an enlarged clitoris, though only one had hypospadias. There was no association between small genitalia and the degree of retardation of bone age, of smallness in height, or of response to ACTH.

Gestation and birth weight. Pregnancy and delivery were said to have been normal in all except 6, of whom 3 were breech deliveries. The length of gestation was 39–41 weeks except for one case each of 37, 38 and 42 weeks. The mean birth weight was 3.4 ± 0.1 kg. When adjusted for gestation and mother's size, the SDS for birth weight ranged from −2.1 to +1.5 and averaged −0.2 ± 0.2.

Familial incidence. The parents' heights averaged 40th centile for fathers and 36th centile for mothers. The mid-parent SDS was −0.2 ± 0.2. There was one pair of brothers represented in our sample. The mother and father are not related and there were three other brothers,

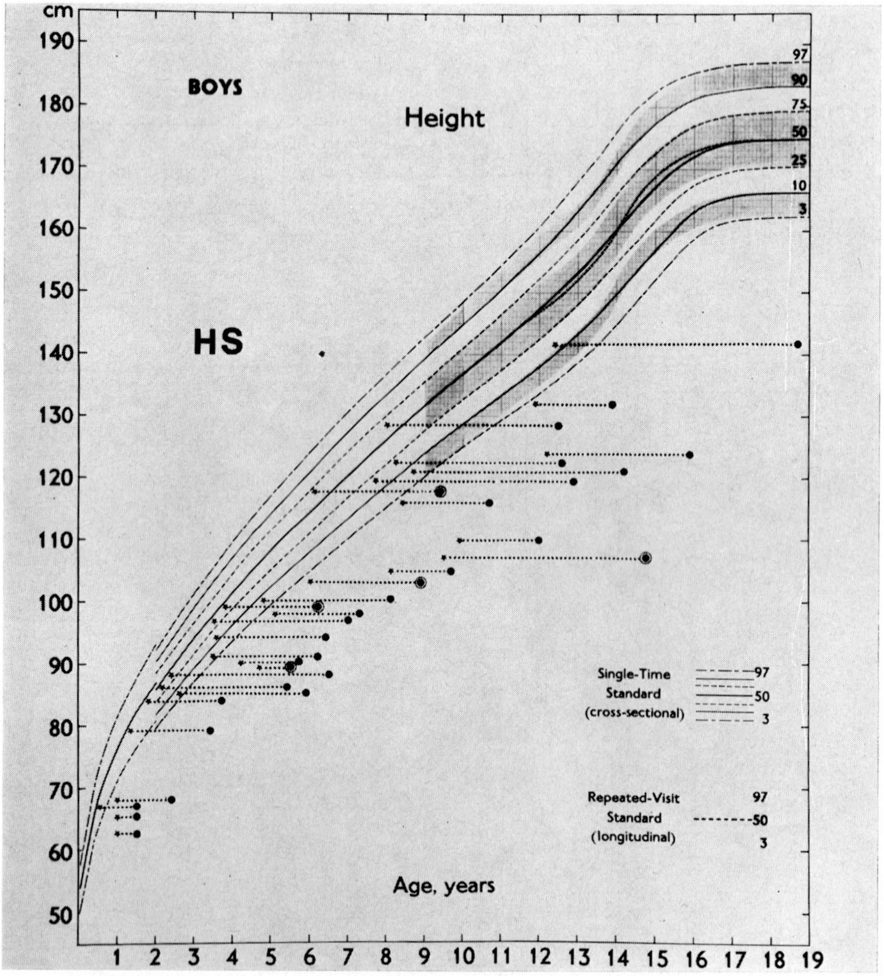

Fig. 1. Height at diagnosis, HS boys, plotted against chronological age, and bone age. The length of the dotted horizontal line indicates the absolute bone age retardation in each patient. Partial deficiency cases: ringed dots.

who were normal. All the other children are unrelated and all the other patients' sibs were said by their parents to be within normal limits for height. In 22 additional HS patients we have seen there are two further sib pairs. Thus approximately 1 in 10 of all our HS patients has an affected sibling.

Height and weight. The absolute heights are plotted, both against chronological age and bone age, in Fig. 1 (boys only). All patients were much below the 3rd height centile for chronological age (CA) though about a half are above it when plotted at bone age (BA). The height SDS for CA ranged from −2.6 to −7.3 with a mean of −4.7.

Since bone age was retarded, height for bone age was within normal limits in many cases. The mean SDS was −1.6, range −5.1 to +1.7.

Weight is plotted for CA in Fig. 2, together with values for low birth weight and some other patients.

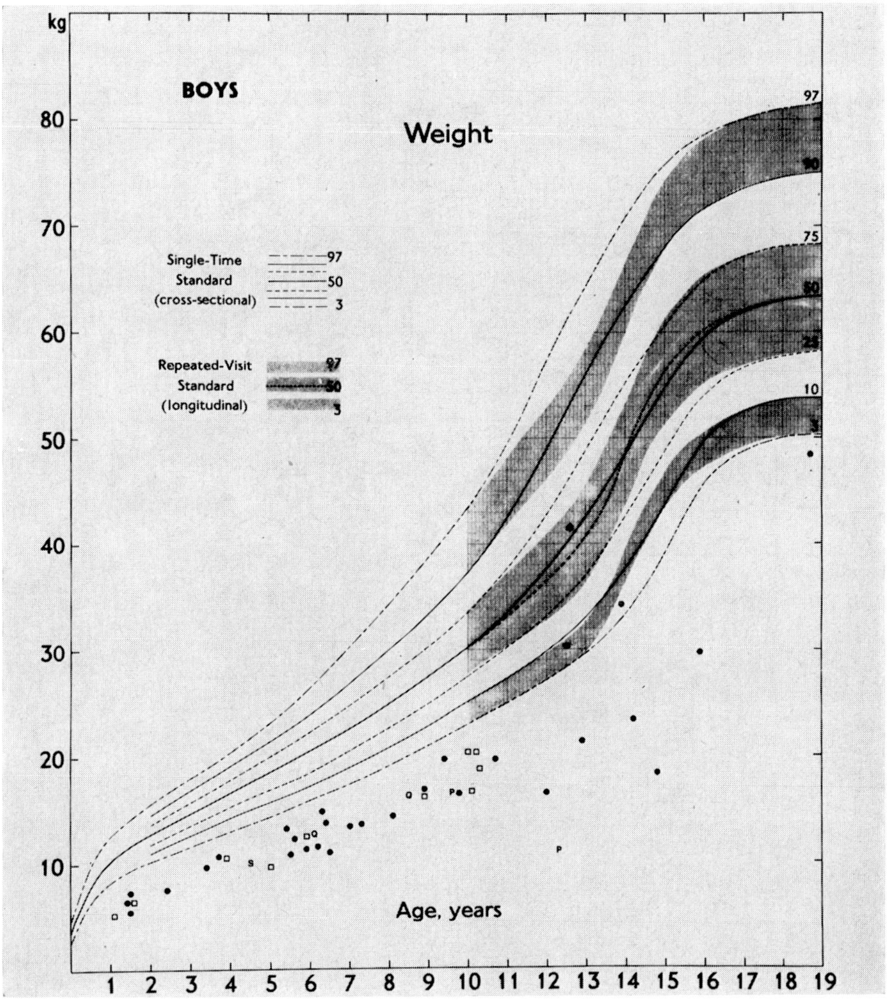

Fig. 2. Weight for CA at diagnosis, HS ●, LBW □, psychosocial (P) and uncertain (Q) patients.

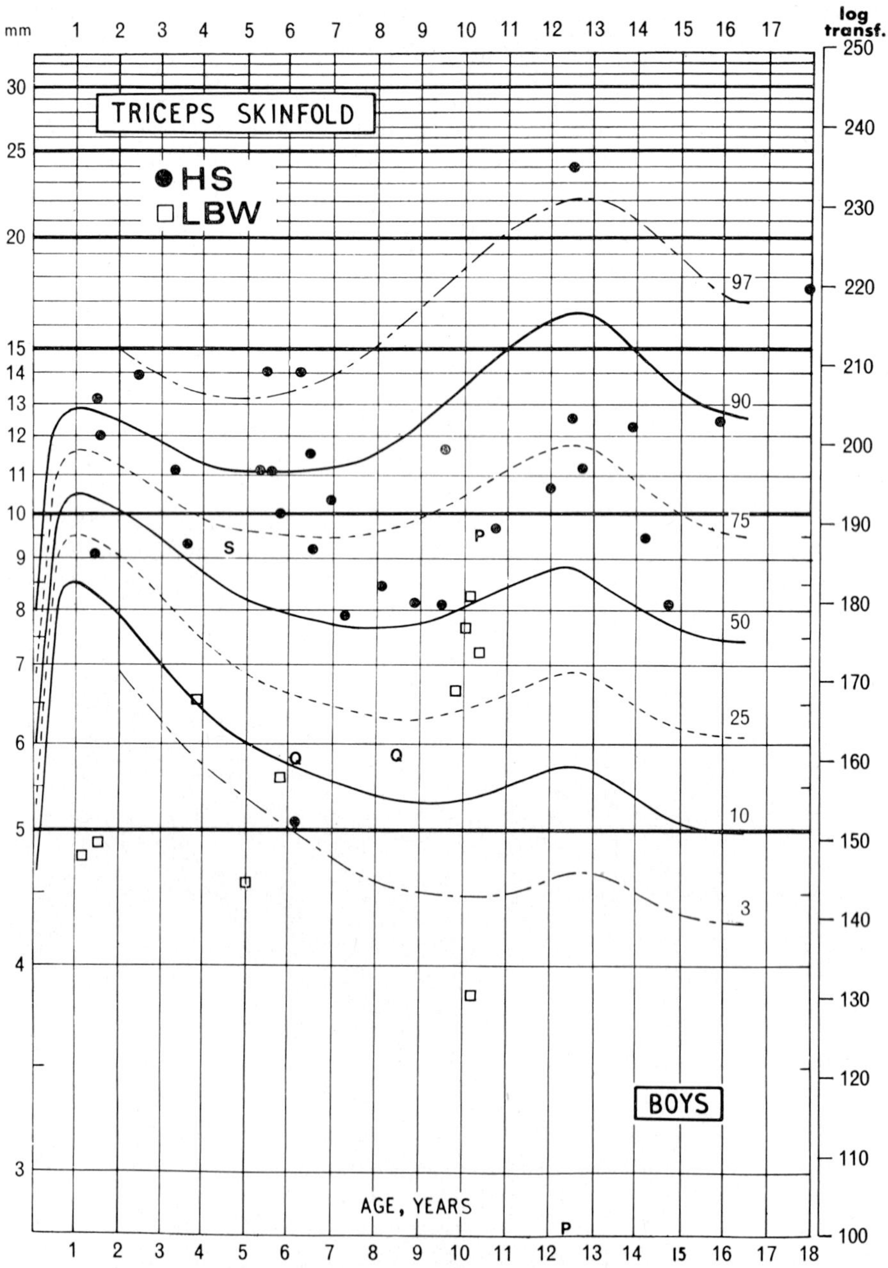

Fig. 3. Triceps skinfolds, HS, LBW and other patients, boys (S, sulphation factor lack; P, psychosocial; Q, uncertain).

Bone age. Bone age was always retarded, as can be seen in Fig. 1 by the length of the lines joining the circles and stars. The average degree of retardation was 3.3 years, but this absolute difference depends on age, negative BA's not being obtainable. The bone age SDS averaged −3.2, with a range of −0.8 to −5.7. It was significantly related to age, those under age 8 averaging −2.7 SDS and those over 8 −3.8 SDS.

Sitting height. The mean SDS for sitting height was −4.6, compared with −4.7 for height. Thus there is no evidence that the trunk-limb proportions of the HS children were different from those of normal children of the same size.

Skinfolds. The triceps skinfolds for boys are plotted in Fig. 3 together with those of the LBW and psychosocial patients. When the transform values of triceps and subscapular folds were averaged nearly all HS patients were above the mean for the normal population.

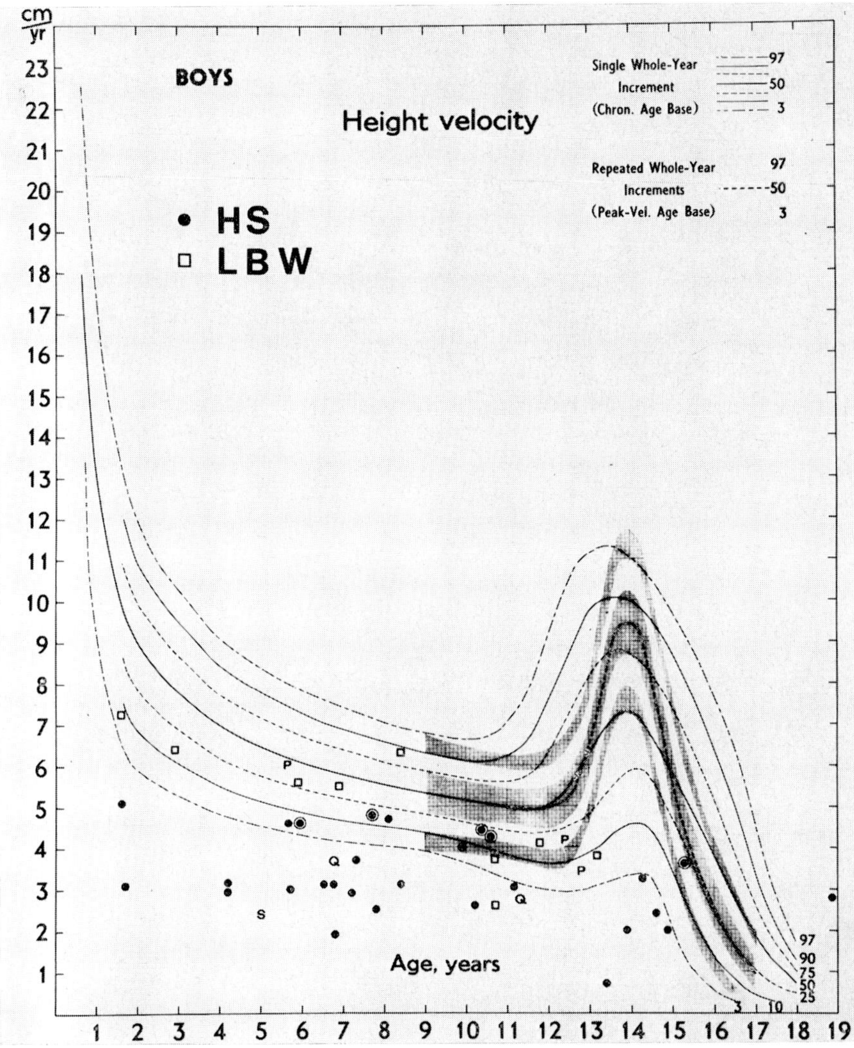

Fig. 4. Height velocity in pre-treatment year, HS ●, LBW □, psychosocial (P) and sulphation factor-lacking (S) boys. Partial deficiency cases: ringed dots.

The range of SDS was −1.0 to +2.5 with a mean of 0.91 ± 0.15. The younger patients were somewhat fatter than the older ones, those under 8 years averaging +1.3, the remainder +0.5.

Limb bone, muscle and fat. The means and standard deviations are given in Table II. In the HS patients both bone and muscle averaged about 3 SD below the mean for chronological age, while fat averaged 2.2 SD above it. Thus muscle growth in this disorder is as limited as growth of bone.

Pre-treatment height velocity. The SDS scores for pre-treatment height velocity ranged from −0.9 to −4.5 with a mean of −2.8 ± 0.2 in the 20 patients in whom they can be calculated.

The absolute velocities are plotted in Fig. 4 together with velocities of the LBW and some other patients. All patients plotted are pre-pubescent, so those over age 11 should be judged in relation to an imagined extension of the pre-pubertal velocity lines. There are 8 HS patients above the 3rd centile (hence within accepted 'normal limits'); we have reason to think that the 5 of them marked with a circle have partial deficiency only (see below).

Pre-treatment velocity is probably quite a good guide to the degree of GH deficiency, provided it covers a full year and is considered in relation to the expected values both for

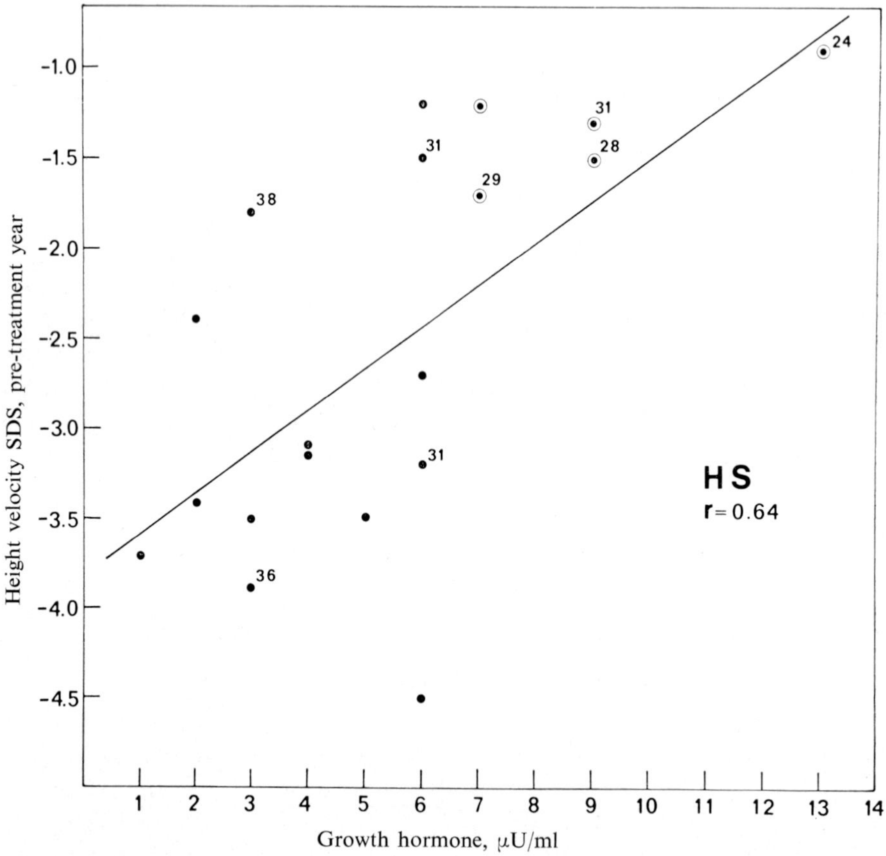

Fig. 5. Relation in HS patients between height velocity SDS for CA in pre-treatment year and peak GH value on insulin hypoglycaemia. The values by some of the points are percent decrease in nitrogen excretion in 3-day metabolic test. Presumed partial deficiency cases are those ringed.

CA and BA. Absolute values may be very biased by age and periods of less than a year can be totally misleading. Most patients with 'complete' deficiency have pre-treatment velocities of less than −2.0 SDS for CA; patients with presumed partial deficiency range from about −1.0 to −2.0 SDS for CA.

Pre-treatment height velocity and GH peak. There was a highly significant relation (r = 0.64, 18 cases) between these two variates, as illustrated in Fig. 5. The regression was 0.24 ± 0.02 (cm/yr on μU./ml). The presumed partial deficiency cases are ringed; they lie in the section of the graph bounded by −2.0 SDS and 6 μU./ml GH peak. The values for the percent decrease in nitrogen retention in the metabolic test, wheie known, are put beside the points. The agreement is good.

Pre-treatment bone age velocity. The bone age velocity in the pre-treatment year was usually less than the norm of 1.0; the average was 0.86 ± 0.1 'years' per year. Height velocity was more depressed than bone age velocity. The fraction *height velocity in % expected/BA velocity in % expected* averaged 0.77 ± 0.7. During the pre-treatment year, therefore, the patients are becoming increasingly short for their bone age. Thus their predicted adult height is becoming steadily lower.

Antibody-developing patients. Four HS patients developed antibodies during treatment

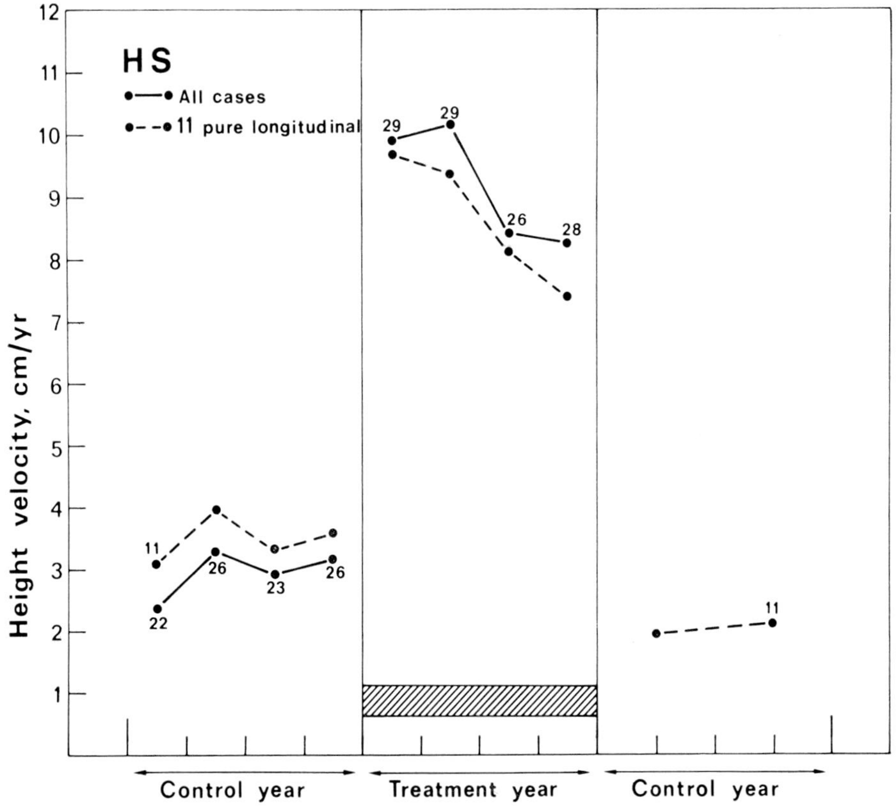

Fig. 6. Mean height velocity during 3-monthly periods before and during, and 6-monthly periods after, treatment; HS patients. Total series (numbers in graph); and 11 patients present on each measuring occasion.

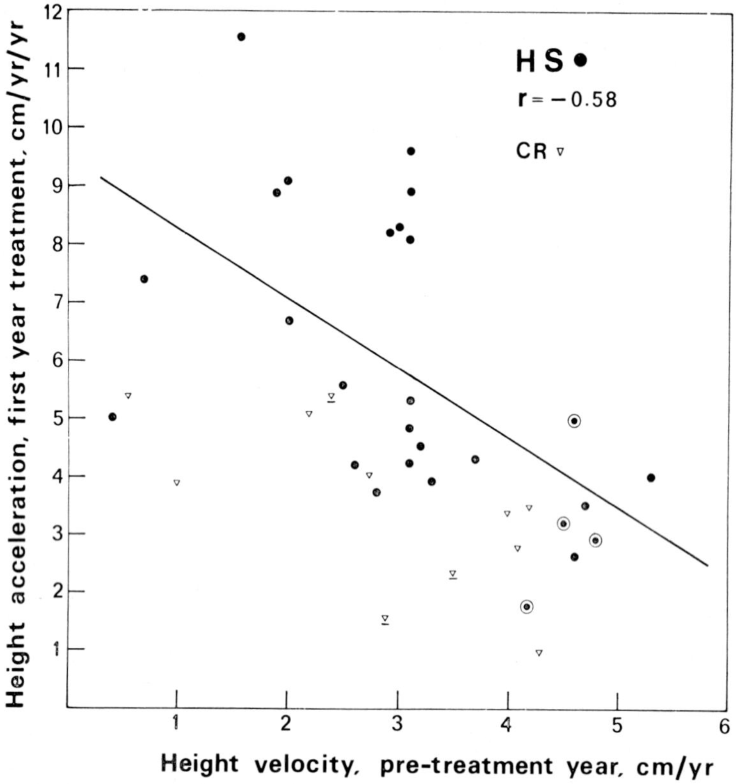

Fig. 7. Relation between height acceleration in first year of treatment (treatment velocity less pre-treatment velocity) and velocity during pre-treatment year. Regression line is for HS patients only, on 18–20 I.U./week HGH. Cranio patients also indicated (▽).

to a degree which stopped their growth response and precluded further HGH treatment (Chalkley and Tanner, 1971). There seemed to be nothing in their pre-treatment investigation which distinguished them from the other HS patients. All were boys; two had normal and two small genitalia. Height and height velocity were similar to the others. One has a rather low birth weight; unfortunately we lack measurements of birth length to compare with the patients of Illig *et al.* (1970).

Craniopharyngioma and other CNS lesion patients

There are 15 patients with craniopharyngiomas, and 3 with other CNS tumours. The means in Table II are for both combined.

Substitution therapy varied. Most were on thyroxine, usually 0.1 or 0.2 mg/day, and cortisone, usually between 5 and 15 mg/day. Some were on thyroxine only and a few on no treatment apart from GH.

Height, weight and bone age. The height SDS varied from −1.2 to −6.5 with a mean of −3.8 ± 0.4. Sitting height averaged the same. Bone age SDS ranged from 0.0 to −5.9 with an average of −3.0 ± 0.4, which is as retarded as the HS patients. Patients on cortisone did not differ from those without it either in height SDS or bone age SDS.

Skinfolds were all above the mean, with an average SDS of 1.3 ± 0.2, which is slightly above even the HS patients. Patients on cortisone did not differ from those who were not.

Pre-treatment height velocity. Five pubescent cases have been excluded from the pre-treatment velocity average, which was 2.7 ± 0.4 cm/yr, range 0.3 to 4.3 cm/yr.

RESPONSE TO HGH TREATMENT (HS PATIENTS)

Height velocity. Treatment began at ages ranging from 2.2 to 20.1 years, bone ages 0.0 to 13.8 'years'. Four patients showed signs of beginning puberty before or during the treatment year, and have been excluded from the means since it is possible that their increase in velocity might have been caused by adolescence rather than HGH.

The mean pre-treatment height velocity was 3.1 ± 0.2 cm/yr and the mean velocity during the first year of treatment was 9.1 ± 0.5 cm/yr, an increase of 6.0 cm/yr. The range of increase was 1.8 to 12.8 cm/yr: all except three patients (two of them partial HS) accelerated by over 3.0 cm/yr.

In Fig. 6 are plotted the mean absolute rates of growth during each of the 3-month periods before and during treatment, and the 6-month periods after treatment. Though in the individual patient the 3-monthly velocity is affected by seasonal variation, the averages of a number of patients are not so much affected, since about half started in April, and half

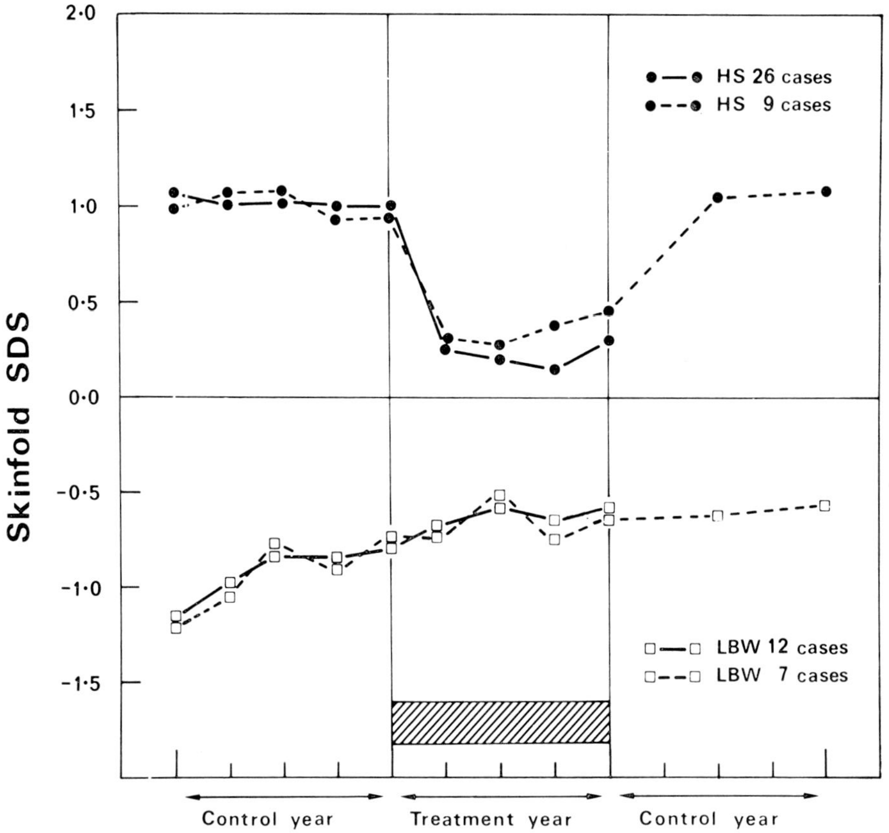

Fig. 8. Mean skinfold SDS (average of triceps and subscapular) at 3-monthly intervals before and during first year of HGH treatment and 6-monthly intervals after treatment. HS and LBW patients.

in October. The solid line shows the total number of cases, varying from 22 to 29; the dashed line the 11 cases followed in pure longitudinal fashion, avoiding bias. There is little difference between the lines. The treatment year curve shows that typical catch-up is achieved (Prader *et al.*, 1963; Tanner, 1963). The lower velocity during the post-treatment year compared with the pre-treatment year in the 11 cases ('regulatory deceleration') is considered below.

Bone age is not so greatly accelerated as height, as the change in the ratio 'height velocity in % expected/bone age velocity in % expected' shows. In the pre-treatment year this averaged 0.77 $\pm$ 0.1; in the treatment year 1.7 $\pm$ 0.1 and in the post-treatment year 0.70 $\pm$ 0.1.

The relation of height acceleration to pre-treatment velocity, GH peak, and age at treatment

Figure 7 shows the relation between the acceleration in the first year of treatment and the velocity in the pre-treatment year. The correlation coefficient for the 27 HS patients (of both sexes but non-pubertal) was $r = -0.58$. The regression had the slope $b = -1.19 \pm 0.3$. Thus the lower the pre-treatment velocity the greater the response, as might be expected. The cranio patients are also indicated in the Figure; for them, interestingly, the correlation is the same, and the regression almost exactly parallel, but with treatment velocity 2 cm less for given age.

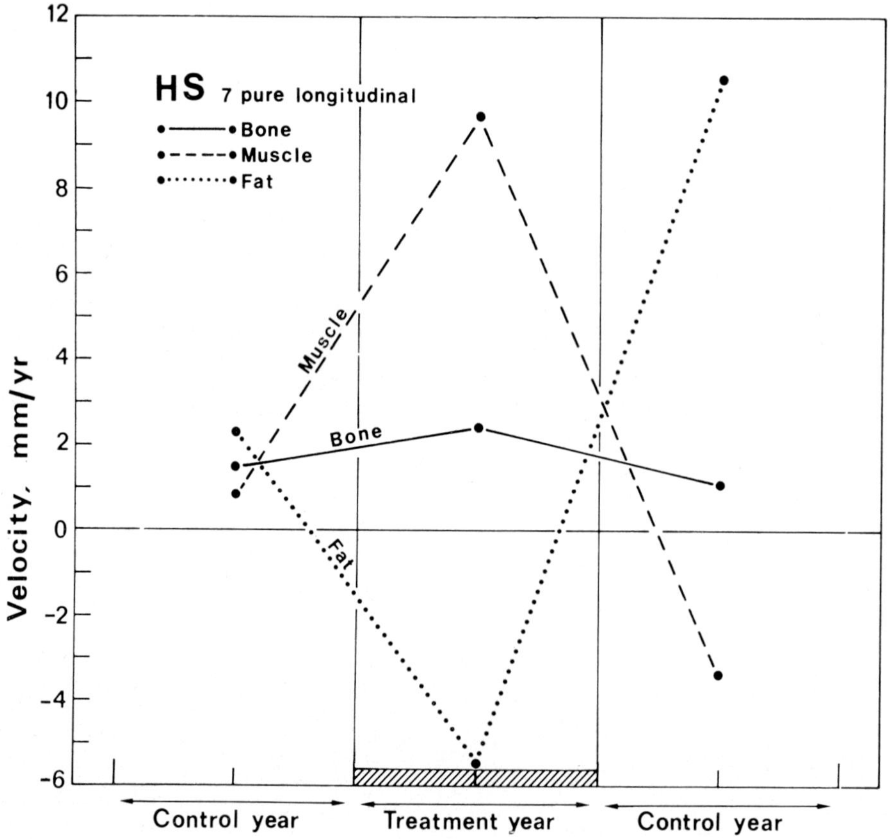

Fig. 9. Mean velocities of widths of bone, muscle and fat measured in radiographs of upper arm and calf, before, during and after HGH treatment. HS patients, pure longitudinal series.

The correlation between height acceleration and the peak GH level on insulin hypoglycaemia, either at diagnosis or during the coasting year, was −0.19, which fails to reach the 5% level of significance. Height acceleration was not much related to chronological age at treatment either (r = 0.25, P > .05). Absolute velocity during treatment did relate significantly to age, however, with r = −0.54. In the pre-treatment year there was still some positive correlation between velocity and age (r = −0.33) even in these GH-lacking patients.

Sitting height was accelerated by treatment to about the same degree as leg length. Before treatment the sitting height SDS averaged −0.66 ± 0.1 and after it −0.60 ± 0.1. At the end of the post-treatment year the average was again −0.66 ± 0.2 (14 cases only).

Skinfolds were decreased on treatment, as shown in Fig. 8 where the means for the HS and for LBW patients are given. Most of the decrease occurred in the first 3 months of treat-

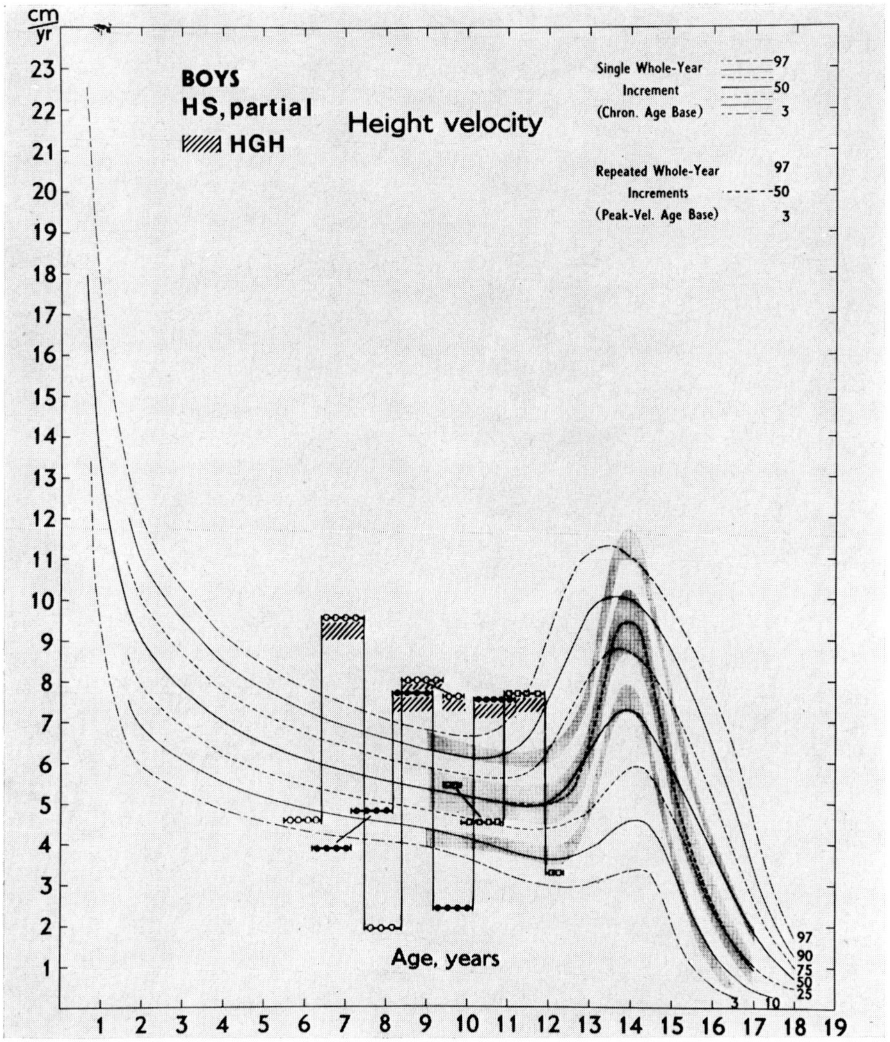

Fig. 10. Height velocity response to HGH in three HS patients, presumed only partially GH deficient. Hatched area signifies treatment with HGH.

ment, and most of the post-treatment rise probably occurred equally quickly. The amount by which the skinfold decreased in the first 6 months was related to the height acceleration. The correlation was 0.41 (24 cases).

Radiographic bone muscle and fat velocities. The means for the HS patients are given in Table II. The increased velocity of bone and muscle on treatment, and the decreased velocity of fat can readily be seen. A better estimate of the changes, however, is given by the 7 patients who were followed in the pure longitudinal way through the 3 years, and it is the means for these which are plotted in Fig. 9.

The changes in muscle width were particularly striking. Examination of the 3-monthly values showed that the increased velocity was greatest in the first 3 months, just as the fat loss

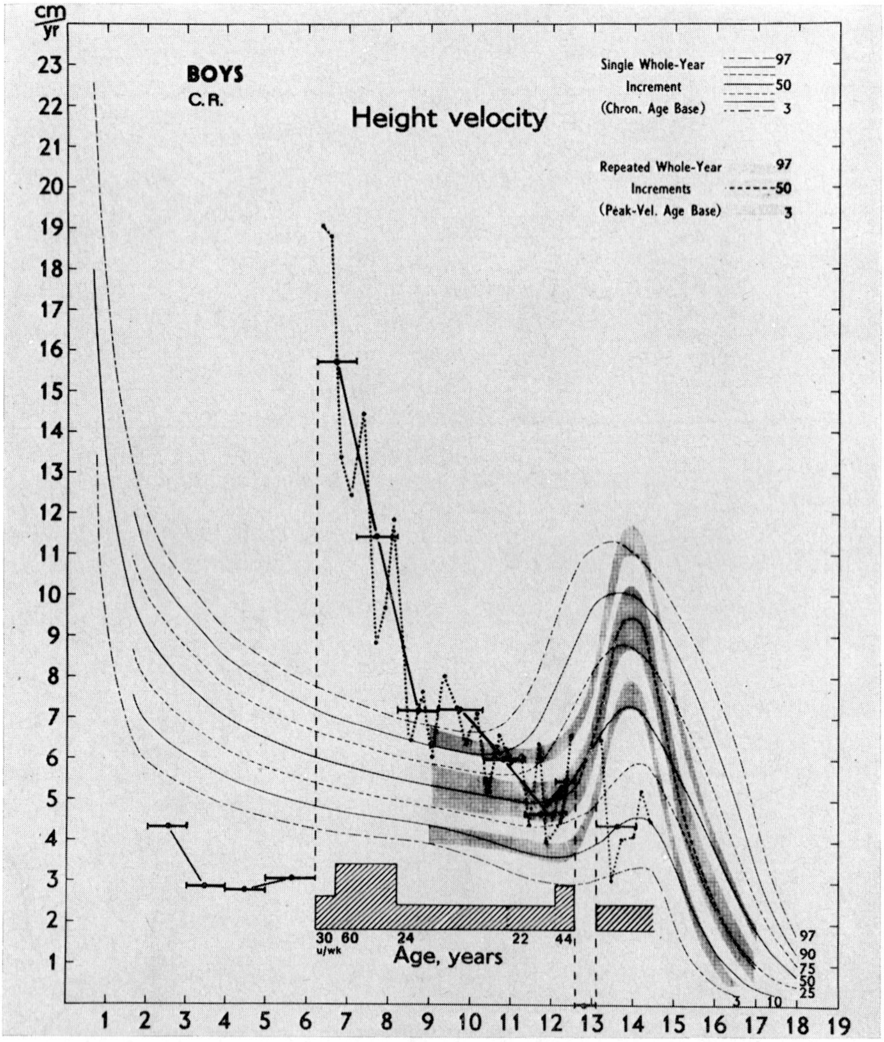

Fig. 11a. Height and height velocity curves of HS patient on long-term treatment. Height plotted at CA, and BA, HGH dose and puberty ratings as indicated. In velocity chart yearly values solid lines, 3-monthly values dotted lines.

was greatest then. Not only did muscle increase proportionately more than bone on treatment, but when treatment ceased muscle width actually diminished, just as fat width did during treatment. Bone, on the other hand, continued to grow in the post-treatment year at about the same rate as before treatment. In the post-treatment year there was a large increase in fat, carrying the absolute fat width to a higher level than it was at the beginning of treatment.

Partial HS deficiency patients

The average height acceleration of the 4 pre-pubescent partial HS patients was 3.2 cm/yr/yr, considerably below the overall HS average of 6.0 cm/yr/yr. However, the partials' response was not out of line with what was predicted from the pre-treatment velocity, which in them averaged 4.5 cm/yr compared with the overall average of 3.2 cm/yr.

The height and height velocity curves of 3 of these patients, Nos. 1.6, 1.12 and 1.18 are

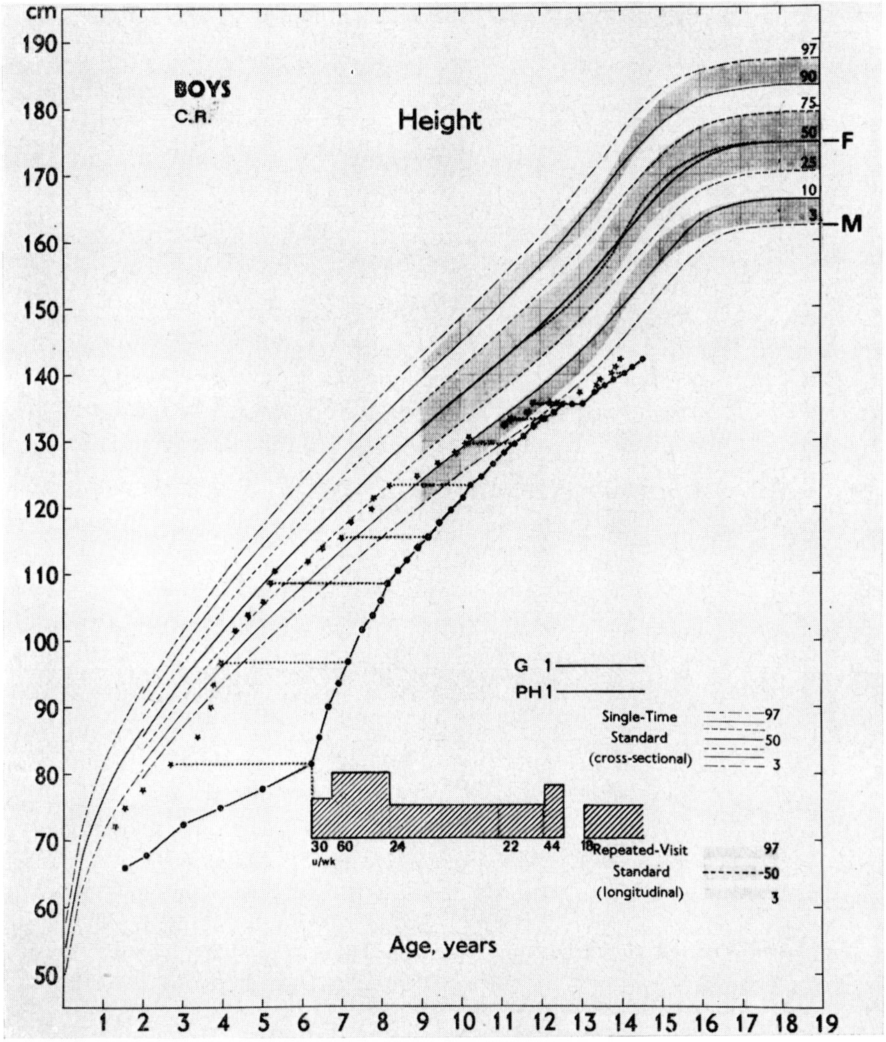

Fig. 11b. See legend to Figure 11a.

plotted. In Fig. 10 the treatment responses are highly significant and large enough to be clinically important. Certainly such patients should have HGH treatment.

HS patients: Long-term treatment

Seven HS patients have been treated continuously, without a coasting year.

The response of our longest-term patient (Fig. 11) is quite typical, though perhaps his very high velocity in the first year was partly due to his doses of HGH being twice as much as that of any other patient. His velocity diminished as he approached the 3rd centile (which is his mother's) despite constancy of the exogenous HGH. When the dose was doubled for 6 months, he did appear to respond somewhat, though this may have been only a seasonal effect. When HGH was stopped, his velocity fell to zero. When restarted, he showed no evidence of a catch-up, his velocity being in line with those of his previous, decelerating, curves. There were no signs of puberty when he was last seen.

The height velocity SDS in each year of treatment is shown in Fig. 12, the coasted and

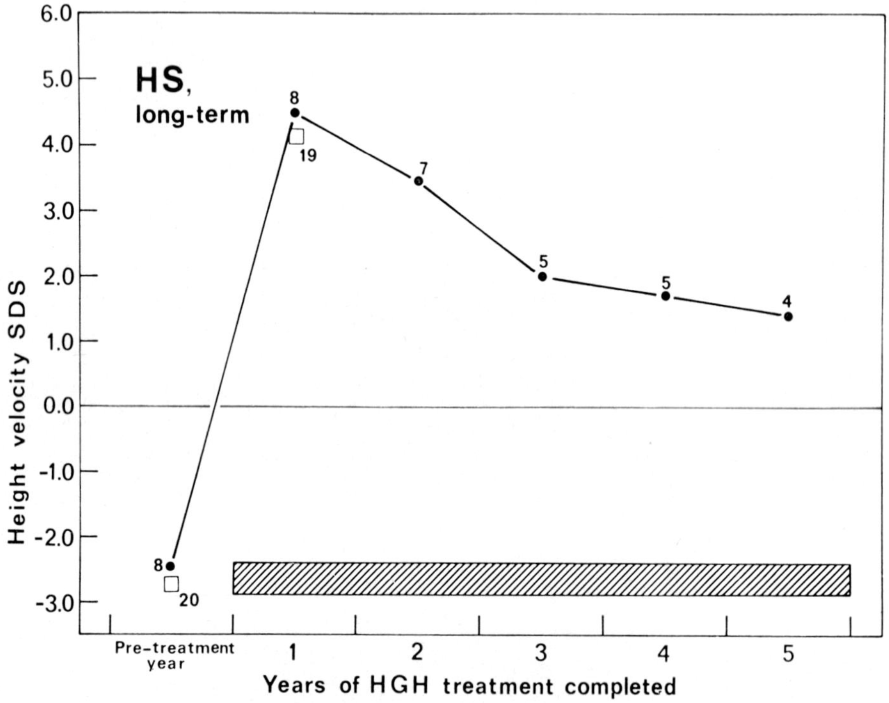

Fig. 12. Height velocity SDS pre-treatment and in first 5 years of treatment. Long-term HS patients. Numbers for each point as indicated are total HS patient group.

the continuously-treated patients being combined. The square points represent the whole HS group. We have no indication that the velocity eventually goes below average, unless antibodies develop or treatment is stopped.

Fig. 13 shows the mean skinfold SDS at the end of each year of treatment. The skinfold reached its minimum level in most cases after six months of treatment (see above); by the end of the first year already some drift back toward the pre-treatment skinfold level was apparent, and this continued in years subsequent to the first. Fig. 13 shows that by the end of 5 years of treatment the skinfolds, on average, were back where they started, at about +1.0 SDS.

The bone age velocity, stimulated in the first year of treatment, likewise drifted back, but by the end of 4 years on average had reached not the pre-treatment figure of 0.9, but the normal average of 1.0. The averages for the first 4 years of treatment were 1.3, 1.3, 1.1 and 1.0 'years' per year.

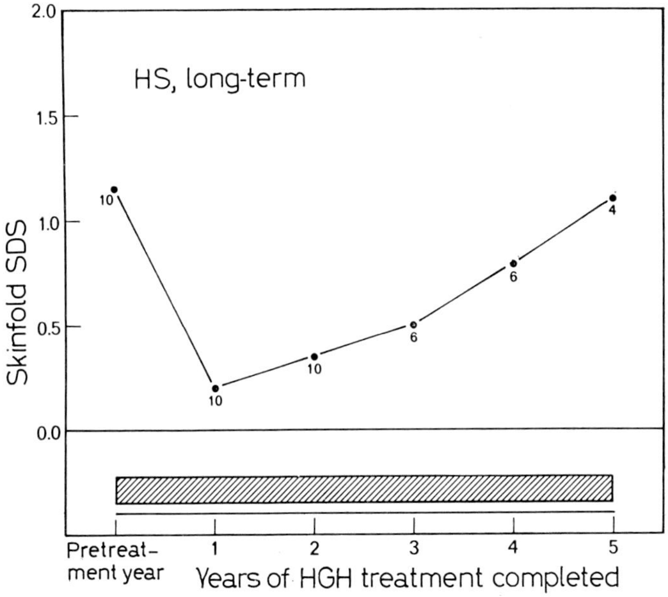

Fig. 13. Skinfold SDS at end of pre-treatment year and first five years of treatment HS patients. Numbers as indicated.

CRANIOPHARYNGIOMAS AND OTHER CNS LESION PATIENTS

In general the response to HGH in height, skinfolds and muscle measurements were similar to those of the HS patients, though less in extent. There were, however, some exceptions.

Height velocity. The pre-treatment velocity of the non-pubescent patients averaged 2.7 ± 0.4 cm/yr, giving a mean acceleration of 3.4 ± 0.4 cm/yr compared with 6.0 cm/yr for the HS patients (who are, of course, younger). Individual patients had accelerations ranging from 1.0 to 5.4 cm/yr. Three were under 2.5 cm/yr.

The treatment acceleration was negatively related to the pre-treatment velocity, just as in the HS patients. In Fig. 7, above, the regression line is for the HS patients only. The open triangles show the cranio patients. All lie below the HS regression line, indicating that their response to HGH is considerably less than that of the HS patients of comparable pre-treatment velocity. The regression line for the cranio patients is nearly parallel to the HS line, and approximately 2 cm/yr below it; the correlations for the two series of cases are similar.

Two of the lowest accelerations (1.0 cm/yr and 2.3 cm/yr) occurred in boys (2.5 and 2.11) whose growth status was not really abnormal. Both were prepubescent; their pre-treatment velocities of 4.3 cm/yr and 3.5 cm/yr were brought up on treatment to 5.3 cm/yr and 5.8 cm/yr, a quite normal pre-adolescent rate.

Of six *pubescent patients* (five girls and one boy) two girls failed to respond at all, for the obvious reason that their period of possible growth was over. Their bone ages were 14.4 and 14.6 'years' (girls approach their final bone age at about 15.5 'years'), their puberty ratings were 3,4 in one case and 5,5 post menarcheal, in the other. Another girl treated at a bone age of 13.7 'years', ratings 3,1 was approaching her pre-spurt height like the boys described above, and responded only slightly.

However, two girls had bone ages of 11.8 'years' and 12.7 'years' with puberty ratings of 3.1 and 3.1 when treatment began, both responded well, with accelerations of 3.2 and 3.6 cm/yr.

Bone age velocity increased only insignificantly on treatment in the pre-pubescent cases, from a mean of 0.6 ± 0.1 to 0.7 ± 0.2 'years'/yr. This marks a contrast with the HS patients.

Limb bone, muscle and fat widths changed comparably with those of the HS patients.

DISCUSSION

HS: diagnosis and aetiology: isolated versus multiple deficiencies

In our clinic, in a paediatric hospital, isolated GH deficiencies outnumber multipel deficiencies. In Goodman *et al.*'s (1968) series, 16 patients were classified as isolated, 7 GH and TSH-deficient, 4 GH- and ACTH-deficient, and 8 GH-, TSH- and ACTH-deficient. ACTH lack was diagnosed on a metopiron test, TSH lack by PBI. In Prader *et al.*'s (1967) series 8 out of 18 patients were given HGH alone, while 10 'needed therapy .... with thyroid preparations in substitution doses' during all or part of the treatment period. But these authors add that this 'simultaneous treatment with thyroid preparations did not enhance growth'. Trygstad (1969) gave thyroid seldom and considered its use should be avoided unless there was a clear indication for it. Evidently different diagnostic criteria are being used for TSH and ACTH deficiency in different centres, and different thresholds are adopted for treatment with thyroid. Goodman *et al.* consider some of their patients developed thyroid deficiency insidiously before and during treatment, but their evidence is by no means conclusive. We have not treated any of our HS patients with thyroxine and their growth responses are as good as or better than others in the literature. Between the isolated and multiple cases neither Goodman nor Prader found any difference in pre-treatment appearance.

We can offer no useful opinion on the question of ACTH deficiency. Possibly as many as a third of the HS patients had it, judged by the test used; but this may not be entirely satisfactory, and these patients responded just as well as the others and have never shown any clinical signs of adrenal deficiency.

A deficiency of gonadotrophins is another matter. At present this cannot be diagnosed till puberty, and in our classification of HS we make no distinction between those who will and those who will not show this deficiency when the time comes. Neither we nor anyone else have reported enough cases of pubertal age to allow a good estimate of the proportion of cases who fail to show spontaneous puberty. Of the 11 HS patients who have so far reached a bone age at which puberty should have started, 7 have developed secondary sex characteristics and 4 have not. (In terms of chronological age puberty may occur late in the HS patient, but that is only to be expected, since maturational age is delayed.)

No less than 13 of our 36 HS boys had an abnormally small penis and ill-developed scrotum, something that has previously been remarked by Laron and Sarel (1970) who found it in an even higher proportion of cases. Goodman *et al.* found it in only 4 out of 22 of their isolated plus multiple deficiencies, and Prader and Trygstad do not remark on its occurrence. We suppose that the maldevelopment is due to lack of androgen stimulation *in utero*, not of GH stimulation, as Laron supposes. Why this should so often occur in GH deficiency is not clear; perhaps there is failure of gonadotrophin secretion also at this time. It will be interesting to see what proportion of our maldeveloped patients have failure of pubertal development; so far one boy with small genitalia has developed a spontaneous puberty and one failed while 4 with normal genitalia have entered puberty and one failed.

The predominance of boys over girls requires an explanation. In our HS cases the ratio of boys to girls was 30/6, in Prader's 11/4, in Trygstad's 14/8 and in Goodman's (isolated plus multiple) 22/12. The average ratio is 67/31, but the variation between series is considerable. No adequate explanation of the difference has been proposed; it seems greater than could be accounted for by a generally higher male susceptibility to foetal and natal traumata.

In our HS patients we have in all 3 pairs of sibs amongst a total of 58 children; thus 10% of our HS patients have affected sibs. Amongst Trygstad's (1969) cases 8 out of 22 (36%) had affected sibs, in Prader *et al.* (1967) 3 out of 15 (20%) were from the same family and Goodman *et al.* (1968) had no affected sibs in 16 cases. The average incidence to date is thus about 15%.

The catch-up response to treatment: pre-treatment and treatment height velocities

The shape of the height velocity curve in response to treatment shows the familiar 'catch-up' phenomenon (Prader, 1963). This is shown in Fig. 6 for the first year, and in Fig. 12 for successive years of treatment. Not only is it true of height response. The fall in skinfolds and the rise in muscle velicity are greater during the first 3 months than subsequently (Figs. 8, 9 and 13), and the skinfolds tend to return to their pre-treatment state after 4 or 5 years of HGH.

One might imagine that the catch-up velocity, or treatment acceleration, would be greater at younger than at older ages. Such is the case, but the correlation is very low ($r = -0.25$). The treatment acceleration is much more closely related to velocity before treatment ($r = 0.58$). Presumably the slowest-growing children have the least capacity to secrete GH as the results of the GH stimulation test indicate.

The cranio patients had a pre-treatment velocity at the level of the 'total' HS patients (2.7 cm/yr, compared with 2.9 cm/yr), which perhaps argues that their doses of cortisone and thyroxine were about right. They showed the same relation between pre-treatment and treatment velocities, but their absolute response was less by an average of 2 cm/yr than that of HS patients of similar pre-treatment velocity. Possibly this reflects limitations of cortisone or thyroxine dosage, or some regulation factor not yet understood. The tissue responses of fat and muscle were also smaller.

Regulatory deceleration after treatment. Children whose growth has been stimulated by these relatively high doses of HGH may sink back in the post-treatment year to levels of height velocity below pre-treatment, or at least to levels below the expected if no treatment had taken place. We have some evidence, not at present conclusive, that this occurs.

Muscle and fat response to HGH. The loss of subcutaneous fat in HS patients given HGH has been described by several authors. The very considerable gain of muscle size, followed by loss when HGH is withdrawn is a newer and striking finding, and confirms the results of Cheek *et al.* (1966) who described an HGH-induced increase in muscle by biopsy in a few HS patients. The gain is expected from much animal work, though the loss is more surprising. It seems that a steady secretion of GH is necessary to maintain muscle at a normal level of development, in the same way that exercise, and in the male, testosterone are required.

We are investigating the possibility of making the radiographic estimate of muscle gain and fat loss into a short-term test for the effectiveness of HGH; it would appear likely that a month of HGH should produce measurable changes.

Prognosis

Our results seem to indicate that in truly isolated GH deficiency HGH brings the patient to within normal height limits at the bone age (more truly, the real developmental point of which bone age is a fallible measure) characteristic of take-off for the adolescent spurt. A normal adolescent spurt then proceeds provided HGH is continued. In cases of added gonadotrophin deficiency, catch-up is the same until the take-off point has been reached (at bone age about 12 in girls and 14 in boys) after which growth proceeds very slowly. Gonadotrophin or sex hormone should then be given, in conjunction with the HGH. A final increase of around 10-12 cm in girls and 15-18 cm in boys may then be expected.

SUMMARY

1. Human growth hormone (HGH) has been given for one whole year or longer to 100 patients, aged 1.5 to 19 years, participating in the Medical Research Council Clinical Trial of HGH. Each patient was measured 3-monthly for a control year before treatment, and the majority for a control year following the first treatment year. All measurements were made by one anthropometrist. Radiographic measurements of widths of bone, muscle and fat in calf and upper arm were made. Methods and standards for assessing the significance of a given height acceleration are given.

2. The characteristics at diagnosis are given of 36 patients with isolated GH deficiency or hyposomatotrophism (18 with craniopharyngiomas and other CNS lesions, and 3 with multiple trophic hormone deficiency).

3. Thirty of the 36 HS patients were boys and 13 had an abnormally small penis and ill-developed scrotum. Only 2 were siblings. Parents averaged 40th centile for height. Four children developed growth-suppressing antibodies, and had to cease treatment. The mean standard deviation score (SDS) for height at diagnosis was −4.7, range −2.6 to −7.3. Bone age SDS averaged −3.2, range −0.8 to −5.7. Skinfold SDS averaged +0.91. Limb muscle width SDS averaged about −3.0. GH peak in insulin hypoglycaemia averaged 4.7 ± 0.7 μU./ml, range 1 to 13.

4. A category of partial growth hormone deficiency patients is defined as those with GH peaks of 7–20 μU./ml inclusive and height velocity SDS in the year before treatment between −1 and −2. Total HS patients have GH peaks of 1 to 6 μU./ml inclusive and height velocity SDS of −2. Partial HS patients are accelerated by HGH and should be treated; but their average acceleration is below that of total HS patients.

5. There is a highly significant relation (r = −.64) between blood GH peak level and pre-treatment height velocity in the HS patients.

6. HGH was given, usually 20 I.U./wk in 2 doses/wk, for between 1 and 7 years to all these patients. The average first year acceleration (treatment year less pre-treatment year velocity) in the HS patients was 6.0 cm/yr (9.1 less 3.1). There was a strong negative correlation of acceleration with pre-treatment velocity, r = −.58. The relation with age was only −.25. In years subsequent to the first the velocity falls, following the familiar catch-up curve. If HGH is stopped after several years of continuous administration, however, zero or near-zero height velocity ensues. Once started, HGH should be given till growth finishes.

7. Skinfolds decreased on treatment and muscle widths increased in the HS and cranio pharyngioma patients. The largest changes were in the first 3 months; after 4 or 5 years treatment skinfolds on average had returned to the high pre-treatment level. When HGH was stopped muscle widths actually decreased. Cranio patients responded like HS patients though to a lesser degree.

ACKNOWLEDGEMENTS

We wish to thank the Medical Research Council for making available the HGH through the Human Pituitary Hormone Subcommittee, Dr. Anne Hartree, Dr. Bangham and Dr. M. Cotes for preparing and supplying the hormone, and Professor Russell Fraser and Dr. Stuart Mason for valuable discussion on procedure. The very exacting work of reducing this great bulk of data to order has been carried out by Mrs. Jacky Willis, to whom we are deeply grateful. For most of the chemical results quoted we are grateful to Professor Barbara Clayton and her staff. Financial support has been provided by the Medical Research Council and the Nuffield Foundation. Lastly we wish to thank all the physicians who allowed us to study their patients.

REFERENCES

CHALKLEY, S. R. and TANNER, J. M. (1971): The incidence and effects on growth of antibodies to human growth hormone. *Arch. Dis. Childh.*, *46*, 160.

CHEEK, D. B., BRASEL, J. A., ELLIOTT, D. and SCOTT, R. (1966): Muscle cell size and number in normal children and in dwarfs (pituitary, cretins and primordial) before and after treatment. *Bull. Johns Hopk. Hosp.*, *119*, 46.

CLAYTON, B. E., TANNER, J. M. and VINCE, F. P. (1971): Short term metabolic response to human growth hormone. *Arch. Dis. Childh.*, *46*, 405.

GOODMAN, H., GRUMBACH, M. M. and KAPLAN, S. L. (1968): Growth and growth hormone. II. A comparison of isolated growth hormone deficiency and multiple pituitary hormone deficiencies in 35 patients with idiopathic hypopituitary dwarfism. *New Engl. J. Med.*, *278*, 57.

HARTREE, A. S. (1966): Separation and partial purification of the protein hormones from human pituitary glands. *Biochem. J.*, *100*, 754.

ILLIG, R., PRADER, A., FERRANDEZ, A. and ZACHMANN, M. (1970): Familiarer pränataler Wachstumshormon-Mangel mit erhöhter Bereitschaft zur Bildung von Wachstumshormon-Antikörpern. *Verh. dtsch. Ges. Endokrinol.*, *16*, 246.

LARON, Z. and SAREL, R. (1970): Penis and testicular size in patients with growth hormone insufficiency. *Acta endocr. (Kbh.)*, *63*, 652.

PRADER, A., TANNER, J. M. and VON HARNACK, G. A. (1963): Catch-up growth. *J. Pediat.*, *62*, 646.

PRADER, A., ZACHMANN, M., POLEY, J. R., ILLIG, R. and SZEKY, J. (1967): Long-term treatment with human growth hormone (Raben) in small doses. *Helv. paed. Acta*, *22*, 423.

TANNER, J. M. (1962): *Growth at Adolescence, 2nd ed.* Blackwell, Oxford.

TANNER, J. M. (1963): Regulation of growth in size in mammals. *Nature (Lond.)*, *199*, 845.

TANNER, J. M. (1964): *The Physique of the Olympic Athlete.* Allen & Unwin, London.

TANNER, J. M. (1969): Growth and endocrinology of the adolescent. In: *Endocrine and Genetic Diseases of Childhood.* Editor: E. L. Gardner. Saunders, Philadelphia-London.

TANNER, J. M. (1972): Physical growth and development. In: *Textbook of Paediatrics.* Editors: J. D. Forfar and G. C. Arneil. Livingstone, Edinburgh. In press.

TANNER, J. M., GOLDSTEIN, H. and WHITEHOUSE, R. H. (1970): Standards for children's height at ages 2 to 9 years, allowing for height of parents. *Arch. Dis. Childh.*, *45*, 755.

TANNER, J. M. and WHITEHOUSE, R. H. (1962): Standards for subcutaneous fat in children. Percentile for thickness of skinfolds over triceps and below scapula. *Brit. med. J.*, *1*, 446.

TANNER, J. M. and WHITEHOUSE, R. H. (1967): The effect of human growth hormone on subcutaneous fat thickness in hyposomatotrophic and panhypopituitary dwarfs. *J. Endocr.*, *39*, 263.

TANNER, J. M., WHITEHOUSE, R. H. and HEALY, M. J. R. (1962): A new system for estimating skeletal maturity from the hand and wrist, with standards derived from a study of 2,600 healthy British children. Centre international de l'Enfance, Paris.

TANNER, J. M., WHITEHOUSE, R. H., HUGHES, P. C. R. and VINCE, F. P. (1971): The effect of human growth hormone treatment on the growth of 100 children with growth hormone deficiency, low birth weight, inherited smallness, Turner syndrome, and other complaints. *Arch. Dis. Childh.*, *46*, 745.

TANNER, J. M., WHITEHOUSE, R. H. and TAKAISHI, M. (1966): Standards from birth to maturity for height, weight, height velocity and weight velocity; British children 1965. *Arch. Dis. Childh.*, *41*, 454.

TRYGSTAD, O. (1969): Human growth hormone and hypopituitary growth retardation. *Acta paediat. scand.*, *58*, 407.

EFFECT OF HGH TREATMENT ON GROWTH, BONE AGE AND SKINFOLD THICKNESS IN 44 CHILDREN WITH GROWTH HORMONE DEFICIENCY

A. PRADER, A. FERRANDEZ, M. ZACHMANN and R. ILLIG

Department of Pediatrics, University of Zürich, Zürich, Switzerland

Patients and methods

Forty-four children with growth hormone (GH) deficiency have been treated with human growth hormone (HGH). Sixteen children had an isolated GH deficiency of the usual type; that is, they had no TSH and ACTH deficiency, although their gonadotropic function was still largely unknown. Six had an isolated GH deficiency of type A as described by Illig *et al.* (1970). Thirteen had combined GH deficiency without tumor and 9 had combined GH deficiency due to craniopharyngioma. Many of the patients with combined GH deficiency have been treated with thyroid and/or cortisol and/or anabolic steroids before and during treatment with HGH. HGH-Raben and HGH-Roos with an average biological activity of approximately 1.3 I.U./mg have been used. The dosage was uniformly 5 mg/sq.m twice weekly. The period of treatment ranged from 9 months to 9 years.

In our experience HGH-Raben and HGH-Roos have the same growth accelerating effect. Furthermore, no differences between our own Roos preparation and the commercially available Roos preparation from the Swedish pharmaceutical firm KABI have been found.

The optimal dosage and whether it should be related to body surface area is arguable. In adults Kowarski *et al.* (1970) have estimated a secretion rate of between less than 500 μg and 1500 μg per day which is about 2 to 3 times less per sq.m per week than our therapeutic dosage. This suggests that a smaller dosage could be used with the same effect. From limited experience the impression is that half our dosage has the same or only a slightly smaller growth promoting effect than the full dosage.

Since the special form of isolated GH deficiency, termed type A, is not widely known, its specific elements should be noted. It is inherited through an autosomal recessivity. At birth the patients have normal weight but are smaller than their healthy siblings. Progressive growth retardation is noted in early infancy and throughout childhood. Dwarfism is therefore more severe than in other cases with isolated GH deficiency. The face is round with a prominent forehead and a small nose with retracted bridge, giving an exaggerated version of the typical face seen in other cases with isolated GH deficiency. The N-retention seen in a standard HGH test and the increase of plasma cortisol seen in a standard insulin tolerance test is greater than in other patients with isolated GH deficiency. The most important biological aspect is the appearance of HGH antibodies in high concentration under HGH treatment which suppresses the growth promoting effect of HGH. The pathogenesis is unknown but it seems reasonable to assume that these children have a hereditary complete deficiency of GH which is effective already before birth and which causes a lack of immune tolerance to homologous HGH (Illig *et al.*, 1970).

Pretreatment values

Table I compares height and bone age with normal values in our four groups of patients before treatment. In order to compare patients of different age the standard score method was used (Churchill, 1966). Height and bone age are given as the difference from the normal mean and the results are not expressed in cm or years but in normal standard deviations (SD). Up to age 12 the normal standards of our Zürich longitudinal growth study could be used (Budliger and Prader, 1971). For height, the standards of Tanner *et al.* (1966) and for bone age, the standards of Greulich and Pyle (1959), were employed.

In isolated GH deficiency using the Zürich standards, height is 5 SD and bone age 2.3 SD below the normal mean (Table I). Thus GH is not only necessary for growth but also

TABLE I

Height and bone age in 44 children with GH deficiency

Type of GH deficiency	Height		Bone age	
	Zürich standard	Tanner standard	Zürich standard	GP* standard
Isolated	–5.0 (13)	–4.4 (16)	–2.3 (13)	–4.2 (16)
Type A	–8.2 (6)	–6.8 (6)	–2.7 (6)	–3.9 (6)
Combined	–4.3 (8)	–4.6 (13)	–2.4 (5)	–5.7 (11)
Craniophar.	–5.8 (2)	–4.0 (8)	–2.3 (2)	–4.4 (7)

Standard score method, number of patients in brackets.
* Greulich-Pyle.

for bone maturation. The discrepancy between the massive growth retardation and the mild bone age retardation indicates a poor prognosis for future adult height. It will be of interest to determine whether the prognosis improves or remains unchanged under HGH treatment and whether treatment beginning early in childhood is better for prognosis than the treatment which begins later. The table also shows that in type A, height is more retarded than in all other types of GH deficiency. Furthermore, it can be seen that in combined GH deficiency, bone age according to the standards of Greulich and Pyle is more retarded than height, whereas it is less retarded than height in isolated GH deficiency. The more pronounced retardation of bone age in combined GH deficiency is probably due to TSH deficiency or to insufficient thyroid replacement therapy. It is not seen in the smaller series of younger patients which were analysed with the Zürich standards.

Effect on growth velocity

Table II gives the growth velocity expressed in cm/year before and during the first year of treatment for all patients, and during the first 3 years of treatment for those who have been under prolonged treatment. In isolated GH deficiency the growth rate increases from 3.4 cm/year before treatment to 9.2 cm/year in the first year of treatment. In the series of Tanner *et al.* (1971) using higher doses the results are very similar; growth rate increases from 3.1 to 9.1 cm/year. After the first year, growth rate decreases slowly and reaches 6.3 cm in the third year of treatment. The abnormally fast growth in the first year, which slows down to a normal growth rate in the following years, gives a growth curve which is typical for catch-up growth and which is seen in many other situations when the cause of growth retardation

TABLE II

Effect of HGH on growth velocity in GH deficiency (cm/year)

	Short-term therapy			Long-term therapy				
	n	before ther.	1st year	n	before ther.	1st year	2nd year	3rd year
Isolated	16	3.4	9.2	4	3.2	9.2	7.4	6.3
Type A	6	3.2	5.2	3	4.0	6.3	2.6	—
Combined	13	3.8	7.3	7	4.0	7.2	5.7	5.4
Craniophar.	9	2.6	7.5	2	1.8	6.6	5.2	6.3

can be corrected or eliminated. In combined GH deficiency the results are similar but not quite as good as in isolated GH deficiency. In type A, height increases much less and falls back to values below the pretreatment level in the second year of treatment because of the development of high antibody titers.

In Table III growth velocity is compared with normal growth velocity as expected for the chronological age of the patients. The comparison of observed to expected velocity is

TABLE III

Effect of HGH on growth velocity in GH deficiency: observed/expected for chronological age

	Short-term therapy			Long-term therapy				
	n	before ther.	1st year	n	before ther.	1st year	2nd year	3rd year
Isolated	16	0.4	1.8	4	0.4	1.5	1.3	1.1
Type A	5	0.9	1.8	2	0.4	0.9	0.3	—
Combined	12	0.7	2.7	4	0.7	1.1	1.1	0.9
Craniophar.	8	0.4	1.6	2	0.3	1.2	0.9	0.9

Velocity standards of Tanner *et al.* (1966).

TABLE IV

Effect of HGH on growth velocity in GH deficiency: observed/expected for bone age

	Short-term therapy			Long-term therapy				
	n	before ther.	1st year	n	before ther.	1st year	2nd year	3rd year
Isolated	4	0.4	1.3	3	—	1.0	0.9	0.8
Type A	6	0.3	0.7	3	0.5	1.0	0.3	—
Combined	7	0.6	1.2	4	0.7	1.3	1.1	0.9
Craniophar.	6	0.4	1.1	2	0.2	1.0	0.9	1.0

Velocity standards of Tanner *et al.* (1966).

made by calculating the ratio of observed velocity to expected velocity using the velocity standards of Tanner *et al.* (1966). This ratio gives the observed velocity in per cent of normal velocity. Before treatment the mean velocity in the four groups varies between 40 and 90% of the normal. During the first year of treatment, it is 180 to 270% and in the third year of treatment it is still 90 to 110%. This is again the typical course of catch-up growth.

In Table IV growth velocity is compared with normal growth velocity as expected from the bone age of the patients. If the results are analysed in this way, the mean velocity before treatment varies in the four groups between 30 and 60% of the normal. During the first year it is, with the exception of type A, 110 to 130% of normal, and in the third year of treatment it is 80 to 100% of normal. These figures show the same trend but are lower than those derived from the comparison with normal velocity for chronological age. The success of HGH therapy is therefore less impressive.

Effect on height, bone age, growth prognosis and skinfold thickness

The next important step in the analysis of growth is to compare the effect of HGH on growth and on bone age. Here the analysis of our group with isolated GH deficiency is the most informative since the results are not influenced by the deficiency of other pituitary hormones or by the problem of antibodies.

In Table V the mean pre- and post-treatment values of height, bone age and skinfold thickness in the isolated GH deficiency group are compared with the normal mean values for chronological age, using the standard score method which expresses the difference from the normal mean in normal standard deviations. Skinfold thickness indicates the amount of subcutaneous fat.

TABLE V

Effect of HGH on height, bone age and skinfold thickness in isolated GH deficiency

	Short-term therapy			Long-term therapy				
	n	start of th.	after 1 year	n	start of th.	after 1 year	after 2 years	after 3 years
Height	13	− 5.0	− 4.0	4	− 5.7	− 4.5	− 3.8	− 3.2
Bone Age	13	− 2.3	− 2.2	4	− 1.6	− 1.5	− 0.7	− 0.4
Skinfolds	12	+ 0.5	− 0.6	5	+ 0.2	− 0.7	− 0.7	—

Standard score method, Zürich standards.

On the left side of Table V the mean values of the whole group before treatment and after the first year of treatment are shown. Before treatment, height is retarded by 5 SD, bone age by 2.3 SD, whereas skinfold thickness is increased by 0.5 SD. These figures compare very well with the series of Tanner *et al.* (1971) who have found −4.7 SD for height, −3.2 SD for bone age and +0.9 SD for skinfold thickness. In the first year of treatment, height catches up by 1 SD, bone age only by 1/10 SD whereas skinfold thickness which is originally above normal mean, decreases by 1 SD. This means that growth is more accelerated than bone age and that the mild obesity which is typical for GH deficiency is disappearing in the first year of treatment. Since growth is faster than bone maturation, the prognosis of the potential future adult height increases also. If future adult height is predicted from chronological height and bone ages using the tables of Bayley and Pinneau (1952) in those patients whose bone age is above 7, an impressive improvement of estimated future adult height under HGH treatment is indeed seen. Such estimations are only of relative value, but they indicate an important trend. It follows that in achieving an optimal adult height, HGH treatment should be started as early as possible, at least in patients with isolated GH deficiency.

The capacity of HGH to accelerate growth more than bone age, at least in patients with GH deficiency, is unique. Testosterone and anabolic steroids, and to a smaller degree also thyroxine, have the opposite effect; they accelerate bone age more than growth, which decreases the potential future adult height. These hormones should therefore not be used in patients with isolated GH deficiency.

On the right side of Table V, the effect of a 3-year treatment period in four patients is shown. Figure 1 gives the same data in graphical form. During the whole period, height and bone age tend towards the normal mean but show characteristic changes in velocity. Height changes most during the first year and less during the second and third year as expected in catch-up growth. In contrast bone age changes in the second and third year more than during the first year. This is an atypical or delayed catch-up growth and tends to reduce part of the first year gain of estimated future adult height. Skinfold thickness remains normal after the first year of treatment and shows no further changes.

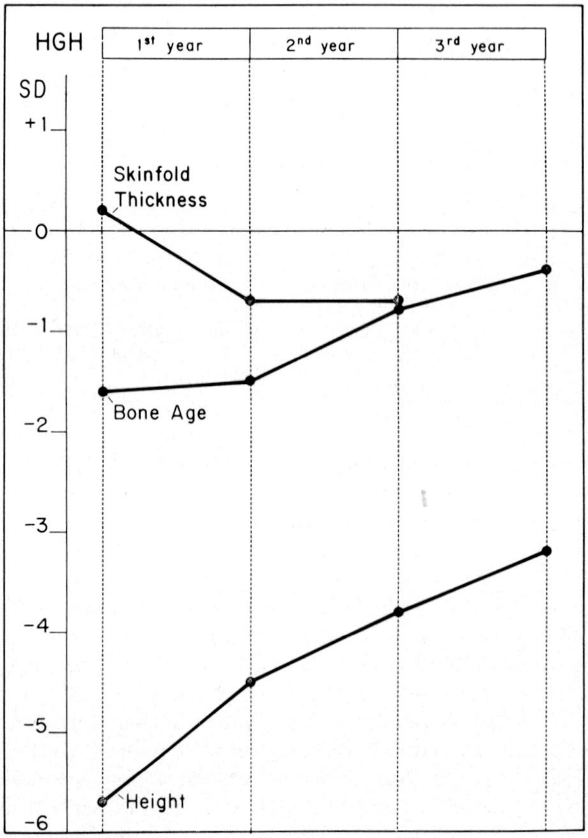

Fig. 1. The effect of HGH in 4 patients with isolated GH deficiency (standard score method, Zürich standards).

SUMMARY

The effect of HGH-Raben and HGH-Roos in a dosage of 5 mg (6.5 I.U.) per sq.m twice weekly has been studied in 44 patients with four different types of GH deficiency (isolated GH deficiency of the usual type, isolated GH deficiency of type A, combined GH deficiency without tumor and isolated GH deficiency with craniopharyngioma). The strongest growth promoting effect is seen in isolated GH deficiency and the weakest in type A, where the therapeutic failure is due to the development of HGH antibodies in high concentrations. The increased growth rate has all the characteristics of catch-up growth with abnormally fast growth in the first year and a reduction to normal growth in the following years. In isolated GH deficiency, the most informative analysis, bone age is less retarded than height and accelerates less than growth. It does not show the typical catch-up characteristics but rather a slow increase in the first year and a stronger increase in the following year. The strong acceleration of growth and the weak acceleration of bone maturation is in contrast to the effect of anabolic steroids and means an improvement, at least theoretically, of expected adult height and indicates that treatment should be started as early as possible. The mild obesity which is typical for GH deficiency disappears at the beginning of treatment.

REFERENCES

BAYLEY, N. and PINNEAU, S. R. (1952): Tables for predicting adult height from skeletal age: revised for use with the Greulich-Pyle hand standards. *J. Pediat.*, *40*, 426.

BUDLIGER, H. and PRADER, A. (1971): Körpermasse und Wachstumsgeschwindigkeit gesunder Kinder in den ersten 12 Jahren. Longitudinale Wachstumsstudie Zürich. In preparation.

CHURCHILL, E. (1966): Statistical considerations. In: *Human Development,* p. 40. Editor: F. Falkner. W. B. Saunders Co., Philadelphia-London.

GREULICH, W. W. and PYLE, S. I. (1959): *Radiographic Atlas of Skeletal Development of the Hand and Wrist, 2nd ed.* Stanford University Press, Stanford, Calif.

ILLIG, R. (1970): Growth hormone antibodies in patients treated with different preparations of human growth hormone. *J. clin. Endocr.*, *31*, 679.

ILLIG, R., PRADER, A., FERRANDEZ, A. and ZACHMANN, M. (1970): Familiärer pränataler Wachstumshormon-Mangel mit erhöhter Bereitschaft zur Bildung von Wachstumshormon-Antikörpern. In: *Endokrinologie der Entwicklung und Reifung, 16. Symposium der Deutschen Gesellschaft für Endokrinologie in Ulm, Februar 1970*, pp. 246–247. Editor: J. Kracht. Springer-Verlag, Berlin-Heidelberg-New York.

KOWARSKI, A., THOMPSON, R. G. and BLIZZARD, R. M. (1970): Integrated concentration of human growth hormone (ICGH) and true secretion rates (SR) in normal and abnormal states. *The Endocrine Society/USA, Abstract No. 155.*

TANNER, J. M., WHITEHOUSE, R. H. and TAKAISHI, M. (1966): Standards from birth to maturity for height, weight, height velocity, and weight velocity: British children, 1965. *Arch. Dis. Childh.*, *41*, 454 (Part I), 613 (Part II).

TANNER, J. M., WHITEHOUSE, R. H., HUGHES, P. C. R. and VINCE, F. (1971): The effects of human growth hormone treatment for 1 to 7 years on growth and body composition in 100 children, with short stature due to GH deficiency, low birth weight, inherited smallness, Turner's syndrome and other complaints. *Arch. Dis. Childh.*, in press.

THE SYNDROME OF FAMILIAL DWARFISM AND HIGH PLASMA IMMUNOREACTIVE HUMAN GROWTH HORMONE (IR-HGH)

Z. LARON, M. KARP, A. PERTZELAN, R. KAULI, R. KERET and M. DORON

Pediatric Metabolic and Endocrine Service, and Rogoff-Wellcome Medical Research Institute, Beilinson Hospital, Petach Tiqva, Tel Aviv University Medical School, Tel Aviv, Israel

In 1966 we described a syndrome of familial dwarfism both clinically and in many laboratory tests indistinguishable from pituitary dwarfism (Laron *et al.*, 1966). There were however abnormally high plasma concentrations of IR-HGH. It was suggested that the circulating GH is biologically inactive in these cases. Additional patients and their family histories were described in a later report (Laron *et al.*, 1968; Pertzelan *et al.*, 1968). More recently we have reported additional clinical and laboratory findings in these and newly discovered patients of Jewish origin (Laron, 1967; Laron and Pertzelan, 1969; Daughaday *et al.*, 1969; Laron and Sarel, 1970; Laron *et al.*, 1971*b*). An increasing awareness of the possibility of this syndrome also led to the finding of several patients who have apparently the same syndrome in other countries.

In the present paper we give a detailed description of the clinical and laboratory studies performed on 26 patients with this syndrome.

PATIENTS

Ethnic origin and genetic aspects

Table I lists the patients who have been studied by us, and mostly followed up for many years. With the possible exception of the family of patient No. 26, all are of Jewish oriental origin. The degree of consanguinity is very high, with 17 patients belonging to 5 family groups. In one instance, the father and two children are affected (patients 15, 16, 17). Analysis of the pedigrees (Figs. 1 and 2) suggests an autosomal recessive mode of inheritance (Pertzelan *et al.*, 1968).

Clinical features

The pertinent perinatal data are shown in Table II. Pregnancy and delivery were normal in all cases. Birth weight was normal with one exception (patient 16). In 7 of the 13 cases in which birth length was known, it was 2 SD below the mean birth length for the respective ethnic group (Zaizov and Laron, 1966). It has not yet been determined whether the incidence of congenital malformations in this syndrome is greater than that in the general communities.

Motor development and skeletal maturation was in general slow (Laron and Pertzelan, 1969; Laron *et al.*, 1971*a*). Many sat up only after the age of one year, and started walking at 1.5 years or more. Teething was late in onset and fontanelle closure occurred between 3 and 7 years. The typical features, the most important of which are shown in Fig. 3, are evident from early childhood. They are indistinguishable from those in children with familial isolated lack of HGH. One of the earliest notable features is a small face and mandible

TABLE I

Pertinent data on the patients' families

No.	Name	Sex	Family group	Country of origin	Hered-itary	Number of normal siblings	Consanguinity
1	R.S.	F	I	Iraq	+	–	1 and 2 are sisters. Parents are first
2	Z.S.	F	I	Iraq	+	–	cousins
3	J.S.	F	I	Iraq	+	–	Parents are second cousins
4	S.B.	F	I	Iraq	+	–	4 and 5 are siblings. Parents are dis-
5	R.B.	M	I	Iraq	+	–	tant relatives
6	Sh.S.	M	II	Yemen	+	6	6, 7 and 8 are siblings. Parents are
7	R.S.	F	II	Yemen	+		third cousins
8	S.S.	M	II	Yemen	+		
9	R.J.	F	II	Yemen	+	4	9 and 10 are siblings. Father is half-
10	G.J.	M	II	Yemen	+		brother of mother's grandmother
11	E.C.	F	III	Afghanistan	+	5	11 and 12 are sisters. Parents are first
12	R.C.	F	III	Afghanistan	+		cousins. Grandparents are first cousins
13	Y.L.	M	IV	Iraq	+	–	13 and 14 are siblings. Parents are
14	M.L.	F	IV	Iraq	+	–	first cousins
15	S.M.	M	V	Iraq	+		Father of 16 and 17
16	A.M.	F	V	Iraq	+	1	16 and 17 are siblings, offspring of 15
17	R.M.	M	V	Iraq	+		Normal sib. is from another mother
18	H.A.	F	VI	Algeria	+	4	Parents are first cousins
19	E.A.	M	VII	Iraq	+	5	Parents are first cousins
20	M.D.	M	VIII	Iran	+	2	Parents are first cousins
21	S.E.	M	IX	Iran	+	8	Parents are first cousins
22	S.C.	F	X	Iran	—	2	
23	P.M.	F	XI	Iran	—	3	
24	M.E.	M	XII	Iran	+	1	Mother is father's niece
25	F.N.	F	XIII	Iran	+	3	Parents related: father and maternal grandmother are first cousins
26	N.Z.	F	XIV	S. America			

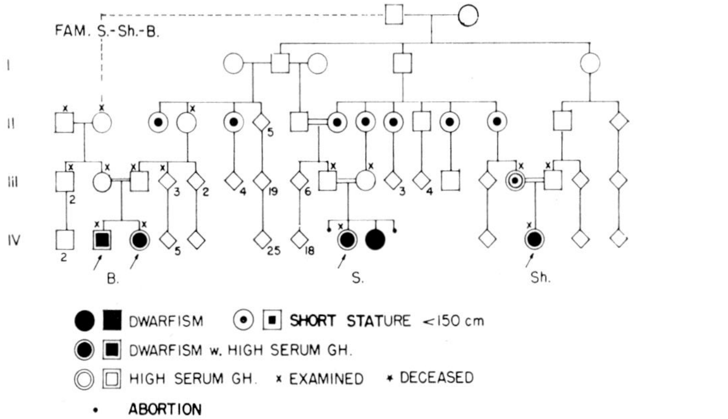

Fig. 1. Genetic pituitary dwarfism with high serum growth hormone. Pedigree of three related families (Fam. Gr. I; patients 1–5). (From Pertzelan *et al.*, 1968, *Israel J. med. Sci.*, *4*, 595, by kind permission of the editors).

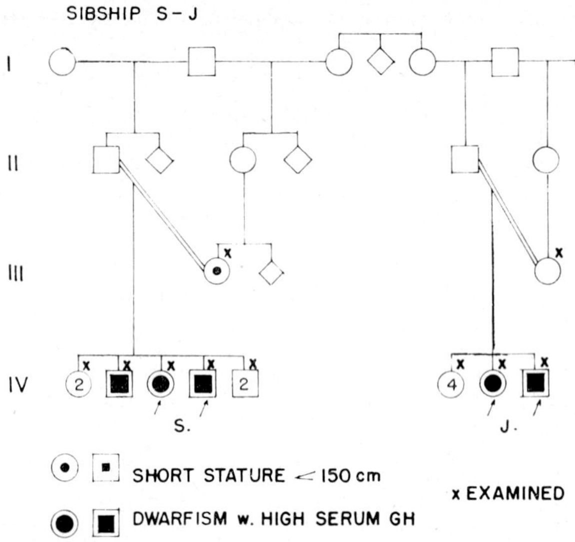

Fig. 2. Genetic pituitary dwarfism with high serum growth hormone. Pedigree of two related families (Fam. Gr. II; patients 6–10). (From Pertzelan *et al.*, 1968, *Israel J. med. Sci.*, *4*, 595, by kind permission of the editors).

TABLE II

Perinatal data of patients with the syndrome of familial dwarfism and high plasma IR-HGH

Pt. No.	Family group	Birth weight (g)	Birth length (cm)	Congenital malformations
1	I	3500		Short fourth metatarsus (bilateral)
2	I	3400		
3	I	3400	46*	Congenital dislocation of hip joints
4	I	3200	49	Brachymesophalanx II, V, left hand
5	I	3740	47	Suspected congenital heart disease (aortic stenosis)
6	II	3200		Clinodactyly V (hands and feet). Feet V fingers over IV.
7	II			Congenital strabismus, left
8	II	3150	47	Clinodactyly V (hands)
9	II	2900	46*	
10	II	2700	45*	Congenital dislocation of hip joints
11	III			
12	III			Congenital(?) cataract
13	IV	3000	49	Nystagmus congenital. Partial syndactyly II–III (feet). Congenita strabismus
14	IV	3100	46*	Congenital strabismus. Partial syndactyly II–III (feet)
15	V			
16	V	2100		
17	V	3200	44*	
18	VI	3050		
19	VII	2600		
20	VIII	3250	46*	
21	IX	3240	45*	
22	X			
23	XI	3200	47	Harelip
24	XII	3500		
25	XIII	3200	50	
26	XIV	3000		

* 2 SD below the mean birth length for respective ethnic community.

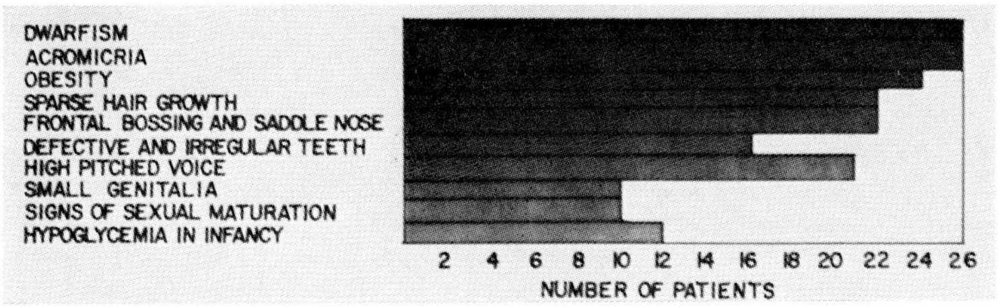

Fig. 3. Main clinical signs and symptoms of the syndrome of familial dwarfism and high plasma IR-HGH.

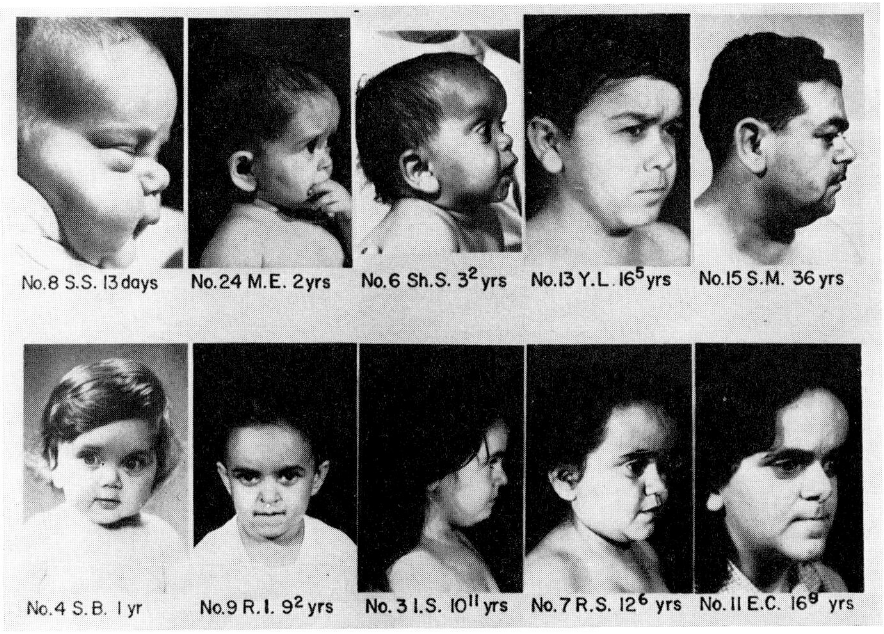

Fig. 4. Typical appearance of face and head in the syndrome of familial dwarfism and high plasma IR-HGH. Note small face, sparse hair, saddle nose.

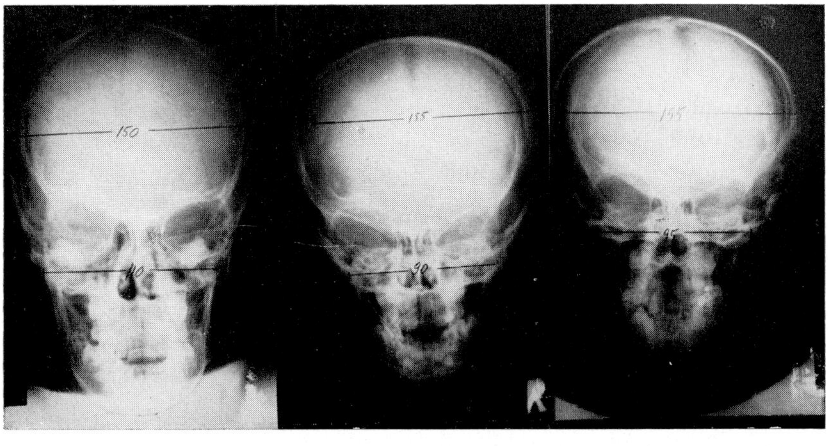

Fig. 5. Antero-posterior X-ray of three 10-year-old children. From left to right: normal healthy child, isolated HGH lack and familial dwarfism with high plasma IR-HGH (patient 6). Upper line = biparietal diameter; lower line = bicondylar diameter.

TABLE III

The bicondylar/biparietal diameter ratio in normal children, and various pituitary disorders

Age group (years)	Sex	Normal controls n		Dwarfism with high plasma IR-HGH n		Isolated HGH lack n	
0 – 3	M	18	66.3 ± 1.2	2	50.5 ± 2.1*		
	F	14	66.9 ± 1.6			1	59
3 – 8	M	16	70.7 ± 2.6	2	52.5 ± 0.2*	4	59.2 ± 1.6*
	F	12	70.5 ± 2.3	3	55.5 ± 2.9*	3	62.6 ± 3.4*
8 – 20	M	22	73.6 ± 3.9	5	63.8 ± 3.0*	5	60.3 ± 2.8*
	F	18	75.4 ± 2.5	9	59.7 ± 2.1*	7	65.0 ± 1.6*

* Statistically significant difference compared with respective control group.

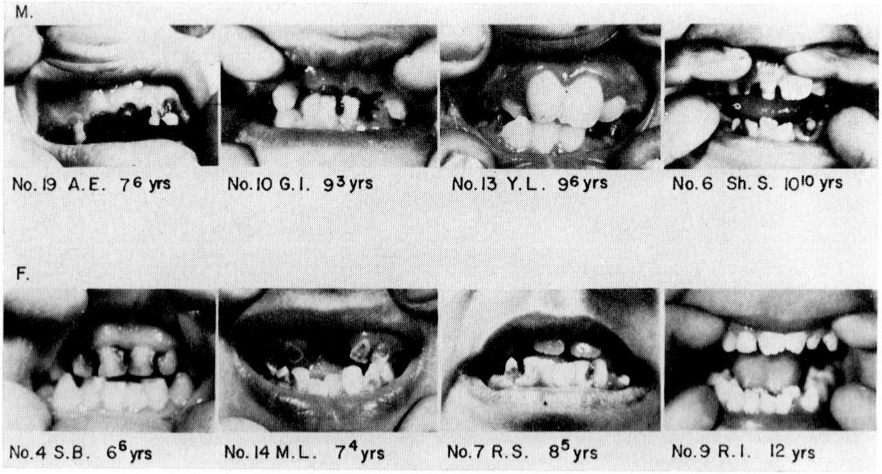

Fig. 6. Appearance of teeth in the syndrome of familial dwarfism and high plasma IR-HGH. Note defects, discoloration and crowding due to small facial bones.

giving the false impression of a large head. The disproportion between face and calvarium results in a saddle nose and the 'sign of the sunset' (Fig. 4). Hair growth is slow and sparse with deep temporal recessions. The small face is due to underdevelopment of the skull base, as seen in the antero-posterior X-ray of the skull (Fig. 5), and the ratio between the bicondylar to biparietal diameters (Scharf and Laron, 1971). This ratio is significantly smaller in these patients than in healthy controls, and is comparable to the ratio in familial isolated HGH deficiency (Table III). Teething is slow, with discolored and defective teeth (Fig. 6*) and, due to the small mandible, crowding develops. The children have a high pitched voice. From early infancy the children grow slowly, their height being below the third percentile (Figs.

* Figures 6, 9, 10, 11, 12, 13, 14, 15, 16, 17, 18, 19, 20, 21, 22, 28, 29, 30 and 31 will be published in a monograph on the syndrome of familial dwarfism and high plasma IR-HGH.

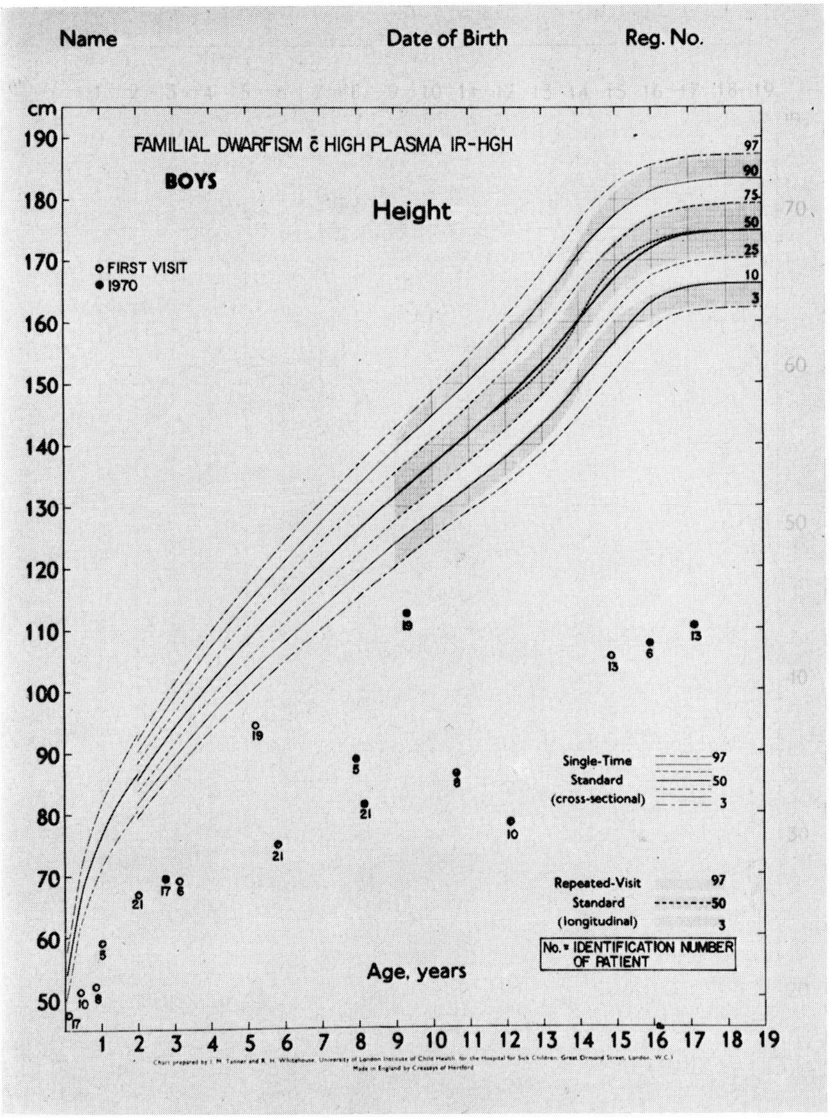

Fig. 7. Linear height of boys with the syndrome of familial dwarfism and high plasma IR-HGH. Patient 15, an adult, is not included.

7 and 8). It would seem that some girls are slightly taller than the boys. This is not so when the height of the parents and normal adult siblings is compared (Fig. 9); there are more mothers and sisters below the third percentile than there are fathers and brothers.

Aside from the short body, the hands and feet are small (acromicria). The body proportions are infantile with a mean upper/lower segment ratio of 1.53 compared to HGH lack: 1.34. In normal boys over 10 years this ratio is 1 or below (Rimoin *et al.*, 1968).

The subcutaneous fat tissue, measured by skinfold thickness, is well developed from early infancy and the patients are obese at all ages (Fig. 10). The fat is mainly truncal and

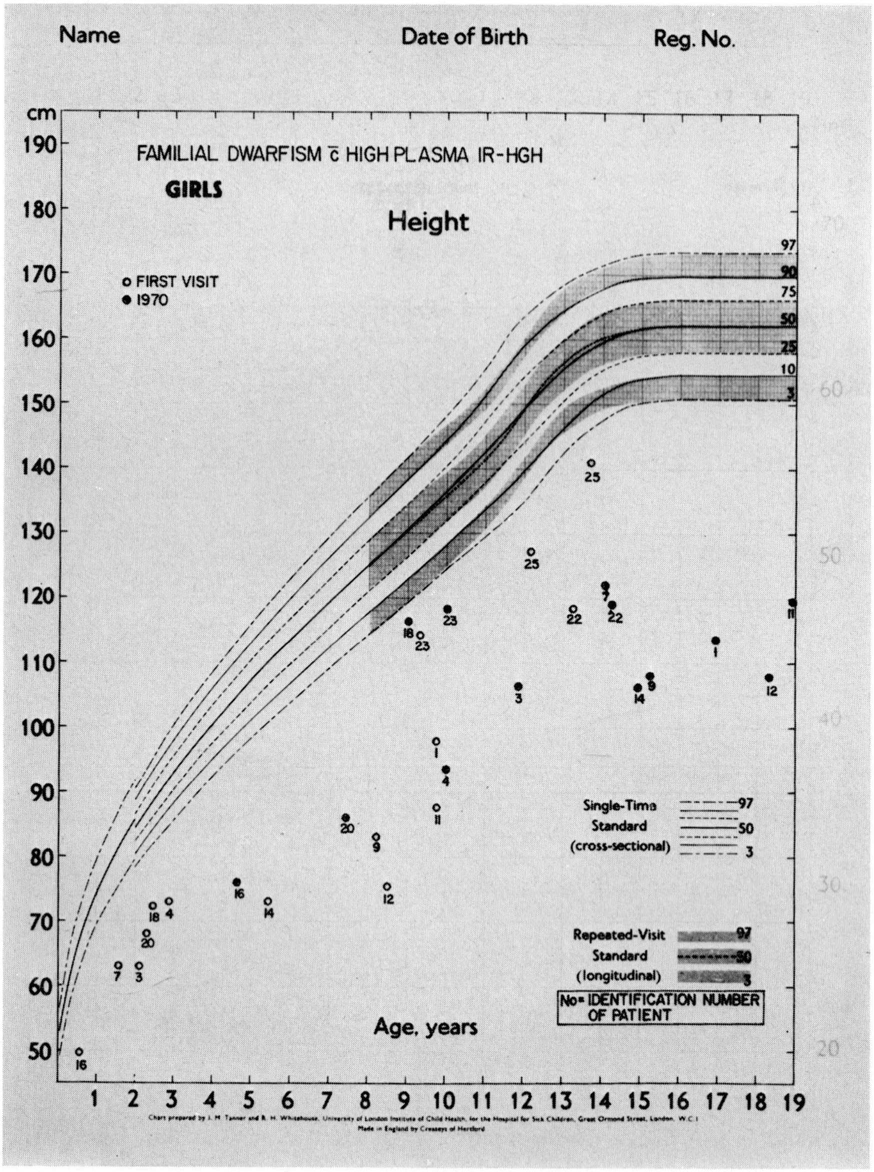

Fig. 8. Linear height of girls with the syndrome of familial dwarfism and high plasma IR-HGH.

the degree of obesity is not always evident by calculating the weight age as the bones are very thin and the muscle tissue underdeveloped. The genitalia are small from early childhood (Fig. 11), a finding which is more easily seen in the boys; nevertheless, the patients do undergo sexual maturation (Fig. 12), seen so far in 10 of our patients, 6 females and 4 males, one of whom (patient 15) is an adult and the father of patients 16 and 17. The boys seem to mature more slowly than the girls.

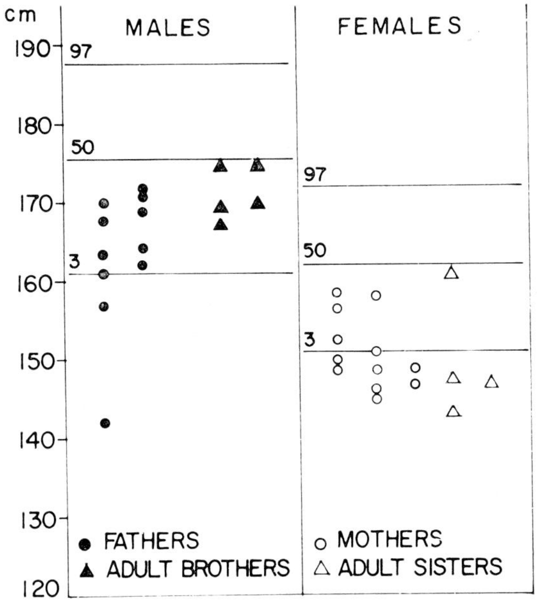

Fig. 9. Linear height of the parents and normal adult siblings of the patients with the syndrome of familial dwarfism and high plasma IR-HGH. (The shortest father is patient 15).

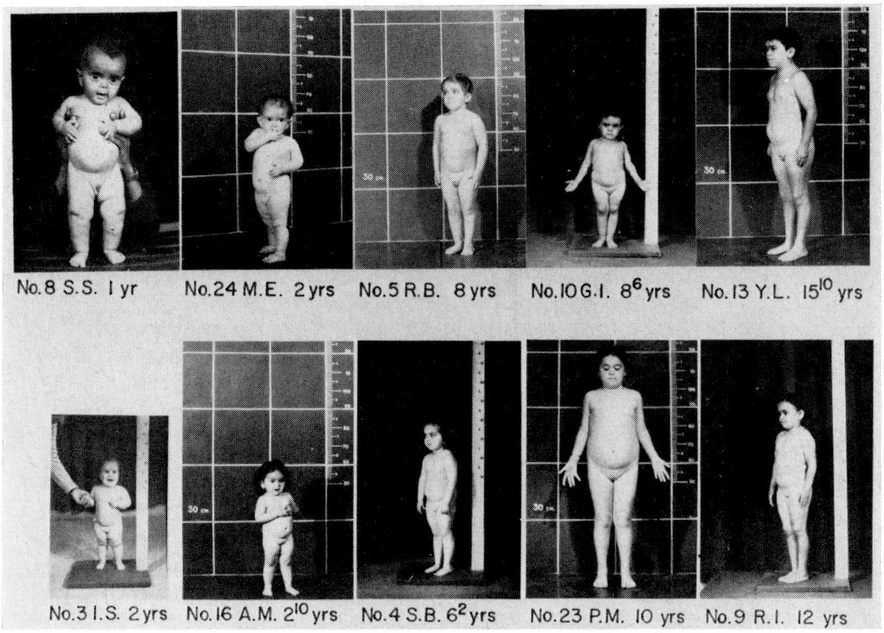

Fig. 10. Obesity in patients with the syndrome of familial dwarfism and high plasma IR-HGH.

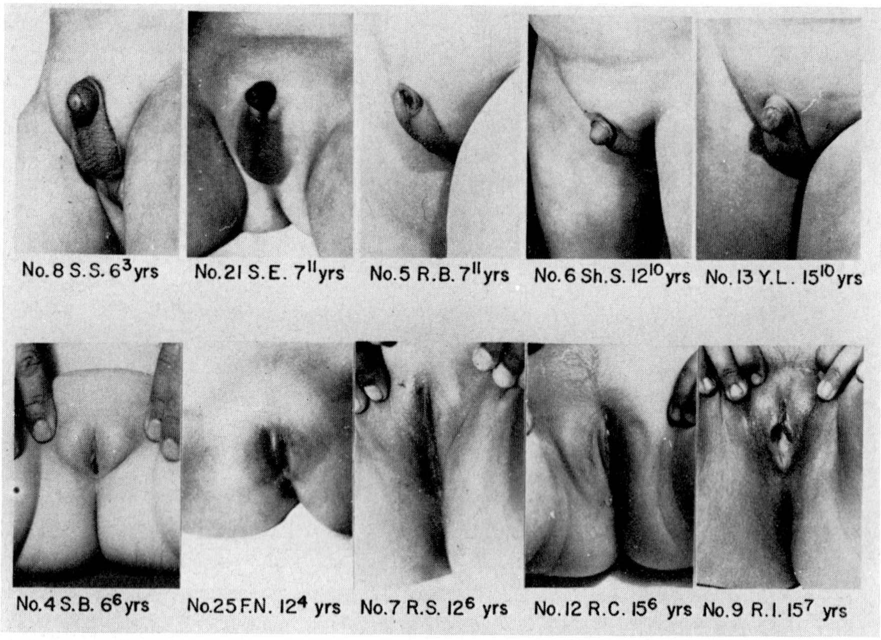

Fig. 11. Genital and male gonadal size in patients with the syndrome of familial dwarfism and high plasma IR-HGH.

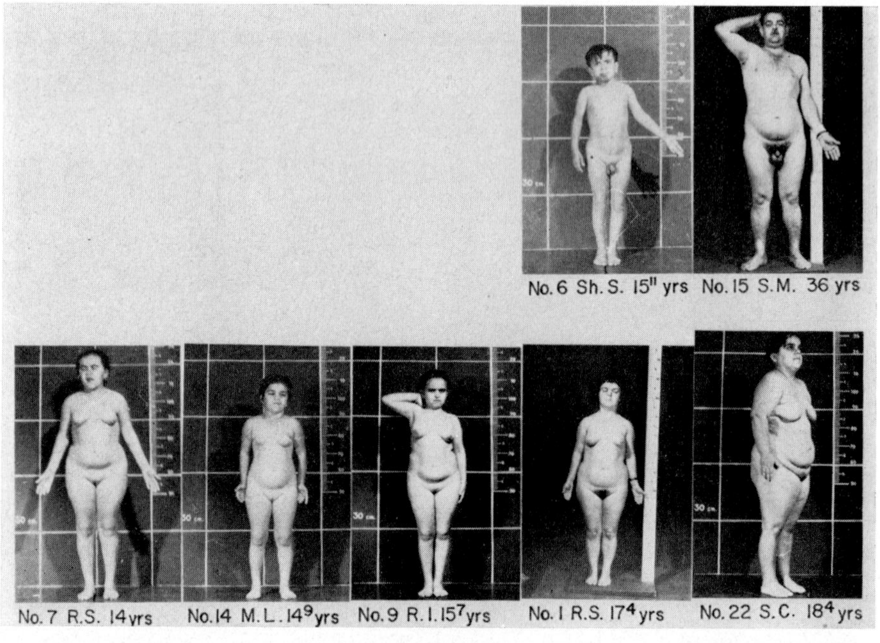

Fig. 12. Sexual maturation in patients with the syndrome of familial dwarfism and high plasma IR-HGH.

History of hypoglycemia

Twelve of the patients gave a history of hypoglycemic spells in infancy.

Skeletal maturation

In all patients with the exception of the one adult patient, bone age was retarded until the late stages of puberty.

Mental development

In 17 of the 26 patients psychological evaluations were carried out. With the exception of 3, all showed some retardation and deficiency in visuomotor functioning (Frankel and Laron, 1968; Laron *et al.*, 1970).

LABORATORY INVESTIGATIONS

General blood and urine examinations with the exception of those discussed below, were normal, as were the assessments of thyroid, adrenal, ACTH, TSH, ADH and gonadotrophin functions.

Fasting plasma HGH

Repeated blood examinations in all the patients revealed HGH concentrations in the acromegalic range (Fig. 13). However, in these same patients there were fluctuations unrelated to age and these have yet to be explained.

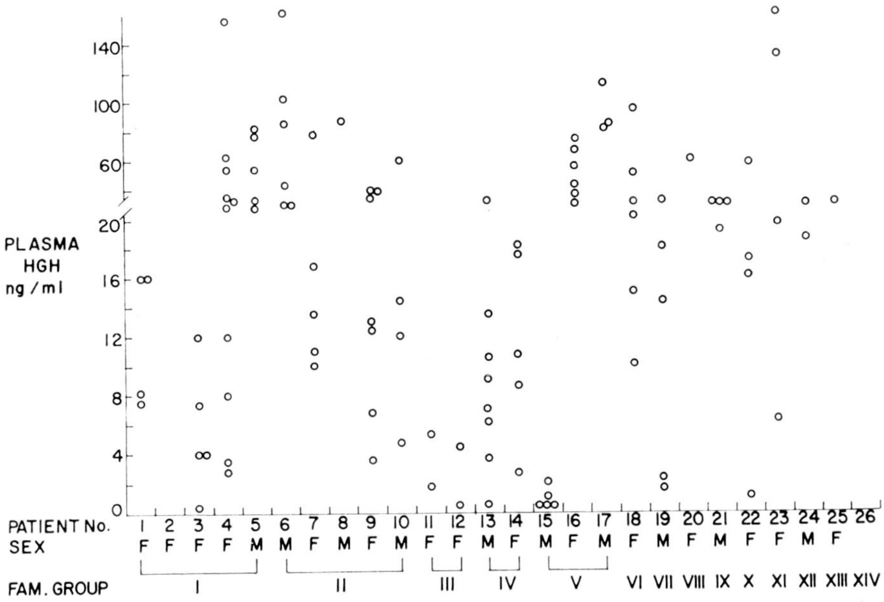

Fig. 13. Fasting plasma HGH in patients with the syndrome of dwarfism with high plasma IR-HGH.

Fasting blood sugar

Blood sugar was measured as total reducing substances. It can be seen that most patients had low blood sugar levels at some time (Fig. 14). The younger the patient the more frequent the incidence of low sugar levels.

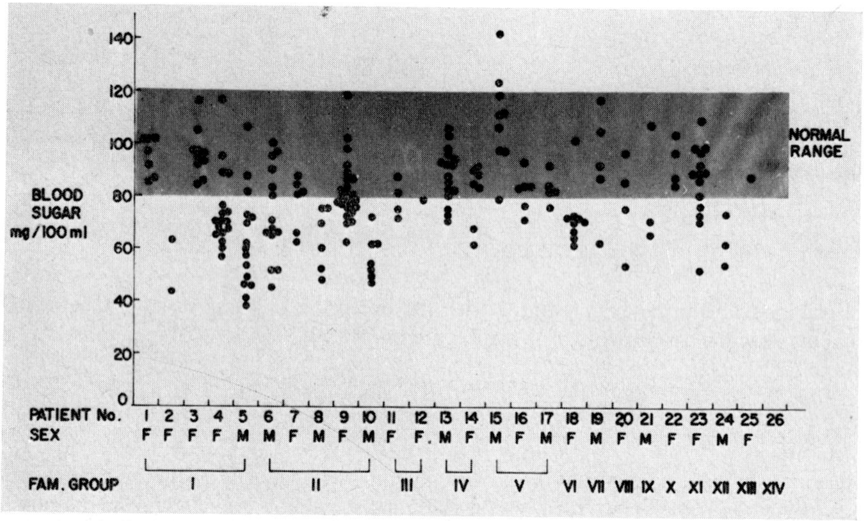

Fig. 14. Fasting blood sugar in patients with the syndrome of dwarfism with high plasma IR-HGH.

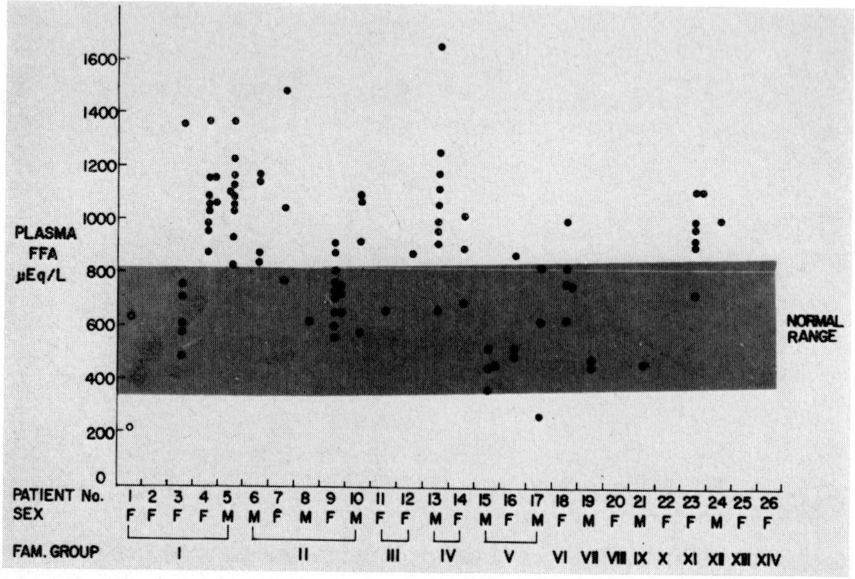

Fig. 15. Fasting plasma FFA in patients with the syndrome of dwarfism with high plasma IR-HGH.

Fasting free fatty acids (FFA) and fasting plasma cortisol (11-OHCS)

In about half of the cases plasma FFA (Fig. 15) and 11-OHCS (Fig. 16) were higher than those in the normal range.

Insulin tolerance test

The patient population was divided into three groups according to age. Below the age of 6 the fasting blood sugar was significantly lower than in older children (Fig. 17, Table IV).

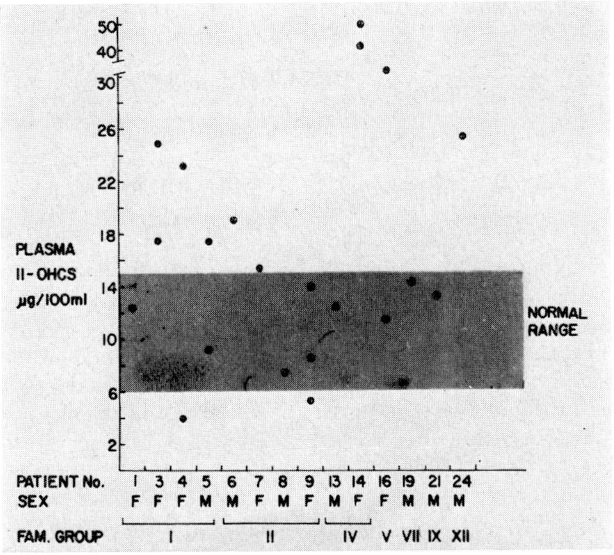

Fig. 16. Fasting plasma 11-OHCS values in patients with the syndrome of dwarfism with high plasma IR-HGH.

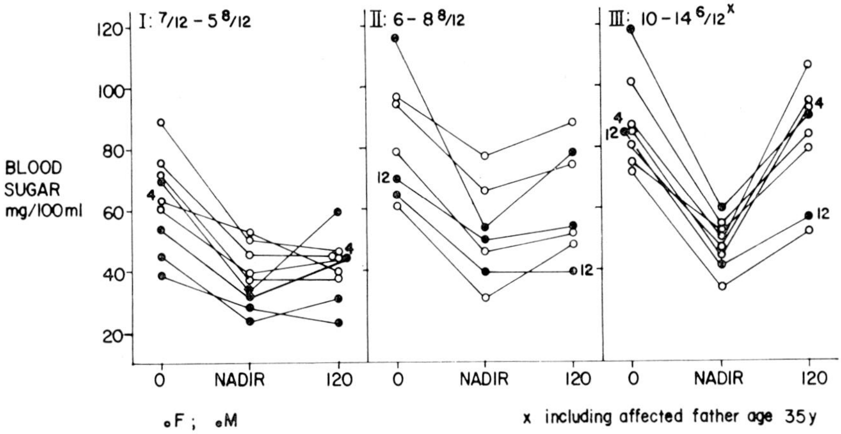

Fig. 17. Insulin tolerance test in dwarfism with high plasma IR-HGH. Age groups in years.

TABLE IV

Blood sugar (TRS) response in insulin tolerance test in patients with the syndrome of familial dwarfism and high plasma IR-HGH

Groups Age (yr)	Gr. I (n = 9) 7/12 – 5 11/12		Gr. II (n = 7) 6 – 8 8/12		Gr. III (n = 8) > 10	
Time (minutes)	TRS mg% mean ±SD	% of 0′	TRS mg% mean ±SD	% of 0′	TRS mg% mean ±SD	% of 0′
0′ (fasting)	63 ± 17		83 ± 21		87 ± 16	
Nadir (15′–30′)	38 ± 9	39 ± 12	52 ± 17	38 ± 13	47 ± 9	45 ± 8
120′ (recovery)	43 ± 12	68 ± 16	61 ± 18	73 ± 13	82 ± 19	94 ± 20

Statistical evaluation: TRS *% fasting value*

Fasting Gr. I *vs* Gr. III $t = 2.9$ $p < 0.02$
Recovery Gr. I *vs* Gr. II $t = 2.28$ $p < 0.05$ – NS
Gr. I *vs* Gr. III $t = 4.9$ $p < 0.001$ – $t = 5.6$ $p < 0.001$
Gr. II *vs* Gr. III $t = 2.19$ $p = 0.05$ – $t = 2.78$ $p < 0.02$

The younger age group showed hypoglycemia nonresponsiveness, which progressively developed into hypoglycemia responsiveness over age 10 years. These changes were proven in the same child by retesting (Fig. 17, see patients 4 and 12).

Plasma HGH response to insulin hypoglycemia was measured in 21 instances (Fig. 18). In 13 patients, concentrations showed a higher than normal rise; in 3 patients the rise was within the normal range, in 2 patients there was no change in the elevated values, and in 3 patients there was a paradoxic decrease.

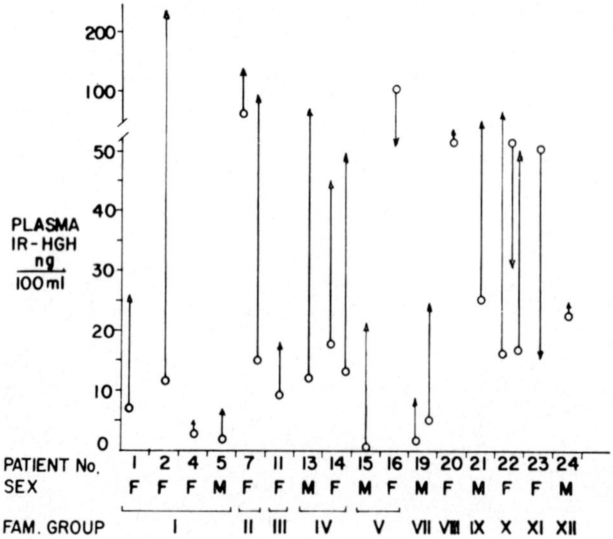

Fig. 18. Plasma HGH response to insulin hypoglycemia in patients with the syndrome of dwarfism with high plasma IR-HGH.

Intravenous arginine stimulation test

Infusion of arginine (Laron, 1969) resulted in a rise of plasma HGH in all cases with the exception of one adult (patient 15) (Fig. 19). It is noteworthy that despite the many high fasting values there was no paradoxic response, *e.g.* a decrease in plasma HGH like that often observed in normal females who start with an elevated HGH concentration. On the other hand we saw 3 paradoxic decreases during insulin hypoglycemia, a finding rarely observed. The many low values of plasma insulin obtained in response to arginine infusion are a characteristic finding with this syndrome. Figure 20 illustrates that almost all peak insulin values are below the mean for normal controls and most of them are well within the range for juvenile diabetics.

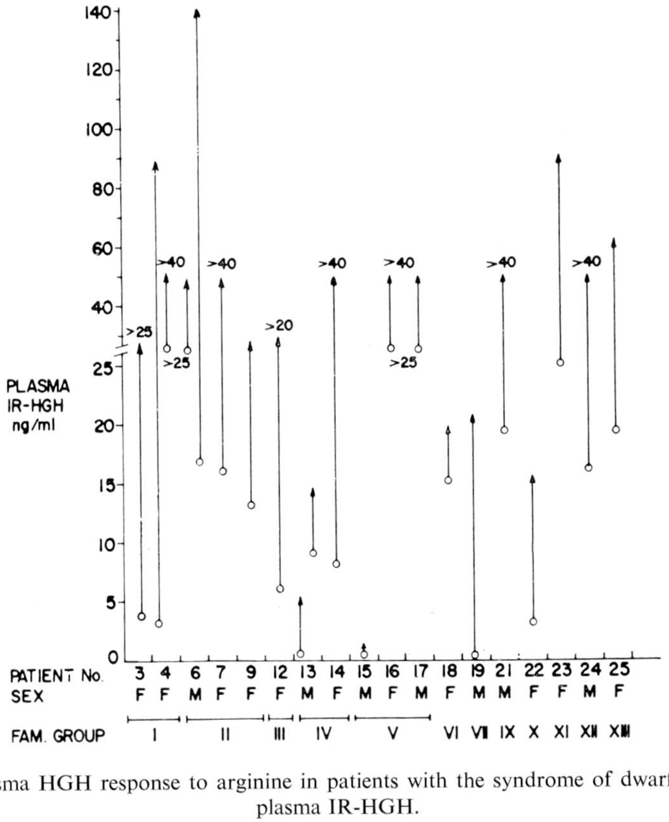

Fig. 19. Plasma HGH response to arginine in patients with the syndrome of dwarfism with high plasma IR-HGH.

Oral glucose tolerance test (OGTT)

Twenty-two complete tests were performed in 15 patients. Six patients showed glucose intolerance (Table V) including the only adult of the group. The plasma insulin response was insufficient in 2 instances (patients 4 and 10), borderline in 3 (patients 12, 13 (1 of 2 tests) and 23), and of a hyperinsulinemia type in 3 patients (Fig. 21). There was no overall correlation between the insulin rise after OGTT and arginine. This may be explained by a difference in the stimulation mechanisms involved or by the basic defects of this disease.

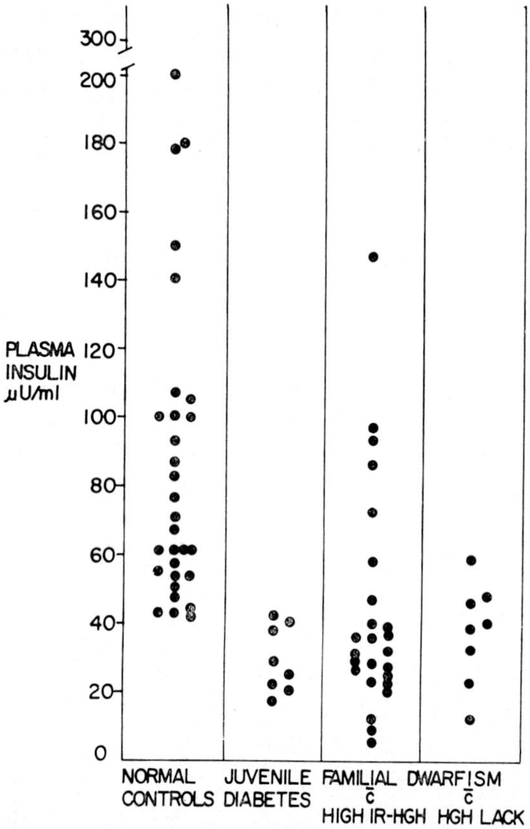

Fig. 20. Peak insulin during arginine stimulation test.

TABLE V

Glucose intolerance during oral glucose tolerance test (OGTT) in 6 patients with the syndrome of dwarfism and high IR-HGH

Family group	Patient No.	Age (yr)	Sex	Blood sugar (mg/100 ml)						
				0	30	60	90 minutes	120	150	180
I	5	8	M	88	143	166	134	122	128	130
II	9	15	F	75	148	188	148	127	116	70
II	10	10	M	46	130	180	185	174	72	42
III	11	19	F	88	128	174	159	142	120	112
III	12	18	F	100	136	174	181	144	121	122
V	15	40	M	97	174	174	255	237	218	200

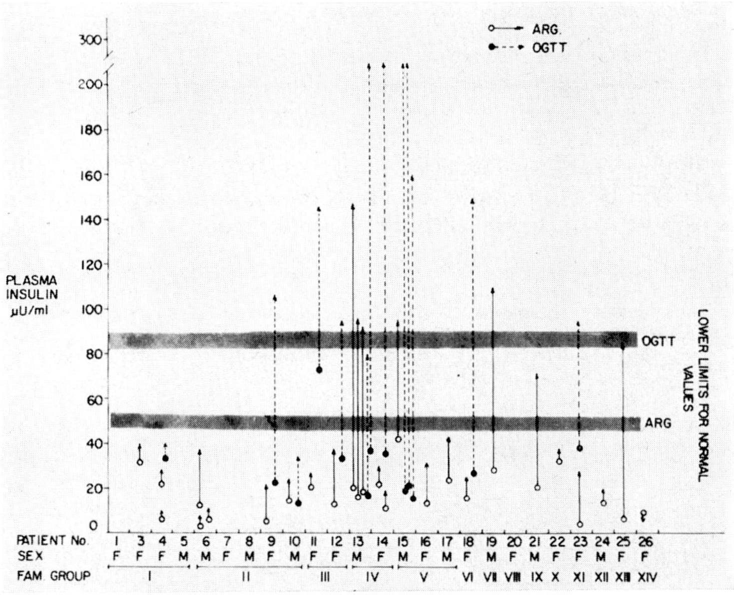

Fig. 21. Plasma insulin response to arginine (○) and OGTT (●) in patients with the syndrome of familial dwarfism with high plasma IR-HGH.

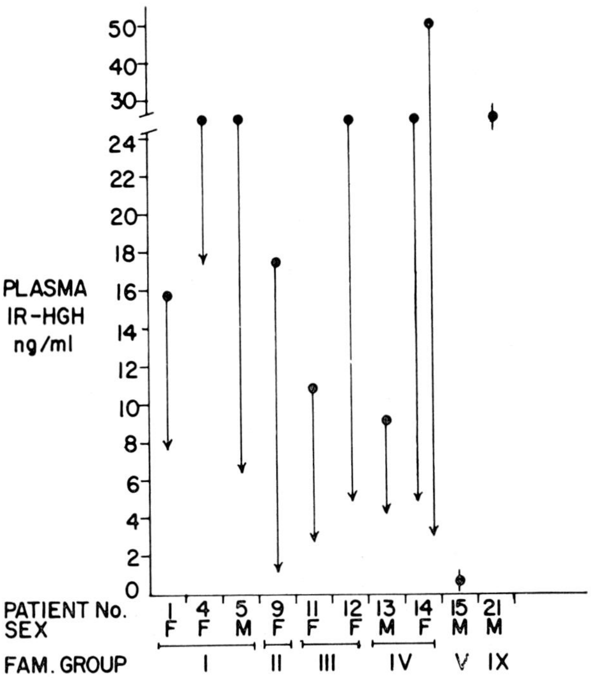

Fig. 22. Plasma growth hormone response to OGTT in patients with the syndrome of dwarfism with high plasma IR-HGH.

Growth hormone suppression tests

In 11 instances plasma IR-HGH was measured during OGTT (Fig. 22). In 9 instances there was a significant suppression; in 1 (patient 15) the starting value was low; and in patient 21 there was no change.

The peroral administration of 6α-methylprednisolone (Medrol®) in a dose of 12 mg/day had some effect in 2 children and caused a paradoxic rise in 3 (Fig. 23). From the comparative data obtained in newborns and active acromegaly, it seems that large amounts of corticosteroids are required in all these conditions in order to suppress GH secretion.

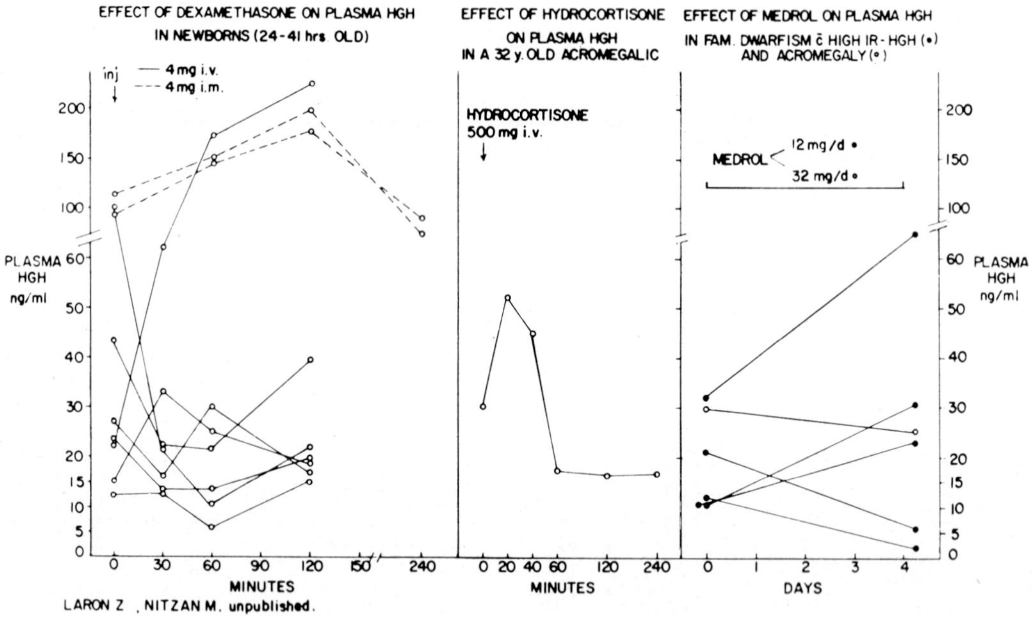

Fig. 23. Effect of dexamethasone, hydrocortisone, and medrol on plasma HGH.

RESPONSE TO EXOGENOUS HGH

Effect of one i.v. dose of HGH on plasma FFA

Intravenous injection of HGH causes, in hypopituitary subjects lacking HGH, a marked rise in the plasma FFA with a peak 3-4 hours after administration (Table VI). Half of the patients with high IR-HGH had a response equivalent to the patients with HGH lack, but their mean response as a group was significantly less. In other studies newborns and active acromegalics were also found to have a reduced lipolytic response to HGH, but this response becomes normal as soon as plasma HGH falls (Laron *et al.*, 1971*b*). The correlation between the fasting HGH concentration at the beginning of the test and the peak FFA concentration after administration of the exogenous HGH, is further illustrated in Figure 24.

Metabolic studies

Metabolic balance studies under controlled conditions were performed in 10 patients.

TABLE VI

Effect of one i.v. dose of HGH on plasma free fatty acids (FFA) (values are expressed as mean ± SD)

Patients	No.	Fasting concentration HGH (ng/ml)	Fasting concentration FFA (μEq/l)	i.v. HGH FFA peak (μEq/l)	i.v. HGH Δ%	
High IR-HGH	15	5–213	935 ± 248	1319 ± 459	43 ± 41	
Isolated HGH lack	11	0	853 ± 266	1791 ± 331	143 ± 97	
Panhypopituitarism	4	0	633 ± 436	1280 ± 495	124 ± 66	
Newborns	3	10–31	1148 ± 465	1099 ± 178	0	
Infant	1	5	315	2300	625	
Acromegaly	1	50	220	300	36	active
(adult)		4	295	839	182	treated

Dose of HGH: infants 3 mg, children 5 mg, acromegalic 10 mg.

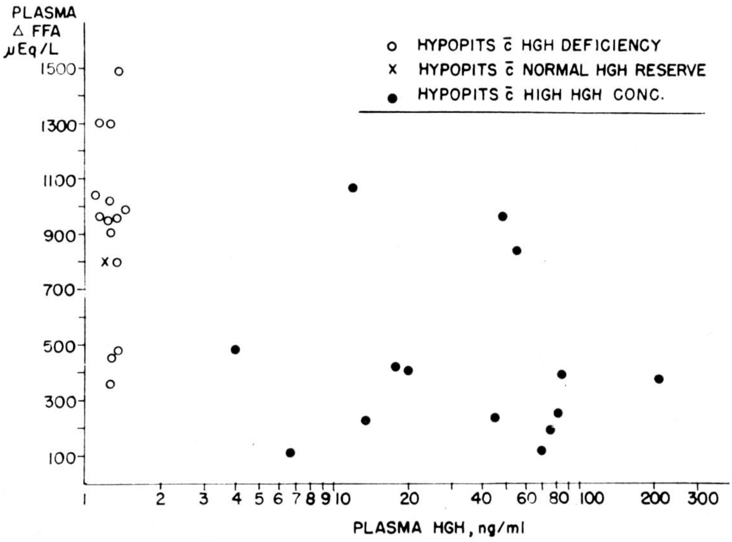

Fig. 24. Correlation between plasma IR-HGH before the test and the increase in plasma FFA after i.v. HGH injection in patients with dwarfism due to HGH lack or the familial syndrome with high plasma IR-HGH. (From Laron *et al.*, 1971*b*, *J. clin. Endocr.*, *33*, 332, by kind permission of the editors).

HGH (prepared in our laboratory or by the National Pituitary Agency) was administered intramuscularly in a dose of 2.5 mg twice a day for 4 days.

The changes in blood sugar, insulin and urea were variable, but there was a significant increase in serum inorganic phosphorus (Fig. 25). A very characteristic finding so far was the lack of response of the serum sulfation factor activity compared to the patients with

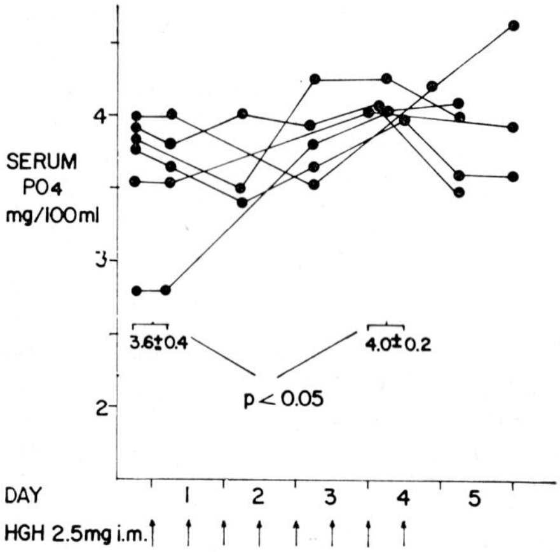

Fig. 25. Serum inorganic phosphorus before and during HGH treatment in dwarfism with high plasma IR-HGH.

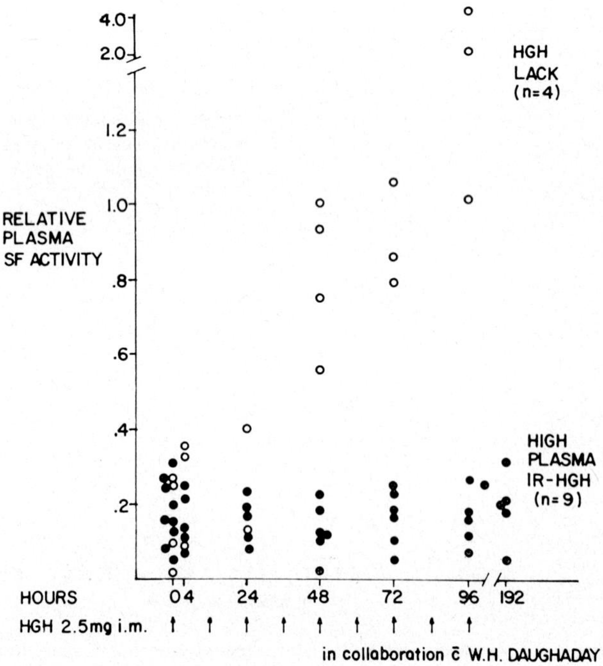

Fig. 26. Plasma sulfation factor (SF) activity before and during HGH treatment in dwarfism due to HGH lack or with high plasma IR-HGH.

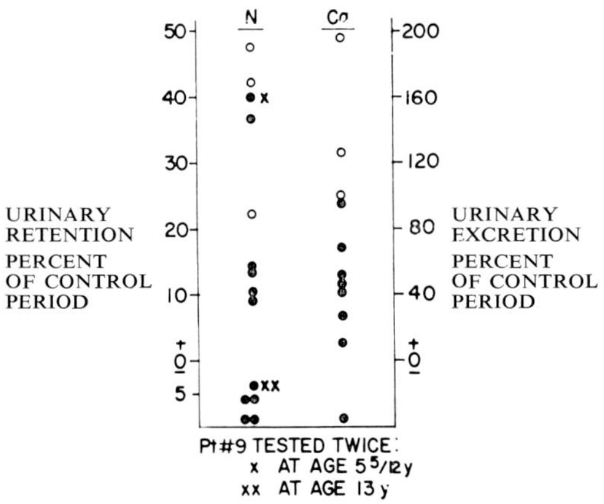

Fig. 27. Effect of HGH (5 mg/day for 4 days) on urinary N and Ca in dwarfism due to HGH lack (○) or high IR-HGH (●).

isolated HGH lack (Fig. 26). These studies were performed in collaboration with Dr. W. H. Daughaday to whom we sent the lyophilized plasma samples.

Urinary nitrogen retention (Fig. 27) varied from none in 5 patients to over 20% of control values in 3 and a slight response in 4 subjects. It is noteworthy that patient 9, tested

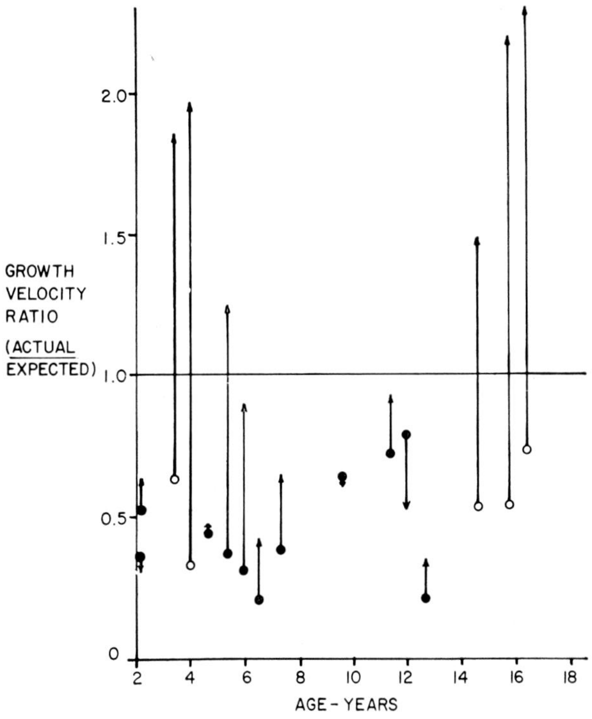

Fig. 28. Effect of HGH treatment on linear growth in pituitary dwarfism.

twice, showed a good urinary retention response to HGH in the first instance, but none in the second, 7.5 years later.

In subjects with hyposomatotropinism, exogenous HGH induces calciuria. In the patients with high plasma IR-HGH this response was reduced (Fig. 27).

Effect of prolonged HGH administration on linear growth

The effect of intramuscular HGH administration three times a week for several months on the linear growth, is shown in Figure 28. The results are expressed as the ratio between the actual growth velocity found, and that expected for the corresponding chronological age (50th percentile for age). This calculation helps to exclude the sex and age growth variability.

It is evident that without treatment the ratio is low in all patients, but whereas the patients lacking HGH show a marked increase in growth velocity ratio during HGH therapy, the patients with familial dwarfism and high IR-HGH show a minor response.

It is noteworthy that patient 9 who received several courses of HGH at various ages (Fig. 29) showed a better response in the initial courses than in the later ones, paralleling nitrogen retention.

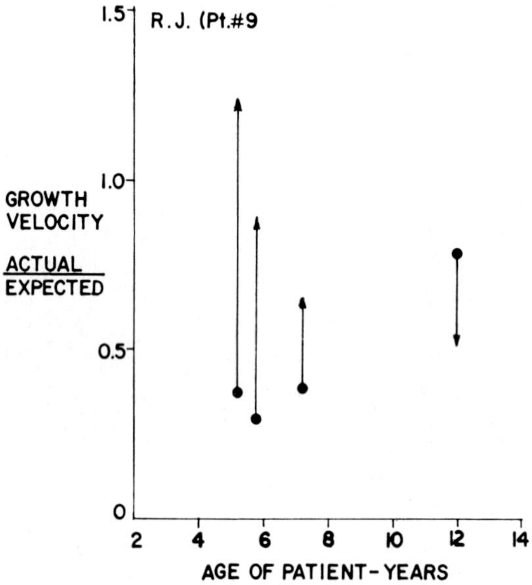

Fig. 29. Effect of repeated courses of HGH therapy on a patient with familial dwarfism and high plasma IR-HGH.

Immunological studies

Using a radioimmunoassay technique (Frasier and Smith, 1966) no antibodies against HGH could be demonstrated at any time.

Ophthalmological investigations

None of the patients, even those with glucose intolerance, including the one adult patient, revealed signs of diabetic retinopathy.

DISCUSSION

As yet, the basic defect in the syndrome of familial dwarfism and high plasma IR-HGH is unknown. However one can assume that there is at least one basic inherited defect which leads to specific clinical and metabolic changes, many of which resemble the patients with familial isolated HGH lack. The basic question is whether the immunoreactive hormone in this syndrome is biologically inactive or whether there is an inherited lack in synthesis of an intermediary substance such as sulfation factor, or whether there is a primary end-organ defect. If we compare various signs and symptoms seen in these patients with different states of either high plasma, biologically active, growth hormone or pituitary dwarfism due to lack of HGH (Fig. 30), it becomes evident that the greatest similarity exists between this syndrome

	HIGH PLASMA HGH STATES: NEWBORNS	NORMAL FEMALES	NORMAL SLEEP	ACROMEGALY	JUV. DIABETES (?)	ASTROCYTOMA	LIVER CIRRHOSIS	PRIM. SF. LACK (?)	FAM. DWARFISM c̄ HIGH IR-HGH	LACK OF HGH: HYPOTH.-PIT. INSUFF.
DWARFISM	—	—		—*	—*			+	+	+
ACROMICRIA	—	—		—*	—			+	+	+
SMALL GENITALIA	—	—		—	—			—?	+	+
OBESITY	—	—		—	—	—		—	+	+
TEND. HYPOGLYCEMIA	—	—		—*	—*			—	+	+
INSULIN SENSIVITY	—	—		—*	—*			—	+	+
INSULINOPENIA (ARG.)	—	—		—*	+		+	—	+	+
GLUCOSE INTOLER.	—	—		+	+		±	—	±	±
LOW PL. SF.	—	—		—*				+	+	+
NO SF. RESP. TO HGH		—		?				+	+	—*
METAB. SUBRESP. TO HGH	+			+				+?	+	—*
INCOMPL. SUPPR. OF HGH BY OGTT	+	+	+	+		+	+	+?	+	

* OPPOSITE EFFECT

Fig. 30. Plasma HGH: clinical and metabolic correlations.

and familial isolated HGH lack. This is true even concerning the glucose intolerance which has been described in sexual ateliotic dwarfism (Merimee *et al.*, 1970; Karp and Laron, 1971), findings which cannot be explained at present by a primary defect in sulfation factor generation.

It is not known whether the regulatory (feed-back) mechanism in these patients is disturbed.

Insulin hypoglycemia and arginine infusion generally further increased the already high IR-HGH levels, indicating sensitivity to stimuli used under normal conditions and of a marked responsiveness of the pituitaries of these patients.

There seems little doubt that there are fluctuations in the levels of IR-HGH in these patients, as well as in their metabolic responses. The changes in carbohydrate metabolism show, with increasing age, a transition from hypoglycemia nonresponsiveness, insulinopenia to glucose intolerance (Fig. 31). This may be due, among other reasons, to the development of compensatory mechanisms.

End-organ refractoriness also does not seem a plausible explanation as many metabolic

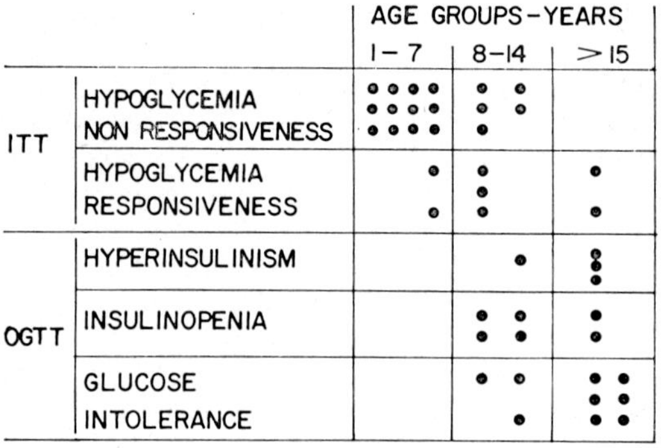

		AGE GROUPS-YEARS		
		1–7	8–14	> 15
ITT	HYPOGLYCEMIA NON RESPONSIVENESS	●●●● ●●●● ●●●●	●● ●● ●	
	HYPOGLYCEMIA RESPONSIVENESS	● ●	● ● ●	● ●
OGTT	HYPERINSULINISM		●	● ● ●
	INSULINOPENIA		●● ●●	● ●
	GLUCOSE INTOLERANCE		●● ●	●● ●● ●●

Fig. 31. The changes in carbohydrate metabolic response in the syndrome of familial dwarfism and high plasma IR-HGH, as related to age of the patients.

systems are involved. At least part of their resistance to exogenous HGH is due to competitive saturation of peripheral tissue receptors of HGH by the endogenous hormone (Laron *et al.*, 1971*b*).

Whatever its etiology, this syndrome does not seem confined to Jews. Awareness of this disease has led to the finding of patients in other countries. Table VII lists 15 such patients. It is noteworthy that half of them are of Dutch and Arab origin.

TABLE VII

Dwarfism with high plasma IR-HGH diagnosed in various countries and not of Jewish origin

No.	Patient Name	Sex	Country Studied	Country Origin	Sibs	Con-sang.	Authors
1	A.V.	M	Holland	Dutch			Van Gemund *et al.*, 1969
2	R.V.	M	Holland	Dutch	1–2	*	
3		M	Holland	Dutch			Van den Brande *et al.*, 1970
4	M	M	Canada	Dutch			Davis, Bailey, Martin, 1968,
5	M	F	Canada	Dutch	4–5	*	1971 p.c.
6		M	Lebanon	Arabian			Najjar, 1969
7		F	Lebanon	Arabian	6–7	*	
8	S.H.	M	USA	Arabian		*	Daughaday, Elders, 1970 p.c.
9	F	F	Iran	Persian		*	Forbes, Noormand, 1968 p.c.
10	T.P.	M	Spain	Spanish			Ferrandez, Collado, Vazquez,
11	J.P.	F	Spain	Spanish	10–11		1970 p.c.
12	S.F.	M	USA	Italian			Merimee *et al.*, 1968 (adult)
13	A.S.	F	Germany	German			Teller *et al.*, 1970 p.c.
14	M.H.	F	Sweden	Swedish			Aronsson, 1970 p.c.
15	D.H.	M	Canada				Ehrlich, 1969 p.c.

p.c. = Personal communication.

ACKNOWLEDGEMENTS

The authors are indebted to our nurses, especially Mrs. Dalia Greenberg who over many years has performed most of the tests in these patients. Mr. Aurel Bratu of Rogoff-Wellcome Medical Research Institute drew the illustrations. Thanks are due to the staff of the Medical Photography Laboratory, especially Mrs. Kay Segal and Mr. Yoshua Schechter for their great help.

Thanks are due to the many colleagues throughout the world who kindly keep communicating any patient who may fit the described syndrome; and last but not least the wonderful cooperation of Dr. W. H. Daughaday who has kindly performed the determinations of sulfation factor activity of our patients

REFERENCES

Daughaday, W. H., Laron, Z., Pertzelan, A. and Heins, J. N. (1969): Defective sulfation factor generation: a possible etiological link in dwarfism. *Trans. Ass. Amer. Phycns*, *82*, 129.

Frankel, J. J. and Laron, Z. (1968): Psychological aspects of pituitary insufficiency in children and adolescents with special reference to growth hormone. *Israel J. med. Sci.*, *4*, 953.

Frasier, S. D. and Smith Jr., P. G. (1966): Antibodies to human growth hormone. *Amer. J. Dis. Child.*, *112*, 383.

Karp, M., Laron, Z. and Pertzelan, A. (1971): Glucose tolerance and insulin secretion in parents and children with familial dwarfism and high plasma IR-HGH and familial isolated HGH deficiency. In: *Proceedings, Second International Symposium on Growth Hormone,* p. 57. ICS 236, Excerpta Medica, Amsterdam.

Laron, Z. (1967): Activity of endogenous growth hormone. *Lancet*, *2*, 1094.

Laron, Z. (1969): The hypothalamus and the pituitary gland (hypophysis). In: *Paediatric Endocrinology* p. 35. Editor: D. Hubble. Blackwell, Oxford-Edinburgh.

Laron, Z., Karp, M., Pertzelan, A. and Frankel, J. J. (1970): ACTH deficiency in children and adolescents (clinical and psychological aspects). In: *Pituitary, Adrenal and The Brain. Progress in Research*, *Vol. 32*, p. 305. Editors: D. De Wied and J. A. W. M. Weijnen. Elsevier, Amsterdam-London-New York.

Laron, Z., Pertzelan, A. and Mannheimer, S. (1966): Genetic pituitary dwarfism with high serum concentration of growth hormone. A new inborn error of metabolism? *Israel J. med. Sci.*, *2*, 152.

Laron, Z., Pertzelan, A. and Karp, M. (1968): Pituitary dwarfism with high serum levels of growth hormone. *Israel J. med. Sci.*, *4*, 883.

Laron, Z. and Pertzelan, A. (1969): Somatotrophin in antenatal and perinatal growth and development. *Lancet*, *1*, 680.

Laron, Z., Pertzelan, A. and Frankel, J. J. (1971*a*): Growth and development in the syndromes of 'familial isolated absence of HGH' or 'pituitary dwarfism with high serum concentration of an immunoreactive but biologically inactive HGH'. In: *Hormones in Development.* Editors: M. Hamburg and E. J. W. Barrington. Appleton-Century-Crofts, New York.

Laron, Z., Pertzelan, A., Karp, M., Kowadlo-Silbergeld, A. and Daughaday, W. H. (1971*b*): Administration of growth hormone to patients with familial dwarfism with high plasma immunoreactive growth hormone: Measurement of sulfation factor, metabolic and linear growth responses. *J. clin. Endocr.*, *33*, 332.

Laron, Z. and Sarel, R. (1970): Penis and testicular size in patients with growth hormone insufficiency. *Acta endocr. (Kbh.)*, *63*, 625.

Merimee, T. J., Fineberg, S. E., McKusick, V. A. and Hall, J. (1970): Diabetes mellitus and sexual ateliotic dwarfism: a comparative study. *J. clin. Invest.*, *49*, 1096.

Merimee, T. J., Hall, J., Rabinowitz, D., Mckusick, V. A. and Rimoin, D. L. (1968): An unusual variety of endocrine dwarfism: subresponsiveness to growth hormone in a sexually mature dwarf. *Lancet*, *2*, 919.

Najjar, S. S. (1969): Pituitary dwarfism with elevated serum HGH. *Acta endocr. (Kbh.)*, *Suppl. 138*, Abstract No. 144.

Pertzelan, A., Adam, A. and Laron, Z. (1968): Genetic aspects of pituitary dwarfism due to absence or biological inactivity of growth hormone. *Israel J. med. Sci.*, *4*, 495.

Rimoin, D. L., Merimee, T. J., Rabinowitz, D. and McKusick, V. A. (1968): Genetic aspects of clinical endocrinology. *Recent Progr. Hormone Res.*, *24*, 365.

SCHARF, A. and LARON, Z. (1971): Skull changes in pituitary dwarfism and the syndrome of familial dwarfism with high plasma immunoreactive growth hormone. A roentgenologic study. *Horm. Met. Res.*, in press.

VAN DEN BRANDE, J. L., DU CAJU, M. V. L., VISSER, H. K. A., SCHOPMAN, W., HACKENG, W. H. L. and DEGENHART, H. J. (1970): Primary plasma growth factor deficiency (sulfation factor and thymidine factor), presenting as hyposomatotropic dwarfism. In: *Abstracts, IX Meeting European Society of Paediatricians, Lyon, 1970*, No. 21.

VAN GEMUND, J. J., LAURENT DE ANGELO, M. S. and VAN GELDEREN, V. V. (1969): Familial prenatal dwarfism with elevated serum immuno-reactive growth hormone levels and end-organ unresponsiveness. *Maandschr. Kindergeneesk.*, *37*, 372.

ZAIZOV, R. and LARON, Z. (1966): Body length and weight at birth and one year of age in different communities. *Acta paediat. (Uppsala)*, *55*, 524.

GROWTH HORMONE LIKE ACTIVITY IN PLASMA AND URINE*

R. M. BALA**, K. A. FERGUSON*** and J. C. BECK

Division of Endocrinology and Metabolism, McGill University Clinic, Royal Victoria Hospital, Montreal, Canada

Human growth hormone (HGH) extracted from pituitary glands has been studied extensively, and the amino acid sequence and subsequent synthesis has been reported by Li and his associates (Li, 1968; Li and Yamashiro, 1970). Human plasma and urine contain pituitary dependent substance(s) with antigenic similarity to extracted pituitary HGH (ep-HGH). Berson and Yalow (1966) and Boucher (1968) have presented evidence that immunoreactive HGH (IR-HGH) in plasma exists in an unassociated monomeric form which resembles ep-HGH, whereas others, including our own laboratories, have presented direct and indirect evidence suggesting that IR-HGH in plasma may occur in different molecular sizes or may be associated with other proteins (Ferguson *et al.*, 1967; Hadden and Prout, 1964, 1965; MacMillan *et al.*, 1967; Irie and Barrett, 1962; Collipp *et al.*, 1964; Bala *et al.*, 1970).

HGH was first identified in urine by immunodiffusion techniques by Salinas *et al.* (1963) and by Geller and Loh (1963) who quantitated HGH in urine using a hemagglutination inhibition method. Subsequently, a number of investigators (Franchimont, 1965; Sakuma *et al.*, 1968; Girard and Greenwood, 1967, 1968) have reported widely varying amounts in both normal and acromegalic subjects. These data are summarized in Table I.

It is generally accepted that physiological, in contrast to pharmacological, amounts of ep-HGH lack biological activity *in vitro* (Salmon and Daughaday, 1957; Kostyo and Knobil,

TABLE I

Urinary growth hormone levels

Authors	Amount (mg/24 hrs)
Salinas *et al.* (1963)	Not quantitated
Geller and Loh (1963)	Normals – 17
	Acromegaly – up to 2,000
Franchimont (1965)	Normals – 737
Sakuma *et al.* (1968)	Normals – 3.7
	Acromegaly – 51.9
Girard and Greenwood (1967, 1968)	Normals – undetectable
	Acromegaly – 0.24

* Supported by the Medical Research Council of Canada, Grant No. MT-631.

** Present address: Division of Medicine, Health Sciences Center, Calgary, Alberta, Canada.

*** Visiting Scientist, C.S.I.R.O., Paramatta, New South Wales, Australia.

1959; Daughaday and Mariz, 1962; Daughaday and Kipnis, 1966; Daughaday and Reeder, 1966; Rigal, 1964; Wettenhall *et al.*, 1969) and that there is a time lag in the expression of certain biological activities usually associated with anabolism when HGH is administered *in vivo*. Plasma contains biologically active substance(s) as measured *in vitro* by stimulation of incorporation of radioactive sulfate, thymidine and proline (Salmon and Daughaday, 1958; Almqvist, 1961; Daughaday and Mariz, 1962; Daughaday and Kipnis, 1966) into chondroitin sulphate, DNA, and hydroxyproline, respectively, by cartilage from hypophysectomized rats. This activity, referred to as sulfation factor (SF), is pituitary dependent and in general reflects the plasma levels of IR-HGH. Table II summarizes some of the argu-

TABLE II

Extracted pituitary HGH : plasma HGH

+ ve			− ve
plasma biologic HGH-like activity		plasma IR-HGH	
↑	acromegaly	↑	– *in vitro* – lack of activity of HGH in physiologic doses
↓	dwarfism	↓	
↓	hypophysectomy	↓	– *in vivo* biologic action time lag
↑	HGH Rx — hypox		– plasma fractionation – peptides with biologic activity > immunologic activity
	HGH Rx — pit. dwarf	↑	– clinical situations:*
			– Laron dwarf – Rimoin pygmy – 'cerebral gigantism' – postop. craniopharyngioma – post ro. Rx pituitary tumors – acromegaly

conclusion

↓ biologic HGH-like activity in plasma is pituitary-dependent

↓ antigenic similarity of plasma substance and HGH extracted from human pituitaries

* Laron *et al.* (1966); Rimoin *et al.* (1968); Daughaday *et al.* (1969).

ments that raise the question of whether plasma and pituitary growth hormone are identical. Some of these data indicate that pituitary-HGH is converted to a biologically active form(s) before or after entering plasma, or that it induces the synthesis of biologically active substance(s) in plasma or in a tissue from which it is secreted into the circulation. Alternatively, plasma HGH may be associated with a larger protein which plays a transport-storage role.

TABLE III

$$Kd = \frac{Ve - Vo}{Vi} \; *$$

Ve = elution volume for the protein
Vo = gel column void volume
Vi = gel imbibed volume

* Andrews (1964, 1966); Laurent and Killander (1964).

We initially studied the distribution and correlation of biologic and IR-HGH like activity in plasma, after fractionation of large volumes by gel filtration on Sephadex. This produced a series of fractions of progressively smaller molecular weight, the latter being estimated from the distribution coefficient (Kd), as depicted in Table III. In view of our preliminary findings of IR-HGH in varying sizes in plasma and the conflicting reports on levels of growth hormone in urine, we attempted further investigation of plasma and urine IR-HGH.

MATERIALS AND METHODS

1. Plasma

An overview of the method is seen in Figure 1. Blood samples from normal and acromegalic patients were collected in anticoagulant ACD solution and centrifuged at 4°C, yielding plasma samples varying from 80 to 350 ml. The plasma samples were kept at 4°C and were used within 24 hours of venipuncture. Lyophilized plasma samples were processed shortly after collection, stored at 4°C, and reconstituted immediately before use. The samples were fractionated by gel filtration, utilizing large closed-end columns 10 × 100 cm, with a minimum gel bed volume of 7000 ml containing Sephadex G-75 superfine. Elution was carried out at 4°C in a reverse flow manner, at pH 8.1, with 0.15 M NH_4HCO_3, which contained 0.02% sodium azide as a bactericidal agent. On the column albumin was eluted with a Kd value of 0.05 and ep-HGH at 0.30. The absorbance (OD) of the eluate at 280 mμ was continuously monitored and the eluate was divided into successive fractions; these were concentrated by ultrafiltration at 4°C through Dialfo UM-2 membranes (Amicon Corporation) specified to retain proteins of greater than 1000 molecular weight. The total amount and concentration of protein was rechecked after ultrafiltration.

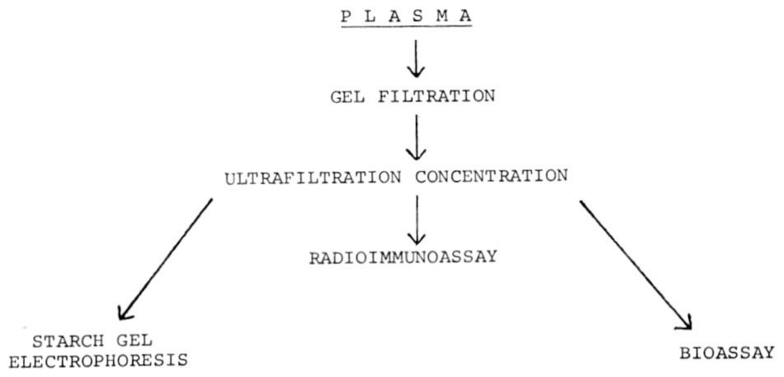

Fig. 1. Flow diagram of procedures for handling plasma.

The amount of IR-HGH in the concentrated fraction was determined by a modified solid phase tube method (Bala *et al.*, 1969), consistently sensitive to less than 0.05 ng HGH and often to values of 0.025 ng HGH. Similarly fractionated bovine plasma was used as a control for non-specific protein effects on the radioimmunoassay.

Sulfation factor assays of the concentrated samples were carried out as described by Daughaday *et al.* (1959) and Kogut *et al.* (1963) with minor modifications. At least two dose-levels were assayed. Pooled normal human plasma was used as a reference standard, a unit of SF activity being arbitrarily defined as the activity of 10 mg protein in the normal pooled reference plasma.

2. Urine

Preliminary measurements of IR-HGH in unconcentrated urine revealed very low to undetectable levels. Therefore in order to avoid changing the nature and amounts of urine proteins by dialysis (Miyasato and Pollak, 1966), lyophilization (Cheever and Lewis, 1969), or extraction procedures (Rohde and Dörner, 1969), urines were concentrated by Diaflo ultrafiltration (Pollak *et al.*, 1968). One or more 24-hour urine specimens were collected at room temperature from five normal, three acromegalic, and six hypopituitary patients (one of whom had received HGH treatment). Sodium azide was added to the urine in a final concentration of 20 mg/100 ml, as an antibacterial agent. All subsequent procedures were carried out at 4°C, and are summarized in Figure 2. The urines were concentrated

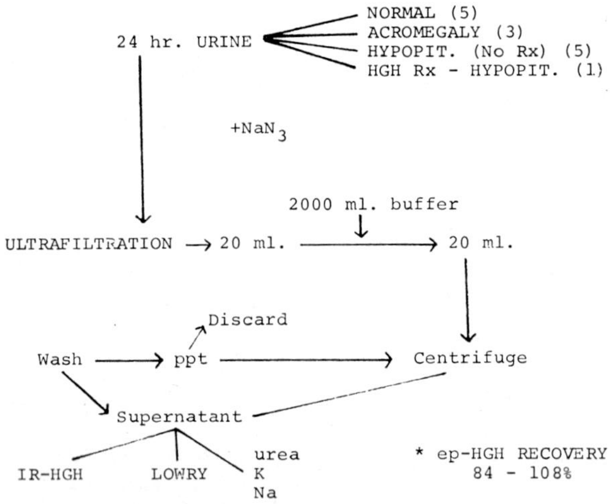

Fig. 2. Flow diagram summarizing methods for handling 24 hour urines from normals, acromegalic and hypopituitary patients (Rx denotes treatment).

to approximately 20 ml by Diaflo ultrafiltration through UM 10 membranes (retention of greater than 10,000 molecular weight). The urine concentrate was diluted to approximately 2000 ml with NH_4HCO_3 buffer, as described above, and reconcentrated to 20 ml. This concentrate, combined with the membrane wash, was centrifuged. The precipitate was washed with NH_4HCO_3 buffer and discarded. The wash was combined with the concentrate supernatant. The concentrations of urea, Na, K and protein in the urine concentrates were determined. Total IR-HGH in the urine concentrates for each 24-hour urine collection was determined using the solid phase tube radioimmunoassay.

The urine concentrates were then combined as indicated in Figure 3, giving one pool from normal subjects, three from acromegalic patients, three from non-treated hypopituitary patients, and two from the hypopituitary patient after treatment with HGH. Total IR-HGH of each pool was determined and the pools were fractionated by gel filtration on columns similar to those described for the handling of the plasma samples and with similar elution techniques. The array of fractions of progressively decreasing molecular weight were concentrated by Diaflo ultrafiltration using UM 2 membranes. The concentrations of protein (absorbance at 280 mμ), Na, K and urea were determined in the fractions, and all fractions had total IR-HGH determined. The effects of varying protein, urine, NaCl and urea concentrations on the solid

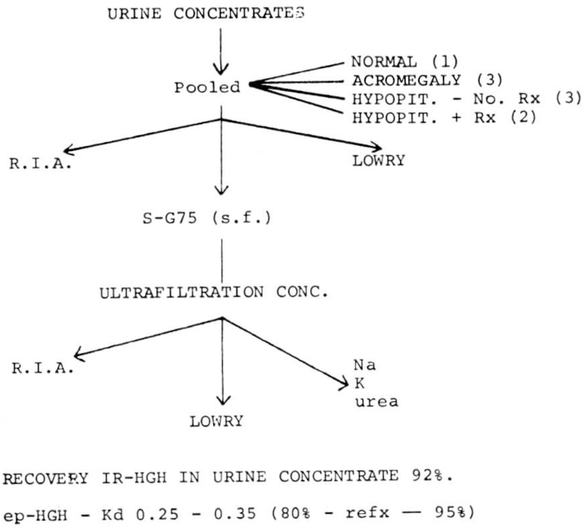

Fig. 3. Flow diagram summarizing methods for handling urine concentrates (Rx denotes treatment; RIA – radioimmunoassay; SG – Sephadex).

phase tube-type radioimmunoassay were first determined, and are reported elsewhere (Bala and Beck, 1971). Within the ranges measured the assay was found to be negligibly affected by these variables.

RESULTS

1. Plasma

Fractionations of two representative samples are shown in Figures 4 and 5. The protein concentration of the eluate is plotted at the mid-sample Kd. The magnified scale reveals minor secondary peaks at Kd greater than 1.0, implying absorption by Sephadex of material with significant absorbance at 280 mμ. This could represent aromatic amino acids alone or could be small peptides. The height of the hatched bars represents the total nanograms (ng) of IR-HGH eluted in the fractions, and the elution volumes are represented by the width of the bars along the Kd scale. The potency of the eluted portions is shown by the height of the solid bars in ng of IR-HGH per mg of protein in these fractions. In all of the samples studied, more than 50% of the total IR-HGH recovered was eluted in the Kd area less than 0.25, that is, in a molecular size range greater than ep-HGH. The maximum potency occurred in the Kd interval corresponding to the molecular size of ep-HGH in all plasma. Some fractions of normal plasma did not have detectable levels of IR-HGH while a fresh acromegalic plasma (Fig. 5) had immunoreactivity in all fractions. The acromegalic plasmas had a greater relative proportion of the total IR-HGH in the ep-HGH molecular size range compared with normal plasma. The lyophilized acromegalic plasmas revealed material of high IR-HGH potency eluted after Kd 1.0 compared with fresh acromegalic plasma. Further evidence derived from the total protein in each Kd interval implied that lyophilization resulted in increased amounts of small-sized molecular material with a high aromatic molecular structure component.

To allow for interexperimental comparison of the plasma fractions, 0.10 Kd intervals were selected semi-arbitrarily. The mean per cent ± SE of the total IR-HGH recovered is

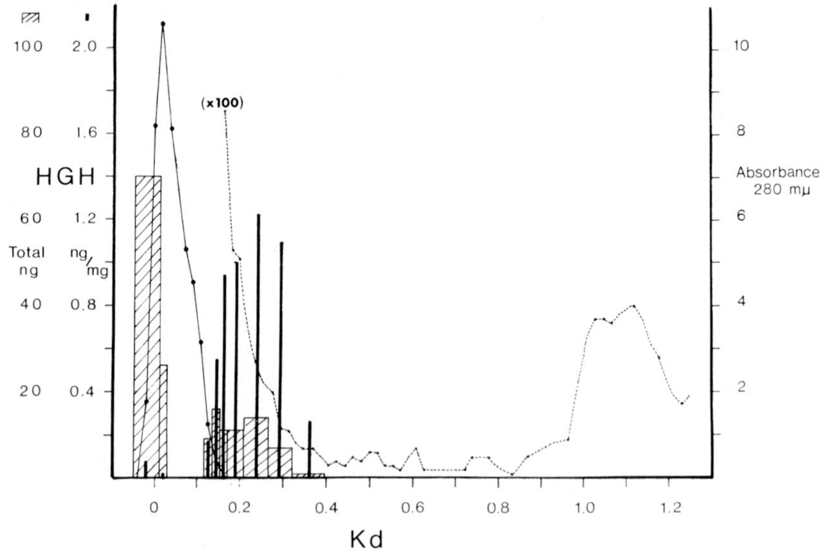

Fig. 4. Fractionation of 115 ml of fresh normal plasma by gel filtration with Sephadex G 75. Protein concentration of the eluate (280 mμ OD) plotted at mid-fraction Kd ●——● and magnified 100 × →.

The height of the hatched bars represents the total ng of IR-HGH in the fractions, while the width of the hatched bars represents the volume of elution corresponding to the Kd area covered. The height of the solid bars represents the potency of the fractions in ng IR-HGH per mg protein. Original plasma IR-HGH 1.9 ng per ml; 0.047 ng per mg protein.

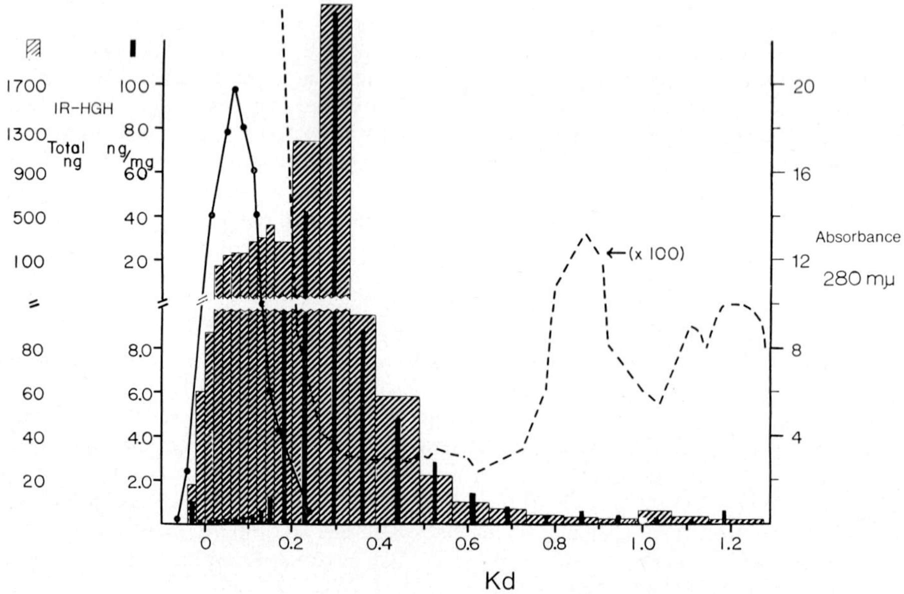

Fig. 5. Fractionation of 272 ml of fresh acromegalic patient plasma. Plotted similarly to Fig. 4. Original plasma IR-HGH 27.8 ng per ml; 0.623 ng per mg protein.

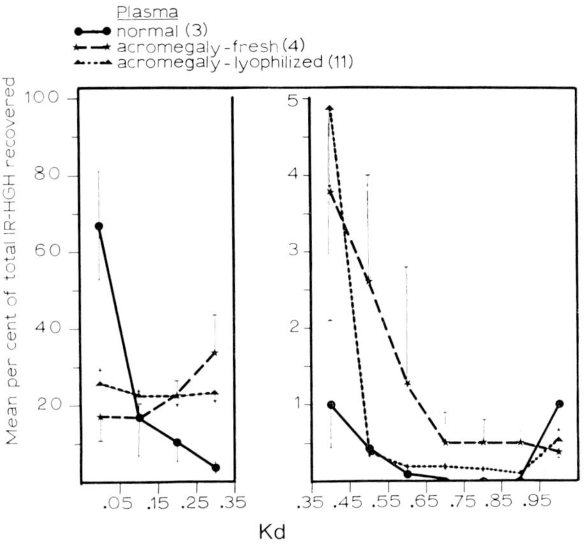

Fig. 6. The total amount of IR-HGH in each Kd interval calculated as a percent of the total IR-HGH recovered in each experiment, the mean percent ± SE is shown for the different types of plasmas. The increased value in the Kd interval greater than 0.95 was mainly due to the one non-fresh normal plasma.

shown for each Kd interval in Figure 6. Pooling of the data broadens the zone of activity compared with the individual experiments. The mean final recovery of IR-HGH in all experiments was 108% (range 60-137). The mean levels of IR-HGH in the normal, fresh and lyophilized acromegalic plasma were 1.7, 27.1, 37.7 ng per ml, respectively. The normal plasmas had a greater proportion of the total IR-HGH eluted before Kd 0.25 areas compared with plasmas from acromegalics. The latter showed a higher percentage of the total IR-HGH in the Kd area corresponding to the molecular size of ep-HGH compared with the normal plasmas. This suggests that in acromegaly, the IR-HGH molecules are of different size or exhibit less aggregation or perhaps less association with other plasma proteins concomitant with the higher levels of initial plasma IR-HGH. A relatively greater proportion of the total IR-HGH occurs in the larger molecular size area in lyophilized plasma compared with fresh acromegalic plasma, indicating aggregation of IR-HGH molecules or association with other protein molecules due to lyophilization. Significant individual variation occurred in the percentages of the total IR-HGH eluted per Kd interval; however, in all plasmas, more than 50% of the total IR-HGH was eluted before Kd 0.25.

The mean potency of the various plasma fractions in ng IR-HGH per mg of protein for each Kd interval is shown in Figure 7. The mean potencies of the original plasmas were 0.028, 0.496 and 0.501 ng IR-HGH per mg of protein for the normal, fresh and lyophilized acromegalic plasmas, respectively. The highest IR-HGH potency of eluted proteins in all plasmas was found in the Kd interval 0.25-0.45 and in most plasmas in the 0.25-0.35 Kd interval, which corresponds to the molecular size interval of ep-HGH. The fractions of acromegalic plasma showed higher potencies than normal plasma fractions throughout. Even in the Kd area corresponding to very large and small molecular sizes, the potency of the acromegalic fractions approximated the peak potency of the normal plasmas in the Kd interval of ep-HGH.

Plasma fractions from Kd 0.15 to 0.85 were assayed for SF activity. To allow interexperi-

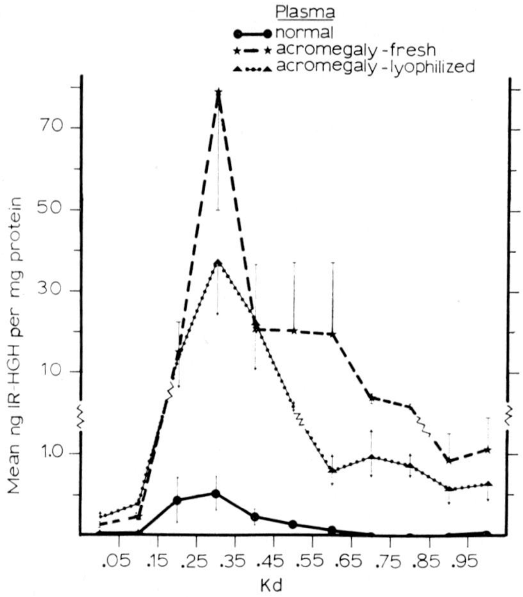

Fig. 7. The potencies of the eluted proteins in each Kd interval for each experiment calculated in ng IR-HGH per mg protein. The mean ± SE potencies are plotted for the different types of plasmas.

mental comparison, the total SF activity in each Kd interval for each plasma was expressed per 100 ml of original plasma. The mean ± SE totals per Kd interval for normal and acromegalic patient plasma are shown in Figure 8. Considerable variation in the SF activity by the various plasmas occurred when expressed per 0.1 Kd intervals. The maximum total SF

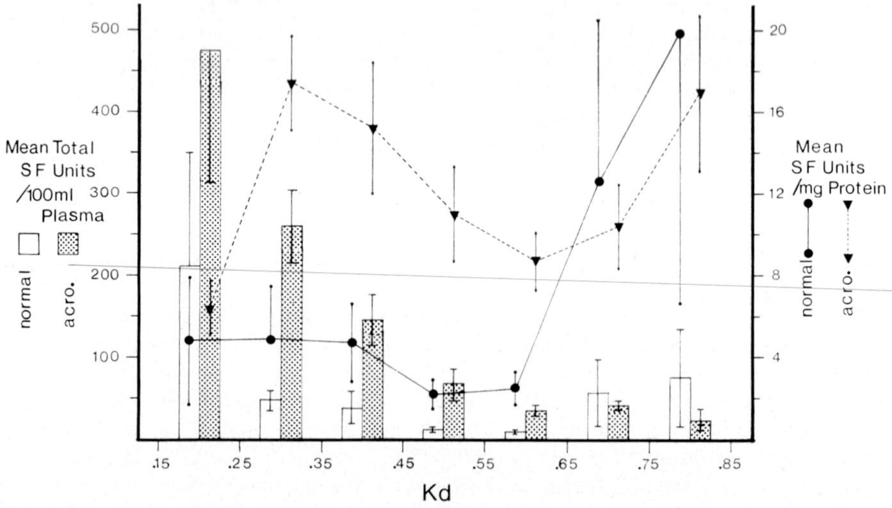

Fig. 8. The total SF units of activity in each Kd interval of each experiment was calculated per 100 ml of original plasma found. The potency in SF units per mg of protein in each Kd interval for each experiment was calculated. The mean ± SE total SF activities and potencies are plotted for acromegalic and normal plasma.

activity occurred in the 0.15-0.25 Kd interval in all plasmas, except in three lyophilized acromegalic plasmas in which the maximal total activity occurred in the 0.25–0.35 Kd area.

In the Kd area less than 0.65, the potency in SF units per mg of eluted protein was greatest in the 0.25-0.35 Kd interval area in most plasmas. The potency of all but two acromegalic plasmas was higher than that of the normal plasmas in the Kd intervals before 0.65. This suggests that acromegalic plasma contains a greater amount of a more potent SF activity stimulating substance than does normal plasma. In the Kd area greater than 0.65, the elution of increased amounts of more potent SF activity stimulating substance suggests a substance of a different nature or a dissociated form.

2. *Urine*

Clinical diagnoses, urine IR-HGH values after ultrafiltration, and percent recovery of IR-HGH post-fractionation are shown in Table IV. If these values are extrapolated to IR-HGH levels in unconcentrated urine, they fall below the sensitivity of most radioimmunoassays. Ep-HGH similarly processed by ultrafiltration resulted in recoveries ranging from 85 to 104%. Although overlap in individual total IR-HGH in 24 hr. urine concentrates occurred in the hypopituitary and normal groups, the mean values are significantly different ($P < .05$; student's t test). The mean total urine IR-HGH in the acromegalic and normal

TABLE IV

Total daily urine HGH

Subjects	Total daily urine IR-HGH (ng/24 hrs)	% Recovery IR-HGH post fractionation
Normals (5)	Mean — 78.8±14.4	77
	Range — 44−121	
Hypopituitarism (6)	Mean — 34.8±4.5	133
	Range — 15−53	
Acromegaly (3)	Mean — 254±73.7	58
	Range — 96−526	
HGH treated	Day 1 — 116	87
Hypopituitary (1)	Day 2 — 58	

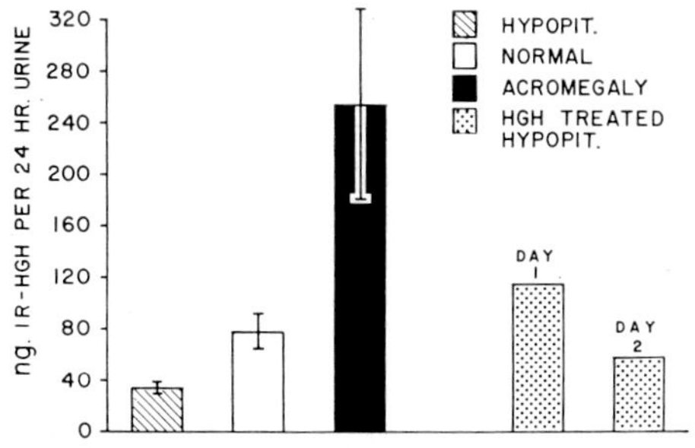

Fig. 9. Comparison of total IR-HGH per 24 hour urine in normal acromegalic and hypopituitary patients. The urine from the treated hypopituitary patients was collected on the day of treatment (Day 1) and the day after treatment (Day 2) with 2.5 mg HGH. Values shown are mean ± SE.

groups are also significantly different, although some overlap did occur with the normal and post-radiotherapy acromegalic patients. These data are depicted in Figure 9 and show the mean ± SEM of the total IR-HGH in the urine of the various groups. The HGH-treated hypopituitary dwarf received 2.5 mg HGH (Medical Research Council of Canada – Clinical grade) every third day. Less than 0.01% of the administered HGH was recovered as urine IR-HGH over the two-day collection.

The urine concentrates were fractionated by gel filtration, and the total IR-HGH was determined in each fraction after concentration. The mean recovery of IR-HGH in the starting material after gel filtration and concentration was 92%. To allow interexperimental comparison, the total IR-HGH in the eluted fractions was calculated per 0.1 Kd interval in each experiment. The total IR-HGH per Kd interval was calculated as a percent of the total IR-HGH recovered in that experiment. The mean of the total IR-HGH recovered per Kd interval for the different groups is shown in Figure 10. Significant variation in total

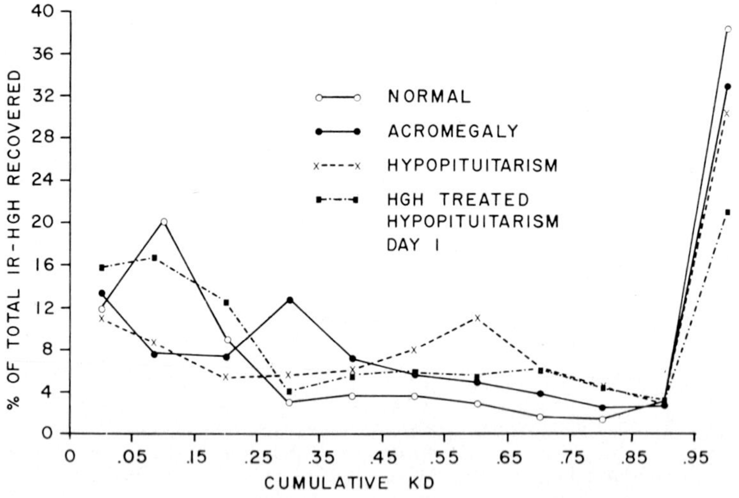

Fig. 10. Comparison of mean % of the total IR-HGH recovered in each experiment per Kd interval of the different groups of patient urine after gel filtration and concentration.

IR-HGH per Kd interval occurred between experiments. Some IR-HGH was detected in all Kd areas even though some of the individual fractions included in these Kd intervals were devoid of IR-HGH activity. The cumulative total IR-HGH after Kd 0.95 includes an extension of elution out to Kd 2.0 or greater. A significant relative amount of the total IR-HGH activity is noted in the Kd area greater than 0.95 in all of the urine groups. The urine from the hypopituitary patient on day 1 of HGH treatment shows relatively less IR-HGH in this Kd area. In the Kd areas smaller than 0.95, the normal urine group shows a peak of IR-HGH activity in the large molecular size area before Kd 0.25. The acromegalic group urine shows a peak value in the Kd area 0.25-0.35, corresponding to the molecular size area of ep-HGH, and is significantly different from the next highest level, the hypopituitary group. The hypopituitary urine group shows a peak at Kd 0.55-0.65, corresponding to a molecular weight area of approximately 8,000. The urine of the hypopituitary patient treated with ep-HGH showed a relatively large amount of IR-HGH activity in the larger molecular size area. The comparison of relative totals of IR-HGH eluted over larger Kd

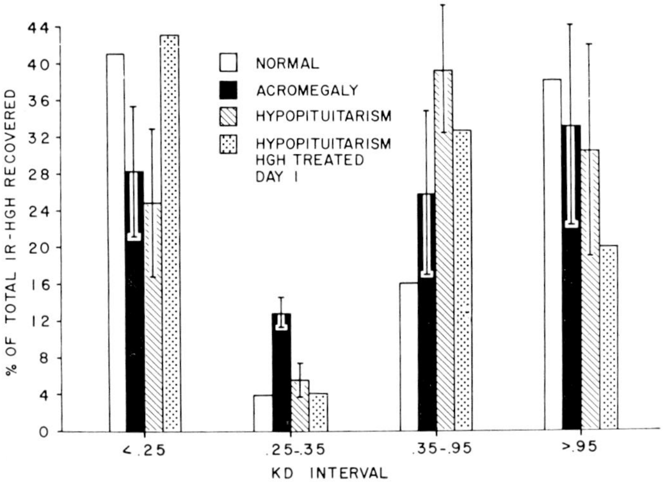

Fig. 11. Comparison of IR-HGH recovered over larger Kd areas expressed similarly to Fig. 10, except that the mean ± SE of the total IR-HGH recovered per Kd interval is shown.

areas in Figure 11 shows the relatively small amount of IR-HGH in urine found in the molecular size area of ep-HGH.

DISCUSSION

We have attempted to investigate the nature of endogenous IR-HGH in plasma and urine. We feel that the technique of gel filtration and the subsequent concentration of eluate fractions by ultrafiltration has allowed the detection of IR-HGH activity in fractions in which it might otherwise not have been apparent. Since electrophoresis might have disruptive effects on protein associations (Franglen and Gosselin, 1958; Ferguson, 1964; Reithel, 1963), we have avoided the technique as an initial separative method for plasma.

Our results show that after fractionation by gel filtration more than 50% of the IR-HGH activity in normal and acromegalic plasmas was present in a molecular size area greater than that of ep-HGH. It is possible that ep-HGH is an altered form or subunit structure of endogenous pituitary HGH and is therefore different from part of the total plasma IR-HGH (Roos *et al.*, 1963; Leaver, 1966; Rohde and Dörner, 1969). If endogenous pituitary HGH is identical to ep-HGH, then either these IR-HGH molecules in plasma are aggregated forms or are associated with other plasma proteins, and these complexes are not disrupted by gel filtration. There is convincing evidence that ep-HGH does aggregate under appropriate conditions (Andrews, 1966; Leaver, 1966; Sluyser, 1964; Hunter, 1963, 1965; Saxena and Henneman, 1966; Squire and Pedersen, 1961; Hanson *et al.*, 1966; Lewis *et al.*, 1969; Cheever and Lewis, 1969; Li *et al.*, 1964); however, evidence with respect to endogenous plasma IR-HGH is lacking.

Our results showing an increased proportion of the total IR-HGH of lyophilized acromegalic plasma in a larger molecular size Kd area may support the finding of others (Squire *et al.*, 1963) that lyophilization does favour aggregation of proteins, but fails to explain our findings with fresh plasmas. Similarly fractionated lyophilized ep-HGH resulted in more than 95% of the IR-HGH being eluted in the same Kd area on refractionation. These results are similar to the findings of others using different buffer systems (Hunter, 1963, 1965; Saxena and Henneman, 1966). This suggests that the conditions in which our plasmas were fractionated did not result in the apparent larger molecular size units for IR-HGH by favoring aggregation of IR-HGH.

The finding that the relative proportions of IR-HGH with a Kd similar to ep-HGH were increased in acromegalic plasma compared to normal plasma, supports the possibility of the association of IR-HGH with plasma proteins, with saturation occurring at higher levels of plasma IR-HGH analogous to the distribution of cortisol in Cushing's syndrome. It is recognized that part of the IR-HGH in the molecular size area larger than ep-HGH may be due to the molecular size zone spreading on gel filtration, in spite of the conditions chosen to minimize this effect. The elution of IR-HGH peptides with Kd values corresponding to molecular sizes smaller than ep-HGH suggests a metabolically changed endogenous plasma HGH or that less molecular weight IR-HGH is produced during storage of the plasma or during gel filtration. Others have shown that degradation of ep-HGH may occur during extraction, storage, freezing or lyophilization (Berson and Yalow, 1966; Leaver, 1966; Rohde and Dörner, 1969; Hunter, 1963; Lewis *et al.*, 1969; Cheever and Lewis, 1969), and this conversion may be prevented to some extent by enzyme inhibitors (Ellis *et al.*, 1968; Lewis and Cheever, 1965).

In general, our data show that the plasmas with the highest initial IR-HGH levels are associated with increased relative amounts of IR-HGH in the small molecular size area, suggesting that part of this IR-HGH may reflect increased levels of metabolically changed IR-HGH in plasma. Our findings do imply that molecules much smaller than ep-HGH can be immunoreactive with it. It is possible that part of the high molecular weight IR-HGH may be due to binding of small IR-HGH molecules with plasma proteins. Fractionation studies of plasma from hypopituitary patients after treatment with ep-HGH revealed a similar distribution of IR-HGH to that found in normal plasma. The relative proportions of IR-HGH with smaller and larger molecular sizes than ep-HGH increased with the time from treatment to withdrawal from plasma. Similarly fractionated plasmas from hypophysectomized patients were devoid of IR-HGH activity.

We have attempted to use the sulfation factor assay as an '*in vitro*' assay for pituitary HGH dependent biological activity in plasma. It is established that SF in plasma is, to a large degree, dependent upon pituitary HGH. That this plasma SF is not due to pituitary HGH directly is shown by the absence of SF activity in physiological amounts of ep-HGH tested *in vitro*, and by the time lag of restoration of SF activity in the plasma of hypophysectomized rats or humans after administration of ep-HGH *in vivo*, as well as the subsequent persistence of the SF in plasma for much longer than IR-HGH (Daughaday and Kipnis, 1966). The greatest total amount of SF in all the plasma fractions assayed over the 0.15-0.85 Kd interval corresponded to a molecular size range larger than that of ep-HGH. In all the Kd areas less than 0.65, all but two acromegalic plasmas contained a greater amount of more potent SF activity than normal plasma. The peak mean potencies in the Kd interval corresponding to ep-HGH molecular size suggests that the most potent SF activity stimulating material has a molecular size similar to ep-HGH. The approximate parallelism, in general, of SF activity and IR-HGH in these plasmas is of interest. Since ep-HGH is biologically inactive *in vitro* in physiological amounts, this could represent an altered pituitary HGH molecule in plasma that has both IR-HGH and *in vitro* biological activity. Alternatively, it may be an *in vitro* biologically active substance(s) in plasma, which is pituitary HGH dependent, that is distinct from IR-HGH but is similar in size. The increase in SF activity after Kd 0.65 suggests an SF of a different nature. Preliminary assays of immunoreactive insulin and insulin-like activity of these fractions show peak activity at Kd 0.65. But the amount of insulin would not fully account for the level of SF activity found, according to our studies and those of others (Salmon and Daughaday, 1957; Salmon *et al.*, 1968; Esanu *et al.*, 1969) on the stimulation of *in vitro* SF activity by insulin. However, insulin could account for part of the activity. We believe that the further isolation and characterization of the *in vitro* biologically active substances in plasma which are pituitary HGH dependent, would be of interest.

The radioimmunoassay used in this study of urine IR-HGH was found acceptably specific. We were unable to find measurable amounts of IR-HGH in non-processed urine as others have (Geller and Loh, 1963; Franchimont, 1965; Sakuma *et al.*, 1968), although urine from acromegalic patients did give values for IR-HGH near the lower limits of sensitivity for the radioimmunoassay. Our findings are in general agreement with Girard and Greenwood (1968) in regard to the total IR-HGH per 24 hours in acromegalic patient urine, but differ from their findings in that we have also detected IR-HGH in the urine concentrates from normal and hypopituitary subjects. The UM 10 Diaflo membranes are specified to filter molecules of a size corresponding to or less than a molecular weight of 10,000; however, our experience suggests that significant amounts of much smaller molecules are retained. It is possible that we also lost variable amounts of small molecular size IR-HGH on ultrafiltration concentration. Studies of UM 10 dialysate concentrated by UM 2 (retention of molecular weight greater than 1000) membrane ultrafiltration did not support this possibility. We believe it is unwise to assume that IR-HGH in urine and ep-HGH are identical on the basis of available evidence. It is possible that urine IR-HGH is a much different molecule to endogenous plasma HGH or ep-HGH, with only the antigenic portion(s) in common. Endogenous HGH has been exposed to metabolic processes and ep-HGH has been subjected to an extraction procedure. The relatively small quantities of IR-HGH found in the urine from a patient treated with ep-HGH suggest that very little IR-HGH is excreted in urine. Bulking concentrates of many urines from normal patients and from acromegalic patients might give sufficient IR-HGH protein for further study of endogenous HGH. This might obviate the need for very large volumes of plasma as the source of IR-HGH in such a study.

Our findings after gel filtration fractionation of these urine concentrates suggest that a significant amount of IR-HGH in urine is a small molecule with a high aromatic amino acid content, since it is absorbed to Sephadex and is eluted after Kd 0.95. The finding of variable amounts of IR-HGH in the intermediate molecular size zones, between the molecular size of ep-HGH and the very small molecules, suggests that the molecular size of IR-HGH varies in urine. The presence of IR-HGH in urine in a molecular size area greater than that of albumin suggests that aggregation of IR-HGH molecules may be occurring or that adsorption of IR-HGH to other proteins in urine has occurred. Relatively greater amounts of IR-HGH occurred in a molecular size area corresponding to ep-HGH on gel filtration in urine from acromegalic patients compared with normal urines. This may indicate a greater relative amount of nonbound HGH associated with the higher plasma IR-HGH levels in acromegaly. Further conclusions as to the character of IR-HGH in urine and its relationship to endogenous HGH metabolism must await the generation of larger amounts of urinary HGH.

SUMMARY

1. The amounts of protein, IR-HGH and SF activity of plasma fractions from normals and acromegalic patients, obtained by gel filtration and concentrated by ultrafiltration, were determined and expressed per 0.1 Kd intervals.

2. More than 50% of the total IR-HGH in these plasma samples was eluted in a molecular size range greater than ep-HGH, indicating plasma IR-HGH in a molecular size larger than ep-HGH. The plasma from acromegalic patients showed a relatively greater amount of IR-HGH in the ep-HGH molecular size area compared with normal plasma.

3. The maximal total SF activity occurred in a molecular size greater than ep-HGH but the maximal SF potency of the eluted proteins occurred in the ep-HGH molecular size area, and plasma from acromegalic patients contained a greater amount of more potent SF activity.

4. Total IR-HGH was measured in 24 hour urine samples collected from normals, acromegalic and hypopituitary subjects, one of whom had received HGH treatment. After concentration by polyionic membrane ultrafiltration, mean levels of total IR-HGH were 79, 35 and 254 mg per 24 hour urine from normal, hypopituitary and acromegalic patients, respectively.

5. Less than 0.01% of the HGH injected into a hypopituitary patient was recovered in the urine over a 2-day collection period.

6. Urine concentrates were fractionated by gel filtration on Sephadex, and the eluted fractions concentrated by ultrafiltration and the total IR-HGH determined. IR-HGH eluted in the fractions corresponded to a wide range of molecular sizes; the greatest relative amount of IR-HGH in all groups was eluted in a very small molecular size range.

7. Acromegalic patient urine concentrates showed more IR-HGH in the molecular size range of ep-HGH than did the others.

8. These data suggest that ep-HGH is different from plasma HGH and from endogenous pituitary HGH. IR-HGH in urine exists in molecular sizes different from ep-HGH.

REFERENCES

ALMQVIST, S. (1961): Studies on sulfation factor (SF) activity of human serum. *Acta endocr. (Kbh.)*, *36*, 31.

ANDREWS, P. (1964): Estimation of the moiecular weights of proteins by Sephadex gel-filtration. *Biochem. J.*, *91*, 222.

ANDREWS, P. (1966): Molecular weights of prolactins and pituitary growth hormones estimated by gel filtration. *Nature (Lond.)*, *209*, 155.

BALA, R. M. and BECK, J. C. (1971): Human growth hormone in urine. *J. clin. Endocr.*, in press.

BALA, R. M., FERGUSON, K. A. and BECK, J. C. (1969): Modified solid-phase (tube) radioimmunoassay of human growth hormone. *Canad. J. Physiol.*, *47*, 803.

BALA, R. M., FERGUSON, K. A. and BECK, J. C. (1970): Plasma biological and immunoreactive human growth hormone-like activity. *Endocrinology*, *87*, 506.

BERSON, S. A. and YALOW, R. S. (1966): State of human growth hormone in plasma and changes in stored solutions of pituitary growth hormone. *J. biol. Chem.*, *241*, 5745.

BOUCHER, B. J. (1968): The molecular weight of radioimmunoassayable growth hormone in human serum. *J. Endocr.*, *42*, 153.

CHEEVER, E. V. and LEWIS, U. J. (1969): Estimation of the molecular weights of the multiple components of growth hormone and prolactin. *Endocrinology*, *85*, 465.

COLLIPP, P. J., KAPLAN, S. A., BOYLE, D. C. and SHIMIZU, C. S. N. (1964): Protein bound human growth hormone. *Metabolism*, *13*, 532.

DAUGHADAY, W. H. and KIPNIS, D. M. (1966): The growth promoting and anti-insulin action of somatotropin. *Recent Progr. Hormone Res.*, *22*, 49.

DAUGHADAY, W. H., LARON, Z. and HEINS, J. N. (1969): Defective sulfation factor generation: a possible etiological link in dwarfism. *Clin. Res.*, *17*, 472 (Abstract).

DAUGHADAY, W. H. and MARIZ, I. K. (1962): Conversion of proline-U-C^{14} to labeled hydroxyproline by rat cartilage in vitro. Effects of hypophysectomy, growth hormone, and cortisol. *J. Lab. clin. Med.*, *59*, 741.

DAUGHADAY, W. H. and REEDER, C. (1966): Synchronous activation of DNA synthesis in hypophysectomized rat cartilage by growth hormone. *J. Lab. clin. Med.*, *68*, 357.

DAUGHADAY, W. H., SALMON, W. D. and ALEXANDER, F. (1959): Sulfation factor activity of sera from patients with pituitary disorders. *J. clin. Endocr.*, *19/7*, 743.

ELLIS, S., MENKE, J. M. and GRINDELAND, K. E. (1968): Identity between growth hormone degrading activity of the pituitary gland and plasmin. *Endocrinology*, *83/5*, 1029.

ESANU, C., MURAKAWA, S., BRAY, G. A. and RABEN, M. S. (1969): DNA synthesis in human adipose tissue in vitro. I. Effect of serum and hormones. *J. clin. Endocr.*, *29/8*, 1027.

FERGUSON, K. A. (1964): Starch-gel electrophoresis – application to the classification of pituitary proteins and polypeptides. *Metabolism*, *13/10*, 985.

FERGUSON, K. A., LAZARUS, L., VANDOOREN, P. and YOUNG, J. D. (1967): The nature of the growth-promoting substances in human plasma. *Acta endocr. (Kbh.)*, *119/Suppl.*, 238 (Abstract).
FRANCHIMONT, P. (1965): Dosage radio-immunologique de l'hormone croissance dans les urines. *Ann. Endocr.*, *26*, 627.
FRANGLEN, G. and GOSSELIN, C. (1958): Separation of metastable polymers by starch gel electrophoresis. *Nature (Lond.)*, *181*, 1152.
GELLER, J. and LOH, A. (1963): Identification and measurement of growth hormone in extracts of human urine. *J. clin. Endocr.*, *23*, 1107.
GIRARD, J. and GREENWOOD, F. C. (1967): The radioimmunoassay of urine for human growth hormone. *J. Endocr.*, *37*, 34.
GIRARD, J. and GREENWOOD, F. C. (1968): The absence of intact growth hormone in urine as judged by radioimmunoassay. *J. Endocr.*, *40*, 493.
HADDEN, D. R. and PROUT, T. E. (1964): A growth hormone binding protein in normal human serum. *Nature (Lond.)*, *202*, 1342.
HADDEN, D. R. and PROUT, T. E. (1965): Studies on human growth hormone. II. The effect of human serum on growth hormone labelled with radioactive iodine. *Bull. Johns Hopk. Hosp.*, *116*, 122.
HANSON, L. A., ROOS, P. and RYMO, L. (1966): Heterogeneity of human growth hormone preparations by immuno-gel filtration and gel filtration electrophoresis. *Nature (Lond.)*, *212*, 948.
HUNTER, W. M. (1963): Proceedings of the Society of Analytical Chemistry. 87th Ordinary Meeting of the Physical Methods Group. *Analyst*, *88*, 251.
HUNTER, W. M. (1965): Homogeneity studies on human growth hormone. *Biochem. J.*, *97*, 199.
IRIE, M. and BARRETT, R. J. (1962): Immunologic studies of human growth hormone. *Endocrinology*, *71*, 277.
KOGUT, M. D., KAPLAN, S. A. and SCHMIZU, C. S. N. (1963): Growth retardation: use of sulfation factor as a bioassay for growth hormone. *Pediatrics*, *31*, 538.
KOSTYO, J. L. and KNOBIL, E. (1959): The stimulation of leucine-Z4-C^{14} incorporation with the protein of isolated rat diaphragm by simian growth hormone added in vitro. *Endocrinology*, *65*, 525.
LARON, Z., PERTZELAN, A. and MANNHEIMER, S. (1966): Genetic pituitary dwarfism with high serum concentration of growth hormone. A new inborn error of metabolism. *Israel J. med. Sci.*, *2/2*, 152.
LAURENT, J. C. and KILLANDER, J. (1964): A theory of gel filtration and its experimental verification. *J. Chromat.*, *14*, 317.
LEAVER, F. W. (1966): Evidence for the existence of human growth hormone-ribonucleic acid complex in the pituitary. *Proc. Soc. exp. Biol. (N.Y.)*, *122*, 188.
LEWIS, U. J. and CHEEVER, E. V. (1965): Evidence for two types of conversion reactions for prolactin and growth hormone. *J. biol. Chem.*, *240*, 247.
LEWIS, U. J., PARKER, D. C., OKERLUND, M. D., BOYER, R. M., LITTERIA, M. and VANDERLAAN, W. P. (1969): Aggregate-free human growth hormone. II. Physicochemical and biological properties. *Endocrinology*, *84*, 332.
LI, C. H. (1968): The chemistry of human pituitary hormone; review 1956–1966. In: *Growth Hormone*, pp. 3–28. Editors: A. Pecile and E. E. Müller. ICS 158, Excerpta Medica, Amsterdam.
LI, C. H., TANAKA, A. and PICKERING, B. T. (1964): Human pituitary hormone. VII. In vitro lipolytic activity. *Acta Endocr. (Kbh.)*, *45*, *Suppl. 90*, 155.
LI, C. H. and YAMASHIRO, D. (1970): The synthesis of a protein possessing growth promoting and lactogenic activities. *J. Amer. chem. Soc.*, *92/26*, 7608.
MACMILLAN, D. R., SCHMID, J. M., EASH, S. A. and READ, C. H. (1967): Studies on the heterogeneity and serum binding of human growth hormone. *J. clin. Endocr.*, *27/8*, 1090.
MIYASATO, F. and POLLAK, V. E. (1966): Serum proteins in urine. An examination of the effects of some methods used to concentrate the urine. *J. Lab. clin. Med.*, *67/6*, 1036.
POLLAK, V. E., GAIZUTIS, M. and REZAIAN, J. (1968): Serum proteins in urine: Examination of new method for concentrating urine. *J. Lab. clin. Med.*, *71/2*, 338.
REITHEL, F. J. (1963): The dissociation and association of protein structures. *Advanc. Protein Chem.*, *18*, 123.
RIGAL, W. M. (1964): Site of action of growth hormone in cartilage. *Proc. Soc. exp. Biol. (N.Y.)*, *117*, 794.
RIMOIN, D. L., MERIMEE, T. J., RABINOWITZ, D., CAVALLI-SFORZA, L. L. and MCKUSICK, V. (1968): Genetic aspects of isolated growth hormone deficiency. In: *Growth Hormone*, pp. 418–432. Editors: A. Pecile and E. E. Müller. ICS 158, Excerpta Medica, Amsterdam.

ROHDE, W. and DÖRNER, G. (1969): Immunochemical studies on the heterogeneity of human growth hormone (HGH). *Acta Endocr. (Kbh.)*, *60*, 101.

ROOS, P., FEVOLD, H. R. and GEMZELL, C. A. (1963): Preparation of human growth hormone by gel filtration. *Biochim. biophys. Acta (Amst.)*, *74*, 525.

SAKUMA, M., IRIE, M., SHIZUME, K., TSUSHIMA, T. and NAKAO, K. (1968): Measurement of urinary human growth hormone. *J. clin. Endocr.*, *28*, 103.

SALINAS, A., MÖNCKEBERG, F. and BEAS, F. (1963): Immunological detection of growth hormone in normal human urine (correspondence). *Lancet*, *2*, 302.

SALMON, W. D. and DAUGHADAY, W. H. (1957): A hormonally-controlled serum factor which stimulates sulfate incorporation by cartilage in vitro. *J. Lab. clin. Med.*, *49/6*, 825.

SALMON, W. D. and DAUGHADAY, W. H. (1958): The importance of amino acids as dialyzable components of rat serum which promotes sulfate uptake by cartilage from hypophysectomized rats in vitro. *J. Lab. clin. Med.*, *51/2*, 167.

SALMON, W. D., DUVALL, R. M. and THOMPSON, E. Y. (1968): Stimulation by insulin in vitro of incorporation of (^{35}S) sulfate and (^{14}C) leucine into protein-polysaccharide complexes, (^{3}H) uridine into RNA, and (^{3}H) thymidine into DNA of costal cartilage from hypophysectomized rats. *Endocrinology*, *82/3*, 493.

SAXENA, B. B. and HENNEMAN, P. H. (1966): Isolation and properties of the electrophoretic components of human growth hormone by sephadex-gel filtration and preparative polyacrylamide-gel electrophoresis. *Biochem. J.*, *100*, 711.

SLUYSER, M. (1964): Possible causes of electrophoretic and chromatographic heterogeneity of pituitary hormones. *Nature (Lond.)*, *204*, 574.

SQUIRE, P. G. and PEDERSEN, K. O. (1961): Sedimentation behavior of human pituitary growth hormone. *J. Amer. chem. Soc.*, *83*, 476.

SQUIRE, P. G., STARMAN, B. and LI, C. H. (1963): Studies of pituitary lactogenic hormone. XII. Analysis of the state of aggregation of the ovine hormone by ultracentrifugation and exclusion chromatography. *J. biol. Chem.*, *238*, 1389.

WETTENHALL, R. E. H., SCHWARTZ, P. L. and BORNSTEIN, J. (1969): Actions of insulin and growth hormone on collagen and chondroitin sulfate synthesis in bone organ cultures. *Diabetes*, *18*, 280.

SUBJECT INDEX

Prepared by L. M. Boot, D.Sc., Amsterdam

INDEX OF AUTHORS